Object-Oriented Systems Analysis And Design Using UML

Object-Oriented Systems Analysis And Design Using UML

Second Edition

SIMON BENNETT
STEVE McROBB
RAY FARMER

The McGraw-Hill Companies

London • Burr Ridge, IL • New York • St Louis • San Francisco • Auckland
Bogotá • Caracas • Lisbon • Madrid • Mexico • Milan • Montreal • New Delhi
Panama • Paris • San Juan • Sào Paulo • Singapore • Sydney • Tokyo • Toronto

Published by McGraw-Hill Education
Shoppenhangers Road
Maidenhead
Berkshire
SL6 2QL
Telephone: +44(0)1628 502500
Fax: +44(0)1628 770224
Web site: http://www.mcgraw-hill.co.uk

British Library Cataloguing in Publication Data
A catalogue record for this book is available from the British Library

Library of Congress Cataloguing in Publication Data
Library of Congress data for this book is available from the Library of Congress, Washington, D.C.

Web site address: http://www.mcgraw-hill.co.uk/textbooks/bennett

Acquisitions Editor: Conor Graham
Editorial Assistant: Sarah Douglas
Senior Marketing Manager: Jackie Harbor
Senior Production Manager: Max Elvey
Produced for McGraw-Hill by Steven Gardiner Ltd
Printed and bound in Great Britain by Bell and Bain Ltd., Glasgow

ISBN 0-07-709864-1

McGraw-Hill books are available at special quantity discounts.
Please contact the Corporate Sales Executive at the above address.

To all our families and friends

Contents

Preface

Background to the Book

The Faculty of Computing Sciences and Engineering at De Montfort University began teaching object-oriented systems analysis and design to first-year students on undergraduate programmes in Business Information Systems and Multimedia Computing in 1997, having taught structured approaches for many years. Object-oriented approaches to analysis, design and programming are now taught on all diploma, undergraduate and postgraduate programmes in Computer Science and Information Systems. At that time, we were all working at De Montfort University and wanted a book to support our teaching, one that put the analysis and design activities in the context of the whole systems life cycle (whatever kind of life cycle that is), and that included generic analysis and design issues, such as fact finding. Most books on object-oriented approaches to analysis and design concentrate on object-orientation and on the notation.

Another concern with many textbooks that we have used is that they do not employ a consistent case study as a source of examples throughout the book. We try to teach or train by means of practical case studies so that students experience something that is close to the development of a real system rather than a series of disjointed exercises. We particularly like the approach taken by Barbara Robinson and Mary Prior in their book *Systems Analysis Techniques* (1995). Throughout the book they use two case studies, one for examples and one for exercises. However, their book teaches a structured approach using the notation of Structured Systems Analysis and Design Method (SSADM).

When McGraw-Hill offered us the opportunity to write our own book, we decided to take a similar approach and have developed two case studies through the course of the text, although not to the same level of detail.

Since the publication of the first edition in 1999, many things have changed. Only one of the authors is still at De Montfort University; another left the University to become an Information Systems Consultant for Ericsson Intracom, and is now working for GEHE UK as a Systems Architect; the third has recently taken up a new appointment as Associate Dean at Coventry University. As well as our careers, our ideas have developed, helped in large part by all the feedback and reviews that we have received. UML itself has also changed, and has become more widely accepted as the common language for systems modelling. But we still believe in the value for teaching and learning of basing the book around a consistent thread of case study material, and have therefore retained this approach in this new edition.

Who Should Read this Book?

The three authors of this book believe that systems analysis and design are activities that should take place in the context of the organizations that will use the information systems that are the result. The examples we use are based on business organizations, but they could be any kind of organization in the public or voluntary (not for profit) sectors, and the approach we adopt is suitable for most kinds of information system, including real-time systems. The book starts with three chapters that set the development of information systems in this context.

We expect most of our readers to be students undertaking a Diploma, Bachelor's or Master's course in a computing or information systems subject. It will also be of relevance to some students of other subjects, such as business studies, who want to understand how business information systems are developed without wanting to be programmers, analysts or designers.

The book is also suitable for professionals in computing and information systems, many of whom began their professional career before the advent of object-oriented development techniques and who want to upgrade their skills by learning about object-oriented analysis and design. We have used the Unified Modelling Language (UML), which is the *de facto* standard notation for object-oriented development.

Case Studies

In our teaching and training we use case studies as the basis of tutorials and practical work. We also use the same case studies to provide examples in taught material and in student assessments. We believe that it is important that students see analysis and design as a coherent process that goes from initial fact-finding through to implementation, and not as a series of disjointed exercises. This book uses two practical case studies. The first of these, Agate Ltd, is an advertising company. Agate is used for examples in most of the chapters of the book that explain techniques and for some exercises. The second case study, FoodCo Ltd, is a grower of fruit and vegetables and a manufacturer of packaged foods. FoodCo is used for most of the exercises that are included in chapters for the reader to complete. In the second edition we have added two more case study chapters.

The two case studies are introduced in short chapters (A1 and B1) that can be found between Chapters 4 and 5. In these first two case study chapters, we provide background material about the two companies and explain some of their requirements for computerized information systems. Chapter A2 (between Chapters 6 and 7) presents examples of the requirements model, while Chapter A3 (between Chapters 7 and 8) brings together some examples from the first version of the analysis model for Agate's new system. Chapter A4 (between Chapters 11 and 12) presents examples from the analysis model after it has been further developed. Chapter A5 (between Chapters 18 and 19) brings together some examples from the design model for the new system. We do not provide full models for the FoodCo case study, but FoodCo forms the basis of most of the practical exercises at the end of chapters. A few partial solutions are provided where they are required as the basis for later exercises.

If you are using this book as a teacher, you are welcome to use these materials and to develop them further as the basis of practical exercises and assessments. Some exercises that you may want to use are provided in each chapter. Further models, solutions and case studies are provided on the book's website, and this will continue to develop over time.

Exercises for Readers

Each chapter contains two kinds of exercises for readers. First we provide Review Questions. The aim of these is to allow you to check that you have understood the material in the chapter that you have just read. Most of these Review Questions should only take a few minutes at most to complete. Solutions to some of these questions are to be found at the back of the book. The answers to many of them are to be found in the text. Some require you to apply the techniques you have learned, and the answers to these are not always provided.

At the end of each chapter are Case Study Work, Exercises and Projects. These are exercises that will take longer to complete. Some are suitable to be used as tutorial exercises or homework, some could be used as assignments, and some are longer projects that could be developed over a matter of weeks. We have provided answer pointers to some of these exercises to help you.

Solutions to more of these exercises are available via the book's website to bona fide university and college teachers who adopt the book as the set text for their courses. Please contact McGraw-Hill for details of the support materials.

Structure of the Book

Although we have not formally divided the book into sections, there are four parts to the book, each of which has a different focus.

Part 1

Chapters 1 to 4 provide the background to information systems analysis and design and to object-orientation. In the first three of these chapters we explain why analysis and design are important in the development of computerized systems and introduce fundamental concepts such as those of systems theory. Chapter 4 introduces some of the ideas of object-orientation that will be developed in the second part.

Part 2

The second part of the book begins with the first two case study chapters (A1 and B1) and includes Chapters 5 to 11. The focus of this part of the book is on the activities of requirements gathering and systems analysis and the basic notation of the Unified Modelling Language (UML). In it we introduce use cases, class diagrams, sequence diagrams, collaboration diagrams, activity diagrams, statechart diagrams and the Object Constraint Language (OCL). Chapter 5, which is new for the second edition, discusses models and diagrams and presents one of the UML diagramming techniques—activity diagrams—as an example. In Chapter 5 we also provide an overview of the way the UML techniques fit together in the iterative development life cycle. We have also added two additional case study chapters (A2 and A3), which illustrate the development of the UML models as the analysis activities progress. This part concludes with Chapter A4, in which we provide further examples from the analysis model for Agate. The purpose of these case study chapters is to illustrate the development of models as the analysis and design progress. We do not have the space in the book to provide all the analysis and design documentation.

Part 3

The third part of the book is about system design. It includes Chapters 12 to 18 and concludes with examples from the design model for the Agate case study (Chapter A5). In this part we develop the use of most of the diagramming techniques introduced in Part 2. We do this by enhancing the analysis models that we have produced to take design decisions into account. This part covers the transition to design, the distinction between system design and object design, system architecture, design patterns, and the design of objects, user interfaces and data storage. The design model at the end of this part serves the same purpose as the one at the end of the analysis chapters.

Part 4

In the final part we cover the later stages of the systems development life cycle and what we have loosely called 'advanced issues'. Originally we intended to include the chapter on design patterns in this part of the book, as some would regard it as an advanced aspect of design. However, we found that we wanted to use patterns in the other design chapters and moved it forwards. This leaves us with only three 'advanced' chapters. The chapter on implementation introduces the last of the UML diagrams, component diagrams and deployment diagrams.

Pathways through the Book

Whatever the formal structure of the book, you the reader are welcome to work through it in whatever order you like.

We have taught two undergraduate analysis and design modules: a first-year module on analysis and design, with the emphasis on the life cycle and on analysis; and a second-year module with an emphasis on design and the use of methodologies. For this purpose, we would expect to cover the following chapters in each module.

Analysis module—Chapters 1, 2, 3, 4, 5, 6, 7, 8, 9, 10, 12, 13, 14. (Statechart diagrams are omitted from this so that students are not overloaded with different diagramming notations in the first year.)

Design module—Chapters 12, 13, 14 (recap), 11, 15, 16, 17, 18, 19, 20, 21, 22.

We have tried to group together techniques under the general headings of analysis and design, even though many of them are used throughout the life cycle. This does not mean that we necessarily advocate following a life cycle model that treats analysis and design as separate phases. We suggest an iterative life cycle in which the models of the system are progressively elaborated and in which analysis and design are interwoven. However, we do believe that analysis and design are separate activities, even if they are not separate stages in a project's life cycle. We also think that it is easier to learn analysis and design as separate activities rather than merged together. (This is like a chef being taught to cook desserts and main courses separately. Later this chef can either follow a structured approach to cooking, in which a dessert is prepared in advance and then chilled and stored until required, and the main course is then cooked separately. Or, when she is experienced, she can take an iterative approach and progressively build the meal, switching from main course to dessert and back again.)

If you plan to use this book for a course that concentrates on UML, then you may want to use the following path through the book—Chapters 5, 6, 7, 8, 9, 10, 11, 17, 18, 19, and including the case study chapters. The other chapters can be read as background.

If you are familiar with the general aspects of information systems development and of object-orientation, and are reading this book in order to gain an understanding of how you can use UML in analysis and design, then you can start at Chapter A1, the first of the case study chapters, and work your way through from there. If you are not familiar with object-oriented approaches, then you should also include Chapter 4.

Transferable Skills

Some of the skills of the systems analyst and designer can be classified as transferable or professional skills. Most employers place a high value on these skills. Many colleges and universities provide special modules on these skills, embed them in other modules or provide self-study packages for students. We have included material on fact finding skills, particularly interviewing and questionnaire design, within the text of the book. We have not included other skills explicitly, but there are opportunities for teachers to use the exercises to develop skills in problem solving, group work, research, report writing and oral presentation.

Web Site and Support Materials

This book is supported by materials that we have placed on the book's website. The website is accessible at www.mcgraw-hill.co.uk/textbooks/bennett/. The website material has been revised for the second edition and the format redesigned. We have added self-test exercises for students, and plan to add more case study material and supplementary articles that provide material that we could not fit into the book. There are also links to other web pages that we think that you the reader will find interesting or that extend the material that we have been able to include in the book. Materials for lecturers also include MS Powerpoint slides to accompany each chapter, solutions to some of the exercises and copies of most of the figures from the book that can be used for teaching. If you use this material in your teaching materials, we ask that you acknowledge our copyright on the material.

We welcome feedback about the book. Some of the changes that we have made in writing the second edition have been based on feedback from lecturers and students around the world who have used our book. You can email us at authors@OOADtext.info or write to us care of McGraw-Hill at the address on p.iv of the book.

Latest Version of UML

This edition is based on UML Version 1.4. Version 2.0 was planned to have been available by now, but is now unlikely to be released until 2002 or 2003. Version 1.5 is being developed. Information on UML updates will be available on the book's website.

Changes in the Second Edition

The second edition has some significant changes from the first. All chapters have been revised and updated to reflect developments in the world of object-oriented analysis and design, the evolution of UML and changes in our own thinking. Chapter 5 on models, diagrams and the iterative life cycle has been added. Two additional case study chapters have been included. Many chapters have been updated with new material, particularly Chapters 7 and 17. Some material has been moved from one chapter to another. A summary of changes is available on the web site.

Simon Bennett, Steve McRobb, Ray Farmer
Leicester, UK
November 2001

Acknowledgments

Our thanks to the reviewers of the original proposal for the book and to everyone who has read part of the book and provided us with feedback. Thanks go to David Howe who reviewed the first edition and to the anonymous reviewers of the second edition. Thanks also to all those readers—students, lecturers, trainers and practitioners—who have emailed us with feedback. Unfortunately, we have not been able to take every single comment on board, or we would have written four different books!

The names of products and of companies mentioned in this book may be claimed as trademarks or registered trademarks of their respective companies. Where those names appear in this book and the authors or publisher were aware of a claim, the name has been printed in all capitals or with initial capitals.

The UML Cube logo is a trademark of Object Management Group, Inc. in the U.S. and other countries.

Extracts from Gamma/Helm/Johnson/Vlissides, DESIGN PATTERNS. (pages 185, 186, 187). © 1995 by Addison-Wesley Publishing Company. Reprinted by permission of Pearson Education, Inc.

Extracts from Checkland/Scholes, SOFT SYSTEMS METHODOLOGY IN ACTION. © 1990 John Wiley and Sons Ltd. Reproduced with permission.

Figure from Allen/Frost, COMPONENT-BASED DEVELOPMENT FOR ENTERPRISE SYSTEMS, © 1998 by Cambridge University Press. Adapted with permission.

Extracts from *Computing* reproduced by permission of *Computing*.

Extracts from *Journal of Information Technology*. © Taylor and Francis Ltd. Reproduced with permission. Journal website: http://www.tandf.co.uk/journals/routledge/02683962.html

Extract from Webster, THEORIES OF INFORMATION SOCIETY. © 1995 Pearson Education Ltd. Reproduced with permission.

Extract from Jacobson/Griss/Jonsson, SOFTWARE REUSE: ARCHITECTURE, PROCESS AND ORGANIZATION FOR BUSINESS SUCCESS, © 1997 by Addison-Wesley Publishing Company. Reprinted by permission of Pearson Education, Inc.

CHAPTER 1

Information Systems—What Are They?

OBJECTIVES

In this chapter you will learn:

- how to define an information system
- some examples and types of information system
- how to apply the basic concepts of systems theory
- how information systems are related to organizations.

1.1 Introduction

Information systems have played an important part in human affairs since our most distant ancestors first became capable of organized collective action. For example, Palaeolithic cave paintings of hunters and animals that date back 30,000 years or more can be seen as a sort of information system, used to capture, store, organize and display information.

The application of information technology (IT) in modern times has brought immense changes to the scope and nature of information systems. So profound are these changes that some writers believe we are living through an information revolution, on a scale that is little short of a second industrial revolution. This has been a recurrent theme since Bell coined the term 'post-industrial society' (1973), and remains an implicit undercurrent in Zuboff's classic *In The Age Of The Smart Machine* (1988).

Not everyone agrees that a direct comparison with the industrial revolution is valid. For example, Webster (1995) argues that contemporary changes in society, while certainly significant, do not represent the kind of radical break with the past implied by a 'revolution'. The picture is also complicated by the fact that, in some countries, the introduction of computers is interwoven with industrialization. Still, it is clear that computers have had a dramatic and pervasive effect on our lives. At the time of writing, it seems that both the spread of computerized information systems and the pace of technological development will continue to accelerate for the foreseeable future. Recent technical developments like the Internet have made possible information systems that would not have been feasible even a few years ago.

Despite the antiquity of some of its subject material, the academic study of information systems is young, even by the standards of the computer age, and owes its importance today chiefly to the rise of the computer. Those who developed the earliest computer systems gave very little thought to the issues with which this book is primarily concerned. Wartime imperatives drove them to apply the new technology to military problems like code-breaking, naval gunnery calculations and other mathematical tasks. Their focus was on the technical difficulties of implementing hardware based on ideas from the very cutting edge of research in electronics and control logic. It was also necessary to invent efficient techniques for controlling these new machines. In due course these techniques became today's computer programming.

By the time that computer scientists had gained a degree of mastery over their hardware, the world was once more at peace. Businesses began to be aware of the potential that computers held for commercial activities, and attention gradually turned to wider questions. Those with which we are concerned are the following.

- How do we establish the business requirements for a new system (often much subtler and more complex than the role of the earliest machines)?
- What effects will the new system have on the organization?
- How do we ensure that the system we build will meet its requirements?

It is primarily with these questions that the field of information systems is concerned, and they are also the main subjects of this book.

Within the relatively new field of information systems, object-oriented analysis and design are even newer, although they are derived from object-oriented programming, which has been around a lot longer. Object-oriented analysis and design were first conceived of only about a decade ago, and today they are still very much a novelty. Only in the last few years has there been a real surge of interest in object-oriented systems development. This relative youth of object-oriented analysis and design might give cause for concern. Are they no more than a passing fashion? We believe they are no mere fad, but on the contrary, for many information system applications—although not all[1]—object-orientation is simply the best way yet found of carrying out the analysis, design and implementation.

So far we have skirted around a question that we must address right at the start. What is an information system? In this chapter we provide the beginnings of an answer (the rest must come through practical experience). We set the scene by describing some information systems, ranging from one that is completely up to date, to others that are considerably older (Section 1.2). We discuss what is meant by 'system' (Section 1.3), by 'information' and finally by 'information system' (Section 1.4). We also put the question in context by raising some business issues, in particular how the managers of an organization decide (or discover) the things that are important for the business to do, and we outline the role of information systems in doing these things successfully (Section 1.5). For business managers, this is the real bottom line. An information system that did not help to ensure the success and prosperity of the organization, its investors and its employees would simply not be worth the cost of its development.

[1] We return to this question in Chapter 4.

1.2 Information Systems in Practice

We set the contemporary scene in Box 1.1 by describing a complex modern information system. Many readers will be familiar with systems that resemble this one, and some may have used them to make purchases over the Internet. This particular example is entirely fictitious, although it is loosely based on elements taken from a number of real companies.

Box 1.1 An on-line retail system

McGregor plc is a chain of retail stores that sells kitchen appliances, mobile phones and electronic home entertainment equipment. The company has recently created an on-line shopping centre, accessed via the World Wide Web. After registering their name and address, shoppers can browse through various products, selecting items and placing them in a virtual trolley. At the end of the trip, shoppers can buy what is in their trolley, remove items or quit without making a purchase. Payment is made by submitting credit card details on-line, or by entering part of the card details and phoning to give the rest. Delivery times are usually within three working days for small items such as mobile phones, but up to three weeks for larger items such as cookers. Goods are despatched direct to the customer's home. Credit cards are debited only on the day of despatch, and between purchase and delivery, customers can use the website to check on the progress of their order.

This is how an on-line shopper interacts with the system, but beneath the surface a great deal more is going on. A whole network of hardware connects the shopper's home PC and modem, through a phone line, a telephone exchange, another phone line and modem, to a computer that acts as a web server. This is connected to other networks at McGregor's head office and shops. Many software applications are also busy processing information captured through the web pages, and feeding various support activities. Some of these are computerized and others are carried out by people.

- Marketing staff keep prices and product details up to date on the electronic product catalogue system. This can also be accessed by touch-screen PCs in the shops.
- Credit card details are stored electronically for relay to the card processing centre when the goods are despatched.
- Robot forklift trucks in the warehouse fetch items to the loading bay when they are due for despatch, and warehouse staff load them onto delivery trucks.
- Delivery drivers follow a schedule linked to an electronic map in the vehicle cab. This is updated by radio every few minutes, helping to avoid traffic jams.
- Out of stock items are re-ordered from the supplier by electronic data interchange (EDI). On delivery, warehouse, delivery and charging begin quite automatically.
- At each significant point in the sequence, a database entry is automatically updated, and this is displayed on the web page, allowing shoppers to discover what stage their order has reached.

There are many users besides the shopper, each with a different view of the overall system. A network manager monitors the number of hits on the web server and network traffic within McGregor, checking for hardware or software failure and breaches of network security (e.g. if hackers try to break in). Her concern is the efficient and secure flow of information; she is not interested in its content. A financial controller uses a linked accounting system to monitor sales transactions and cash flow. A despatch clerk regularly checks forklift schedules on her PC, and compares them to delivery truck schedules faxed daily from the courier company. She smoothes the flow of goods through the loading bay, and spends long hours on the phone sorting out delays. A market researcher uses a specialized statistical package on a portable PC to analyse on-line sales, assessing the success of the web pages' presentation styles. Registration allows individual customers to be tracked as they browse the site; using information about their preferences, the design can be finely tuned to attract high-spending customers.

The staff described at McGregor all rely on computer systems for the information they need to do their jobs. But only recently have information systems begun to use information technology in such a sophisticated way. Our next example, described in Box 1.2, is almost exactly contemporary with the birth of the electronic digital computer, yet it contains little that we would recognize today as information technology. Checkland and Holwell (1998) describe it aptly as 'the information system that won the war.'

Box 1.2 Fighter Command's 1940 information system

This vital communications and control system was used to monitor and control Spitfire and Hurricane fighter squadrons as they flew against Luftwaffe bombing raids during the Battle of Britain. It is quite certain that, without it, the RAF could not have succeeded in the defence of Britain in 1940. The main 'hardware' was a large map kept at Command headquarters at Bentley Priory. Coloured counters were used to mark the current position of various aircraft formations, and were moved around the map by hand as reports came in via radar, telephones, teleprinters and radio. Reports of enemy raids were collected both by radar and a network of observer posts throughout the country, while information about RAF deployment was sent in via Group and Sector control rooms. A special filter room correlated and checked all reports before passing them through to the main control room for display. Other displays included a blackboard showing at a glance the status of all current Luftwaffe raids, and a row of coloured lights flagged the readiness of RAF fighter squadrons. Carefully co-ordinated duplicates of these displays were based at other control rooms, including those at Sector level where the controller spoke directly by radio to the tense young pilots in their aeroplanes. Using this basic, but effective technology, human operators performed communication, filtering, processing and display tasks that would today be carried out by computers.

Some accounts of the Battle of Britain have claimed that it was won chiefly due to the courage of British pilots or to the RAF's use of radar. But these miss a most important point. The new radar technology was certainly important, but the British variety was still technically far inferior to its German equivalent, already in use for naval and anti-aircraft gunnery. Equally, there could have been no victory without the bravery of the RAF pilots, but was this really greater than that of their German opponents, who also faced extreme danger each time they flew over the English coast?

Important as these factors were, it was only through Fighter Command's information system that the British human and technical resources were deployed so effectively.

With a little imagination, it is easy to see how a modern system could use computers to fulfil similar functions to those described in Box 1.2, and indeed this is the situation in most airforces today. Box 1.3 looks even further back in time, introducing briefly an information system that dates back over a century, yet still operates effectively today with similar technology in many parts of the world.

The distinguishing characteristics of the pre-computer information systems described in Boxes 1.2 and 1.3 would change relatively little if they were to be re-implemented today using computers. But the system described in Box 1.1 would not be feasible at all without the use of modern information technology. Thus, while the computer is not a necessary precondition for the existence of many information systems, in some cases today computers have made possible information systems that could not exist without them.

Box 1.3 A railway signalling system

This example is particularly interesting because, according to most classifications of work and social status, an old-fashioned railway signalman would be regarded as essentially a manual worker of the early industrial age. Yet on closer examination he can be seen clearly to be an information worker, differing from many modern computer users only in that the technology available to him was much less sophisticated than would now be the case. According to Webster (1995) '. . . The railway signalman must have a stock of knowledge about tracks and timetables, about roles and routines; he needs to communicate with other signalmen down the line, with station personnel and engine drivers, is required to "know the block" of his own and other cabins, must keep a precise and comprehensive ledger of all the traffic that moves through his area. . .'. In other words, the signalman operates an information system that comprises his ledgers, the levers, signals and lamps and what he knows in his mind.

Nowadays many railway signalling systems are largely automated[2]. However, whether the work is done by networked computers linked electronically to electric motors that switch points and operate signal lights, or by a traditional signalman who pulls levers connected by steel cables to the points and signals, and keeps careful records in a hand-written ledger, the tasks that are actually carried out are little different. All that has changed is the technology that implements the system.

1.3 General Systems Theory

We now introduce some basic concepts of general systems theory. This is essentially a way of understanding a system in terms of those components and characteristics that are common to all systems. We use the term *system* here in a very specific sense, although one that has much broader application than just computer systems. In everyday speech, people may refer to the legal system, a tropical storm system, the system of parliamentary democracy, an eco-system, a system for winning at roulette, a computer system in someone's office, a system for shelving books in a library, a system-built house, and many more. Some of these certainly meet our definition of a system, and others do not. Probably the only thing that they all have in common is that they have some kind of organization. But in general systems theory, a system is rather more than just anything that shows a degree of organization.

[2] This raises questions about the ethics and efficacy of the computer simulation of human judgement in safety-critical situations, but such arguments are beyond the scope of this book.

1.3.1 Characteristics of a system

So what are systems? For our purposes, a system has the following characteristics.

- A system exists in an environment.
- A system is separated from its environment by some kind of boundary.
- Systems have inputs and outputs. They receive inputs from their environment, and send outputs into their environment.
- Systems have interfaces. An interface allows communication between two systems.
- A system may have sub-systems. A sub-system is also a system, and may have further sub-systems of its own.
- Systems that endure have a control mechanism.
- System control relies on feedback (and sometimes feed-forward). These comprise information about the system's operations or its environment, that is passed to the control mechanism.
- A system has some properties that are not directly dependent on the properties of its parts. These are called emergent properties, as they only emerge at the level of the system as a whole.

Figure 1.1 presents the relationships between these concepts schematically.

Boundary and environment

An understanding of the related concepts of *boundary* and *environment* is essential to making sense of any system. In fact the first step in understanding a system is to choose the system that you wish to understand, and this largely means choosing its boundary.

For example, a cell biologist may think about a single human cell as a system. Her interest, and therefore the system she thinks about in detail, is bounded by the cell membrane. A biochemist might be interested in a slightly larger system, perhaps a chemical reaction that takes place in a particular group of cells. A specialist physician attempting to diagnose an illness may consider a patient's kidney as a system. The boundary of this system may coincide with the organ, or it may be wider still, depending on the particular

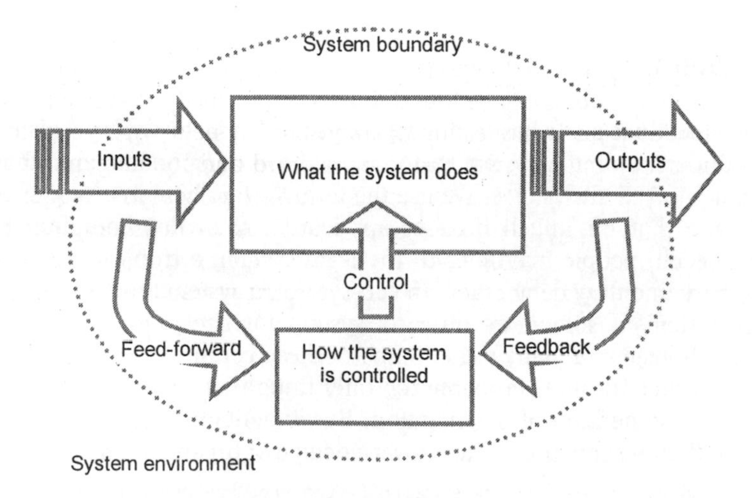

Figure 1.1 Parts of a system, and their relationship to each other.

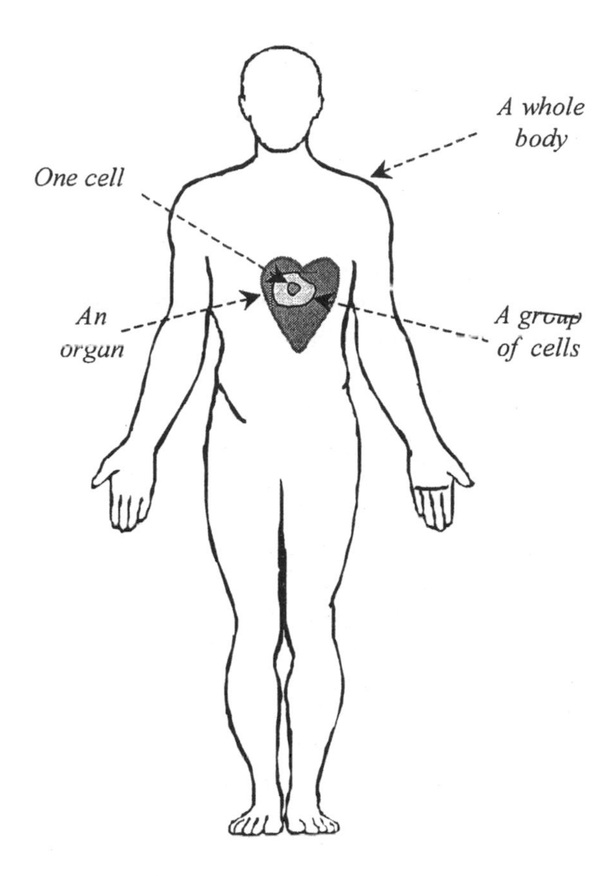

Figure 1.2 Systems at different scales.

illness. For a doctor in general practice, a person's whole body may be considered as a system bounded by its skin. These different boundaries are illustrated in Figure 1.2. What is readily apparent is that we can continue zooming out in this way, perhaps until we reach a cosmologist (whose interest might be bounded by the limits of the physical universe) or even a theologian (whose interests are even wider).

But the choice of a system that corresponds to a subject of interest is not simply a matter of scale. Systems can overlap with each other, and this is also an issue for boundary setting. Let us think again about a whole person as a system. Within this image, many other systems can be found. For example, medical specialisms each have their own view of what is interesting or important. A neurologist may focus on the nervous system, consisting of the brain, spinal cord and the network of nerves that spreads through the body to just beneath the surface of the skin. While its physical boundary is almost identical to that of the body as a whole, the nervous system contains only various specialized nerve cells. The interests of a haematologist thinking about the circulatory system have a similar boundary, yet this system contains instead blood cells, blood vessels and the heart. A person can also be considered as a variety of non-physical systems. A psychologist might study an individual's cognitive system or emotional system, or may consider a child's intellect as a learning system. If we move along the scale again, a social psychologist may think of a family as many overlapping systems: a child-rearing system, an economic system, a house-maintenance system, etc. These views of the family as a system have purely conceptual boundaries, since family members remain part of their system no matter how distant they are in space.

The boundaries of different systems can overlap or coincide. Indeed, two systems may be closely related, may have identical boundaries, and yet still be distinct. This is potentially a tricky part of the theory. At what point does one system end and another begin? Also, can a thing (e.g. a human organ) simultaneously be part of one system and of another, at one and the same time?

The answer to this lies in the fact that the usefulness of systems is in helping us to understand something about how the world works. They do this by representing selected aspects of the world in an abstract way. In most cases, it is not important whether or not the system corresponds in every detail to the thing it represents. Checkland and Scholes (1990) explain this as follows.

> . . . it is perfectly legitimate for an investigator to say 'I will treat education provision *as if it were* a system,' but that is very different from saying that it *is* a system . . . Choosing to think about the world as if it were a system can be helpful. But this is a very different stance from arguing that the world *is* a system, a position that pretends to knowledge that no human being can have.[3]

Figure 1.3 illustrates the same point graphically.

This does not mean to suggest that no system has a basis in reality. On the contrary, many systems are made of real components. For example, it is quite possible to touch physically all the parts of a central heating system. But it is a matter of choice, based on our interest at a given moment, whether we choose to *think* about it as a system. Any system that we think about necessarily exists in our thoughts, rather than in the world, and such a system, however much it corresponds to the world, is still a subjective view of reality, not the reality itself.

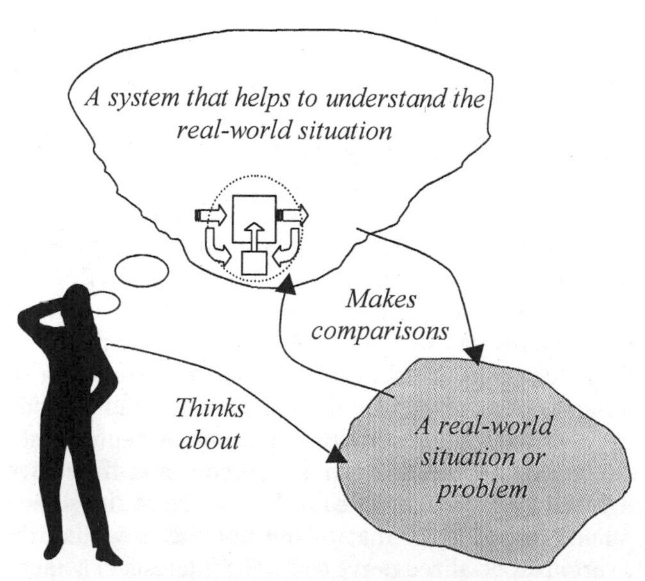

Figure 1.3 The relationship between system and reality (loosely adapted from Checkland and Scholes, 1990).

[3] Italics in the original.

System	Inputs	Outputs
A student	Information Exercises Guidance	New knowledge New ideas Solutions
A family	Money Social standards and norms (e.g. laws) Purchases Daily news	New citizens (i.e. children) Products of family members' work Social influence Votes in elections
A business	Raw materials and labour Investment Information (e.g. customer orders)	Profit and taxes Finished products Information (e.g. the company report)

Figure 1.4 System inputs and outputs (note that there is not necessarily a one-to-one correspondence between inputs and outputs).

Figure 1.5 A black box view of a system.

Input, output and interface

Systems have interactions with their environment. For example, human cells take in food and oxygen and excrete waste materials. Some cells produce things of use to the body as a whole. For example some white blood cells produce the antibodies that fight infection. The nervous system receives sensory information in the form of light, sound, touch, etc. and transforms these ultimately into signals that move the body's muscles and generate speech. Each interaction is based on a set of inputs and outputs. Inputs originate outside the system, and are taken in to be used in some way. Outputs are created by the system, and sent into the environment in order to have an effect somewhere else, but always to achieve a purpose of the system. Figure 1.4 shows some inputs and outputs for three systems.

An important characteristic of purposeful systems, such as a business, is the way they transform inputs into outputs, since this is how they fulfil their objectives. For some purposes, it is enough to know the relationship between inputs and outputs. For example, a customer of the McGregor on-line shopping system, described in Box 1.1, is probably not interested in what goes on behind the scenes. Her main focus is on the interaction with the system-in other words, the inputs she must provide (information, money) and the outputs she receives (information, a new fridge/freezer, customer service). This is called a *black box approach* since it treats the system as an opaque box whose internal workings are completely hidden. All that can be seen is what goes into the system, and what comes out. This is illustrated in Figure 1.5.

An output from one system can simultaneously be an input to another. The two systems share part of their boundary, across which inputs and outputs pass between them. The shared boundary is an interface. For example, tiny little air sacs in your lungs act as an interface between the circulatory and respiratory systems. Here, freshly inhaled oxygen is output by the lungs and input to the bloodstream for distribution, while spent carbon dioxide is output by the blood and input to the lungs for exhalation. We shall see later, in many places throughout this book, that the identification and understanding of interfaces is of particular importance in the development of information systems.

Sub-systems

Sub-systems are a natural consequence of other systems ideas introduced above. For example, a white blood cell making an antibody can be seen as a system on its own, or as a sub-system that contributes to the larger system. Figure 1.6 illustrates some sub-systems that can be found in the description of the Agate case study in Case Study A1 (this case study is used to provide many of the illustrative examples throughout the book).

Sub-systems are at once part of a larger system, and also coherent systems in their own right[4]. Communication between sub-systems is, by definition, through interfaces. This particular kind of diagram is sometimes called a system map.

Another way of arranging systems and sub-systems is as a hierarchy. Hierarchies are a very important aspect of systems theory, to which we will return frequently. For example, in Chapter 4 we shall see the importance of the concept of hierarchy to understanding object-orientation.

Control in systems

Many systems have a specialist sub-system whose function is to control the operation of the system as a whole. In fact, general systems theory sprang partly from the science of cybernetics, which studies control in natural and artificial systems. A familiar type of cybernetic control is the family of thermostatic devices that control central heating, hot water and air conditioning systems, and also temperature-controlled appliances such as freezers and ovens. Many are based on a simple feedback loop, as shown in Figure 1.7. System control is usually based on a comparison between two or more input values, whose similarity or difference guides a decision about whether any controlling action is required.

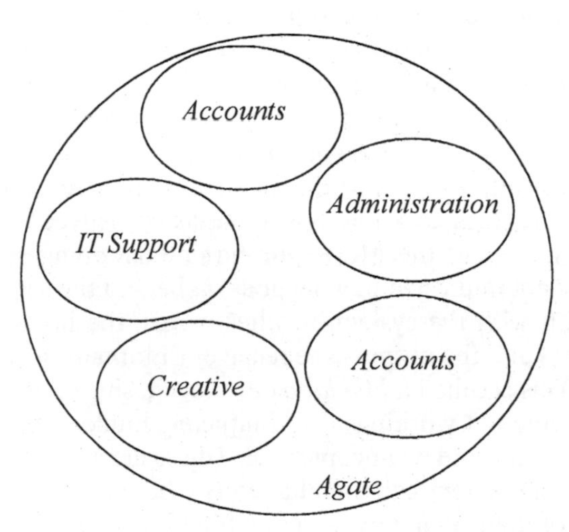

Figure 1.6 Sub-systems of Agate.

[4] Koestler (1967) coined the word 'holon' to express this duality. A holon is something that, seen from one perspective, looks like a whole thing, but, viewed in another way, looks like a part of something else.

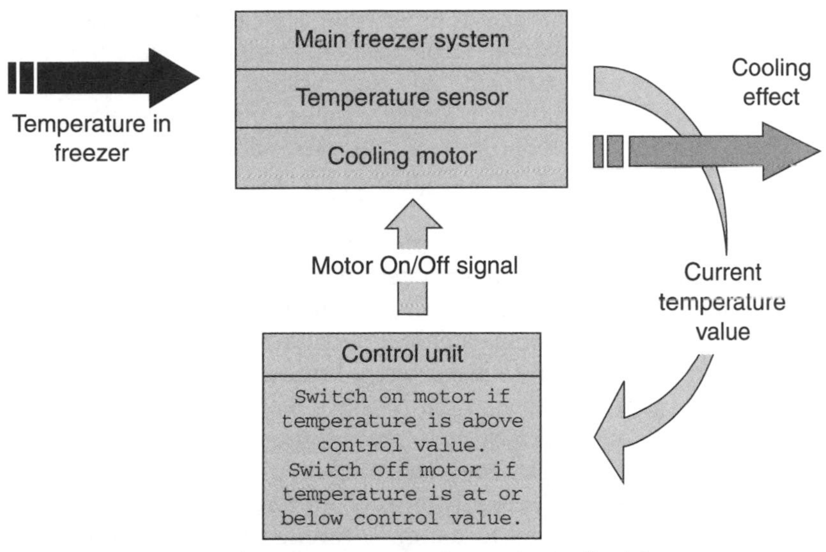

Figure 1.7 Thermostatic control in a freezer—a simple negative feedback loop.

Feedback

The term *feedback* indicates that one or more outputs of the system is sampled and literally fed back to the control unit. In Figure 1.7, a sensor samples the temperature in the main freezer compartment (an output measure) and this is fed back to the control unit. Here a logic mechanism (this could be either mechanical or electronic) decides whether action is required. Possible actions are turning the cooling motor on and turning it off. Both are effected by a signal sent to the motor by the control unit, thus 'closing' the 'feedback loop'. In this way, the control unit monitors the current temperature, compares it to a reference value (in this case probably set by the owner using a dial or switch), and if necessary adjusts the system's operation to bring the difference between the two values within pre-defined limits.

This is *negative* feedback, because it aims to maintain the system's equilibrium by opposing deviations from some norm. Negative feedback is used widely in physical systems, such as electronic equipment, manufacturing systems, etc. By contrast, *positive* feedback works by reinforcing deviations instead of opposing them and therefore tends to increase movements away from equilibrium. Systems governed by positive feedback are inherently unstable, which is undesirable when the deviation that is reinforced is itself undesirable. For example, the ear-splitting howl sometimes heard from public address systems is the result of a positive feedback loop. It usually occurs when sound output from a loudspeaker is caught by a microphone for re-input to the amplifier. Under certain conditions the signal resonates round and round the loop, growing louder at each turn. The volume then builds quickly until the amplifier reaches its limit.

Both negative and positive feedback occur in social and business systems, but here they are rarely so clear-cut or simple. For example, the Bank of England regularly reviews the performance of the UK economy to determine whether or not it meets the Government target for monetary inflation. When the economy deviates from the target, action is taken to bring it back on track, usually by altering the bank's base interest rate. This would be a simple example of negative feedback, though of course, in reality the situation is much more complex. Another example of negative feedback is when a business adjusts its manufacturing output to maintain a constant inventory.

If sales rise, stock falls, and the rate of production is accordingly increased. If sales fall, stock tends to rise, and the rate of production is reduced. Here the stock level acts as a *buffer* between production and sales sub-systems.

Positive feedback is beneficial when a steady state is undesirable. For example, in a soccer game, equilibrium would tend always to result in a goal-less draw. Any competitive edge that one team gains over its opponents is exploited to retain and maximize the advantage. However, this picture is greatly complicated by the fact that both teams continually strive to get the upper hand in a dynamic game, and so it is rare to see the cycle build to an extreme. Instead, the system (i.e. the game) usually tends to continue in a state of dynamic disequilibrium. Negative feedback also plays a role, as teams try to maintain any equilibrium that acts in their favour. A complex mix is seen also in business competition, where firms capitalize on any competitive advantage they can gain over their rivals (positive feedback, aiming to disrupt an unhelpful equilibrium), while also trying to prevent their rivals from gaining an edge (negative feedback, maintaining a beneficial equilibrium).

Positive feedback does not necessarily mean that every deviation from the norm is reinforced. Some may be allowed to die away through lack of encouragement. In audio amplifiers all frequencies are amplified, but only resonant frequencies are disastrously reinforced by the feedback cycle, while others fade naturally without ill-effect. Often a sound engineer need only adjust the amplifier's tone control, reducing slightly the volume in the troublesome frequency band, to cure a feedback problem. A similar type of control can be discerned in businesses that directly link productive output to sales, perhaps because there is a very short-term seasonal market (say, toys at Christmas). In this case, when sales rise, production is increased as far as possible. But when sales fall below a critical level, a product may be dropped altogether.

Effective feedback is an essential part of all learning. No one can develop a new skill, without receiving appropriate feedback that highlights which aspects of their performance are satisfactory and which parts need improvement. This applies equally to the activity of software development. Most professionals continue to learn how to do the job throughout their careers. This is partly because techniques and technologies evolve continuously. But also, every project is unique, poses a new set of challenges and demands new approaches from the developers. Feedback on what worked in the past, and what did not, helps to guide the developer's choices in future.

Feed-forward

Feed-forward information relies on sampling the system's inputs rather than its outputs. For example, the Christmas toy business mentioned above may find that the use of feedback from sales (a measure of output) does not allow them to react quickly enough to changing market conditions. This may leave them with unsold stock, bought from manufacturers before they realized that demand had slumped. Ideally, they should adjust their manufacturing to suit the level of demand, and they may be able to use market research to forecast which toys will be popular among children this year. This would allow the firm to avoid buying products for which there is no market. Another way to use feed-forward would be to find out which toys are being manufactured in large numbers, and then advertise these aggressively in order to stimulate demand.

While feed-forward control information can help a system to be more responsive to environmental fluctuations, it is not always easy to implement or manage in a business organization. Difficulties still arise if the rate at which conditions change in the environment is faster than the rate at which the business can adapt. The effects of this are apparent to anyone who visits the kind of specialist book shop where remaindered titles are sold cheaply.

In the Agate case study, the agency must employ and train sufficient staff to cope with the anticipated workload. If there is a serious slump in orders for new work, it may not be possible to reduce the number of staff quickly enough to avoid bankruptcy, due to the need to give staff a period of notice before they can be laid off. The company may also not be able to respond quickly enough to a sudden surge of orders, due to the lead time for recruiting and training new staff. Forecasting the level of demand for a service such as Agate's is an important role of information systems in business.

Emergent properties

One of the truly distinctive characteristics of a system is that it is more than just the sum of its parts. The system possesses some feature or ability of its own that is not present in any of its components. These are called emergent properties. For example, a car is only a form of transport if there is enough of it to drive. It then has the property of being a vehicle, but the wheels, windscreen, motor, etc. do not have this property until they are correctly assembled.

This recognition that systems have emergent properties is the main reason why the systems approach is described as *holistic*. This means thinking about each system as a whole, with important aspects that will be overlooked if we think only about its parts in isolation from each other. The holistic world view of the systems approach is manifest in techniques like the rich picture. This aims at capturing in one picture or diagram all that is essential to an initial understanding of a system[5]. An opposing view is taken by reductionists, who begin with the assumption that complex phenomena can be explained by reducing them to their component parts. This approach has a very important place in the physical sciences (e.g. physics and chemistry, although there are dissenters even in those disciplines). We believe it is thoroughly misapplied when used to explain complicated human situations.

1.3.2 Systems of interest to this book

We have used a mixture of natural and artificial systems as examples. General systems theory is applicable to both, but in this book we are principally interested in the artificial kind. Even some of these are more relevant than others. For example, we will probably only be professionally interested in the system of parliamentary democracy if we intend to develop an information system that supports it. Two types of system are of particular importance to us. The first is the *human activity system* (Checkland, 1981). A key feature of this type is that it centres on a purposeful activity, such as we find when we look at a business, a club, a hospital, or any organized activity. There is often wide disagreement among the participants about exactly what that purpose is, and this can be a significant problem for the analysis of information system requirements. Figure 1.8 shows how some of the people who work at Agate see the purpose of one system in which they are engaged.

[5] See Checkland and Scholes (1990) for a full explanation of rich pictures with many examples of their use in practical situations.

System	Purpose of system	As seen from the perspective of . . .
Agate (a business system)	To become a successful advertising agency on the international stage, thus providing both wealth and prestige for its directors	A director
	To provide varied and interesting work with a good salary, and also to be a useful stepping stone towards the next career move	A copy-writer
	To provide a pleasant and comfortable life until retirement (five years away), without the need to make too much effort	Another director

Figure 1.8 A human activity system with multiple purposes.

The second kind of system in which we have an interest is, naturally, the information system. Information systems are constructed to help people in a human activity system to achieve their goals, whatever these might happen to be. Thus human activity systems are the context of, and provide the meaning for, the activity of information systems development. Anyone who does not understand the meaning and purpose of a human activity system will find that it is impossible to specify, still less to build, an information system that supports it in any useful way.

We shall discuss many examples of information systems throughout the main part of the book. It is also worth mentioning here that the activity of systems development can be usefully regarded as a human activity system. This system usually has the purpose of developing effective software solutions to business problems, or something similar to this. Its environment is typically the organization in which the developers work, including users of the software and their managers. Its subsystems include individual members of the project team, and also the various analysis and design models that describe the software. Inputs will include information from users about how they want the software to work, while the main outputs are the software itself, together with manuals and other documentation. Control will be exercised by a team leader or project manager, who receives regular feedback on progress and problems. Suitable feed-forward will help to alert the manager to anticipated problems, as well as those that have already happened. One advantage of taking a systems view of any activity is that it encourages those involved to think about the sorts of feedback and control that are needed for everything to run smoothly. This applies just as much to software development as to anything else.

1.4 Information and Information Systems

Some computing professionals see the development of information systems as essentially a matter of designing and building computer technology (including software) that meets a set of clearly understood needs. While this may be the ideal situation, in practice it is often a simplistic view that misses much of importance. It happens that information technology is now the normal technology used to implement an information system, but, as we saw in Section 1.2, information systems were constructed in many other ways prior to the development of digital computers.

Designing and building technology can be the easy part of the job—at least the easy part to understand—while the hard part is often determining the needs that the technology must serve. This involves identifying ways that an information system can support the purposes of a human activity system. An understanding of the information that will be useful to the human actors is an important ingredient in this, as is an understanding of how the information can be used effectively. As a result of these many different concerns, Information Systems has become a multidisciplinary subject that bridges many other fields, in particular computer science and business management, but also psychology, social theory, philosophy and linguistics, among others. In the following sections, we discuss the relationship between information, information systems and the human activity systems they are intended to assist.

1.4.1 Information

Information is conveyed by messages and has meaning. Meaning always depends on the perspective of the person who receives a message. We are always surrounded by a vast mass of potential information, but only some of this ever comes to our attention, and only some of that is actually meaningful in our present context. Checkland and Holwell (1998) describe the process by which raw facts become useful (i.e. become information) through a sequence of stages.

For an example of how this happens, consider three people watching the evening sky. A plume of smoke is rising in the middle distance. For the first, the smoke is just part of the general view, and she does not consciously notice it. The second sees it, and it evokes a memory of a camping trip long ago. But he is aware that the only connection between past and present smoke is a coincidence of shape and colour, so he moves on to look at something else. The third is thrown into consternation, when she realizes that the smoke is rising from her house, which is on fire. She runs to phone the Fire Service before doing whatever else she can to save her house.

The sight of the smoke is, on the face of it, a single message available to be received by all three, yet its meaning is different in each case. As the first person does not even notice it, Checkland and Holwell call this *data* (from the Latin for 'given'), meaning that this is a fact, but one that has not been selected for attention. The second person notices the smoke but does not relate it to any present context. Checkland and Holwell call this *capta* (from the Latin for 'taken'), meaning a fact that has been selected for attention. The third person not only notices the smoke, but gives it a meaning derived from the context in which she sees it (recognizing her house, understanding the implications of the smoke, etc.). This can be called *information* since it has a meaning within the context. The meaning of a fact is thus entirely dependent on its relevance to the observer.

There is a final step where information becomes *knowledge*, by being structured into more complex meanings related by a context. We can see this in the third person's response to the smoke. She integrates information from many sources: the link between smoke and fire, the effect of fire on houses, the existence and purpose of a Fire Service and the location of nearby phones. All this information comes together in a single framework of knowledge that is relevant to the context. In a word, she *knows* what to do.

1.4.2 Information systems

In the past, it has been usual for systems analysis and design textbooks to include a section that broadly classifies the variety of information systems an analyst may encounter in practice. Such a classification is less useful today than it was once, since information systems have tended to become ever more integrated with each other, and thus the boundaries have blurred between categories that, not long ago, were quite distinct. Thus, while it is still helpful to present a brief overview of some of the general types of application in organizations, the following is intended to introduce some aspects of the role of information systems, rather than to depict actual systems.

Operational systems

Operational systems automate the routine, day-to-day record-keeping tasks in an organization. The earliest commercial information systems were operational ones, because routine, repetitive tasks that involve little judgement in their execution are the easiest to automate. Accounting systems are one example. These derive from the need in all organizations to keep track of money-the amount coming in, the amount going out, the cash available to be spent and the credit that is currently available. Few modern organizations could survive long without a computerized accounting system. Sensible organizations have a 'disaster recovery plan' that details how they intend to cope with an emergency that destroys data or renders computer systems inoperable.

The flow of information through an accounting system is based on thousands, even millions of similar *transactions*, each of which is an exchange of a quantity of something, usually a money value. For example, when you buy a tube of toothpaste in a supermarket, two separate records are made. One records that a tube of toothpaste was sold, and the other records the money you paid in exchange for it. As this repeats day after day for each item, customer, checkout and branch, an overall picture is built up that allows the company's accountants to compare total income with total costs and determine whether a profit has been made. Real accounting systems are more complicated than this, often with sub-systems to handle wages, taxation, transport, budget planning and major investments.

Organizations use personnel systems to keep track of their employees, partly so that managers can control the money spent on wages, but also for many other reasons. The simplest personnel systems provide information that keeps the internal phone directory up to date, or invites employees to retirement planning seminars when they reach a certain age. More complex applications calculate each employee's working hours for the next few months, based on the company's current level of orders, how many employees are currently sick, how many have booked holidays and the mix of skills needed for each production team to be effective. ERP (Enterprise Resource Planning) systems produced by companies such as SAP and Oracle, integrate personnel applications like this with other systems that monitor, forecast and control virtually all of the processes and resources of a business. In recent years, the boundary has become very blurred between 'operational' systems like this and the management support systems described in the next section.

Management support systems

Information systems intended to support management usually work at a much higher level of complexity than operational systems. But the blurring of boundaries in recent years has had particular effect here. For example, the personnel planning application discussed at the end of the preceding section could be considered as management support rather than operational. Moreover, much of the information used by management to make decisions is derived directly from information stored at the operational level. In practice, many management support systems are built on top of operational systems, while other systems combine elements of the two, meeting a complex set of needs at different levels of the organization.

Many of the earliest management support systems were developed simply by adding a set of programs (known as a management information system, or MIS) to extract data from existing operational systems, and analyse or combine it to give managers information about the part of the organization for which they were responsible. We can easily see how this happened with an accounting system. Once all routine sales transactions had been stored on a computer, it was a short step to the realization that if this data were analysed appropriately, it could tell managers at a glance which products were not selling well, which checkout operators took too long dealing with a customer, which store had the lowest volume of trade, and so on.

This information is useful to managers because they have a responsibility to maximize the performance of an organizational sub-system. An important part of this is identifying and resolving problems as they occur. Thus, the crucial aspect of a management support system is the feedback or feed-forward that it provides, alerting managers to problems and opportunities, and assisting them in the process of tuning the organization's performance. Let us return for a moment to the diagram in Figure 1.1, and consider this as a human activity system that represents an organization as a whole. The chief difference, then, between operational and management support systems is that they fit into different parts of this diagram. Operational systems either are located in the central box (labelled 'what the system does'), or, alternatively, they assist its work by supporting the flow of inputs or outputs. Management support systems either are located in the box in the lower part of the diagram (labelled 'how the system is controlled'), or, alternatively, they assist its work by supporting the flow of feedback to, or control information from, the control unit.

Real-time control systems

Real-time systems are explicitly concerned with the direct control of a system's operations, often physical in nature. Examples include lift control systems, aircraft guidance systems, manufacturing systems and the robot forklifts in the McGregor system described in Box 1.1. For this reason, they are perhaps best considered as a control sub-system of a physical processing system. Their role is thus very different from both operational and management support systems. Real-time systems usually have human operators (to date, few are completely independent of human supervision, though this may become common in the future), but they are generally insulated from the surrounding human activity system. In fact, many authors would not agree that real-time systems are information systems at all. We do not regard this as an important issue. The techniques used for the analysis, design and implementation of real-time systems are very similar to those used for other computer systems, so in practical terms any distinction is artificial.

1.4.3 Information technology

We have left information technology until last because decisions about information technology should ideally be made last in the cycle of development. Only when the human activity system has been understood, the need for an information system has been identified, and the information system requirements have been defined—only then should the emphasis turn to the information technology that will implement it. Sadly, we cannot pretend that this is how things always happen in the real world. Indeed it is partly for this reason that so many systems in the past have been so unsuccessful. But it is how they should happen, wherever possible. Section 1.5 of this chapter examines some of the reasons for this in more detail.

Here we offer only a broad definition of information technology, and this only for the sake of clarity. We understand it to include all the varieties of hardware familiarly known as, based upon, or that include within them, a computer or its peripheral devices. For example, obvious things like desktop PCs, pocket electronic organizers, modems, network cabling, file-servers, printers and computer-controlled machinery in factories and airliners, and also less obvious things like digital mobile phones, the electronic circuits in some cars that calculate fuel consumption, the microchips in some cameras that set the aperture and shutter speed—in other words, everything misleadingly described in marketing literature as 'intelligent'. Due to the unprecedented rate of technical progress, the range of devices that can be described as information technology increases almost daily, and the boundaries between them blur. As digital devices continue to advance in speed and processing power, manufacturers exploit these advances to develop and market new products. For example, consider the communication devices that combine a mobile phone, modem, e-mail software and a web browser, or the add-on cards that let a PC display broadcast TV, or the new digital TV sets that behave in many ways like network computers. Interface technologies such as voice-activation may make it easier to interact with computers without needing to sit at a desk and press keys or click mouse buttons. Mobile commerce (for example, using WAP mobile phones) has had a disappointing start from the supplier's point of view, but it is expected by some commentators to begin to have a serious impact around 2004. This is also likely to change the way that many people access information and communicate with companies and other organizations. In particular, it will remove the physical restriction that requires a user to be in the same place as a bulky PC when he or she wants to access the Internet. On the whole, it appears likely that over the next few years computers will progressively disappear from view, while their effects will paradoxically be felt in more and more areas of everyday life.

All the examples of information technology that have just been mentioned are really just tools that, like any tool, can be used for many different tasks—and not only those for which they were intended. There is a saying that, if your only tool is a hammer, the whole world looks like a nail. The corollary is also true: if you can see only nails, you will be inclined to use any tool that comes to hand as a hammer, regardless of whether it is actually a wrench, a book or a can of beans. So how a tool will *actually* be used matters much more than how it is *meant* to be used. A modern word-processing package provides a skilled user with the facilities to automate many complex tasks by using macro programs, mailing lists and embedded objects like spreadsheets, sound clips and hyperlinks to the Web. Yet many users have no need of all this and would be just as happy with an electronic typewriter. The question is, then, if an electronic typewriter is all that is required, why install a powerful PC running all the latest software?

1.5 Strategies for Success

Business strategy is an essential part of the context for this book, since all information systems analysis and design projects begin with the identification of a business issue that can be aided or a problem that can be solved by the use of an information system. The underlying assumption is that information systems are only worthwhile if they meet the needs of the organization in which they are installed. In this section, we consider some ways that business needs can be identified at a very high level, suggesting possible application areas for information systems and information technology.

1.5.1 Identifying a business strategy

The development of a business strategy essentially begins with the question: 'Where would we like our organization to be in (say) ten years' time?' Answers to this question identify a set of goals for the organization. Logically, the next question is then: 'How do we get from where we are now to where we want to be?' In effect, how do we meet those business goals? Answers to this question will ideally suggest some practical steps that can be taken towards achieving the strategic goals. A number of different techniques have been developed to help with this process. One example is the well-known SWOT approach[6], but many others have been used by organizations to develop their business strategies, including the Value Chain Model, which we introduce in the next section.

The detailed contents of any particular strategy (i.e. the actual goals and steps it contains) depend very much on the characteristics of the organization, its environment, the skills of its workforce, and many other factors. In the Agate case study, we are told that the strategy is 'to continue to grow slowly and to develop an international market' (these are the goals). The directors also have a view about how to achieve these objectives: they want to get more business from multinationals, and they hope to succeed in this through the quality of their work and by developing 'campaigns that have a global theme but are localized for different markets around the world' (these are the steps). We can safely assume that these elements have been included in the strategy because the directors are confident of certain other things. For example, they may believe that the technical quality of Agate's work and the creativity of their staff are both strengths of the company, and will meet the demands placed on them. They probably also believe that their current client base and contact list is extensive enough for them to win the kind of work they are seeking.

1.5.2 The contribution of information systems

Information systems can contribute to the achievement of business goals in so many different ways that it can seem daunting to decide which are the systems that really matter. Many techniques have been developed to help arrive at useful answers, some adapted from their original purpose as tools for developing a business strategy. One example is Porter's Value Chain Model (1985). While not the only technique for information system planning, and not necessarily the best for all situations, it is a useful way of structuring this discussion because of the systemic view of an organization that it presents.

The model is illustrated in Figure 1.9, and depicts an end-to-end 'chain' or flow of materials through the organization. The flow is transformed at each stage, beginning with raw inputs that become a product through operations applied to them, then are shipped and sold to a customer who finally receives an after-sales service. In practical applications, a group of managers and other staff allocate activities and departments to appropriate compartments of the model, and identify the value each adds to the overall product or service.

Porter used the chain metaphor to reinforce the point that all primary activities are essential, but any weak link negates the value of work done at every other stage. For example, if a business is good at selling its products, but the products themselves

[6] This stands for Strengths, Weaknesses, Opportunities, Threats. Usually carried out as a group brainstorming exercise, it involves identifying and categorizing everything about the organization's current circumstances that falls into these categories. The resulting strategy is based on finding ways of exploiting the strengths and opportunities, while counteracting the weaknesses and threats.

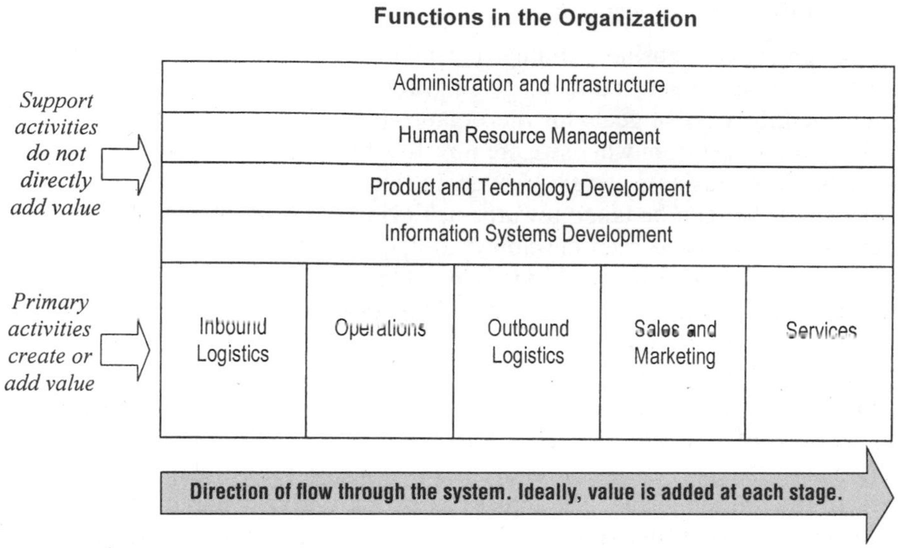

Functions in the Organization

Support activities do not directly add value

Administration and Infrastructure				
Human Resource Management				
Product and Technology Development				
Information Systems Development				

Primary activities create or add value

Inbound Logistics	Operations	Outbound Logistics	Sales and Marketing	Services

Direction of flow through the system. Ideally, value is added at each stage.

Figure 1.9 The Value Chain Model (adapted from Porter, 1985).

are of poor quality, it is unlikely to be successful. Alternatively, a business that makes excellent products but has very poor arrangements for obtaining its raw materials will also not be successful. In a successful organization, each primary activity (the lower part of the diagram) adds value to the products (i.e. it benefits the company more than it costs). Activities in the upper part of the diagram provide services, but do not directly add value to products. They are not essential, and are therefore only worth doing if they make a contribution to the efficiency or effectiveness of primary activities. Their role must therefore be tuned to support primary activities.

This tool can be useful in information systems planning since it focuses attention on activities that are critical to a business, either because they are currently a problem, or because they represent a major source of profit or competitive edge. Information system development projects can then be targeted at assisting those operations that can directly contribute to the success of the organization as a whole.

1.5.3 Information systems and information technology strategies

The best managed organizations separate their strategic thinking into the three layers illustrated in Figure 1.10. The key idea is that the development of new information systems should only be considered in the context of a well-thought-out business strategy, while purchases of information technology hardware should only be considered in the context of specific information systems that are planned for development. Thus, the business strategy drives the information system strategy, which in turn drives the information technology strategy. Information flows in the diagram are not only in one direction. In formulating a business strategy, managers need to be advised on those areas of the business where the information system can contribute to fulfilling business goals. Thus the planning cycle is iterative. A similar two-way communication takes place between the information system and information technology strategy planning functions. The role of the information technology strategy is to enable the successful implementation of the systems defined in the information system strategy, while also informing the information system strategists about what is feasible.

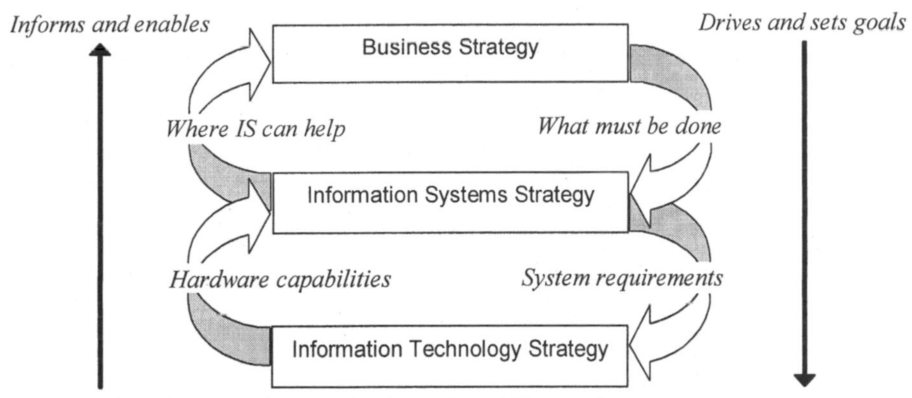

Figure 1.10 The relationship between business, IS and IT strategies.

The importance of the relationships shown in Figure 1.10 cannot be overstated. For example, consider the McGregor system described earlier in Box 1.1. One objective of the company's current business strategy is to capture a share of the lucrative on-line market. This may even be a business imperative. If other on-line retailers have already taken a significant share of the overall market, they may pose a serious threat to McGregor's continued survival. In order to fulfil this business objective, McGregor's managers must identify, define and then develop an appropriate set of software systems—this includes the on-line customer order system, the robot warehouse system, the stock control system, the purchase order system, and so on.

It is the main purpose of the IS strategy correctly to identify which of the many possible systems should be chosen for development, and also how these need to interface with other systems. Those chosen for development will then become systems development projects. Selection of the wrong systems for development wastes time and resources, and can also distract attention from the business priorities. In this context, 'wrong' may mean systems that are unimportant, but it can also mean systems that are not compatible with other vital systems. In this case, it could cause major business problems for McGregor if the on-line customer order system cannot correctly transfer details of the items ordered to the warehouse system, resulting in the wrong goods being delivered to the customer. This kind of difficulty could occur simply because a project team is not sufficiently aware of the need for the software they are developing to interface with another system.

Another critical issue might be poor design of the on-line catalogue, resulting in frustrated customers who buy instead from a competitor's website. While the actual design of the website would not form part of the IS strategy, major business concerns such as the need for clear navigation and interaction on a customer website, are often spelt out as part of the overall IS strategy.

The IT strategy is responsible for identifying the hardware components and configurations that will allow the software to operate effectively. In McGregor's case, this will include specifying the web servers to ensure that the response time is always fast enough to satisfy customers. Slow response times could also frustrate customers and result in lost sales for the company. The detailed specification of the servers (operating system, number of incoming lines, processors, RAM, etc.) will not be worked out until later, but the strategy will identify this as a business concern, and will also explicitly relate the various hardware components to the software systems they must run.

For many businesses, success depends on finding an appropriate fit between the overall business goals, the information systems that help to fulfil those goals and the IT on which the information systems run. This question of strategy alignment has even greater importance when the business is engaged in e-commerce, whether this is business-to-customer (B2C) or business-to-business (B2B). In either case, for customers, suppliers, partners, collaborators—indeed for any interaction that occurs electronically—the information systems *are* the company, since the website is really all that can be seen. Moreover, an Internet presence can be seen—and judged—by the whole world. Either an inappropriate strategy, at any of the three levels, or a poor implementation can bring swift business failure, as a number of dot.com companies discovered to their cost over the first year of the new millennium. In Chapter 2, we explore in more detail the problems that can occur during information systems development, and even *because of* information systems, while in Chapter 3 we introduce some ways that these problems can be resolved.

1.6 Summary

In this chapter we have introduced the main concepts that underpin information systems. These include some important ideas from general systems theory, such as the ideas of control, communication and emergent properties. We have also discussed the relationship between information and meaning, the central point here being that information only makes any sense within a specific context. For systems analysts and designers this context is usually the human activity system of an organization. This leads to a necessary set of relationships between the goals of an organization, the steps it must take in order to fulfil them, the information that its staff must have to carry out their activities, the information systems that provide the information, and finally the information technology that supports the information systems. While we recognize that there are many real organizations where this does not happen in the way we describe, we can only wonder how much more could be achieved, and express the hope that readers of this book will start work from firmer foundations.

The idea of information systems is far from new, yet information technology has changed the field almost beyond recognition in the last few years. This process will undoubtedly continue in the future, but however much information systems may change, valuable lessons may still be learned from historical information systems, both those developed since the introduction of the computer and those that pre-date it.

Review Questions

1.1 What is the difference between an *information system* and *information technology*?

1.2 Identify some things that a computerized *information system* can do, which are difficult or impossible for a non-computerized equivalent.

1.3 Why does it not matter whether a system is real, or exists only in someone's mind?

1.4 Why are boundary and environment important for understanding a system?

1.5 What is the difference between feedback and feed-forward?

1.6 Why has a human activity system more than one purpose?

1.7 What is the purpose of a management support system?

1.8 What is meant by disaster recovery? Why is it important for a business organization?

1.9 What are the relationships between business goals, information systems strategy and information technology strategy?

1.10 Define information. How does it differ from data?

1.11 Describe how knowledge differs from information.

1.12 Give an example of some knowledge that you possess. What is its purpose?

Case Study Work, Exercises and Projects

1.A Think of three or four information systems that are not computerized (either historical or contemporary). Identify (or imagine) a computerized equivalent. For each pair, write a brief description of the boundary and main inputs and outputs. What are the main differences between the computerized and non-computerized versions?

1.B Re-read the description of the on-line shopping system in Box 1.1, and assume that everything described (computer software, hardware, human activities, etc.) is a single system. Identify its main sub-systems and control mechanisms. What feedback and feed-forward information do you think they would use? Don't be constrained by the description in Box 1.1—use your imagination too. And remember that some control may not be computerized.

1.C Read the first part of the FoodCo case study in Chapter B1, up to and including Section B1.2.1. What do you think are FoodCo's business goals for the next ten years? Make any assumptions that you feel are justified.

1.D Roughly copy out the diagram from Figure 1.9 (make it large, with plenty of space in each compartment). Try to map FoodCo to the Value Chain Model by identifying people, activities, sections, etc. that fit in each compartment. What flows through your value chain model?

1.E Identify the value that you think is added by each primary activity to FoodCo's products. Which do you think are the weak links?

Further Reading

- Checkland and Holwell (1998) is an accessible and up-to-date account of the related subjects of systems, information and information systems.
- Zuboff (1988) is a personal view of the way that information technology and information systems are changing the nature of work and management based on research in many American companies during the 1980s. One of Zuboff's conclusions is that managers who do not understand how information technology changes an organization, may undermine all its potential benefits.
- Webster (1995) is a scholarly debunking of many of the more exaggerated claims about how the 'information revolution' is changing social relationships.
- Turban and Aronson (2001) gives an up-to-date and broad-ranging review of modern software technology for the support of managers at all levels.
- Fidler and Rogerson (1996), although now dated in its coverage of software applications, is still a useful introduction to the role of information systems in providing support to managers, and to Strategic Information Systems Planning. There are many practical case studies throughout the book, although the approach taken is not an object-oriented one.
- Timmers (2000) is one of the few books on e-commerce that give serious attention to the relationship between e-commerce activity, business strategy and marketing theory. This may sound rather dry and academic, but the book is also very readable and contains a number of detailed real-life case studies.
- Koestler (1967) is a classic text that applies systems concepts to many aspects of life. This book ranges widely over human history and society, and presents some early speculations on the systemic role of evolution in modern social behaviour.

CHAPTER

2

Problems in Information Systems Development

OBJECTIVES

In this chapter you will learn

- who the main players are in an information systems project
- the problems in information systems development
- the underlying causes of these problems
- how the concept of a stakeholder helps identify ethical issues in information systems development
- the costs of problems and ethical issues.

2.1 Introduction

It is only by understanding what can go wrong during a system development project that we can hope to avoid failure. Some authors use the metaphor of a journey to describe the process of information systems development (Connor, 1985), highlighting the fact that the main concern in any project is to avoid wrong turns on the way to the destination. At the start of a project, many different routes are available. Some lead to the planned destination, while others lead to a destination that may surprise everyone by turning out to be satisfactory, but is nevertheless unintended. Both are acceptable, although the latter can be nerve-wracking along the way. There are also routes that arrive at the wrong destination, while others are dead ends that lead nowhere. We must recognize and avoid these. This is the explicit purpose of carrying out analysis and design before developing a new system. Analysis is a way of understanding what needs doing before we begin trying to do it, and design is a way of checking that the planned action or solution really meets the needs of the situation, before it is put into practice. An understanding of potential problems is an essential precursor to systems analysis and design.

Projects fail for many different reasons, and in varying degrees. As we shall see, some failures are more visibly catastrophic than others. In this chapter we look at the question of project failure from the perspective of each of the main players (Section 2.2). We discuss the causes that can lead to failure. These are broadly categorized into those that relate to the quality of what is delivered, and those that relate to the productivity of the project (Section 2.3). While not usually a direct cause of project failure, we also consider some ethical issues associated with systems development (Section 2.4). Some of these may act as a delay or an obstacle for the developer, while others only appear as problems when viewed from a wider context. Finally, we mention the costs of the problems, whether or not they result in outright system failure (Section 2.5).

2.2 What are the Problems?

Many information systems are very successful, and failure is the exception rather than the rule. But the consequences of failure can be severe. One report estimated that failed information systems projects in the USA accounted for $81bn annually—almost a third of total information systems development expenditure (Standish Group, 1995). A similar picture probably prevails in the UK, where total IS/IT expenditure for 1995 was estimated at £33.6bn (Willcocks and Lester, 1996). The bill for one aborted project (the London Stock Exchange Taurus system) is estimated at £480m. All potential causes of failure are at least to some extent under the control of the developers. A professional must take the possibility of failure seriously, and work hard to avoid it, even if this sometimes is limited to an awareness of the risks, followed by a damage-reduction exercise.

One difficulty with the question 'What can go wrong?' is that the answer depends partly on who gives it. But this is not surprising, as information systems development is a complex activity that always involves people. This is not to say that all people are difficult—this only applies to a very few! But, in any organization, people have varying perspectives that influence their view of a situation, and what, if anything, is to be done about it. Although some attitude differences derive from psychology and personal history, which are beyond the scope of this book, others relate to an individual's position within the organization and their link to an information system project. It is useful to discriminate between three categories of people with important relationships to a project. The first is the group of employees who will become end-users of the information system when it is completed. The second is the group of managers, here called 'clients', who have control (or at least influence) over the initiation, direction or progress of a project. The third is the group of professionals responsible for the development of the information system. For the moment, we ignore differences within each group.

2.2.1 An end-user's perspective

End-users come in many varieties, and can have varying relationships to an information system. The examples that follow concentrate on the experiences and frustrations of those who either consume the products of an information system (i.e. they use its outputs to help achieve a task, such as despatching an ambulance to an emergency), or are responsible for entering raw data into an information system.

'What system? I haven't seen a new system'

One problem that can be experienced by an end-user is vividly expressed by a term that gained widespread usage in the 1980s. *Vapourware* describes a software product that is much talked about, but never released to its intended users. In other words, instead of arriving, it evaporates. While many businesses are reluctant to talk about information system project failures in public, vapourware may be surprisingly common.

Some surveys have found that an astounding proportion of information systems development projects fails to deliver any product at the end. In the US it has been suggested that 'as many as 25% of all projects in large MIS departments are never finished' (Yourdon, 1989). In the UK, total costs on a Regional Information System for the Wessex Regional Health Authority amounted to £63m, yet no system was delivered (Collins, 1998b). This example indirectly hurt patients, as the aim of the system was to help manage hospital resources more effectively throughout the region, and thereby deliver a better, more responsive service. When a project is not completed, the expected benefits to users and other beneficiaries are not achieved.

'It might work, but it's dreadful to use!'

This relates to systems that are unpleasant or difficult to use. Systems may fail to meet the criterion of usability in a number of ways, including: poor interface design, inappropriate or illogical sequence of data entry, incomprehensible error messages, unhelpful 'help', poor response times and unreliability in operation. Figure 2.1 gives some examples (what these mean, and how to avoid them, is explained more in later chapters).

Not long ago one of the authors bought a pair of shoes in a local shoe shop, and saw staff struggling to register the sale correctly with a new cash register system. The difficulty arose indirectly from a promotional offer. Any customer buying this style of shoe was also entitled to a free pair of socks. Since the socks were a normal stock item, correct records of stock had to be maintained. This meant that the socks had to be 'sold' through the till, even though they were actually being given away for free.

System characteristic	Example
Poor interface design	A web page with yellow text on a white background.
Inappropriate data entry	A system where the backspace key sometimes deletes whole words.
Incomprehensible error messages	A system message that says 'error #13452'.
Unhelpful 'help'	A system message that says 'wrong date format—try again'.
Poor response times	Nurses in an Intensive Care Unit complained that their new computerized patient chart system took longer to store and retrieve data, compared with the manual system. This took them away from patients for longer[1].
Unreliability in operation	A national motor vehicle insurance company lost most of its digital records of customers' policies due to a system error. Staff were unable to send renewal notices, but were compelled instead to write to customers asking them to phone in with their policy details[2].

Figure 2.1 Aspects of poor system usability with examples.

[1] Source: Goss *et al.* (1995).

[2] One of the authors is a customer of this company.

A simple way to handle this would have been for the assistant to over-ride the price with zero at the time of sale. The assistant tried this, but the software specifically prevented the 'sale' of a stock item at a zero price. The assistant called the manager. After some experimentation, it appeared that the only way to deal with this transaction was to reduce the price of the shoes by 1p, and to sell the socks at a cost of 1p, thus giving the correct total for the sale. Now that staff know how to do this, it will not cause so much difficulty in future. But it will always be an unnecessarily awkward way of handling a routine task. There are many examples of this sort of poor design, and they cause a great deal of frustration and lost time for users.

'It's very pretty—but does it do anything useful?'

A system may appear well designed and easy to use, but still not do the 'right' things. This may be a question of the tasks that should be carried out by the system. For example, a library catalogue enquiry system would be of limited use if it could only retrieve shelving information about a book when provided with the title and the author's name, in full and spelt correctly. Readers often do not know the title of the book for which they are searching. Even if the author's name is known, it may be spelt incorrectly. Another way that a system may fail to meet its users' needs is through poor performance (this overlaps with the question of usability, discussed in the previous sub-section).

A system may also be of doubtful value to its users because it requires them to work in a way that seems nonsensical. One report describes a warehouse management system, designed partly to increase managers' control over the routine activities of warehouse workers. The workers found that the new system removed much of their discretion in their work, and this prevented them being able to maximize the use of scarce storage space.

> . . . because they could see how improvements in these areas would save money for the company they found ways of working around the system.
> . . . They were reproved by management for their bad attitude, and yet, it was their commitment to the company as a major employer in their local community which led to their frustration with what they regarded as unnecessarily wasteful rules and procedures.
>
> (Symons, 1990)

It is much more worrying when software errors and failures present a hazard to life. An extreme example was the London Ambulance Service emergency despatch system, abandoned shortly after delivery in 1992, after incurring a total development cost estimated at £43m.

> Far from speeding up the dispatch of London's ambulances and paramedic teams, response times became slower than ever.
> After a series of high-profile failures, the system was canned amid claims—never officially substantiated—that it was responsible for several needless deaths.
>
> (Barker, 1998)

Controversy also continues to this day about whether it could have been software errors that caused the crash on the Mull of Kintyre of a Royal Air Force Chinook helicopter in 1994. All 29 people on board were killed, including a number of high-ranking police and military intelligence officers. The official verdict, still accepted by the Government, stated that the helicopter crashed because its pilots were grossly negligent, but an earlier Board of Enquiry concluded that the crew may have been distracted by a major technical (i.e. software) malfunction. A series of reports in

Computer Weekly and an item on British TV's Channel 4 News claim that internal Ministry of Defence reports have raised concerns about the reliability under certain conditions of the engine control software in this type of helicopter. While there is no evidence that a technical malfunction caused the crash, the House of Lords is sufficiently concerned that it recently agreed to form a select committee to investigate the circumstances (Collins, 2001).

2.2.2 A client's perspective

By client we mean that person, or group of people, responsible for paying for the development of a new information system. A client usually has influence over whether or not approval is given to a project before it starts. Some clients (but not all) also have the power to stop a project once it is underway. A client may also be a user. If so, we can assume that they share the user's perspective on the kind of things that can be a problem. They may make only indirect use of the system's outputs, which insulates them from the immediate experience of a badly designed interface, for example. While the concerns of a client may overlap with those of an end-user, they also include distinct matters relating to payment, ownership and value-for-money.

'If I'd known the real price, I'd never have agreed'

It is almost routine in many organizations for information systems projects to exceed their budget. A survey of approximately 200 large US organizations found that 'the typical project was one year late and 100% over budget' (Capers Jones, 1986). Any project that overruns its budget can reach a point where the total costs outweigh the benefits that will be provided on completion. This point is not always recognized when it is reached; this may result in the expensive completion of a system that might have been better cancelled. In other cases the project is cancelled, either because it is recognized that costs are escalating out of control, or because it becomes apparent the benefits will not be as great as originally promised. The rationale is summed up in the saying: 'Don't throw good money after bad'. One of the most striking examples in recent years was the London Stock Exchange electronic trading system known as Taurus, cancelled after three years' development.

> The City of London was shocked by the sudden cancellation of Project Taurus. The London Stock Exchange had spent [. . .] £80 million in the venture. The securities industry had invested over £400 million developing their own system in readiness for Taurus. On Wednesday 11 March 1993 the Stock Exchange publicly admitted it had all been for nothing.
>
> (Drummond, 1996)

'It's no use delivering it now—we needed it last April!'

A project that is completed late may no longer be of any use. For example, a bricks-and-mortar retailer, threatened by rivals who sell at a lower price on the Internet, may have little use for an e-commerce website if it is not operational until all the customers have defected and the company has been declared bankrupt.

Many other kinds of project are time-critical. This can be due to new legislation that affects the organization's environment. An example of this was the de-regulation of the UK electricity supply market in April 1998. This required electricity companies to make extensive modifications to their computer systems so that they would be able to handle customers' new freedom to switch between suppliers. A few years earlier, all

Local Authorities in the UK faced a similar challenge twice in three years, when Central Government changed the basis for local tax calculations. Each change required hundreds of councils to specify, develop (or purchase) and successfully install new computer systems that allowed them to produce accurate invoices and record income collected from local tax-payers. Failure to implement the new systems in time risked a massive cashflow problem at the beginning of the new tax year.

Commercial pressures can also have an effect. This sometimes translates into whether a business succeeds in being the first to market a new product or service, although the advantage is not always permanent. For some time the continuing success of the Internet bookstore Amazon.com derived, at least in part, from the considerable competitive advantage of being the first of its kind. Some competitors (notably the established US bookseller Barnes and Noble) felt obliged to follow Amazon's lead. For the followers, there is not the same need to take risks with new technology. But attracting customers away from a leader may mean differentiating yourself in some way, perhaps offering new services, or perhaps by being even better at what the leader already does well. At the time of writing (Spring 2001), and in the wake of the many dot.com crashes during the year 2000, some press reports suggest that Amazon's commanding lead may be faltering, but it is still far from certain how this particular contest will end.

'OK, so it works—but the installation was such a mess my staff will never trust it'

Once a new system gets a bad press it can be very difficult to overcome the resistance of the people who are expected to use it, particularly if there is an existing alternative available to them. The following scenario is based on a real situation, observed at first hand by one of the authors.

> A small company introduced a local area network (LAN) to connect the PCs in its office. Staff were encouraged to store files on the LAN disk-drive, where other staff could also access them (previously, all data was stored on local hard drives, accessible only from one PC). Most saw the mutual benefit of sharing information, and complied. Management claimed that the daily back-up of the LAN drive onto tapes was a further benefit, since there was no longer a need to keep personal back-ups on floppy disks. Then a mechanical fault occurred. This erased all data on the LAN drive, and when the engineer tried to restore it from the tape, it emerged that the tape drive had not operated correctly for several weeks. All tapes recorded over the previous six weeks were useless. Staff were told that all data stored in that time was permanently lost. Re-entering it all took many person-days. The faulty disk and tape drives were replaced, and tapes are now checked after every back-up, but many staff reverted to keeping all important data on their local hard drives. Most keep personal back-ups on floppy disks too. Perhaps nothing will persuade them to trust the LAN again.

'I didn't want it in the first place'

Organizations are complex and political by nature. The politics with which we are concerned here are to do with conflicting ideals and ambitions, and the play of power within the organization. There can be disagreement between management and workers, as in the case of the warehouse management system mentioned earlier in this section. There can also be contention between individual managers, and between groups of managers. One result can be that a manager is sometimes an unwilling client in relation to a project. The following scenario is based on another real-life situation recently observed by one of the authors.

The head office of a multinational company decided to standardize on a single sales order processing system in all its subsidiaries throughout the world. But the Hong Kong office already had information systems that linked electronically with customers in Singapore, Taiwan and other places in South East Asia. It became apparent that the existing links would not work with the new system. For the Hong Kong management, this meant extra costs to make the new system work, and disruption to established relationships which, in their view, already worked smoothly and did not need to be changed. They therefore had little desire to see the project succeed in their region, but felt they had no other choice. Had they been less scrupulous, they might have tried to find ways of sabotaging its progress, either in the hope that it would be abandoned altogether, or at least that they might be exempted from the global rule.

'Everything's changed now—we need a completely different system'

It is almost inevitable for any information system project that, by the time the system is completed, the requirements are no longer the same as they were thought to be at the beginning. Requirements can change for many reasons.

- Project timescales can be very long (the Taurus project ran for three years), and business needs may change in the meantime.
- Users naturally tend to ask for more, as they learn more about what is available.
- External events can have a dramatic impact—for example a currency crisis that led to the collapse of a significant market might render a new information system pointless.

This does not apply only to new systems currently under development. Systems that have been in operation for some time may also be affected. This is part of the natural, ongoing process of maintenance, modification, upgrading and eventual replacement that all information systems undergo, and a topic to which we return in Chapters 3 and 19. From a client's perspective, the motivation is usually to make an information system fit better with the business, and therefore provide better support for business activities.

2.2.3 A developer's perspective

The perspective of the developer is quite different both from that of an end-user and from that of a client. This is because the developer adopts the role of 'supplier' to the 'customer' (i.e. client or end-user). For this reason, when problems occur the developer may feel forced into a defensive position of justifying the approach taken during the project. Since at this stage we are discussing only problems, many of the problems identified by a developer tend to centre on blame and its avoidance.

'We built what they said *they wanted'*

Changes to the requirements for a system, based on sound business reasons, always seem perfectly reasonable from a client's point of view. However, for a developer, given the responsibility for building a system to meet those requirements, they can be a real headache. If we were able to distil the essence of how many developers feel about this, it would read something like the following.

No matter how skilled you are, you can't achieve anything until the users, clients, etc., tell you what they want, and at the start they don't even agree with each other. Eventually, with skill and perseverance, you produce a specification with which everyone is reasonably happy.

> You work for months to produce a system that meets the specification, and you install it. In no time at all, users complain that it doesn't do what they need it to do. You check the software against the specification, and you find that it does exactly what it was supposed to do. The problem is that the users have changed their minds. They just don't realize that it's not possible to change your mind late in a project. By then, everything you have done depends on everything else, and to change anything you would almost have to start all over again. Or it turns out that they didn't understand the specification when they accepted it. Or there is some ambiguity about what it meant, and you've interpreted it differently from them. Whatever the reason, it's always your fault, even though all you ever tried to do was to give them what they wanted.

In reality, of course, analysts, programmers, etc. often do understand why users and clients change their minds during a project, or after delivery. But this doesn't always make it less frustrating when it happens.

'There wasn't enough time to do it any better'

In every project, there are pressures from outside that limit the ability of the development team to achieve excellence. In the first place, projects almost invariably have a finite budget, which translates directly into a finite amount of time to do the work. There may also be an externally-imposed deadline (in a Year 2000 project for example). Another external pressure results from the impatience of users and clients to see tangible results. This, too, is often understandable, since they are not so much concerned with the information system itself, as with the benefits it can bring them—an easier way to do a tedious job, a quicker way to get vital information, and so on. But it can be very counter-productive, if it becomes a pressure within the project team to cut short the analysis and get on with building something (anything!) quickly to keep the users happy. The result of haste in these circumstances is usually a poor product that meets few of the needs of its users. Developers know this, but they don't always have the power to resist the pressure when it is applied.

'Don't blame me—I've never done object-oriented analysis before!'

In a successful information system development team, the members must possess a harmonious blend of skills that are appropriate to the needs of the project. These may include the use of techniques (such as object-oriented analysis), knowledge of methodologies (such as the Unified Software Development Process), skill in programming languages (such as Java) or detailed knowledge of hardware performance (such as networking devices). There must be a complementary set of skills within the team for a project to succeed. For example, the analysts must all use the same or related techniques otherwise the results could not be coherent (e.g. the products of object-oriented analysis are entirely incompatible with those of structured analysis). Similarly, all programmers must be able to interpret the designs or else they will not know what the class operations are meant to do. Problems occur when the available staff do not possess adequate expertise in the particular skills required for a project. Many highly skilled and experienced systems analysts may know SSADM very well (a structured methodology used widely in the UK), but have little experience of object-oriented analysis. Many excellent programmers know COBOL inside out, but (even today) have little experience in Java or Smalltalk. As a consequence, some projects with highly skilled staff are still carried out rather poorly, because the staff are inexperienced with the particular techniques they must use.

'How can I fix it?—I don't know how it's supposed to work'

This complaint is often heard from programmers who have been asked to modify an existing program, and who have then discovered that there is no explanation of what it does or how it works. To modify or repair any artefact, whether a computer system or a bicycle, it is usually necessary to understand how it was intended to work, and thus what the consequences are of changing this or that feature. Anyone who has ever tried to repair anything electronic or mechanical, such as a motor vehicle, washing machine, or VCR, will know that much of the time is spent trying to understand what the various parts do, and how they interact with each other. This is true even when a maintenance manual is to hand. The situation is no different for computer software. While software may be more intangible in form than a VCR, it is no less mechanistic in its operation.

'We said it was impossible, but no-one listened'

Just like client managers, systems developers can sometimes be overwhelmed by organizational politics. At times this means that a project is forced on an unwilling team, who do not believe that it is technically feasible to achieve the project's goals. Alternatively, the team may not believe the project can be completed within the time made available. But if opposing views prevail, the team may find itself committed to trying to achieve what it said could not be done. In these circumstances, it can be very hard for team members to become enthusiastic about the project.

'The system's fine—the users are the problem'

A few information systems professionals, usually those who understand least about business and organizations, are sometimes prone to blame 'the user' for everything. Those who hold this attitude believe that problems that occur in the use of software inevitably result from the fact that most users are far too stupid or ignorant to make proper use of the system. On the other hand, they believe that the design and execution of the software is not open to serious question. Many of these technocrats are undoubtedly very talented, but this view is patently absurd since it assumes that the answer to a problem is known before the situation has even been investigated. In a word, it is a prejudice. We will say no more about it, other than to comment that anyone who hopes to learn the truth about a situation must also be prepared to examine critically their own preconceptions.

2.3 Why Things Go Wrong

Flynn (1998) proposed an analytical framework to categorize project failures, and this is illustrated in Figure 2.2.

Complete failure is the most extreme manifestation of a problem, but Flynn's framework can also be applied to less catastrophic problems. In Flynn's view, projects generally fail on grounds of either unacceptable quality, or poor productivity. In either case, the proposed system may never be delivered, it may be rejected by its users or it may be accepted yet still fail to meet its requirements.

These categories are what are sometimes called 'ideal types'. This means that they are intended to help explain what is found in reality, but that does not imply that any

Type of failure	Reason for failure	Comment
Quality problems	The wrong problem is addressed	System conflicts with business strategy
	Wider influences are neglected	Organization culture may be ignored
	Analysis[3] is carried out incorrectly	Team is poorly skilled, or inadequately resourced
	Project undertaken for wrong reason	Technology pull or political push
Productivity problems	Users change their minds	
	External events change the environment	New legislation
	Implementation is not feasible	May not be known until the project has started
	Poor project control	Inexperienced project manager

Figure 2.2 Causes of IS project failure (adapted from Flynn, 1998).

real example, when examined in all its detail, will precisely match any one category. Real projects are complex, and their problems can seldom be reduced to one single cause. Many of the examples in the following sections show some of the characteristics of more than one category of cause.

2.3.1 Quality problems

One of the most widespread definitions of the quality of a product is in terms of its 'fitness for purpose' (see any standard text on software quality assurance, e.g. Ince, 1994 or Sanders and Curran, 1994). In order to apply this to the quality of a computer system, clearly it is necessary to know (i) for what purpose the system is intended, and (ii) how to measure its fitness. As we shall see, both parts of this can be problematic at times.

The wrong problem

Here the emphasis is on the purpose for which a new system is intended. The argument goes like this: if an information system does not contribute anything worthwhile to the aims of the organization, then it is, at best, a waste of resources and a distraction from those things that really matter. At worst, it may do real harm if the objectives of an information system are in direct conflict with the organization's business strategy. The difficulty is in knowing the right viewpoint to take when defining the aims of a project. If the aims of the organization as a whole are unclear, or are not communicated to those responsible for planning information system projects, there is always a risk that a project will fall into this error. It may then be regarded as a complete success by its developers and users, yet appear a failure when seen in a wider frame of reference.

A primary cause of system failures in general is that some projects are started with no clear idea about exactly what are the nature and goals of the client organization. This means that failure, or at least lack of success, is almost inevitable. If an organization itself is not understood, then it is extremely difficult to identify and develop information systems that support it in fulfilling its aims.

[3] To this category, we would add design and implementation. Even when analysis is carried out correctly, this is still no guarantee that the software will be well designed, nor that it will be correctly programmed.

Neglect of the context

This emphasizes the fitness of an information system to fulfil its purpose. This can take the form of a system that is too difficult to use, since the designers have taken insufficient account of the environment in which its users work, or the way that they like to work. Some examples given earlier in this chapter can be interpreted in this way, depending on assumptions about the situation. For example, in one case cited earlier (Section 2.2.1) workers and managers held different views about the purpose of the warehouse management system. Managers believed they needed to control the activities of workers more closely. Yet the system designed to do this had also the side effect of obstructing the workers from carrying out their work in an efficient way, to the detriment of the whole company.

Incorrect requirements analysis

We prefer to interpret this category as including, not only analysis, but also design and implementation. Here, too, the focus is on a system's fitness to fulfil its purpose, rather than the purpose itself. Even if the aims are clear at the outset, many pitfalls lie along the route, particularly if the development team does not have the right skills, the right resources or enough time to do a good job. However, even when none of these present a difficulty, the project can still fail if the techniques being applied by the team are inappropriate to the project.

The results of this category of failure are generally among the most visible systems problems, when seen from a user's point of view. This may be because they directly affect the external design of the system (e.g. the content or layout of its screens), or the selection of tasks that the system performs (e.g. an essential function may not be included), or the operation of the software (the system as coded may not work in the way that its analysts and designers intended).

Project carried out for the wrong reason

Here, the emphasis is once more on the intended purpose of the system. To give a recent example, over the last few years many organizations seem to have rushed into some sort of e-commerce activity. While this is clearly a great success for some, others have derived little or no benefit. During the year 2000 there seemed to be so many 'dot.com' crashes that the *Guardian* newspaper's website ran a column called 'Dot.com deathwatch', which featured only stories about troubled and failing Internet companies.

It has been pointed out that many organizations did not think carefully enough about some key questions (McBride, 1997).

- What does the business aim to achieve by a presence on the Internet?
- How must the business re-organize to exploit the opportunities offered by the Internet?
- How can the business ensure that its presence on the Internet is effective?

The sensible conclusion must be that there is a great deal more to successful trading on the Internet than just writing a few web pages in HTML, and placing them on a web server where surfers can find them. Some organizations, in moving onto the World Wide Web, were simply following a trend. They do so, not because they understand what it can do for their business, but precisely because they do not understand, and therefore fear the consequences of being left out of something good.

Two possible underlying reasons explain why this can happen. One of these is political push within the organization. For example, a powerful group of managers may wish the business to look modern, even when no clear benefit has been identified. The other is the pull of new technology. An organization is very vulnerable to this if the most senior managers have little understanding of information technology, and therefore no rational basis for evaluating the exaggerated claims that vendors are prone to make about their newest products. In practice these two reasons often combine into a single force that can be irresistible.

Other companies, typically Internet start-up businesses with no history of trading in the physical world, seem simply to have believed that the Internet represented a 'new economy' where established business rules no longer applied. However, while some aspects of business on the Internet are clearly different—in particular, the speed at which things happen, including business failures—the business fundamentals have not really changed. It is still just as important to plan and design with care, to pay attention to costs and income and to ensure that proper controls are in place.

2.3.2 Productivity problems

Productivity problems relate to the rate of progress of a project, and the resources (including time and money) that it consumes along the way. If quality is the first concern of users and clients, then productivity is their other vital concern. The questions that are likely to be asked about productivity are as follows.

- Will the product be delivered?
- Will it be delivered in time to be useful?
- Will it be affordable?

Requirements drift

Requirements almost always drift if they are allowed to do so. In the simplest case, this just means that users' requests change over time. In principle, it would be unreasonable to prevent this from happening, but, in extreme cases, change requests can bedevil a project and even prevent its completion. The longer a project proceeds, the more complex both its products and its documentation become. To compound this, each part of the final system probably depends on many others, and the inter-dependencies grow more numerous and complex. It becomes progressively more difficult over time to make changes to a system under development, because any change to one part requires changes to many other parts so that they will continue to work together. A limit is reached when a project is stalled by an escalating workload of making all the changes required as a consequence of other changes that were requested by the users. At this point, management have only two choices. They can cancel the project and write off the money spent so far (this happened at the London Stock Exchange in 1993). Alternatively, an effort can be made to bring the project back on track. This is almost always both difficult and expensive, and requires highly skilled management.

External events

This is one cause of failure that is normally beyond the control of both project team and higher management. Depending on the environment in which the organization operates, decisive external events can even be impossible to anticipate. Nevertheless, it is prudent on any project at least to assess the vulnerability of the project to external events, since some are much more at risk than others. For example, a project to build a distributed information system that is to operate on new, state-of-the-art computers communicating over public telephone circuits may be sensitive to external factors such as the reliability of the telephone network and call pricing. By contrast, a project to build an information system that will operate on existing, tried and tested hardware within one building can safely ignore these factors.

Poor project management

The manager of a project is ultimately responsible for its successful completion, and it could therefore be argued that any project failure is also a failure of the project management. To some extent this is true, but there are also some cases where the only identifiable cause of failure overall is a failure of management. This is almost always due to either poor planning at the start, or a lack of care in monitoring progress against the plan. As a result, the manager allows the project to falter, or permits its costs to grow in an uncontrolled way.

Implementation not feasible

Some projects are over-ambitious in their technical aims, and this may not become evident until an attempt is made to implement the system. This is particularly the case when a system is intended to work together with other systems, whether or not these are already in use. The problems of testing and debugging a new system can grow steadily more complex as attention is focused on larger and larger sub-systems. Sometimes the task of interfacing several large, complex software systems, written in different programming languages, installed at different sites and running on different makes of computer hardware, can turn out to be impossible to achieve.

An example of the kind of complex installation routinely tackled in many modern projects is a £25m system called Caseman, which has computerized many administrative tasks in the 250 County Courts of England and Wales. This system is based on an Oracle database combined with the WordPerfect word-processing package, running on the UNIX operating system on DEC computers. Although each of these components, taken independently, is thoroughly tried and tested, nevertheless implementation was delayed because the company responsible for the project encountered problems in making the different parts work together (Schneider, 1997). A further difficulty occurs when each supplier insists that their product is performing exactly as specified, and any problem must lie elsewhere. It can then be very hard to locate the exact origins of the problem.

Technical problems with the implementation do not always become evident until after the system is implemented. One example of this was with the LAS ambulance despatch system. A practical difficulty that staff encountered in its use was with the deliberately high-tech design that used a digital on-screen map:

> . . . which showed the location of the ambulances. The staff hated it and said it was dangerous [. . .] The problem with a map system is accuracy.
>
> (Barker, 1998)

This was a contributory factor to the system's overall failure to speedily despatch ambulances to emergencies, resulting from the application of a new technology to a critical task that was not sufficiently understood.

Another recent implementation problem led to the crash of the online sportswear retailer Boo.com in May 2000. The software for their website was much delayed in development, but proved a disaster even when delivered. It turned out that very few home PCs were sufficiently advanced to run the sophisticated 3D visualizations without crashing. Even when the software ran without crashing, most images were very slow to download, adding to the users' frustration. As a result, too few customers bought from the website. The company called in the liquidators after reportedly spending £80m over six months, with no realistic prospect of sales increasing to the point where the business would become viable (Hyde, 2000).

Problems of this kind can usually be avoided by sound design practices. Beginning with Chapter 12, we consider later what the software designer can do to avoid costly and damaging failures.

2.4 The Ethical Dimension

Ethics can be loosely defined as a branch of philosophy that is concerned with the rightness or wrongness of human character and conduct, and the establishment of moral rules or principles to guide our behaviour. In practical terms, thinking about ethics normally means that we are trying to establish a way of judging the effects that one person's behaviour has on other people.

Given that all computer-based information systems have a direct effect on someone's life, it is hard indeed to think of one that does not have a significant ethical dimension to its design, construction or use. As with any system of rules, it is necessary to consider the consequences if they are broken. It is not overstating the case to say that a breach of the ethical rules that apply to any aspect of an information system project is necessarily a problem for someone. Sometimes these problems are noticeable at the time and must be handled within the project, otherwise they can lead to a failure of the project overall. For example, a system intended to automate some activities of a business may result in job losses among the staff. At other times there is hidden damage to a project, an organization, the information systems profession as a whole or even in extreme cases to society at large. For example, there is a debate about whether individual users of e-mail systems have a right to privacy from surveillance. Some evidence in the prosecution case that led to the impeachment proceedings against President Clinton was gathered from personal e-mails. Investigating officers were apparently able to retrieve copies of certain sensitive messages, even though their author believed that all copies had been safely destroyed (*Wall Street Journal*, 1998).

One of the difficulties in assessing the ethical issues in a project is that the person who may have a problem is not necessarily the developer of the system, its user, its client—or indeed anyone at all who is obviously connected with the project. In fact, one of the first problems that must be faced in this area is the identification of all the people who may be affected by the system. These are often called *stakeholders*, since, in their different ways, each has a stake in the outcome of the project.

To illustrate the diversity of stakeholders who may be associated with a project, consider the introduction of a network of bank ATM machines in the branches of a supermarket chain. Figure 2.3 shows a preliminary tracing out of the possible effects of this system, together with the groups of people affected.

You may not feel that every group identified in this analysis is equally affected by the project, and in some cases you may feel that the effect is quite minor. However, until such an analysis has been carried out, it is not possible to speak with any authority about the nature or the extent of impact that a new information system will have.

Stakeholder affected	Possible consequence of system	Nature of effect on stakeholder
Bank clerks	Automation of bank activities currently carried out manually	Reduced need for staff—redeployment or redundancies
Bank customers	More convenient access to bank services	Improved service
Supermarket customers	More people using supermarket car park	Reduction in service
Bank shareholders	More people attracted to use bank, so greater commercial success	Increased dividends
Supermarket shareholders	More people attracted to use supermarket, so greater commercial success	Increased dividends
Local citizens	More journeys to supermarket to use ATM	Increased pollution

Figure 2.3 Possible stakeholders in a bank ATM network.

2.4.1 Ethical issues within a project

The issue of professionalism is at the forefront of any discussion of ethics within information systems development. By their very nature, information systems projects have deep consequences for the lives and work of many people who will either use the software, or will be affected in some way by its use. There are several reasons for this. First, information systems projects are often major investments for the client organization, and money spent on these is necessarily money that now cannot be spent on other worthwhile projects. Second, they are often concerned with the way that important business activities are carried out, and they can therefore have a direct effect on the overall success or failure of the organization. Third, the introduction or modification of information systems often causes radical changes in the way that employees carry out their work, and how they relate to their managers, colleagues and customers. Given these responsibilities, it is important that project team members behave in a professional manner. To some extent this is just a matter of being aware of the rules for behaving in a professional way, and then following them.

Behaving ethically is usually not just a matter of applying a straightforward rule in a mechanical way. We often find ourselves confronting ethical dilemmas, where no one course of action seems entirely right, yet we must find some basis for making a decision. This is compounded when our actions affect different stakeholders in different ways.

Some ethical effects arise from aspects of a project that are not under the direct control of an individual developer. For example, Sachs (1995) describes a 'Trouble Ticketing System' that was intended to improve the work efficiency of testers (telephone repair staff) by tracking the progress of repairs and allocating the next step in a job to any tester that was available at the time. In practice, the system disrupted informal communication between employees, thus damaging their efficiency instead of improving it. According to Sachs, the fault lay in a failure to develop an adequate understanding of the testers' working practices, in particular the way that their informal communications helped in troubleshooting difficult problems. But individual developers are rarely free to choose which aspects of a situation should be analysed.

Legislation adds a further ethical dimension for the members of an information system project team. In the UK the relevant legislation includes the Data Protection Act 1998, the Computer Misuse Act 1990 and the Health and Safety (Display Screen)

Regulations 1992. The increasing use of the Internet as a medium for information and exchange also brings cross-border complications. One recent example was the widely reported 'Internet adoption' case, which involved a British couple who adopted twin baby girls advertised on a website based in the US. A British court later declared the adoption to be invalid and the twins were returned to the care of their natural father in the US. In many cases it is currently far from clear which laws apply when information and services are provided via the Internet to residents of one country, but either the service provider or the information content are hosted within a different country.

2.4.2 Wider ethical issues

Among the wider ethical issues associated with information systems development, one of the longest-running debates is over the effect that IT and IS have had on levels of unemployment throughout the world. Some authors claim that, within the foreseeable future, one effect of the spread of information technology will be an unprecedented, sustained rise in global unemployment that will leave countless millions idle and impoverished (Rifkin, 1995). Others argue that, while IT has certainly destroyed some jobs (tens of thousands of jobs lost in the UK banking sector over the last decade), it has also created many new jobs to replace them, often bringing new opportunities to neglected backwaters in the world economy.

For many years, an ethical debate has raged between the proponents of freedom (some would say anarchy) on the Internet, and the big business interests who already own its infrastructure and may soon dominate most of its content. In its brief history, the Internet has been at the centre of many debates revolving around freedom of access, including the widespread concerns that have been expressed about the ready availability of pornography, political materials and other contentious content. Yet for many others, this chaotic situation has brought unprecedented opportunities to meet (in a virtual sense) and communicate with similarly minded people all over the world.

One issue that is likely to receive more attention in the next few years is the way that some companies use new applications of information technology to gather unprecedented quantities of detailed personal information. For example, at the time that the first edition of this book was being prepared, a South African bank had recently begun to issue its customers with mobile phones that displayed their current account balance each morning. The phones also enabled the bank to monitor all phone numbers dialled, and this added to the bank's profile of data about individual customers. Customers are categorized, and those that the bank feels are least profitable are 'encouraged' to leave the bank, e.g. by higher service charges. The bank was said to be considering plans to add a geographic information system capability, which would also allow customers' daily movements to be tracked, and a link with Internet providers, which would allow the bank to ascertain which websites are visited by their customers (Collins, 1998a).

Questions of access to computers and computerized information have wide implications. There have been arguments over the years about whether the use or the availability of computers have tended to favour men over women, the middle class over the working class, and those in the affluent north over those from the poor south of the globe. Some even suggest that computers and the Internet have already begun to create a new division in the world's population, dividing people everywhere into the information-rich and the information-poor.

So many things are changing so fast at present that it is not possible to be sure how these questions will seem in a few years time. What is clear is that future generations will look back on our time as one of great change. Such an era is inevitably one that raises profound ethical questions about the effects on our fellow citizens of the way that we design and apply technology.

2.5 Costs of Failure

Many projects discussed in this chapter, particularly the more famous ones (Taurus, the LAS despatch system and Boo.com), have very large price tags attached. Some companies are widely thought to be reluctant to admit their failures, since this can reflect badly on the business as a whole, perhaps damaging customers' or investors' confidence. For this reason, it is thought that the known, high profile failures represent only the visible tip of a much larger iceberg. But these are not the only costs associated with project problems. Many costs are associated with projects that have not failed outright, but which, for a variety of reasons, do not fully meet their requirements.

A system that is poorly designed or functionally limited has many consequences for its users. If we take interface design as an example, an ill-considered screen layout (for example one that compels users to switch back and forth between two screens as they enter data) can have effects that range from mild irritation, through increased error rates, absence due to sickness, to greater staff turnover. Each has an associated cost—but it may be hard to measure accurately, and in some cases may not be recognized at all. If an employee leaves to take another job one reason may be his or her frustration with a poorly designed computer system.

System reliability can be important in determining overall costs. Recall the motor insurance company mentioned above in Figure 2.1, where customers were asked to resubmit personal details, as a computer system crash had caused most customer data to be lost. This is not a reassuring picture for any insurance company to give its customers, and it is a safe conclusion that many will have switched to another insurer as a result. Some will also have told friends and colleagues, who in turn will be less likely to use this company in the future. It is very unlikely that anyone could accurately determine the full cost to this business in terms of lost customers. In the case of Boo.com, a technically inappropriate implementation resulted in the complete failure of the business.

Some of the more routine effects that can occur are summarized in Figure 2.4.

The list of effects in Figure 2.4 is by no means exhaustive. But, in showing how poor

Design aspect	Example	Immediate effects	Other consequences
User Interface	Illogical screen layout. Difficult to read screens. Unhelpful help messages.	Wasted time. Increased frustration. Increased error-rate.	Loss of confidence in system. Increased sickness. Increased absenteeism. Greater staff turnover.
Program Execution	System response is slow.	As above.	Increased operating costs.
Data Storage	Lost data. Inaccurate outputs.	Extra work re-entering data. Extra work checking outputs.	Reduced income. Loss of customer confidence. Lost sales.

Figure 2.4 Some hidden costs of poor design.

design and operational characteristics of an information system can affect the organization as a whole, it reinforces the importance of getting things right before the problems have a chance to appear. This is true even if we ignore the ethical consequences of some information systems applications under development or already in use today. The full social cost of some ethical issues outlined in Section 2.4 is probably incalculable.

2.6 Summary

In this chapter we have looked at the issue of project failure from many different perspectives, including those of the people who use information systems, the people who purchase them and the people who build them. We have also outlined some of the deeper causes of project and system failure, and considered the costs and wider ethical issues.

The failures in information systems development can teach valuable lessons, and, moreover, ignoring a difficulty does not make it go away. Rather, it increases the likelihood of repeating past mistakes. So it is important to understand as much as possible about what can go wrong with an information system development project, the better to be able to avoid it. But after so much concentrated attention on the many dead ends that our unluckier (or more careless) predecessors have encountered, it seems appropriate to end the chapter on a final upbeat note. Two of the failures discussed earlier have since become success stories.

The London Stock Exchange has successfully installed an online share trading system. The new system is much simpler than Taurus—and not all users are entirely happy with its functionality—but it does meet its basic requirements. It was introduced in April 1997, on time and within budget (Philips, 1997).

The London Ambulance Service also has a new ambulance despatch system, introduced five years after its disastrous predecessor was scrapped in 1992.

> Now the service's emergency call room plays host to visitors from around the world who come to see the successful implementation of an emergency response system in one of the biggest emergency call centres in the world. Indeed, the new system won the prestigious BCS Elite Group Information Systems Management Award in May [1997].
>
> (Barker, 1998)

We can also remind ourselves of the many positive contributions that modern information systems make to our lives, enabling so many things that were not possible for earlier generations. It is difficult to imagine an aspect of life in modern society that is not facilitated in some way by a computerized information system. Our goal, then, should be to ensure that tomorrow's systems build on the successes of the past, while avoiding the problems and failures, as far as it is in our power so to do.

Review Questions

2.1 Why do users, clients and developers disagree on the nature and causes of the problems in information systems development?

2.2 What are the main underlying causes of problems in information systems development?

2.3 Define quality.

2.4 What are the main differences between quality problems and productivity problems?

2.5 Why do the requirements drift once a project is underway?

2.6 What can be the results of ignoring the organizational context of an information system?

2.7 Define stakeholder.

2.8 What ethical issues might be involved in setting up an online shopping system that has links to an organization's management information systems?

Exercises, Case Study Work and Projects

2.A Do some research in computing trade journals, and find examples of recent projects that failed or ran into difficulties. Draw up a table with four separate columns. Give these titles like: 'Nature of problem', 'Who sees it as a problem', 'Probable cause' and 'Flynn category'. Enter the projects into your table, using your own intuition to complete column 3. Then complete the 4th column using Flynn's categories (summarized in Figure 2.2). How do your causes compare with Flynn's categories?

2.B The British Computing Society (BCS) publishes a Code of Conduct for its members, who include thousands of computing and information systems professionals in the UK. In the US, similar codes are published by the Association for Computing Machinery (ACM) and the Institute of Electrical and Electronic Engineers (IEEE). In many other countries around the world, including India, Zimbabwe and Singapore, there is a national professional society with an equivalent code of professional ethics. Write down some ethical problems associated with the development of an information system to support staff who despatch ambulances to medical emergencies, and use this to identify a list of issues you would expect to be covered in a professional code of conduct for information systems developers. Then obtain a copy of the BCS Code (or, if you are a reader in another country, your nearest local equivalent). Compare it to your list of issues. What are the main differences?

2.C Write down all the stakeholders who you think are associated with an emergency ambulance despatch system. How are they affected?

2.D Review the ethical issues you identified in Exercise 2.B, and identify one or more issues that appear as a problem from the perspective of one stakeholder, but do not appear as a problem from the perspective of another stakeholder.

2.E Find out what legislation applies to information systems development activity in your country, and what implications it has for developers. Would the South African bank customer profiling system described in Section 2.4.2 be fully legal under these laws? Which particular aspects of the system make it ethically questionable, and from whose perspective?

Further Reading

- Trade magazines such as *Computing* and *Computer Weekly* regularly publish articles that report on problems in current projects as they occur.
- Sauer (1993) discusses reasons for project failures in terms of the project's organizational environment, and includes several practical case studies that illustrate the problems in practice.
- Ince (1994) is one of many books that deal with software quality assurance (SQA), a discipline that endeavours to ensure that software systems fully meet their defined requirements. The latest edition of Pressman's standard software engineering text (2000) also has a good chapter on SQA.
- De Montfort University's Centre for Computing and Social Responsibility runs a website that focuses on ethical issues in information systems development. This has many links to other sites, and also carries the full text of a number of academic papers in this field. It can be found at www.ccsr.cse.dmu.ac.uk.

CHAPTER 3

Avoiding the Problems

OBJECTIVES

In this chapter you will learn

- the stages in the waterfall life cycle
- about prototyping and incremental life cycles
- the importance of project management
- how users may be involved in a project
- the role of CASE tools in systems development.

3.1 Introduction

Problems that commonly affect the development of computerized information systems have been described in Chapter 2. These problems can significantly reduce the success of a software development project, and it is clearly important to adopt strategies and procedures that will minimize their occurrence. In this chapter we consider the steps that can be taken to make a computerized information system development project as successful as possible.

The problems identified in Chapter 2 fall into two major categories, those that are concerned with the management of the project and those that relate to poor quality in the delivered system. These are the two major areas on which we should focus if we wish to produce information systems within budget, on time and providing the required functionality. We discuss project lifecycles in Section 3.2 and consider some of the issues regarding the management of information systems development in Section 3.3.

There is much debate about which approach to information systems development is most likely to produce a quality system. One major source of difficulty is the inherent complexity of software development. In an object-oriented approach complexity is managed in such a way that it helps to address this problem.

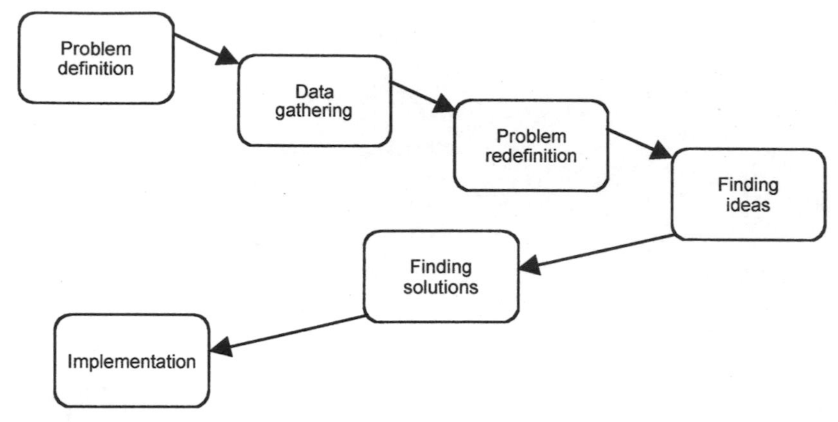

Figure 3.1 General problem solving model (adapted from Hicks, 1991).

From a more general perspective, building computerized information systems can be viewed as a form of problem solving. Figure 3.1 shows a general problem solving model adapted from Hicks (1991). The phases *Data Gathering* and *Problem Redefinition* are concerned with understanding what the problem is about; the *Finding Ideas* phase attempts to identify ideas that help us to understand more about the nature of the problem and possible solutions. *Finding Solutions* is concerned with providing a solution to the problem and *Implementation* puts the solution into practice. This approach to problem solving divides a task into sub-tasks, each with a particular focus and a particular objective.

The information systems development process may be subdivided simply into three main tasks, identifying what is required, planning how to deliver what is required and delivering what is required. There are many other ways of subdividing an information systems project but they all include an analysis activity that identifies what the system should do, a design activity that determines how best to do it and some construction that builds the system according to the design. The phases that contain these activities are given various names but the core activities remain the same. For example, Larman (1998) suggests a development process that has the three phases *Plan and Elaborate*, *Build* and *Deploy*.

An important consequence of subdividing the development process is that techniques and skills specific to the different phases can be identified. Teams of developers with these specialized skills can be allocated to the particular phases or activities, maximizing the chance that the activities are completed as well as possible. The importance of user involvement is considered in Section 3.4. Subdividing the development process results in smaller tasks that can be managed more easily to achieve the appropriate quality standards and to stay within the allocated resource budget. The importance of methodological approaches to systems develoment are discussed in Section 3.5.

A further source of help to overcome the problems the software developer faces is computer support. Software development is an activity that, like many others, may benefit from computerized information systems. It is interesting that computerized support for software development did not became realistic in terms of cost and technology until the mid-1980s. It is now accepted that computerized support is an important component of a development project (Allen, 1991). This is covered in more detail in Section 3.6.

We have already alluded to the benefits to be gained from managing the software development process effectively and have explicitly identified poor project management as a source of many of the problems. Building a software system is very different from building almost any other human artefact. Software is intangible, it cannot be weighed, its strength cannot be measured, its durability cannot be assessed, its resistance to physical stress cannot be gauged. Of course we try to find (and with some success) measures of a software system that enable us to make judgements about its size, its complexity, the resource required to build it, and so on. But these measures are much less well understood than their counterparts in the design and construction of such tangible artefacts as buildings and cars.

3.2 Project Life Cycles

Subdividing the process of software development produces what is known as a life cycle model. Just as an animal goes through a series of developmental stages from its conception to its demise so, it is argued, does a computerized information system. Various project life cycles can be applied to computerized information systems development. We will discuss some of the most commonly used life cycle models. Arguably two activities precede the information systems development project, strategic information systems planning and business modelling. The successful completion of these activities should ensure that the information system that is developed is appropriate to the organization. It can be argued that these are part of the information systems development life cycle. However, their focus is not on computerization in itself but rather the identification of organizational requirements. Whether they are viewed as part of the life cycle or not, their importance is almost universally accepted for commercially oriented computer systems development. If a system is not commercially oriented (e.g. an embedded control system) strategic information systems planning and business modelling are not precursors to systems development. There is a distinction to be made between systems development, where a system may incorporate human, software and hardware elements, and software development which focuses primarily on software construction, although it involves the human users and the hardware upon which it executes. It is perhaps a matter of perspective. Thus a software development project is, by definition, focused solely on producing a software system that will satisfy the user requirements whereas strictly speaking a systems development project has a wider scope and may not even include software as part of the solution.

Strategic Information Systems Planning. As we saw in Chapter 1, information systems work within the context of an organization and must satisfy its current requirements as well as providing a basis from which future needs can be addressed. In order to do this, strategic plans are developed for the organization as a whole and within their context a strategic view of information systems needs can be formed. For example, in the Agate case study a strategic decision may be made to target multinational companies for international advertising campaigns. This has consequences for campaign management and its supporting information systems.

Business modelling. In order to determine how an information system can support a particular business activity it is important to understand how the activity is performed and how it contributes to the objectives of the organization. Campaign management is an important business function for Agate and it should be modelled in order to determine how it is carried out, thus providing some of the parameters for subsequent information systems development.

3.2.1 Traditional life cycle

A diagrammatic view of the traditional life cycle (TLC) for information systems development is shown in Figure 3.2. This model is also known as the waterfall life cycle model (the difficulty of returning to an earlier phase once it is completed has been compared to the difficulty of swimming up a waterfall). The individual phases of a TLC are described in the following sections.

System Engineering. An information system involves human, software and hardware elements. The first stage of an information systems project is to identify the major requirements for the whole system and then to identify those parts of the system that are best implemented in software, those parts that are best implemented in hardware and those components that should be allocated to human participants. This phase produces a high-level architectural specification that defines how these major parts of the system will interact with each other.

Requirements Analysis. All the system's requirements need to be defined clearly and requirements gathering becomes more intensive. If the project is concerned largely with the development of elements that will be implemented in software then these are the main focus of requirements analysis. The objective is then to identify what the users would require from the software elements of the system. Requirements gathering for large or complex systems benefits from fact-finding techniques that help to identify the requirements, and further techniques to document them appropriately and to analyse them[1]. User requirements can also be used to develop the criteria for user acceptance tests.

Design. Once it is known what is required from the system, the design process determines how best to construct a system that delivers these requirements. Design is first concerned with the specification of a software architecture that defines the

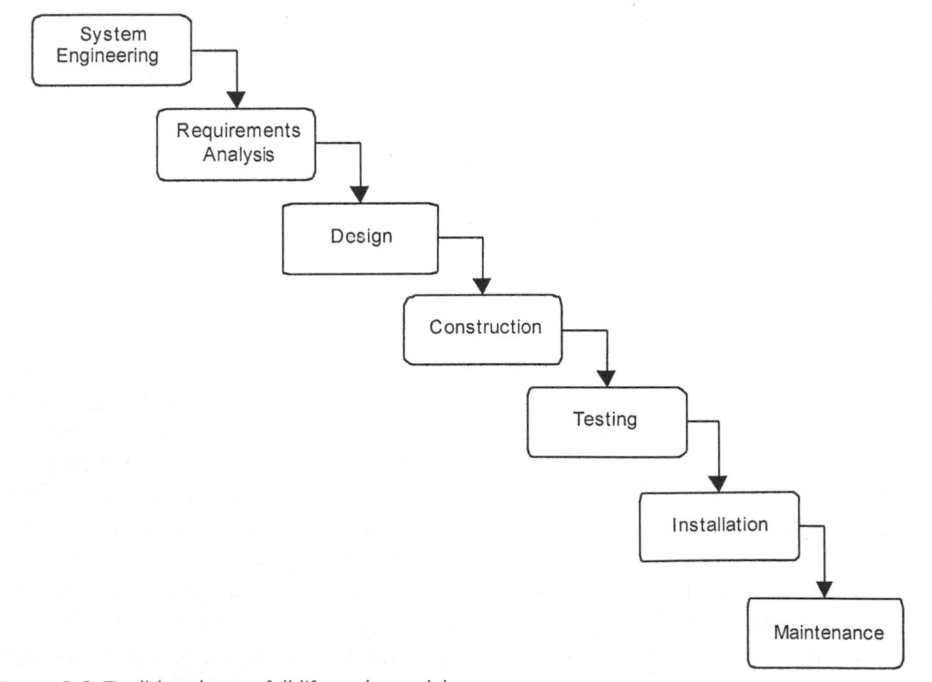

Figure 3.2 Traditional waterfall life cycle model.

[1] We describe the techniques provided by UML for these purposes in Chapters 5–8.

major software components and their relationships. Design involves reaching a balance between requirements that conflict with each other (e.g. maintainability and performance) within implementation environment constraints. Human activities may also be designed at this stage.

Construction. The completed design is now translated into program code. Depending on how the design phase has been completed, part of the construction may be automated. Construction may utilize different programming languages and database management systems for different parts of the system.

Testing. The system is tested to ensure that it satisfies the user requirements accurately and completely. Typically, several levels of testing are performed. Individual components are tested independently, then are tested together as a sub-system and then the sub-systems are tested together as a whole system. Often users perform some form of acceptance testing before the system is finally accepted as complete.

Installation. Once the system has been tested satisfactorily it is delivered to the customer and installed for use. The introduction of the system has to be managed carefully so as not to cause unnecessary disruption and to minimize the attendant risk of change. For example, one approach is to run both the old and new systems in parallel to ensure that the new system operates effectively before discontinuing the operation of the old system. This is an expensive strategy and it may be impractical to adopt such a fail-safe approach. Nonetheless a contingency plan appropriate to the level of risk should be in place. These aspects of project management are discussed further in Chapter 21.

Maintenance. It is most likely that the system will be subject to change during its operating life. The delivered system may operate erroneously and corrections may have to be made to the software (corrective maintenance). Certain aspects of the system's behaviour may not have been fully implemented (because of cost or time constraints for instance), but are then completed during the maintenance phase (perfective maintenance). The operating environment may also change in various ways causing requirements changes that have to be accommodated (adaptive maintenance).

There are many variations of the waterfall model (e.g. Pressman, 1997 and Sommerville, 1992), differing chiefly in the number and names of phases, and the activities allocated to them. A major justification for life cycle models like this is that their phases have explicitly defined products or *deliverables*. Sommerville (1992) suggests a series of deliverables produced by different phases of development, shown in Figure 3.3.

These products can be used to monitor productivity and the quality of the activity performed. Several phases have more than one deliverable. If we need to show a finer level of detail to assist in the monitoring and control of the project, phases can be split so that each sub-phase has only one deliverable. Alternatively, a phase may be viewed as comprising a series of activities each of which has a single deliverable and can be managed individually. Project management issues are discussed in Chapter 21. An organization that embarks upon systems development should specify phases in the systems development process and their associated deliverables in a way that is applicable to the organizational context and the nature of the systems being developed.

The traditional life cycle has been used for many years but is the subject of several criticisms.

- Real projects rarely follow such a simple sequential life cycle. Project phases overlap and activities may have to be repeated.
- Iterations are almost inevitable, because inadequacies in the requirements analysis may become evident during design, construction or testing.

Phase	Output deliverables
System Engineering	High level architectural specification
Requirements Analysis	Requirements specification Functional specification Acceptance test specification
Design	Software architecture specification System test specification Design specification Sub-system test specification Unit test specification
Construction	Program code
Testing	Unit test report Sub-system test report System test report Acceptance test report Completed system
Installation	Installed system
Maintenance	Change requests Change request report

Figure 3.3 Life cycle deliverables (adapted from Sommerville, 1992).

- A great deal of time may elapse between the initial systems engineering and the final installation. Requirements will almost inevitably have changed in the meantime and users find little use in a system that satisfies yesterday's requirements but may hamper current operations.
- The TLC tends to be unresponsive to changes in client requirements or technology during the project. Once architectural decisions have been made (e.g. during systems engineering) they are difficult to change. A technological innovation that may make it feasible to automate different parts of the whole system may become available after the project has been running for some time. It may not be possible to incorporate the new technology without re-doing much of the analysis and design work already completed.

The diagram in Figure 3.4 shows possible paths for iteration within the TLC but these iterations can be very costly. For example, if a major deficiency in the requirements analysis is discovered during construction iterating back to requirements analysis may result in significant redesign making some of the construction so far completed inappropriate. If problems such as this do occur, project constraints may result in the missed requirements being addressed without a suitable redesign but instead using an *ad hoc* coding solution. This approach tends to produce unmaintainable systems. However, despite its drawbacks the TLC does provide a very structured approach to systems development that has the following advantages.

- The tasks in a particular stage may be assigned to specialized teams. For example, some teams may specialize in analysis, others in design and yet others in testing.
- The progress of the project can be evaluated at the end of each phase and an assessment made as to whether the project should proceed.
- The controlled approach can be effective for managing the risks on large projects with potentially high levels of risk.

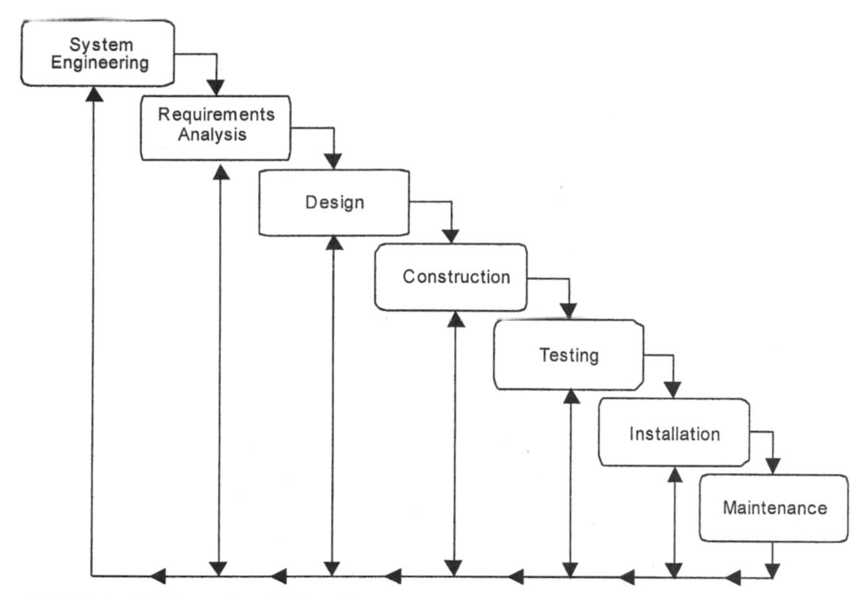

Figure 3.4 Waterfall life cycle with iteration.

3.2.2 Prototyping

Many approaches to systems development incorporate some iteration. For instance, analysis may involve a series of tasks that are repeated iteratively until the analysis models are deemed complete. The waterfall model discussed so far aims to deliver a final working system as the finished product. However such approaches have a significant difficulty, in that the user only actually experiences how the system operates once it is delivered. Sometimes users find it difficult to imagine how their requirements will be translated into a working system and it is certainly the case that different implementations can be produced from the same set of requirements. The prototyping approach overcomes many of the potential misunderstandings and ambiguities that may exist in the requirements.

In software development a prototype is a system or a partially complete system that is built quickly to explore some aspect of the system requirements and that is not intended as the final working system. A prototype system is differentiated from the final production system by some initial incompleteness and perhaps by a less resilient construction. If the prototype is to be discarded once it has fulfilled its objectives the effort required to build a resilient prototype would be wasted. A prototype will typically lack full functionality. It may have limited data processing capacity, it may exhibit poor performance characteristics or may have been developed with limited quality assurance. Prototype development commonly uses rapid development tools, though such tools are also used for the development of production systems.

Prototypes may be constructed with various objectives in mind. A prototype may be used to investigate user requirements as described in Chapter 5. For example, a prototype may be focused on the human–computer interface in order to determine what data should be presented to the user and what data should be captured from the user. A prototype might also be used to investigate the most suitable form of interface. A prototype may be constructed to determine whether a particular implementation platform can support certain processing requirements. A prototype might be concerned with determining the efficacy of a particular language, a database management system or a communications infrastructure.[2] A life cycle for prototyping is shown in Figure 3.5.

The main stages required to prepare a prototype are as follows:

- perform an initial analysis
- define prototype objectives
- specify prototype
- construct prototype
- evaluate prototype and recommend changes.

These are described in more detail below.

Perform an initial analysis. All software development activity utilizes valuable resources. Embarking upon a prototyping exercise without some initial analysis is likely to result in an ill-focused and unstructured activity producing poorly designed software.

Define prototype objectives. Prototyping should have clearly stated objectives. A prototyping exercise may involve many iterations, each iteration resulting in some improvement to the prototype. This may make it difficult for the participants in a prototyping exercise to determine if there is sufficient value to continue the prototyping. However, with clearly defined objectives it should be possible to decide if they have been achieved.

Specify prototype. Although the prototype is not intended for extended operation it is important that it embodies the requisite behaviour. It is almost certainly the case that the prototype will be subject to modification and this will be easier if the software is built according to sound design principles.

Construct prototype. Since it is important that prototype development is rapid, the use of a rapid development environment is appropriate. For example, if an interactive system is being prototyped, environments such as Delphi™ or Visual Basic® can be most effective.

Evaluate prototype and recommend changes. The purpose of the prototype is to test or explore some aspect of the proposed system. The prototype should be

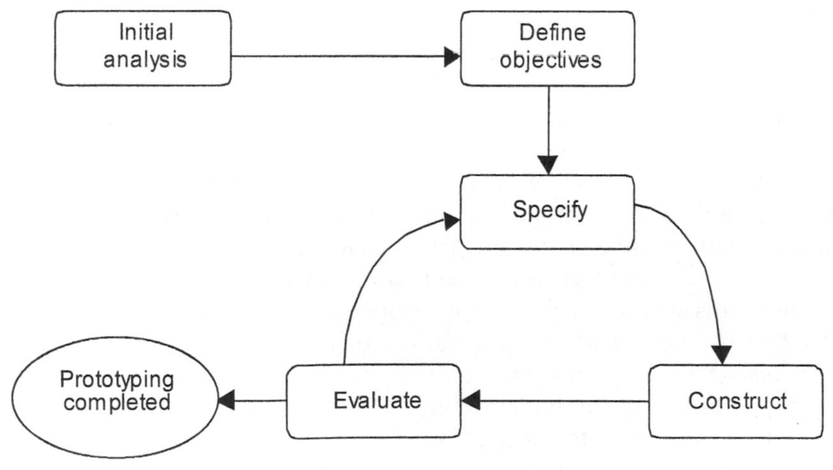

Figure 3.5 A prototyping life cycle

[2] For example, an Object Request Broker (these are described in Chapter 18).

evaluated with respect to the objectives identified at the beginning of the exercise. If the objectives have not been met then the evaluation should specify modifications to the prototype so that it may achieve its objectives. The last three stages are repeated until the objectives of the prototyping exercise are achieved.

Prototyping has the following advantages:

- early demonstrations of system functionality help identify any misunderstandings between developer and client;
- client requirements that have been missed are identified;
- difficulties in the interface can be identified;
- the feasibility and usefulness of the system can be tested, even though, by its very nature, the prototype is incomplete.

Prototyping also has several problems and their impact on a particular project should be estimated before engaging in prototyping:

- the client may perceive the prototype as part of the final system, may not understand the effort that will be required to produce a working production system and may expect delivery soon;
- the prototype may divert attention from functional to solely interface issues;
- prototyping requires significant user involvement;
- managing the prototyping life cycle requires careful decision making.

3.2.3 Incremental development

Several suggestions on how to overcome the problems with the TLC have been made. Gilb (1988) suggests that successful large systems start out as successful small systems that grow incrementally. An incremental approach performs some initial analysis to scope the problem and identify major requirements. Those requirements that will deliver most benefit to the client are selected to be the focus of a first increment of development and delivery. The installation of each increment provides feedback to the development team and informs the development of subsequent increments.

Boehm's (1988) spiral model can be viewed as supporting incremental delivery. However, Gilb (1988) argues that the spiral model does not fully support his view of incremental development, as there are aspects of systems development that it does not emphasize or include. These are as follows:

- the production of high-value to low-cost increments;
- the delivery of usable increments of 1% to 5% of total project budget;
- a limit to the duration of each cycle (e.g. one month);
- a measure of productivity in terms of delivered functionality or quality improvements;
- an open-ended architecture that is a basis for further evolutionary development.

Figure 3.6 shows how Boehm's spiral model can be adapted to suit incremental delivery. Note that prototyping may be used either during the risk analysis or during the software development part of the development cycle.

3.2.4 The Unified Software Development Process

The Unified Software Development Process (USDP) (Jacobson et al., 1999) reflects the current emphasis on iterative and incremental life cycles. It builds upon previous approaches by Jacobson (1992), Booch (1994) and Rumbaugh (1991). The USDP incorporates the UML and comprises much good advice on software development. The USDP will be discussed in more detail later (Chapter 22); however an overview of its lifecycle is presented here.

A development cycle for the USDP is illustrated in Figure 3.7 and comprises four phases:

- *Inception* is concerned with determining the scope and purpose of the project;
- *Elaboration* focuses requirements capture and determining the structure of the system;
- *Construction*'s main aim is to build the software system;
- *Transition* deals with product installation and rollout.

A particular development cycle may be made up of many iterations. In Figure 3.7 there are two iterations in the inception phase and four in the construction phase. This is purely illustrative and the number of iterations in each phase is determined on a project by project basis. At the end of each iteration an increment is delivered and its composition may range from elements of a requirements model to working program code for a portion of the system. In the USDP an increment is not necessarily additive, it may be a reworked version of a previous increment.

The diagram also illustrates that a phase may involve a series of different activities or workflows. This is different from the waterfall life cycle in which each phase largely comprises a single activity. The inception phase may include elements of all of the workflows, though it is likely that design, implementation (i.e. constructing the software) and test would be focused on any necessary exploratory protyping. However, most commonly inception would primarily involve the requirements and analysis workflows.

Texel and Williams (1997) suggest a detailed object-oriented life cycle model that includes 17 stages. A life cycle with such a large number of delineated stages is probably more suited to large projects. If used for small projects the management overhead is likely to be excessive.

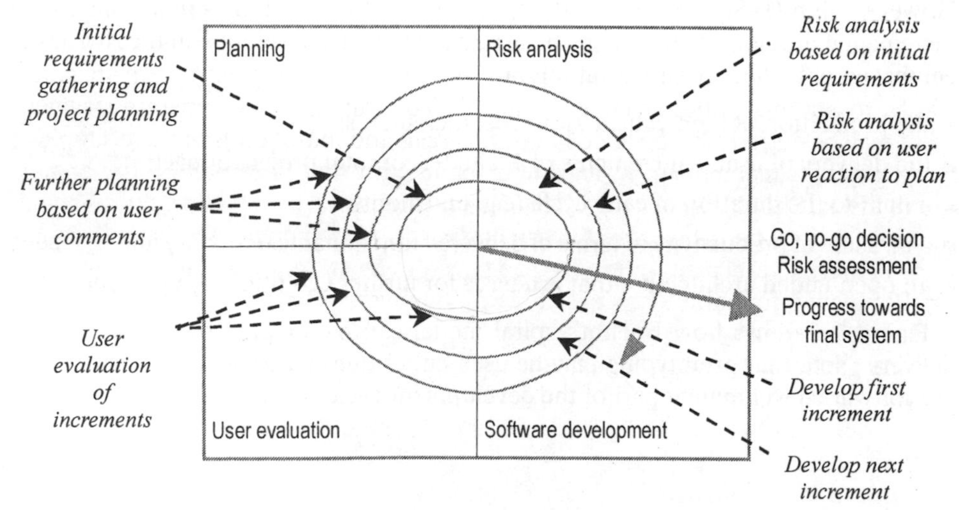

Figure 3.6 Spiral model for incremental delivery (adapted from Boehm, 1988).

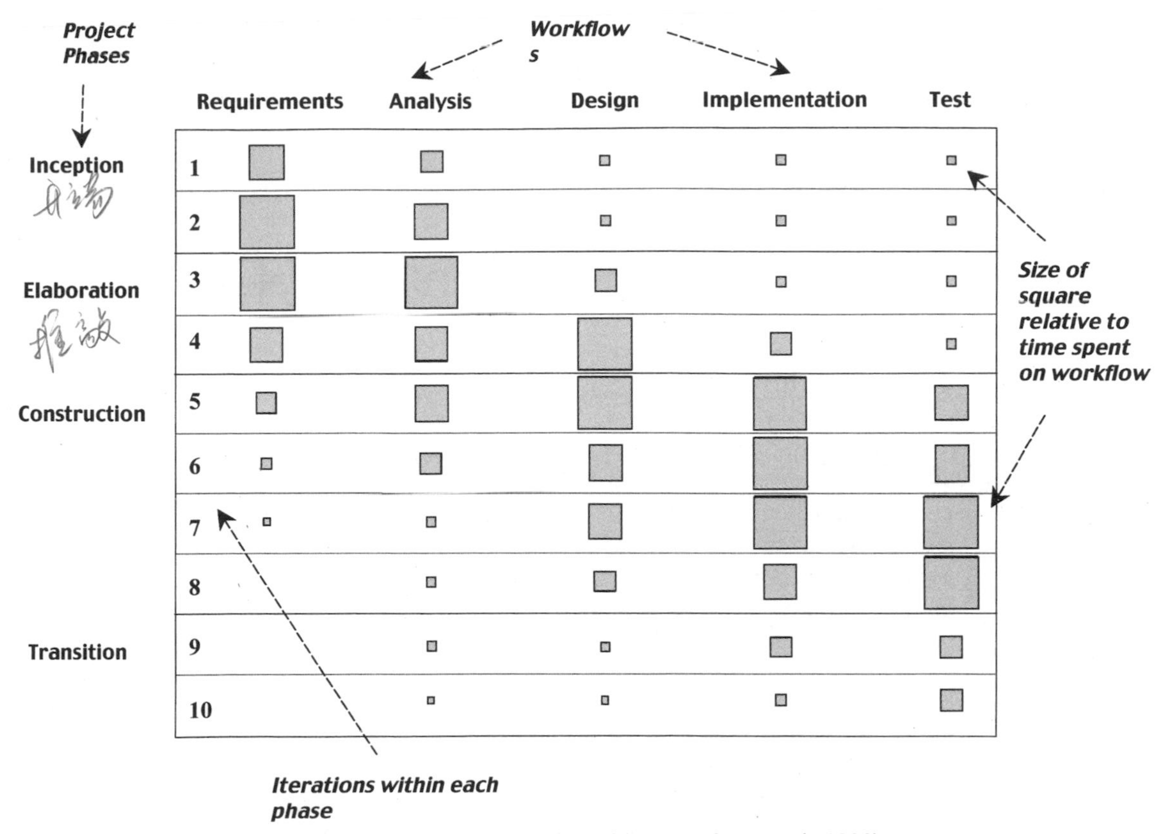

Figure 3.7 The Unified Software Development Process (adapted from Jacobson et al., 1999).

3.3 Managing Information Systems Development

An information system development project has many attendant risks. Late delivery may result in lost income and missed opportunities for the client, and punitive legal action against the software developers. The continuing ability of an enterprise to compete effectively may be dependent upon a new information system. Failure to deliver a new system may even cause a company to fail. There are also risks associated with adopting a new information system. For example, suppose an organization chooses to automate the payment of all invoices. An automatic invoice payment system may be designed to balance the commercial and moral duties to pay suppliers promptly with a requirement to maximize the interest that is earned on the organization's assets. Should the system fail to achieve this balance in practice, the company may suffer serious consequences. On the one hand, late payment of an invoice might lead to damaging legal action. Of course, one way to avoid this particular risk is not to adopt the new information system but the alternatives will also have attendant risks and costs. For instance, a manual system for processing invoices may have higher costs and greater risk of error. On the other hand, if invoices are paid too quickly the company's interest earnings will be reduced. Use of any new information system must be controlled and monitored to ensure that it continues to operate within the parameters defined by the company.

Organizations are increasingly dependent on the operation of computerized information systems to manage the business, to the point where a new system that interferes with the successful operation of existing systems may prevent the company from being able to operate at all. As a project proceeds, more detail becomes available as to what the proposed system should do and how it will operate. It is important at each stage to identify risks to the organization and determine their significance so that judgements can be made on the continuing viability of the project. If a new system is thought likely to place the organization at risk of a catastrophic failure then either the system should be redesigned to avoid this (perhaps at significant extra cost) or the project should be cancelled. Note that in Figure 3.6 each iteration includes a risk analysis that determines whether the next cycle should proceed.

3.4 User Involvement

A key factor in maximizing the chance of success is ensuring that there is continued and effective user involvement throughout the project. The traditional waterfall life cycle is less amenable to user involvement throughout the whole project and hence is less likely to stay in line with the changing user requirements. A prototyping approach is normally dependent upon continuing user involvement and by its very nature encourages user involvement. However, care has to be taken to ensure that the users have sufficient time to perform their roles effectively. In prototyping, the evaluation of the prototype requires significant time from the users.

Users can be involved in projects at various levels and can play many different roles. Some approaches to software development (e.g. DSDM[3]) directly involve users in the development team, so that they then have a significant opportunity to influence the way the project proceeds, perhaps by identifying difficulties and suggesting more acceptable alternatives. It is important that users who have such a significant influence on the direction of the project should understand the organizational requirements and the needs of fellow users. Direct involvement of users is more likely to be successful if they are considered as full members of the project team and if they are genuinely empowered to represent the organization and make decisions within clearly defined operating parameters. There is always a danger that users who become members of a project team cease over time to represent the user perspective effectively, as they begin to view themselves more as a team member and less as a user representative. One way of overcoming this tendency is to rotate membership of the development team among a group of users. However, this can result in a loss of continuity. A more satisfactory approach is to scope each activity so that a user team member can see it through to completion in a reasonably short time, say within three months.

At the other end of the spectrum are users whose only involvement in the project is as participants in fact gathering. Such a user may provide information about current working practice but they have little or no influence on the design of the new system. In this situation users are likely to be concerned about the project and may fear the effect it will have on their jobs. As a result, they may be less co-operative and will be less willing to take ownership of the new system when it is installed.

Even when users are not invited to join the project team, effective participation can still be encouraged through a consultative approach. Procedures are set up so that users are able to review certain aspects of the systems development, provide feedback, give their views of the system and then be able to see the response of the project team to their feedback. In some circumstances a large part of the task of requirements

[3] DSDM is introduced in Chapter 21.

capture may be delegated to users. They are likely to feel a strong affinity to the delivered system, provided that it satisfies the requirements they specified.

Whatever form of involvement users have with the project it is important that their role is carefully explained and that training is given as required. Some large organizations have gone to the trouble of training users so that they understand the terminology and models used by the systems developers. Furthermore, users must be given the time to participate. It is no good expecting users to review requirements documents effectively if the only time they have available is during their lunch break.

Users who participate in a systems development project can be selected in various ways. They can be designated by management as being the most appropriate representatives or they can be selected by their peers. In either circumstance they must be genuine representatives of the user perspective.

3.5 Methodological Approaches

One of the major influences on the quality of the systems developed is the software development approach adopted. If the approach used is not appropriate for a particular type of application then it may limit the quality of the system being produced. In part at least, the object-oriented approach provides a mechanism for mapping from real-world problems to abstractions from which software can be developed effectively. Furthermore object-orientation provides conceptual structures that help to deal with modelling complex information systems. As information systems requirements are becoming increasingly complex the use of an object-oriented approach is more necessary. It is a sensible strategy to transform the development of a large, complex system into the development of a set of less complicated sub-systems. Object-orientation offers conceptual structures that support this sub-division. Object-orientation also aims to provide a mechanism to support the reuse of program code, design and analysis models.

A methodology consists of an approach to software development (e.g. object-orientation), a set of techniques and notations (e.g. the Unified Modelling Language–UML) that support the approach, a life cycle model (e.g. spiral incremental) to structure the development process and a unifying set of procedures and philosophy. In this text we do not espouse a particular methodology but apply object-oriented techniques in a co-ordinated fashion using UML. Methodologies are discussed in more detail in Chapter 22.

3.6 CASE

Computer Aided Software Engineering (CASE) tools have been widely available since the mid-1980s and now provide support for many of the tasks the software developer must perform. CASE tools have been categorized in various ways according to the phase in the life cycle in which they should be used. Upper-CASE tools provide support for the analysis and design while lower-CASE tools are concerned with the construction and maintenance of software. These two categories of tools directly support the overall development process. Other computer tools address the tasks involved in the management of information system development projects.

Early CASE tools could be categorized clearly as upper-CASE or lower-CASE but this distinction is much less true these days when CASE tools offer facilities that support many life cycle phases. Typically CASE tool vendors offer a series of products that are integrated to offer extensive life cycle coverage. Such CASE tools are known as I-CASE (the I stands for Integrated). CASE tools tend to be designed for use with a particular approach to software development, particular notations and implementation environments. In the 1980s and early 1990s CASE tools were being developed to support an apparently ever increasing range of different development approaches. The growth in popularity of object-oriented approaches has exacerbated the situation with the emergence of many new methodologies with different notations. Although it is acknowledged that no single approach to software development is appropriate for all projects it has become accepted that a common notation set can be used for requirements and design modelling. The move to a common notation has resulted in the development of the Unified Modelling Language, which it must be emphasized, is not a methodology. Modern CASE[4] tool-sets provide an increasingly wide range of facilities and cover most life cycle activities. These are discussed in turn below.

3.6.1 Diagram support

Most approaches to information systems development utilize diagrams to represent the relationships between different elements of the requirements and to document the design structure before the system is constructed. Analysis and design diagrams are models of the system and of its requirements. Diagrammatic analysis and design techniques were used in information systems development for many years before computerized support became available. However the difficulties in preparing diagrams by hand and then ensuring that they remain up to date whenever the system is modified meant that in practice they quickly became inconsistent. The automated diagram support offered by CASE tools appears to overcome this problem. However, although the process of creating and maintaining diagrammatic models of the system are much eased by CASE tools, they must still be enforced by quality assurance procedures.

Computer support for preparing and maintaining diagrams requires various features which include:

- checks for syntactic correctness;
- repository support;
- checks for consistency and completeness;
- navigation to linked diagrams;
- layering;
- traceability;
- report generation;
- system simulation;
- performance analysis.

These features are described in turn below.

Syntactic correctness. The CASE tool checks that the correct symbols are being used on the diagrams and that they are being linked in permissible ways. This ensures that the correct vocabulary is being used but does not ensure that it is meaningful or relevant to client requirements.

[4] These facilities are similar to those suggested for Integrated Project Support Environments (IPSE) in the 1980s.

Repository. Originally repositories were only designed to hold the definitions of the data attributes of a system. They were then termed data dictionaries because in the same way that a language dictionary provides clear definitions of the words in a language, the data dictionary contains definitions of all the elements in an information system. However, increasingly it became clear that all elements in a system development project are open to ambiguous interpretation unless clearly defined in a repository of information about the system. A repository may contain diagrams, descriptions of diagrams and specifications of all the elements in the system. Some CASE vendors use the term encyclopaedia instead of repository.

Consistency and completeness. Most CASE tools support various diagrammatic models that capture different aspects of the system. As all relate to the same system it is important that any one element that appears on several diagrams or models (perhaps viewed from different perspectives) should be consistent with itself. Most approaches to analysis and design stipulate that certain diagrams must be completed and that the elements referred to in those diagrams must all be documented in the repository. To manually check the consistency and completeness of a system of any significant size is a task that is very onerous, time-consuming and error-prone. A good CASE tool may check the consistency and completeness of a large model in seconds and provide the developer with a comprehensive report on any inconsistencies found or omissions identified.

Navigation to linked diagrams. A complex system is likely to require many diagrammatic models to describe its requirements and its design. For a CASE tool to be usable, easy navigation between linked diagrams is essential. For example, double-clicking on a component at one level of abstraction may automatically open up a diagram that describes it at a more detailed level. It is also helpful to be able to move directly from one view that contains a particular element to another view that contains the same element.

Layering. An information system of any significant size is by nature complex and it is unlikely that all relationships between its components can be shown on a single diagram. Just as maps are drawn at different scales with different levels of detail, system models are produced at various levels of abstraction. A high-level diagram may represent the relationships between large components such as sub-systems. A diagram drawn at a lower level of abstraction may describe the elements within a particular component in detail. In order to cope with complexity, we divide the system into manageable chunks and link them in layers. A good CASE tool provides a capability to layer the models of the system at different levels of abstraction. The consistency and completeness checking discussed earlier should also check that the representations of one element at different levels of abstraction are consistent.

Traceability. Most of the elements created during the development of an information system are derived from other elements, and the connections between them should be maintained. It must be possible to trace through from the repository entries that describe a particular requirement to the program code that provides the functionality that satisfies the requirement. If a requirement changes, the maintenance activity is easier if all the code that implements that requirement can be readily identified. It should be possible to trace all requirements from the analysis documentation, through the design documentation to the implemented code. This feature is known as requirements traceability.

Report generation. Complex systems involve modelling many elements. Comprehensive reporting capabilities improve the usability of a CASE tool by ensuring that the developer can easily obtain information about the models for a system in suitable formats. In fact a CASE tool would be of little use if the information it held about a project were not readily available, no matter how effective it was in other respects.

System simulation. When a CASE tool has been populated with models of an application it should be possible to simulate some aspects of system behaviour. For example, how does the system respond to a particular event? Some CASE tools provide capabilities that enable a software developer to examine the consequences of a design decision without the need to actually build the software.

Performance analysis. The performance of a system is an important ingredient in its success. For example, a system that supports staff who deal directly with customer enquiries should be able to respond quickly to a query about the availability of a certain product. If customers are kept waiting for too long, this will probably result in lost sales. The analysis of performance is particularly difficult for an application that runs on multiple processors and uses a complex communications infrastructure. Some CASE tools even provide the capability to perform a 'what if' analysis to examine the implications of alternative implementation architectures.

3.6.2 Software construction

CASE tools can offer a range of features to support software construction and maintenance. These include code generation and maintenance tools.

Code generators. The capability to generate code directly from a design model is a major benefit to the developer for several reasons. First, a working software system is likely to be produced more quickly. Second, one source of error is largely removed when the code is produced automatically and consistent with the design. Third, when requirements change, a consequent change to the design documentation can be followed by automatic code generation. If the application logic is defined completely and precisely in the design model, full code generation is possible. If a design model contains detailed operation specifications[5] then it is likely that a code framework can be generated to which further code can be added. In order to reduce the level of detail required for the design model, code generators may make certain assumptions concerning the implementation. Code generators are available for many different languages and development environments and are likely to include the capability to generate database schemata for the major proprietary database management systems.

Maintenance tools. Software maintenance is a major issue. All systems are subject to change as the enterprise changes, perhaps in response to legislative change. For example, the introduction of the Euro in Europe has necessitated significant change to many software systems. Various tools are available to help with systems maintenance. For some programming languages, reverse engineering tools may also be available that can generate design documentation directly from program code (although if the program code is poorly structured the resulting design documentation may be of little use). Tools are also available that can analyse program code and identify those parts that are most likely to be subject to change.

3.6.3 Benefits and difficulties of using CASE

CASE tools can bring many benefits to the development activity. They help to standardize the notation and diagramming standards used within a project, and this aids communication among the team members. They can perform automatic checks on many aspects of the quality of the models produced by analysts and designers. The

[5] These define how the system will function.

report generation capabilities of a CASE tool reduce the time and effort that needs to be spent by analysts and designers in retrieving data about the system upon which they are working. Where a CASE tool can carry out automatic code generation, this further reduces the time and effort that is required to produce a final system. Finally, the electronic storage of models is essential to the reuse of models, or components of them, on other projects that address similar analysis or design problems.

Like any other technology, CASE tools also have their disadvantages. These include limitations in the flexibility of the documentation that they can provide. However, some case tools now include the capability to specify and tailor documentation templates to suit particular reporting requirements. The development approach may also be limited by the need to work in a particular way in order to fit in with the capabilities of the CASE tool. The ability of a CASE tool to check all models for their consistency, completeness and syntactic correctness can in itself give rise to a danger. Developers may make the erroneous assumption that, because their models are correct in those specific senses, they are therefore also necessarily relevant to user requirements. There are also certain costs attached to the installation of a CASE tool. Aside from the cost of the software and manuals, there is also likely to be a significant cost in additional training for developers who will be expected to use the CASE tools.

On balance CASE tools can provide useful and effective support for the software development activity, but it requires appropriate management for this to be achieved without any damaging side-effects.

3.7 Summary

We have considered how to avoid the problems that typically arise during information systems development. Several strategies have been discussed. Life cycle models are used to provide structure and management leverage for the development process itself. User involvement is crucial to ensure relevance and fitness for purpose of the delivered system. Furthermore, many of the difficulties that occur during installation are limited if ownership of the proposed system has been fostered by effective participation. The modelling requirements of the activities involved in the software development are well supported by object-orientation. Finally we considered the importance of CASE tool support for the software developer.

Review Questions

3.1 What are the advantages of the traditional waterfall life cycle?

3.2 What are the disadvantages of the traditional waterfall life cycle?

3.3 How are some of the disadvantages listed in your answer to Question 3.2 overcome?

3.4 What is prototyping?

3.5 How does prototyping differ from incremental development?

3.6 What are the different ways of involving users in the systems development activity? What are potential problems with each of these?

3.7 How do 'syntactic correctness', 'consistency' and 'completeness' differ from each other?

3.8 What does requirements traceability mean?

3.9 Why is it not enough for a diagram to be syntactically correct, consistent and complete?

3.10 What is the purpose of a repository?

Case Study Work, Exercises and Projects

3.A Read the Case Study Chapter B1. What life cycle model would you recommend for the development of the production control system for FoodCo? Justify your decision.

3.B For a CASE tool with which you are familiar, explore and critically assess the consistency and completeness checking facilities available.

3.C For a CASE tool with which you are familiar, explore and critically assess its system generation capabilities.

3.D In your library find references for three life cycle models not discussed in this chapter. Briefly review each of these life cycle models.

Further Reading

- Hicks (1991) provides a comprehensive introduction to problem solving skills that are valuable to the software developer. Sommerville (1992) and Pressman (1994) provide good discussions of life cycle issues. Gilb (1988) contains much good advice concerning software development and is well worth reading.
- Texel and Williams (1997) offer a detailed object-oriented methodology specifying deliverables and tasks clearly.
- Jacobson, Booch and Rumbaugh (1999) provide a description of the USDP and further information the Rational variant of the USDP can be found at www.rational.com.
- Many CASE tools are in widespread use and increasingly are providing support for the UML standard, though the styles of implementation do vary. One of the commonly used tools is Rational Rose™ from Rational Corporation, the company that developed UML. More information can be obtained from the website: www.rational.com.

CHAPTER 4

What Is Object-Orientation?

OBJECTIVES

In this chapter you will learn

- the fundamental concepts of object-orientation
- the justifications for an object-oriented approach
- how object-orientation is used in practice.

4.1 Introduction

It is a major premise of this book that the object-oriented approach to systems development helps to avoid many of the problems and pitfalls described in earlier chapters. In this chapter we lay the foundations for understanding object-orientation by presenting an explanation of the fundamental concepts. This is important for all readers who are new to object-orientation, but particularly so for those who are already familiar with other approaches to systems analysis and design, for example, structured analysis. There are important differences between the object-oriented approach and earlier approaches. It is necessary to have a sound grasp of the basic concepts in order to be able to apply the techniques of object-orientation in an effective way.

At this stage the most important concepts to grasp are object, class, instance, generalization and specialization, message passing and polymorphism. We explain each of these concepts in turn, showing both what the parts of an object-oriented system are and how they use message passing to isolate different parts of a system effectively from each other, thus controlling the complexity of the interfaces between sub-systems (Section 4.2). The discussion is necessarily conceptual in places, but practical examples and analogies are used to illustrate the theoretical points wherever this is appropriate. We then place object-orientation in its historical context, and look at some of the reasons why it has been able to make a contribution to successful information systems development (Section 4.3). Finally, we conclude by summarizing the state of evolution that object-oriented languages have reached today (Section 4.4).

4.2 Basic Concepts

The most important concept addressed in this section is the *object* itself, and it is to this that we first pay attention. The other concepts explained in this section are strongly dependent on each other, and all contribute to an adequate understanding of the way that objects interact, and thus to their significance for information systems. The following explanations concentrate on giving an understanding of the territory that later chapters will explore in greater depth.

4.2.1 Objects

In an early book on object-oriented analysis and design, Coad and Yourdon (1990) define *object* as follows:

> Object. An *abstraction* of something in a problem domain, reflecting the capabilities of the system to keep information about it, interact with it, or both.

This may not immediately appear to help very much, as parts of the definition are themselves a little obscure and raise further questions. In particular, what exactly do we mean by 'abstraction', and which 'system' are we talking about?

A useful definition of abstraction in this context might be: 'A form of representation that includes only what is important or interesting from a particular viewpoint.' To give a common example, a map is an abstract representation. No map shows every detail of the territory it covers (impossible, in any case, unless it were as large as the territory, and made from similar materials!). The intended purpose of the map guides the choice of which details to show, and which to suppress. Road maps concentrate on showing roads and places, and often omit landscape features unless they help with navigation. Geological maps show rocks and other sub-surface strata, but usually ignore towns and roads. Different projections and scales are also used to emphasize parts of the territory or features that have greater significance. Each map is an abstraction, partly because of the relevant features it reveals (or emphasizes), and also because of the irrelevant features it hides (or de-emphasizes). Objects are abstractions in much the same way. An object represents only those features of a thing that are deemed relevant to the current purpose, and hides those features that are not relevant.

This brings us to our second question. To which system are Coad and Yourdon referring? The context of the quotation makes it reasonably clear that their answer would be something like this: 'The proposed object-oriented software system, whose development is under consideration'. This seems appropriate—after all, this book is about the development of object-oriented software—but we should note that other systems are also involved. The most important of these is the human activity system that we must understand before beginning the construction of any software.

We should also note that we are using one tool to serve two purposes. As we shall see particularly in Chapter 7, objects are used to model an understanding of the application domain (essentially part of a human activity system), yet we shall also see in later chapters that objects are also understood as parts of the resulting software system. These are distinct purposes and there will be some occasions when we may need to take care to be clear about which meaning is intended.

Rumbaugh et al. (1991) take a slightly different slant on defining an object. These authors also explicitly recognize the dual purpose noted in the last paragraph.

> We define an *object* as a concept, abstraction, or thing with crisp boundaries and meaning for the problem at hand. Objects serve two purposes: They promote understanding of the real world and provide a practical basis for computer implementation.

This definition contains a repetition (although this does no harm)—by 'concept' we actually mean an abstraction, something that is logical rather than physical. A concept may be intangible, even imaginary (in the sense that not every participant agrees on its meaning, or sometimes even its existence). In the Agate case study, one concept is the 'campaign'. While they are clearly important, campaigns are intangible and difficult to define with precision. They really exist only as a relationship between a client (say Yellow Partridge, a jewellery company), the Account Manager, some other staff, some advertisements and various tasks and components that go into creating advertisements.

Concepts include many kinds of relationship between people, organizations and things. In information systems development, it is often necessary to recognize relationships such as contracts, sales or agreements. While intangible, some of these relationships can be quite long-lasting, and can have a complex influence on how people and other things in the application domain are able to act.

Let us take a simple transaction as an example. Imagine buying a tube of toothpaste in your local supermarket. On one level, this is just a sale, an exchange of money for goods. On a deeper level you might be entering into a complicated relationship with the shop and the manufacturer. This probably depends on other factors, e.g. the warranty may vary depending on the country that you are in at the time of the purchase, and perhaps the sale will earn points for you on a loyalty card. Perhaps the packaging includes a money-off coupon for your next purchase, or a contest entry form which must be accompanied by a valid proof of purchase. Now suppose you find something wrong with the toothpaste—you might be able to claim a refund or replacement. Maybe you can even sue the shop for damages. The point here is that we cannot understand the business without understanding these possible consequences of the sale in some appropriate way. In this case, the real world 'sale' will almost certainly be modelled as an object in the system.

At a fairly abstract level, when choosing the objects we wish to model—in fact, at the level that corresponds to a mapmaker—we need to ask: 'What sort of map is this; what details should it show, and what should it suppress?' In the real world, probably the only common characteristic of all objects is that they exist. However, all objects in a model or in an information system have certain similarities to all other objects, summarized by Booch in the statement that an object 'has state, behaviour and identity' (Booch, 1994). Here, 'state' represents the particular condition that an object is in at a given moment, 'behaviour' stands for the things that the object can do (and that are relevant to the model) and 'identity' means simply that every object is unique.

For example, Figure 4.1 lists some characteristics of a person, a shirt, a sale and a bottle of ketchup. Supposing that we wished to model these as objects, we can identify some possible identities, behaviours and states (but note that these are for illustration only, and do not indicate any particular system perspective).

In some texts (e.g. Wirfs-Brock et al., 1990), objects are deliberately characterized as if each were a person, with a role in the system that is based on its answers to three questions.

- Who am I?
- What can I do?
- What do I know?

Object	Identity	Behaviour	States
A person.	'Hussain Pervez'.	Speak, walk, read.	Studying, resting, qualified.
A shirt.	'My favourite button-down white denim shirt'.	Shrink, stain, rip.	Pressed, dirty, worn.
A sale.	'Sale no 0015, 15/06/02'.	Earn loyalty points.	Invoiced, cancelled.
A bottle of ketchup.	'This bottle of ketchup'.	Spill in transit.	Unsold, opened, empty.

Figure 4.1 Characteristics of some objects.

This approach defines an object in terms of its responsibilities and its knowledge, themes that we will encounter again in later chapters.

4.2.2 Class and instance

Let us begin by being clear that we are here dealing with objects as abstractions within an information system—either a model or the resulting software—and not with the real world objects that these represent. The UML specification (OMG, 2001) gives a definition that illustrates the strength of the connection between the three concepts of object, instance and class.

> An object represents a particular instance of a class.

Objects that are sufficiently similar to each other are said to belong to the same *class*. *Instance* is another word for a single object, but it also carries connotations of the class to which that object belongs: every object is an instance of some class. So, like an object, an instance represents a single person, thing or concept in the application domain. A class is an abstract descriptor for a set of instances with certain logical similarities to each other.

Figure 4.2 shows some classes that might be identified from the Agate case study (Chapter 7 describes a practical approach to identifying classes).

A class and its instances are related in the following manner. For staff at Agate, the idea of 'a campaign' is an abstraction that could represent any one of several specific campaigns. In an object-oriented software system, the class `Campaign` represents the relevant features that all campaigns have in common. There is one instance of the class to represent each real-world campaign. Some examples of campaigns are: a series of magazine adverts for various Yellow Partridge jewellery products placed throughout Europe in January 2002; a national series of TV, cinema and magazine adverts for the launch of the Soong Motor Co's *Granda* estate in Summer 2002.

Each instance of a class is unique, just as every living person is unique, however closely they resemble someone else. This is true even when two instances have identical characteristics. For example, there could be two members of staff at Agate

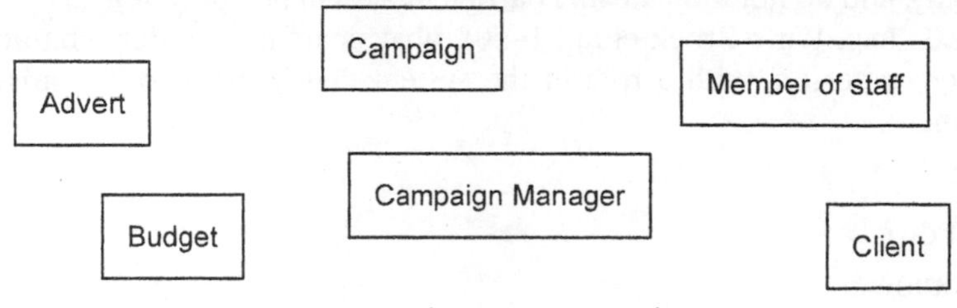

Figure 4.2 Some possible object classes in the Agate case study.

with the same name—say 'Ashok Patel', as this is a fairly common name. It is even possible (though unlikely) that both joined the company on the same date and work in the same section. Yet each remains a separate individual, and would therefore be represented by a separate instance of the class `StaffMember`.

4.2.3 Class membership

The idea that instances belong to a class logically implies that there must be a test that determines to which class an instance belongs. Since membership is based on similarity, such a test will also be capable of determining whether any two instances belong to the same class. There are two distinct types of logical similarity which must be tested. First, all objects in a class share a common set of descriptive characteristics.

For example, the staff at Agate record a client's company name, address, telephone number, fax number, e-mail address and so on. Each item is included in the list because it is useful in some way to the users of the system[1], and the full list gives a complete description of a client. The value of each item (e.g. the actual company name) will vary from one client to another. But, while the values may differ, this information *structure* is the same for every client.

To take another example, users of the Agate system also need to know about the member of staff assigned as staff contact to a client. A staff member might be described by a name, staff number and start date. Again, the full list gives a complete description of a staff member. Here, too, the value of each item (e.g. the staff name) will normally vary from one person to another, but again the structure is the same for all staff members.

Now compare the two descriptions, summarized in Figure 4.3.

Both staff members and clients have a name, but otherwise there is little in common. The information structure used to describe a client would not be capable of describing a member of staff, and vice versa. Even the sorts of name that would be valid in each case are likely to be different (apart from the relatively unusual cases where a company is known by the name of a person)[2]. It is reasonable to consider all clients as members of one class, and all staff as members of one class, but it would not be justified to consider them as members of the *same* class. When two objects cannot be described by a single set of features, they cannot belong to the same class.

A second logical similarity must also be tested. All objects in a class share a common set of valid behaviours. For example, clients may initiate a campaign, may be assigned a staff contact, may pay for a campaign, terminate a current campaign, and so on. Perhaps

Class	Characteristics	Class	Characteristics
Staff Member	Name Staff number Start date	Client	Name Address Telephone number Fax number E-mail address

Figure 4.3 Information structures for two classes.

[1] What appears in the list depends on the needs of the application. Another team may be developing a system to monitor whether client companies comply with environmental laws. Their list may share some characteristics with ours, but may also add others of no interest to our model.

[2] The values that an attribute can take are said to be contained within the *namespace* of the class. This means that each attribute value (for example, a company name) must be unique for its class. An attribute of another class might take the same value (for example, a staff member's name) but, since this is in a separate namespace, there is no conflict between the two.

no one client will actually do all this in reality, but that does not matter. The point is that any client could do any of these things, and the information system must reflect this.

If we consider staff members, we find a different set of valid behaviours. Staff can be assigned to work on a campaign, be assigned to a campaign as staff contact, change grade and maybe other things we do not yet know about. It may be more likely for staff members than for clients that they will actually go through the same sequence of behaviours, but this, too, does not matter. The point again is that a member of staff could do these things.

All clients, then, have a similar set of possible behaviours, and so do all staff members. But clients can do some things that a member of staff cannot do, and vice versa. On this count, too, we would consider clients as a class, and staff as a class, but we cannot make a case for considering clients and members of staff as instances of the same class. To summarize the examples, `Client` is a valid class, and `StaffMember` is a valid class. We can also note that, while the discussion here has been rather formal, an informal version of this dual test also works. In fact, we have already met this in Section 4.2.1: all members of a class give the same answers to the questions 'What do I know?' and 'What can I do?'

4.2.4 Generalization

In this section, we look at the concept of generalization and its application within object-orientation. In the UML Specification (OMG, 2001), generalization is formally defined as:

> . . . a taxonomic relationship between a more general element and a more specific element. The more specific element is fully consistent with the more general element (it has all of its properties, members and relationships) and may contain additional information.

This is perhaps a little much to grasp in one take, so in the sections that follow we will dismantle the definition into its main components, and examine each part on its own. We will highlight the general principles by looking first at an example of species classification (note that this is intended only to illustrate the concept of generalization, and is not a technical presentation of modern biological taxonomy). Later, in Chapter 8, we present an extended practical example from the Agate case study.

Taxonomic relationship

'Taxonomy' means a scheme of hierarchic classification—either an applied set of classifications, or the principles by which that set is constructed. The word was originally used for the hierarchic system of classification of plant and animal species, hence the example shown in Figure 4.4.

The taxonomic relationship between the two elements in this hierarchy labelled 'cat' and 'mammal' can be simply rephrased as 'the cat belongs to the class of mammals', or even more simply 'a cat *is a kind* of mammal'. Many other relationships are also identified in the diagram. For example, a domestic cat is a cat, as is a tiger, and both are also animals and living things. We can summarize this by saying that in each case the common relationship is that one element 'is a kind of' the other element.

Box 4.1 Class and object

Different books use the same terminology in different ways, and this can be confusing for the novice. For example, some authors use 'object' and 'class' interchangeably to mean a group of similar objects. But 'object' can also mean a single individual object, with 'class' reserved for the definition of a group of similar objects.

A further confusion arises if it is not clear whether 'object' refers to application domain things and concepts, or to their representation in an object-oriented model. The map is not the territory, even when the same words are used in both.

A third source of confusion can occur when the project focus shifts from analysis to design. During this transition, what was a model of the application domain is transformed into a model of software components. However, as we have not yet reached that point in the life cycle, we defer further discussion until later.

The UML Specification (OMG, 2001) defines the terms as follows.

A **class** is *'A description of a set of objects that share the same attributes, operations, methods, relationships and semantics.'* Moreover *'The purpose of a class is to declare a collection of methods, operations and attributes that fully describe the structure and behavior of objects.'*

An **object** is *'an instance that originates from a class, it is structured and behaves according to its class.'*

We follow these definitions. From now on 'object' is used consistently and interchangeably to mean a single instance of a class in a model. 'Class' is used consistently to mean the descriptor of (specification of) a set of similar object instances in a model. When we need to refer explicitly to people, things or concepts in the application domain, we will say so explicitly, as in this sentence.

A further distinction should be made between a class and a type. The UML Specification defines a **type** as *'A domain of objects together with operations applicable to the objects without defining the physical implementation of those objects. A Type may not contain any methods, but it may contain behavioural specifications of its operations.'*

What does this rather obscure definition mean? Cook and Daniels (1994) express the distinction with more clarity, noting that they prefer to use the term type rather than class to denote a collection of similar real-world objects. A class has connotations of a software implementation. Hence the UML constraint that a type may not contain any methods, for a method is an implementation of an operation. An operation is a logical specification of behaviour, while a method is a program fragment that implements the specification. In the same way, a type is a specification for a set of similar objects, while a class is an implementation of a type.

We prefer to take a pragmatic approach and use 'class' for the most part as interchangeable with 'type'. Wherever the context does not make it sufficiently clear which meaning is intended, further explanation will be given.

A more general element

The mammal is a more general element than the cat, which in turn is more general than the domestic cat or the tiger. Thus any description of a mammal must apply to many different animals: domestic cats, tigers, dogs, whales, etc. A description of a general mammal really means just a description of common characteristics shared by all mammals. This may be rather brief, perhaps consisting of only one characteristic, e.g. that all mammals suckle their young. In the tree diagram, the more general an element is, the nearer it is to the root of the tree.

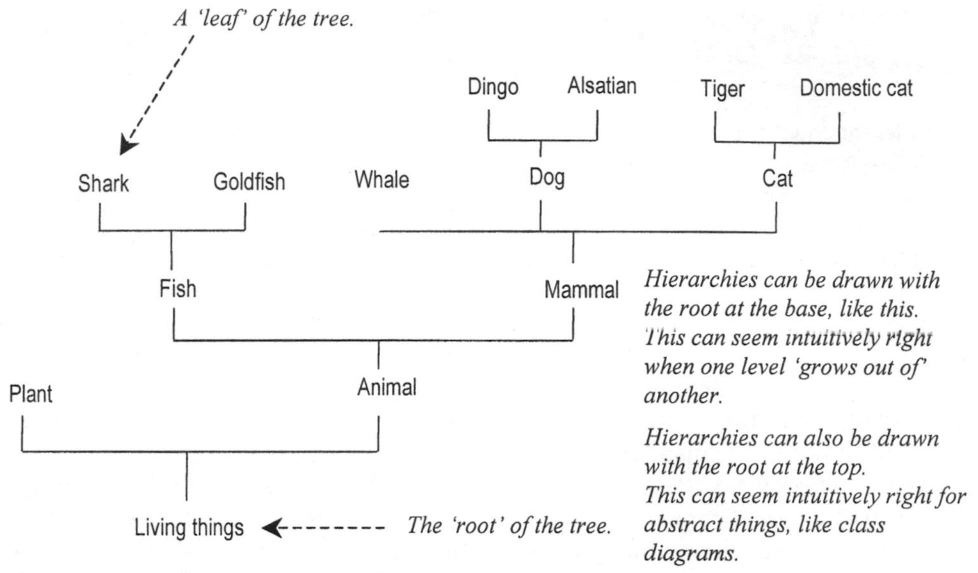

Figure 4.4 A simple taxonomy of species.

A more specific element

As an element of the hierarchy, 'cat' is more specific than 'mammal' and 'domestic cat' is still more specific. In other words, the former word in each pair conveys more information. Knowing that an animal is a cat, we can guess something about its diet, general body shape, size (within certain limits), number of legs (barring accidents, etc.), and so on. If we know only that it is a mammal, we can guess little, if anything at all, of its description. The more specialized elements of a hierarchy are those which are further from the root of the tree and closer to the leaves. The most specialized elements of all are those that actually form the leaves of the tree. In Figure 4.4, the leaves are individual species[3].

Fully consistent with the first element

Whatever is true for a mammal is also true for a domestic cat. If the defining characteristic of a mammal is that it suckles its young, then a domestic cat also suckles its young. This is true for every mammal, whether a tiger, dog or whale. This is an important feature of any hierarchic taxonomy. As an illustration, we can think about what would happen if a zoologist discovers that a defining mammalian characteristic does not apply to an animal previously thought to be a mammal. For example, suppose a research project found conclusive evidence that common field mice lay eggs instead of bearing live young. This alone would give enough grounds either to re-classify the field mouse, or to re-define what distinguishes a mammal from other animals. Perhaps a brand new classification would be invented specifically to accommodate egg-laying mice (as occurred following the discovery of the duck-billed platypus, which in most respects resembles a mammal, except that it lays eggs instead of bearing live young).

[3] If 'species' is considered as a class, then this in turn is another sort of abstraction of its members.

Adds additional information

A full description of the domestic cat would contain a great deal more information than that for a general member of the mammal class. For example, we might define a domestic cat by saying that it suckles its young, has a certain skeletal structure, a particular arrangement of internal organs, carnivorous teeth and habit, the ability to purr, and so on. Apart from suckling its young, none of these characteristics apply to *all* other mammals. A full zoological description of any species contains at least one characteristic (or a unique combination of characteristics) that differentiates it from all other species. Otherwise it would not make sense to consider it a species in the first place.

Using generalization

The main application for *generalization* in object-orientation is to describe relationships of similarity between classes. Object classes can be arranged into hierarchies much the same as the species example. This has two main benefits.

The first results from the use of object classes to represent different aspects of a real-world situation that we wish to understand. Using generalization, we can build logical structures that make explicit the degree of similarity or difference between classes. This is an important aspect of the *semantics* of a model—in other words, it helps to convey its meaning. For example, to know what hourly-paid and monthly-salaried employees in a business have in common with each other may be just as important as to know how they differ. The former may help to understand that some types of information must be recorded in identical ways for both types of employee. Figure 4.5 illustrates this with an example that might be suitable for a payroll system.

In this model, each employee is represented by their date of appointment, date of birth, department, employee number, line manager and name, together with some details that depend on whether they are paid weekly or monthly (these are the only significant differences modelled). A hierarchically structured model allows the close similarity to be shown clearly.

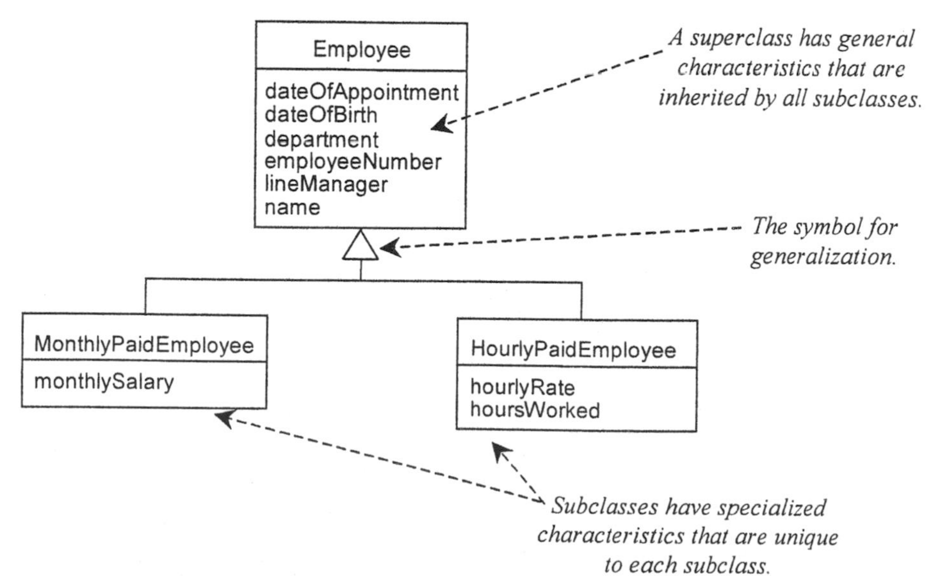

Figure 4.5 Hierarchy of employee types.

A second benefit comes from the relative ease with which a hierarchy can be extended to fit a changing picture. If this company were to decide that a new, weekly-paid type of employee is required, it is a simple matter to add a new subclass to the hierarchy to cater for this, as shown in Figure 4.6.

Other features of generalization

Three features of generalization that are not mentioned explicitly in the definition given earlier are sufficiently important to deserve some discussion at this point. These are: the mechanism of *inheritance*, the *transitive* operation of inheritance and the *disjoint* nature of generalization hierarchies.

Inheritance. This is a mechanism for implementing generalization and specialization in an object-oriented programming language. When two classes are related by the mechanism of inheritance[4], the more general class is called a *superclass* in relation to the other, and the more specialized is called its *subclass*. At a first approximation, the rules of object-oriented inheritance work more or less as shown below, although this is very simplified and will need to be refined shortly.

1. A subclass inherits all the characteristics of its superclass[5].
2. The definition of a subclass always includes at least one detail not derived from its superclass.

Inheritance is very closely associated with generalization. Generalization describes the logical relationship between elements that share some characteristics, while inheritance describes an object-oriented mechanism that allows the sharing to occur.

Transitive operation. This means that the relationship between two elements at adjacent levels of a hierarchy 'carries over' to all more specialized levels. Thus, in Figure 4.4, the definition of an animal applies in turn to all mammals, and thus by a series of logical steps to a domestic cat. So we can rewrite the rules of inheritance given above, as follows.

1. A subclass always inherits all the characteristics of *all* its superclasses.
2. The definition of a subclass always includes at least one detail not derived from *any* of its superclasses.

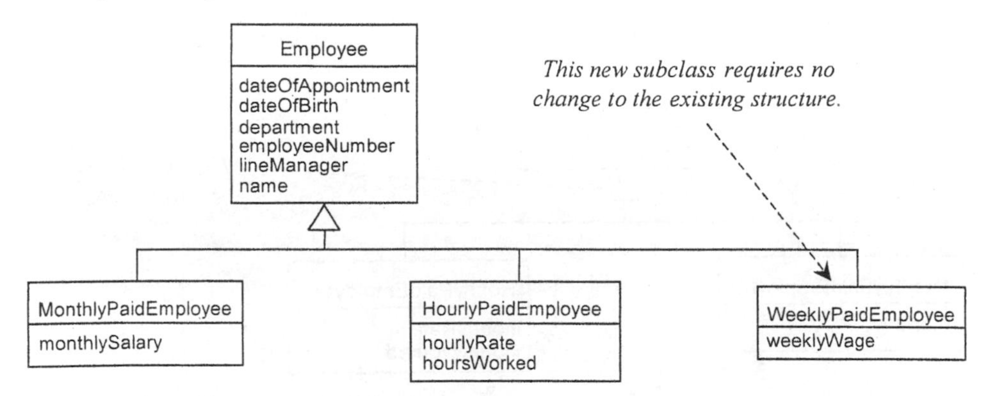

Figure 4.6 Hierarchies are easy to extend.

[4] The name comes from a superficial similarity to biological inheritance, which also takes the form of a hierarchy. But object-oriented inheritance is very different from inheritance in living things, and care should be taken not to confuse the two.

[5] We shall see in Chapter 8 that inherited characteristics are sometimes over-ridden in a subclass.

Disjoint nature. In a hierarchic system, the branches of the tree diverge as they get further away from the root and closer to the leaves. They are not permitted to converge. This means, for example, that a cat cannot be both a mammal and a reptile. In other words, each element in a hierarchy can only be a member of one classification at any given level of the hierarchy (although, of course, it can be a member of other classifications at other levels of the hierarchy, due to the transitive nature of the relationship).

The disjoint aspect of generalization means that we sometimes need to be careful about the characteristics chosen to express a generalization. For example, we could not use 'Has four feet' as the only defining characteristic of a mammal, even supposing that it were true of all mammals—because many lizards also have four feet, and this would make it possible to classify a lizard as a mammal. A class must be defined in terms of a unique set of characteristics that differentiate it from all other classes at its level of the hierarchy.

It is worth stressing here that generalization structures are abstractions that we choose to apply, since they express our understanding of some aspects of an application domain. This means that we can also choose to apply more than one generalization structure to the same domain, if it expresses a relevant aspect of the situation. Thus for example, a human might be simultaneously classified as a creature (*Homo sapiens*), as a citizen (a voter in a city electoral division) and as an employee (an Account Manager in Agate's Creative Department). If each of these were represented as a hierarchy in an object-oriented model, the position of a human might be an example of multiple inheritance, which means that a subclass is at once a member of more than one hierarchy, and inherits characteristics from every superclass in each hierarchy.

We should also note that real-world structures are not compelled to follow the logical rules applied in object-oriented modelling. Sometimes they are not disjoint or transitive, and therefore not strictly hierarchic. This does not detract from the usefulness of hierarchic structures in object-oriented development.

4.2.5 Message passing

In an object-oriented system, objects communicate with each other by sending messages. This might seem a little obvious if we think of the system as a kind of simulation, with objects that represent things in a real-world system collaborating to carry out a collective task. After all, this is how people communicate with each other: we send messages too. In the case of human communication, the means by which we transmit our messages are diverse, and sometimes it is not obvious who is the sender or receiver for a particular message, but all human communication can be seen as made up of messages. For example, everything we say to our friends and family, the e-mails we read when we log onto the network, advertising posters on the bus, games shows and cartoons on TV, even the clothes we wear, our tone of voice and our posture—these are all messages of one sort or another. If this is so, then just what is special about saying that objects only communicate via messages?

What makes it special is the fact that software was not constructed in this way until quite recently. Earlier approaches to systems development tended to separate data in a system from the processes that act on the data. This was done for sound analytical reasons, and is still appropriate for some applications, but it can give rise to difficulties. Chief among these is the need for a process to understand the organization of the data that it uses; for such a system, processes are said to be dependent on the structure of the data.

In itself, this is not necessarily a problem—in fact, it is to a certain extent inescapable. But if taken too far, and for insufficient reason, dependency of process upon data structure can cause many problems. For example, if the data structure were changed for any reason (a not uncommon occurrence in business), those processes that use the data may also need to be changed. This is an example of coupling between sub-systems, and we must minimize coupling if we are to construct systems that are reliable, that can be upgraded or modified, and that can be easily repaired if they break down (coupling is discussed in Chapters 12 and 13). Object-oriented systems avoid these problems by locating each process, as far as is practicable, with the data it uses. This is another way of describing an object: it is really little more than a bundle of data together with some processes that act on the data. These processes are called *operations*, and each has a specific *signature*. An operation signature is a definition of its interface. In order to invoke an operation, its signature must be given (signatures are sometimes also called *message protocols*).

In practice, it is not usually possible for all processes to be located with all the data that they must access, and data and processes are distributed among many different objects. Message passing is a way of insulating each object from needing to know any of the internal details of other objects. Essentially, an object knows only its own data, and its own operations. But in order for collaboration to be possible, the 'knowledge' of some objects includes knowing how to request services from other objects, which may include the retrieval of data. In this case, an object knows which other object to ask, and how to formulate the question. But it is unnecessary for an object to know anything about the way that another object will deliver the service. Such knowledge would introduce an undesirable and unnecessary degree of coupling into the relationship between the two objects.

When an object receives a message it can tell instantly whether the message is relevant to it. If the message includes a valid signature to one of its operations, the object can respond. If it does not, the object cannot respond. So we can think of operations as residing within objects, only able to be invoked by a message that gives a valid operation signature. Meanwhile, the object's data lies even deeper inside, since the data can only be accessed by an operation of that object. Clearly, then, if the signatures to an object's operations are not changed, it makes no difference what changes are made to the way that the operations run or the way that the data is stored. Neither of these changes would be visible from the outside. For this reason, the complete set of signatures for an object are sometimes known as its *interface*. This is called *encapsulation*, and is illustrated in Figure 4.7.

Consider a simple system to print pay cheques for employees in a business. One way of designing this would be to have a class `Employee`, with an instance that represents each person on the payroll. Each `Employee` object is responsible for knowing about the salary earned by the real employee it represents. Next, there might be a `PaySlip` object that is responsible for printing each employee's payslip each month. In order to print a payslip, each `PaySlip` object must know how much that employee has earned. The object-oriented approach to this is for each `PaySlip` object to send a message to the `Employee` object with which it is associated, asking how much salary to pay. The `PaySlip` object need not know how the `Employee` object works out the salary, nor what data it stores. All it needs to know is that if it asks an `Employee` object for a salary figure, an appropriate response is given. Message passing allows objects to encapsulate (i.e. hide) their internal details from other parts of the system, thus minimizing the knock-on effects of any changes to the design or implementation.

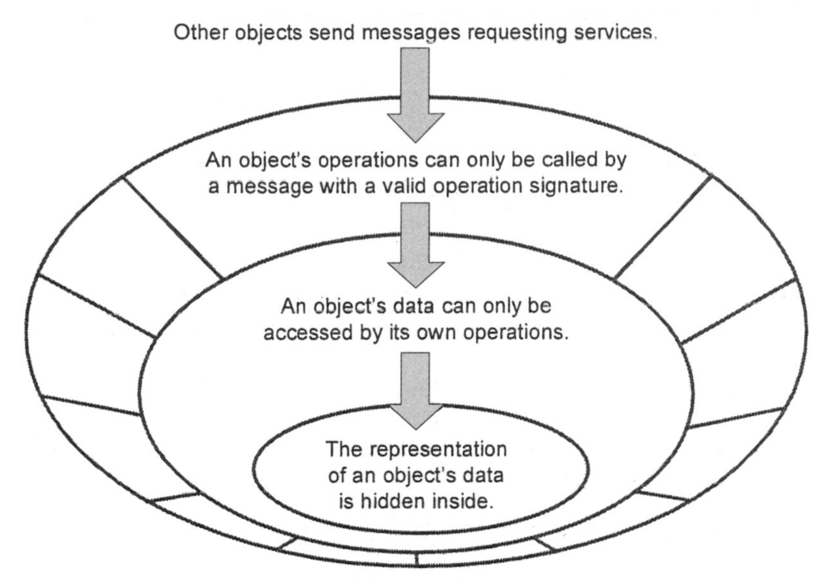
Other objects send messages requesting services.
An object's operations can only be called by a message with a valid operation signature.
An object's data can only be accessed by its own operations.
The representation of an object's data is hidden inside.

Figure 4.7 Encapsulation: the layers of protection that surround an object.

4.2.6 Polymorphism

When one person sends a message to another, it is often convenient to ignore many of the differences that exist between the various people that might receive the message. For example, in some situations a mother may use the same phrasing regardless of the age or gender of her child. 'Go to bed now!' carries much the same import whether the child is a five-year-old boy or a thirteen-year-old girl. But the precise tasks to be carried out by each child in going to bed may be very different. The five-year-old may set off towards bed by himself, but perhaps then requires help with washing his face, brushing his teeth, putting on his pyjamas and he may also expect to be read a bedtime story. The thirteen-year-old may not require any further help, once convinced that it really is bedtime.

This is rather like *polymorphism*, which is an important element in the way that object-oriented approaches encourage the decoupling of sub-systems. Polymorphism literally means 'an ability to appear as many forms', and it refers to the possibility of identical messages being sent to objects of different classes, each of which responds to the message in a different, yet still appropriate way. This means the originating object does not need to know which class is going to receive the message on any particular occasion. The key to this is that a receiving object is responsible for knowing how to respond to messages.

Figure 4.8 uses a collaboration diagram to illustrate polymorphism in a business scenario (the notation for collaboration diagrams is described fully in Chapter 9). The diagram assumes that there are different ways of calculating an employee's pay. Full-time employees are paid a salary that depends only on his or her grade; part-time staff are paid a salary that depends in a similar way on grade, but must also take into account the number of hours worked; temporary staff differ in that no deductions are made for the company pension scheme, but the salary calculation is otherwise the same as for a full-time employee. An object-oriented system to calculate pay for these employees might include a separate class for each type of employee, each able to perform the appropriate pay calculation. However, following the principle of polymorphism, the message signature for all `calculatePay` operations is the same. Suppose one of the

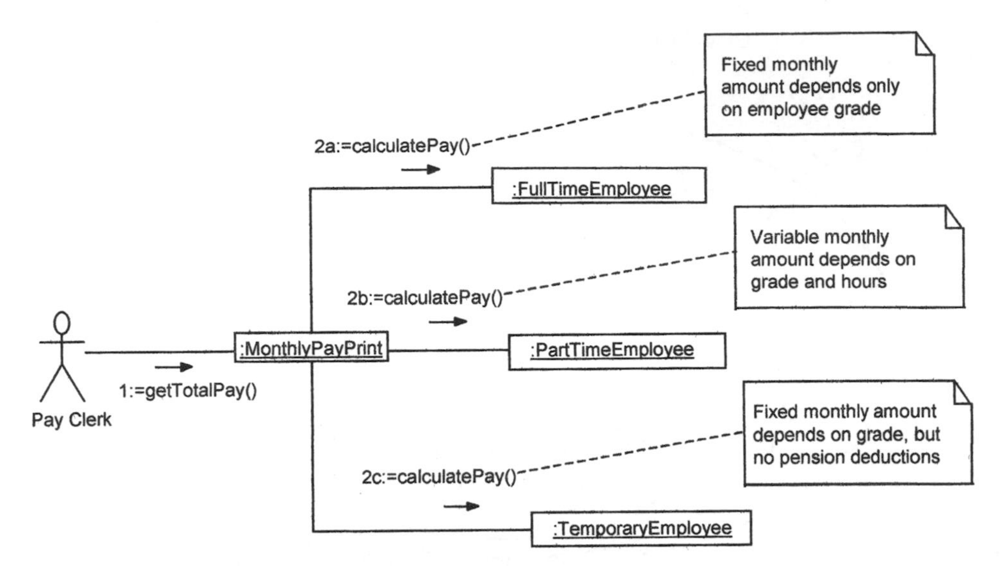

Figure 4.8 Polymorphism allows a message to achieve the same result even when the mechanism for achieving it differs between different objects.

outputs from this system is a print-out showing the total pay for the current month: to assemble the total, a message is sent to each employee object, asking it to calculate its pay. Since the message signature is the same in each case, the requesting object (here called `MonthlyPayPrint`) need not know the class of each receiving object, still less how each calculation is carried out.

Polymorphism is a powerful concept for the information systems developer. It permits a clear separation between different sub-systems that handle superficially similar tasks in a different manner. This means that a system can easily be modified or extended to include extra features, since only the interfaces between classes need to be known. The way that each part of the system is implemented (its internal structure and behaviour) is hidden from all parts of the system that do not absolutely need this knowledge.

Program designers and programmers have struggled for many years to achieve such a high level of modularity in software. Object-orientation gives the greatest promise yet of practical success.

4.2.7 Object state

In the real world, people and objects do not always behave in exactly the same way in response to similar stimuli. For example, if you have just eaten a good lunch, you will be much more likely to refuse the offer of a large, sticky gateau. However, if you happen to be very hungry at the time, this offer may be much more tempting. A simple way to represent this difference is to say that at any given time you can be in either of two states: well-fed or hungry. Each state is characterized by data that describes your condition—in this case your level of hunger. Each state is also characterised by a difference in behaviour. In other words, your response to certain messages differs according to the current value of your data—when well-fed, you refuse food, while when hungry you accept it. Your behaviour in response to a stimulus can also change your state. After eating a large gateau, your state changes from hungry to well-fed (perhaps for some people more than one would be required).

Objects can also occupy different states, and this affects the way that they respond to messages. Each state is represented by the current values of data within the object, which can in turn be changed by the object's behaviour in response to messages. Booch et al. (1999) define state as

> . . . a condition or situation during the life of an object during which it satisfies some condition, performs some activity, or waits for some event.

The concept of object state is fundamental to an understanding of the way that the behaviour of an object-oriented software system is controlled, so that the system responds in an appropriate way when an external event occurs. The significance of control is perhaps most apparent for safety-critical real-time systems. For example, consider the engine and flight controls in a modern airliner. Normally, for the greater part of the flight, an onboard computer flies the aircraft entirely automatically. During take-off and landing the pilot and flight crew take control, but even then it is still the software that directly operates the engine throttles, elevators, and so on. It could be disastrous if all the engines of an airliner were shut down during the final approach to land, and the software is designed to prevent this occurring. However, it may sometimes be necessary for the pilot to override this constraint during an emergency. The software must also be designed to permit this to occur, and to correctly discriminate between the different situations. In order to ensure safety, the control software must be designed to take account of all possible states of the aircraft (parked, climbing, flying level on auto-pilot, landing), the appropriate control behaviours for each state (shut down engine, full throttle, climb, descend, turn), and the external events that can trigger a change in behaviour (pilot operates throttle, turbulence causes course deviation, loss of cabin pressure).

The identification of object states can also be critical to the correct operation of business information systems. Here, the consequences of error are not so often life threatening, but they can threaten the survival of the organization and thus have an impact on the lives and livelihoods of customers, workers, investors and others associated with the enterprise. We return to this topic again in Chapter 7, and it is explained in detail in Chapter 11.

4.3 The Origins of Object-Orientation

Object-orientation is only one stage, and perhaps not the last, in a long development path. We review below some strands in the history of computing that have led to object-oriented analysis and design.

Increasing abstraction

From the very earliest days of digital computers, there has been a steady increase in the level of abstraction at which programmers have been able to operate. The increase in abstraction applies both to the activity of programming itself, and to the tasks that computer programs are expected to perform. In a relatively short time, the purposes to which computers are applied have become enormously more complex and demanding, thus greatly increasing the complexity of the systems themselves. Some of the waymarks in this progress are summarized in Figure 4.9.

Period	Programming languages	Interface technology	Typical applications
1940s	Machine code	Hardwired input	Military and scientific
1950s	Assembly language. Early third-generation languages (3GLs)	Punched tape and cards	Simple commercial software, such as early project planning programs
1960s	More mature 3GLs, e.g. Fortran and COBOL	Punched cards, teleprinters	Commercial systems such as accounts and payroll spread; scientific applications grow more complex
1970s	Non-procedural languages: Simula, ADA	VDU terminals	Mainframe networks begin to support early management information systems
1980s	Object-oriented languages, e.g. Smalltalk, Eiffel, C++	PCs. Early GUIs	Spread of office automation and spreadsheets; e-mail introduced in some businesses
1990s	Java, Python	GUIs, Web applets	Enterprise resource planning, data warehouses, multimedia applications

Figure 4.9 The increasing abstraction of programming (greatly simplified).

Event-driven programming

Early work on computer simulation led directly to the object-oriented paradigm of independent, collaborating objects that communicate via messages. A typical simulation task is to model the loading of vehicles onto a large ferry ship, in order to determine the quickest safe way to do this. This simulation would be run many times under differing assumptions, for example, the sequence of loading, the arrangement of vehicles on the storage decks, the speed at which vehicles are driven onto the ship, the time separation between one vehicle and the next, and so on. The real-world situation that is being modelled is very complex, and consists of a number of independent agents, each responding to events in ways that are easy to model at the level of individual agents, but very difficult to predict in the large, with many agents interacting with each other all at once. This kind of task is very difficult to program effectively in a procedural 3GL, since designs for these languages are based on the underlying assumption that the program structure controls the flow of execution. Thus for a program to tackle the simulation task described above in a 3GL, it must have separate routines that test for, and respond to, a vast number of alternative conditions.

One solution to this problem, that evolved into early simulation languages such as Simula-67, was to structure the program in a similar way to the problem situation itself: as a set of independent software agents, each of which represents an agent in the real-world system that is to be simulated. This remains one of the key ideas in object-oriented software development: that the structure of the software should mirror the structure of the problem it is intended to solve. In this way, the tension between the model of the application domain and the model of the software (mentioned in Section 4.2.1) is resolved, turning a potential weakness in fact into a strength.

The spread of GUIs

The rapid spread of graphical user interfaces (GUIs) in the 1980s and 1990s posed particular difficulties for contemporary development methods. GUIs brought some of the problems encountered earlier in simulation programming into the world of mainstream business applications. The reason for this is that users of a GUI are presented on their computer screen with a highly visual interface, that offers many alternative actions all at once, each one mouse-click away. Many other options can be reached in two or three more clicks via pull-down menus, list boxes and other dialogue techniques. Interface developers naturally responded by exploiting the opportunities offered by this new technology, with the result that it is now almost impossible for a system designer to anticipate every possible route that a user might take through a system's interface. This means that the majority of desktop applications are now very difficult to design or control in a procedural way. The object-oriented paradigm offers a natural way to design software, each of whose components offers clear services that can be used by other parts of the system quite independently of the sequence of tasks or the flow of control.

Modular software

In an object-oriented system, classes have two kinds of definition. From an external perspective, a class is defined in terms of its interface, which means that other objects (and their programmers) need only know the services that are offered by objects of that class and the signature used to request each service. From an internal perspective, a class is defined in terms of what it knows and what it can do—but only objects of that class need to know anything about this internal definition. It follows that an object-oriented system can be constructed so that the implementation of each part is largely independent of the implementation of other parts, which is what modularity means. This contributes to solving some of the most intractable problems in information systems development.

- It is easier to maintain a system built in a modular way, as changes to a sub-system are much less likely to 'ripple' through the rest of the system.
- For the same reason, it is easier to upgrade a modular system. As long as replacement modules adhere to the interface specifications of their predecessors, other parts of the system are not affected.
- It is easier to build a system that is reliable in use. This is because sub-systems can be tested more thoroughly in isolation, leaving fewer problems to be addressed later when the whole system is assembled.
- A modular system can be implemented in small, manageable increments. Provided each module is designed to provide a useful and coherent package of functionality, they can be introduced one at a time.

Life cycle problems

Most structured methodologies of the 1980s and early 1990s were based on the waterfall life cycle model originally designed for large engineering projects, such as constructing new jet aircraft and building road bridges. The difficulties with this life cycle model have already been discussed in Chapter 3. Object-orientation addresses them by encouraging a cyclic development approach. In this, there is little difficulty in revisiting earlier stages and revising earlier products in an iterative process that can repeat, if necessary, until everyone is satisfied with the quality of the software—subject, of course, to time and budget constraints.

This aspect is strongly linked to the highly modular character of an object-oriented system, described in the previous section, and also to the 'seamless' development of models throughout an object-oriented life cycle. The latter point is discussed in the next section.

Model transitions

In structured approaches to information systems development, the models of process that are developed during the analysis phase (e.g. data flow diagrams) have only an indirect relationship to the process models developed during the design phase (e.g. structure charts or, in SSADM, update process models). This has meant that designs for new systems, however good in their own right, have been hard to trace back to the original requirements. Yet, what makes a design successful is that it meets the requirements in a way that is functional, efficient, economic, and so on (see Chapter 12). This is particularly frustrating when one considers that a major strength of the structured methods was their concentration of effort on achieving a good understanding of users' requirements.

Object-oriented analysis and design avoid these transition problems by using a core set of models throughout analysis and design, adding more detail at each stage, and avoiding the awkward discontinuities that arise when one model must be discarded to be replaced by another with a different, incompatible structure. In UML, the fundamental analysis models are the use case and the class diagram (described in Chapters 6 and 7, respectively), and these continue as the backbone of the design, with other design models derived directly or indirectly from them.

Reusable software

Information systems are very expensive, yet in the past their developers have tended to reinvent new solutions to old problems, many times over. This wastes time and money, and has led to the demand for reusable software components, which can eliminate the need to keep reinventing the wheel. Object-oriented development methods offer great potential, not yet fully realized, for developing software components that are reusable in other systems, for which they were not originally designed. This is a result of the highly modular nature of object-oriented software, and also of the way that object-oriented models are organized. Inheritance is of particular importance in this context, and we say more about this in Chapter 8.

4.4 Object-Oriented Languages Today

A number of object-oriented programming languages are available today, with some significant differences between their capabilities and the extent to which they adhere to the object-oriented paradigm. Figure 4.10 lists some of the main characteristics of the most widely used of these languages. This is not a programming textbook, so we offer only a very brief description of each feature, as follows.

Strong typing refers to the degree of discipline that a language enforces on the programmer when declaring variables. In a strongly typed language (most modern languages are strongly typed), every data value and object that is used must belong to an

Feature	Smalltalk	C++	Eiffel	Java
Strong typing	✓	optional	✓	✓
Static/dynamic typing (S/D)	D	S	S	S+D
Garbage collection	✓	✗	✓	✓
Multiple inheritance	✗	✓	✓	✗
Pure objects	✓	✗	✓	✗
Dynamic loading	✓	✗	✗	✓
Standardized class libraries	✓	✗	✓	✓
Correctness constructs	✗	✗	✓	✗

Figure 4.10 Characteristics of some widely used object-oriented languages.

appropriate type for its context. Static typed languages enforce this with type-checking at compile time. Dynamic typed languages check types at run-time, but some languages offer a hybrid approach that allows the flexibility of loading classes at run-time. Garbage collection is important for memory management in systems that create and delete many objects during their execution. If objects are not removed from memory when they are deleted, the system may run out of memory in which to execute. When this is provided automatically, it removes the responsibility for this task from the programmer. Multiple inheritance refers to the capacity of an object to acquire features from more than one hierarchy. This is important as it minimizes the amount of code duplication and hence reduces inconsistencies that can cause maintenance problems. In static typed languages, multiple inheritance can allow a new class to stand in for any of its superclasses, and this reduces the amount of special-case programming required elsewhere in the system. Languages in which all constructs are implemented as classes or objects are said to be 'pure' object-oriented languages. Some languages permit data values that are not objects, but this introduces extra complexities for the programmer. Other languages allow un-encapsulated types, but this gives the sloppy programmer opportunities to bypass the safe encapsulation of classes. Both of these circumstances can cause a system to be difficult to maintain and extend. Dynamic loading refers to the ability of a language to load new classes at run-time. This can be used to allow software to reconfigure itself, for example, to cope with hardware or environment changes. It can also be used in a client-server environment to allow the client software to evolve by loading it with different classes from the server. Dynamic loading can help to propagate improvements and bug fixes, by concentrating maintenance efforts on the server side. Standardized class libraries give the programmer access to classes that are known to run successfully on a variety of hardware platforms, and under a variety of operating systems. When these are not available, it can be difficult to modify an application so that it will run on another platform, or in conjunction with applications that have used a different library. Finally, correctness constructs include pre-conditions and post-conditions on methods, forming a contract between any client-supplier pair. Contracts are important to the development of robust software and are discussed in Chapter 10.

Limitations of object-orientation

Some applications are not ideally suited to object-oriented development, and in this section we make a few comments about these. There are two main examples. The first kind includes systems that are strongly database-oriented. By this we mean both that they have a record-based structure of data that is appropriate to a relational database management system (RDBMS), and also that their main processing requirements centre on the storage and retrieval of the data (e.g. a management information system used mainly for querying data in the database). Such applications cannot easily be adapted to an object-oriented implementation without losing the many benefits of using a RDBMS for data storage. Commercial RDBMS are a very mature form of technology that organize their data according to sound mathematical principles. This ensures a good balance of efficiency of retrieval, resilience to change and flexibility in use. However, RDBMS are limited in their capabilities for storing and retrieving certain kinds of complex data structure, such as those that represent multimedia data. Graham et al. (1998) mention the spatial (map-based) data that forms the basis of a geographic information system (GIS) as a particular example of data structures to which RDBMS are not well suited, but which are ideally suited to object-oriented development. We return to the relative advantages and disadvantages of RDBMS as compared to object-oriented database systems in Chapter 18.

Applications that are strongly algorithmic in their operation are less suited to an object-oriented development approach. For some scientific applications that involve large and complex calculations (for example, satellite orbit calculations) it may be neither feasible nor desirable to split the calculation down into smaller parts. Such a system, if developed in an object-oriented manner, might contain very few objects, but each would be extremely complex. This would not be sound object-oriented design, and so either a procedural or a functional approach (these are alternative styles of programming) is recommended instead.

4.5 Summary

In this chapter we have introduced the most important concepts in object-orientation, in particular, object and class, generalization and specialization, message passing, object state and polymorphism. Understanding these gives an essential foundation for later chapters that deal with the practical application of object-oriented analysis and design techniques. There is a great deal of synergy in the way that the different fundamental concepts contribute to the success of object-orientation. For example, message passing and polymorphism both play a significant role in achieving sound modularity in a system. But there is no clean break with the past; instead the characteristics of object-orientation are best seen as the result of a gradual process of evolution that can be traced back to the earliest days of electronic digital computers. This evolutionary process is by no means finished yet, but rushes onwards into the future. As applications and computing environments grow ever more complex, there is a continuing need for reliable, maintainable, modifiable information systems.

Review Questions

4.1 Define object, class and instance.

4.2 What do you think is meant by 'semantics'?

4.3 How does the object-oriented concept of message passing help to encapsulate the implementation of an object, including its data?

4.4 What is polymorphism?

4.5 What is the difference between generalization and specialization?

4.6 What rules describe the relationship between a subclass and its superclass?

4.7 What does it mean to say that an object-oriented system is highly modular?

4.8 Why is it particularly hard for a designer to anticipate a user's sequence of tasks when using a GUI application?

4.9 What does 'object state' mean?

4.10 What is an operation signature?

Case Study Work, Exercises and Projects

4.A Section 4.2.1 mentions the human activity system and the proposed software system as particularly important systems to consider, but these are not the only systems that an analyst will encounter or work with. Make a list of any other systems you can think of that might be involved in the process of software development. What interfaces exist between them?

4.B Re-read the description of generalization given in Section 4.2.4. How does object-oriented inheritance differ from inheritance between a parent and a child: (i) in biology, and (ii) in law?

4.C Arrange the following into a hierarchy that depends on their relative generalization or specialization: person, thing, green, shape, Maria, cub, polar bear, square, law, child, colour, animal. Add more classifications as necessary so that it is clear what is generalized or specialized at each level.

4.D Read the first section of the case study material for FoodCo (Section B1.1), and identify classes that represent FoodCo's whole business environment.

4.E List all FoodCo's products that are identified in the case study material in Case Study B1 and arrange these into a hierarchy. Imagine some more products that make your hierarchy more interesting, and add these to your diagram.

Further Reading

- Most standard texts on object-oriented analysis and design contain a section that introduces the fundamental concepts of object-orientation. Although predating the development of UML, Jacobson et al. (1992) and Rumbaugh et al. (1991) remain good general introductions.
- Most more recent books use UML notation. Naturally, the most authoritative are those from the Rational stable, in particular Jacobson et al. (1999).

CHAPTER

Agate Ltd Case Study Introduction

Agate Ltd

A1.1 Introduction to Agate

Agate is an advertising agency in Birmingham, UK. Agate was formed as a partnership in 1982 by three advertising executives, Amarjeet Grewal, Gordon Anderson and Tim Eng (the name is a combination of their initials). Amarjeet and Gordon had previously worked for one of the UK's largest and most successful advertising companies in London, but felt frustrated at the lack of control they had over the direction of the campaigns they worked on. As a result, they moved to the West Midlands region of the UK and set up on their own business in 1981. Shortly afterwards they were joined by Tim Eng, with whom they had worked on a project in Hong Kong, and Agate was formed.

In 1987, the three partners formed a UK limited company and between them own all the shares in it. Gordon Anderson is Managing Director, Amarjeet Grewal is Finance Director and Tim Eng is Creative Director. They now employ about fifty staff at their office in the centre of Birmingham (see Figure A1.1) and a further hundred or so at seven offices around the world. Each of these other offices is set up locally as a company with the shares owned jointly by Agate and the local directors.

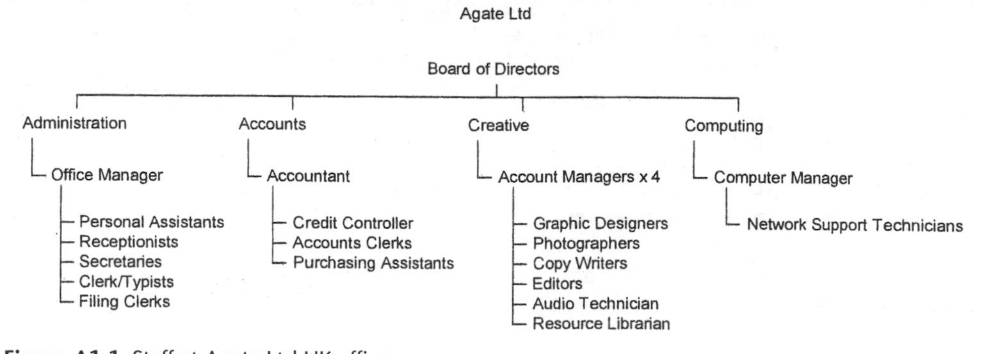

Figure A1.1 Staff at Agate Ltd UK office.

Initially the company concentrated on work for the UK motor industry, much of which is located in the West Midlands region. However, as it has expanded and internationalized, the type of work it takes on has changed, and it now has clients across a wide range of manufacturing and service industries.

The company strategy is to continue to grow slowly and to develop an international market. The directors would like to obtain business from more large multinational companies. They feel that they can offer a high standard of service in designing advertising campaigns that have a global theme but are localized for different markets around the world.

The company's Information Systems strategy has a focus on developing systems that can support this international business. Currently, as well as considering the development of a new business information system, the directors are deciding whether to invest in hardware and software to support digital video editing. They have until now sub-contracted video editing work, but recently have invested in ISDN connections for video-conferencing with other offices, and this means that they have the capability for fast file-transfer of digital video between offices.

A1.2 Existing computer systems

Agate uses computers extensively already. Like most companies in the world of design and creativity, Agate uses Apple Macintosh computers for its graphic designers and other design-oriented staff. The secretaries and personal assistants also use Apple Macs. However, the company also uses PCs to run accounts software in Windows.

Last year, Agate had a basic business system for the UK office developed in Delphi for Windows NT. However, after the system was developed, the directors of Agate decided that it should have a system developed in Java, the object-oriented language originated by Sun Microsystems Inc., which is designed to run over the Internet. One of the reasons for the choice of Java was that it is portable across different hardware platforms, and the company wants software that could run both on the PCs and on the Macs. Another reason is that Agate foresees the possibility of using Java to deliver information to clients via the World Wide Web. Unfortunately, the person who developed the Delphi software for the company (and was going to rewrite it in Java) was headhunted by an American software house, because of her skills in Java. She has moved to the USA, and is now earning twice what she could earn as a self-employed system developer in the UK! Fortunately, this developer, Mandy Botnick, was methodical in her work and has left Agate with some object-oriented system documentation for the system she designed and developed.

This existing system is limited in its scope: it only covers core business information requirements within Agate. It was intended that it would be extended to cover most of Agate's activities and to deal with the international way in which the business operates.

A1.3 Business activities in the current system

Agate deals with other companies that it calls clients. A record is kept of each client company, and each client company has one person who is the main contact person within that company. His or her name and contact details are kept in the client record. Similarly, Agate nominates a member of staff—a director, an account manager or a member of the creative team—to be the contact for each client.

Clients have advertising campaigns, and a record is kept of every campaign. One member of Agate's staff, again either a director or an account manager, manages each campaign. Other staff may work on a campaign, and Agate operates a project-based management structure, which means that staff may be working on more than one project at a time. For each project they work on, they are answerable to the manager of that project, who may or may not be their own line manager.

When a campaign starts, the manager responsible estimates the likely cost of the campaign, and agrees it with the client. A finish date may be set for a campaign at any time, and may be changed. When the campaign is completed, an actual completion date and the actual cost are recorded. When the client pays, the payment date is recorded. Each campaign includes one or more adverts. Adverts can be one of several types:

- newspaper advert—including written copy, graphics and photographs;
- magazine advert—including written copy, graphics and photographs;
- TV advert—using video, library film, actors, voice-overs, music etc;
- radio advert—using audio, actors, voice-overs, music etc;
- poster advert—using graphics, photographs, actors;
- leaflet—including written copy, graphics and photographs.

Purchasing assistants are responsible for buying space in newspapers and magazines, space on advertising hoardings, and TV or radio air-time. The actual cost of a campaign is calculated from a range of information. This includes:

- cost of staff time for graphics, copy-writing etc;
- cost of studio time and actors;
- cost of copyright material—photographs, music, library film;
- cost of space in newspapers, air-time and advertising hoardings;
- Agate's margin on services and products bought in.

This information is held in a paper-based filing system, but the total estimated cost and the final actual cost of a campaign are held on the new computer system.

The new system also holds the salary grades and pay rates for the staff, so that the cost of staff time on projects can be calculated from the timesheets that they fill out. It has been partially implemented and is not used in the existing system.

A1.4 Summary of requirements

This section summarizes the requirements for the new system.

1. **To record details of Agate's clients and the advertising campaigns for those clients.**
   - 1.1 To record names, address and contact details for each client.
   - 1.2 To record the details of each campaign for each client. This will include the title of the campaign, planned start and finish dates, estimated costs, budgets, actual costs and dates, and the current state of completion.
   - 1.3 To provide information that can be used in the separate accounts system for invoicing clients for campaigns.

1.4 To record payments for campaigns that are also recorded in the separate accounts system.

1.5 To record which staff are working on which campaigns, including the campaign manager for each campaign.

1.6 To record which staff are assigned as staff contacts to clients.

1.7 To check on the status of campaigns and whether they are within budget.

2. To provide creative staff with a means for recording details of adverts and the products of creative process that leads to the development of concepts for campaigns and adverts.

2.1 To allow creative staff to record notes of ideas for campaigns and adverts.

2.2 To provide other staff with access to these concept notes.

2.3 To record details of adverts, including the progress on their production.

2.4 To schedule the dates when adverts will be run.

3. To record details of all staff in the company.

3.1 To maintain staff records for creative and administrative staff.

3.2 To maintain details of staff grades and the pay for those grades.

3.3 To record which staff are on which grade.

3.4 To calculate the annual bonus for all staff.

4. Non-functional requirements.

4.1 To enable data about clients, campaigns, adverts and staff to be shared between offices.

4.2 To allow the system to be modified to work in different languages.

B1

CHAPTER

FoodCo Ltd Case Study Introduction

FoodCo Ltd

B1.1 Introduction to FoodCo

FoodCo produces a range of perishable foods for supermarkets, and is based in the flat agricultural lands of the East Anglian region of the UK. John Evans, the present Chairman, started the company when he was demobilized from the Royal Air Force at the end of World War II. He borrowed money to buy 200 acres of arable farmland, but his ambition was to be more than a farmer. As soon as Home Farm was running he opened a factory in a converted barn.

The first product was a pickle made to a traditional family recipe. It sold well, and success financed expansion. Two years later, John acquired derelict land next to the farm and the company moved into a larger, purpose-built factory. The product range soon extended to pre-packed vegetables and salads, and later a wide range of sauces, pickles and sandwich toppings, in fact almost anything made of vegetables that can be sold in jars. FoodCo's traditional customers are major UK supermarket chains. Some lines (e.g. washed salads) sell to all customers, while others (most of the cooked products) are produced for a specific supermarket chain. Most are packaged under the supermarket's 'own brand' label.

The pickle started a company tradition that, as far as possible, ingredients are grown on the company's own farm. This now covers 1,500 acres, and includes a market garden growing tomatoes, peppers, courgettes, chillis and other exotic vegetables under glass, and an extensive herb garden. Ingredients which do not grow well in the UK climate are supplied from carefully selected farms abroad, in Mediterranean Europe, East Africa, the United States and the Far East.

The company's annual turnover and employee numbers are summarized in Figure B1.1.

There are now three factories on the site. Beechfield is the oldest, and this is where raw vegetables are prepared. This work is relatively unskilled. The newer Coppice and Watermead factories concentrate on the more complex cooking processes involved in making sauces, pickles and the like. These need more skilled and experienced staff.

FoodCo Limited	
Number of employees (actual)	278
Number of employees (full-time equivalent)	213
Annual turnover (current projection for 2001/2002)	£6.5m

Figure B1.1 FoodCo's current staff complement and annual turnover.

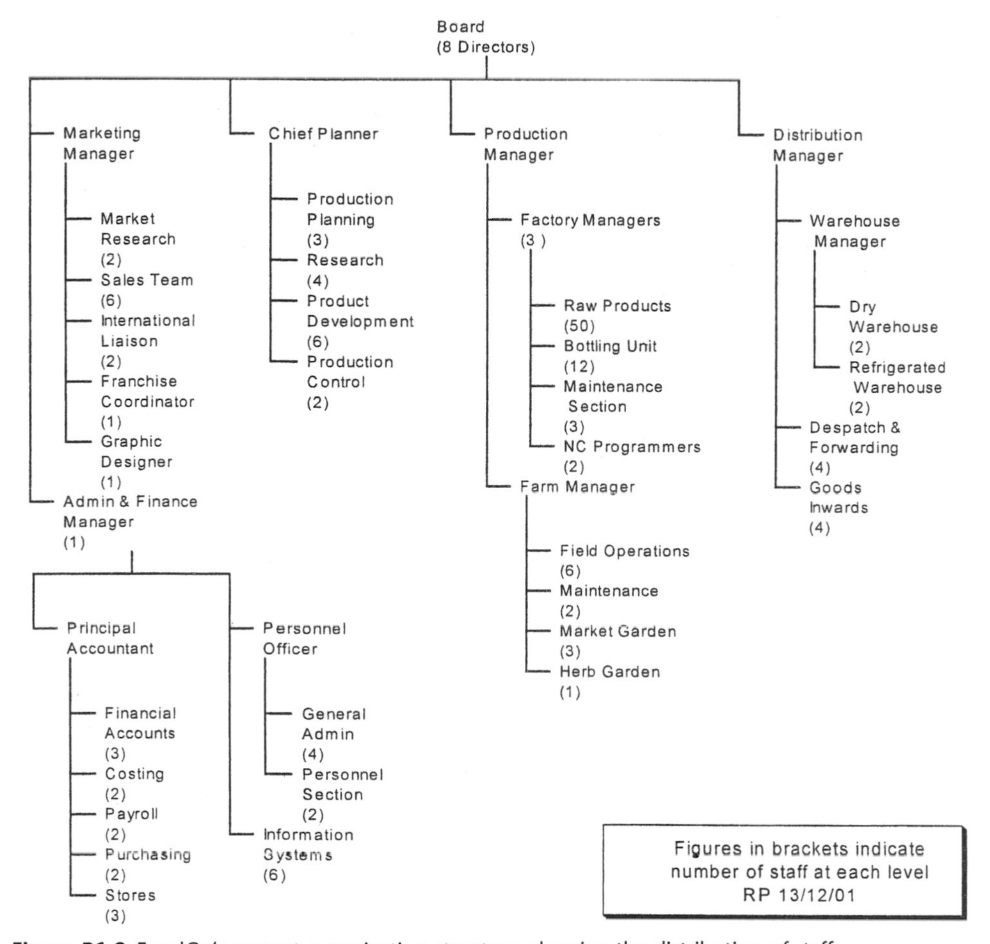

Figure B1.2 FoodCo's current organization structure, showing the distribution of staff.

A bottling plant is also located in Watermead, and there are two warehouses in separate buildings. One is refrigerated and stores fresh vegetable and salad packs, while the other stores dry and bottled products. Figure B1.2 shows a recent organization structure chart.

The company is still privately owned and managed. In 1993 John made way for his elder son Harold to become Managing Director. At the time, it was generally agreed by the Board that the company must improve profitability but there was no consensus on how to achieve this. In May 1996, John persuaded the Board that FoodCo must bypass its traditional supermarket customers and market directly to consumers. Partly as a result of John's speech to the Board (Box B1.1 gives an edited version), the *World Tradition* range was launched in 1997. This now sells successfully at the quality end of the UK market. Helped by the growing reputation of British food and cooking, the range has also begun to penetrate continental European and North American markets.

Box B1.1 Scene: the Board Room, May 1996. John Evans speaks to the Directors about Foodco's problems and a possible solution

'Fellow directors, we all know the company faces great difficulties. This year's profits will be the lowest ever. If we don't do something drastic, we'll be bankrupt in another year. But if we are to turn the situation round, we must know why things got so bad.

'I believe the reason goes right back to our beginnings. Those of you who were with me in 1947, when I started this company, will remember we had a degree of control that seems incredible now. Everything ran the way we wanted: farm, production, sales, distribution. We made consistently high quality goods, and by 1962 the new supermarkets were clamouring to buy. That was all a long time ago, but I think our early success is a direct cause of our present predicament. Let me explain.

'Remember 1968? When we borrowed heavily to finance expansion to meet that demand? Those loan repayments squeezed our profits hard. And then in 1974? When the TrustMart chain emerged as our dominant customer, and began driving down prices? We simply hadn't the financial muscle to fight them. We were still paying off a huge loan! That's why for ten years or more, TrustMart has dictated prices that are so low they have crippled us. And we've been unable to do a thing, because we've simply been scared they'll buy elsewhere. So last year TrustMart bought 65% of our total production—altogether over £3m in sales—and we'll be lucky to clear £100,000 profit on it!

'That's also why TrustMart call all the shots on new products. We don't have the cash to develop products for other customers. Now, I know we've grown in spite of that. It's not all been bad, but let's not kid ourselves it's good. We haven't really run the game since 1980. We all know it! We've been towed along behind TrustMart—and the supermarket sector—like a child dragged along by its father. We've only survived this long because TrustMart had no other source of supply. Now that's changing. We have serious new rivals for the supermarket supply business, and TrustMart have driven our prices still lower, to the point where we may make no profit at all this year.

'We can beat off this attack, but only if we develop new products and sell in a wider market. There is no argument about that, but there is a problem. Our *real* customers are not the supermarkets, but *their* shoppers. And they don't know we exist, because our name is hidden behind all the TrustMart own brand labels on all our packs and jars.

'The answer is to reach the consumers directly. Our market can only expand if they know our name, if they ask for our products. So here's what we do. We're going to launch our own brand name, and promote it so well that everyone knows about us. Customers will begin to insist on our brand, and TrustMart will have to pay our price, for a change.

'It won't be cheap. We'll need serious market research. We'll need more staff in the Product Development team, and we'll need time. We'll need a new corporate image. We'll need TV advertising. But it will be worth it. There's a vast market out there, and I'm not just thinking of the UK.

'So can we finance it? Certainly! It means heavy borrowing again, but we can repay the loans out of profits on increased sales. Sure, it's a risk, but we'll sink if we don't take it. There are many details to work out, but this plan can succeed. It *will* succeed! When we started, we were the best in the business. I believe we can be the best again.

'Thank you. Are there any questions?'

B1.2 FoodCo Today

B1.2.1 Current thinking

John Evans still believes that the company's major difficulty was over-reliance on one customer, and that this will be solved over time as the *World Tradition* market expands. His son, Harold, feels that management procedures are now the main problem, particularly management information. He sees the systems as hopelessly inadequate and thinks the company has simply outgrown them. For him, this is an extremely serious issue since it will inevitably worsen as the company grows. But father and son are each as stubborn as the other, so they have never settled their differences on this vital point.

Late in 1995 a new Finance Director, Clare Smythe, was appointed. Less than one year later she achieved a compromise that averted open war in the family. First, she championed the *World Tradition* brand that successfully met John's concerns. This is a range of international condiments, prepared to traditional recipes from many cultures, and using only the finest ingredients. Growing numbers of people in the affluent world want to be able to prepare authentic dishes from world cuisine, ranging from aloo brinjal (Indian potato and eggplant curry) to Yucatan-style cod (Mexican fish cooked in orange, lime and coriander). The new range allowed the company to reposition itself in a new international market, where growth has been highly profitable. It also helps FoodCo to free itself from dependence on TrustMart, still by far their largest customer.

Second, Clare recently helped Harold to persuade the Board that the introduction of a new product range compelled the company to manage its information more effectively and efficiently. The Board agreed to undertake a major review and updating of all information systems, and a national firm of consultants was commissioned to recommend a strategy.

B1.2.2 Information systems

The current systems are a mixed set of applications, some dating back to the late 1960s, that run on diverse hardware platforms. An ageing mini-computer runs an inflexible suite of accounting programs, a sales order processing system and a stock control system. The stock control system also generates product barcodes for the jar and bottle labels. The mini-computer is accessible from VT100 character-only terminals dotted throughout the factories and offices. Payroll is run off-site on a local computer bureau's mainframe.

Some managers and other staff have networked PCs with standard office software. A handful of proprietary packages include the Computer Aided Design programme used to design production line layouts. The PC network is not linked to the mini-computer or to the outside world for Internet, e-mail, or Electronic Data Interchange.

Some production is automated, including some washing and chopping operations in the Beechfield factory. The automated machines are of the numerical-control type, and are now obsolescent. Although they still do a reasonable job, by modern standards they are awkward to re-program, and maintaining them requires particular specialist skills.

The consultants' report

After some months of investigation the consultants submitted their report early in 2001. This identified serious failings in a number of areas, and recommended a phased approach to change. The top priority was to develop new product costing and production planning systems that would interface to a new in-house payroll package. The improved product costing and production information would give tighter control of production costs. Price negotiations with customers could be conducted on a more realistic basis, and better management information would help managers respond to the volatile international market for *World Tradition* products.

The second main recommendation in the report was for a substantial investment in upgraded hardware. Many more PCs were to be installed, networked to each other and to the mini-computer. These needed to be in place prior to phase 2 of the information systems plan, which called for a rapid spread of automation through the production and distribution departments. Finally, all new software development was to fit in with a medium-term plan to make the most of new technology opportunities. For this reason, a move to an object-oriented development method was seen as a critical aspect. This would help later with building an integrated set of systems, ultimately to include factory automation, management information, electronic links with suppliers and customers and an exploration of on-line Internet marketing. The Board accepted all the recommendations, and a detailed investigation into the requirements for the first systems was begun by FoodCo's in-house IS team, who had recently been trained in object-oriented development methods. Two staff members were seconded from the consultants to act as mentors to the first two or three projects.

Product costing: current operations

This section describes the way that product costing activities are currently carried out at FoodCo. It concentrates particularly on Beechfield factory, as this was an area identified by the consultants' report as a priority for action. Further information is given within chapters, where necessary, as part of case study exercises or review questions. Some information that is not strictly necessary to the completion of the exercises in this book has been included, in order to give a broader view of the overall operations.

Line operations. The nature of production control varies between the various factory and farm departments, depending on the operations undertaken and the nature of the product. At Beechfield, the main products are packs of washed salads and prepared raw vegetables, and some uncooked products such as coleslaw and Waldorf salad. There are three production lines. Each can be adapted to produce different products as the need arises, but only one at a time. Operatives' pay and the overall production costs for these lines are based on the entire batch produced during a single run, which often, although not always, equates to a single eight-hour shift. The line is switched on at the beginning of the run and temporarily halted for coffee breaks and lunch, or when a problem occurs. When a line is switched to a different product, this is treated as a separate run. If operatives are required to wait while the line is changed over to another product, or while a problem with the line is sorted out, they are paid a standing rate to compensate them for lost earnings.

Payroll and costing. For workers on the older lines at Beechfield, earnings are calculated using an algorithm that has as its input variables: the piecework rate for each item, the quantity of that item produced, the number of productive hours spent on the line by each employee and the employee's grade. For each run, the line

Daily Production Record Sheet No:

Line: .... Supervisor: .... Date: ....

Product: .... Job No: .... Circle Day: M T W T F S S

Run Start: .... Run Finish: ....

Problem Report When & where problem occurred: ....

Brief Description:

Fault Log No: .... Downtime (H/m): .... (Only if applicable)

Production Control use only

Signed: .... (Supervisor)

Total Qty Produced: ....

Checked by: ....

PR5/1.3

Figure B1.3 Daily production record sheet for the Beechfield factory.

supervisor completes a *daily production record sheet* (see Figure B1.3). These are sent to the costing section for analysis before being passed on to the payroll section.

The supervisors also complete a *weekly timesheet* (see Figure B1.4) for each employee. These are passed direct to payroll section. Each Tuesday, the entire week's production record sheets and timesheets are batched up and sent to the computer bureau. Data from the production sheets and timesheets is input to a piecework system at the bureau to produce a weekly earnings figure for the payroll calculation. After the payroll has been run all paperwork is returned to FoodCo's costing section for analysis. In practice, however, only a limited amount of analysis is carried out.

Some parts of the overall product costing function are outside the scope of this initial project, and will either be included in a later increment, or possibly in phase 2 of the plan. These are Coppice and Watermead factories, where the problems with product costing are not as significant as at Beechfield, and Home Farm, where the operations are very different in nature.

Problems in product costing

The mini-computer accounting system includes a product costing module. This meets only some of the information needs of the Finance Director and very few of the requests of any other managers, least of all the factory and farm managers who have direct control of most operations. Since the existing product costing system cannot answer most of the queries that are put to it, the costing clerks attempt to provide additional reports by using a spreadsheet application. But the sheer volume of data available for input each week is impossible to process accurately for all products. Due to ongoing staff shortages in the office, it has only usually been possible to produce actual costs for one production line each week. Making the best of a bad job, each line is costed accurately every fifth week and estimates produced for the other four weeks in between. As a result, the 'actual' costs quoted in management reports are often really no more than estimates derived from samples of the data available.

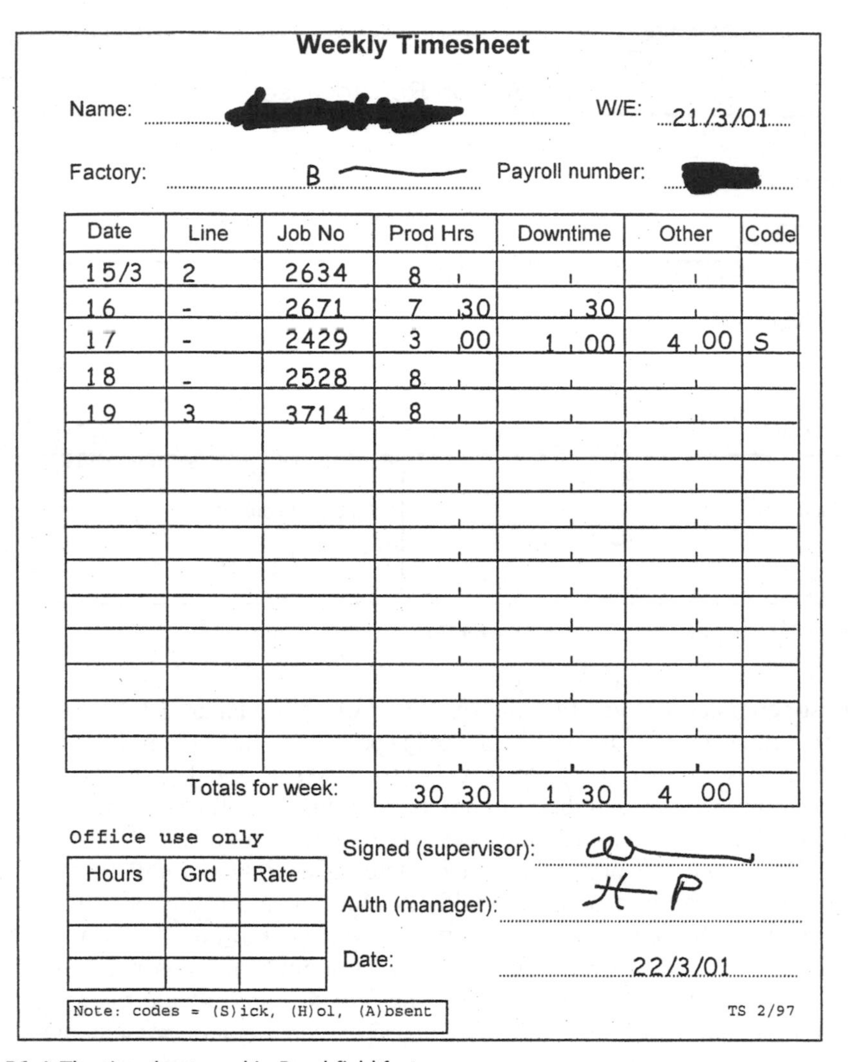

Weekly Timesheet

Name: [redacted] W/E: 21/3/01

Factory: B—— Payroll number: [redacted]

Date	Line	Job No	Prod Hrs	Downtime	Other	Code
15/3	2	2634	8			
16	-	2671	7 30	30		
17	-	2429	3 00	1 00	4 00	S
18	-	2528	8			
19	3	3714	8			
		Totals for week:	30 30	1 30	4 00	

Office use only

Hours	Grd	Rate

Signed (supervisor): [signature]

Auth (manager): [signature]

Date: 22/3/01

Note: codes = (S)ick, (H)ol, (A)bsent

TS 2/97

Figure B1.4 The timesheet used in Beechfield factory.

Both Harold Evans and Clare Smythe are convinced that their inability to get accurate costs is a major contributory factor to the company's decline in profitability. In effect, it means that senior management cannot say with confidence which operations are profitable and which are not. Until they have better information, they cannot even tell where their real problems are.

B1.3 The Proposal

The first system proposed for development is one to automate product costing. In Box B1.2, some staff from FoodCo's IS team are heard during a meeting early in the project. Louise Lee is the project manager, Ken Ong and Rosanne Martel are both analyst/developers on the team, and Chris Pelagrini is a consultant in object-oriented development (he works for the consultancy firm that produced the IS report).

Box B1.2 Scene: FoodCo's IS Team Meeting Room

Louise Lee: I'll begin by welcoming Chris Pelagrini. Chris is a consultant on object-oriented development, and he will be working closely with us on this project.

Chris Pelagrini: Thank you, Louise. Yes, that's right, I've been seconded to you for the duration. Provided, that is, we complete in six months (laughs).

LL: Don't worry, we can hit that deadline. OK, let's get started. Today I mainly want to set the scope for the project. Rosanne, you had a meeting with the Beechfield factory manager, Hari Patel. What did you find out?

Rosanne Martel: Yes, I met Hari on Thursday. He's the principal client for this system, and he'll be a user too. He confirmed the reasons why we picked this as our first object-oriented project. It's strategically important, but not so critical that the whole company will fold if we mess up. It's tightly scoped, and only really affects operations in Beechfield and the costing office. But it does need access to a lot of data currently on the mini-computer and it's a feeder system for payroll and production planning. If we develop a system that integrates with these, we'll have a sound basis for re-engineering the entire IS provision.

LL: Good. This confirms the consultants' report too. Did you get any idea of the main functionality of the system? We'll need this to estimate timescales with any confidence.

RM: Ken, you've done some work on this. How far did you get?

Ken Ong: Well, it's too early to be precise, but I've sketched out some use cases and a rough class diagram. Users include Hari, his line supervisors, the sales team, production control and the costing office. The main system inputs are staff timesheets and production record sheets, and data we can import from payroll records and the existing costing system. The main system outputs will be production run costs. One obvious problem is that we don't hold any payroll data electronically, so we'll need access to the bureau's files at some point. I would say that as a whole it is not highly complex. My first class diagram has about a dozen classes. There are a few interactions that involve most classes—for example, producing the final cost for a production line run, but most are simpler.

LL: So this is a fairly small system with relatively few users, but lots of interfaces to other systems. Can you show us some of this on the whiteboard?

KO: Yes, of course. Just give me a few minutes (goes to whiteboard and starts to draw).

LL: (While Ken draws) What do you think so far, Chris? Perhaps you could say a little about how you see your role.

CP: My task is to help you apply the object-oriented techniques that you have all learned on the training courses to this project. You all know there is a big difference between knowing the techniques, and understanding how they fit together. I'm here to help when you're unsure about anything. Rosanne's summary suggests this project is an ideal start and I'm confident we will make it a complete success.

LL: That's great, Chris, coming from the expert. OK Ken, now let's take a look at your diagrams.

5 CHAPTER

Modelling Concepts

OBJECTIVES

In this chapter you will learn

- what is meant by a model
- the distinction between a model and a diagram
- the UML concept of a model
- how to draw *activity diagrams* to model processes
- the approach to system development that we have adopted in this book.

5.1 Introduction

Systems analysts and designers produce models of systems. A business analyst will start by producing a model of how an organization works; a systems analyst will produce a more abstract model of the objects in that business and how they interact with one another; a designer will produce a model of how a new computerized system will work within the organization. In Section 5.2 we explain what is meant by a model and the relationship between models and diagrams. In UML, the term 'model' has a specific meaning, and we explain the UML concept of a model and how it relates to other UML concepts, such as the idea of a package.

The best way to understand what we mean by a model is to look at an example. In the Unified Process (the method of developing systems that is promoted by the developers of UML) activity diagrams are used to model the development process itself. We introduce the basic notation of activity diagrams in Section 5.3. Activity diagrams are one of the techniques that can be used to model the dynamic view of a system, and their use in systems analysis and design is explained in more detail in Chapter 10 where they are used as one way of specifying operations. Finally, in Section 5.4 we give an overview of the approach to system development that we have adopted in this book. We should stress that we are not presenting this development approach as a method in its own right, but only introducing it as a way of providing some structure to the process that we follow in the book.

5.2 Models and Diagrams

In any development project that aims at producing useful artefacts, the main focus of both analysis and design activities is on models. This is equally true for projects to build highways, space shuttles, television sets or software systems. In the days before computer modelling, aircraft designers built wooden or metal scale models of new aircraft to test their characteristics in a wind tunnel. A skilled furniture designer may use a mental model, visualizing a new piece of furniture without drawing a single line.

In IS development, models are usually both abstract and visible. On the one hand many of the products are themselves abstract in nature. Most software is not tangible for the user. On the other hand, software is usually constructed by teams of people who need to see each other's models. However, even in the case of a single developer working alone, it is still advisable to construct visible models. Software development is a complex activity, and it is extremely difficult to carry all the necessary details in one person's memory.

5.2.1 What is a model?

Like any map, models represent something else. They are useful in several different ways, precisely because they differ from the things that they represent.

- A model is quicker and easier to build.
- A model can be used in simulations, to learn more about the thing it represents.
- A model can evolve as we learn more about a task or problem.
- We can choose which details to represent in a model, and which to ignore. It is an abstraction.
- A model can represent real or imaginary things from any domain.

Many different kinds of thing can be modelled. Civil engineers model bridges, city planners model traffic flow, economists model the effects of government policy and composers model their music. This book is a model of the activity of object-oriented analysis and design.

A useful model has just the right amount of detail and structure, and represents only what is important for the task at hand. This point was not well understood by at least one character in *The Restaurant at the End of the Universe* by Douglas Adams (1980). Here, a group of space colonists are trying to re-invent things they need after crash-landing on a strange planet.

> 'And the wheel,' said the Captain, 'What about this wheel thingy? It sounds a terribly interesting project.'
> 'Ah,' said the marketing girl, 'Well, we're having a little difficulty there.'
> 'Difficulty?' exclaimed Ford. 'Difficulty? What do you mean, difficulty? It's the single simplest machine in the entire Universe!'
> The marketing girl soured him with a look.
> 'Alright, Mr. Wiseguy,' she said, 'you're so clever, you tell us what colour it should be.'

Real projects do get bogged down in this kind of unnecessary detail if insufficient care is taken to exclude irrelevant considerations (though this example is a little extreme).

What IS developers must usually model is a complex situation, frequently within a human activity system. We may need to model what different stakeholders think about the situation, so our models need to be rich in meaning. We must represent functional and non-functional requirements (see Section 6.2.2). The whole requirements model must be accurate, complete and unambiguous. Without this, the work of designers and programmers later in the project would be much more difficult. At the same time, it must not include premature decisions about how the new system is going to fulfil its users' requests, otherwise designers and programmers may later find their freedom of action too restricted. Most system development models today are in the form of diagrams, with supporting textual descriptions and logical or mathematical specifications of processes and data[1].

5.2.2 What is a diagram?

Analysts and designers use diagrams to build models of systems in the same way as architects use drawings and diagrams to model buildings. Diagrammatical models are used extensively by systems analysts and designers in order to:

- communicate ideas;
- generate new ideas and possibilities;
- test ideas and make predictions;
- understand structures and relationships.

The models may be of existing business systems or they may be of new computerized systems. If a system is very simple it may be possible to model it with a single diagram and supporting textual descriptions. Most systems are more complex and may require many diagrams fully to capture that complexity.

Figure 5.1 shows an example of a diagram (a UML activity diagram) used to show part of the process of producing a book. This diagram alone is not a complete model. A model of book production would include other activity diagrams to show other parts of the overall system such as negotiating contracts and marketing the book. This diagram does not even show all the detail of the activities carried out by authors and the other participants in the process. Many of the activities, shown as rectangles with rounded ends in Figure 5.1, could be expanded into more detail. For example, the activity `Write Chapter` could be broken down into other activities such as those shown in Figure 5.2.

We might break some of the activities shown in Figure 5.2 down into more detail, though it will be difficult to show the detail at a lower level, as activities like `Write a Paragraph`, `Add a Figure`, `Revise a Paragraph` and `Move a Figure` do not lend themselves to being represented in the flowchart notation of the activity diagram. There is also a limit to what we want to show in such a diagram. There are many activities such as `Make Coffee`, `Change CD` and `Stare out of Window` that are part of the process of writing, but like the colour of the wheel in the quote from *The Restaurant at the End of the Universe*, they represent unnecessary detail.

The diagrams of Figures 5.1 and 5.2 are typical of the kind of diagrams used in systems analysis and design. Abstract shapes are used to represent things or actions from the real world. The choice of what shapes to use is determined by a set of rules

[1] Some approaches rely primarily on formal logic techniques and rigorous mathematical specification. These are most often applied to real-time and safety-critical systems, such as those that control aircraft in flight or manage nuclear power plants, and are not covered in this book.

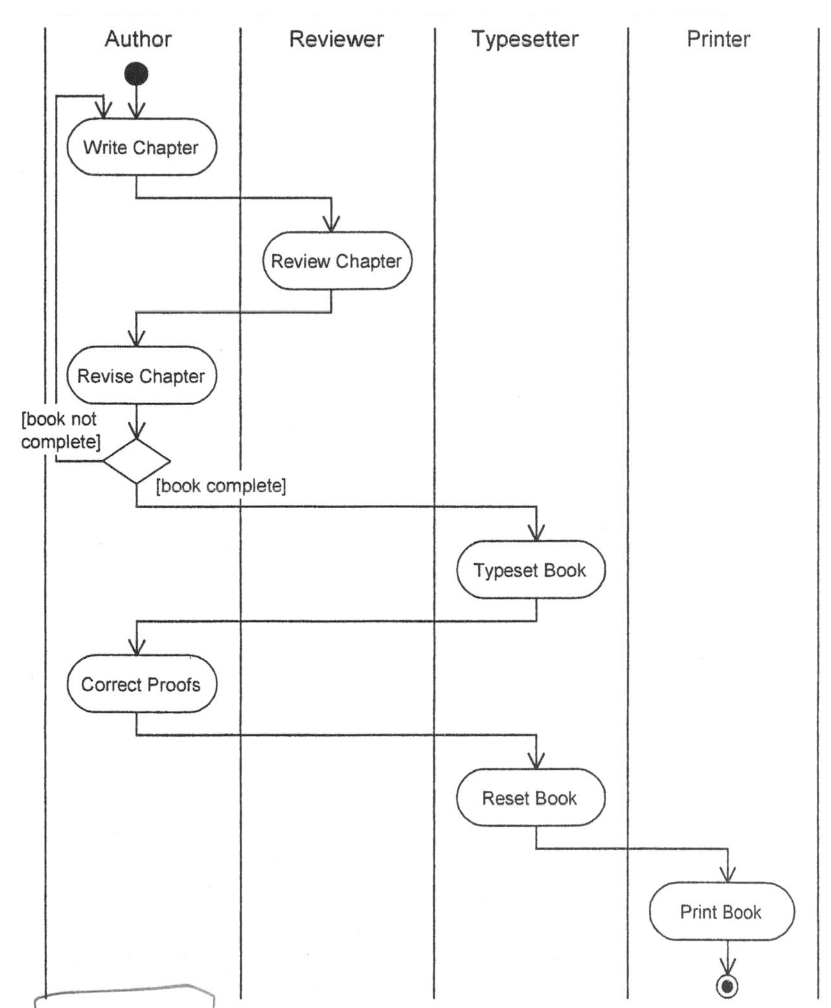

Figure 5.1 Activity diagram for producing a book.

that are laid down for the particular type of diagram. In UML, these rules are laid down in the *OMG Unified Modeling Language Specification* (OMG, 2001). It is important that we follow the rules about diagrams, otherwise the diagrams may not make sense, or other people may not understand them. Standards are important as they promote communication in the same way as a common language. They enable communication between members of the development team if they all document the information in the same standard formats. They promote communication over time, as other people come to work on the system, even several years after it has been implemented in order to carry out maintenance. They also promote communication of good practice, as experience of what should be recorded and how best to do that recording builds up over time and is reflected in the techniques that are used.

Modelling techniques are refined and evolve over time. The diagrams and how they map to things in the real world or in a new system change as users gain experience of how well they work. However, for the designers of modelling techniques, some general rules are that the techniques should aid (and enforce):

- simplicity of representation—only showing what needs to be shown;
- internal consistency—within a set of diagrams;

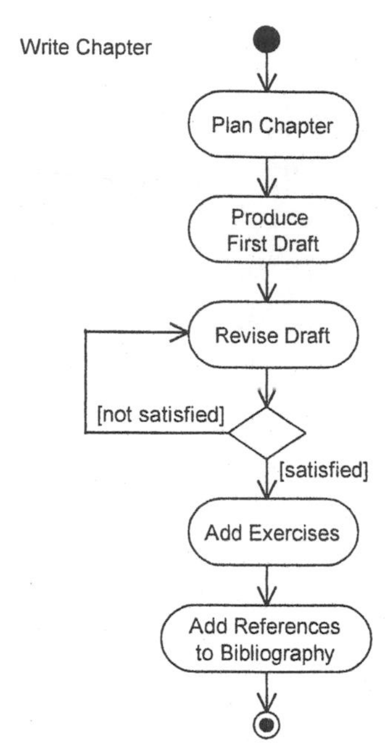

Figure 5.2 Activity diagram for the activity `Write Chapter`.

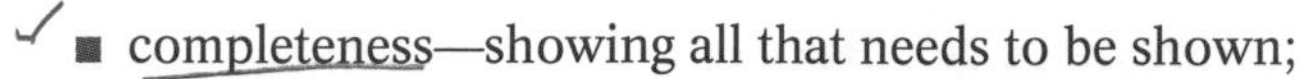

- completeness—showing all that needs to be shown;
- hierarchical representation—breaking the system down and showing more detail at lower levels.

Figure 5.3 shows some symbols from a label in an item of clothing. These icons belong to a standard that allows a manufacturer of clothing in Argentina to convey to a purchaser in Sweden that the item should be washed at no more than 40°C, should not be bleached and can be tumble dried on a low setting.

While not following the UML standards will not cause your T-shirts to shrink, it will cause you problems in communicating with other analysts and designers—at least if they are using UML as well. We have chosen to use UML in this book, as we believe that it is developing into an industry standard.

UML consists mainly of a graphical language to represent the concepts that we require in the development of an object-oriented information system. UML diagrams are made up of four elements:

- icons,
- two-dimensional symbols,
- paths and
- strings.

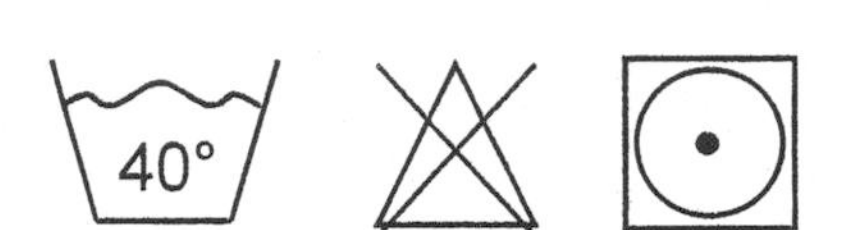

Figure 5.3 Example of a diagram standard.

UML diagrams are *graphs*—composed of various kinds of shapes, known as *nodes*, joined together by lines, known as *paths*. The activity diagrams in Figures 5.1 and 5.2 illustrate this. Both are made up of *two-dimensional symbols* that represent activities, linked by arrows that represent the transitions from one activity to another and the flow of control through the process that is being modelled. The start and finish of each activity graph is marked by special symbols—*icons*—the dot for the initial state and the dot in a circle for the final state. The activities are labelled with *strings*, and strings are also used at the decision points (the diamond shapes) to show the conditions that are being tested. (Two-dimensional symbols have compartments that can contain other symbols or strings. Icons do not contain other symbols.)

The UML Semantics section of the UML Specification provides the more formal grammar of UML—the syntax—and the meaning of the elements and of the rules about how elements can be combined—the semantics. The UML Notation Guide section of the UML Specification explains the different diagrams in more detail and provides examples of their construction and use. Box 5.1 explains with the aid of an example how the UML specification defines the syntax and semantics of UML. It may be difficult to understand at this stage in your understanding of UML, so feel free to skip it and come back to it when you understand more about UML.

5.2.3 The difference between a model and a diagram

We have seen an example of a diagram in the previous section. A single diagram can illustrate or document some aspect of a system. However, a model provides a complete view of a system at a particular stage and from a particular perspective.

Box 5.1 The UML metamodel

The metaclass diagrams use the notation of UML class diagrams (see Chapter 7) to define the *metamodel*—the model of how the elements of UML fit together. Rules in a special formal language called Object Constraint Language (OCL) formally define constraints on how these elements can be used. The textual part of the UML Specification explains the meaning of the elements and how they work in English text.

As an example, Figure 5.4 shows a simplified extract from the diagram in the UML Semantics section that defines the rules about activity diagrams. (Activity diagrams are, in fact, a special kind of *statechart diagram*, at least in UML Version 1.4.)

The diagram shows us that a `FinalState` is a subtype of `State`, which is a subtype of `StateVertex`, and that `StateVertex` has two associations with `Transition`. We can tell from this that each `Transition` goes from one and only one `StateVertex` (its `source`) to another (its `target`). However, each StateVertex can have many (depicted by the asterisks) `incoming` and `outgoing` transitions.

Because `FinalState` is a subtype of `StateVertex`, it appears that it can also have many incoming and outgoing transitions. However, it does not make any sense for a final state in a statechart or an activity diagram to have any outgoing transitions. If it has outgoing transitions, then there must be a state after it, so it is not a final state. OCL can be used to add a constraint to the metaclass `FinalState` that stops it from having outgoing transitions. This is written as follows.

```
context FinalState inv:
        self.outgoing->size = 0
```

This means that this OCL statement is an *invariant* (something that cannot be changed) of `FinalState`, and says that the size of the set of `Transitions` linked to any `FinalState` by the `outgoing` association must be zero. That is, it cannot have any outgoing transitions.

This is also explained in text in the UML Specification as follows.

> *FinalState*
>
> A special kind of state signifying that the enclosing composite state is completed. If the enclosing state is the top state, then it means that the entire state machine has completed.
>
> A final state cannot have any outgoing transitions.
>
> FinalState is a child of State.
>
> (OMG, 2001, p.2-150)

You do not need to worry about the implications of the first part of this definition, but if you want to understand it, take a look at Figures 5.1 and 5.2 again. All the activities in Figure 5.2 take place within the `Write a Chapter` activity of Figure 5.1. When the activities of Figure 5.2 have been completed, then the enclosing activity, `Write a Chapter`, has been completed. At the end of Figure 5.1 there is a final state icon, and since this activity chart is not enclosed within some other activity, it means that the entire activity diagram has completed.

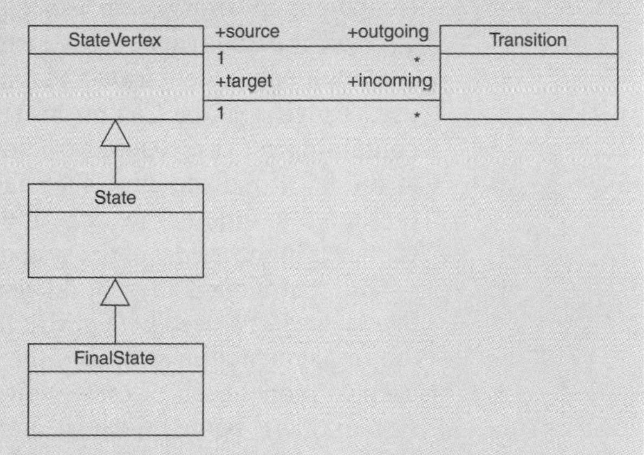

Figure 5.4 Extract from UML metamodel for statechart and activity diagrams.

For example, a Requirements Model of a system will give a complete view of the requirements for that system. It may use one or more types of diagram and will most likely contain sets of diagrams to cover all aspects of the requirements. These diagrams may be grouped together in models in their own right. In a project that uses UML, a Requirements Model would probably consist of a Use Case Model, which comprises use case diagrams, use case descriptions and prototypes of some use cases (see Chapter 6) and an initial Architectural Model showing sub-systems (see Section 5.2.4). On the other hand a Behavioural Model of a system will show those aspects of a system that are concerned with its behaviour—how it responds to events in the outside world and to the passage of time. At the end of the analysis stage of a project, the Behavioural Model may be quite simple, using Collaboration Diagrams to show which classes collaborate to respond to external events and with informally defined messages passing between them. By the end of the design stage of a project, the Behavioural Model will be considerably more detailed, using Interaction Sequence Diagrams to show in detail the interaction between classes within a collaboration, and with every message defined as an event or an operation of a class.

A model may consist of a single diagram, if what is being modelled is simple enough to be modelled that way, but most models consist of many diagrams—related to one another in some way—and of supporting data and documentation. Most models consist of many diagrams because it is necessary to simplify complex systems to a level that people can understand and take in. For example, the class libraries for Java are made up of hundreds of classes, but books that present information about these classes rarely show more than about twenty on any one diagram, and each diagram groups together classes that are conceptually related.

5.2.4 Models in UML

In UML there are a number of concepts that are used to describe systems and the ways in which they can be broken down and modelled. A *system* is the overall thing that is being modelled, such as the Agate system for dealing with clients and their advertising campaigns. A *sub-system* is a part of a system, consisting of related elements, for example the Campaigns sub-system of the Agate system. A *model* is an abstraction of a system or sub-system from a particular perspective or *view*. An example would be the use case view of the Campaigns sub-system, which would be represented by a model containing use case diagrams, among other things. A model is complete and consistent at the level of abstraction that has been chosen. Different views of a system can be presented in different models, and a *diagram* is a graphical representation of a set of elements in the model of the system.

Different models present different views of the system. Booch et al. (1999) suggest five views to be used with UML: the use case view, the design view, the process view, the implementation view and the deployment view. The choice of diagrams that are used to model each of these views will depend on the nature and complexity of the system that is being modelled. Indeed, you may not need all these views of a system. If the system that you are developing runs on a single machine, then the implementation and deployment views are unnecessary, as they are concerned with which components must be installed on which different machines.

UML provides a notation for modelling sub-systems and models that uses an extension of the notation for *packages* in UML. Packages are a way of organizing model elements and grouping them together. They do not represent things in the system that is being modelled, but are a convenience for packaging together elements that do represent things in the system. They are used particularly in CASE tools as a way of managing the models that are produced. For example, the use cases can be grouped together into a Use Case Package. Figure 5.5 shows the notation for packages, sub-systems and models. In diagrams we can show how packages, sub-systems and models contain other packages, sub-systems and models. This can be done either by containing model elements within larger ones or by using a special icon, which indicates containment. Figures 5.6 and 5.7 show the two notations for an example of a system containing two sub-systems.

5.2.5 Developing Models

The models that we produce during the development of a system change as the project progresses. They change along three main dimensions:

- abstraction,
- formality and
- level of detail.

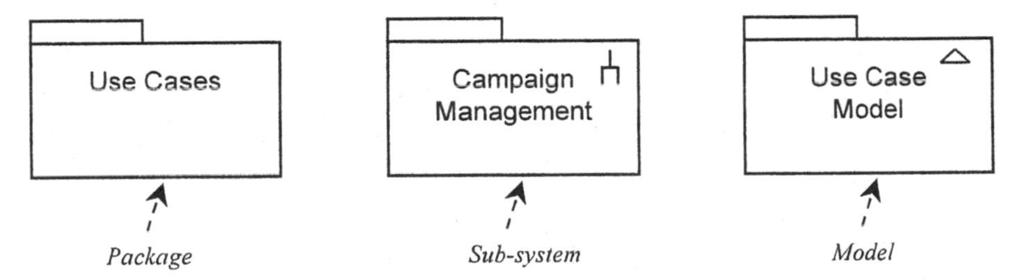

Figure 5.5 UML notation for packages, sub-systems and models.

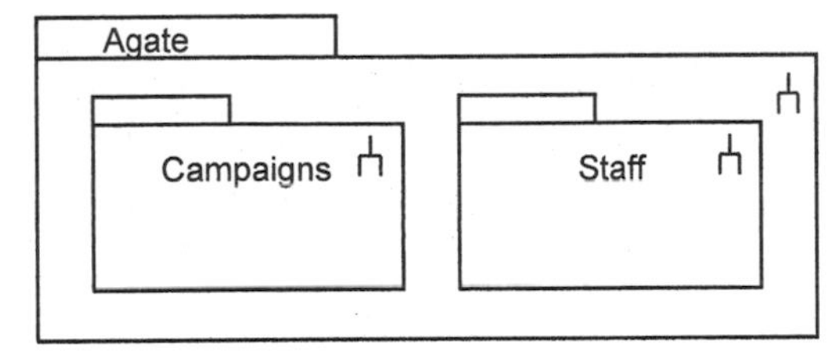

Figure 5.6 UML notation for a system containing sub-systems, shown by containment.

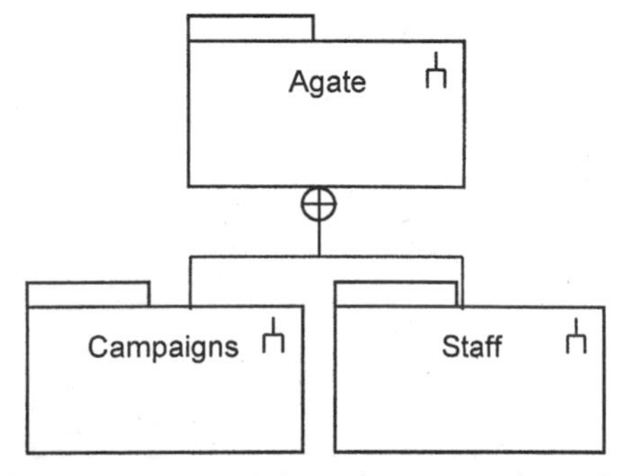

Figure 5.7 UML notation for a system containing sub-systems, shown by relationship.

During a particular phase of a project we may extend and elaborate a model as we increase our understanding of the system that is to be built. At the end of each phase we hope to have a model that is complete and consistent, within the limitations of that phase of the project. At that point, that model represents a view of our understanding of the system at that point in the project.

In a system development project that uses an iterative life cycle, different models that represent the same view may be developed at different levels of detail as the project progresses. For example, the first use case model of a system may show only the obvious use cases that are apparent from the first iteration of requirements capture. After a second iteration, the use case model may be elaborated with more detail and additional use cases that emerge from discussion of the requirements. Some prototypes may be added to try out ideas about how users will interact with the system. After a third iteration, the model will be extended to include more structured descriptions of how the users will interact with the use cases and with relationships among use cases. (Use cases are explained in Chapter 6.) Figure 5.8 illustrates this process of adding detail to a model through successive iterations. The number of iterations is not set at three. Any phase in a project will consist of a number of iterations, and that number will depend on the complexity of the system being developed.

It is also possible to produce a model that contains a lot of detail, but to hide or suppress some of that detail in order to get a simplified overview of some aspect of the system. For example class diagrams (explained in Chapter 7) can be shown with the compartments that contain attributes and operations suppressed. This is often useful for showing the structural relationships between classes, using just the name of each class, without the distracting detail of all the attributes and operations. This is the case in the diagrams that show the classes in the Java class libraries (referred to in Section 5.2.3), where the intention is to show structural relationships between classes rather than the detail.

As we progress through analysis and design of a system, elements in the model will become less abstract and more concrete. For example, we may start off with classes

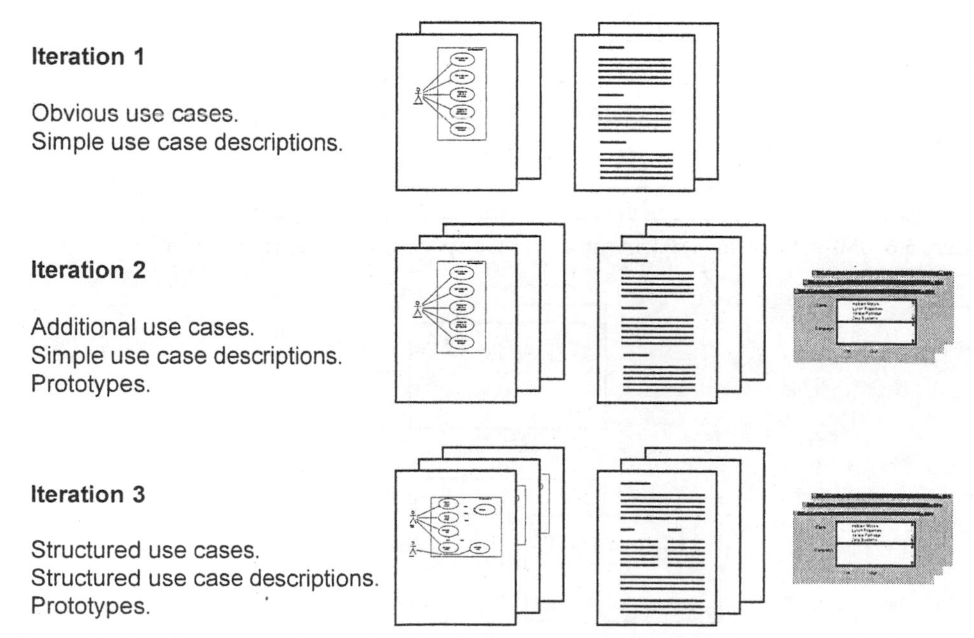

Figure 5.8 Development of the Use Case Model through successive iterations.

that represent the kinds of objects that we find in the business, `Campaigns`, `Clients` etc., that are defined in terms of the responsibilities that they have. By the time that we get to the end of design and are ready to implement the classes, we will have a set of more concrete classes with attributes and operations, and the classes from the domain will have been supplemented by additional classes such as collection classes, caches, brokers and proxies that are required to implement mechanisms for storing the domain classes (see Chapter 18).

In the same way, the degree of formality with which operations, attributes and constraints are defined will increase as the project progresses. Initially, classes will have responsibilities that are loosely defined and named in English (or whatever language the project is being developed in). By the time that we reach the end of design and are ready to implement the classes, they will have operations defined using Object Constraint Language (OCL) (see Chapter 10), with pre-conditions and post-conditions for each operation.

This iterative approach, in which models are successively elaborated as the project progresses, has advantages over the waterfall model, but it also has shortcomings. First, it is sometimes difficult to know when to stop elaborating a model, and second, it raises the question of whether to go back and update earlier models with additional information that emerges in later stages of the project. Issues like these are addressed either as part of a methodology (Chapter 22) or of a project management approach (Chapter 21). For now, we shall look at a first example of a UML diagram and see how it is developed.

5.3 Drawing Activity Diagrams

We have used activity diagrams earlier in this chapter to illustrate what is meant by a diagram. In this section we explain the basic notation of activity diagrams in UML and give examples of how they are used.

5.3.1 Purpose of activity diagrams

Activity diagrams can be used to model different aspects of a system. At a high level, they can be used to model business activities in an existing or potential system. For this purpose they may be used early in the system development lifecycle. They can be used to model a system function represented by a use case, possibly using object flows to show which objects are involved in a use case. This would be done during the stage of the lifecycle when requirements are being elaborated. They can also be used at a low level to model the detail of how a particular operation is carried out, and are likely to be used for this purpose late in analysis or during the design stage of a project. Activity diagrams are also used within the Unified Software Development Process (USDP) (Jacobson et al., 1999) to model the way in which the activities of the USDP are organized and relate to one another in the software development lifecycle. We use them for a similar purpose in later chapters to show how the activities of the simplified process that we have adopted for this book fit together. (This process is described in Section 5.4.)

In summary, activity diagrams are used for the following purposes:

- to model a task (in business modelling for instance);
- to describe a system function that is represented by a use case;

- in operation specifications, to describe the logic of an operation;
- in USDP to model the activities that make up the lifecycle.

Fashions change in systems analysis and design—new approaches such as structured analysis and design and object-oriented analysis and design replace old approaches and introduce new diagrams and notation. One diagram type that is always dismissed by the inventors of new approaches but always creeps back in again is the flowchart[2]. Activity diagrams are essentially flowcharts clothed in the notation and semantics of statechart diagrams (which are explained in more detail in Chapter 11).

Activity diagrams are really most useful to model business activities in the early stages of a project. For modelling operations, interaction sequence diagrams are closer to the spirit of object-orientation. However, there may be occasions when the analyst wants to model the activities that must be carried out, but has not yet identified the objects or classes that are involved or assigned responsibilities to them. In such circumstances, activity diagrams may be an appropriate tool to use.

5.3.2 Notation of activity diagrams

Activity diagrams at their simplest consist of a set of activities linked together by transitions from one activity to the next. Each activity is shown as a rectangle with rounded ends. The name of the activity is written inside this two-dimensional symbol. It should be meaningful and summarize the activity. Figure 5.9 shows an example of two activities joined by a transition.

Activities exist to carry out some task. In the example of Figure 5.9, the first activity is to add a new client into the Agate system described in Chapter A1. The transition to the second activity implies that as soon as the first activity is complete, the next activity is started. Sometimes there is more than one possible transition from an activity.

In this example from the Agate system, the flow of work is summarized by this brief statement from an interview with one of the directors of Agate.

> When we add a new client, we always assign a member of staff as a contact for the client straightaway. If it's an important client, then that person is likely to be one of our directors or a senior member of staff. The normal reason for adding a new client is because we have agreed a campaign with them, so we then add details of the new campaign. But that's not always the case—sometimes we add a client before the details of the campaign have been firmed up, so in that case, once we have added the client the task is complete. if we are adding the campaign, then we would record its details, and if we know which members of staff will be working on the campaign, we would assign each of them to work on the campaign.

This transcript from an interview describes some choices that can be made, and these choices will affect the activities that are undertaken. We can show these in an activity diagram with an explicit decision point, represented by a diamond shaped icon, as in Figure 5.10.

However, it is not necessary to use an explicit decision point like this. The diagram can be simplified by just showing the alternative transitions out of the activity `Assign Staff Contact`, as in Figure 5.11.

[2] Flowcharts are useful because they model the way that people perform tasks as a sequence of activities with choice points where they take one of a set of alternative paths and in which some activities are repeated either a number of times or until some condition is true.

The alternative transitions in both cases are each labelled with a *guard condition*. The guard condition is shown inside square brackets and must evaluate to either true or false. Alternative guard conditions from a single activity or decision point do not have to be mutually exclusive, but if they are not, you should specify the order of evaluation in some way, otherwise the results will be unpredictable. We would recommend that they should be.

Figures 5.10 and 5.11 illustrate another element of the notation of activity diagrams: when an activity has completed that ends the sequence of activities within a particular diagram, there must be a transition to a final state, shown as a black circle within a white circle with a black border. Each activity diagram should also begin with another special icon, a black circle, which represents the start of the diagram. Figure 5.12 shows the addition of the start state into the diagram of Figure 5.11. It also shows an additional activity-to assign a member of staff to work on a campaign-and additional guarded transitions.

Note that there is a loop or iteration created at the bottom of this diagram, where the activity `Assign Staff to Campaign` is repeated until there are no more staff to assign to this particular campaign.

Activity diagrams make it possible to represent the three structural components of all procedural programming languages: sequences, selections and iterations. This ability to model processes in this way is particularly useful for modelling business procedures, but can also be helpful in modelling the operations of classes.

In an object-oriented system, however, the focus is on objects carrying out the processing necessary for the overall system to achieve its objectives. There are two ways in which objects can be shown in activity diagrams:

- the operation signature of an object can be used as the name of an activity;
- an object can be shown as providing the input to or output of an activity.

Figure 5.13 shows an example of the first of these uses of objects in activity diagrams. In this example, the total cost of a campaign is calculated from the cost of all the individual adverts in the campaign added to the campaign overheads.

The second way that objects are shown in activity diagrams is by using *object flows*. An object flow is a dependency between an object and an activity that results in a change to the state of that object. The state of the object can be shown in square brackets within the symbol for the object. Figure 5.14 shows an example of this for the activity `Record Completion of a Campaign`, which changes the state of a `Campaign` object from `Active` to `Completed`. (Objects and classes are covered in much more detail in Chapters 7 and 8, and the idea of 'state' is covered in more detail in Chapter 11, where we explain statechart diagrams.)

Figure 5.9 Example of two activities joined by a transition.

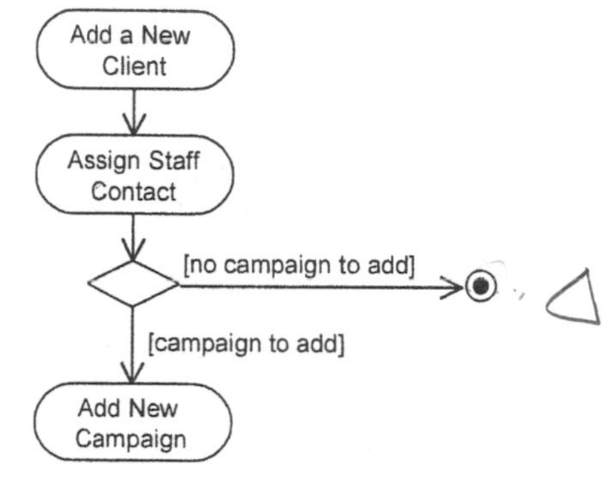

Figure 5.10 Activities with a decision point.

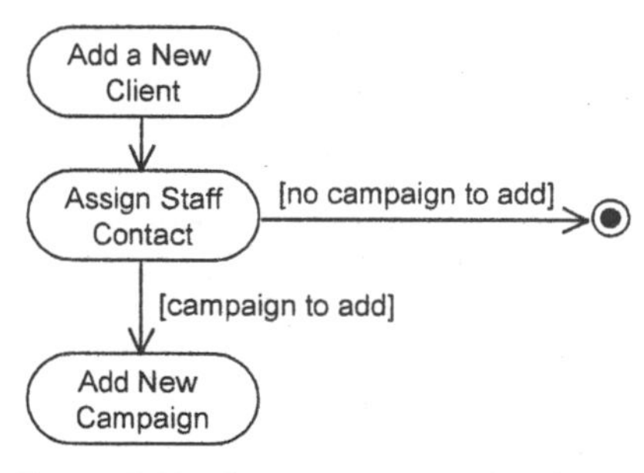

Figure 5.11 Choice represented without an explicit decision point.

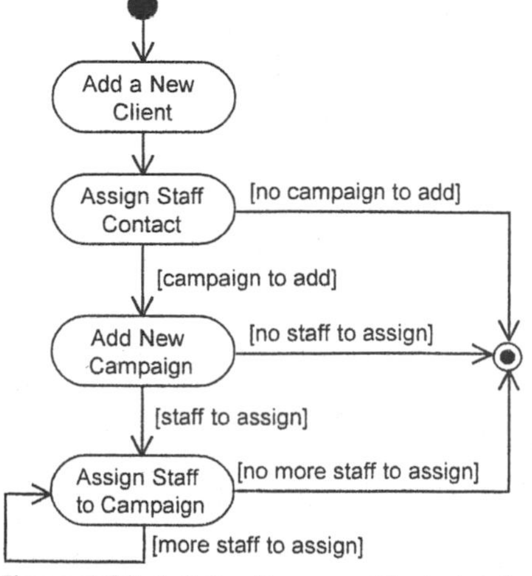

Figure 5.12 Activity diagram with start state.

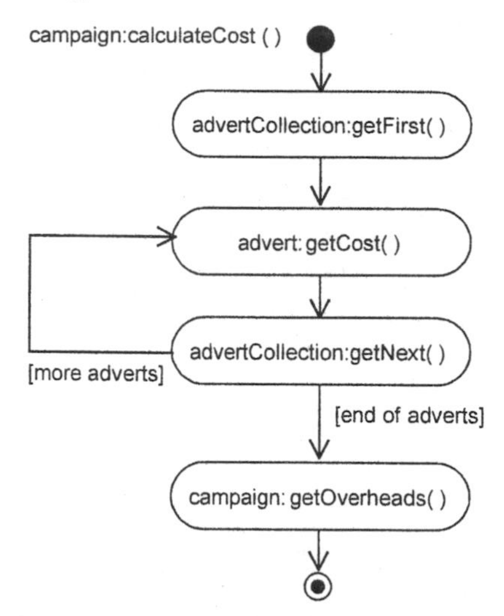

Figure 5.13 Activity diagram with object operation signatures as activities.

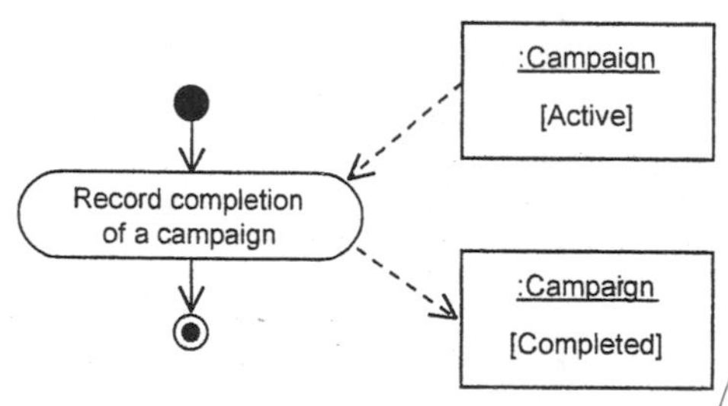

Figure 5.14 Activity diagram with object flows.

A final element of the notation of activity diagrams that it is useful to understand at this stage is the idea of *swimlanes*. Swimlanes are particularly useful when modelling how things happen in an existing system, and can be used to show where activities take place or who carries out the activities. The diagram is divided into columns for each location and the name of the location is shown at the top of each column.

In the Agate case study, when an advertising campaign is completed, the campaign manager for that advertising campaign records that it is completed. This triggers off the sending of a record of completion form to the company accountant. An invoice is then sent to the client, and when the client pays the invoice, the payment is recorded. (Some of these activities are documented as requirements in Section A1.4.

In order to model the way that the system works at the moment, we might draw an activity diagram like the one in Figure 5.15 in order to show these activities taking place. The brief for this project is to concern ourselves with the campaign management side of the business, as there is an existing accounts system in the company. However, the act of drawing this diagram raises the question of what happens to the payment from the client.

- Does the payment go to the accountant, and is there some way in which the campaign manager is notified?
- Does the payment go to the campaign manager, and does he or she record the payment and then pass it on to the accountant?

Clarifying points like these are part of the process of requirements capture, which is covered in detail in Chapter 6.

One of the reasons for introducing activity diagrams at this point is that they are used in the Unified Software Development Process to document the activities of the software development lifecycle. In USDP, the diagrams are *stereotyped*—the standard UML symbols are replaced with special icons to represent activities and the products of those activities. In the next section, we describe the simplified process model that we have adopted in this book. We use activity diagrams to summarize this process in the case study chapters later in the book.

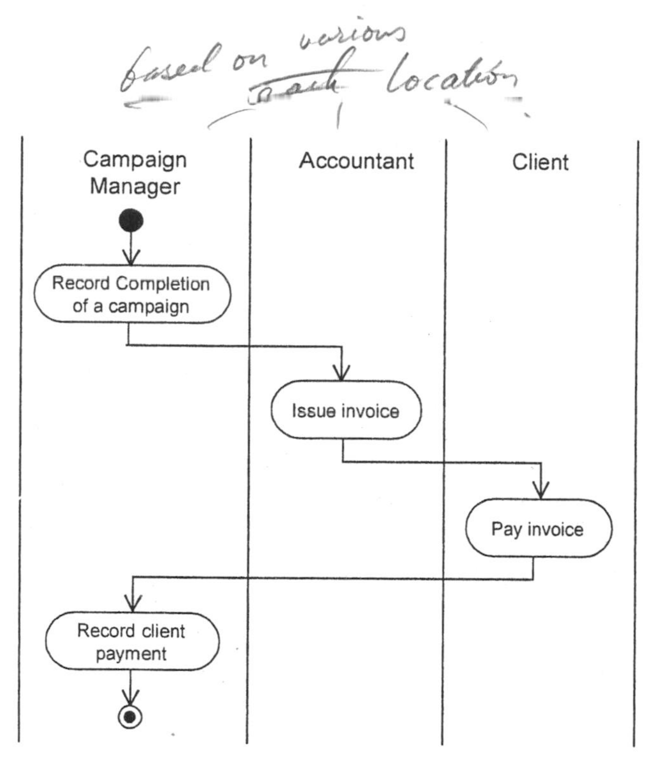

Figure 5.15 Activity diagram with swimlanes.

5.4 A Development Process

A development process should specify what has to be done, when it has to be done, how it should be done and by whom in order to achieve the required goal. Project Management techniques (see Chapter 21) are used to manage and control the process for individual projects. One of the software development processes currently in wide use is the Unified Software Development Process (USDP) (Jacobson et al., 1999). The USDP has been developed by the team that created UML. It is claimed that the USDP embodies much of the currently accepted best practice in information systems development. These include:

- iterative and incremental development;
- component-based development;
- requirements driven development;
- configurability;
- architecture centrism;
- visual modelling techniques.

The USDP is explained in more detail in Chapter 22 on System Development Methodologies. USDP is often referred to as the *Unified Process*.

The USDP does not follow the Traditional Life Cycle shown in Figure 3.2 but adopts an iterative approach within four main *phases*. These phases reflect the different emphasis on tasks that are necessary as systems development proceeds (Figure 5.16). These differences are captured in a series of *workflows* that run through the development process. Each workflow defines a series of activities that are to be carried out as part of the workflow and specifies the roles of the people who will carry out those activities. The important fact to bear in mind is that in the waterfall lifecycle activities and phases are one and the same, in iterative lifecycles like the USDP the activities are independent of the phases, and it is the mix of activities that changes as the project proceeds. Figure 5.17 illustrates how a simplified waterfall lifecycle would look using the same style of diagram as Figure 5.16.

5.4.1 Underlying Principles

In order to place the techniques and models described in this book in context we have assumed an underlying system development process. We are not attempting to invent yet another methodology. The main activities that we describe here appear in one form or another in most system development methodologies. The system development process that we adopt is largely consistent with USDP though it incorporates ideas from other sources. This approach incorporates the following characteristics. It is

- iterative;
- incremental;
- requirements driven;
- component-based;
- architectural.

These principles are embodied in many commonly used methodologies and are viewed as elements of best practice.

5.4.2 Main activities

The systems development process embodies the following main activities:

- Requirements Capture and Modelling;
- Requirements Analysis;
- System Design;
- Class Design;
- Interface Design;
- Data Management Design;
- Construction;
- Testing;
- Implementation.

These activities are interrelated and dependent upon each other. In a waterfall development process they would be performed in a sequence (as in Figure 5.17). This is not the case in an iterative development process, although some activities clearly precede others. For example, at least some requirements capture and modelling must take place before any requirements analysis can be undertaken. Various UML techniques and notations are used, as well as other techniques, and these are summarized in the table in Figure 5.18.

Only the key deliverables are listed in the table and are likely to be produced in a series of iterations and delivered incrementally. A brief summary of each activity follows. The models that are produced and the activities necessary to produce them are explained in more detail in subsequent chapters.

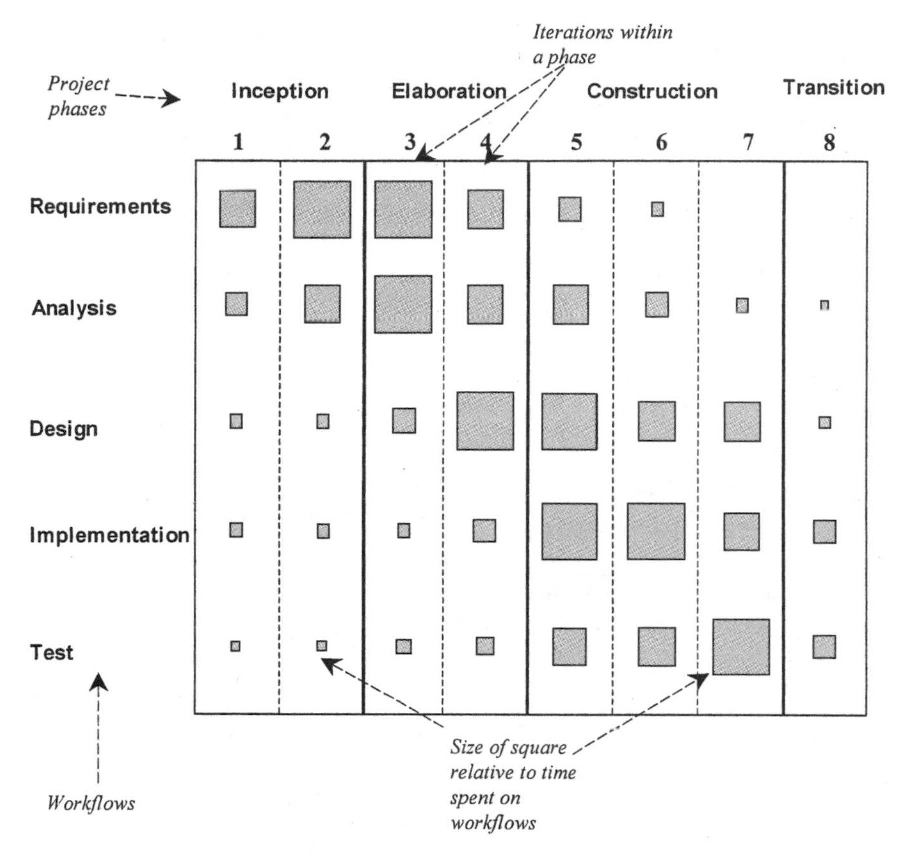

Figure 5.16 Phases and workflows in the Unified Software Development Process.

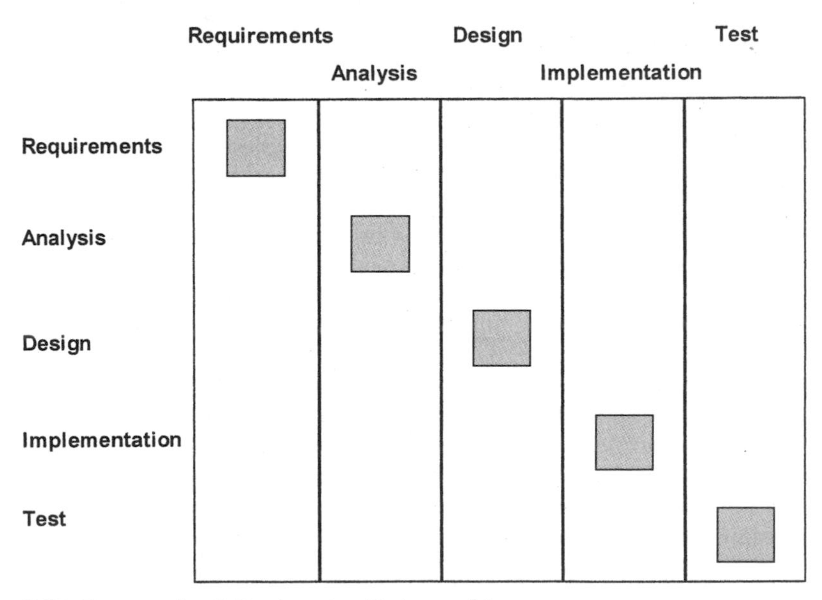

Figure 5.17 Phases and activities in a simplifed waterfall process.

Activity	Techniques	Key Deliverables
Requirements Capture and Modelling	Requirements elicitation Use case modelling Prototyping	Use case model Requirements list Initial architecture Prototypes Glossary
Requirements Analysis	Collaboration Diagrams Class and Object Models Analysis Modelling	Analysis models
System Design	Deployment Modelling Component Modelling Package Modelling Architectural Modelling	Overview design and implementation architecture
Class Design	Class and Object Modelling Interaction Modelling State Modelling Design Patterns	Design models
Interface Design	Class and Object Modelling Interaction Modelling State Modelling Package Modelling Prototyping Design Patterns	Design models with interface specification
Data Management Design	Class and Object Modelling Interaction Modelling State Modelling Package Modelling Design Patterns	Design models with database specification
Construction	Programming Component Reuse Database DDL Programming Idioms	Constructed System Documentation
Testing	Programming Test Procedures	Tested System
Implementation		Installed System

Figure 5.18 Table of system development process activities.

5.4.2.1 Requirements Capture and Modelling

Various fact-finding techniques are used to identify requirements. These are discussed in Chapter 6. Requirements are documented in use cases. A use case captures an element of functionality and the requirements model may include many use cases. For example, in the Agate case study the requirement that the accountant should be able to record the details of a new member of staff on the system is an example of a use case. It would be described initially as follows.

Use Case: Add a new staff member

When a new member of staff joins Agate, his or her details are recorded. He or she is assigned a staff number, and the start date is recorded. Start date defaults to today's date. The starting grade is recorded.

The use cases can also be modelled graphically. The use case model is refined to identify common procedures and dependencies between use cases. The objective of this refinement is to produce a succinct but complete description of requirements.

Prototypes of some key user interfaces may be produced in order to help to understand the requirements that the users have for the system.

An initial system architecture (see Figure 5.19 for part of the Agate system) may be developed to help guide subsequent steps during the development process. This initial architecture will be refined and adjusted as the development proceeds.

5.4.2.2 Requirements Analysis

Essentially each use case describes one or more requirement. Each use case is analysed separately to identify the objects that are required to support it. The use case is also analysed to determine how these objects interact and what responsibilities each of the objects has in order to support the use case. Collaboration diagrams (Figure 5.20) are used to model the object interaction. The models for each use case are then integrated to produce an analysis class diagram, as described in Chapters 7 and 8. Figure 5.21 shows an example of an analysis class. The initial system architecture may be refined as a result of these activities.

5.4.2.3 System Design

At this stage various decisions concerning the design process are made including the further specification of a suitable systems architecture. For example, a possible architecture for the system in the Agate case study is shown in Figure 5.22. This architecture

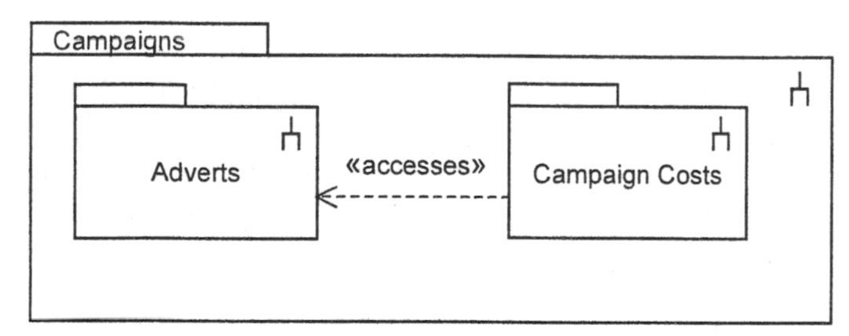

Figure 5.19 Part of the initial system architecture for the Agate system.

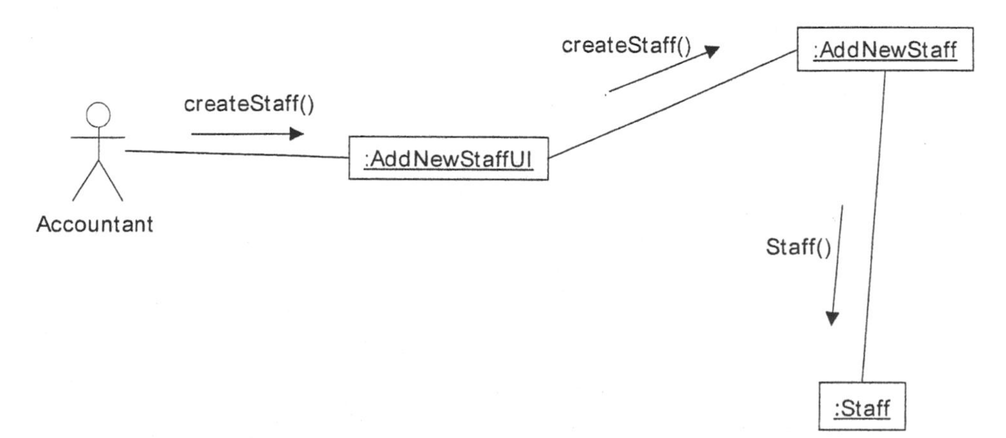

Figure 5.20 Part of a collaboration diagram for the use case `Add New Staff`.

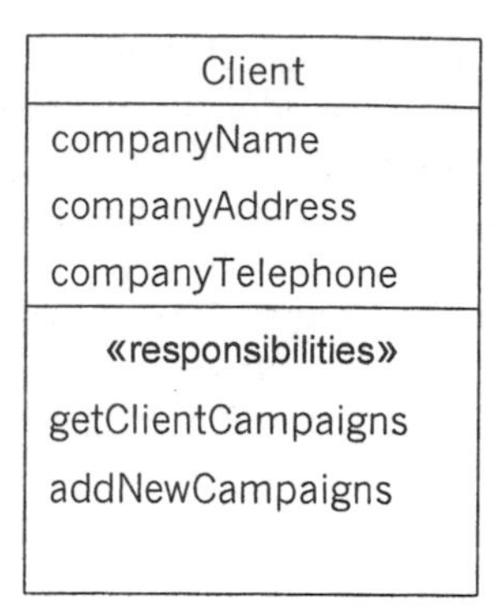

Figure 5.21 Partly completed sample analysis class.

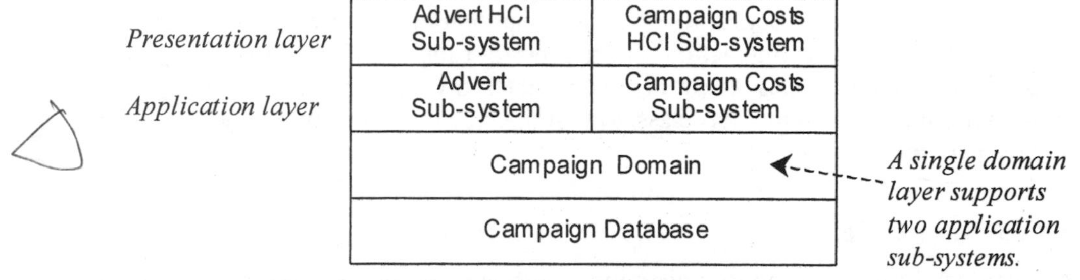

Figure 5.22 Possible architecture for part of the Agate system.

has four layers. The two bottom layers provide common functionality and database access for the staffing and advert planning sub-systems. Part of the architectural specification may include the identification of particular technologies to be used. In this case it may be decided to use a client–server architecture with the sub-system interfaces operating through a web-browser to give maximum operational flexibility.

System design is also concerned with identifying and documenting suitable development standards (e.g. interface design standards, coding standards) for the remainder of the project. System Design is explained in Chapter 13.

5.4.2.4 Class Design

Each of the use case analysis models is now elaborated separately to include relevant design detail. Interaction sequence diagrams may be drawn to show detailed object communication (Chapter 9) and state diagrams may be prepared for objects with complex state behaviour (Chapter 10). The separate models are then integrated to produce a detailed design class diagram. Design classes have attributes and operations specified (Figure 5.23) to replace the less specific responsibilities that may have been identified by the analysis activity (Figure 5.21). The detailed design of the classes normally necessitates the addition of further classes to support, for example, the user interface and access to the data storage system (typically a database management system). Class design is explained in Chapter 14.

Client
-companyName:String -companyAddress:String -companyTelephone:Phone
+getClientCampaigns():campaignList +addNewCampaign(newCampaign: Campaign)

Figure 5.23 Partly completed sample design class.

5.4.2.5 Interface Design

The nature of the functionality offered via each use case has been defined in requirements analysis. Interface design produces a detailed specification as to how the required functionality can be realized. Interface design gives a system its look and feel and determines the style of interaction the user will have. It includes the positioning and colour of buttons and fields, the mode of navigation used between different parts of the system and the nature of online help. Interface design is explained in Chapter 17 and is very dependent on class design.

5.4.2.6 Data Management Design

Data management design focuses on the specification of the mechanisms suitable for implementation of the database management system being used (see Chapter 18). Techniques such as normalization and entity-relationship modelling may be particularly useful if a relational database management system is being used. Data management design and class design are inter-dependent.

5.4.2.7 Construction

Construction is concerned with building the application using appropriate development technologies. Different parts of the system may be built using different languages. Java might be used to construct the user interface, while a database management system such as Oracle would manage data storage and handle commonly used processing routines.

5.4.2.8 Testing

Before the system can be delivered to the client it must be thoroughly tested. Testing scripts should be derived from the use case descriptions that were previously agreed with the client. Testing should be performed as elements of the system are developed.

5.4.2.9 Implementation

The final implementation of the system will include its installation on the various computers that will be used. It will also include managing the transition from the old systems to the new systems for the client. This will involve careful risk management and staff training.

5.5 Summary

As in many kinds of development projects, we use models to represent things and ideas that we want to document and to test out without having to actually build a system. Of course our aim ultimately is to build a system, and the models help us to achieve that. Models allow us to create different views of a system from different perspectives, and in an information system development project, most models are graphical representations of things in the real world and the software artefacts that will be used in the information system.

These graphical representations are diagrams, which can be used to model objects and processes. In UML a number of diagrams are defined and the rules for how they are to be drawn are documented. UML diagrams are made up of four graphical elements: icons, two-dimensional symbols, paths and strings. Diagrams are also supported with textual material, some of which may be informal, for example in plain English, while some may be formal, for example written in Object Constraint Language.

As a project progresses a variety of models are produced in order to represent different aspects of the system that is being built. A model is a complete and consistent view of a system from a particular perspective, typically produced using diagrams. An example of a diagram notation that is used in UML is the activity diagram. Activity diagrams model activities that are carried out in a system and include sequences of activities, alternative paths and repeated activities. As well as being used in system development projects, activity diagrams are also used in the Unified Software Development Process to document the sequence of activities in a workflow.

The Unified Software Development Process provides a specification of a process that can be used to develop software systems. It is made up of phases, within which models of the system are elaborated through successive iterations in which additional detail is added to the models until the system can be constructed in software and implemented. For the purpose of this book, we have broken the software development process into a number of activities that must be undertaken in order to develop a system. These activities are described in more detail in subsequent chapters.

Review Questions

5.1 What is the difference between a diagram and a model?

5.2 What are the four elements of a UML diagram?

5.3 Why do we use models in developing computerized information systems and other artefacts?

5.4 Why do we need standards for the graphical elements of diagrams?

5.5 What is the UML notation for each of the following: package, sub-system and model?

5.6 In what two ways can we show in UML that something is contained within something else, for example a sub-system within another sub-system?

5.7 What is the notation used for an activity in a UML activity diagram?

5.8 What links activities in an activity diagram?

5.9 In what two ways can a decision be represented in a UML activity diagram?

5.10 What are the two special states shown in an activity diagram?

5.11 What is meant by a guard condition?

5.12 What is an object flow?

5.13 What is the notation for an object flow?

5.14 What is the difference between the USDP and the waterfall life cycle in the relationship between activities and phases?

Case Study Work, Exercises and Projects

5.A Some people suggest that information systems are models or simulations of the real world. What are the advantages and disadvantages of thinking of information systems in this way?

5.B Think of other kinds of development project in which models are used. For each kind of project list the different kinds of models that you think are used.

5.C Choose a task that you carry out and that you understand, for example preparing an assignment at college or university, or a task at work. Draw an activity diagram to summarize the activities that make up this task. Use swimlanes if the task involves activities that are carried out by other people.

5.D Choose some of the activities in your activity diagram and break them down into more detail in separate diagrams.

5.E Read about the Rational Unified Process (RUP) (see references in the Further Reading section). Identify some of the differences between RUP and USDP.

Further Reading

- Booch et al. (1999) discuss the purpose of modelling and the differences between models and diagrams. They also describe the notation of activity diagrams.
- Jacobson et al. (1999) describe the Unified Software Development Process and explain the notation of the stereotyped activity diagrams that they use to model the workflows in the USDP.
- An alternative to the USDP is the Rational Unified Process, see Kruchten (1999) or the Rational Corporation website (www.rational.com).

6 CHAPTER

Requirements Capture

OBJECTIVES

In this chapter you will learn

- the distinction between the current and required systems
- when and how to apply the five major fact finding techniques
- the need to document requirements
- how to draw *use case diagrams* to document requirements
- how to write *use case descriptions.*

6.1 Introduction

Part of the job of the systems analyst is to find out from users what they require in a new information system. Indeed, identifying what a new system should be able to do is one of the first steps in its development, whether you are developing some simple programs for your own use or embarking on the development of a large-scale system for a commercial client. The user requirements can be classified in different ways (Section 6.2), and analysts use a range of techniques to identify and document the requirements (Sections 6.3 and 6.5). Each of the main fact finding techniques has advantages and disadvantages and is appropriate for different situations. Stakeholders are the people who have an interest in the new system and whose needs must be considered (Section 6.4). UML provides a diagramming technique that can be used to document the stakeholders' requirements. This is the use case diagram, a relatively simple diagram that is supported by written information in the form of use case descriptions (Section 6.6).

6.2 User Requirements

The aim of developing a new information system must be to produce something that meets the needs of the people who will be using it. In order to do this, we must have a clear understanding both of the overall objectives of the business and of what it is that the individual users of the system are trying to achieve in their jobs. Unless you are in the rare position of developing a system for a new organization, you will need to understand how the business is operating at present and how people are working now. Many aspects of the current system will need to be carried forward into the new system, so it is important that information about what people are doing is gathered and documented. These are the requirements that are derived from the 'current system'. The motivation for the development of a new information system is usually problems with and inadequacies of the current system, so it is also essential to capture what it is that the users require of the new system that they cannot do with their existing system. These are the 'new requirements'.

6.2.1 Current system

The existing system may be a manual one, based on paper documents, forms and files; it may already be computerized; or it may be a combination of both manual and computerized elements. Whichever it is, it is reasonably certain that large parts of the existing system meet the needs of the people who use it, that it has to some extent evolved over time to meet business needs and that users are familiar and comfortable with it. It is almost equally certain that there are sections of the system that no longer meet the needs of the business, and that there are aspects of the business that are not dealt with in the existing system.

It is important that the analyst, gathering information as one of the first steps in developing a new system, gains a clear understanding of how the existing system works: parts of the existing system will be carried forward into the new one. It is also important, because the existing system will have shortcomings and defects, which must be avoided or overcome in the new system. It is not always easy or possible to replace existing systems. So-called *legacy systems* may have been developed some time ago and may contain millions of lines of program code, which have been added to and amended over a period of time. One approach to dealing with such systems is to create new front-ends, typically using modern graphical user interfaces and object-oriented languages, and *wrap* the legacy systems up in new software. If this is the case, then it is also necessary to understand the interfaces to the legacy systems that the new *wrappers* will have to communicate with.

It is not always possible to leave legacy systems as they are and simply wrap them in new code. It was not possible to ignore the problems that faced companies at the turn of the century when it was realized that many systems were in danger of catastrophic collapse as a result of the decision to use two decimal digits to store the year. However, the process of changing the program code in such systems is a matter of understanding the internal working of existing systems rather than gathering information about the way the organization works and the way that people do their jobs.

Not everyone agrees that a detailed understanding of the current system is necessary. Ed Yourdon (1989) argues that it is a waste of time to model the current system in great detail. Yourdon points out that so much time can be spent investigating and modelling the current system that the analysts lose sight of their objective, and impatient users cancel the project. He makes the case for concentrating on the behaviour that is required of the new system. The opposite position is taken by SSADM (Structured Systems Analysis and Design Method), which expends considerable time on the investigation and modelling of the current system. This is done in order to refine it to its logical essence and to be able to merge it with the new requirements to produce a model of the required system that includes the essentials of the existing system.

We believe that a case can be made for investigating the existing system.

- Some of the functionality of the existing system will be required in the new system.
- Some of the data in the existing system is of value and must be migrated into the new system.
- Technical documentation of existing computer systems may provide details of processing algorithms that will be needed in the new system.
- The existing system may have defects that we should avoid in the new system.
- Studying the existing system will help us to understand the organization in general.
- Parts of the existing system may be retained. Information systems projects are now rarely 'green field' projects in which manual systems are replaced by new computerized systems; more often there will be existing systems with which interfaces must be established.
- We are seeking to understand how people do their jobs at present in order to characterize people who will be users of the new system.
- We may need to gather baseline information against which we can set and measure performance targets for the new system.

For all these reasons, an understanding of the current system should be part of the analysis process. However, the analyst should not lose sight of the objective of developing a new system. In the sections on functional, non-functional and usability requirements below, we shall explain what kind of information we are gathering.

6.2.2 New requirements

Most organizations now operate in an environment that is rapidly changing. The relative strength of national economies around the world can change dramatically and at short notice; the fortunes of large companies, which may be an organization's suppliers, customers or competitors, can be transformed overnight; new technologies are introduced which change production processes, distribution networks and the relationship with the consumer; governments and (particularly in Europe) supra-governmental organizations introduce legislation that has an impact on the way that business is conducted. Some authors make the case for developing business strategies to cope with this turmoil. Tom Peters in 'Thriving on Chaos' (1988) argues that we must learn to love change and develop flexible and responsive organizations to cope with the dynamic business environment. A clear result of responding to a dynamic environment is that organizations change their products and services and change the way they do business. The effect of this is to change their need for information. Even in less responsive organizations, information systems become outdated and need enhancing and extending. Mergers and demergers create the need for systems to be replaced. The process of replacement offers an opportunity to extend the capabilities of systems to take advantage of new technological developments, or to enhance their usefulness to management and workforce. Many organizations are driven by internal factors to grow and change the ways in which they operate, and this too provides a motivation for the development of new information systems.

Whether you are investigating the working of the existing system or the requirements for the new system, the information you gather will fall into one of three categories: 'functional requirements', 'non-functional requirements' and 'usability requirements'. Functional and non-functional requirements are conventional categories in systems analysis and design, while usability is often ignored in systems development projects. In many university courses, issues surrounding the usability of systems are taught under the separate heading of Human Factors or Human–Computer Interaction, or are only considered in the design stage of the development process. However, the lesson of human factors research is that usability considerations should be integral to the systems development lifecycle, and so they are included here.

Functional requirements

Functional requirements describe what a system does or is expected to do, often referred to as its *functionality*. In the object-oriented approach, which we are taking here, we shall initially employ use cases to document the functionality of the system. As we progress into the analysis stage, the detail of the functionality will be recorded in the data that we hold about objects, their attributes and operations. In structured methods, such as SSADM, the function may be the unit around which we structure the system, and the function is described in progressively greater detail as we move through analysis into design and implementation.

At this stage, we are setting out to establish what the system must do, and functional requirements include the following.

- Descriptions of the processing that the system will be required to carry out.
- Details of the inputs into the system from paper forms and documents, from interactions between people, such as telephone calls, and from other systems.
- Details of the outputs that are expected from the system in the form of printed documents and reports, screen displays and transfers to other systems.
- Details of data that must be held in the system.

Non-functional requirements

Non-functional requirements are those that describe aspects of the system that are concerned with how well it provides the functional requirements. These include the following.

- Performance criteria such as desired response times for updating data in the system or retrieving data from the system.
- Anticipated volumes of data, either in terms of throughput or of what must be stored.
- Security considerations.

Usability requirements

Usability requirements are those that will enable us to ensure that there is a good match between the system that is developed and both the users of that system and the tasks that they will undertake when using it. The International Standards Organization (ISO) has defined the usability of a product as 'the degree to which specific users can achieve specific goals within a particular environment; effectively, efficiently, comfortably and in an acceptable manner'. Usability can be specified in terms of measurable objectives, and these are covered in more detail in Chapter 16 on Human–Computer Interaction. In order to build usability into the system from the outset, we need to gather the following types of information.

- Characteristics of the users who will use the system.
- The tasks that the users undertake, including the goals that they are trying to achieve.
- Situational factors that describe the situations that could arise during system use.
- Acceptance criteria by which the user will judge the delivered system.

Paul Booth (1989) describes the issues surrounding system usability in more detail.

6.3 Fact Finding Techniques

There are five main fact finding techniques that are used by analysts to investigate requirements. Here we describe each of them in the order that they might be applied in a system development project, and for each one we explain the kind of information that you would expect to gain from its use, its advantages and disadvantages, and the situations in which it is appropriate to use it.

6.3.1 Background reading

If an analyst is employed within the organization that is the subject of the fact-gathering exercise, then it is likely that he or she will already have a good understanding of the organization and its business objectives. If, however, he or she is going in as an outside consultant, then one of the first tasks is to try to gain an understanding of the organization. Background reading or research is part of that process. The kind of documents that are suitable sources of information include the following:

- company reports,
- organization charts,
- policy manuals,
- job descriptions,
- reports and
- documentation of existing systems.

Although reading company reports may provide the analyst with information about the organization's mission, and so possibly some indication of future requirements, this technique mainly provides information about the current system.

Advantages and disadvantages

+ Background reading helps the analyst to get an understanding of the organization before meeting the people who work there.
+ It also allows the analyst to prepare for other types of fact finding, for example, by being aware of the business objectives of the organization.

+ Documentation on the existing system may provide formally defined information requirements for the current system.
- Written documents often do not match up to reality; they may be out of date or they may reflect the official policy on matters that are dealt with differently in practice.

Appropriate situations

Background reading is appropriate for projects where the analyst is not familiar with the organization being investigated. It is useful in the initial stages of investigation.

6.3.2 Interviewing

Interviewing is probably the most widely used fact finding technique; it is also the one that requires the most skill and sensitivity. Because of this, we have included a set of guidelines on interviewing that includes some suggestions about etiquette in Box 6.1.

Box 6.1 Guidelines on Interviewing

Conducting an interview requires good planning, good interpersonal skills and an alert and responsive frame of mind. These guidelines cover the points you should bear in mind when planning and conducting an interview.

Before the interview

You should always make appointments for interviews in advance. You should give the interviewee information about the likely duration of the interview and the subject of the interview.

Being interviewed takes people away from their normal work. Make sure that they feel that it is time well spent.

It is conventional to obtain permission from an interviewee's line manager before interviewing them. Often the analyst interviews the manager first and uses the opportunity to get this permission.

In large projects, an interview schedule should be drawn up showing who is to be interviewed, how often and for how long. Initially this will be in terms of the job roles of interviewees rather than named individuals. It may be the manager who decides which individual you interview in a particular role.

Have a clear set of objectives for the interview. Plan your questions and write them down. Some people write the questions with space between them for the replies.

Make sure your questions are relevant to the interviewee and his or her job.

At the start of the interview

Introduce yourself and the purpose of the interview.

Arrive on time for interviews and stick to the planned timetable—do not over-run.

Ask the interviewee if he or she minds you taking notes or tape-recording the interview. Even if you tape-record an interview, you are advised to take notes. Machines can fail! Your notes also allow you to refer back to what has been said during the course of the interview and follow up points of interest.

Remember that people can be suspicious of outside consultants who come in with clipboards and stopwatches. The cost-benefit analyses of many information systems justify the investment in terms of savings in jobs!

During the interview

Take responsibility for the agenda. You should control the direction of the interview. This should be done in a sensitive way. If the interviewee is getting away from the subject, bring them back to the point. If what they are telling you is important, then say that you will come back to it later and make a note to remind yourself to do so.

Use different kinds of question to get different types of information. Questions can be open-ended—'Can you explain how you complete a timesheet?'—or closed—'How many staff use this system?'. Do not, however, ask very open-ended questions such as 'Could you tell me what you do?'

Listen to what the interviewee says and encourage him or her to expand on key points.

Keep the focus positive if possible. Make sure you have understood answers by summarizing them back to the interviewee. Avoid allowing the interview to degenerate into a session in which the interviewee complains about everyone and everything.

You may be aware of possible problems in the existing system, but you should avoid prejudging issues by asking questions that focus too much on problems. Gather facts.

Be sensitive about how you use information from other interviews that you or your colleagues have already conducted, particularly if comments were negative or critical.

Use the opportunity to collect examples of documents that people use in their work, ask if they mind you having samples of blank forms and photocopies of completed paperwork.

After the interview

Thank the interviewee for their time. Make an appointment for a further interview if it is necessary. Offer to provide them with a copy of your notes of the interview for them to check that you have accurately recorded what they told you.

Transcribe your tape or write up your notes as soon as possible after the interview while the content is still fresh in your mind.

If you said that you would provide a copy of your notes for checking then send it to the interviewee as soon as possible. Update your notes to reflect their comments.

A systems analysis interview is a structured meeting between the analyst and an interviewee who is usually a member of staff of the organization being investigated. The interview may be one of a series of interviews that range across different areas of the interviewee's work or that probe in progressively greater depth about the tasks undertaken by the interviewee. The degree of structure may vary: some interviews are planned with a fixed set of questions that the interviewer works through, while others are designed to cover certain topics but will be open-ended enough to allow the interviewer to pursue interesting facts as they emerge. Thc ability to respond flexibly to the interviewee's responses is one of the reasons why interviews are so widely used.

Interviews can be used to gather information from management about their objectives for the organization and for the new information system, from staff about their existing jobs and their information needs, and from customers and members of the public as possible users of systems. While conducting an interview, the analyst can also use the opportunity to gather documents that the interviewee uses in his or her work.

It is usually assumed that questionnaires are used as a substitute for interviews when potential interviewees are geographically dispersed in branches and offices around the world. The widespread use of desktop video conferencing may change this and make it possible to interview staff wherever they are. Even then, questionnaires can reach more people.

Interviewing different potential users of a system separately can mean that the analyst is given different information by different people. Resolving these differences later can be difficult and time-consuming. One alternative is to use group interviews in order to get the users to reach a consensus on issues. Dynamic Systems Development Method (DSDM) is a method of carrying out systems development in which group discussions are used (Stapleton, 1997). These discussions are run as workshops for knowledgeable users with a facilitator who aims to get the users to pool their knowledge and to reach a consensus on the priorities of the development project.

Advantages and disadvantages

+ Personal contact allows the analyst to be responsive and adapt to what the user says. Because of this, interviews produce high quality information.
+ The analyst can probe in greater depth about the person's work than can be achieved with other methods.
+ If the interviewee has nothing to say, the interview can be terminated.
- Interviews are time-consuming and can be the most costly form of fact gathering.
- Interview results require the analyst to work on them after the interview: the transcription of tape recordings or writing up of notes.
- Interviews can be subject to bias if the interviewer has a closed mind about the problem.
- If different interviewees provide conflicting information, it can be difficult to resolve later.

Appropriate situations

Interviews are appropriate in most projects. They can provide information in depth about the existing system and about people's requirements from a new system.

6.3.3 Observation

Watching people carrying out their work in a natural setting can provide the analyst with a better understanding of the job than interviews, in which the interviewee will often concentrate on the normal aspects of the job and forget the exceptional situations and interruptions which occur and which the system will need to cope with. Observation also allows the analyst to see what information people use to carry out their job. This can tell you about the documents they refer to, whether they have to get up from their desks to get information, how well the existing system handles their needs. One of the authors has observed staff using a tele-sales system where there was no link between the enquiry screens for checking the availability of stock and the data entry screens for entering an order. These tele-sales staff kept a pad of scrap paper on the desk and wrote down the product codes for all the items they had looked up on the enquiry screens so that they could enter them into the order-processing screens. This kind of information does not always emerge from interviews.

People are not good at estimating quantitative data, such as how long they take to deal with certain tasks, and observation with a stopwatch can give the analyst plentiful quantitative data, not just about typical times to perform a task but also about the statistical distribution of those times.

In some cases where information or items are moving through a system and being dealt with by many people along the way, observation can allow the analyst to follow the entire process through from start to finish. This type of observation might be used in an organization where orders are taken over the telephone, passed to a warehouse for picking, packed and despatched to the customer. The analyst may want to follow a series of transactions through the system to obtain an overview of the processes involved.

Observation can be an open-ended process in which the analyst simply sets out to observe what happens and to note it down, or it can be a closed process in which the analyst wishes to observe specific aspects of the job and draws up an observation schedule or form on which to record data. This can include the time it takes to carry out a task, the types of task the person is performing or factors such as the number of errors they make in using the existing system as a baseline for usability design.

Advantages and disadvantages

+ Observation of people at work provides first hand experience of the way that the current system operates.
+ Data are collected in real time and can have a high level of validity if care is taken in how the technique is used.
+ Observation can be used to verify information from other sources or to look for exceptions to the standard procedure.
+ Baseline data about the performance of the existing system and of users can be collected.
– Most people do not like being observed and are likely to behave differently from the way in which they would normally behave. This can distort findings and affect the validity.
– Observation requires a trained and skilled observer for it to be most effective.
– There may be logistical problems for the analyst, for example, if the staff to be observed work shifts or travel long distances in order to do their job.
– There may also be ethical problems if the person being observed deals with sensitive private or personal data or directly with members of the public, for example in a doctor's surgery.

Appropriate situations

Observation is essential for gathering quantitative data about people's jobs. It can verify or disprove assertions made by interviewees, and is often useful in situations where different interviewees have provided conflicting information about the way the system works. Observation may be the best way to follow items through some kind of process from start to finish.

6.3.4 Document sampling

Document sampling can be used in two different ways. First, the analyst will collect copies of blank and completed documents during the course of interviews and observation sessions. These will be used to determine the information that is used by people in their work, and the inputs to and outputs from processes which they carry out, either manually or using an existing computer system. Ideally, where there is an existing system, screen shots should also be collected in order to understand the inputs and outputs of the existing system. Figure 6.1 shows a sample document collected from Agate, our case study company.

Agate
Campaign Summary

Date 23rd February 2002

Client Yellow Partridge
Park Road Workshops
Jewellery Quarter
Birmingham B2 3DT
U.K.

Campaign Spring Collection 2002

Billing Currency GBP £

Item	Curr	Amount	Rate	Billing amount
Advert preparation: photography, artwork, layout etc.	GBP £	15,000.00	1	15,000.00
Placement French Vogue	EUR €	6 500,00	1.61	4,037.27
Placement Portuguese Vogue	EUR €	5 500,00	1.61	3,416.15
Placement US Vogue	USD $	17,000.00	1.44	11,805.56
Total				34,258.98

This is not a VAT Invoice. A detailed VAT Invoice will be provided separately.

Figure 6.1 Sample document from the AGATE case study.

Second, the analyst may carry out a statistical analysis of documents in order to find out about patterns of data. For example, many documents such as order forms contain a header section and a number of lines of detail. (The sample document in Figure 6.1 shows this kind of structure.) The analyst may want to know the distribution of the number of lines in an order. This will help later in estimating volumes of data to be held in the system and in deciding how many lines should be displayed on screen at one time. While this kind of statistical sampling can give a picture of data volumes, the analyst should be alert to seasonal patterns of activity, which may mean that there are peaks and troughs in the amount of data being processed.

Advantages and disadvantages

+ Can be used to gather quantitative data, such as the average number of lines on an invoice.
+ Can be used to find out about error rates in paper documents.
- If the system is going to change dramatically, existing documents may not reflect how it will be in future.

Appropriate situations

The first type of document sampling is almost always appropriate. Paper-based documents give a good idea of what is happening in the current system. They also provide supporting evidence for the information gathered from interviews or observation.

The statistical approach is appropriate in situations where large volumes of data are being processed, and particularly where error rates are high, and a reduction in errors is one of the criteria for usability.

6.3.5 Questionnaires

Questionnaires are a research instrument that can be applied to fact finding in system development projects. They consist of a series of written questions. The questionnaire designer usually limits the range of replies that respondents can make by giving them a choice of options. (Figure 6.2 shows some of the types of question.) YES/NO questions only give the respondent two options. (Sometimes a DON'T KNOW option is needed as well.) If there are more options, the multiple choice type of question is often used when the answer is factual, whereas scaled questions are used if the answer involves an element of subjectivity. Some questions do not have a fixed number of responses, and must be left open-ended for the respondent to enter what they like. Where the respondent has a limited number of choices, these are usually coded with a number, which speeds up data entry if the responses are to be analysed by computer software. If you plan to use questionnaires for requirements gathering, they need very careful design. Box 6.2 lists some of the issues that need to be addressed if you are thinking of using questionnaires.

Advantages and disadvantages

+ An economical way of gathering data from a large number of people.
+ If the questionnaire is well designed, then the results can be analysed easily, possibly by computer.
- Good questionnaires are difficult to construct.
- There is no automatic mechanism for follow up or probing more deeply, although it is possible to follow up with an interview by telephone or in person if necessary.
- Postal questionnaires suffer from low response rates.

YES/NO Questions

Do you print reports from the existing system? (Please circle the appropriate answer.)	YES	NO	10

Multiple Choice Questions

How many new clients do you obtain in a year? (Please tick one box only.)	a) 1–10	☐	11
	b) 11–20	☐	
	c) 21–30	☐	
	d) 31 +	☐	

Scaled Questions

How satisfied are you with the response time of the stock update?
(Please circle one option.)

1. Very satisfied	2. Satisfied	3. Dissatisfied	4. Very dissatisfied	12

Open-ended Questions

What additional reports would you require from the system?

____

____

____

Figure 6.2 Types of question used in questionnaires.

Box 6.2 Guidelines on Questionnaires

Using questionnaires requires good planning. If you send out 100 questionnaires and they do not work, it is difficult to get respondents to fill in a second version. These guidelines cover the points you should bear in mind when using questionnaires.

Coding

How will you code the results? If you plan to use an optical mark reader, then the response to every question must be capable of being coded as a mark in a box. If you expect the results to be keyed into a database for analysis, then you need to decide on the codes for each possible response. If the questions are open-ended, how will you collate and analyse different kinds of responses?

Analysis

Whatever analysis you plan should be decided in advance. If you expect to carry out a statistical analysis of the responses, you should consult a statistician before you finalize the questions. Statistical techniques are difficult to apply to responses to poorly designed questions.

You can use a special statistical software package, a database or even a spreadsheet to analyse the data.

Piloting

You should try out your questionnaire on a small pilot group or sample of your respondents. This enables you to find out if there are questions they do not understand, they misinterpret or they cannot answer.

If you plan to analyse the data using statistical software, a database or a spreadsheet, you can create a set of trial data to test your analysis technique.

Sample size and structure

If you plan to use serious statistical techniques, then those techniques may place lower limits on your sample size. If you want to be sure of getting a representative sample, by age, gender, department, geographical location, job grade or experience of existing systems, then that will help to determine how many people to include. Otherwise it may be down to you to choose a sensible percentage of all the possible respondents.

Delivery
How will you get the questionnaires to your respondents, and how will they get their replies back to you?

You can post them, or use internal mail in a large organization, fax them, e-mail them or create a web-based form on the company intranet and notify your target group by e-mail. If you use the intranet, you may want to give each respondent a special code, so that only they can complete their own questionnaire.

Your respondents can then post, fax or e-mail their responses back to you.

Respondent information
What information about the respondents do you want to gather at the same time as you collect their views and requirements? If you want to analyse responses by age, job type or location, then you need to include questions that ask for that information.

You can make questionnaires anonymous, or you can ask respondents for their name. If the questionnaire is not anonymous, you need to think about confidentiality. People will be more honest in their replies if they can respond anonymously or in confidence.

If you ask for respondents' names and you store that information, then in the UK you should consider the provisions of the Data Protection Act (1998). (See also Chapter 12.) There are similar requirements in other countries.

Covering letter
In a covering letter you should explain the purpose and state that the questionnaire has management support. Give an estimate of the time required to fill in the questionnaire and a deadline for its return. Thank the respondents for taking part.

Structure
Structure the questionnaire carefully. Give it a title, and start with explanatory material and notes on how to complete it. Follow this with questions about the respondent (if required). Group questions together by subject. Avoid lots of instructions like 'If you answered YES to Q. 7a, now go to Q. 13.' Keep it reasonably short.

Return rate
Not everyone will necessarily respond. You need to plan for this and either use a larger sample than you need or follow up with reminders. If you use a form on the Intranet, you should be able to identify who has not responded and e-mail them reminders. Equally, you can e-mail a thank you to those who do respond.

Feedback
This needs to be handled carefully—telling everyone that 90% of the company cannot use the existing system may not go down well—but people do like to know what use was made of the response they made. They may have spent half an hour filling in your questionnaire, and they will expect to be informed of the outcome. A summary of the report can be sent out to branches, distributed to departments, sent to named respondents or placed on the company intranet.

Appropriate situations

Questionnaires are most useful when the views or knowledge of a large number of people need to be obtained or when the people are geographically dispersed, for example, in a company with many branches or offices around the country or around the world. Questionnaires are also appropriate for information systems that will be used by the general public, and where the analyst needs to get a picture of the types of user and usage that the system will need to handle.

6.3.6 Remembering the techniques

For those who like mnemonics, these techniques are sometimes referred to as SQIRO—Sampling, Questionnaires, Interviewing, Reading (or Research) and Observation. This order has been chosen to make it possible to pronounce the mnemonic. However, this is not the order in which they are most likely to be used. This will depend on the situation and the organization in which the techniques are being used.

6.3.7 Other techniques

Some kinds of systems require special fact finding techniques. *Expert systems* are computer systems that are designed to embody the expertise of a human expert in solving problems. Examples include systems for medical diagnosis, stock market trading and geological analysis for mineral prospecting. The process of capturing the knowledge of the expert is called *knowledge acquisition* and, as it differs from establishing the requirements for a conventional information system, a number of specific techniques are applied. Some of these are used in conjunction with computer-based tools.

6.4 User Involvement

The success of a systems development project depends not just on the skills of the team of analysts, designers and programmers who work on it, or on the project management skills of the project manager, but on the effective involvement of users in the project at various stages of the life cycle. The term *stakeholders* was introduced in Chapter 2 to describe all those people who have an interest in the successful development of the system. Stakeholders include all people who stand to gain (or lose) from the implementation of the new system: users, managers and budget-holders. Analysts deal with people at all levels of the organization. In large projects it is likely that a steering committee with delegated powers will be set up to manage the project from the users' side. This will include the following categories of people:

- senior management—with overall responsibility for running the organization,
- financial managers with budgetary control over the project,
- managers of the user department(s) and
- representatives of users.

Users will be involved in different roles during the course of the project as:

- subjects of interviews to establish requirements,
- representatives on project committees,
- those involved in evaluating prototypes,
- those involved in testing,
- subjects of training courses and
- end-users of the new system.

Case Study Example

The section that follows applies to what has been covered in this chapter so far to the case study.

One of the first tasks in fact finding is to draw up a plan that outlines what information is being sought, which techniques will be used, who is involved and how long the fact finding will take. A draft plan for fact finding at Agate is shown below. The jobs of the subjects are those shown in Figure A1.1 in the Agate case study.

Objective	Technique	Subject(s)	Time commitment
To get background on the company and the advertising industry	Background reading	Company reports, trade journals	0.5 day
To establish business objectives. Agree likely scope of new system. Check out involvement of non-UK offices	Interview	Two directors	2 x 1 hour each
To gain understanding of roles of each department. Check out line management and team structure in the Creative Department. To agree likely interviewees among staff	Interview	Department heads (only 1 account manager)	2 x 1 hour each
To find out how the core business operates	Interview	1 account manager 1 graphic designer 1 copy writer 1 editor	1.5 hours each
To follow up development of business understanding	Observation	2 creative staff	0.5 day each
To determine role of support/admin staff and relationship to core business	Interview	2 admin staff (based on experience with the company)	1.5 hours each
To establish what records and resources are kept	Interview/ document sampling	Filing clerk Resource librarian	2 x 1 hour each
To determine what use is made of current computer system. To determine functionality of current system	Interview	Computer manager	2 x 1 hour
To establish additional requirements for new system	Interview	2 account managers 3 staff from Creative Department	3 x 1 hour each
To establish accounting requirements for new system	Interview	Accountant Credit controller 1 purchasing assistant 1 accounts clerk	1.5 hours each

6.5 Documenting Requirements

Information systems professionals need to record facts about the organization they are studying and its requirements. As soon as the analysts start gathering facts, they will need some means of documenting them. In the past the emphasis was on paper forms, but now it is rare for a large-scale project to depend on paper-based documentation. As we have explained in Chapter 5, systems analysts and designers model the new system in a mixture of diagrams and text. The important thing to bear in mind is that within a project some set of standards should be adhered to. These may be the agreed standards of the organization carrying out the analysis and design project or they may be a requirement of the organization that is having the work done. For example, government and military projects usually require that developers conform to a specific set of standards. We are using UML to produce models of the system from different perspectives. Computer Aided Software Engineering (CASE) tools are normally used to draw the diagrammatic models and to maintain in a repository the associated data about the various things that are shown in the diagrams.

However, there will also be other kinds of documents, not all of which fit into the UML framework. In large-scale projects a librarian or configuration manager may be required to keep track of these documents and ensure that they are stored safely and in a way that enables them to be retrieved when required. Such documents include the following:

- records of interviews and observations,
- details of problems,
- copies of existing documents and where they are used,
- details of requirements,
- details of users and
- minutes of meetings.

Even in smaller projects which cannot justify a librarian, a filing system with an agreed set of conventions on how material is to be filed, and for recording who has taken items from the filing system is good practice.

In many projects, these documents will be stored digitally, using a document management system or a version control system. In this case, many people can access the same document simultaneously. The system enforces control over whether a document can be updated, and ensures that no more than one person at a time is able to 'check out' a document in order to amend it.

Not all of the documents listed above represent requirements, and it is necessary to maintain some kind of list or database of requirements. There are software tools available to hold requirements in a database, and some can be linked to CASE tools and testing tools. This makes it possible to trace from an initial requirement through the analysis and design models to where it has been implemented and to the test cases that test whether the requirement has been met.

Use cases, which are explained in the next section, can be used to model requirements, but because they focus on the functionality of the system, are not good for documenting non-functional requirements. Jacobson et al. (1999) suggest that the use case model should be used to document functional requirements and a separate list of 'supplementary requirements' (those not provided by a use case) should be kept. They say that together, the use case model and the list of supplementary requirements constitute a traditional requirements specification. Rosenberg and Scott (1999) argue that use cases are not the same as requirements: use cases describe units of system behaviour, whereas requirements are rules that govern the behaviour of the system; one requirement may be met by more than one use case, and one use case may meet more than one requirement; some non-functional requirements are difficult to attribute to any particular use case.

Some people try to document requirements in use cases by writing long use case descriptions using templates that enable them to include non-functional requirements as well as functional requirements. One of the authors has also come across developers who use the process of brainstorming for use cases as a way of eliciting requirements. However, this tends to produce some very odd use cases.

We favour the view that use cases can be used to model functional requirements, but a separate list of requirements should be kept, containing all requirements—functional and non-functional—for the system. Where there is a relationship between a particular use case and a particular requirement, this should be recorded. Moreover, some requirements describe very high-level units of behaviour and may need to be broken down into low-level requirements that describe more precisely what is to be done. Any database of requirements should make it possible to hold this kind of hierarchical structure of requirements.

Sometimes the process of requirement gathering throws up more requirements than can be met in a particular project. They may be outside the scope of the project, over-ambitious, too expensive to implement or just not really necessary at this point in time. The process of building a requirements model for a system involves going through all the candidate requirements to produce a list of those that will be part of the current project. Figure 6.3 shows this as an activity diagram.

6.6 Use Cases

Use cases are descriptions of the functionality of the system from the users' perspective. Use case diagrams are used to show the functionality that the system will provide and to show which users will communicate with the system in some way to use that functionality. Figure 6.4 shows an example of a use case diagram. This is a relatively simple diagramming technique, and its notation is explained below in Section 6.6.2.

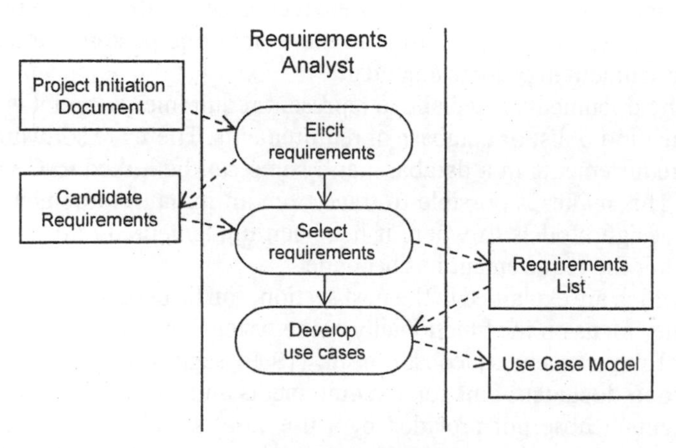

Figure 6.3 Activity diagram to show the activities involved in capturing requirements.

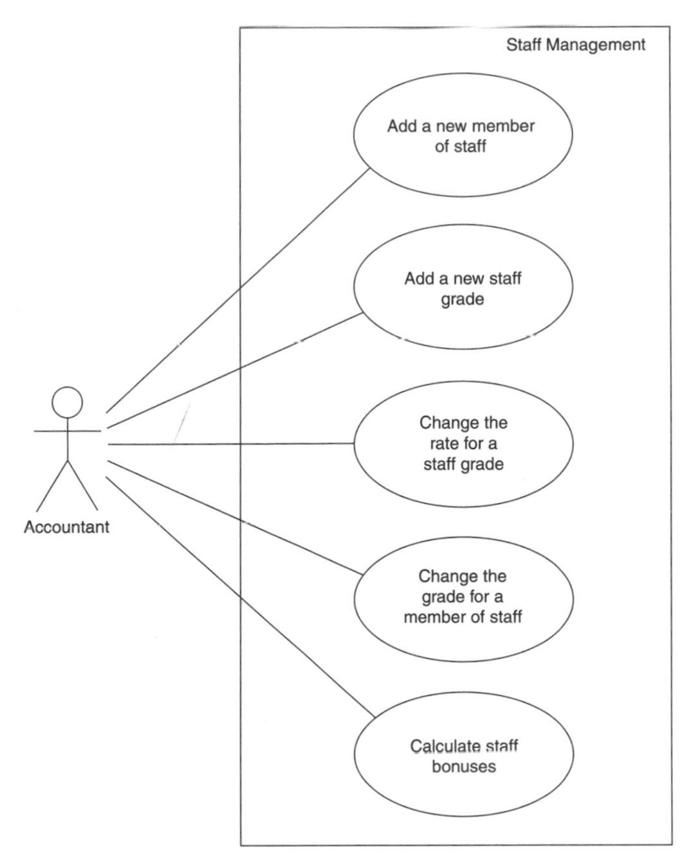

Figure 6.4 Example use case diagram.

Use case diagrams were developed by Jacobson et al. (1992), and the subtitle of the book in which they are presented is *A Use Case Driven Approach*. Jacobson and his co-authors offer a complete approach to the development of object-oriented software systems, but use case diagrams are the starting point for much of what follows in their approach.

6.6.1 Purpose

The use case model is part of what Jacobson et al. (1992) call the requirements model; they also include a problem domain object model and user interface descriptions in this requirements model. Use cases specify the functionality that the system will offer from the users' perspective. They are used to document the scope of the system and the developer's understanding of what it is that the users require.

Use cases are supported by *behaviour specifications*. These specify the behaviour of each use case either using UML diagrams, such as *collaboration diagrams* or *sequence diagrams* (see Chapter 9), or in text form as *use case descriptions*.

Textual *use case descriptions* provide a description of the interaction between the users of the system, termed *actors*, and the high level functions within the system, the use cases. These descriptions can be in summary form or in a more detailed form in which the interaction between actor and use case is described in a step-by-step way. Whichever approach is used, it should be remembered that the use case describes the interaction as the user sees it, and is not a definition of the internal processes within the system, or some kind of program specification.

6.6.2 Notation

Use case diagrams show three aspects of the system: actors, use cases and the system or sub-system boundary. Figure 6.5 shows the elements of the notation.

Actors represent the roles that people, other systems or devices take on when communicating with the particular use cases in the system. Figure 6.5 shows the actor `Staff Contact` in a diagram for the Agate case study. In Agate, there is no job title `Staff Contact`: a director, an account manager or a member of the creative team can take on the role of being staff contact for a particular client company, so one actor can represent several people or job titles. Equally, a particular person or job title may be represented by more than one actor on use case diagrams. This is shown in Figures 6.5 and 6.6 together. A director or an account manager may be the `Campaign Manager` for a particular client campaign, as well as being the `Staff Contact` for one or more clients.

The use case description associated with each use case can be brief:

Assign staff to work on a campaign

The campaign manager selects a particular campaign. A list of staff not already working on that campaign is displayed, and he or she selects those to be assigned to this campaign.

Alternatively, it can provide a step-by-step breakdown of the interaction between the user and the system for the particular use case. An example of this extended approach is provided below.

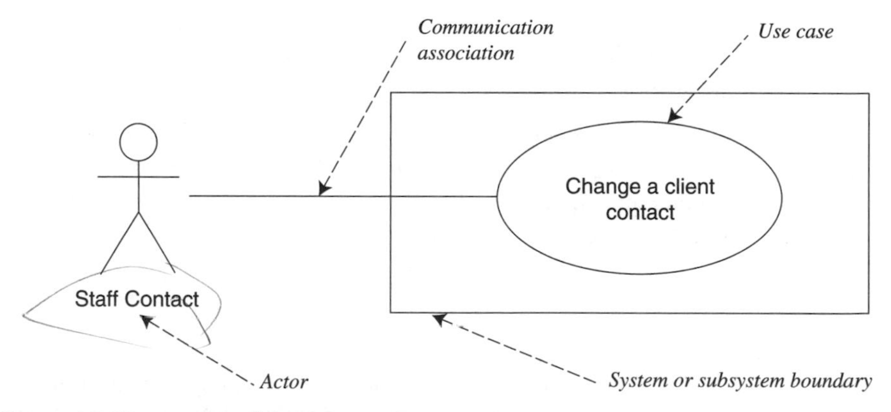

Figure 6.5 The notation of the use case diagram.

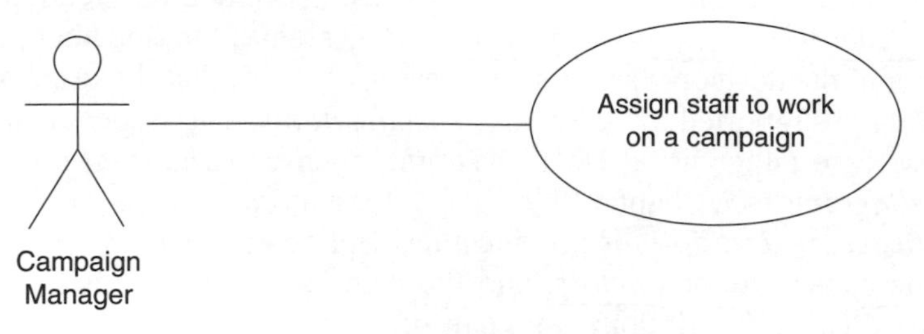

Figure 6.6 Use case showing `Campaign Manager` actor.

Assign staff to work on a campaign

	Actor Action		**System Response**
1.	The actor enters the client name.	2.	Lists all campaigns for that client.
3.	Selects the relevant campaign.	4.	Displays a list of all staff members not already allocated to this campaign.
5.	Highlights the staff members to be assigned to this campaign.	6.	Presents a message confirming that staff have been allocated.

Alternative Courses

Steps 1-3. The actor knows the campaign name and enters it directly.

Constantine (1997) makes the distinction between *essential* and *real* use cases. Essential use cases describe the 'essence' of the use case in terms that are free of any technological or implementation details, whereas real use cases describe the concrete detail of the use case in terms of its design. During the analysis stage, use cases are almost always essential, as the design has not yet been decided upon. In a real use case, Step 2 in the use case description for `Assign staff to work on a campaign` could be described as 'Lists all campaigns for the client in a list box, sorted into alphabetical order by campaign title.'

Each use case description represents the usual way in which the actor will go through the particular transaction or function from end to end. Possible major alternative routes that could be taken are listed as *alternative courses*. The term *scenario* is used to describe use cases in which an alternative course is worked through in detail, including possible responses to errors. The use case represents the generic case, while the scenarios represent specific paths through the use case.

As well as the description of the use case itself, the documentation should include the purpose or intent of the use case, that is to say details of the task that the user is trying to achieve through the means of this use case, for example:

The campaign manager wishes to record which staff are working on a particular campaign. This information is used to validate timesheets and to calculate staff year-end bonuses.

One way of documenting use cases is to use a template (a blank form or word-processing document to be filled in). This might include the following sections:

- name of use case,
- pre-conditions (things that must be true before the use case can take place), Entry Condition
- post-conditions (things that must be true after the use case has taken place), Exit condition.
- purpose (what the use case is intended to achieve) and
- description (in summary or in the format above).

Two further kinds of relationships can be shown on the use case diagram itself. These are the *Extend* and *Include* relationships. They are shown on the diagram using two pieces of UML notation that you will come across in other diagrams: dependencies and stereotypes.

Dependencies—a dependency is a relationship between two modelling elements where a change to one will probably require a change to the other because the one is dependent in some way on the other. A dependency is shown by a dashed line with an open arrowhead pointing at the element on which the other is dependent. There are many kinds of dependencies in UML, and they are distinguished from one another using stereotypes.

Stereotypes—a stereotype is a special use of a model element that is constrained to behave in a particular way. Stereotypes can be shown by using a keyword, such as 'extend' or 'include' in matched *guillemets*, like «extend». (Guillemets are used as quotation marks in French and some other languages.) Stereotypes can also be represented using special icons. The actor symbol in use case diagrams is a stereotyped icon—an actor is a stereotyped class and could also be shown as a class rectangle (see Chapter 7) with the stereotype «actor» above the name of the actor. So by stereotyping classes as «actor» we are indicating that they are a special kind of class that interacts with the system's use cases. Note, however, that actors are external to the system, unlike use cases and classes.

The Extend and Include relationships are easy to confuse. «extend» is used when you wish to show that a use case provides additional functionality that may be required in another use case. In Figure 6.7, the use case `Print campaign summary` extends `Check campaign budget`. This means that at a particular point in `Check Campaign Budget` the user can optionally invoke the behaviour of `Print campaign summary`, which does something over and above what is done in `Check campaign budget` (print out the information in this case). There may be more than one way of extending a particular use case, and these possibilities may represent significant variations on the way the user uses the system. Rather than trying to capture all these variations in one use case, you would document the core functionality in one and then extend it in others. Extension points can be shown in the diagram, as in `Check campaign budget` in Figure 6.7. They are shown in a separate compartment in the use case ellipse, headed `Extension points`. The name of the extension point is given and a description of the point in the use case where it occurs. A condition can be shown next to the dependency relationship. This condition must be true for the extension to take place in a particular instance of the use case.

«include» applies when there is a sequence of behaviour that is used frequently in a number of use cases, and you want to avoid copying the same description of it into each use case in which it is used. Figure 6.8 shows that the use case `Assign staff to work on a campaign` has an «include» relationship with `Find campaign`. This means that when an actor uses `Assign staff to work on a campaign` the behaviour of `Find campaign` will also be included in order to select the relevant Campaign.

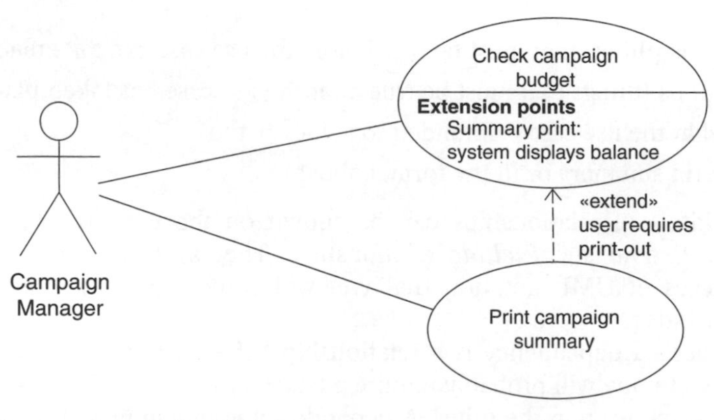

Figure 6.7 Use case diagram showing «extend»

As well as describing the use cases, it is worth describing who the actors are in terms of job titles or the way in which they interact with the system. Although at the moment, we are concentrating on requirements, later we shall need to know who the actual users are for each high level function that is represented by a use case. This may help in specifying the security for different functions or in assessing the usability of the functions.

Bear in mind that actors need not be human users of the system. They can also be other systems that communicate with the one that is the subject of the systems development project, for example, other computers or automated machinery or equipment.

Figure 6.9 shows a use case diagram for the Campaign Management subsystem with both Extend and Include relationships. Note that you do not have to show all the detail of the extension points on a diagram: the `Extension points` compartment in the use case can be suppressed. Of course, if you are using a CASE tool to draw and manage the diagrams, you may be able to toggle the display of this compartment on and off, and even if the information is not shown on a particular diagram, it will still be held in the CASE tool's repository.

In Chapter 4, the concepts of generalization, specialization and inheritance were introduced. They are explained in more detail in Chapter 8. However, generalization and specialization can be applied to actors and use cases. For example, suppose that we have two actors, `Staff Contact` and `Campaign Manager`, and a `Campaign`

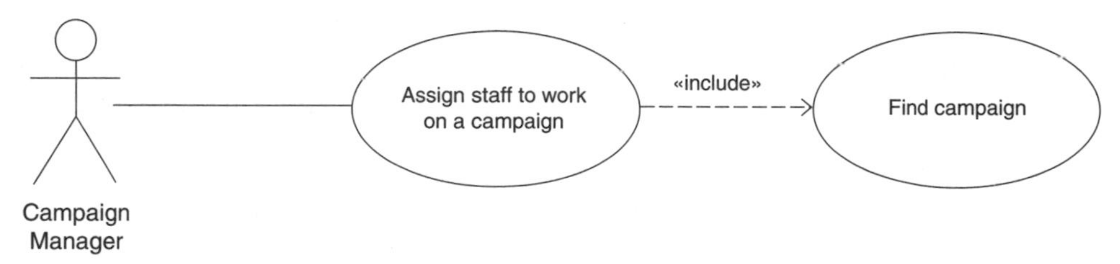

Figure 6.8 Use case diagram showing «`include`»

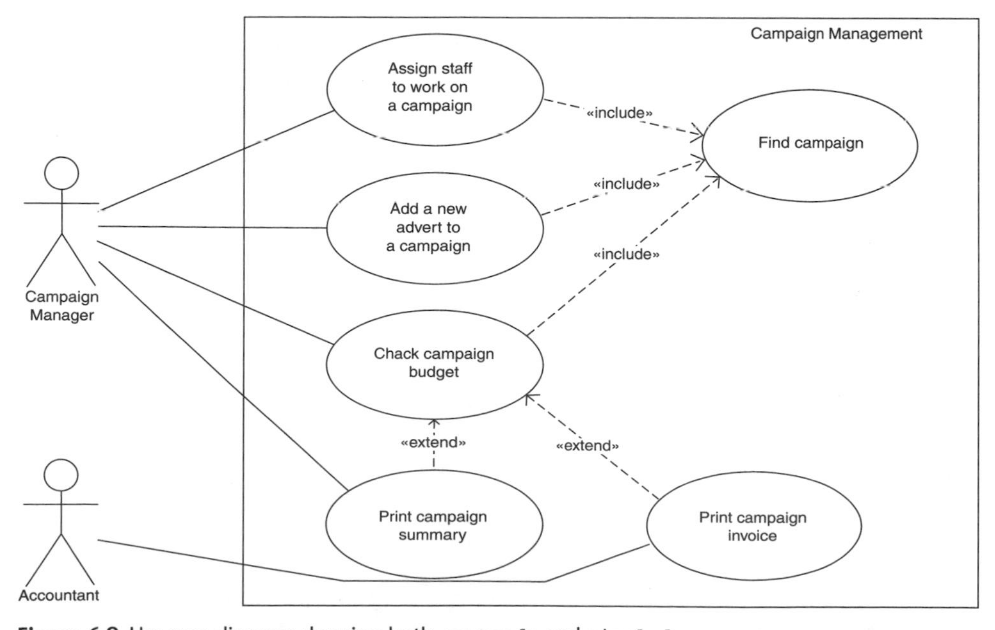

Figure 6.9 Use case diagram showing both «`extend`» and «`include`».

Manager can do everything that a Staff Contact can do, and more. Rather than showing communication associations between Campaign Manager and all the use cases that Staff Contact can use, we can show Campaign Manager as a specialization of Staff Contact, as in Figure 6.10. Similarly, there may be similar use cases where the common functionality is best represented by generalizing out that functionality into a 'super-use case' and showing separate. For example, we may find that there are two use cases at Agate Assign individual staff to work on a campaign, and Assign team of staff to work on a campaign, which are similar in the functionality they offer. We might abstract out the commonality into a use case Assign staff to work on a campaign, but this will be an abstract use case. It helps us to define the functionality of the other two use cases, but no instance of this use case will ever exist in its own right. This is also shown in Figure 6.10.

6.6.3 Supporting use cases with prototyping

As the requirements for a system emerge in the form of use cases, it is sometimes helpful to build simple prototypes of how some of the use cases will work. A prototype is a working model of part of the system—usually a program with limited functionality that is built to test out some aspect of how the system will work. (Prototypes were discussed in Section 3.2.2 and are explained in more detail in Chapter 17 on the design of the user interface.)

Prototypes can be used to help elicit requirements. Showing users how the system might provide some of the use cases often produces a stronger reaction than showing them a series of abstract diagrams. Their reaction may contain useful information about requirements.

For example, there are a number of use cases in the Campaign Management subsystem for Agate that require the user to select a campaign in order to carry out some

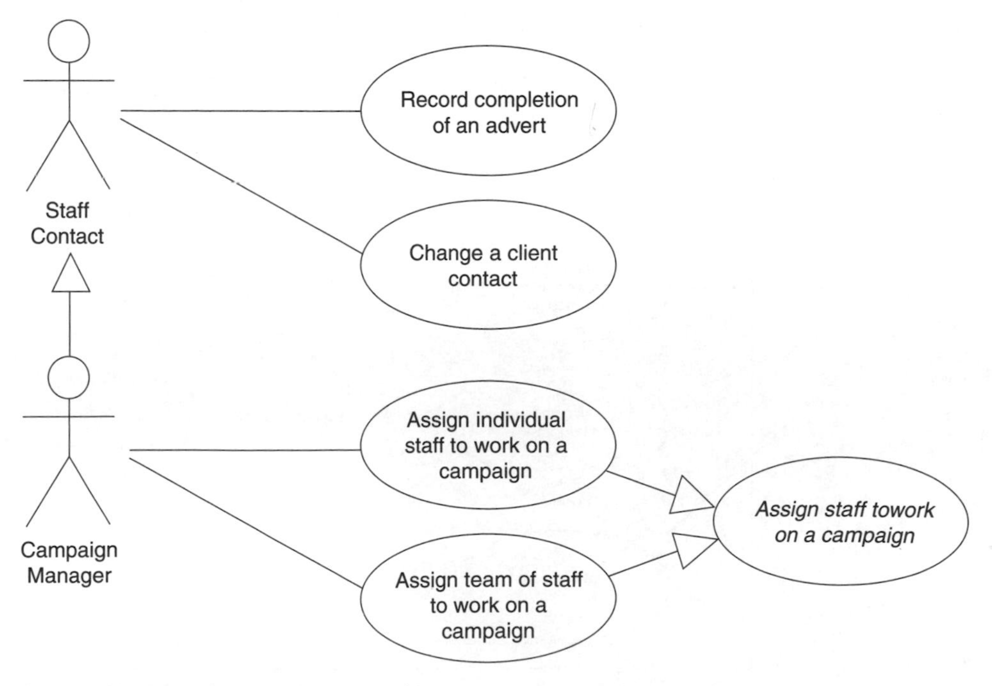

Figure 6.10 Generalization of actors and use cases.

business function. The use case diagram in Figure 6.9 reflects this in the «include» relationships with the use case `Find campaign`. The use case `Find campaign` will clearly be used a great deal, and it is worth making sure that we have the requirements right. A prototype could be produced that provides a list of all the campaigns in the system. A possible version of this is shown in Figure 6.11.

Showing this prototype interface design to the users may well produce the response that this way of finding a campaign will not work. There may be hundreds of campaigns in the system, and scrolling through them would be tedious. Different clients may have campaigns with similar names, and it would be easy to make a mistake and choose the wrong campaign if the user does not know which client it belongs to. For these reasons, the users might suggest that the first step is to find the right client and then only display the campaigns that belong to that client. This leads to a different user interface—shown in Figure 6.12.

The information from this prototyping exercise forms part of the requirements for the system. This particular requirement is about usability, but it can also contribute to meeting other, non-functional requirements concerned with speed and the error rate: it might be quicker to select first the client and then the campaign from a short-list than it is to search through hundreds of campaigns; and it might reduce the number of errors made by users in selecting the right campaign to carry out some function on.

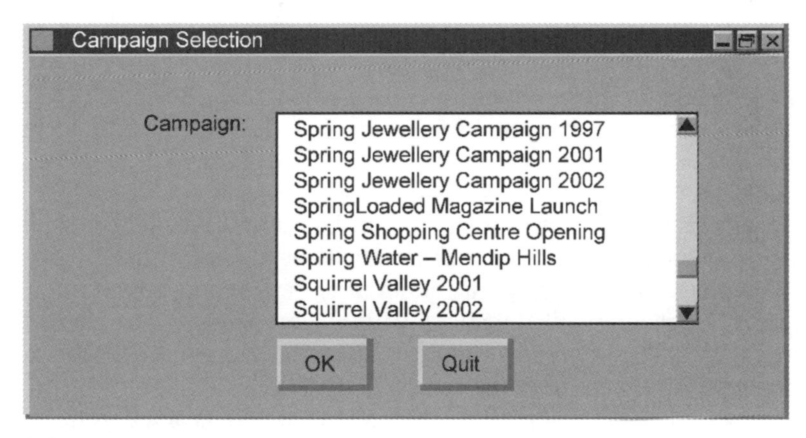

Figure 6.11 Prototype interface for the `Find campaign` use case.

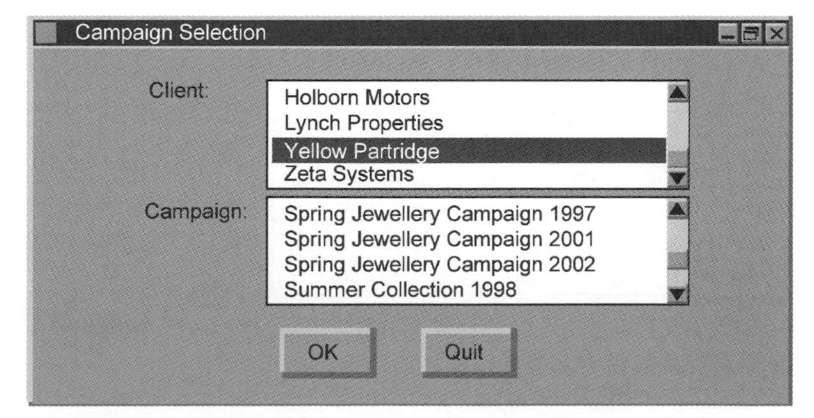

Figure 6.12 Revised prototype interface for the `Find campaign` use case.

Prototypes can be produced with visual programming tools, with scripting languages like TCL/TK, with a package like Microsoft PowerPoint® or even as web pages using HTML.

Prototypes do not have to be developed as programs. Screen and window designs can be sketched out on paper and shown to the users, either formally or informally. A series of possible screen layouts showing the steps that the user would take to interact with a particular use case can be strung together in a storyboard, as in Figure 6.13.

6.6.4 CASE tool support

Drawing any diagram and maintaining the associated documentation is made easier by a CASE tool, as described in Section 3.6.

As well as allowing the analyst to produce diagrams showing all the use cases in appropriate subsystems, a CASE tool should also provide facilities to maintain the repository associated with the diagram elements, and to produce reports. Automatically generated reports can be merged into documents that are produced for the client organization. The behaviour specification of each use case forms part of the requirements model or requirements specification, which it is necessary to get the client to agree to.

6.6.5 Business modelling with use case diagrams

We have used use case diagrams here to model the requirements for a system. They can also be used earlier in the life of a project to model an organization and how it operates. Business modelling is sometimes used when a new business is being set up, when an existing business is being 'reengineered', or in a complex project to ensure that the business operation is correctly understood before starting to elicit the requirements.

In the examples that we have shown above, the actors have all been employees of the company interacting with what will eventually be at least in part a computerized system. In business modelling, the actors are the people and organizations outside the company, interacting with functions within the company. For example, Figure 6.14 shows the `Client` as an actor and use cases that represent the functions of the business rather than functions of the computer system.

A full business model of Agate would show all the functions of the company, and the actors would be the other people and organizations with which Agate interacts, for example the media companies (TV stations and magazine and newspaper publishers) from which Agate buys advertising time and space, and the subcontractors that Agate uses to do design work and printing.

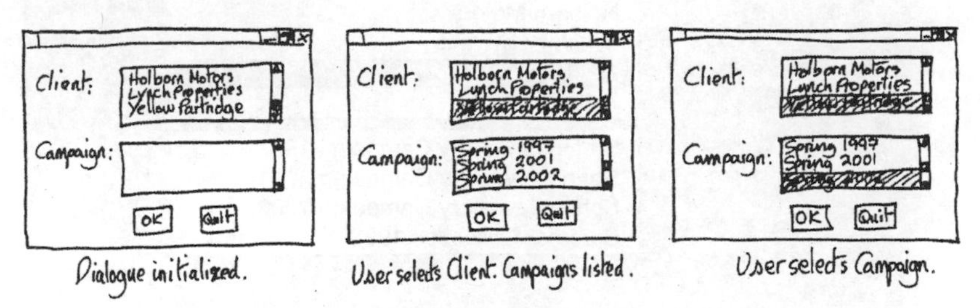

Figure 6.13 Prototype storyboard.

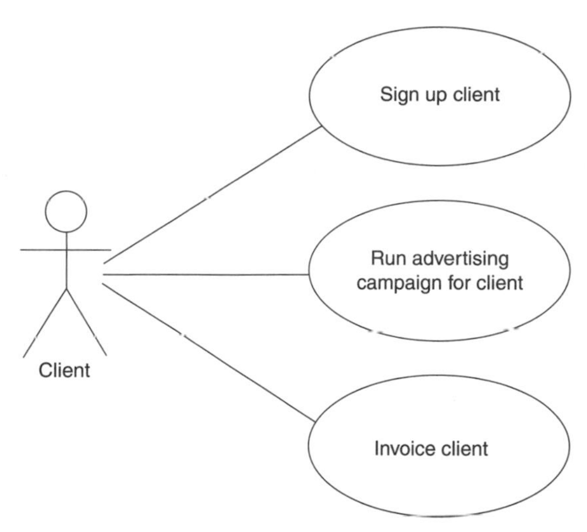

Figure 6.14 Example of business modelling with use cases.

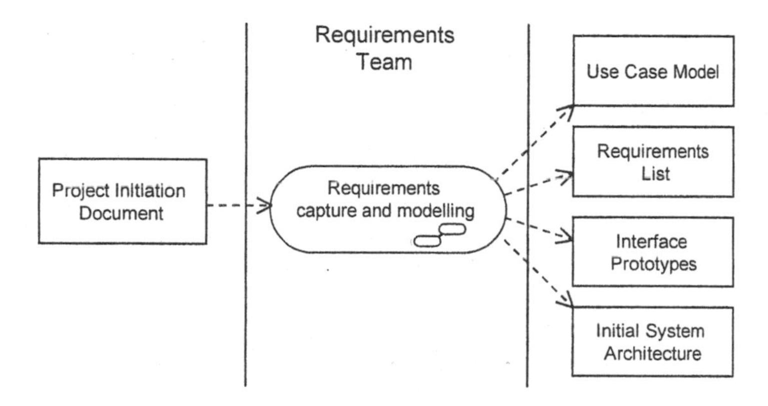

Figure 6.15 Activity diagram for `Requirements capture and modelling`.

6.7 Requirements Capture and Modelling

The first stage of most projects is one of capturing and modelling the requirements for the system. As we progress through the book, we shall include activity diagrams to illustrate the main activities in and products of each phase. These diagrams link back to the table in Figure 5.18, which summarizes the approach that we are taking in this book. Figure 6.15 shows the first such diagram.

In this case we have not broken the activity `Requirements capture and modelling` down into more detail, though it could potentially be broken down into separate activities for the capture of the requirements (interviewing, observation, etc.) and for the modelling of the requirements (use case modelling, prototyping, etc.).

We have used object flows to show the documents and models that are the inputs to and outputs from activities, and swimlanes to show the role that is responsible for the activities. In this case, one or more people in the role of `Requirements Team` will carry out this activity. In a small project, this may be one person, who carries out many other analysis and design activities; in a large project or organization, this may be a team of requirements analysis specialists taking more specialist roles.

The Case Study Chapter A2, which follows this one, provides more extended examples of the outputs of the `Requirements capture and modelling` activity, and the book website provides a full use case model.

6.8 Summary

Analysts investigating an organization's requirements for a new information system may use five main fact finding techniques—background reading, interviews, observation, document sampling and questionnaires. They use these to gain an understanding of the current system and its operation, of the enhancements the users require to the current system and of the new requirements that users have for the new system.

Using agreed standards to document requirements allows the analysts to communicate these requirements to other professionals and to the users. Use case diagrams are one diagramming technique that is used to summarize the users' functional requirements in a high level overview of the way that the new system will be used.

Case Study Example

You have already seen several examples from the case study in this section. The use cases are determined by the analyst from the documentation that is gathered from the fact-finding process. What follows is a short excerpt from an interview transcript, which has been annotated to show the points which the analyst would pick up on and use to draw the use case diagrams and produce the associated documentation. The interview is between Dave Harris, a systems analyst, and Peter Bywater, an Account Manager at Agate. It is from one of the interviews with the objective 'To establish additional requirements for new system' in the fact finding plan in the earlier case study section in this chapter.

Dave Harris: You were telling me about concept notes. What do you mean by this?
Peter Bywater: At present, when we come up with an idea for a campaign we use a word-processor to create what we call a concept note. We keep all the note files in one directory for a particular campaign, but it's often difficult to go back and find a particular one.
DH: So is this something you'd want in the new system?
PB: Yes. We need some means to enter a concept note and to find it again.
(This sounds like two possible use cases. Who are the actors?)
DH: So who would you want to be able to do this?
PB: I guess that the staff working on a campaign should be able to create a new note in the system.
DH: Only them?
(Any other actors?)
PB: Yes, only the staff actually working on a campaign.
DH: What about finding them again? Is this just to view them or could people modify them?
PB: Well, we don't change them now. We just add to them. It's important to see how a concept has developed. So we would only want to view them. But we need some easy way of browsing through them until we find the right one.
(Who are the actors for this?)

DH: Can anyone read the concept notes?
PB: Yes, any of the staff might need to have a look.
DH: Would you need any other information apart from the text of the concept itself?
(Thinking ahead to Chapter 7!)
PB: Yes. It would be good to be able to give each one a title. Could we use the titles then when we browse through them? Oh, and the date, time and whoever created that concept note.
DH: Right, so you'd want to select a campaign and then see all the titles of notes that are associated with that campaign, so you could select one to view it?
(Thinking about the interaction between the user and the system.)
PB: Yes, that sounds about right.
...
From this information, Dave Harris is going to be able to develop the use case descriptions for two use cases:

Create concept note,
Browse concept notes.

The use case diagram is shown in Figure 6.16. The use case descriptions will be as follows.

Create concept note.
A member of staff working on a campaign can create a concept note, which records ideas, concepts and themes that will be used in an advertising campaign. The note is in text form. Each note has a title. The person who created the note, the date and time are also recorded.

Browse concept notes.
Any member of staff may view concept notes for a campaign. The campaign must be selected first. The titles of all notes associated with that campaign will be displayed. The user will be able to select a note and view the text on screen. Having viewed one note, others can be selected and viewed.

The interaction here is quite straightforward, so we shall not need a more detailed breakdown of the interaction between user and system.

Note that in Figure 6.16, because Campaign Staff is a specialization of Staff, we do not need to show a communication association between the Campaign Staff actor and the Browse concept notes use case.

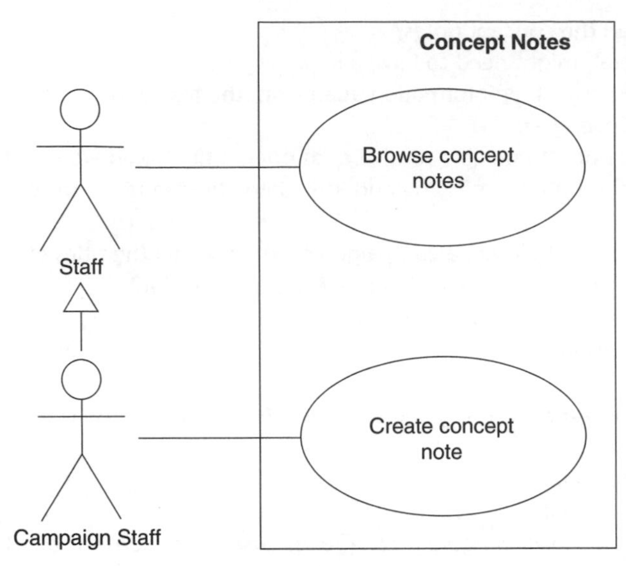

Figure 6.16 Use cases for `Concept Notes` subsystem.

Review Questions

6.1 Read the following description of a requirement for FoodCo, and decide which parts of it are functional requirements and which are non-functional requirements.

The allocation of staff to production lines should be mostly automated. A process will be run once a week to carry out the allocation based on the skills and experience of operatives. Details of holidays and sick leave will also be taken into account. A first draft Allocation List will be printed off by 12.00 noon on Friday for the following week. Only staff in Production Planning will be able to amend the automatic allocation to fine-tune the list. Once the amendments have been made, the final Allocation List must be printed out by 5.00 pm. The system must be able to handle allocation of 100 operatives at present, and should be capable of expansion to handle double that number.

6.2 Name the five main fact finding techniques and list one advantage and one disadvantage of each.

6.3 Imagine that you will be interviewing one of the three staff in Production Planning at FoodCo. Draw up ten questions that you would want to ask him or her.

6.4 What is the purpose of producing use cases?

6.5 Describe in your own words the difference between the «extend» and «include» relationships in use case diagrams.

6.6 What is the difference between an 'essential' and a 'real' use case?

6.7 Write a use case description in the extended form, used for the `Assign staff to work on a campaign` example in Section 6.6.2, for either `Create concept note` or `Browse concept notes`.

6.8 Think of the different possible uses you could make of a library computer system and draw a use case diagram to represent these use cases.

6.9 List some non-functional requirements a library computer system (as in Question 6.8) that you would not model using use cases.

6.10 In what way are use case diagrams different when used for business modelling?

Case Study Work, Exercises and Projects

6.A Refer to the material for the second case study—FoodCo (introduced in Case Study B1). Draw up your initial fact finding plan along the lines of the plan given above.

6.B Read the following excerpt from a transcript of an interview with one of the production planners at FoodCo. Draw a use case diagram and create use case descriptions for the use cases that you can find in this information.

Ken Ong: So what happens when you start planning the next week's allocation?
Rik Sharma: Well, the first thing to do is to check which staff are unavailable.
KO: Would that be because they are on holiday?
RS: Yes, they could be on holiday, or they could be off sick. Because staff are handling raw food, we have to be very careful with any illness. So factory staff often have to stay off work longer than they would if they were office workers.
KO: So how do you know who's off sick and who's on holiday?
RS: They have to complete a holiday form if they want a holiday. They send it to the Factory Manager, who authorizes it and sends it to us. We take a copy, and enter the details into our system. We then return the form to the member of staff.
KO: What details do you enter?
RS: Who it is, the start date of the holiday and the first date they are available for work again.
KO: What about illness?
RS: The first day someone is off sick they have to ring in and notify us. We have to find someone to fill in for them for that day if we can.
KO: Right. Let's come back to that in a minute. How do you record the fact that they're off sick for your next week's production plan?
RS: We make an entry in the system. We record which member of staff it is, when they went off sick, the reason and an estimate of how many days they're likely to be off.
KO: Right, so how do you get at that information when you come to plan next week's allocation?
RS: Well, we run off three lists. We enter Monday's date, and it prints us off one list showing who is available all week, a second list showing who is not available all week, and a third list showing who is likely to be available for part of the week.
KO: Then what?
RS: Then we start with the people who are available all week and work round them. We pull each operative's record up on the screen and look at two main factors—first their skills and experience, and second which line they're working on at the moment and how long they've been on that line. Then we allocate them to a line and a session in one of the three factories.
KO: So you can allocate them to any one of the three factories. Do you enter the same data for each one?
RS: No, there are slight variations in the allocation screen for each of the factories—mainly for historical reasons.
...

Further Reading

- Booth (1989) Chapter 5 describes the issues surrounding the usability of systems in more detail than we can here, and explains the process of Task Analysis.
- Oppenheim's book (2000) provides a very detailed coverage of questionnaire design for survey purposes. It is aimed mainly at social science and psychology students, but has some relevant chapters on how to formulate effective questions. Many books for students on how to carry out a research project cover fact-gathering techniques such as interviewing and questionnaire design. Allison et al. (1996) is an example, but most university libraries and bookshops will have a selection of similar books.
- Hart (1997) gives a detailed explanation of the techniques that are specific to the development of expert systems.
- Roberts (1989) addresses the role of users in a systems development project. This book is one of a series of guides written for civil servants in the UK government service, and is relatively bureaucratic in its outlook. However it ranges widely over the issues that users may face. Yourdon (1989) discusses users and their roles in Chapter 3.
- Jacobson et al. (1992) present the original ideas behind use cases as an analysis technique. For a more recent view, look at Rosenberg and Scott (1999) or Cockburn (2000).
- Examples of use cases from the Agate case study are available on the book website. This includes references to examples of templates for use case descriptions.

CHAPTER A2

Agate Ltd Case Study Requirements Model

Agate Ltd

A2.1 Introduction

In this chapter we bring together the models (diagrams and supporting textual information) that constitute the Requirements Model. In Chapters 5 and 6 we have introduced the following UML diagrams:

- use case diagram,
- activity diagram and
- package diagram.

There is not the space in this book to produce a complete Requirements Model. However, in this chapter we have included a sample of the diagrams and other information. This is done to illustrate the kind of material that should be brought together in a Requirements Model. We have also tried to illustrate how iteration of the model will produce versions of the model that are elaborated with more detail.

A2.2 Requirements List

The Requirements List includes a column to show which use cases provide the functionality of each requirement. This Requirements List includes some use cases not in the first iteration through the use case model.

No.	Requirement	Use Case(s)
1	To record names, address and contact details for each client.	Add a new client
2	To record the details of each campaign for each client. This will include the title of the campaign, planned start and finish dates, estimated costs, budgets, actual costs and dates, and the current state of completion.	Add a new campaign

No.	Requirement	Use Case(s)
3	To provide information that can be used in the separate accounts system for invoicing clients for campaigns.	Record completion of a campaign
4	To record payments for campaigns that are also recorded in the separate accounts system.	Record client payment
5	To record which staff are working on which campaigns, including the campaign manager for each campaign.	Assign staff to work on a campaign
6	To record which staff are assigned as staff contacts to clients.	Assign a staff contact
7	To check on the status of campaigns and whether they are within budget.	Check campaign budget
8	To allow creative staff to record notes of ideas for campaigns and adverts (concept notes).	Create concept note
9	To provide other staff with access to these concept notes	Browse concept notes
10	To record details of adverts, including the progress on their production.	Add a new advert to a campaign; Record completion of an advert
11	To schedule the dates when adverts will be run.	Add a new advert to a campaign
12	To maintain staff records for creative and administrative staff.	Add a new member of staff
13	To maintain details of staff grades and the pay for those grades.	Add a new staff grade; Change the rate for a staff grade
14	To record which staff are on which grade.	Change the grade for a member of staff
15	To calculate the annual bonus for all staff.	Calculate staff bonuses
16	To enable data about clients, campaigns, adverts and staff to be shared between offices.	Not applicable
17	To allow the system to be modified to work in different languages.	Not applicable

A2.3 Actors and Use Cases

Actor	Description
Accountant	The accountant works in the Accounts departments and is responsible for the major resourcing issues for campaigns including staffing and related financial matters.
Campaign Manager	Either a Director or an Account Manager (job titles), who is responsible for estimating the campaign cost and agreeing it with the client. They are responsible for assigning staff to the team and supervising their work, managing the progress of the campaign, conducting any further budget negotiations and authorizing the final invoices.
Staff Contact	Member of staff who is the contact for a particular client. They provide a first point of contact for the client when the client wants to contact Agate.
Staff	Any member of staff in Agate.
Campaign Staff	Member of staff working on a particular campaign.

Figures A2.1–A2.3 shows the use cases from the first iteration, with use case descriptions in the tables.

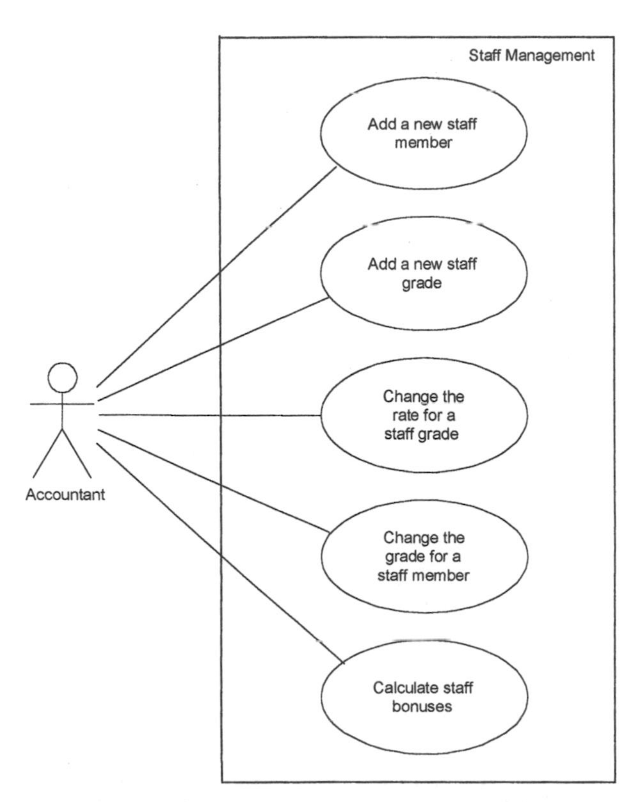

Figure A2.1 `Staff Management` use cases.

Use Case	Description
Add a new staff member	When a new member of staff joins Agate, his or her details are recorded. He or she is assigned a staff number, and the start date is entered. Start date defaults to today's date. The starting grade is entered.
Add a new staff grade	Occasionally a new grade for a member of staff must be added. The name of the grade is entered. At the same time, the rate for that grade and the rate start date are entered; the date defaults to today's date.
Change the rate for a staff grade	Annually the rates for grades are changed. The new rate for each grade is entered, and the rate start date set (no default). The old grade rate is retrieved and the rate finish date for that grade rate set to the day before the start of the new rate.
Change the grade for a staff member	When a member of staff is promoted, the new grade and the date on which they start on that grade are entered. The old staff grade is retrieved and the finish date set to the day before the start of the new one.
Calculate staff bonuses	At the end of each month staff bonuses are calculated. This involves calculating the bonus due on each campaign a member of staff is working on. These are summed to give the total staff bonus.

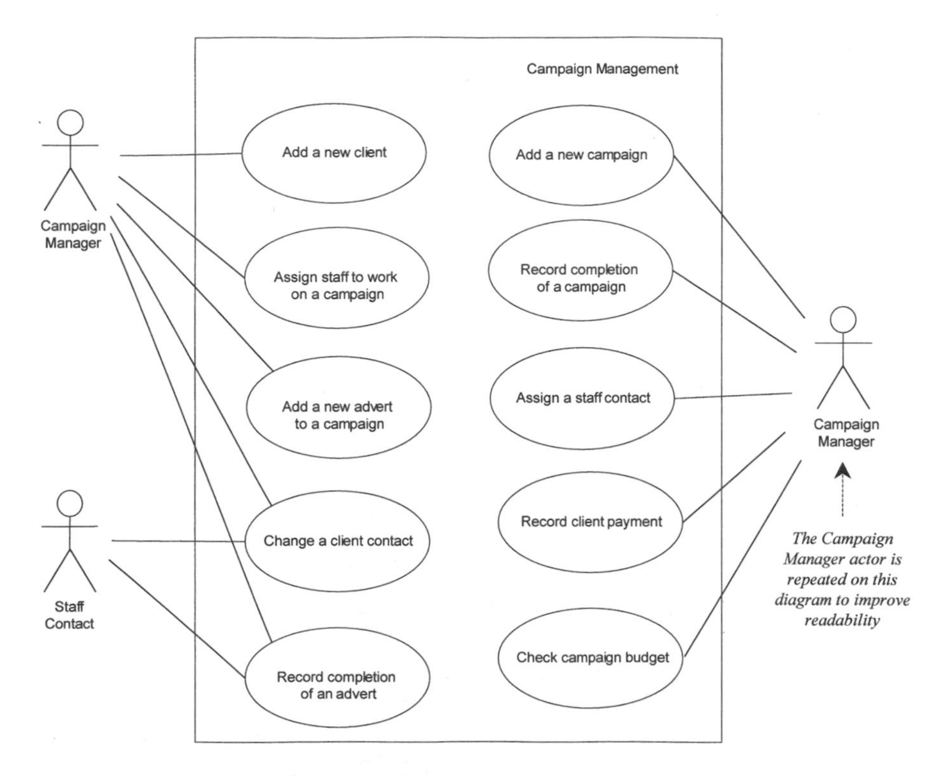

Figure A2.2 `Campaign Management` use cases.

Use Case	Description
Add a new client	When Agate obtains a new client, the full details of the client are entered. Typically this will be because of a new campaign, and therefore the new campaign will be added straight away.
Assign staff to work on a campaign	The campaign manager selects a particular campaign. A list of staff not already working on that campaign is displayed, and he or she selects those to be assigned to this campaign.
Add a new advert to a campaign	A campaign can consist of many adverts. Details of each advert are entered into the system with a target completion date.
Change a client contact	Records when the client's contact person with Agate is changed.
Record completion of an advert	The actor selects the relevant client, campaign and advert. The selected advert is then completed by setting its completion date.
Add a new campaign	When Agate gets the business for a new campaign, details of the campaign are entered, including the intended finish date and the estimated cost. The manager for that campaign is the person who enters it.
Record completion of a campaign	When a campaign is completed, the actual completion date and cost are entered. A record of completion form is printed out for the Accountant as the basis for invoicing the client.

Use Case	Description
Assign a staff contact	Clients have a member of staff assigned to them as their particular contact person.
Record client payment	When a client pays for a campaign, the payment amount is checked against the actual cost and the date paid is entered.
Check campaign budget	The campaign budget may be checked to ensure that it has not been exceeded. The current campaign cost is determined by the total cost of all the adverts and the campaign overheads.

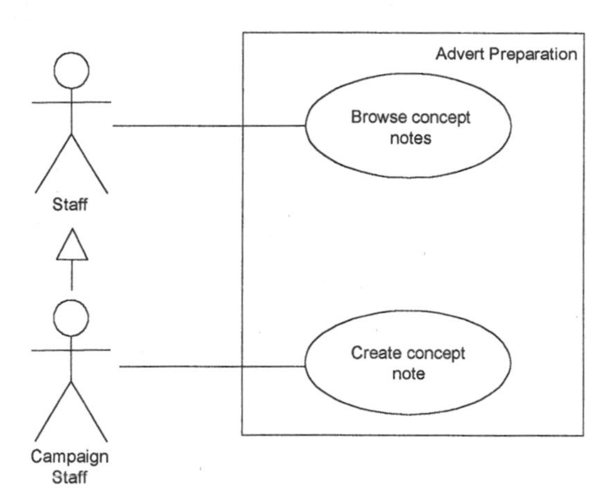

Figure A2.3 `Advert Preparation` use cases.

Use Case	Description
Browse concept notes	Any member of staff may view concept notes for a campaign. The campaign must be selected first. The titles of all notes associated with that campaign will be displayed. The user will be able to select a note and view the text on screen. Having viewed one note, others can be selected and viewed.
Create concept note	A member of staff working on a campaign can create a concept note, which records ideas, concepts and themes that will be used in an advertising campaign. The note is in text form. Each note has a title. The person who created the note, the date and time are also recorded.

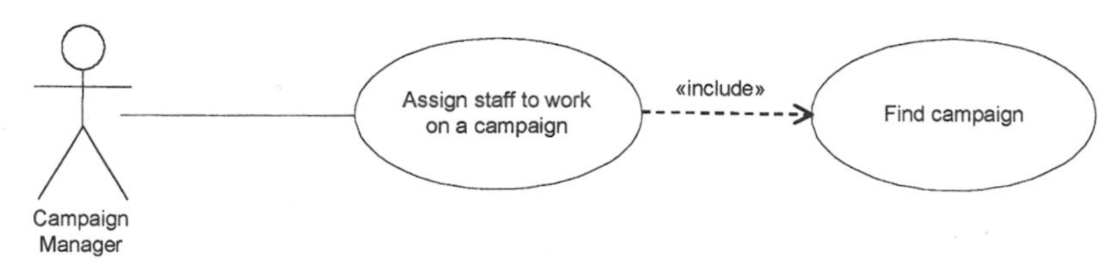

Figure A2.4 Inclusion of `Find campaign` use case.

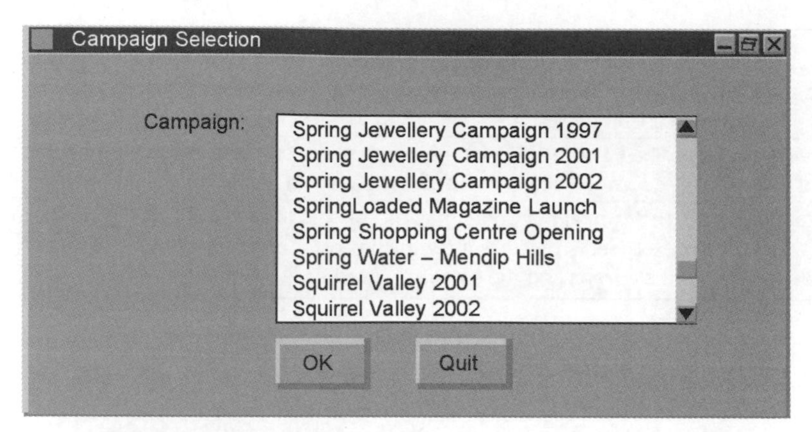

Figure A2.5 Prototype interface for the `Find campaign` use case.

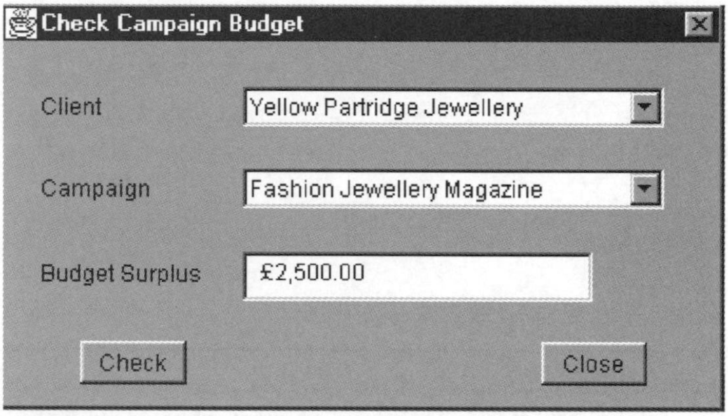

Figure A2.6 Prototype interface for the use case `Check campaign budget`.

As part of the second iteration of use case modelling, it is suggested that all the use cases that require the user to select a client, a campaign or an advert should have include relationships with use cases called `Find client`, `Find campaign` and `Find advert`. An example of this is shown in Figure A2.4.

In order to test out this idea, prototypes of the user interface were produced in the second iteration. The first prototypes used a separate user interface for these included use cases, as shown in Figure A2.5.

However, feedback from the users indicated that this approach was not acceptable. They did not want to have to keep opening extra windows to find clients, campaigns and adverts. The users expressed the view that they should be able to select these from listboxes or dropdown lists that were part of the interface for whatever use case they were in at the time.

In the third iteration of use case modelling, a set of prototypes was produced that uses listboxes. Figure A2.6 shows an example.

In the third iteration, some additional functionality was identified and added to the use case diagrams. As an example of this, Figure A2.7 shows the use case `Check`

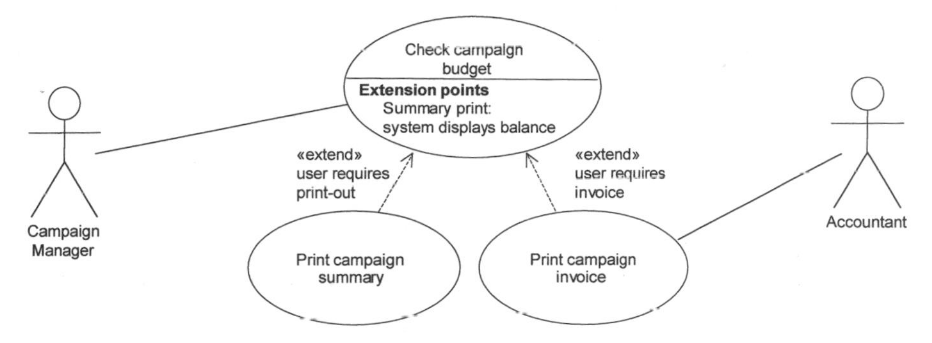

Figure A2.7 Modified use case `Check campaign budget` with extensions.

`campaign budget` extended by the use case `Print campaign summary` and `Print campaign invoice`. This additional functionality will also require a change to the prototype interface in Figure A2.6. Two additional buttons, Print Summary and Print Invoice, need to be added to the row of buttons at the bottom of the window.

Also in the third iteration, the use case descriptions are elaborated to provide more detail about the interaction between the actors and the system. Two examples of these use case descriptions are provided below.

Use case description: Check campaign budget

Actor Action	System Response
1. None	2. Lists the names of all clients.
3. The actor selects the client name.	4. Lists the titles of all campaigns for that client.
5. Selects the relevant campaign. Requests budget check.	6. Displays the budget surplus for that campaign.

Extensions
After step 6, the campaign manager prints a campaign summary.
After step 6, the campaign manager prints a campaign invoice.

Use case description: Assign staff to work on a campaign

Actor Action	System Response
1. None	2. Displays list of client names.
3. The actor selects the client name.	4. Lists the titles of all campaigns for that client.
5. Selects the relevant campaign.	6. Displays a list of all staff members not already allocated to this campaign.
7. Highlights the staff members to be assigned to this campaign.	8. Presents a message confirming that staff have been allocated.

Alternative Courses
Step 4. The advert will exceed the budget and a budget extension request is generated.

A2.4 Glossary

A glossary of terms has been drawn up, which lists the specialist terms that apply to the domain of this project-advertising campaigns.

Term	Description
Admin Staff	Staff within Agate whose role is to provide administrative support that enables the work of the creative staff to take place, for example secretaries, accounts clerks and the office manager.
Advert	An advertisement designed by Agate as part of a campaign. Adverts can be for TV, cinema, web-sites, newspapers, magazines, advertising hoardings, brochures or leaflets. Synonym: Advertisement.
Agate	An advertising agency based in Birmingham, UK, but with offices around the world. The customer for this project.
Campaign	An advertising campaign. Adverts are organised into campaigns in order to achieve a particular objective, for example a campaign to launch a new product or service, a campaign to rebrand a company or product, or a campaign to promote an existing product in order to take market share from competitors.
Campaign Staff	Member of staff working on a particular campaign.
Client	A customer of Agate. A company or organization that wishes to obtain the services of Agate to develop and manage an advertising campaign, and design and produce adverts for the campaign.
Concept Note	A textual note about an idea for a campaign or advert. This is where creative staff record their ideas during the process of deciding the themes of campaigns and adverts. Synonym: Note.
Creative Staff	Staff with a creative role in the company, such as designers, editors and copy-writers; those who are engaged in the work of the company to develop and manage campaigns and design and produce adverts.
Grade	A job grade. Each member of staff is on a particular grade, for example 'Graphic Artist 2' or 'Copywriter 1'.
Grade Rate	The rate of pay for a particular grade, for example the Grade 'Graphic Artist 2' is paid £18,500 per year in the UK from 1/1/2001 to 31/12/2001.
Staff	Any member of staff in Agate. Synonyms: Staff member, member of staff.

A2.5 Initial Architecture

The initial architecture of the system is based on the packages into which the use cases are grouped. These use cases have been grouped into three sub-system packages: `Campaign Management`, `Staff Management` and `Advert Preparation`. Figure A2.8 shows the initial architecture of these three packages, and a package which will provide the mechanisms for the distribution of the application. At this early stage in the project, it is not clear what this will be, but something will be necessary to meet Requirement 16.

A2.6 Activities of Requirements Modelling

In Chapter 5, we outlined the phases and activities of the iterative life cycle, and in Chapter 6 we included an activity diagram to show the activity `Requirements capture and modelling`. Figure A2.9 shows the same diagram. This activity can be broken down into other activities, and these are shown in Figures A2.10, A2.11 and A2.12.

It is important to remember that in a project that adopts an iterative life cycle, these activities may take place over a series of iterations. In the first iteration, the emphasis will be on requirements capture and modelling, in the second, it will shift to analysis, but some requirements capture and modelling activities may still take place. Refer back to Figure 5.16. You may also want to look at Figure 5.8, which illustrates the development of the use case model through successive iterations.

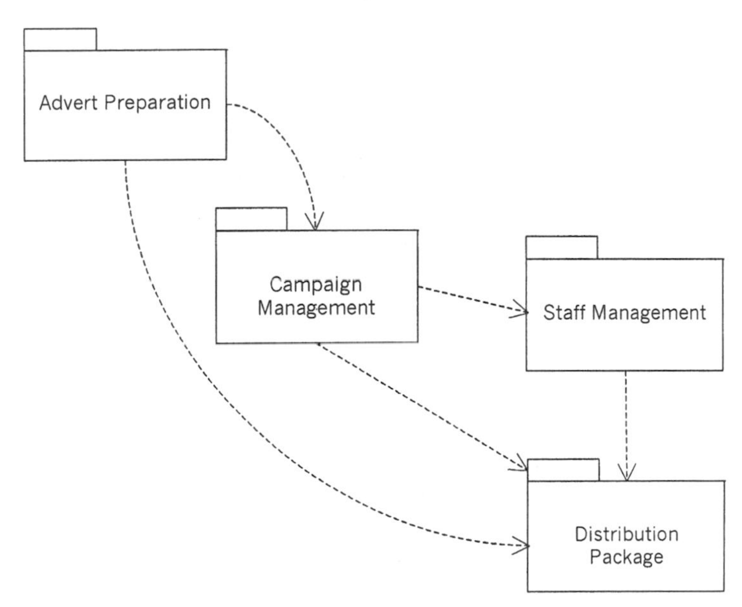

Figure A2.8 Initial package architecture.

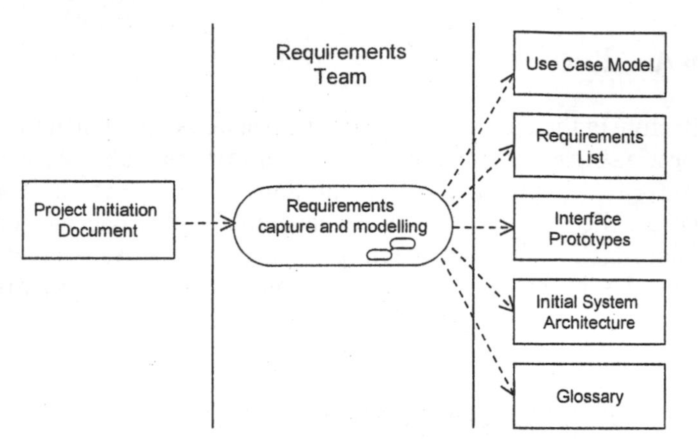

Figure A2.9 Activity diagram for `Requirements capture and modelling`.

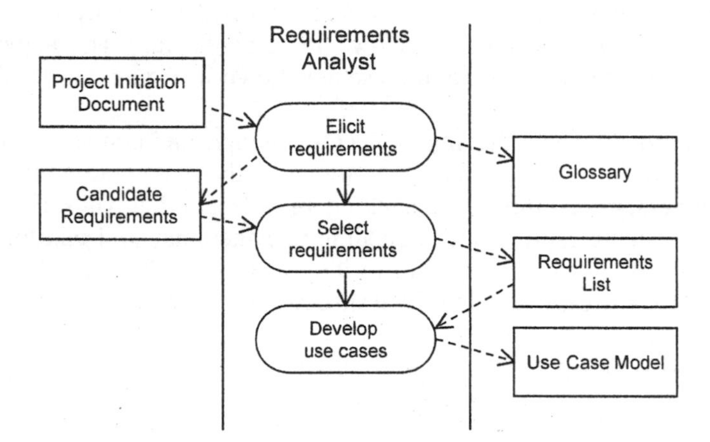

Figure A2.10 Activity diagram to show the activities involved in capturing requirements.

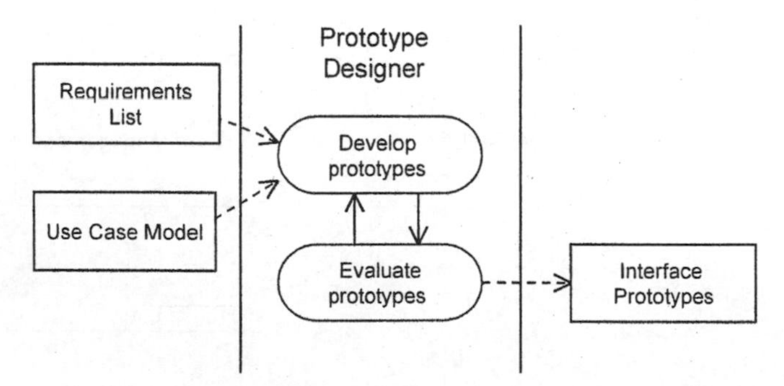

Figure A2.11 Activity diagram to show the activities involved in developing prototypes.

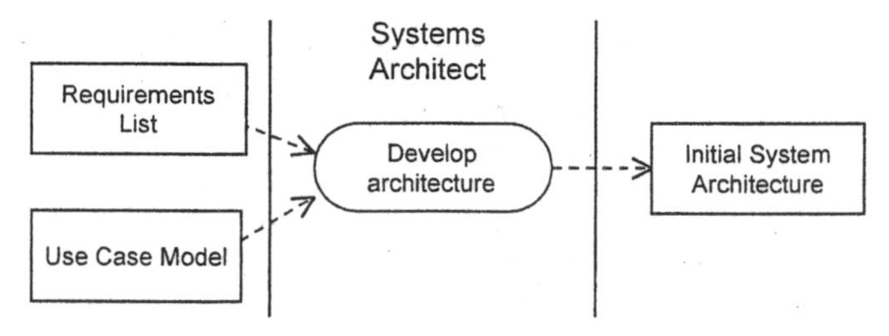

Figure A2.12 Activity diagram to show the activities in developing an initial architecture.

7 CHAPTER

Requirements Analysis

OBJECTIVES

In this chapter you will learn

- why we analyse requirements
- the technical terms used when working with class diagrams
- how the UML class diagram expresses a detailed model of user requirements
- how to realize use cases with collaboration diagrams and class diagrams
- how the CRC technique helps to identify classes and allocate their responsibilities.

7.1 Introduction

In Chapter 6 we saw how use case diagrams are used to build an initial model based on users' requirements for a new system. In this chapter we analyse those requirements in more detail. In Section 7.2, we begin by considering what a requirements model must do, and how it differs from models used at other stages in the lifecycle. Then, in Section 7.3, we discuss how to move from use cases, which are rather high level and concentrate on a user-centred view of the system, to the analysis class diagram, which is the primary model for describing the internal structure and behaviour of the proposed system. This class diagram also forms a basis for the later design class diagram, from which the executable code will be developed. Use cases and class diagrams are linked by an activity called use case realization, which uses collaboration diagrams to help with the transition from use cases to class diagrams. In this section we briefly introduce the notation for collaborations and collaboration diagrams, but the topic is covered in more depth in Chapter 9. In Section 7.4, we illustrate the notation of the class diagram at some length, in order to show just how it helps to analyse in greater detail the model of requirements that is initially expressed as use cases. This involves elaborating the class definitions using attributes and operations, and analysing the logical structure and relationships between classes using associations. In Section 7.5, we give practical guidance on one approach to realizing a use case as an analysis class diagram. In Section 7.6, we introduce a non-UML technique called Class–Responsibility–Collaboration (CRC) cards that predates the UML by many years, but is still widely used to complement UML techniques in requirements analysis. Finally, in Section 7.7 we briefly discuss the derivation of an overall analysis class diagram from the various partial class diagrams that result from the CRC and use case realization techniques.

7.2 What Must a Requirements Model Do?

The most influential factor for the success of an IS project is whether the software product fulfils its users' requirements. Models constructed from an analysis perspective aim at determining these requirements. This means much more than just gathering and documenting facts and requests. The use case model introduced in Chapter 6 gives a perspective on many user requirements and models them in terms of what the software system can do to help the user perform their work (or play). Before we can begin to design software that will deliver this help, we must analyse both the logical structure of the problem situation, and also the ways that its logical elements interact with each other. This is so that we can be sure that we understand what is really happening. We must also go beneath the surface and examine the way in which different, possibly conflicting, requirements affect each other. Then we must communicate this understanding clearly and unambiguously to those who will design and build the software. We do all of this by building further models, and these must meet several objectives.

- They must contain an overall description of what the software should do.
- They must represent any people, physical things and concepts that are important to the analyst's understanding of what is going on in the application domain.
- They must show connections and interactions among these people, things and concepts.
- They must show the business situation in enough detail to evaluate possible designs.
- Ideally, they should also be organized in such a way that they will be useful later for designing the software.

In fact, we can reverse the last point to give a general measure of usefulness for a requirements model. The software developed to help meet a need should have a structure that reflects that of the situation in which the need arises. Thus our description of the application domain must include its conceptual structure. This statement is one of the conceptual foundations of the object-oriented approach, and the UML class diagram is designed to do just this. It communicates an analysis of requirements in such a way that the model's structure can later be directly translated into software components.

7.3 Use Case Realization

7.3.1 From use case to collaboration and class diagrams

To move from an initial use case ultimately to the implementation of software that adequately fulfils the requirements identified by the use case involves at least one iteration through all of the development activities, from requirements modelling to implementation. In relation to a single use case, this activity is known as *use case*

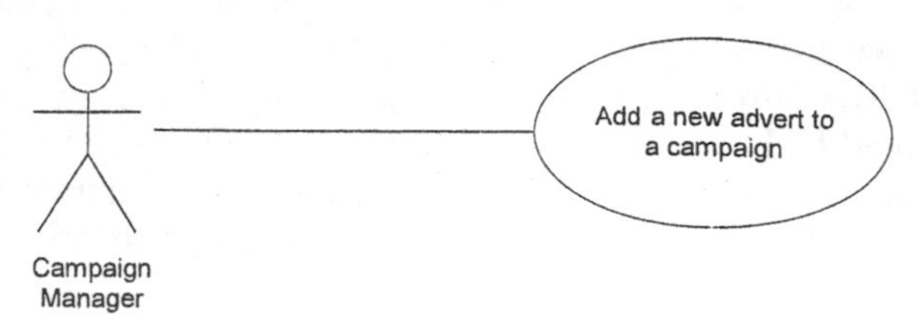

Figure 7.1 Use case diagram for `Add a new advert to a campaign.`

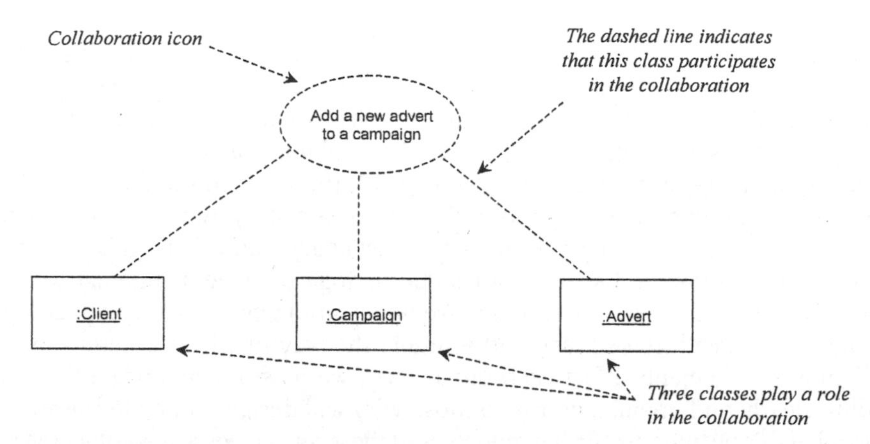

Figure 7.2 Collaboration for `Add a new advert to a campaign.`

realization, and in this section we look at the first steps in this activity. We will illustrate this with the use case `Add a new advert to a campaign`, introduced in Figure 6.9. Figure 7.1 shows the use case diagram, and the series of figures that follow show some of the alternative ways that the use case can be represented, viewing it both from different perspectives and at different levels of abstraction. Some of the notation used in the diagrams may not mean much at this stage, but do not worry about this. It will be explained step by step later in the chapter.

Among other things, use case realization involves the identification of a possible set of classes, together with an understanding of how those classes might interact to deliver the functionality of the use case. The set of classes is known as a *collaboration*. The simplest representation of a collaboration as a set of classes is shown in Figure 7.2.

You can see immediately that this tells us only that these three classes participate in the collaboration—in other words, they can interact, when implemented as software, in such a way as to achieve the result described by the use case. It doesn't tell us how they interact nor how they relate to other parts of the model, but this is just one view of a collaboration. Another view looks at the collaboration 'from the outside'. In the example in Figure 7.3, this is used to show the relationship between a collaboration and the use case that it realizes. Note the deliberate similarity between the two icons: a collaboration appears to differ from a use case in this view only in being drawn with a dashed line instead of a solid one. However, the collaboration is not the same thing thing as the use case. Instead, it has a relationship with the use case that is shown here by the *dependency* arrow. Here, these mean that the definitions of these classes must maintain a reference to the collaboration, which in turn must maintain a reference to the use case. Later, we will see many other

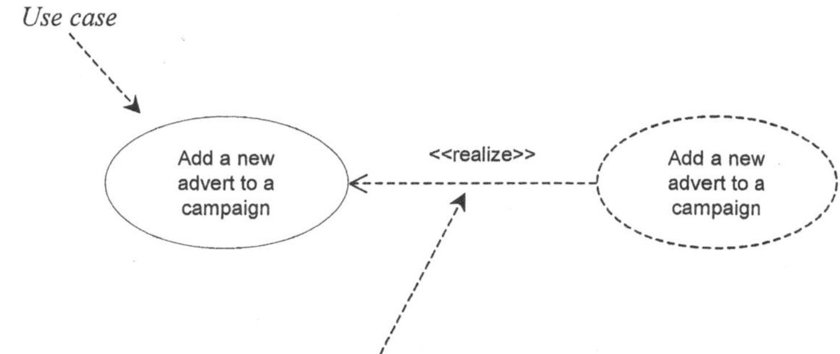

Figure 7.3 A collaboration realizes a specific use case.

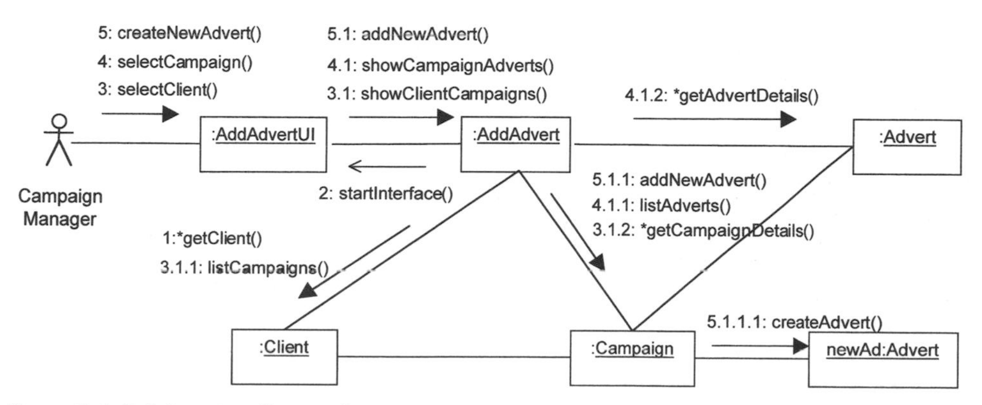

Figure 7.4 Collaboration diagram for `Add a new advert to a campaign`.

examples of dependencies between one model element and another (this notation is particularly useful when the two elements are in different packages).

Collaborations can also be represented in various ways that reveal their internal details. The *collaboration diagram* is probably the most useful of these for our present purpose. A collaboration diagram for this example is shown in Figure 7.4. Don't be too concerned about making full sense of this diagram for the moment. The important thing to notice is that it shows some of the structure that exists between the objects that take part in the collaboration.

Next, a collaboration can be represented as an *object* (or *instance*) diagram. The corresponding object diagram is shown in Figure 7.5. Note the strong structural and notational similarity to the collaboration diagram in Figure 7.4. This is because both show object instances and links, although only the collaboration diagram shows messages between the objects.

Finally (for the time being, at any rate), a collaboration can be represented as a class diagram. Figure 7.6 shows a class diagram for this example (and, during a first iteration through requirements analysis, this is as far as we need to go).

Again, note that there are both structural and notational similarities to the collaboration diagram in Figure 7.4 and the object diagram in Figure 7.5. There is a class for each object, and some classes have associations that correspond to the links between objects.

In this example, the obvious differences are perhaps more apparent than real. For example, this class diagram shows a lot of the internal detail of the classes. However,

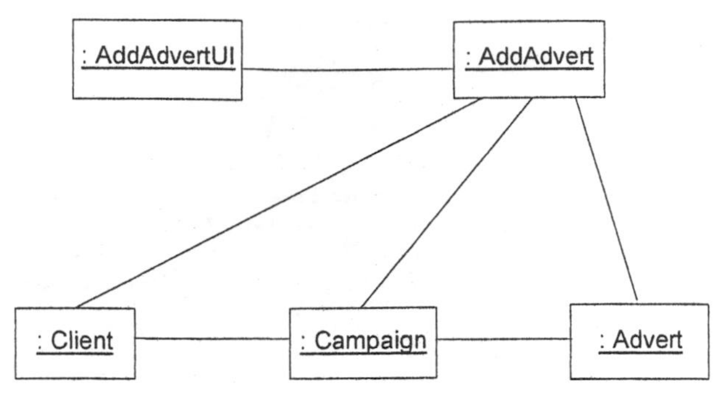

Figure 7.5 Object instance diagram for `Add a new advert to a campaign.`

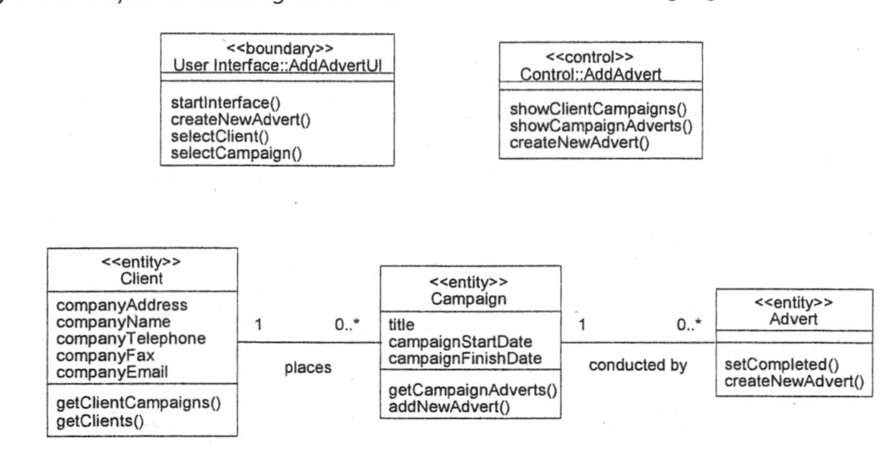

Figure 7.6 Analysis class diagram for `Add a new advert to a campaign.`

both object diagrams and collaboration diagrams can show some (though not all) of these details if this is desired. Also, some links between objects in the object diagram have no corresponding association in the class diagram. The reason for this will be explained a little later in this chapter.

The distinction between these different representations can be a little confusing at first, so it will probably help to say a little about the main purpose of each of the diagrams.

The collaboration icon in Figure 7.2 is in itself simply a high-level abstraction that can stand for any of the other forms. The diagrams in Figures 7.3–7.6 show some of the intermediate forms that realize a use case during the progressive and iterative development of the resulting software. Each form in this series is, in a sense, one step closer to a design model, and thus ultimately to executable code. Each also serves a particular modelling perspective. For instance, a collaboration diagram highlights the interaction among a group of collaborating objects, but it ignores internal details of classes and some aspects of the structure. It is also not easy to read the sequence of messages on a collaboration diagram. A class diagram ignores the interaction altogether, but shows the structure in more detail and can show a lot of the internal features of the classes. Collaborations can also be expressed in ways that do not concern us so much from a requirements analysis perspective—for example, interaction sequence diagrams hide most of the structure but display the sequence of messages with greater clarity (we introduce interaction sequence diagrams in Chapter 9). For the most part, these other representations correspond to the design, test or implementation perspectives of the system model.

7.3.2 Analysis class stereotypes

Before we go any further, this is a suitable point to introduce the concept of *analysis class stereotypes*. These represent three particular kinds of class that will be encountered again and again when carrying out requirements modelling. We have already encountered the idea of stereotypes in Chapters 5 and 6, but it may be useful here to give the UML definition (OMG, 2001).

> **stereotype** A new type of modelling element that extends the semantics of the metamodel. Stereotypes must be based on certain existing types or classes in the metamodel. Stereotypes may extend the semantics but not the structure of pre-existing classes. Certain stereotypes are defined in the UML, others may be user defined . . .

Essentially, this means that instances of a class stereotype have a shared focus on certain kinds of task, that this distinguishes them in a significant way from classes that are instances of another stereotype and, finally, that it is useful to identify this in our models. If you think of the everyday use of the word stereotype, this is not so very different. For example, if a friend says that the roles played by Arnold Schwarzenegger in his movies are quite stereotyped, you would probably understand that they think his characters are all similar to each other in certain ways, even though the plot and context may vary a lot from one film to another. Thus, if you know that Schwarzenegger is the star of a film that you are going to see, you already have some idea of what the film will be like, and also of some of the ways that it might differ from a Leonardo DiCaprio film.

Since UML is designed to be capable of extension, developers can add new stereotypes where there is a clear need to do so. But this is only done when it is absolutely necessary, and we need only concern ourselves here with the three analysis class stereotypes that are defined in an appendix to the UML: *boundary*, *control* and *entity* classes. What these mean, how they are used and how they are shown on diagrams is described in the following sections.

Note, however, that it is not always necessary to stereotype the classes in a model. Nor, if classes are stereotyped, is it always necessary to show the stereotype of a class on diagrams. Stereotypes are shown where they add useful meaning to a model, but their use is not obligatory. In many diagrams shown later in this book, the stereotype is omitted, either because it can be assumed from the context or because it has no specific relevance to the purpose of the diagram.

Boundary classes

Boundary classes 'model interaction between the system and its actors' (Jacobson et al, 1999). Since they are part of the requirements model, boundary classes are relatively abstract. They do not directly represent all the different sorts of interface widget[1] that will be used in the implementation language. The design model may well do this later, but from an analysis perspective we are interested only in identifying the main logical interfaces with users and other systems. This may include interfaces with other software and also with physical devices such as printers, motors and sensors. Stereotyping these as boundary classes emphasizes that their main task is to manage the transfer of information across system boundaries. It also helps to partition the system, so that any changes to the interface or communication aspects of the system can be isolated from those parts of the system that provide the information storage or business logic.

[1] For instance, buttons, windows, list boxes and so on that will appear on the computer screen.

The class `User Interface::AddAdvertUI` (shown in the collaboration diagram in Figure 7.4 and in the class diagram in Figure 7.6) is a typical boundary class. This style of writing the name shows that the class is `AddAdvertUI` (the UI is just an abbreviation for user interface) and it belongs to the `User Interface` package (the concept of packages was introduced in Chapter 5). When we write the package name in this way before the class name, it means that this class is imported from a different package from the one with which we are currently working. In this case, the current package is the `Agate application` package, which contains the application requirements model, and thus consists only of domain objects and classes.

We do not yet necessarily know what this user interface will look like or how it will behave. We may not have chosen the programming language or application package in which the software will be written, and this could restrict the choices of user interface. However, we know that some sort of interface will be needed to manage communication between the actor and the computerized information system, and we can already identify its main responsibilities, modelled here as operations. On class diagrams and collaboration diagrams, the stereotype of a class can be shown in different ways. Figure 7.7 shows the different symbols that are used to represent a boundary class.

Entity classes

The second analysis class stereotype is the entity[2] class, illustrated in Figure 7.6 by the three classes `Client`, `Campaign` and `Advert` (note that these exist within the current package, and so the package name does not need to be made explicit). Entity classes are used to model 'information and associated behaviour of some phenomenon or concept such as an individual, a real-life object, or a real-life event' (Jacobson et al., 1999). As a general rule, entity classes represent something within the application domain, but external to the software system, about which the system must store some information. This thing might be quite abstract, for example, a campaign, or it may be quite concrete, for example a member of staff.

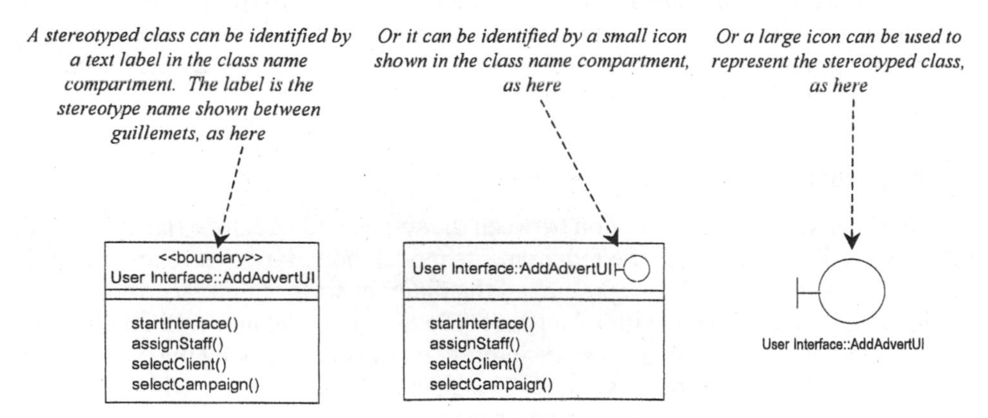

Figure 7.7 Alternative notations for boundary class stereotype.

[2] Some readers may be tempted to confuse the concept of an entity class with the similarly named 'entity' in relational data modelling. There are similarities (e.g., entity classes may show the same kind of logical data structure that is revealed by relational data modelling), but there is also a crucial difference. Entity classes may have complex behaviour related to their information, whereas relational entities represent pure data structures with no behavioural aspect.

Instances of an entity class will often require persistent storage of information about the things that they represent. This can sometimes help to decide whether an entity class is the appropriate modelling construct. For example, an actor is often not represented as an entity class (although they *can* be when it is appropriate). This is in spite of the fact that all actors are within the application domain, external to the software system and important to its operation. But most systems have no need to store information about their users nor to model their behaviour. While there are some obvious exceptions to this (consider a system that monitors user access for security purposes), these are typically separate, specialist applications in their own right. In such a context, an actor would be modelled appropriately as an entity class, since the essential requirements for such a system would include storing information about users, monitoring their access to computer systems and tracking their actions while logged on to a network. But it is more commonly the case that the software we develop does not need to know anything about the people that use it, and so actors are not normally modelled as classes.

Figure 7.8 shows the symbols used for an entity class.

Control classes

The third of the analysis class stereotypes is the control class, illustrated in Figure 7.6 by the class `Control::AddAdvert` (note the explicit package name). Control classes 'represent coordination, sequencing, transactions and control of other objects' (Jacobson et al, 1999). In the USDP, as in the earlier methodology Objectory, it is generally recommended that there should be a control class for each use case (there are exceptions but these need not concern us here). In a sense, then, the control class represents the calculation and scheduling aspects of the logic of the use case—at any rate, those parts that are not specific to the behaviour of a particular entity class, and that *are* specific to the use case. Meanwhile the boundary class represents interaction with the user and the entity classes represent the behaviour of things in the application domain and storage of information that is directly associated with those things (possibly including some elements of calculation and scheduling).

Figure 7.9 shows the symbols for a control class.

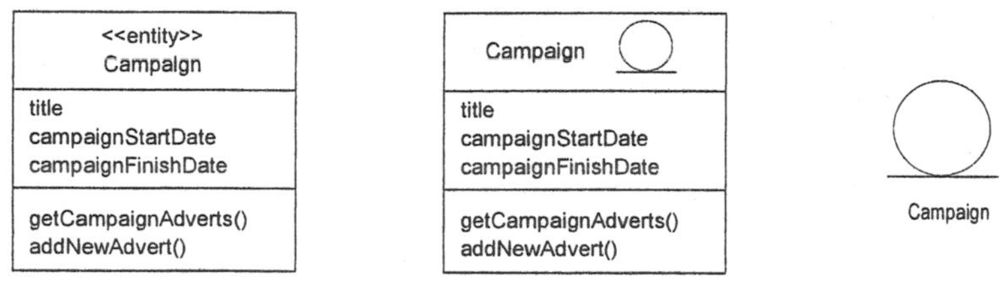

Figure 7.8 Alternative notations for an entity class.

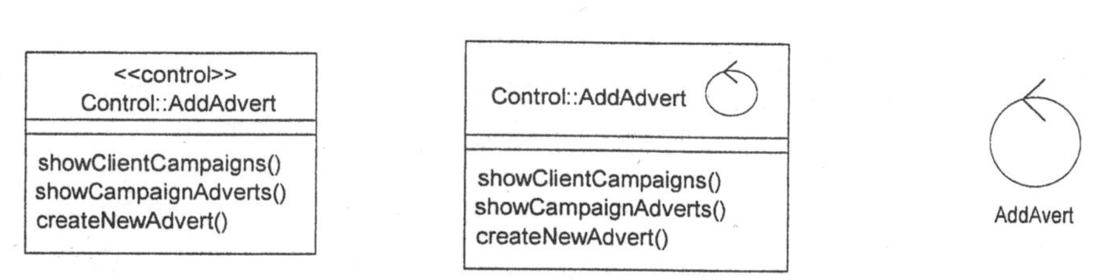

Figure 7.9 Alternative notations for a control class.

7.4 The Class Diagram

7.4.1 Relative stability of classes and instances

Entity classes often represent the more permanent aspects of an application domain, while boundary and control classes represent relatively stable aspects of the way that the software is intended to operate. For example, as long as Agate continues to operate in the advertising business, its business activities are likely to involve campaigns, budgets and adverts, creative staff and campaign managers. And as long as the user requirements for the system do not change, the same boundary and control classes can model the operation of the software and its interaction with users. Thus the description of each class (its data and behaviour—i.e. what it knows, and what it can do) is also relatively stable, and will probably not change frequently.

By contrast, object instances often change frequently, reflecting the need for the system to maintain an up-to-date picture of a dynamic business environment. Instances are subject to three main types of change during system execution.

First, they are created. For example, when Agate undertakes a new campaign, details are stored in a new `Campaign` object. When a new member of staff is recruited, a corresponding `StaffMember` object is created.

Second, they can be destroyed. For example, after a campaign is completed and all invoices are paid, eventually there comes a time when it is no longer of interest to the company[3]. All information relating to the campaign is then deleted by destroying the relevant `Campaign` instance.

Instances of boundary and control classes are particularly volatile—that is, they have short lifetimes and are subject to frequent creation and destruction. For example, a control object and one or more boundary objects are typically instantiated at the start of each execution of a use case, and then destroyed again as soon as the execution is completed.

Finally, objects can be updated, which means a change to the recorded values of one or more characteristics. This is typically done to keep each object in step with the thing that it represents. For example, a client may increase the budget for a campaign, in order to cover a longer run of a TV commercial than was originally planned. To reflect this, the budget value set in the corresponding `Campaign` object must also be changed. Many entity objects are relatively long-lived in comparison with boundary and control objects, and some are updated frequently during their lifetime. Both boundary and control objects may be subject to frequent updates, but for the most part these updates record transient changes of state during the execution of the software, rather than lasting changes to stored information about the application domain.

7.4.2 Attributes

Attributes are part of the essential description of a class. They belong to the class, unlike objects, which instantiate the class. Attributes are the common structure of what a member of the class can 'know'. Each object will have its own, possibly unique, value for each attribute (or values, if the attribute is an array).

Figure 7.10 shows some possible attributes of `Client` and `StaffMember` in the Agate case study. Note that this symbol for a class is sub-divided into three compart-

[3] For example, in the UK many financial records must be kept for six years for tax purposes before they may legally be destroyed.

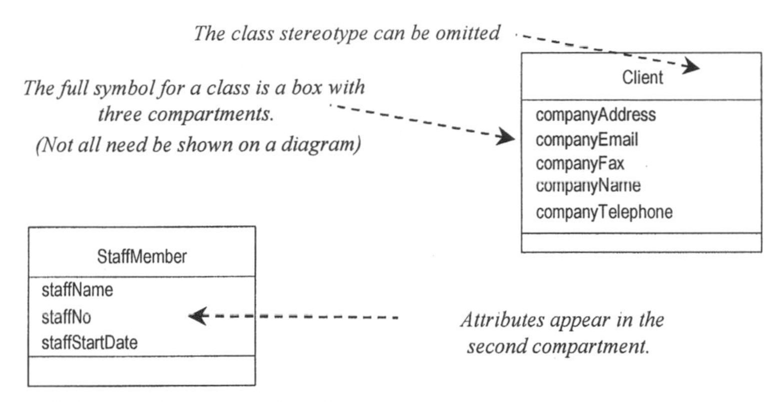

Figure 7.10 Including attributes on a class diagram.

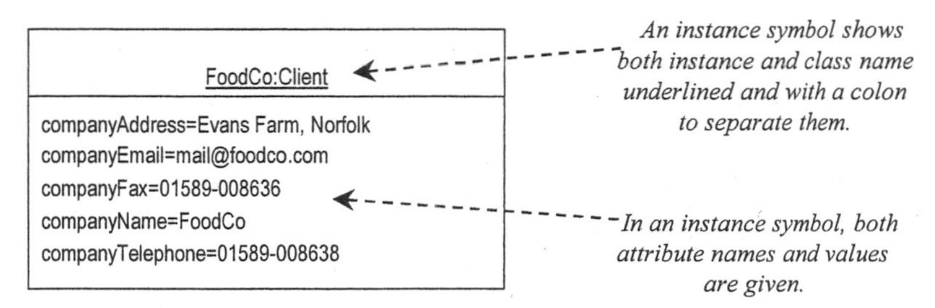

Figure 7.11 Including attribute values on an instance diagram.

ments. The top compartment contains the class name while the second contains the attribute names (which begin with a lower case letter). The third compartment remains empty for the moment.

In much the same way that a class is partly defined by its attribute structure, instances are described by the values of their attributes. To describe a client we give an appropriate value to each attribute. 'FoodCo' is the value given to the company name attribute for the instance of `Client` that represents the real-world client FoodCo. To describe an instance completely, we give a value to all its attributes, as in Figure 7.11.

Some attribute values change during the life of an object. For example, FoodCo may change their e-mail address, their telephone number or their address. They may even decide to change the company's name. In each case, the corresponding attribute value should be updated in the object that represents FoodCo. Other attribute values do not change. For example, a client reference number may be assigned to each client. As long as this is correctly recorded in the first place, there is probably no need for its value ever to change.

7.4.3 Attributes and state

We are now in a position to introduce the idea of object state. The current state of an object is partly described by the instance values of its attributes. When an attribute value changes, the object itself may change state. Some state changes are not significant, which simply means they do not affect the overall behaviour of the object (i.e. the way that it responds to events), and hence of the system as a whole. Others have important implications for object and system behaviour. An example that may be familiar (to some people!) is when your cash card will not allow further cash withdrawals because you have reached your weekly limit. We can understand this in terms of object states, as follows.

Within the bank's computer system, an object <u>yourAccount</u> has an attribute withdrawnDuringWeek that stores your total withdrawals over the last week, and states CanWithdraw and WithdrawalBarred (this is a greatly simplified picture—a real bank system is much more complex). Each week, the value of withdrawnDuringWeek is reset to zero. The object is in its CanWithdraw state. Each time you withdraw cash, withdrawnDuringWeek is updated to take account of the transaction. Once the value of withdrawnDuringWeek reaches a critical threshold set by the bank, <u>yourAccount</u> changes state to WithdrawalBarred. Figure 7.12 shows this as a UML state transition diagram.

Significant changes to the state of an object that affect how they respond to events are modelled using statechart diagrams (see Chapter 11). This short section here is not intended as a presentation of state transition diagrams, but the notation is useful to illustrate what can happen when an object changes state. At the instance level this is merely an update to one attribute value, yet the consequences extend beyond the boundaries of the software system into the user's daily life—as some bank customers know from bitter experience.

7.4.4 Link and association

The concepts of link and association, like those of object and class, are very closely related. A link is a logical connection between two or more objects[4]. In requirements analysis, this expresses a logical relationship in the application domain. An example for Agate is the connection between FoodCo and the 'World Tradition' TV campaign, described by: 'FoodCo is the client for the World Tradition campaign.' This is shown in Figure 7.13.

Linked instances may be from different classes (as with <u>:Client</u>[5] and <u>:Campaign</u>) or from the same class. An example of the latter might be the link supervises between a manager and another staff member who are both instances of Staff

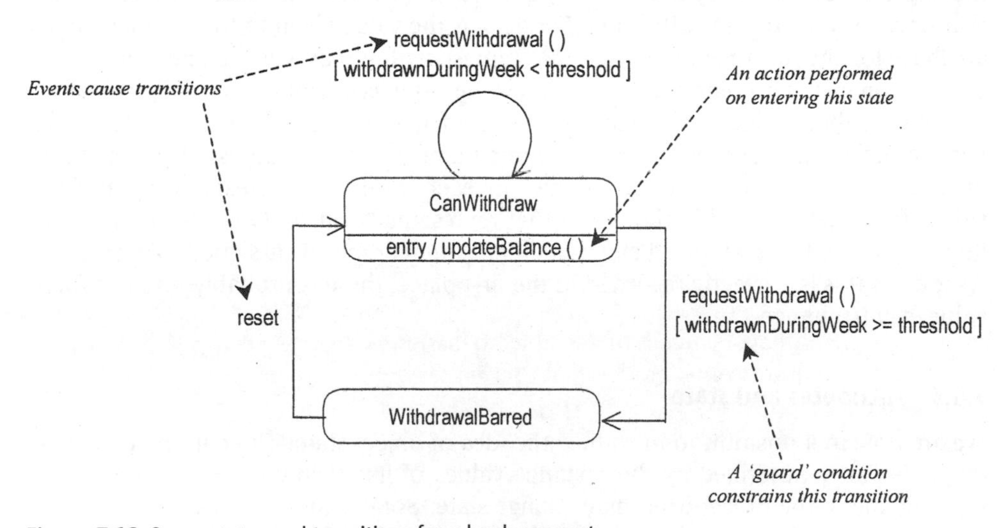

Figure 7.12 Some states and transitions for a bank account.

[4] In most modelling situations, a link connects only two instances, as here. Rarely, a link connects three or more instances, but this is not described here.

[5] This form of the notation refers to an anonymous instance of the class, e.g. :<u>Client</u> = *any* client.

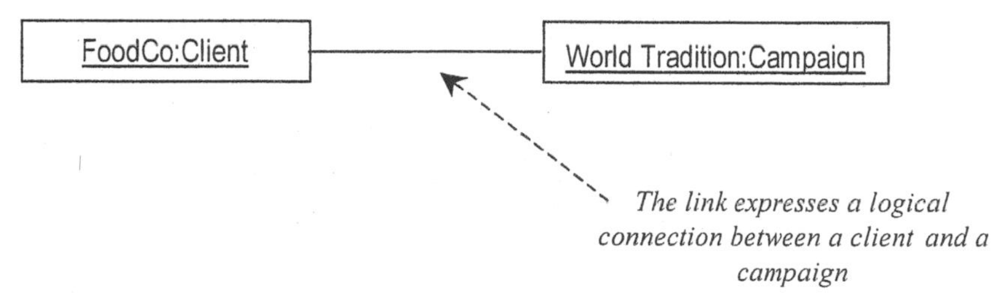
FoodCo:Client
World Tradition:Campaign
The link expresses a logical connection between a client and a campaign

Figure 7.13 A link between instances.

Member. Although this is uncommon, a link can even connect an instance to itself. An example might arise for the captain and players in a hockey team where the captain selects the players. Assuming that the captain is also a player, she may have a link to herself—for example 'the captain selects herself for the team'.

Just as a link connects two instances, an association connects two classes, and just as a class describes a set of similar instances, an association describes a set of similar links (links are called *association instances* by some authors). The similarity is that those links that exist are all between objects of the associated classes. An association is a connection between two classes that represents the possibility of their instances participating in a link.

Some associations can be recognized easily without first being aware of any specific links. For example, it is probably obvious that there is an association between Client and Campaign. Staff at Agate need only record information about clients because they have won (or hope to win) business in the form of campaigns, and it is very unlikely that a campaign would be undertaken except on behalf of a specific client. Other associations are identified through the existence of links that are modelled on collaboration diagrams, as part of the activity of use case realization. As a general rule, wherever a link exists between two entity objects, there will be a corresponding association between their classes.

Like objects, links are not often modelled explicitly on class diagrams at this stage, although they do appear on collaboration diagrams.

7.4.5 Significance of associations

While a link between two objects represents a real-world connection, an association between two classes represents the possibility of links. For example, at Agate a member of staff is assigned to each client as a staff contact. This can be mirrored by links between each :Client and a corresponding :StaffMember, but a link is modelled only if it supports a specific requirement. We know from the Agate use cases that campaign managers need to be able to assign and change a client contact. Therefore the requirements model must permit this link, otherwise it will not be possible to design software that meets these needs.

Figure 7.14 shows some links, but this is not a very economic way of modelling this. To show every link would be unnecessarily complex. There may be many staff and hundreds of clients, not to mention many thousands of links between instances of other classes.

Figure 7.15 shows an association that includes all possible links between clients and members of staff. Rather than modelling every link, this models a general ability to identify a link between a member of staff and a client[6].

[6] But knowing that two classes have an association does not tell us which instances are linked, if any. An association is abstract and general, not particular.

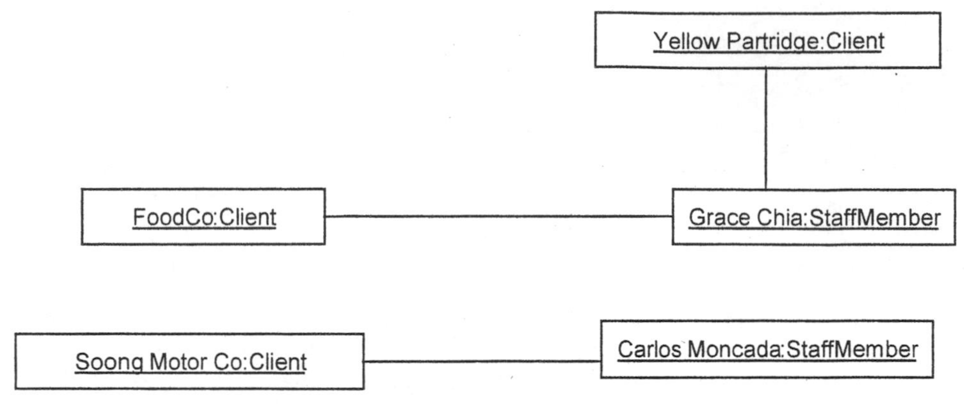

Figure 7.14 Links between instances of `StaffMember` and `Client`.

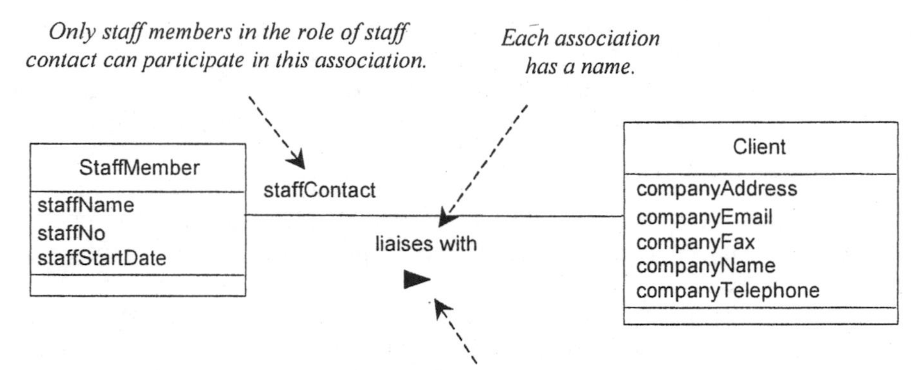

Figure 7.15 `Liaises with` association between `StaffMember` and `Client`.

7.4.6 Associations and state

We can now return briefly to our discussion of object state. In Section 7.4.3 we saw that state is partly defined by an object's current attribute values. The second aspect of an object's state is defined by its current set of links to other objects. Whenever a link to another object is created or destroyed, the object changes state. As with state changes that depend on attribute updates, some of these are significant and others are less so. An example of link creation is when a campaign manager assigns a member of staff to be staff contact for a new client. If, instead, an existing staff contact is changed, this involves destroying one link and creating another in its place.

7.4.7 Multiplicity

Associations represent the possible existence of links between objects. It is often important to define a limit on the number of links with objects of another specific class in which a single object can participate. Multiplicity (the range of allowed *cardinalities*) is the term used to describe constraints on the number of participating objects. Multiplicity reflects *enterprise* (or *business*) *rules*, which are constraints on the way that business activities can take place.

A familiar example is the number of people allowed to be the designated account-holder for a bank account. A sole account has *one and only one* account-holder, a joint account may have *exactly two* account-holders and a business partnership account may have *two or more* account-holders. It is important to model these constraints correctly, as they may determine which operations will be permitted by the software. A badly specified system might incorrectly allow an unauthorized second person to draw money from a sole account. Alternatively it might prevent a legitimate customer from being able to draw money from a joint account. In practice, the multiplicity of an association defines upper and lower limits on the number of other instances to which any one object may be linked.

As we saw in the previous section, it is not usually necessary for a requirements model to show the exact names of the members of staff who are staff contacts for every client. This is inappropriate, not only because it may make the diagram so complex it would become unreadable, but also because links may change frequently. By the time we reached the stage of designing our software, such a detailed model would already be out of date. But a requirements model should be able to tell us something about the variety of ways that individual instances can be linked, and in particular any constraints that apply.

We already know that in the Agate case study, each client has one member of staff assigned as `staffContact`, while each member of staff may be assigned to zero or more clients. This is modelled by adding a little more information to the simple class diagram in Figure 7.15. Figure 7.16 shows this, with each end of the association now qualified by its multiplicity. Multiplicity must be read separately in the opposite directions for each association. Thus the possible number of clients allocated to a staff member ranges from 'zero' to 'any number', while the possible number of staff members allocated to a client must be exactly one[7].

To be useful, the notation should be able to show all multiplicities, not just 'exactly one' and 'zero or more', and Figures 7.17–7.19 show some of the variations. These examples are not exhaustive, and there are many variations on the form, allowing any range of values to be specified, for example `0..3`, `1..5`, `2..10`, `3..*`, or discrete

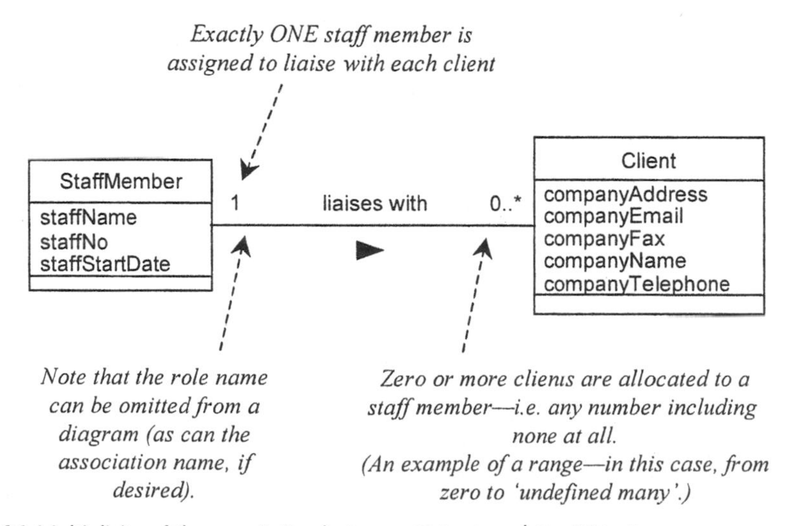

Figure 7.16 Multiplicity of the association between `Client` and `StaffMember`.

[7] This kind of knowledge always emerges, directly or indirectly, from users.

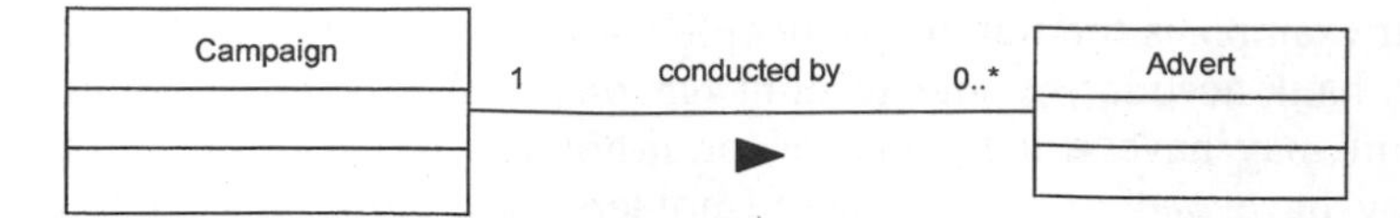

Figure 7.17 A `Campaign` is conducted by zero or more `Adverts` while each `Advert` belongs to exactly one `Campaign`.

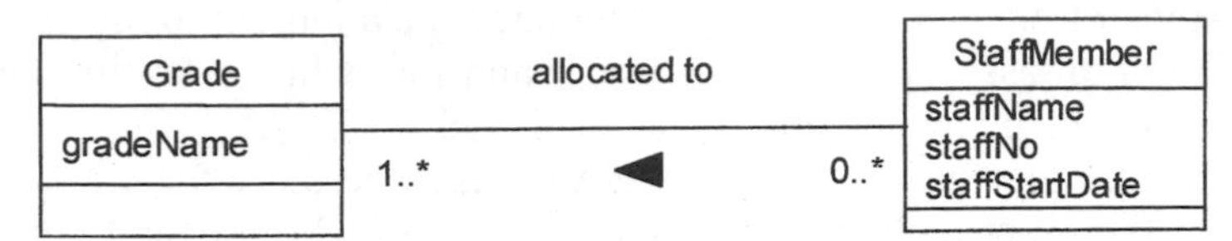

Figure 7.18 Every `StaffMember` must be allocated to one or more `Grades`, while a `Grade` may have zero, one or more staff allocated to it.

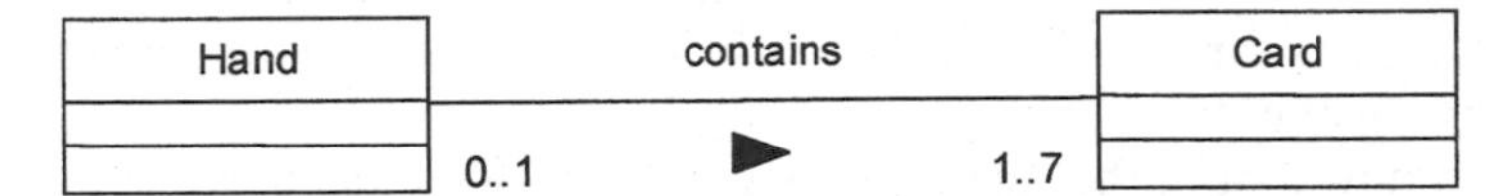

Figure 7.19 A Poker `Hand` contains up to 7 `Cards`. Each `Card` dealt must be in only one `Hand` (although a card may still be undealt in the pack). This assumes no cheating!

values, such as `3`, `5` or `19`, or combinations of the two, for example `1,3,7..*`. Normally, however, it is best not to restrict the multiplicity unnecessarily.

To summarize, association multiplicity conveys important information about the structure of the problem domain. Different assumptions about multiplicity have significant effects on the software design. If association multiplicities are modelled incorrectly, this may later make it impossible for the software to do things that users want it to do.

7.4.8 Operations

Operations are the elements of common behaviour shared by all instances of a class. They are actions that can be carried out by, or on, an object. The classes modelled during requirements analysis represent real-world things and concepts, so their operations can be said to represent aspects of the behaviour of the same things and concepts. However, as the basic idea of an object-oriented system is that it should consist of independent, collaborating objects, it is probably easier to understand operations as aspects of behaviour required to *simulate* the way that the application domain works (see Box 7.1). Another way of putting this is that operations are services that objects may be asked to perform by other objects. For example, in the Agate case study, `StaffMember` has an operation that calculates the amount of bonus pay due for a staff member. And, since staff bonus is partly based on the profit of each campaign a member of staff has worked on, Campaign has an operation to calculate the profit for each campaign.

An operation is a specification for some aspect of the behaviour of a class. Operations are thus defined for the whole class, and are valid for every instance of the class. Figure 7.20 shows some examples of operations for the Agate case study. At this stage we do not need to be concerned with the details of how each operation will work, nor with when they are permitted to work or constrained from working.

Box 7.1 Operations as responsibilities

There is a world of difference between saying, on the one hand, that aStaffMember has an operation to calculate its own bonus, and, on the other hand, saying that a real member of staff has responsibility for calculating her own bonus. The former is an appropriate way of simulating the real-world operation, but there is certainly no implication of identity between the two. Nor do we wish to imply that real (but abstract) campaigns are capable of calculating their own profit—or anything else, for that matter. What is implied is, first, that the ability to carry out these tasks is a requirement of the system, and second, that we must make a judgement about where to locate these operations within the model, and our judgement should be based on the idea of object responsibilities. Later in this Chapter we introduce CRC cards and in Chapter 9 we introduce interaction diagrams, two techniques that can be used to help with the allocation of responsibilities to classes in a coherent manner.

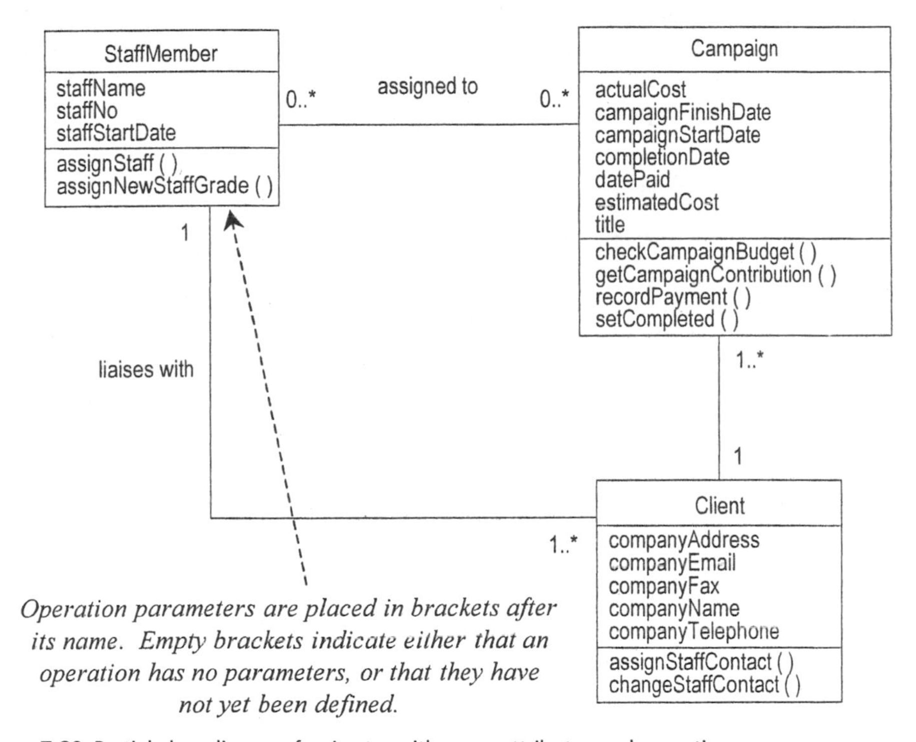

Figure 7.20 Partial class diagram for Agate, with some attributes and operations.

But we do need to know which operations to include in the requirements model, and we need to make a first guess at their appropriate location within classes. Operation names are always placed in the third compartment of the rectangular class symbol. As for attributes, operations are written beginning with a lower case letter. There is no separate notation for showing the operations of an object instance. In this respect, operations are unlike attributes and associations, in that they have exactly the same significance for instances as they do for classes.

Operations are eventually implemented by *methods*, and what a method actually does on any given occasion may be constrained by the value of object attributes and links when the method is invoked. In the bank cash card example described earlier, yourAccount has the operation makeWithdrawal() (written as yourAccount.makeWithdrawal()) and the attribute withdrawnThisWeek (yourAccount.withdrawnThisWeek). The value of yourAccount.withdrawnThisWeek determines how the operation responds. If your limit has not yet been reached, the withdrawal is made. Otherwise, the operation may cause a message to be returned that says you have exceeded your limit. Alternative courses of action, and the basis for choosing between them, are included in the detailed definition of an operation (we describe this in Chapter 10).

The effects that an operation can have include changing the characteristics of an object, for example, updating an attribute value. Another effect may be to change an object's links with other objects, for example assigning :StaffMember to work on :Campaign. The creation of this particular link is necessary for that member of staff to earn a bonus from the campaign profits. Some operations delegate part of their work to other objects: aStaffMember.calculateBonus() provides an example, as this operation must call aCampaign.getCampaignContribution() for each associated campaign before it can arrive at an answer. Thus each individual operation may represent only a small part of a larger task that is distributed among many objects. A relatively even distribution of computational effort is regarded as highly desirable—it is a large step on the way to building a modular system.

7.4.9 Operations and state

Perhaps it is already evident that there is a relationship between an object's operations and its states. An object can only change its state through the execution of an operation. In fact, we are saying no more here than that attributes cannot store or update their own values, and links cannot make or break themselves. But this illustrates an important point about the way that object encapsulation assists modularity in a system. The services that an object provides can only be accessed through the object's interface, which consists entirely of operation signatures defined at the class level. In order to get an object to do anything at all—change its data, create or destroy links, even respond to simple queries, another object must send a message that includes a valid call on an operation.

7.5 Drawing a Class Diagram

In this section we look at some of the practical aspects of drawing a class diagram. This involves suggesting where to look for the necessary information, and also a recommended sequence for carrying out the various tasks. It should be stressed that neither the information sources nor the sequence of tasks are intended to be prescriptive. Nor do we cover every eventuality on every project. In practice, one project differs greatly from another, and some steps may be omitted altogether or carried out at a completely different stage of the life cycle. Experienced analysts will always use their own judgement on how to proceed in a given situation.

7.5.1 Identifying classes

The class diagram is fundamental to object-oriented analysis. Through successive iterations, it provides both a high-level basis for systems architecture, and a low-level basis for the allocation of data and behaviour to individual classes and object instances, and ultimately for the design of the program code that implements the system. It is important to identify classes correctly. However, given the iterative nature of the object-oriented approach, it is not essential to get this right on the first attempt.

As we saw in Section 7.3.1, one of the first steps in creating a class diagram is to derive from a use case, via a collaboration (or collaboration diagram), those classes that participate in realizing the use case. Through further analysis, a class diagram is developed for each use case and the various use case class diagrams are then usually assembled into a larger analysis class diagram. This can be drawn first for a single subsystem or increment, but class diagrams can be drawn at any scale that is appropriate, from a single use case instance to a large, complex system.

In parallel with analysing the static structure shown on class diagrams, we are also interested in the dynamic interaction among classes. This, too, can be derived from the use cases and is shown on the collaboration diagrams. Interaction is normally further explored using sequence diagrams, and, in practice, these are often developed side-by-side with the class diagrams. But in this chapter we will concentrate mainly on the notation and development of the class diagram (object interaction is covered in Chapter 9).

Identifying the objects involved in a collaboration can be difficult at first, and takes some practice before the analyst can feel really comfortable with the process. In our view, some texts—including the authoritative text on the USDP (Jacobson et al., 1999)—do not give much guidance that would help the novice learn how to carry out the task. These authors suggest that a collaboration (i.e. the set of classes that it comprises) can be identified directly for a use case, and that, once the classes are known, the next step is to consider the interaction among the classes and so build a collaboration diagram. We believe that it is usually easier to identify classes through considering their interaction together with their static structure. But it is still worth stressing that first-cut models are often tentative, and may be refined and modified more than once during later iterations.

Our starting point is the use case, and an extended version of a use case description is repeated below (for simplicity, we ignore alternative courses).

Use case description: Assign staff to work on a campaign

Actor Action	**System Response**
1. None	2. Displays list of client names.
3. The actor (a campaign manager) selects the client name	4. Lists the titles of all campaigns for that client
5. Selects the relevant campaign	6. Displays a list of all staff members not already allocated to this campaign
7. Highlights the staff members to be assigned to this campaign	8. Presents a message confirming that staff have been allocated

The task is to find a set of classes that can interact to realize the use case. This means, above all, thinking about those things and concepts in the application domain that are important to the goals of the system that is being developed. In this case, we know from the use case diagram that the campaign manager is the actor for this use case. The use case description tells us that the manager selects a client name, the system responds by displaying all campaigns for that client, the manager selects a campaign, the system displays all staff not yet allocated to it, the manager selects staff to assign to the campaign and the system confirms the result. The objective of the use case is to allow the manager to assign staff to a campaign.

Let's begin by picking out from the description all the important things or concepts in the application domain. (Finding objects and classes is discussed further in Section 7.5.3.) Our first list might include: campaign manager, client name, campaigns, client, staff. But we are only interested in those about which the system must store some information or knowledge in order to achieve its objectives. The campaign manager will be modelled as an actor. For the purposes of this particular use case, it is unlikely the system will need to encapsulate any further knowledge about the actor. We can also eliminate client name, since this is really just part of the description of a client. This leaves `Client`, `Campaign` and `StaffMember` in the collaboration.

Next, we can try to draw a collaboration diagram using these classes. (Note that collaboration diagrams are covered in more depth in Chapter 9). This will help us to see whether they are all needed, and also whether any less obvious classes are needed as well. It will also help us to identify their structure.

Figure 7.21 shows an initial collaboration diagram for this use case. It should be stressed that this is only a very rough first-cut diagram (hence the differences in both detail and structure from the more highly-developed diagram for `Add advert to a campaign` shown earlier in Figure 7.4. This, as we shall see over the next few pages, was the result of further analysis[8]). So far, this does no more than identify the participating entity objects, their classes and an outline sequence of messages that might achieve the goals of the use case.

A collaboration is between individual object instances, not between classes. This is shown in the diagram by the convention of writing a colon before the class name, which indicates that this is an anonymous instance of the class, rather than the class itself. Messages between classes are shown by arrows, and their sequence is indicated by the number alongside. In this example, these are not yet labelled, although some—those that can be most easily related to the use case description—will probably soon be given names that correspond to responsibilities of the class to which the message is addressed.

Class names are always written in the singular, although we know there are many staff, campaigns, adverts, etc. This convention reinforces the view of a class as a *descriptor* for a collection of objects, rather than the collection itself. Another convention (derived from object-oriented programming style) is that most names are written in lower case, but classes are capitalized at the beginning of the name. Multiple words are run together, punctuated by upper case letters at the start of each new word to improve readability, for example, the control class `AssignStaff`. Note, however, that this does not apply to use case names, such as `Assign staff to work on a campaign`, nor to operation names, such as `getClients`.

In Figure 7.21, we have used entity object icons and each class has only a name. This simplest form is appropriate at this stage, since we do not yet know any of the other details that will be added later. Once a class diagram has been derived from this relatively simple collaboration diagram, it will eventually grow quite complex. Nevertheless it is a good start at abstracting some useful details from a description that may be cluttered with many irrelevant facts. As more detail is added to the class diagram, it will soon be much easier to absorb and less ambiguous than the corresponding text—subject, of course, to the modeller's skill and comprehension. In

[8] Hand-drawn diagrams are not necessarily rough-cut or transient, although it happens that this one is both. Most UML diagrams are drawn on a CASE tool, but this does not automatically lend authority. As long as a diagram adheres to the standard, any medium is as good as any other.

real life, of course, even first-cut models are often more complicated than this, and a great deal more effort may be needed to arrive at a preliminary understanding.

The collaboration diagram in Figure 7.21 does not yet show any boundary or control objects, and these must be added. It is also based on certain assumptions about how the interaction between objects would take place, and we must make these assumptions explicit and question them. Although the messages are not yet labelled, the numbers alongside the message arrows indicates their sequence. The diagram implies a linear flow of messages, along the following lines. An initial message is directed to a `Client`, which is assumed to know its `Campaigns` and returns the list (responses to a message are not usually shown explicitly, so the arrow only points in one direction). Each `Campaign` is also assumed to know which `StaffMembers` are currently assigned to it, and which are not. The `Client` object asks the selected `Campaign` object to return a list of unassigned `StaffMembers`, and the `Client` passes this on to the interface. The `Client` object then asks the `Campaign` object to tell the selected `StaffMember` object to assign itself to the `Campaign`, by creating a link between them.

Although we are primarily concerned with analysis questions at present, this scenario raises some serious design issues. In particular, it effectively locates control of the use case within the client object, which would give this class responsibility for tasks that are not directly relevant to the responsibilities of a `Client`. The introduction of a control object allows this responsibility to be encapsulated separately from the application domain knowledge that the entity classes represent. Figure 7.22 shows the collaboration diagram after this refinement. A boundary object has also been added; this will be responsible for the capture of input from the user and the presentation and display of results. In a limited sense, we have begun to design a software architecture that will allow a great deal of flexibility as to how the system will be implemented.

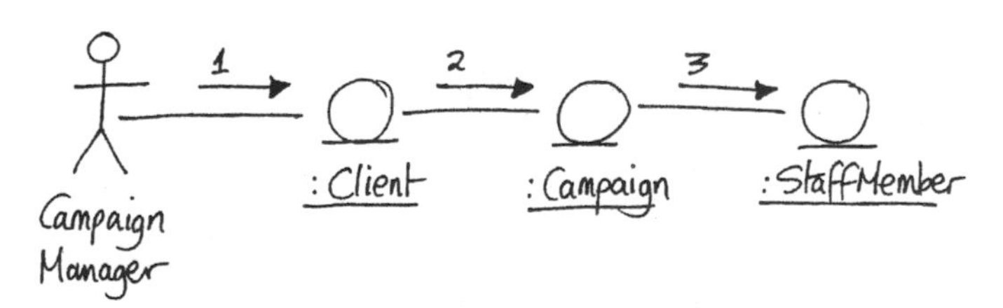

Figure 7.21 Initial collaboration diagram for `Assign staff to work on a campaign.`

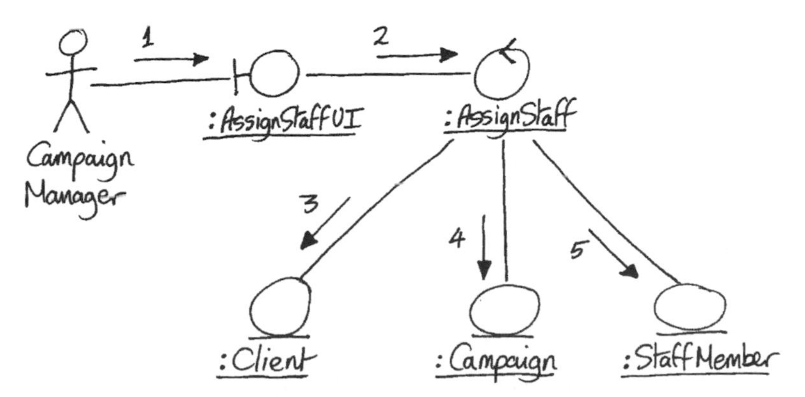

Figure 7.22 Boundary and control objects added, giving a different view on how the object interaction might work.

All messages are now routed centrally through the control object. This means that no entity class needs to know anything about any other entity class unless this is directly relevant to its own responsibilities. For example, a `Client` object is now assumed to know (and to tell the control object) which are its `Campaigns`, but it is no longer responsible for knowing (or finding out) which `StaffMembers` are assigned to its `Campaigns`.

This diagram also uses the icon symbols for objects, but it could equally well be represented using rectangular object symbols, as shown in Figure 7.23 (which also adds labels to some of the more significant messages).

The version in Figures 7.22 and 7.23 has addressed one major issue, but we must now raise some other questions. It seems reasonable to assume that a `Client` is responsible for knowing its own `Campaigns`. But the collaboration diagram now shows no communication between `Clients` and `Campaigns`, so it is not clear how this knowledge will be maintained. A similar consideration applies to the question of how a `Campaign` object maintains its knowledge of the `StaffMembers` that have been assigned to work on it. The first version (Figure 7.21) relied too much on delegation of responsibility through the `Client`. It now seems that we have taken too much responsibility away from the entity objects.

Figure 7.24 shows another way that the interaction might work, and we could perhaps expand this as follows. The control object obtains a list of `Clients` and asks the boundary object to display them. It then asks a `Client` for a list of its `Campaigns`.

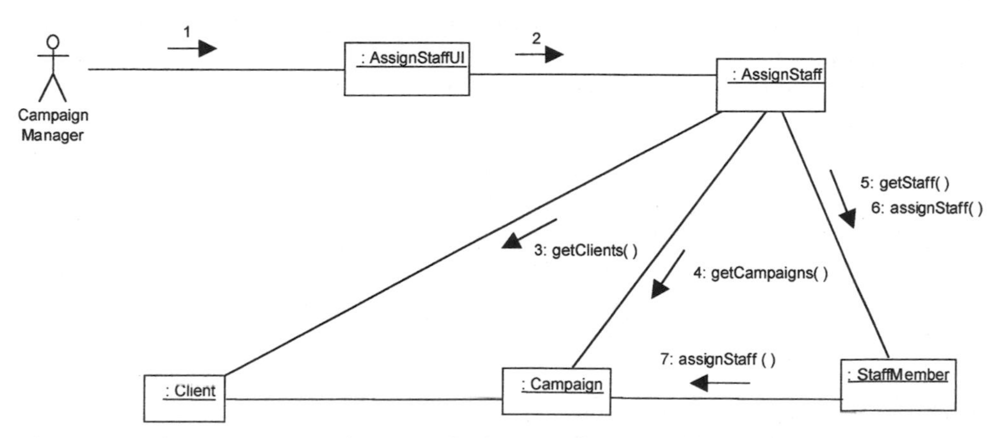

Figure 7.23 Alternative notation for essentially the same diagram as Figure 7.22.

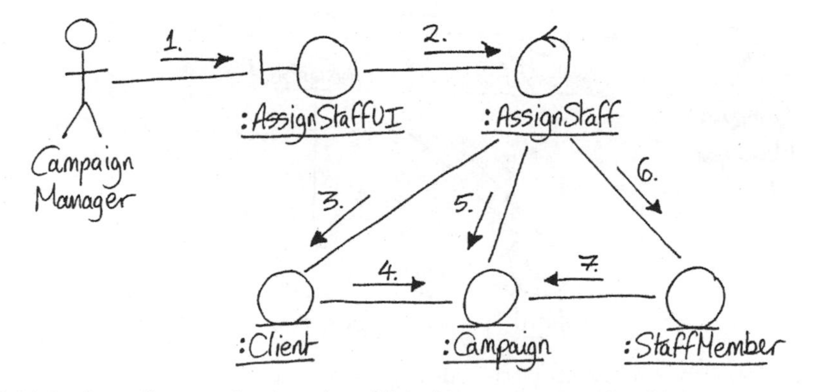

Figure 7.24 Further refinement has introduced links between some entity objects.

The Client may obtain some information directly from the Campaign objects themselves (perhaps their names or status). The control object then directly asks the selected Campaign for information regarding which staff are currently assigned to it, and next asks unassigned StaffMembers for their details, passing each bundle of information to the boundary object for display. The control object then instructs the selected StaffMember to assign itself, which it does by sending a message to the Campaign.

The final diagram in this set is shown in Figure 7.25. We have added associations between Client and Campaign and between Campaign and StaffMember. We have also arrived at an initial judgement about how to distribute the responsibility for this use case among the various collaborating objects (this is shown by the sequence and labelling of the messages). This is still a simplified version of the full interaction, with many details left to be determined by further analysis. For the moment, the link between :Client and :Campaign has been left without any message, since it is not clear whether any messages need to be exchanged between the two objects for this particular interaction, although it is likely to be needed by other use cases, or possibly by alternative courses through this use case. But, while some issues remain to be clarified, this approaches the level of understanding that we need in order to develop a robust class model capable of fully supporting the use case. It should be stressed that no decisions made at this stage are necessarily final, and we may well need to make several iterations through this activity before we achieve a full understanding.

7.5.2 From collaboration diagram to class diagram

The next step in the development of a requirements model is usually to produce a class diagram that corresponds to each of the collaboration diagrams. The class diagram that corresponds to the use case Assign staff to work on a campaign is shown in Figure 7.26.

Provided that the collaboration diagrams are themselves the result of reasonably careful analysis, the transition is not usually too difficult. For a start, there are some obvious similarities between the two diagrams, although there are also some important differences.

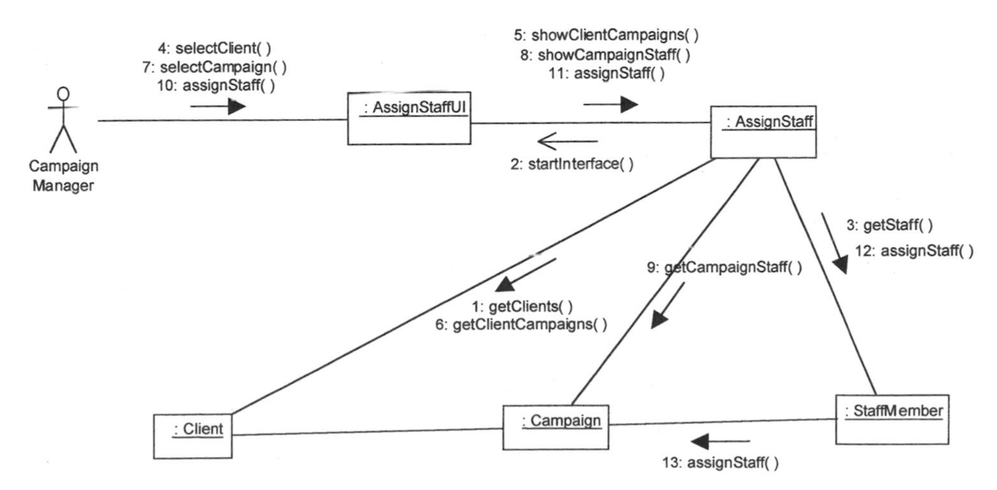

Figure 7.25 Near-final collaboration diagram for Assign staff to work on a campaign.

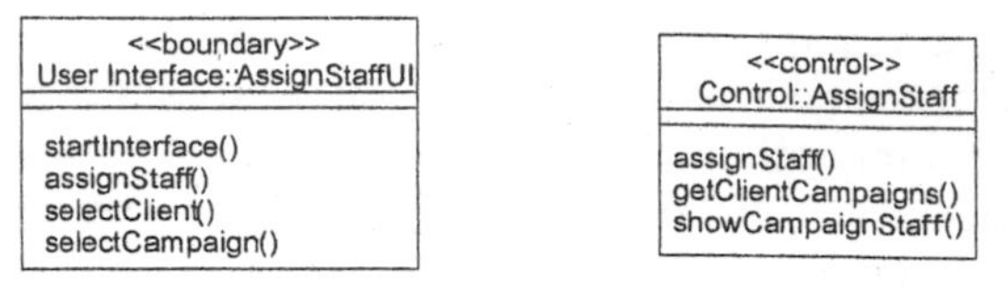

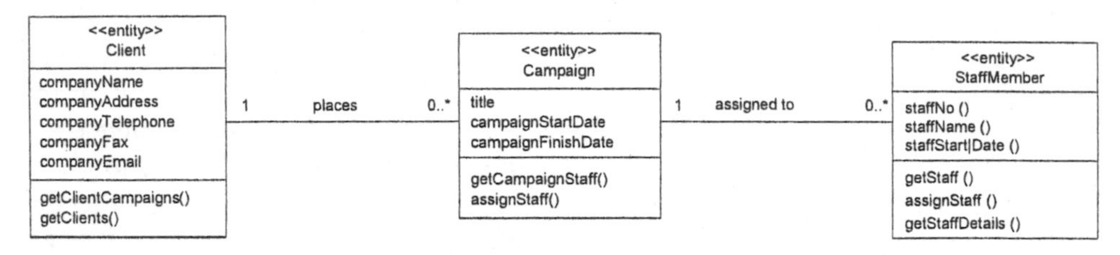

Figure 7.26 Class diagram for the use case `Assign staff to work on a campaign`.

First, we consider the similarities. Both show class or object symbols joined by connecting lines. In general, a class diagram has more or less the same structure as the corresponding collaboration diagram. In particular, both should show classes or objects of the same types. Any of the three analysis stereotype notations for a class can be used on either diagram, and stereotype labels (if used) can also be omitted from individual classes, or from an entire diagram.

Next, we examine the differences, some of which are less obvious than others. Perhaps the most obvious difference is that an actor is almost always shown on a collaboration diagram, but not usually shown on a class diagram[9]. This is because the collaboration diagram represents a particular interaction (for example, one that supports a specific path through a single use case) and the actor is an important part of this interaction. However, a class diagram shows the more enduring structure of associations among the classes, and frequently supports a number of different interactions that may represent several different use cases.

Some subtler details are associated with this change in conceptual emphasis. First, a collaboration diagram usually contains only object instances, while a class diagram (as its name suggests) usually contains only classes[10]. This is visible (but by no means obvious) in the names given to the class and object symbols on each diagram. Another difference is that the connections between the object symbols on a collaboration diagram symbolize links between objects, while on a class diagram the corresponding connections stand for associations between classes. This is why some links—for example, those between `User Interface::AssignStaffUI` and `Control::AssignStaff` and between `Control::AssignStaff` and `Client`—have not been shown on the class diagram. Transient boundary and control objects like these are created only when needed during the execution of the software, while entity objects and their links normally endure beyond one cycle of execution, and probably therefore require persistent storage. But the classes that boundary and control objects instantiate are important aspects of the requirements, and so they are still included in the class diagram. As the model develops, we can anticipate that these classes will be located in separate packages. But, since the class diagram is essentially a model of

[9] However, actors can be shown on a class or object diagram when necessary. They usually are shown if the actor is to be represented by a class—we mentioned this possibility earlier, in the discussion about entity classes in Section 7.3.2.

[10] Actually, a class diagram can contain instances too, but this is relatively unusual in practice.

static structure, we take the view that their transient links do not need to be modelled in an analysis class diagram, hence their disappearance at this point. In contrast to this, a collaboration diagram shows the dynamic interaction of a group of objects and thus every link needed for message passing is shown. The labelled arrows alongside the links represent messages between objects. On a class diagram, the associations themselves are usually labelled, but messages are not shown.

Finally, although any of the three stereotype symbols can be used on either diagram, there are also differences in this notation. When the rectangular box variant of the notation is used in a collaboration diagram it represents object instances rather than classes, is normally undivided and contains only the class name (optionally, together with the object name). On a class diagram, the symbol is usually divided into three compartments that contain in turn the class name (optionally, together with its stereotype), its attributes and its operations (but all except the class name can be omitted if desired).

7.5.3 Other approaches to finding objects and classes

So far, we have said that use cases are the best place to look for entity objects, and the best way to find them is through thinking about interactions between them that support the use case. Some authors follow different approaches. One approach is to first develop a *domain model*—an analysis class model that is independent of any particular use cases. For example, the domain model is a significant feature of the ICONIX method (Rosenberg with Scott, 1999). In the approach that we follow in this book, the development of a domain model is considered primarily as a step that follows the development of class diagrams for each use case (see Section 7.7 and Chapter 8). However, this is largely for pedagogic reasons. We do not believe that any one approach is necessarily the best for all situations (nor, indeed, do most other writers, including Jacobson and Rosenberg). Sometimes a domain model will already exist, and in other cases it will make sense to produce a domain model before producing any use cases. The key to success is iterative refinement of the models, however they are produced in the first place.

It is worth reviewing any background documentation gathered during the fact-finding stage. A second reading, after an initial attempt at class modelling, can discover more classes, due to your clearer understanding of the problem.

Ideally, user representatives will be closely involved in discussing and developing the class diagram. Nowadays users often work alongside professional analysts as part of the project team. Most projects are a learning experience for everyone involved, so it is not unusual for users' understanding of their own business activity to grow and develop, and it is likely that users will identify a number of additional classes that were not apparent at first.

Your own intuition is another useful source, together with that of colleagues. And you can look for analysis patterns, but this is a more advanced technique that will not be covered in this chapter.

With experience these can all give guidance, but always check your intuitions with someone who knows the business intimately. If you let yourself be overwhelmed by similarities to other projects, you may overlook important differences. As an analyst, you should remember at all times that users are the experts on the business and what the software system should do. Your role is to make users aware of the possibilities offered by modern information systems, and to help translate their requests into a system that meets as many of their needs as possible.

However you approach the identification of classes, it helps to have a general idea of what you are looking for. Some pointers have been developed over the years that help to discriminate between likely classes and unlikely ones. Rumbaugh et al. (1995) usefully categorized the kinds of things and concepts that are more likely than others to need representation as an entity object or class.

The main categories shown in Figure 7.27 are based on their categories. It is best to keep a list of potential classes, with a brief description for each (a rough list is fine; it will grow over time, but many items will also be crossed out and removed). When you enter your models into a CASE tool *repository* (a term for the model storage within a CASE tool environment), these textual descriptions and definitions will be an important supplement to the diagrams. Check your list carefully as it grows. Even the most experienced analyst will probably include at first some potential classes that might cause confusion later if they are retained in the model.

Next, there are some guidelines to help you to prune out unsuitable candidate classes. For each item on the list, ask yourself the following questions.

Is it beyond the scope of the system?

You may include people, things or concepts that are not strictly necessary to describe the application domain that you are investigating. Remove these from your list. They may become clear from use case descriptions or from collaborations, but do not worry if the odd one slips through. There will be lots of opportunities to catch them later on. Remember, too, that only the users can finally set the system boundary.

Beginners often include classes that represent the people who operate the current system, perhaps because their names or job titles appear in a use case description. It is rarely necessary to model the operators of the system as classes. But the exception is when they are themselves the subjects of one of the business activities. An example might be an office worker handling a company's pension scheme, who is also a member of the scheme. In this case, you may need to model them as a member of the scheme (i.e. a potential object), as well as an operator of the system (i.e. an actor). On the whole, however, actors do not need to be represented as objects.

Category	Examples
People	Mr Harmsworth (a campaign manager), Dilip (a copywriter)
Organizations	Jones & Co (a forklift truck distributor), the Soong Motor Company, Agate's Creative Department
Structures	Team, project, campaign, assembly
Physical things	Fork-lift truck, electric drill, tube of toothpaste
Abstractions of people	Employee, supervisor, customer, client
Abstractions of physical things	Wheeled vehicle, hand tool, retail goods
Conceptual things	Campaign, employee, rule, team, project, customer, qualification
Enduring relationships between members of other categories	Sale, purchase, contract, campaign, agreement, assembly, employment

Figure 7.27 Looking for objects.

Does it refer to the system as a whole?

You may include an item that refers to the system you are modelling, or to the department or organization that contains it. It is not usually necessary for a model to contain classes that represent the entire system.

Does it duplicate another class?

You may include two items that are really synonyms. If you are not sure, check with your users exactly what they understand by each item on the list. This should become clearer as you write the descriptions for each class.

Is it too vague?

Eliminate any potential classes for which you are unable to write a clear description, unless you are sure this is only because of a temporary lack of information.

Is it too specific?

Unless you are modelling a specific interaction (for example, when drawing an initial collaboration diagram), it is usually better to model classes, rather than instances. Think carefully about any items on your list that are unique. For example, a company may currently have only one supplier, tempting you to model the specific supplier. But a supplier might be replaced tomorrow for business reasons. A class named `Supplier` would be unaffected by this, whereas one modelled too closely on the specific company might require modification.

Is it too tied up with physical inputs and outputs?

Avoid modelling things that depend closely on the physical way system inputs and outputs are currently handled. For example, printed order forms and telephone enquiries may be how the current system operates, but it is much too soon to make a decision on whether they will play the same role in the new system. Try to think of names that express a logical meaning rather than a physical implementation: `Order` and `Enquiry` would be acceptable alternatives.

On the other hand, physical objects that are an essential part of the business activity should be included. This can depend a lot on context—`Truck` may be an acceptable class in a system to co-ordinate vehicles used for parcel deliveries, but irrelevant in another system that records customer payments for the deliveries, even though invoices and payments might travel on the same truck.

Is it really an attribute?

An attribute is a characteristic of a class. What makes this a problem is that an item that is an attribute in one domain may be a class in another, depending on the requirements. So some items on your potential class list may be better modelled as attributes. The primary test is this: does the item only have meaning as a description or qualification of another item? To illustrate this, we will look at examples that show how the significance of a date can vary between two different application domains.

In the Agate case study, the significance of a staff member's start date is to allow appropriate salary, bonus and grading calculations to be carried out. It would therefore be appropriate to model `staffStartDate` as a single attribute of `StaffMember`. But now consider a weather forecasting agency, keeping daily records of atmospheric conditions, and producing analyses for different weeks, months and years. Each date may be described by many other variables, e.g. maximum, minimum and average temperature, hours of sunshine, total precipitation, average windspeed, etc. These analyses might also require separate attributes for day of the week, month and year. We might then choose to model a `Date` class, with the other variables as its attributes.

Is it really an operation?

An operation is an action, a responsibility of a class. This can also be confusing, as some actions can be better modelled as classes. It is particularly easy to confuse the two if the use case descriptions are ambiguous. For an example of an action that can be considered as a class, consider a *sale* transaction. Whenever you buy something in a shop (a new CD, say), some sort of record is kept of the sale. The nature of this record depends on how the shop intends to use the information. This, in turn, determines whether we should model the sale as a class or as an operation. There are two considerations that might make a sale transaction a class rather than an operation.

- A sale may have characteristics of its own, that would be best modelled as attributes of a class, e.g. value, date, etc.
- There may be a requirement for the system to remember particular sales over a period of time, e.g. in order to respond to warranty claims, or to audit the shop's accounts.

If there is no requirement to record a history made up of individual sales, or to describe sales in terms of their value, date, etc. it may make more sense to model them as an operation of another class, perhaps as `StockItem.sell()`. This would probably be quite adequate if the shopkeeper was only interested in knowing the total value or quantity sold for each item. For each action on your preliminary class list, consider whether these criteria apply. If they do not, it may be an operation rather than a class.

Is it really an association?

An association is a relationship of some kind between two classes. But this too can be confusing as we may prefer to represent some relationships as classes. The sales transaction can also be counted as an example of this (a sale is both an action and a relationship). How do we decide which relationships to represent as associations, and which as classes? This can sometimes be a difficult and complex problem. You can apply a similar test to those described above for attributes and operations. If an association is something we need to describe in terms of further characteristics—if it is apparent that it has attributes of its own—then it should be modelled as a class. If it only has meaning as a relationship between two classes, leave it as an association.

But the best answer at this stage is not to spend too long on making the distinction. The important thing during requirements analysis is to make sure all significant relationships are modelled, whether as classes or associations. We can review our judgements later when we understand more about the situation. Indeed, following the transition from requirements modelling to software design, it is often the case that certain types of association may be changed into classes, or further classes may be added to help implement the association effectively (this is covered in Chapter 14).

7.5.4 Adding and locating attributes

Many attributes will already appear in the use case descriptions. Others will become apparent as you think about your model in more detail. The simple rule is that attributes should be placed in the class they describe. This usually presents few problems. For example, the attributes `staffNo`, `staffName`, and `staffStartDate` all clearly describe a member of staff, so should be placed in the Staff class.

Sometimes it is more difficult to identify the correct class for an attribute. The attribute may not properly belong to any of the classes you have already identified. An example will help to illustrate this. Consider this extract from an interview with Amarjeet Grewal (Agate Finance Director):

Amarjeet Grewal: Agate's pay structure is based on distinct grades. Directors and managers negotiate their own salaries, but other staff are placed on a grade that determines their basic salary. You can only change grade as a result of an appraisal with your line manager.

A member of staff has one grade at a time, but it sounds like they may have several previous grades, and several members of staff may be on the same grade at the same time. Staff and Grade are probably classes with an association between them.

The basic salary for each grade is fixed, usually for a year at a time. Every year after the final accounts are closed, I review the grade rates with the Managing Director, and we increase them roughly in line with inflation.

A grade has only one rate at a time, though it can change, and each rate has a money value. Grade may have a rate attribute.

If the company has performed well, we increase the rates by more than the rate of inflation. In case there are any queries, either from an employee or from the Tax Office, it is most important that we keep accurate records of every employee's grades; that is, the rates for all present and all past grades, and the dates these came into force.

There's quite a lot in this bit. A grade may have several previous rates, which suggests either that Grade has multiple rate attributes, or that Rate and Grade are distinct classes. If the latter, then Rate must have a date attribute, since we need to know when it took effect. We must also record when a member of staff changes to a grade, and possibly also when they change from a grade, which suggests one or two more date attributes. Each grade has a date it came into force—another attribute.

It's actually quite complicated, because you can have an employee who changes to several different grades, one after the other, and then the rate for each grade also changes each year. So, for each employee, I have to be able to tell exactly what grade they were on for every day they have worked for the company, and also what the rate for each grade was when they were on it. This is all quite separate from bonus, which is calculated independently each year. For creative staff, bonus is based on the profits from each campaign they have worked on, and for other staff we use an average profit figure for all campaigns.

This is necessarily tentative, but a preliminary analysis yields the following list of classes and attributes:

- classes: StaffMember, Grade, Rate;
- attributes: gradeStartDate, gradeFinishDate, rateStartDate, rateFinishDate, rateValue.

In order to be reasonably lifelike, we can assume some other attributes not given above, such as staffName and gradeDescription, and also some other operations, such as assignNewStaffGrade and assignLatestGradeRate. An initial, though incomplete, class diagram might then look like the one in Figure 7.28.

One problem is where to put the attributes gradeStartDate and gradeFinishDate. These could be placed in Grade, but this would commit it to recording multiple start and finish dates. There may be also many members of staff associated with a grade. The computer system must be able to identify the member of staff to which each date applies, so the structure of dates that might need to be stored could grow quite complex. A similar problem occurs if date attributes are placed in Staff. The explanation for this difficulty is that these attributes do not describe either a member of staff or a grade in isolation. They only have meaning as a description of the link between a specific member of staff and a specific grade. Thus, the clearest answer is to create an additional class (called an association class) specifically to provide these attributes with a home. This is shown in Figure 7.29.

Some readers may be familiar with the relational database technique known as normalization, a technique that provides a rigorous guide to placing attributes in tables (or relations) and ensures minimum redundancy of data. The case illustrated is an example of normalization in practice, but a full treatment of the underlying theory is beyond the scope of this book. Normalization is used in object-oriented design (this is discussed in Chapter 18) but on the whole object-oriented approaches concentrate on capturing the structure of the world as perceived by the system's users. Un-normalized relations are therefore acceptable in an object model, provided that they correspond accurately to users' intuitions about how their business activities are organized.

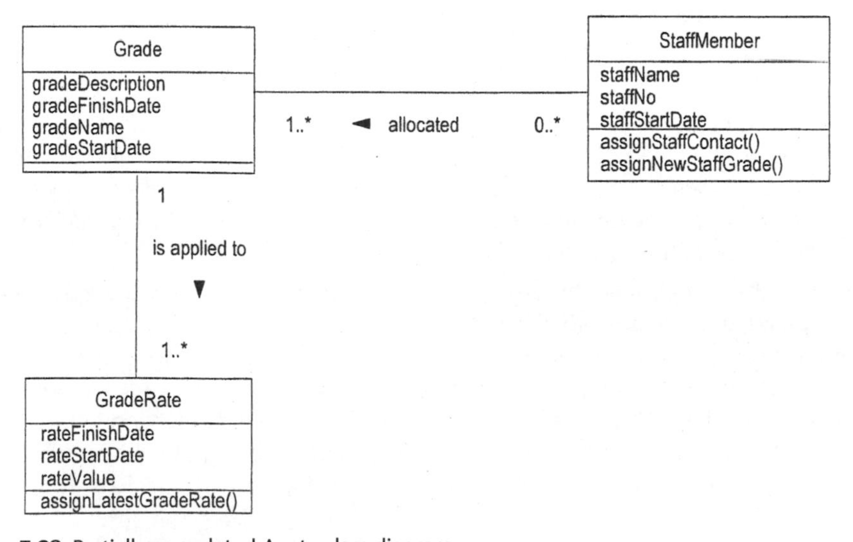

Figure 7.28 Partially completed Agate class diagram.

7.5.6 Adding associations

Find associations by considering logical relationships among the classes in the model. Associations may be found in use case descriptions, and other text descriptions of the application domain, as stative verbs (which express a permanent or enduring relationship), or as actions that need to be remembered by the system. 'Customers *are responsible for* the conduct of their account' is an example of the first, while 'purchasers *place* orders' is an example of the second.

But this is not a very reliable way of finding associations. Some will not be mentioned at all, while others may be too easily confused with classes, attributes or operations. A full understanding of the associations in a class model can only be reached later by analysing the interaction between different classes. With practice, the most important ones will be found fairly easily, and for the moment it is not important if some are missed.

7.5.7 Determining multiplicity

Since association multiplicities represent constraints on the way users carry out their business activities, it is important to get these right, and the only way to do this is to question users about each association in turn. This is true even when the existence and character of the association have been inferred from user documents. An analyst should always check what is said in documents, in case it is ambiguous, erroneous, or out of date.

Rosanne Martel: So let me be clear about this. A client must have exactly one staff contact, but a member of staff can be contact for no clients, one client or several clients. Is there an upper limit on that? A campaign must have one client, but a client can have many campaigns. Can you have a client with no campaigns—say, a new client who hasn't given you any business yet?

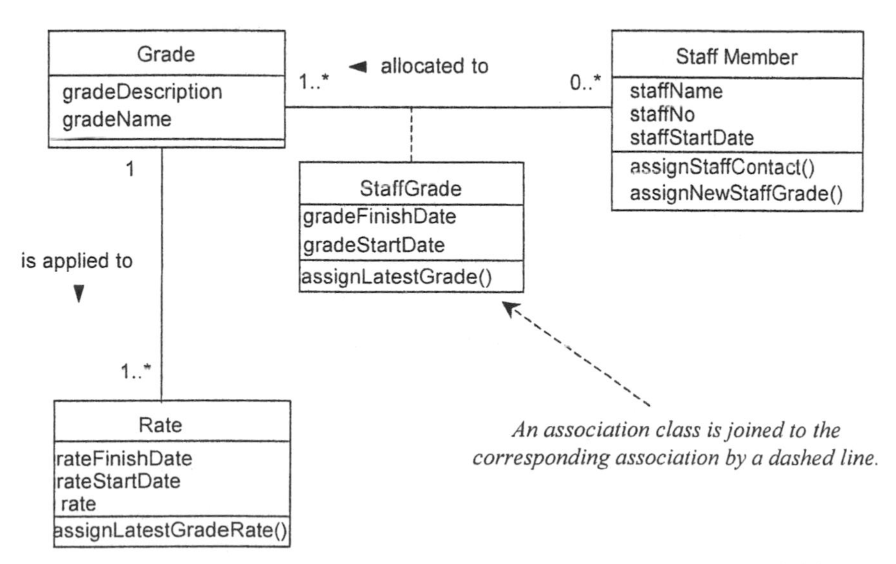

Figure 7.29 An association class gives a home to attributes that properly belong to a link between two objects.

7.5.8 Finding operations

Operations are really a more detailed breakdown of the high-level system responsibilities already modelled as use cases. An operation can be thought of as a small contribution of one class to achieving the larger task represented by a whole use case. They are sometimes found as action verbs in use case descriptions, but this picture is likely to be fairly incomplete until the interaction between classes has been understood in more depth. Chapter 8 describes how to model class interaction. So, as with associations, do not worry if your first attempt has gaps.

7.5.9 Preliminary allocation of operations

Before attempting to allocate operations to specific classes, it is worth remembering that each entity class is only a representation of something in the application domain. As an analyst, you are trying to build a logical model that helps to understand the domain, not necessarily a replica that is perfect in every detail. Two guidelines help in deciding which class to locate each operation in, but there is not a single answer—only a satisfactory fit.

1. Imagine each class as an independent actor, responsible for doing or knowing certain things. For example, we might ask 'What does a staff member need to know or need to be able to do in this system?'
2. Locate each operation in the same class as the data it needs to update or access. However, this is often problematic, as you may not have identified all the attributes yet.

As a general comment on this stage, the most important thing is not to expect to get things right at the first attempt. You will always need to revise your assumptions and models as your understanding grows.

7.6 CRC (Class Responsibility Collaboration) Cards

We have recommended use cases as the starting point for the identification of classes. A further examination of the use cases also helps in identifying operations and the messages that classes need to exchange. However, it is often easier to think first in terms of the overall responsibilities of a class rather than its individual operations. A *responsibility* is a high level description of something a class can do. It reflects the knowledge or information that is available to that class, either stored within its own attributes or requested via collaboration with other classes, and also the services that it can offer to other objects. A responsibility may correspond to one or more operations. It can be difficult to determine the most appropriate choice of responsibilities for each class as there may be many alternatives that all appear to be equally justified.

Class Responsibility Collaboration (CRC) cards provide an effective technique for exploring the possible ways of allocating responsibilities to classes and the collaborations that are necessary to fulfil the responsibilities.[11] CRC cards can be used at several different stages of a project for different purposes. For example, they can be used early in a project to aid the production of an initial class diagram and to

[11] CRC cards were invented by Kent Beck and Ward Cunningham (1989) while they were working together on a Smalltalk development project for Tektronix.

develop a shared understanding of user requirements among the members of the team. Here we concentrate on their use in modelling object interaction. The format of a typical CRC card is shown in Figure 7.30.

CRC cards are an aid to a group role-playing activity that is often fun to do[12]. Index cards are used in preference to pieces of paper due to their robustness and to the limitations that their size (approx. 15cm × 8cm) imposes on the number of responsibilities and collaborations that can be effectively allocated to each class. A class name is entered at the top of each card and responsibilities and collaborations are listed underneath as they become apparent. For the sake of clarity, each collaboration is normally listed next to the corresponding responsibility.

Wirfs-Brock et al. (1990) and others recommend the use of CRC cards to enact a system's response to particular scenarios. From a UML perspective, this corresponds to the use of CRC cards in analysing the object interaction that is triggered by a particular use case scenario. The process of using CRC cards is usually structured as follows.

- Conduct a brainstorming session to identify which objects are involved in the use case.
- Allocate each object to a team member who will play the role of that object.
- Act out the use case. This involves a series of negotiations among the objects (played by team members) to explore how responsibility can be allocated and to identify how the objects can collaborate with each other.
- Identify and record any missing or redundant objects.

Before beginning a CRC session it is important that all team members are briefed on the organization of the session. Some authors (Bellin and Simone, 1997) recommend that a CRC session should be preceded by a separate exercise that identifies all the classes for that part of the application to be analysed. The team members to whom these classes are allocated can then prepare for the role-playing exercise by considering in advance a first-cut allocation of responsibilities and identification of collaborations. Others prefer to combine all four steps into a single session and perform them for each use case in turn. Whatever approach is adopted, it is important to ensure that the environment in which the sessions take place is free from interruptions and conducive to the free flow of ideas (Hicks, 1991).

During a CRC card session, there must be an explicit strategy that helps to achieve an appropriate distribution of responsibilities among the classes. One simple but

Class Name:	
Responsibilities	Collaborations
Responsibilities of a class are listed in this section	*Collaborations with other classes are listed here, together with a brief description of the purpose of the collaboration*

Figure 7.30 Format of a CRC card.

[12] A useful spin-off is that this can support team building and help a team identity to emerge.

effective approach is to apply the rule that each object (or role-playing team member) should be as lazy as possible, refusing to take on any additional responsibility unless persuaded to do so by its fellow objects (the other role-playing team members). During a session conducted according to this rule, each role-player identifies the object that they feel is the most appropriate to take on each responsibility, and attempts to persuade that object to accept the responsibility. For each responsibility that must be allocated, one object (one of the role-players) is eventually persuaded by the weight of rational argument to accept it.[13] This process can help to highlight missing objects that are not explicitly referred to by the use case description. When responsibilities can be allocated in several different ways it is useful to role-play each allocation separately to determine which is the most appropriate. The aim normally is to minimize the number of messages that must be passed and their complexity, while also producing class definitions that are cohesive and well focused.

We illustrate how a CRC exercise might proceed by considering the use case `Add a new advert to a campaign`. The use description is repeated below for ease of reference.

The campaign manager selects the required campaign for the client concerned and adds a new advert to the existing list of adverts for that campaign. The details of the advert are completed by the campaign manager.

This use case involves instances of `Client`, `Campaign` and `Advert`, each role played by a team member. The resulting CRC cards are shown in Figure 7.31, and in the discussion that follows we explain how they can be derived from the use case.

The first issue is how to identify which client is involved. In order to find the correct `Client` the Campaign Manager (an actor and therefore outside the system boundary from the perspective of this use case) needs access to the client's name. Providing a client name and any other details for that client is clearly a responsibility of the `Client` object.

Next, the Campaign Manager needs a list of the campaigns that are being run for that client. This list should include the title, start date and finish date for each campaign. Although a `Campaign` object holds details of the campaign, it is not clear which object (and hence which class) should be responsible for providing a list of campaigns for a client. The team member playing the `Campaign` object argues that although it knows which `Client` object commissioned it, it does not know which other `Campaign` objects have been commissioned by the same `Client`. After some discussion, the `Client` object is persuaded to accept responsibility for providing a list of its campaigns and the `Campaign` object is persuaded that it should provide the information for this list. Once the Campaign Manager has obtained details of the campaigns for that client she requests that the `Campaign` object provide a list of its adverts, to which list the new advert will be added. Since the Campaign object already has responsibility for looking after the list of adverts it is reasonable for it to add the new advert to its list. In order to do this it must collaborate with the `Advert` class which, by definition, has responsibility for creating a new `Advert` object. This completes the analysis of the use case interaction, and the new responsibilities and collaborations that have been identified are added to the cards, as shown in Figure 7.31. We have already seen a preliminary collaboration diagram earlier in the chapter, in Figure 7.4 and a class diagram developed from this use case in Figure 7.6. The reader is recommended to refer back to these to see how CRC cards relate to the development of a requirements model.

[13] An alternative strategy is for each object to be equally keen to take on a responsibility, with the final choice determined by negotiation. Irrespective of the strategy chosen, it is important that all team members understand the need for an effective distribution of responsibilities.

Class Name *Client*	
Responsibilities	Collaborations
Provide client information.	
Provide list of campaigns.	*Campaign provides campaign details.*

Class Name *Campaign*	
Responsibilities	Collaborations
Provide campaign information.	
Provide list of adverts.	*Advert provides advert details.*
Add a new advert.	*Advert constructs new object.*

Class Name *Advert*	
Responsibilities	Collaborations
Provide advert details.	
Construct adverts.	

Figure 7.31 CRC cards for the use case `Add a new advert to a campaign.`

During a CRC session, the team can keep track of the relationships between classes by sticking the index cards on a large board and attaching pieces of thread or string to represent collaborations. This is particularly useful when CRC cards are used early in the development cycle to help produce a class diagram. The cards and pieces of thread can be a very effective prototype of the class diagram[14]. CRC cards can also be extended in various ways. For example, superclasses and subclasses can be shown beneath the class name and some users of the technique also like to list attributes on the back of each card.

[14] This is sometimes known as *paper CASE* technology!

7.7 Assembling the Analysis Class Diagram

The final step that we look at in this chapter is to assemble the various class diagrams that result from use case realization into a single analysis class diagram. This may consist of a single package of entity classes (the domain model), with boundary and control classes typically located in separate packages. With large systems, the domain model alone may comprise several distinct packages, each representing a different functional subsystem of the overall system.

There is usually little conceptual or technical difficulty in this step. All we really have to do is to place the various entity classes into a single class diagram. Where we find that we have defined the same class in different ways to meet the needs of different use cases, we simply assemble all of the operations and attributes into a single class definition. For example, consider the Campaign class as seen in relation to `Add a new advert to a campaign` and `Assign staff to work on a campaign`. Different use cases have suggested different operations. Putting these together results in a class that is capable of meeting the needs of both use cases. When we consider other use cases too, a more complete picture of the class emerges. The stages are illustrated in Figure 7.32.

Integrating the various associations derived from different use cases can seem a little more problematic, but it is actually quite straightforward. The general rule is that if *any* use case requires an association, it should be included. Where there is an

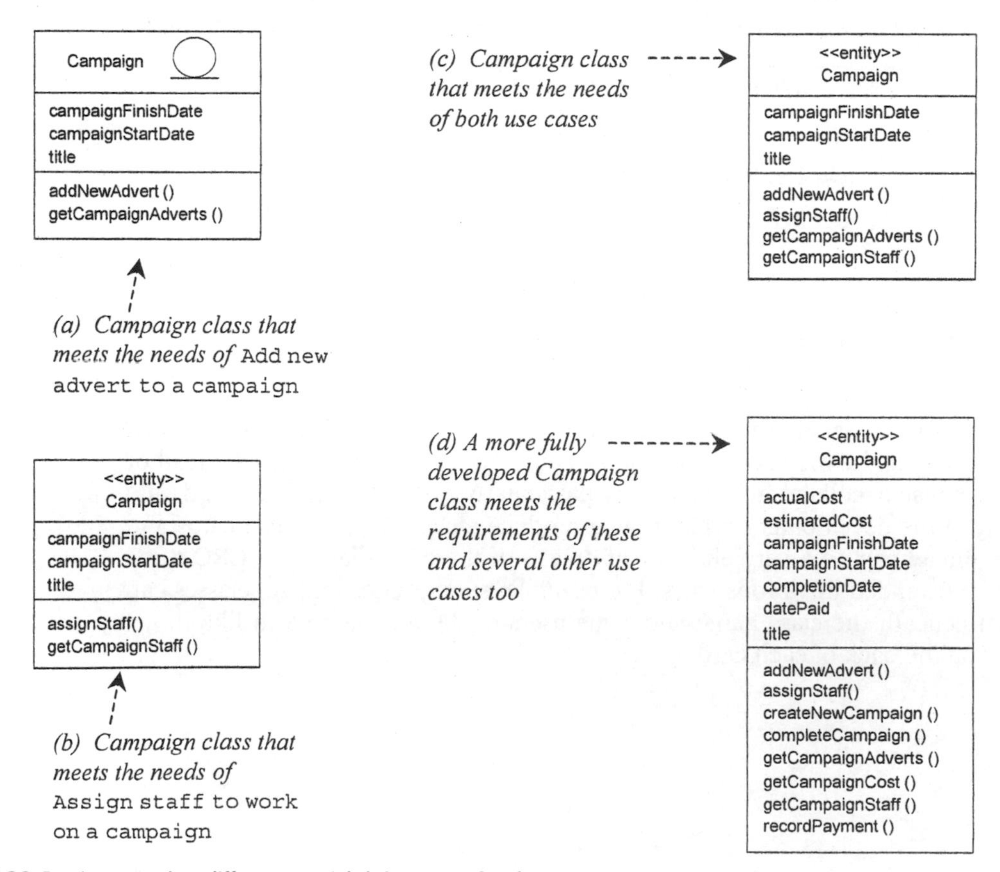

Figure 7.32 Putting together different partial definitions of a class.

apparent conflict in the multiplicity values for an association, then clarification should be sought from users who understand the business rules of the organization.

Figure 7.32 illustrates the process of consolidation. A full analysis class diagram can be found in case study Chapter A3. This includes the requirements identified from many of the use cases modelled in Chapter 6.

7.8 Summary

In this chapter we have seen how to build an initial version of the class diagram for a requirements model, following the process known as use case realization. We have also seen how the CRC technique can help with the preliminary allocation of attributes and operations to classes. The important elements of the analysis model at this stage are classes, with their attributes and operations, and associations, which show the relationships between classes, together with any multiplicity constraints. Once this task has been completed to a reasonable first-cut standard, the model embodies a good understanding of the main functional requirements for the system, in terms of its responsibilities for providing services. It has also defined a logical architecture that is a basis for the design work that follows. However, before design can be successfully undertaken, we must refine the requirements model by identifying any latent generalization or aggregation associations, together with opportunities for the application of patterns. This can simplify the structure, and may also suggest opportunities for reusing some of the requirements modelling work that has already been carried out on other projects. In the next chapter, we look at the most important concepts and techniques involved in refining the requirements model.

Review questions

7.1 Explain what is meant by 'use case realization'.

7.2 Why are requirements models in IS development neither entirely graphical nor entirely textual?

7.3 Distinguish between attribute and value.

7.4 In what sense are classes generally more stable than their instances, and why is this usually the case?

7.5 Distinguish between link and association.

7.6 What is multiplicity, and why can it be called a constraint?

7.7 What is an operation?

7.8 How are operations related to messages?

7.9 What is an attribute?

7.10 Explain why Section 7.4.6 makes no mention of updating a link when it is changed, but instead only discusses the creation and destruction of links.

7.11 What is a collaboration?

7.12 How does a collaboration diagram differ from a class diagram?

7.13 Outline the main steps in developing a class diagram for a use case.

7.14 What are the advantages of team members acting the parts of objects when they are developing a set of CRC cards?

Case Study Work, Exercises and Projects

The following transcript gives the first part of an interview that Rosanne Martel conducted with Hari Patel, the Factory Manager in charge of FoodCo's Beechfield Factory. Read this through carefully, and then carry out the exercises that follow.

Rosanne Martel: Hari, for the benefit of the tape, I'd be grateful if you could confirm that you're the manager responsible for all production at Beechfield.

Hari Patel: Yes, that's right.

RM: Good. Now the purpose of this interview is for me to find out about operations on the production lines. Can you tell me how this is organized?

HP: Sure. How much detail do you want?

RM: Can we start with those aspects that are common to all lines? That will give me a general feel, then if there are differences we can go into more detail later.

HP: OK, there are quite a few similarities. First, there are two main grades of shop-floor staff: operatives and supervisors. Different operatives have a range of skills, of course, but that doesn't affect the way the line works.

RM: How many operatives work on a line, and what do they actually do?

HP: There might be anything from around six operatives to over twenty, depending on the product. They really do all the actual work on the line, either by hand or operating a machine. This could be a semi-skilled labourer feeding in the different kinds of lettuce for salad packs, or a more skilled operator running one of the automatic mixing machines. In this factory, unlike Coppice and Watermead, the work is mostly quite unskilled.

RM: How many supervisors are there to each line?

HP: Just one. They are on full-time supervision duties, and they each look after one production line.

RM: Always the same line?

Rosanne is trying to find out what possible classes there are. What else do you think her questions seek to discover?

HP: Well, let's just say nobody has changed line in the last couple of years.

RM: How about the operatives—are they always on the same line too?

HP: No, we swap them around quite a bit. But it doesn't really matter what line an operative works on. They get paid piecework rates depending on the production run, and the rates are based on the job numbers that appear on their timesheets. There's a separate job number for each run.

RM: I'd like a copy of a timesheet please—preferably a real one with some data, if that's all right. We can blot out the name and staff number on the copy for confidentiality.

A sensible request. Real documents with live data are an invaluable source of information. Figure B1.4 shows the timesheet that Rosanne collected.

HP: Sure. Remind me when we finish, and I'll get you one.

RM: Thanks. Now, does one line always produce the same product?

HP: No, that changes from one day to the next. The production planners produce a new schedule every Friday, and this lists all the production runs for each line for the following week.

RM: I'll take a copy of a production schedule too please. So the supervisor finds out on Friday what their line is working on over the next week?

Here Rosanne is checking where the inputs come from, as well as what they contain.

HP: That's right.

RM: Good, I think I've got that clear. Now let's talk about what happens when people come in to work. Do all the lines start up first thing in the morning?

HP: Usually. Production runs generally last for a whole day if possible, or sometimes a half-day. Production Planning try to keep the change-overs simple, so they tend to schedule changes during breaks to avoid wasting productive time.

RM: The lines don't keep running all the time?

HP: No, they stop for coffee and meal breaks.

RM: What role does the line supervisor play in this?

HP: Well, they make sure the lines have enough raw materials, and they deal with minor emergencies. They also monitor output, liaise with production control, keep track of employee absences, and so on.

RM: Can we go through what a supervisor does on a typical run, please, step-by-step?

Another sensible request. Asking someone to go over things again in more detail will often reveal aspects of the situation that are not obvious from a brief description.

HP: First, they make sure everything is ready before the run starts. They check the storage area to see there is enough of each ingredient. If a long run is planned, you don't need all the ingredients ready at the beginning, but there has to be enough to keep the line running smoothly until the next supply drop. They also have to check if the staff allocated to that run have turned up. A line can usually run for a little while with one or two staff missing, but it's best to have everyone there from the start.

RM: How does a supervisor know what ingredients are required, and how many staff?

A good analyst always probes to find out how, what, why, when, where and who.

HP: Every run has a job card, with this information on it. The warehouse gets a copy of the job card too, so in theory they know what supplies to deliver, to which line and when they will be needed.

RM: Does that usually work?

HP: (Laughs). Sometimes!

RM: What if there aren't enough staff?

HP: Sometimes the supervisor can find a spare body on another line. Or they can run the line slower. You can manage with fewer staff if necessary, but productivity is a lot lower.

RM: Let's say the ingredients are all ready, and all the staff are there waiting to go. What next?

HP: The supervisor switches on the line, and then it's mostly troubleshooting and paperwork.

RM: What does the paperwork involve?

HP: Well, they start by taking the names of all the staff at the start of the run. They copy the job number from the job card to the production record sheet and all the timesheets. If it is the first time that operative has worked that week, then the supervisor makes out a new timesheet. When they start the line, they note the time on the production record sheet. Then they keep a rough note of anyone who leaves the line during a run, and how long they're absent.

RM: What kind of problems does the supervisor deal with?

HP: The main problem is if something goes wrong with the run. Say the line breaks down. They would have to call in maintenance, record the downtime while the line's not running, and try to find useful things for the staff to do while they're waiting for it to be repaired. If an ingredient runs out this could also halt the line, and might mean chasing the warehouse, or contacting the farm or an outside supplier. Sometimes people go missing, or leave early because they're sick. The supervisor has to find a replacement as quickly as possible.

RM: Right, now let's go to the end of a run. What information is formally recorded, and by whom?

HP: First the supervisor notes the finish time on the production record sheet.

RM: I'll have one of those too please.

HP: OK, no problem.

Figure B1.3 shows a blank production record sheet.

HP: Next the supervisor phones for someone to come over from Production Control to verify the quantity produced and note this on the production record sheet. Then the supervisor totals all the absences, because if anyone has more than 15 minutes absence, it's deducted from their total unless they have a good reason, say a medical certificate. Then they work out the total hours for each operative. If someone joined the line in mid-session they might not have a timesheet, so one is made out now and their hours are added in. By the time all that has been done, Production Control has usually checked out the total quantity produced, and this goes on the production record sheet. After that, it's just returning unused ingredients to the warehouse, tidying up the line ready for the next run, that kind of thing.

RM: Thanks, that was really helpful. Now I'd like to ask about how the piecework formula works. Can you tell me what the calculation is?

HP: To be honest, I can never remember the exact formula. You'd do better asking a supervisor or someone from payroll . . .

Now carry out the following exercises, based on the information given in the interview transcript.

7.A Write descriptions for the following use cases:

- `Start line run`
- `Record employee joining the line`
- `Record employee leaving the line`
- `Stop line`
- `Record line problem`
- `End line run`

7.B From your use case descriptions, produce use case realizations in the form of collaboration diagrams and then class diagrams.

7.C Produce a draft analysis class diagram, initially showing only classes and associations.

7.D Review your analysis class diagram together with the various intermediate models, and add any attributes and operations that you think are justified by your use cases. Make reasonable assumptions and add others that you think might be justified by other use cases not directly derived from the transcript.

Further Reading

- The natural source for all readers interested in this subject is the Three Amigos book on the USDP (Jacobson et al., 1999). But this text, while authoritative, is not (in our view, at any rate) ideally suited to the novice requirements analyst.
- Rosenberg and Scott (1999) describe in a very accessible way a process that uses UML in object-oriented requirements modelling. Rosenberg and Scott's process differs in many respects from the one followed in this book, but is very much in sympathy with our aim of producing a robust class model.
- Larman (1998) also describes a process for using UML in object-oriented requirements modelling. Larman's approach is very different again from the one taken in this book, and also from that recommended by Rosenberg.

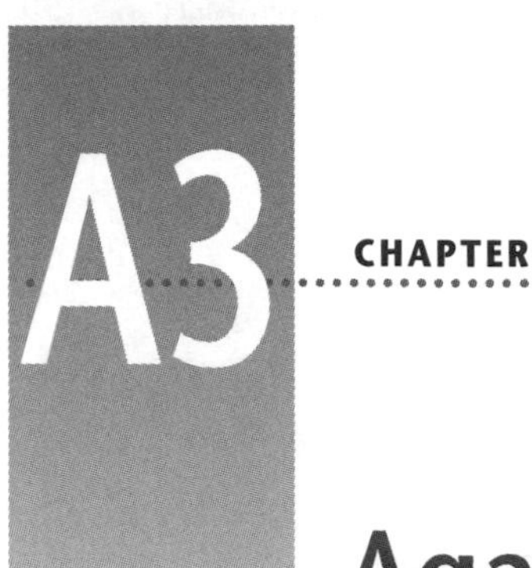

Agate Ltd Case Study Requirements Analysis

Agate Ltd

A3.1 Introduction

In this chapter we analyse the Requirements Model described in Chapter A2 and produce a number of use case realizations. The stages involved in the preparation of a use case realization are described in Chapter 7 and involve the production of the following UML diagrams:

- collaboration diagram,
- use case realization and
- class diagram.

After the use case realizations have been developed, a combined analysis class diagram is produced from them. A more detailed analysis class diagram is also included to indicate how the model develops as the use cases are analysed.

A3.2 Use Case Realizations

The first use case that is analysed here is `Add a new campaign` (all the use cases are specified in Chapter A2). Figure A3.1 shows the classes that collaborate to realise the use case. Figure A3.2 shows the collaboration diagram. Note that the initiation of the dialogue described by the use case is not modelled explicitly. Detail such as this will be added later for this system though in some projects it may be important to model these features early on. The class diagram supports this use case and its collaboration diagram is shown in Figure A3.3. Notice that the class `Campaign` includes only the attributes that are required for the use case. The requirements analyst may identify the need for additional attributes (or functionality) for the use case as it is being analysed but it is important to confirm any changes with the user before they are made. In these models we have named the constructor operation `createCampaign()` to make it clear

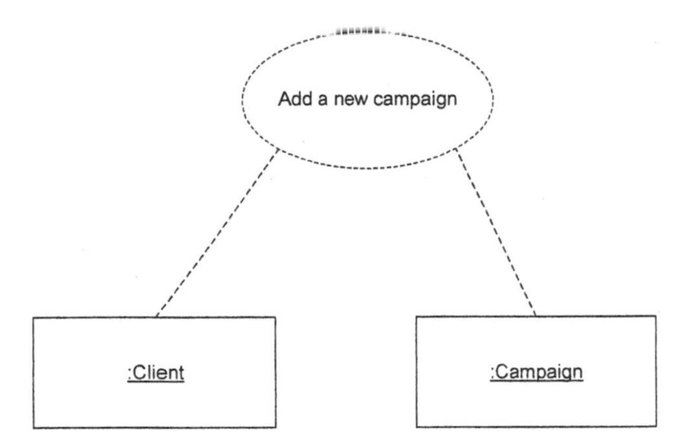

Figure A3.1 Collaboration for the use case `Add a new campaign`.

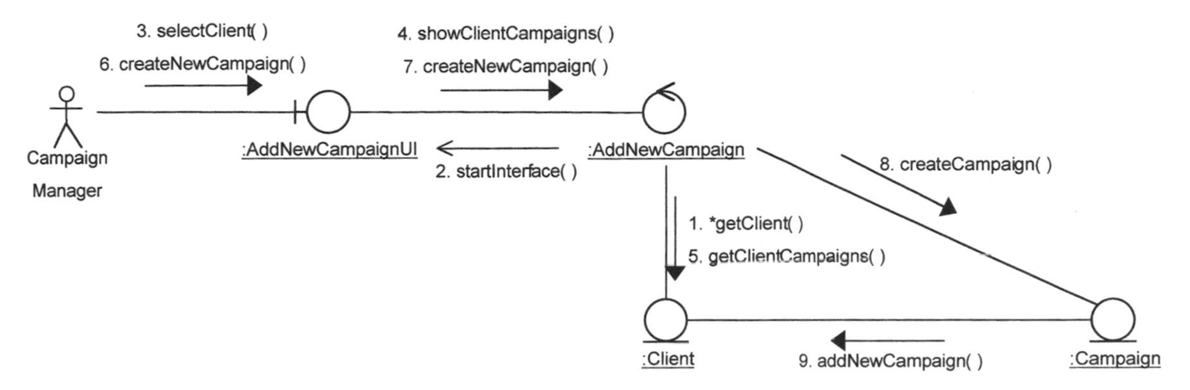

Figure A3.2 Collaboration diagram for the use case `Add a new campaign`.

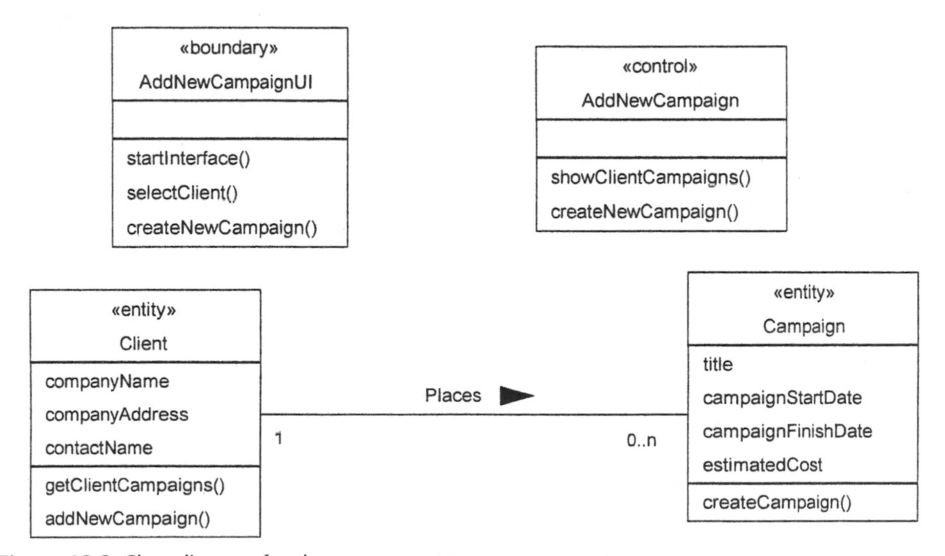

Figure A3.3 Class diagram for the use case `Add a new campaign`.

where in the interaction the new campaign object is being created. When preparing design oriented models the naming conventions used in object-oriented programming languages are more appropriate.

Figures A3.4 to A3.12 show the development of the use case realizations for the use cases `Assign staff contact`, `Check campaign budget` and `Record completion of a campaign`. The use case `Record completion of a campaign` involves the production of a completion note. The boundary class `CompletedCampaignPI` (we use the suffix PI to stand for Printer Interface) is responsible for printing the completion note.

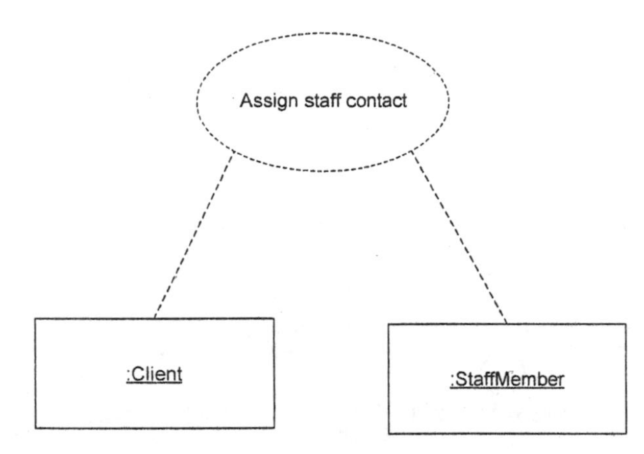

Figure A3.4 Collaboration for the use case `Assign staff contact`.

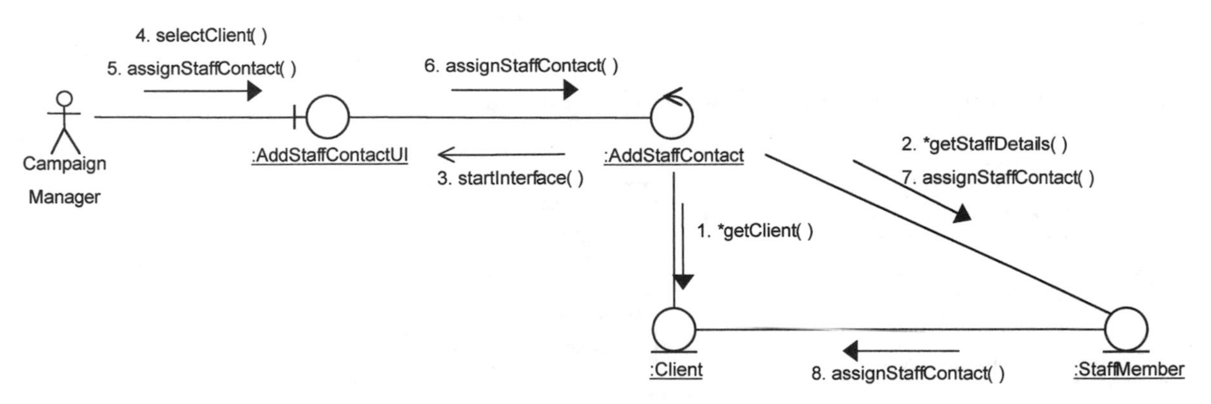

Figure A3.5 Collaboration diagram for the use case `Assign staff contact`.

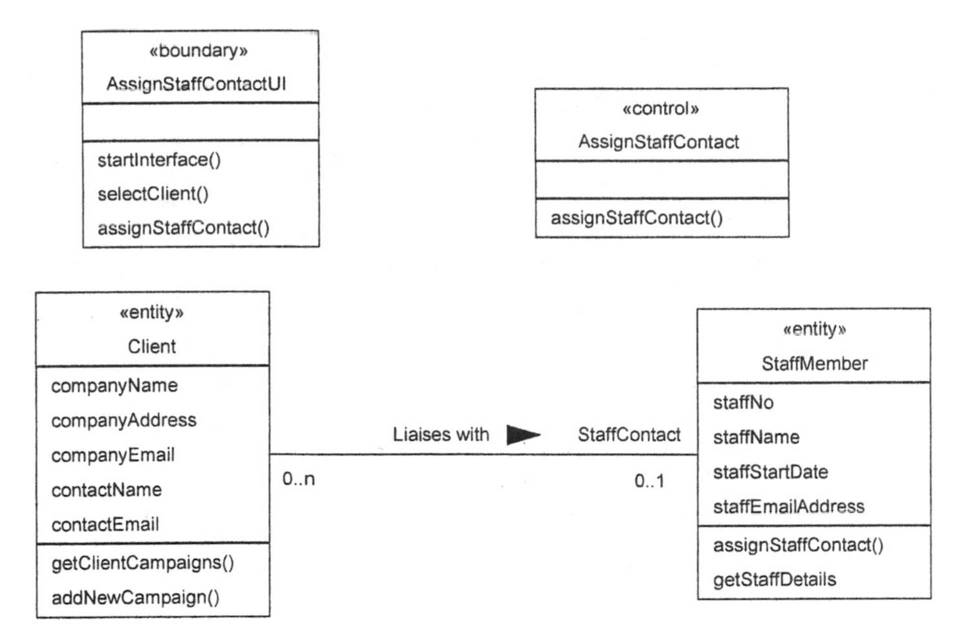

Figure A3.6 Class diagram for the use case `Assign staff contact`.

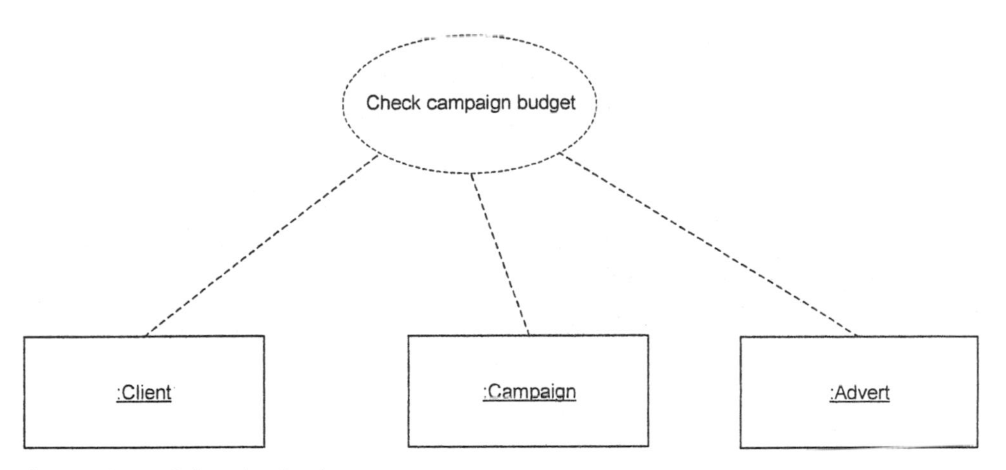

Figure A3.7 Collaboration for the use case `Check campaign budget`.

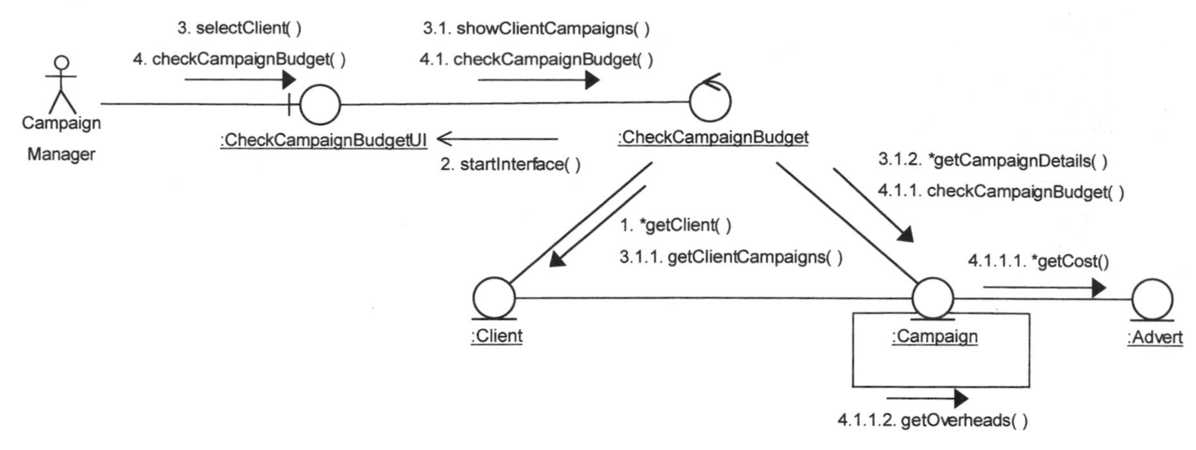

Figure A3.8 Collaboration diagram for the use case `Check campaign budget`.

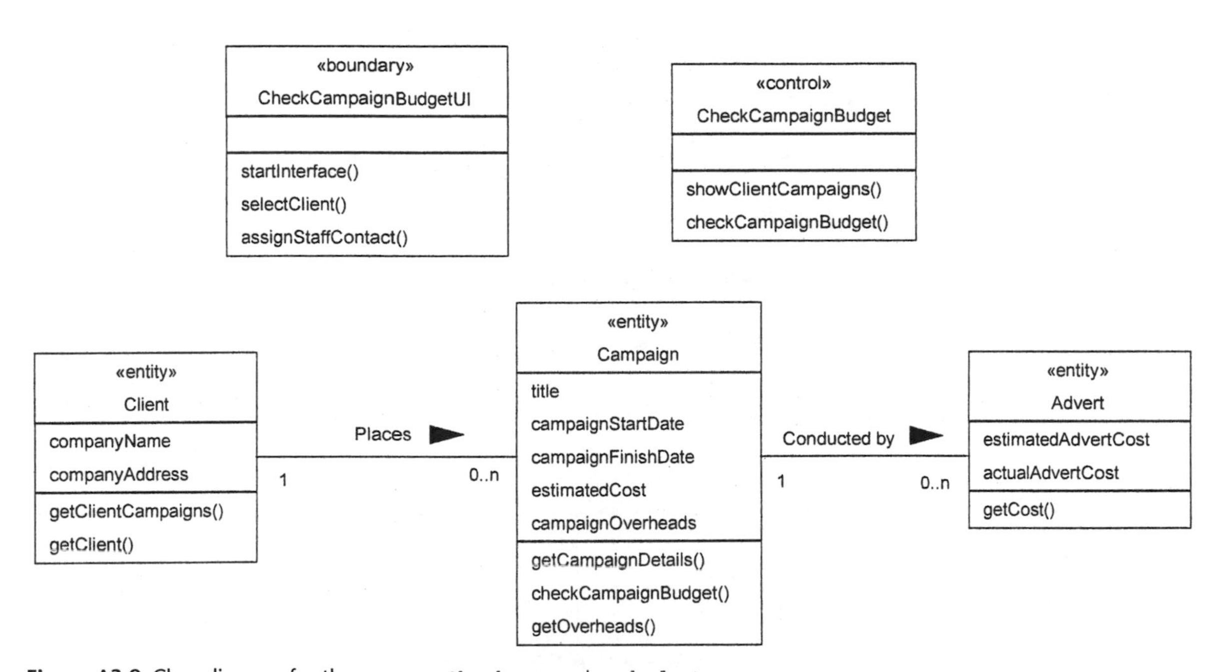

Figure A3.9 Class diagram for the use case `Check campaign budget`.

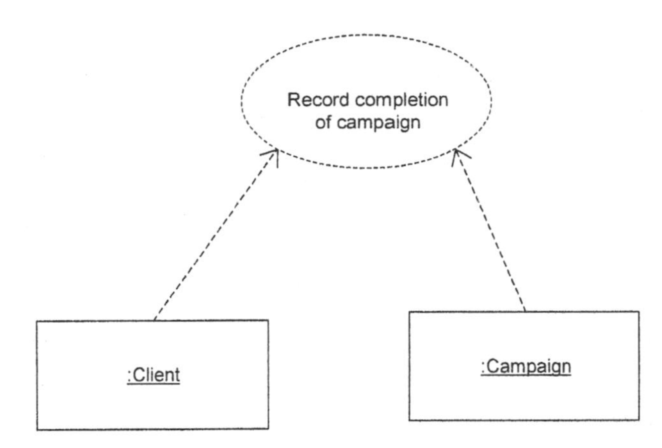

Figure A3.10 Collaboration for the use case `Check campaign budget`.

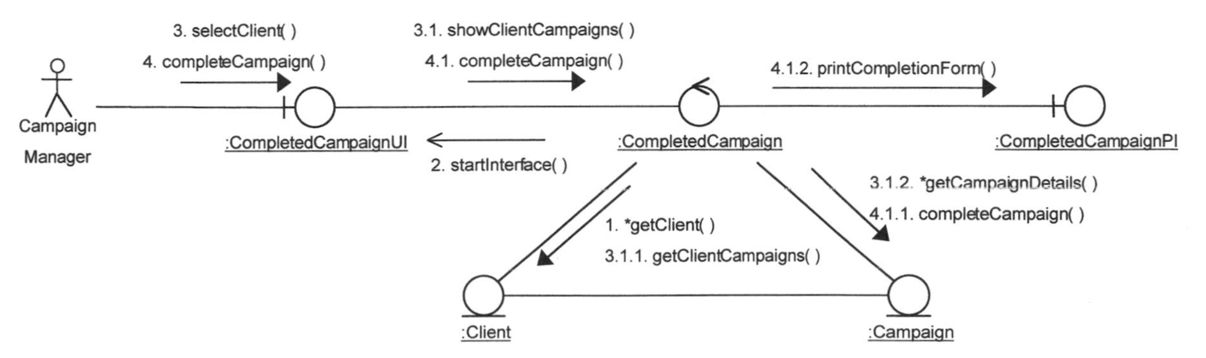

Figure A3.11 Collaboration diagram for the use case `Record completion of a campaign`.

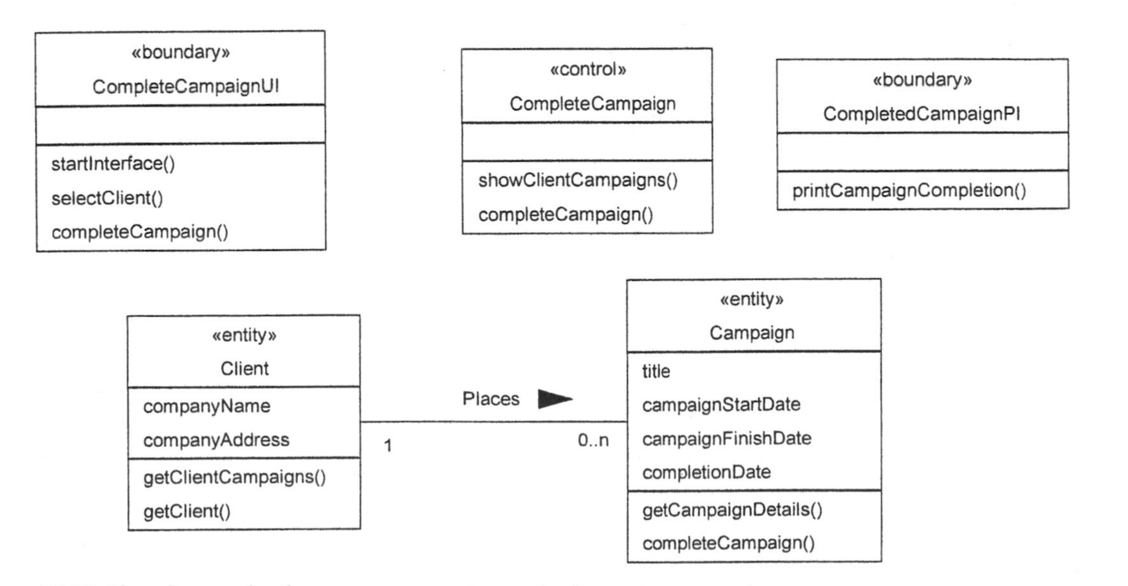

Figure A3.12 Class diagram for the use case `Record completion of a campaign`.

A3.3 Assembling the analysis class diagram

The class diagram in Figure A3.13 has been assembled from use case realizations for the use cases `Add a new campaign`, `Assign staff contact`, `Check campaign budget` and `Record completion of a campaign`. Figure A3.14 shows a more fully developed class diagram that includes the classes, attributes, operation and associations that have been identified from the other use cases in the `Campaign Management` package. This illustrates how a more detailed and complete picture of the analysis model is developed as the use cases are analysed. The use cases `Add a new advert to a campaign` and `Assign staff to work on a campaign` are analysed in Chapter 7. The collaborations are shown in Figures 7.4 and 7.25 respectively.

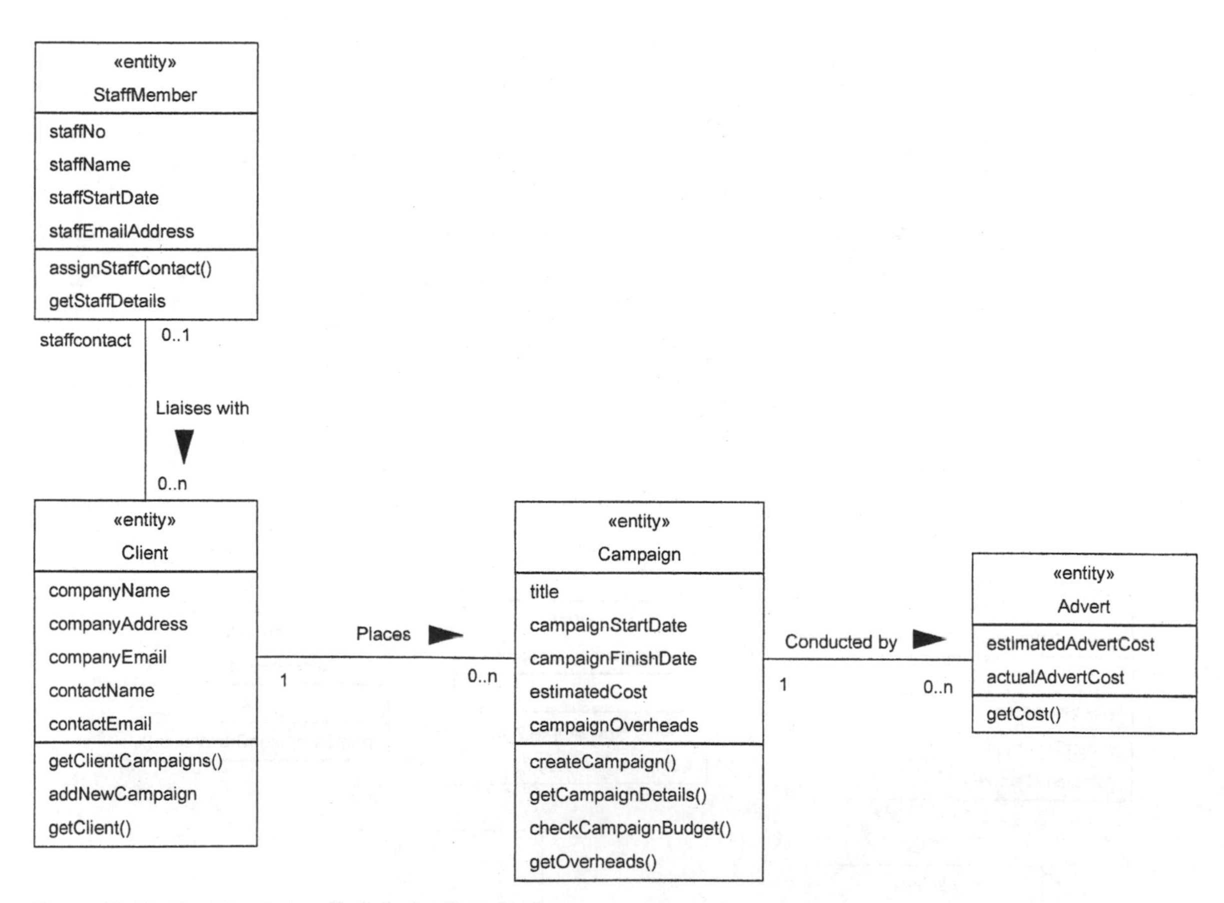

Figure A3.13 Combined class diagram for four use cases.

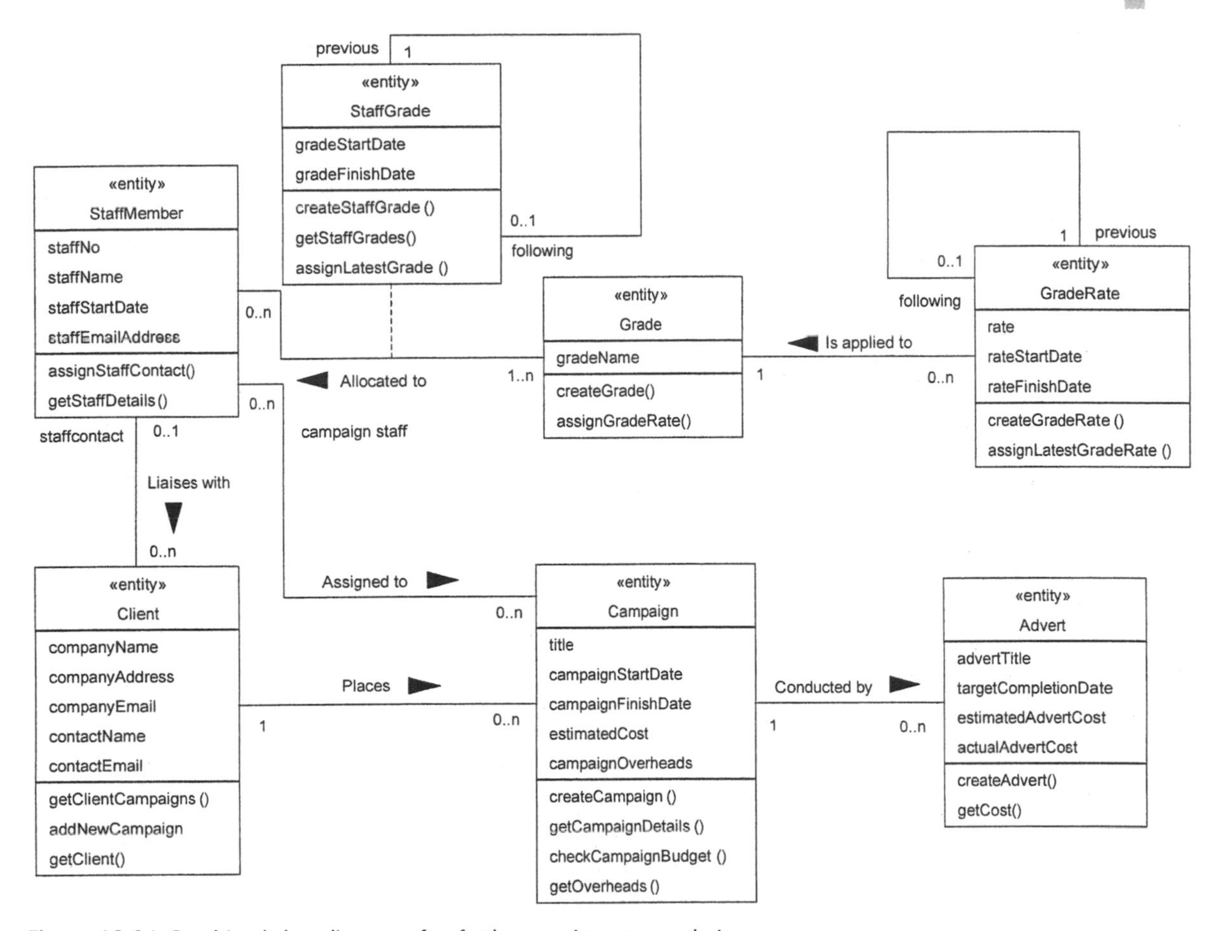

Figure A3.14 Combined class diagram after further requirements analysis.

A3.4 Activities of Requirements Analysis

Figure A3.15 shows an activity diagram illustrating the relationship between the requirements models and the products of use case realization. The activity diagram in Figure A3.16 shows the main activities involved in requirements analysis.

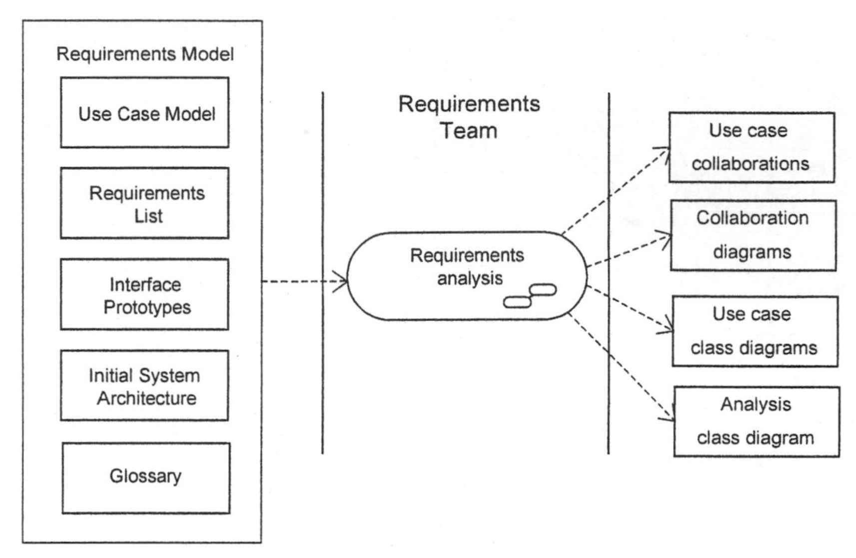

Figure A3.15 High level activity diagram for `Requirements analysis`.

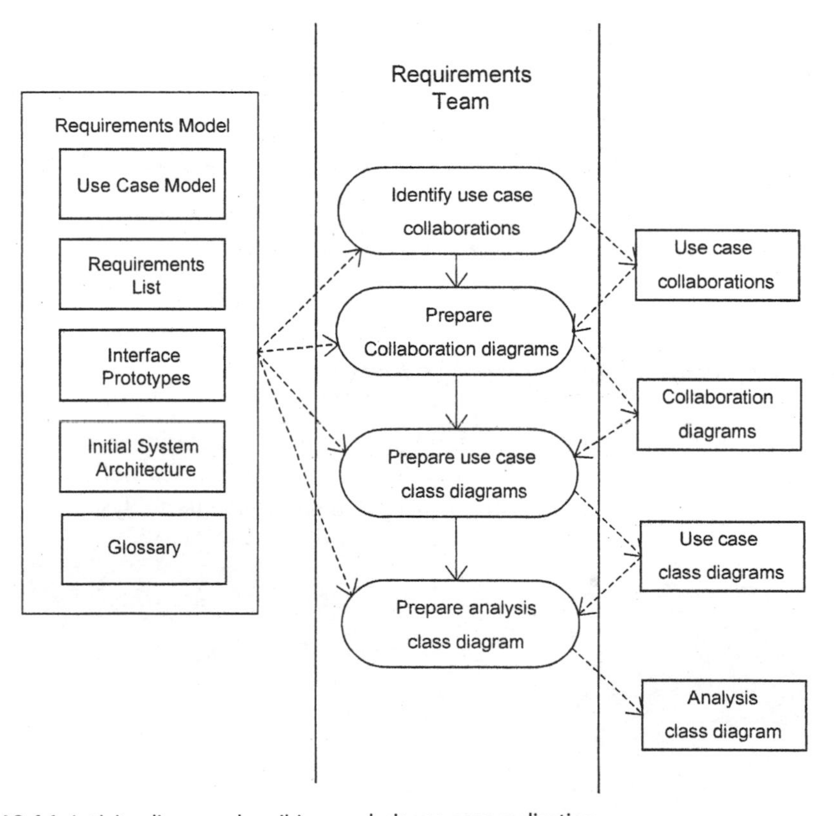

Figure A3.16 Activity diagram describing analysis use case realization.

CHAPTER 8

Refining the Requirements Model

OBJECTIVES

In this chapter you will learn

- what is meant by a component
- how generalization and aggregation help develop reusable components
- how to identify generalization and composition
- how to model generalization and composition
- what is meant by the term pattern
- what types of patterns have been identified in software development.

8.1 Introduction

Once the requirements have been identified and the various use case class diagrams have been assembled into a single analysis diagram that shows all relevant classes and associations, the next step is to refine this model. This is undertaken with a particular view to creating the conditions for developing components that can be reused either within the current project, or on future projects. In this chapter we introduce the concept of a component and discuss the role of abstraction in an object-oriented approach to identifying and building reusable components (Section 8.2). The principal abstraction mechanisms used for this purpose are the class hierarchy, which we have already met in Chapter 4, and a very different type of abstraction known as composition, which we encounter for the first time in this chapter. We show how these and other refinements can be used to encourage reuse (Section 8.3). In practice, most reuse actually occurs later in the life cycle than requirements analysis, so the benefits are often not gained directly by analysts. This is a situation that requires careful management, and we discuss this further in later chapters. Here we concentrate on laying the groundwork. Over the last few years, the 'Patterns' movement has offered new ways of capturing and communicating generalizable knowledge at many stages of the life cycle, particularly during the analysis and design activities. We introduce and explain the concept of patterns and illustrate this with some examples of analysis patterns (Section 8.4). Later (in Chapters 13 and 15 respectively), we discuss architecture and design patterns.

8.2 Component-based Development

It is quite hard nowadays to think of an industry where the use of standard components, and even standard designs, is not common. To take a familiar example, an architect may design many different houses, each unique in its own particular ways. But it would be very unusual to design every component from scratch. The architect typically chooses components from a catalogue, each house being assembled from a standard set of bricks, roofing timbers, tiles, doors, window frames, electrical circuits, water pipes, carpet tiles, paint, etc. Thus, while one house may be completely unlike another in overall appearance, floorplan and number of rooms, the differences between them lie in the way standard components have been assembled.

In this way a house can be designed to meet the needs of an individual customer, without the architect needing to think about 'low-level' details, such as which clay to use for the bricks, how long they should be fired in the kiln, or at what temperature. These problems have already been understood and solved by the brick manufacturer and can therefore be ignored. The architect need concentrate only on those aspects of the bricks relevant to the overall design of the house, such as colour, texture and size. In fact, many aspects of a brick's design, such as its precise ingredients and details of the manufacturing process, will not be described at all in the catalogue. In effect, these details are hidden from the architect—in other words, abstracted out of the description. Other aspects relevant to the architect, such as colour and size, are described in the catalogue.

8.2.1 Software components: the problems

It sometimes seems that many information systems development organizations believe that the analysis of requirements should begin from scratch on every new project. In one sense, this is necessary, since at first we know nothing about what the requirements for a new system are. It is also advantageous if it encourages analysts to take account of the unique characteristics of the proposed system and its environment. But starting from a position of knowing nothing can also have significant disadvantages. Effort may be wasted on finding new solutions to old problems that have already been adequately solved by others in the past. While it is important that real differences between two projects should not become blurred, this should not prevent the team from capitalizing on successful past work, provided that it is relevant to the current problem.

In the majority of cases, those analysts, designers and programmers who seem unnecessarily to reinvent the wheel do not do so deliberately. Good professionals have always tried to learn as much as possible from experience, both their own and that of their colleagues. Programmers have built up extensive libraries that range from personal collections of useful subroutines, to commercially distributed products that contain large numbers of industry-standard components. One example of this is the use of code libraries such as .DLL (Dynamic Link Library) files in Microsoft Windows. So why do some software developers carry on reinventing so many wheels? One reason for this is the NIH ('Not Invented Here') syndrome, which mainly seems to afflict programmers. Another reason is the functionally-based decomposition of the modelling techniques in structured analysis, which primarily affects analysts, and to a lesser extent designers.

The NIH syndrome

In spite of all the library resources available today, some professional programmers (even occasionally whole departments) still fall prey to the NIH syndrome. This is the attitude of one who thinks: 'I don't trust other people's widgets—even those that appear to work, suit my purpose and are affordable—I want to invent my own widgets anyway.' This is understandable in someone who enjoys a technical challenge, or has reasons not to trust the work of others, but it usually does not make good commercial sense. After all, if you want a new lightbulb for your room, it does not seem very sensible to invent and build your own. Even if the knowledge and equipment are available to do so, the cost would be prohibitive.

One remedy can be found in object-orientation, partly because of the different attitude to program development it engenders, but also partly because object-orientation actually makes it easier to use library components.

Model organization

Analysts suffer from the NIH syndrome too, but the biggest obstacle to reuse of successful analysis work has been the way that structured models are organized. It is difficult enough to create a model that is useful at other stages of one single project. It is even harder to create a structured model that is useful on a completely different project. This is partly because structured models are partitioned according to functions, which are a particularly volatile aspect of a business domain. There is also little in structured analysis models such as data flow diagrams that explicitly encourages encapsulation. This is an area where object-orientation can make a distinctive contribution to the reuse of requirements analysis.

8.2.2 How object-orientation contributes to reuse

Object-oriented software development tackles the problem of achieving reuse in ways that resemble the practice of other industries that use standard components. The aim is to develop components that are easy to use in systems for which they were not specifically developed. Ideally, software analysts should be in the position described for an architect, free to think about how their client intends to use the system, without needing to worrying about how individual components are built. The productivity gains are potentially enormous.

One of the keys is the encapsulation of internal details of components, so that other components requesting their service need not know how the request will be met. This allows different parts of the software to be effectively isolated in operation and greatly reduces the problems in getting different sub-systems to interact with each other, even when the sub-systems have been developed at different times or in different languages, and even when they execute on different hardware platforms. Sub-systems that have been constructed in this way are said to be *decoupled* from each other. Later in the book we describe an application of this approach that would enable one sub-system to interact with others over a network (the Agate security sub-system described in Chapter 20).

The effect can be scaled up to the level of complex sub-systems, by applying the same principle of encapsulation to larger groups of objects. Any part of a software system—or, by extension, a model of one—can be considered for reuse in other contexts, provided certain criteria are met.

- A component should meet a clear-cut but general need (in other words, it delivers a coherent service).
- A component should have a simple, well-defined external interface.

In theory, reusable components can be designed within any development approach, but object-orientation is particularly suited to the task. Well-chosen objects meet both of the criteria above, since an object requesting a service need only know the message protocol, and the identity of an object that can provide it[1]. Another important aspect of object-orientation is the way that models, and hence also code, are organized. For example, Coleman et al. (1994) point out that generalization hierarchies are a very practical way of organizing a catalogue of components. This is because they encourage the searcher to begin first with a general category, then progressively refine their search to more and more specialized levels. This mirrors the way that the Dewey decimal system is used to catalogue books in a library.

Requirements reuse

When we write of a 'reusable requirement', we really mean a reusable *model* of a requirement. To date, this is one of the least developed areas of software reuse. In any case, only parts of any model are likely to be reusable. But the key to reuse in requirements modelling is that models are organized so that they abstract out (hide) those features of a requirement that are not necessary for a valid comparison with a similar requirement on another project. Second, the whole point of reuse is to save work, so it should also not be necessary to develop a full model of the second requirement in order to make the comparison. Finally, any relevant differences between the two requirements being compared should be clearly visible—and it should not be necessary to develop a full model of the second requirement in order to see these either.

Generalization

As a simple example of generalization that enables reuse of a component, consider the load-bearing characteristics of a common brick. Depending on the number of floors, type of roofing and other features of the design, different houses place very different loads on the bricks of which their structural walls are built. Yet, within broad parameters, an architect need not be over-precise in calculating the anticipated load, since standard bricks are suitable for use under a range of different loads. This aspect of the design of a brick, which makes it suitable for use in many different situations, rather than being restricted to one narrow set of circumstances, is an example of abstraction by generalization.

Object-oriented developers have one significant advantage over the architect. Using inheritance, a 'software architect' has a way of spawning new products from old ones with minimal effort. There is nothing analogous to this in house-building. For example, each time a new kind of brick is made, it takes much the same effort as

[1] Some recent writers have suggested that useful components are actually at a larger scale than individual objects. This point is discussed further in Chapter 20.

any other brick. But inheritance provides a way of designing and building the larger part of a new software component in advance, leaving only the specialized details to be completed at a later stage. This is because, in a class hierarchy, those characteristics that are shared by subclasses are maintained at the superclass level, and are instantly available to any subclass when required.

Composition

Composition is a type of abstraction that encapsulates groups of classes that collectively have the capacity to be a reusable sub-assembly. Unlike generalization, the relationship is that between a whole and its parts. The essential idea is that a complex whole is made of simpler components. These, while less complex than the whole, may themselves be made of still less complex sub-assemblies, elementary components or a mixture of the two.

A simple example of the usefulness of the idea can be seen in a house-builder fitting a window frame to a new house. Like many other house components, window frames are delivered to site as ready-assembled units. All internal details of the sub-assembly are 'hidden' from both architect and builder, in the sense that they do not need to think about them. In some other part of the building industry, specialist designers tackle the problems of what materials, structure, etc. make a good window-frame, and specialist constructors are busy building window-frames to the designers' specifications. For this all to happen, it is first necessary that the window-frame be identified as a component that can be encapsulated, and thus treated by the house-builder as a single, simple thing, even though really it is not.

8.2.3 Composition in UML

Composition (or *composite aggregation*) is based on the rather less precise concept of *aggregation*, which is a feature of many object-oriented programming languages. The UML Specification (OMG, 2001) is deliberately rather vague about aggregation, but says the following about the relationship between aggregation and composition.

> Composite aggregation is a 'strong form of aggregation which requires that a part instance be included in at most one composite at a time and that the composite object has sole responsibility for the disposition of its parts. This means that the composite object is responsible for the creation and destruction of its parts.'

While composition and aggregation may be identified during requirements analysis, their main application is during design and implementation activities, where they can be used to encapsulate a group of objects as a potentially re-usable sub-assembly. This is more than just a matter of labelling the group of objects with a single name. The fact that they are encapsulated is much more important. The external interface for a composition is actually the interface of the single object at the 'whole' end of the association. Details of the internal structure of the composition—that is, what other objects and associations it contains—remain hidden from the client.

On a practical level, composition is familiar to users of most common computer drawing packages. For example, many of the drawings in this text were prepared or edited using a widely used drawing package. Within this application, several drawing objects can be selected and grouped. They then behave exactly like a single object, and can be sized, rotated, copied, moved or deleted with a single command. Figures 8.1 and 8.2 show this type of composition as both objects and classes. Notice that this

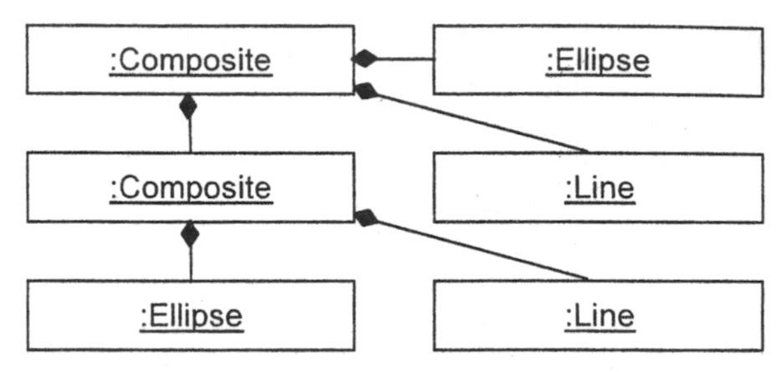

Figure 8.1 Some objects in a computer drawing package.

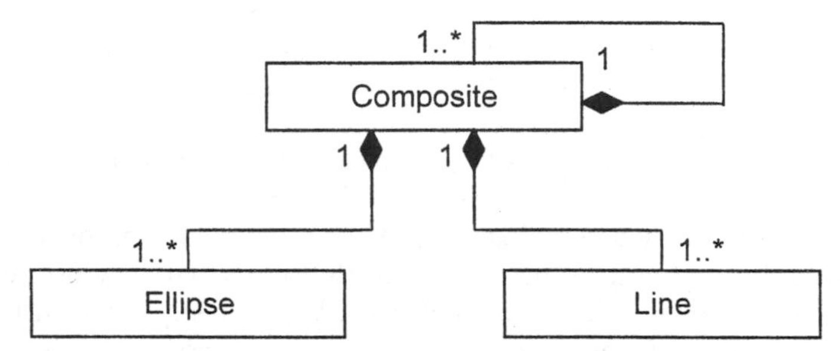

Figures 8.2 Composition used in class diagram to represent composite objects.

particular example is nested—the composition itself contains a further composition. In just the same way that a composite drawing object can only be handled as a single drawing component, the 'part' objects in a composition structure can usually not be directly accessed, and the whole presents a single interface to other parts of the system.

This notation is similar to a simple association, but with a diamond at the 'whole' end. The diamond is filled with solid colour to indicate composition, and left unfilled for aggregation.

Use of composition and aggregation in business-oriented applications is more problematic than in this example, but it is still worth modelling where it conveys useful information about the structure of the business domain. An example from the Agate case study is used later to illustrate aggregation in Section 8.3.2.

8.3 Adding Further Structure

This section shows how to add structure to the class diagram that will help with reuse at later stages of development. First, in Section 8.3.1 we concentrate on generalization, since it is the more useful of the two concepts for this purpose. Then in Section 8.3.2 we show how to model a structure that combines generalization and aggregation.

8.3.1 Modelling generalization

Figure 8.3 shows a note of an interview carried out by an analyst in the Agate case study. Her main objective was to understand more about different types of staff. In her haste, the analyst gathered only a handful of facts, but these highlight some useful information that must be modelled appropriately:

```
17 March - brief interview with Amarjeet Grewal (Finance Director)
Purpose - clarification of points from last Thursday's interview

Asked about staff types
        - only two types seem relevant to system -
                creative staff (C) and  admin staff (A)
How do they differ?
        - main difference is bonus payment...
                1. (C) bonus calculated on basis of campaign profits
                        (only those campaigns they worked on)
                2. (A) paid rate based on average of all campaign profits
Any other diffs?  Amarjeet says -
        - C qualifications need to be recorded
        - C can be assigned as contact for a client
        - A are not assigned to specific campaigns
No other significant differences.
                (NOTE - at next interview, get details of both algorithms)
```

Figure 8.3 Analyst's note of the differences between Agate staff types.

- there are two types of staff,
- bonuses are calculated differently and
- different data should be recorded for each type of staff.

Figure 8.4 shows a fragmentary class diagram that corresponds to this (for clarity, only affected classes are shown). The diagram includes a generalization association between `StaffMember`, `AdminStaff` and `CreativeStaff`. Of these, `StaffMember` is the superclass, while `AdminStaff` and `CreativeStaff` are subclasses. The generalization symbol states that all characteristics of the superclass `StaffMember` (its attributes, operations and associations) are automatically inherited by `AdminStaff` and `CreativeStaff`. There is no need to repeat superclass characteristics within a subclass definition.

For example, as `StaffMember` has an attribute `staffName`, we can take it for granted that `AdminStaff` and `CreativeStaff` also have this attribute. Since `StaffMember` has an operation `getStaffDetails()`, `AdminStaff` and `CreativeStaff` inherit it too. Finally, both `AdminStaff` and `CreativeStaff` inherit from `StaffMember` its association with `Grade`.

From a subclass perspective, inherited features are actually features that belong to the subclasses but have been removed to a higher level of abstraction. These are the shared characteristics that justify the generalization hierarchy. Generalization saves the analyst from the need to show these characteristics explicitly for each subclass to which they apply. Common attributes, operations and associations thus may be shared by several subclasses, but need be shown only once, in the superclass. This aids the general consistency of the model, and can also considerably reduce its complexity.

Each subclass also requires one or more characteristics that differentiate it from other subclasses in the hierarchy. In this example, `CreativeStaff` has the attribute `qualification` and the operation `assignStaffContact()`, neither of which is shared by `AdminStaff`.

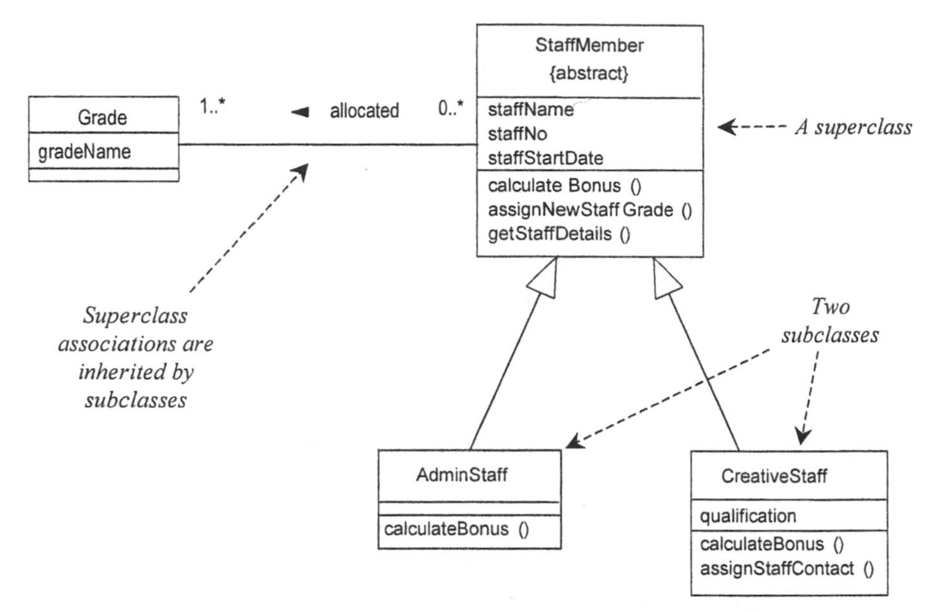

Figure 8.4 A generalization hierarchy describes a relationship among Agate staff types.

Redefined operations

What are we to make of the fact that all three classes involved in the hierarchy in Figure 8.4 have an operation `calculateBonus()`? Has the analyst made a mistake? Or has she failed to take advantage of the economy of representation offered by the generalization notation?

The explanation lies in the fact that, while both `AdminStaff` and `CreativeStaff` require an operation `calculateBonus()`, it works differently in each case. Since the precise logic for the calculation differs between these two groups of staff, the two operations will need to be treated separately later when each algorithm is designed, and also when programme code is written to implement the algorithm. This difference in logic justifies the separate appearance of a superficially similar operation in both subclasses. This sort of fine distinction is not always recognized during analysis.

Why, then, has the operation `calculateBonus()` been included in the superclass `StaffMember`? The answer is that it is an attempt at 'future-proofing'. One of the consequences of identifying a superclass is that it may later acquire other subclasses, that are as yet unknown. In this case the analyst has recognized—or assumed—that objects belonging to *all* subclasses of `StaffMember` are likely to need an operation of some kind to calculate bonus. For this reason, at least a 'skeleton' of the operation is included in the superclass. This may consist of no more than the signature, but since the interface is all that matters to other classes, this alone justifies its inclusion in the superclass definition. Even if the superclass operation is defined in full, some subclasses may choose not to use it because the logic for their version of the operation is different. For all these cases, the operation is said to be *redefined* or *overridden* in the subclass.

Abstract and concrete classes

We now turn to the `{abstract}` annotation below the `StaffMember` class name in Figure 8.4. This may at first be puzzling if we recall learning in Chapter 4 that *all* classes are abstract, in the sense that a class is an abstraction of its members. However, `StaffMember` is abstract in the still more compelling sense that it has no instances. This is shown by the `{abstract}` *property* (an alternative notation for this is to write the class name in italics).

The `{abstract}` property can only be applied to a superclass in a generalization hierarchy. All other classes have at least one instance, and are said to be *concrete* or *instantiated*. Applying the `{abstract}` property to `StaffMember` means that no staff exist at Agate who are 'general' members of staff, and not members of a particular sub-group. All staff members encountered so far (among those that are relevant to the model) are defined as either `AdminStaff` or `CreativeStaff`. Should we later discover another group of staff that is distinct in behaviour, data or associations, and if we need to model this new group, it should be represented in the diagram by a new subclass. The whole reason for the existence of a superclass is that it sits at a higher level of abstraction than its subclasses. This generality allows it to be adapted for use in other systems. While in itself this is not always enough for it to be declared as an abstract class, this is usually the case, and for the moment we can safely ignore exceptions to the rule.

The usefulness of generalization

We can now consider the contribution of generalization hierarchies to reuse. Imagine that the Agate system is completed and in regular use. Some time after installation, the Directors decide that they want to reorganize the company, and one of the results is that Account Managers are to be paid bonuses related to campaign profits. Their bonus is to be calculated in a different way from both administrative and other creative staff, and is to include an element from campaigns that they supervise, and an element from the general profitability of the company. This can be seen in Figure 8.5.

Note also the alternative notation style. In Figure 8.4 each subclass is joined to the superclass by its own generalization association, while in Figure 8.5 the three subclasses are organized into a tree structure with a single triangle joining this to the superclass. The UML Specification (OMG, 2001) calls these, respectively, the separate target and shared target styles. Both are acceptable.

Adding a new subclass requires relatively little change to the existing class model, essentially just adding a new subclass `AccountManager` to the staff hierarchy. In practice, some judgement would be needed as to whether this is better modelled as a subclass of `StaffMember`, or of `CreativeStaff`, and this would be based on assumptions about difficulty of implementation and likely future benefits. In either case, the impact is minimal.

The reuse lies in the fact that the existing abstract class `StaffMember` has been used as a basis for `AccountManager`. The latter can inherit all attributes and operations that are not part of its own specialism—in this case roughly 85% of its entire specification. This is for only one class; over the many classes of a large system, the saving in design and coding effort can clearly be significant. But the opportunity is only available because we have first taken the trouble to identify the generalized aspects of a staff member. This is the main benefit of generalization, that hierarchies can usually be extended without significant effects on existing structures. Were it not for this, there

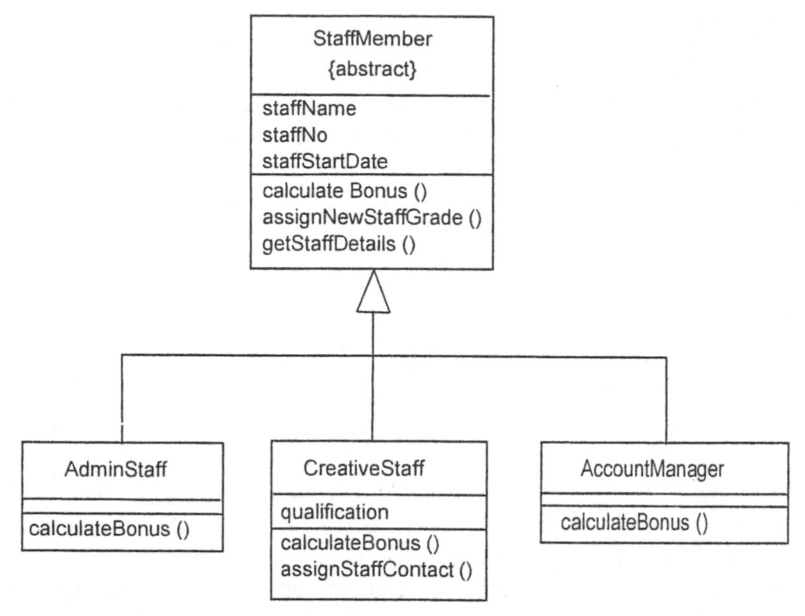

Figure 8.5 A new subclass is easy to add.

would be little justification for the time taken to organize a class diagram using generalization, when it would be so much quicker to leave it alone once the classes and simple associations are defined.

Identifying generalization (a top-down approach)

It is relatively easy to discover generalization where this exists between classes that have already been identified. The rule of thumb is very straightforward. If an association can be described by the expression *is a kind of*, then it can usually be modelled as generalization. Sometimes this is so obvious, that you may wonder if that is all there is to it, for example, 'administrative staff are a kind of staff'. More often, it is not quite so obvious, but still straightforward. For example, 'a helicopter is a type of aircraft and so is a jumbo-jet' and 'a truck is a type of vehicle and so is a buffalo cart' imply generalizations with similar structures, as shown in Figure 8.6.

It is not uncommon to find multiple levels of generalization. This simply means that a superclass in one relationship may be a subclass in another. As Figure 8.6 suggests, `Aircraft` is both a superclass of `Helicopter` and a subclass of `Vehicle`. In practice, more than about four or five levels of generalization in a class model is too many (Rumbaugh et al., 1991), but this is primarily for design reasons.

Adding generalization (a bottom-up approach)

An alternative approach to adding generalization is to look for similarities among the classes in your model, and consider whether the model can be 'tidied up' or simplified by introducing superclasses that abstract the similarities. This needs to be done with some care. The purpose of doing this is quite explicitly to increase the level of abstraction of the model, but any further abstraction introduced should still be 'useful'. The guiding principle is still that any new generalization must meet all the tests described in the previous section. Above all, it must be possible to represent the association with the phrase 'is a kind of'.

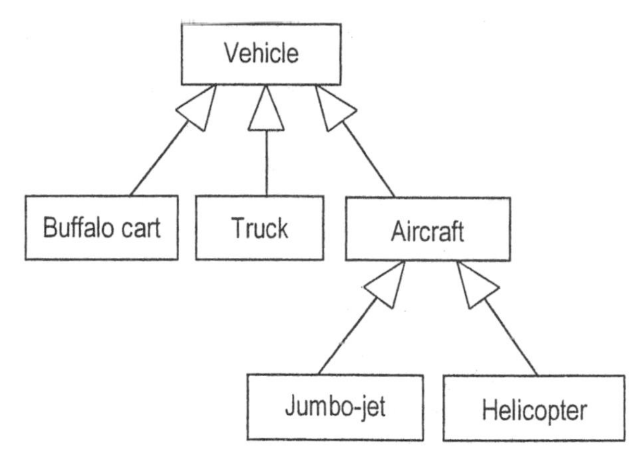

Figure 8.6 A hierarchy discovered top-down.

Some judgement is needed to determine the likely future usefulness of generalization, and this is discussed in the next section.

When not to use generalization

Generalization can be over-used, so it is worth highlighting some implicit judgements made in deciding to model the generalization structure developed in Figures 8.4–8.5. First, as a rule we only model a class as a superclass of another if we are confident that what we know about the former (i.e. its attributes, operations and associations) applies *completely* to the latter. In this example, the analyst had to be reasonably sure that everything she knew about `StaffMember` applied also to both `AdminStaff` and `CreativeStaff`. This is part of the UML definition of generalization: a subclass is 'fully consistent with' its superclass. Even when this is true, if the differences between two potential subclasses are too great, the forced creation of a superclass can give rise to confusion rather than clarity.

We can illustrate this best with an example of inappropriate modelling. At Agate, staff and (some) clients are people. This is perhaps stretching a point, since many clients are in fact companies, but for the sake of illustration we will ignore this and assume all clients are individuals. Following the rule of thumb without reflection, an inexperienced analyst might feel that this justifies the creation of a `Person` superclass, to contain any common attributes and operations of `Client` and `StaffMember`. But it may quickly become apparent that the new class definition contains little but the attribute `personName`. This is really an attempt to force a generalization hierarchy to include subclasses that are too dissimilar.

Second, we should not anticipate subclasses that are not justified by currently known requirements. For example, at Agate `AdminStaff` and `CreativeStaff` are distinct classes based on differences in their attributes and operations. We also know about other kinds of staff in the organization, e.g. the directors. But we have no information that suggests a need to model directors as part of the system, and thus we should not create another subclass of `StaffMember` called `DirectorStaff`. Even if it were to turn out that we do need to model Directors, there is no reason yet to suppose they must be a distinct class. They might turn out to be members of an existing class. This, too, is part of the UML standard definition: a subclass must 'add additional information' (that, we might add, 'is relevant to the system under study').

There is a tension here. On the one hand, generalization is modelled in order to permit future subclassing in situations that the analyst inevitably cannot fully anticipate. Indeed the ability to take advantage of this is one of the main benefits of constructing a generalization hierarchy. Yet, on the other hand, if generalization is overdone, it just adds needlessly to the complexity of the model for no return. There is at present no simple answer to this problem, save the judgement that comes with experience. But it seems probable that over the next few years, organizations will grow much better at managing the activity of reuse, and part of this will involve the communication of clear guidance to project teams on this tricky point.

Multiple inheritance

It is quite possible, and often appropriate, for a class to be simultaneously the subclass of more than one superclass. This is not an unfamiliar concept in terms of everyday classification. For example, if we were to classify some household items according to their use, a coffee mug is a drinking vessel. If we classify the same items according to their aesthetic qualities, the mug might be an attractive craft item. If we classify the same items according to their health risk, the mug might be a hazard (because it is cracked). The mug can belong at one time to various categories derived from different classification schemes without any logical conflict.

In object-oriented modelling, especially during design, it can be useful to define classes that inherit features from more than one superclass. In each case, all features are inherited from every superclass. We do not cover this topic further at this point, but Chapter 17 illustrates the application of multiple inheritance to the design of user interface classes.

8.3.2 Combining generalization with composition or aggregation

At this point it is worth further examining the Agate case study material for further generalization and composition. One possibility is revealed in the statement in Section A1.3 that 'adverts can be one of several types'. For each type, it is certainly true that, for example, a newspaper advert *is a kind of* advert. For simplicity, we assume that 'advert' refers to a *design*, rather than an *insertion*—so an advert that appears five times in one newspaper is one advert appearing five times, and not five adverts each appearing once. This suggests that advert could be a superclass, with newspaper advert, etc. as its subclasses. We need to ask if it meets the tests first described in Section 4.2.4. Figure 8.7 shows the partial class model that results from this analysis, but how was this arrived at?

The process of thought goes something like this. First, is *everything* that is true of `Advert` also true of `NewspaperAdvert`? The answer appears to be yes. All actual adverts must be one of the specific types, with no such thing in reality as a 'general' advert. `Advert` is thus a sound generalization of common features shared by its specialized subclasses. Second, does `NewspaperAdvert` include some details (attributes, operations or associations) that are *not* derived from `Advert`? An authoritative answer would require a detailed examination of attributes and operations for each class, but in general terms the answer appears to be yes. A newspaper advert consists of a particular set of parts. The precise composition of each type of advert is different, and so this structure of associations could not be defined at the superclass level (the attributes, operations and composition structure of Television adverts may resemble newspaper adverts in some respects, but will certainly differ in others). Finally, there is no reason to suppose that a newspaper advert has any other potential ancestor besides advert, so we do not need to consider this rule at higher levels of recursion.

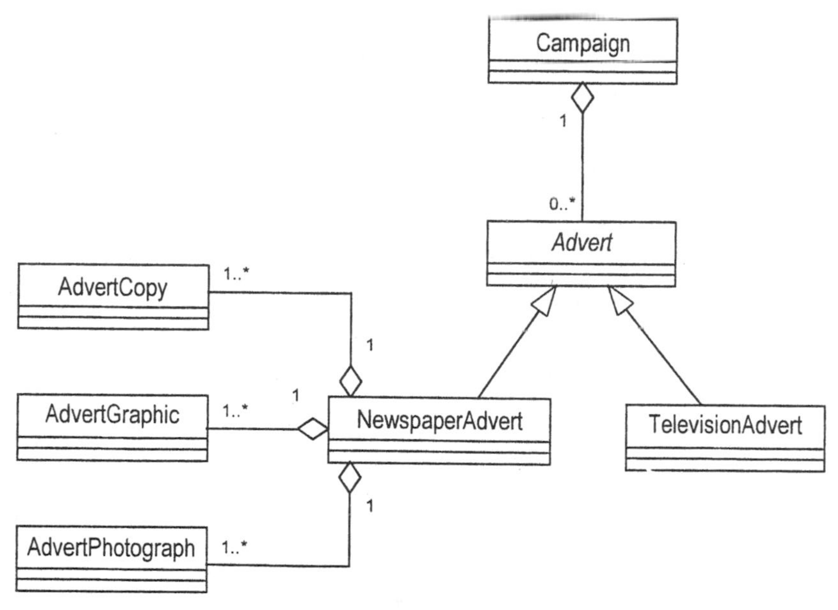

Figure 8.7 Aggregation and generalization in the Agate model (composition does not appear fully justified in any of these cases).

Next, we can see possible composition in the association between `Campaign` and `Advert`, and in turn between `Advert` and its associated parts. A campaign includes one or more adverts. A newspaper advert includes written copy, graphics and photographs.

What kind of composition is involved? This can be analysed by reviewing the UML definitions given in Section 8.2.3. First, can an advert belong to more than one campaign? This is not stated in the case study, but it seems unlikely that an advert can simultaneously be part of more than one campaign. Second, has each `Advert` a coincident lifetime with its `Campaign`? Again, this is not given explicitly, but a client might wish to use an expensive television advert again on another campaign. This point needs to be clarified, but in the meantime it does not appear justified to model this as composition. Thirdly, can copy, graphics or photographs belong to more than one newspaper advert? This seems unlikely, but should really be clarified. Finally, has each of these components a coincident lifetime with the advert? Probably some do, and some do not. It is hard to imagine that advertising copy would be used again on another advert, but photographs and graphics seem less certain. This is another point to be clarified, but in the meantime composition does not seem justified in any of these cases, and aggregation has therefore been used. A similar process of analysis can be applied to each of the other types of advert.

8.3.3 Analysis packages and dependencies

It requires both skill and judgement on the part of the analyst to create a model that will be robust in the face of changing requirements. To some extent this depends on defining analysis packages that are relatively independent of each other while still internally highly cohesive (see Chapter 14). The UML package is a tool for managing the complexity of a model, and they are also a useful way of identifying subsystems that can stand alone as components. Packages are the means by which a developer can 'factor out' classes or structures that have potential use in a wider context than this project alone. But when a model is partitioned into packages (or when pre-existing components are used to support a current project) it is very important to keep track of the dependencies between different classes and packages.

The Agate case study suggests two related but distinct application areas: campaign management and staff management. If this model proves to be only one part of a larger domain, it will probably make sense to model these as two separate analysis packages: `Campaign Management` and `Staff Management`. If this is done, it is quite likely that some entity objects will prove to be common to both packages.

Based on our preliminary analysis so far, the `StaffMember` entity class plays a role in both packages. This leads to an architectural decision. Figure 8.8 illustrates some of the variations described below.

- We could decide to locate this class within the `Staff Management` package. In this case, we need to model a dependency from `Campaign Management` to `Staff Management`, since the `Client` and `Campaign` classes need an association with `StaffMember` (diagram variation (i)).
- We could remove `StaffMember` to a separate package. This would be justified if it appears to have more widespread use. For example, there may also be wages, personnel, welfare and pension applications that need this class. In this case, we need to model dependencies from all the corresponding packages to the package that contains the `StaffMember` class (diagram variation (ii)).
- Further analysis may reveal that in fact `StaffMember` is not a single class. In this case, the derivative classes may remain within their respective packages, but it is

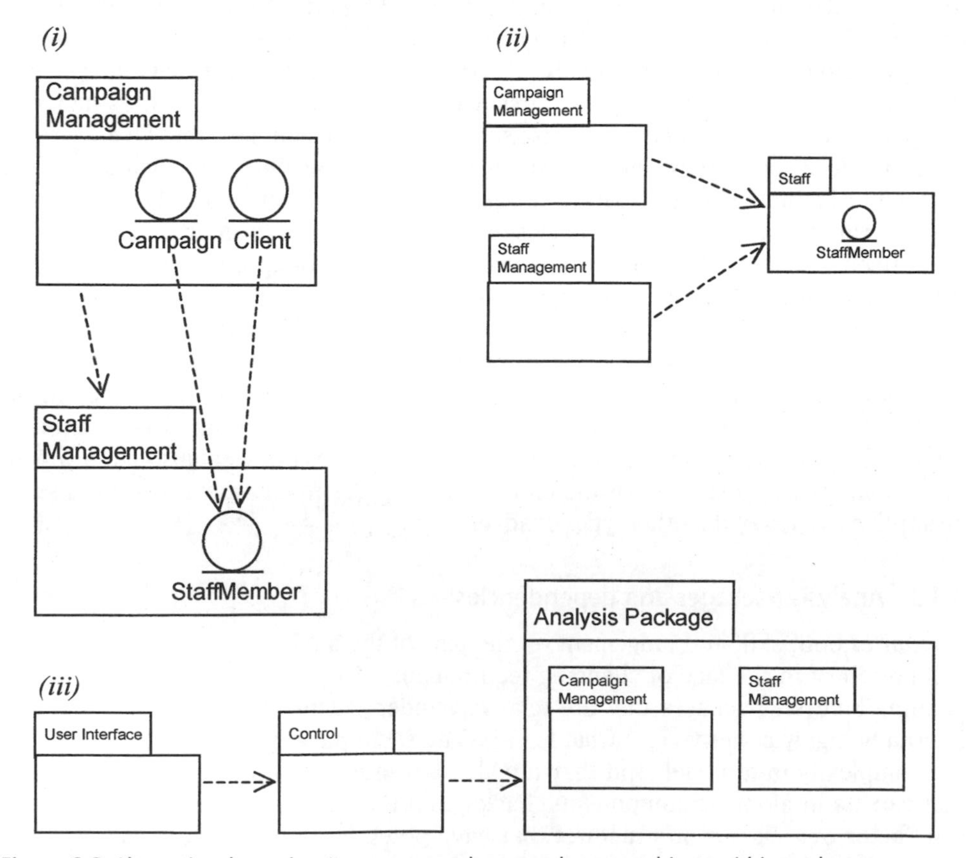

Figure 8.8 Alternative dependencies among packages and among objects within packages.

likely that there will be an association between them and the dependency must still be documented.

- In addition, we have already made an implicit judgement (in Chapter 7) to separate all boundary objects into a `User Interface` package and all control objects into a `Control` package. Objects in these specialized packages will certainly have dependencies on objects in other packages (diagram variation (iii)).

The significance of documenting package dependencies is perhaps not obvious when dealing, as we are here, with a relatively simple model. But it will become more important when the models are large, complex and, above all, when there is a substantial element of reuse involved, whether this occurs at the level of individual classes or at a component scale.

8.4 Software Development Patterns

8.4.1 What is a pattern?

In everyday speech, a pattern often refers to a kind of design that is used to reproduce images in a repetitive manner (on wallpaper or fabric, for example). In software development, the term is related to this idea, but has a much more specific meaning. This particular usage is traced back to the architect Christopher Alexander, who first used the term pattern to describe solutions to recurring problems in architecture. Alexander identified many related patterns for the development of effective and harmonious architectural forms in buildings. Alexander's patterns address many architectural issues—for example the best place to site a door in a room, or how to organize and structure a waiting area in a building so that waiting can become a positive experience. Alexander argued that his patterns became a design language within which solutions to recurring architectural problems could be developed and described. Alexander's definition of a pattern is as follows.

> Each pattern describes a problem which occurs over and over again in our environment, and then describes the core of a solution to that problem, in such a way that you can use this solution a million times over, without ever doing it the same way twice.[2]
>
> Alexander et al. (1977)

A pattern provides a solution that may be applied in different ways depending upon the specific problem to which it is being applied. Alexander's definition makes no reference to buildings or architecture but only to 'our environment'. Although Alexander intended 'environment' to be interpreted as the physical environment in which we live there is clearly an analogous concept appropriate to software development. One definition of a pattern that is appropriate to software systems development is this:

> A pattern is the abstraction from a concrete form which keeps recurring in specific non-arbitrary contexts.
>
> Riehle and Zullighoven (1996)

In fact, this definition could be applied to any domain of human endeavour in which there is some recurring form from which abstractions may be developed. It is not clear how tangible 'concrete' is intended to be but a reasonable interpretation might be 'specific' or 'particular'. Gabriel (1996) gives an alternative and more detailed definition that has some insight into the structure of a pattern.

[2] Quoted by permission of Oxford University Press Inc, New York.

> Each pattern is a three-part rule, which expresses a relation between a certain context, a certain system of forces which occurs repeatedly in that context, and a certain software configuration which allows these forces to resolve themselves.[3]
>
> Gabriel (1996)

This definition focuses on three elements—a *context* that can be understood as a set of circumstances or preconditions, *forces* that are issues that have to be addressed and a software configuration that addresses and resolves the forces.

In this definition the term 'software configuration' might suggest that patterns are limited to software design and construction. In fact, patterns are applied much more widely in systems development. The Boundary, Control and Entity object architecture introduced in Chapter 7 is in fact a pattern widely applied during requirements analysis and systems design. The Layer, Broker and Model–View–Controller architectures that we will consider later, in Chapter 13, are also examples of the application of patterns from Buschmann et al. (1996) to the activity of systems design.

Coad et al. (1997) make the distinction between a *strategy* which they describe as a plan of action intended to achieve some defined purpose and a *pattern* which they describe as a template that embodies an example worth emulating. This view of a pattern is slightly different from the views described earlier as it does not emphasize contextual aspects to the same extent. An example of a Coad et al. strategy is 'Organize and Prioritize Features' and relates to the need to prioritize requirements (discussed in Chapter 3). A simple example of an analysis pattern from Coad et al. (1997) is the Transaction–Transaction Line Item pattern (Figure 8.9).

Figure 8.10 shows the pattern as it might be applied to a sales order processing system. Here the Transaction suggests a `SalesOrder` class and the Transaction Line Item suggests a `SalesOrderLine` class.

Very similar structures are used in a wide variety of other circumstances (e.g. shipment and shipment line item, payment and payment line item). A novice software developer has to learn this structure, or to reinvent it. The latter is much less efficient. The act of describing it as a pattern highlights it as a useful piece of development expertise and makes it readily available for the novice. An example of a principle that makes this pattern advantageous is the desirability of low interaction coupling (see Chapter 14).

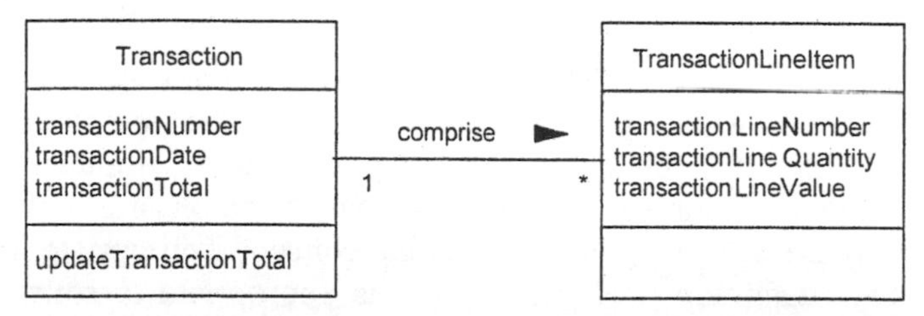

Figure 8.9 Transaction–Transaction Line Item pattern (adapted from Coad et al., 1997).

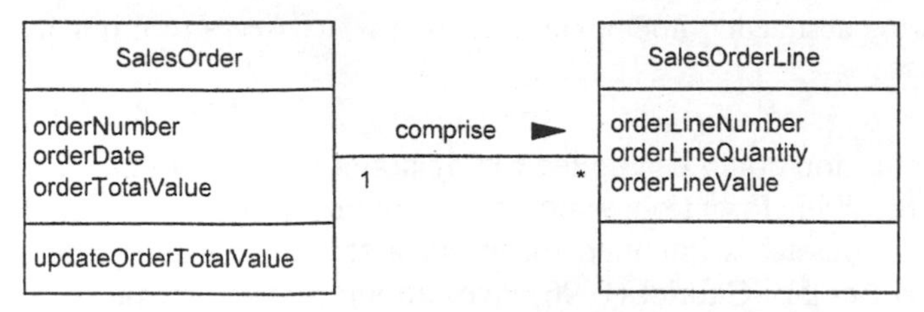

Figure 8.10 Simple application of the Transaction–Transaction Line Item pattern.

[3] Quoted by permission of New York Oxford University Press.

Coplien (1996) identifies the critical aspects of a pattern as follows.

- It solves a problem.
- It is a proven concept.
- The solution is not obvious.
- It describes a relationship.
- The pattern has a significant human component.

The human component of Christopher Alexander's patterns relates to the aesthetic qualities and the sense of harmony that a building may engender in its users. The analogous concept in software is harder to describe. We do not inhabit software systems—we use them as aids to other activities. But the human component Coplien refers to is not simply a good user interface to a working application, it is concerned with the nature of the software constructs used to build the application. Software patterns should result in structures that are sympathetic to the human perspective. It is argued that a good software pattern not only offers a solution that works but a solution that has an aesthetic quality, that is in some way elegant. This aesthetic quality of patterns is sometimes termed Quality Without A Name (QWAN for short) to avoid the overloading of current terminology. As you might imagine, QWAN is the subject of much controversy. Our discussion of patterns will not address issues of elegance though the reader may judge for themselves whether the solutions developed from the patterns have a sense of elegance.

In the same way that a pattern captures and documents proven good practice, *anti-patterns* capture practice that is demonstrably bad. It is sensible to do this. We should ensure not only that a software system embodies good practice but also that it avoids known pitfalls. Antipatterns are a way of documenting attempted solutions to recurring problems that proved unsuccessful. An antipattern can also include reworked solutions that proved effective (Brown et al., 1998).

For example, Mushroom Management (Brown et al., 1998) is an example of an antipattern in the domain of software development organizations. It describes a situation where there is an explicit policy to isolate systems developers from users in an attempt to limit requirements drift. In such an organization, requirements are passed through an intermediary such as the project manager or a requirements analyst. The negative consequence of this pattern of development organization is that inevitable inadequacies in the analysis documentation are not resolved. Furthermore design decisions are made without user involvement, and the delivered system may not address users' requirements. The reworked solution suggested by Brown et al. (1998) is to use a form of spiral process development model (see Chapter 3). Other reworked solutions include the involvement of domain experts in the development team, as recommended by the Dynamics Systems Development Method (DSDM) (we introduce DSDM in Chapter 21).

Patterns have been applied to many different aspects of software development. Beck and Cunningham (1989) documented some of the earliest software patterns in order to describe aspects of interface design in Smalltalk environments. Subsequently Coplien (1992) catalogued a set of patterns specifically for use in C++ programming (patterns that are related to constructs in a specific programming language are now known as *idioms*). Coad et al. (1997) greatly widened the field by describing a series of analysis and design patterns. Patterns have been applied to software development approaches other than object-orientated ones. For example, Hay (1996) identified a series of analysis patterns for data modelling. Hay discusses patterns relating to concepts such as Party and Contract, which appear commonly in information systems.

Design patterns were popularized by Gamma et al. (1995) in their book 'Design Patterns: Elements of Reusable Object-Oriented Software'[2]. The publication of this book gave significant impetus to the use of patterns in software development, but other authors have identified patterns that are concerned with analysis (Fowler, 1997), organizational issues (Coplien, 1996) and systems architecture using CORBA[3] (OMG, 1995).

Architectural patterns have also been identified that address some of the issues concerning the structural organization of software systems. Definitions have been suggested for some of these pattern categories (Buschmann et al., 1996):

Architectural patterns describe the structure and relationship of major components of a software system. An architectural pattern identifies subsystems, their responsibilities and their interrelationships.

Design patterns identify the interrelationships among a group of software components describing their responsibilites, collaborations and structural relationships.

Idioms describe how to implement particular aspects of a software system in a given programming language.

Analysis patterns (Fowler, 1997) are defined as describing groups of concepts that represent common constructions in domain modelling. These patterns may be applicable in one domain or in many domains.

A particular pattern may belong to more than one category. For example, the MVC architecture can be viewed as an architectural pattern (Buschmann et al., 1996) when it is applied to subsystems, as a design pattern when it is concerned with smaller components or individual classes, and as a Smalltalk idiom.

8.4.2 Analysis patterns

In this section we offer a brief introduction to analysis patterns. The use of analysis patterns is an advanced approach that is principally of use to experienced analysts, and we leave the interested reader to follow this up through further reading. They are closely related to design patterns, which we cover in some detail in Chapter 15.

An analysis pattern is essentially a structure of classes and associations that is found to occur over and over again in many different modelling situations. Each pattern can be used to communicate a general understanding about how to model a particular set of requirements, and therefore the model need not be invented from scratch every time a similar situation occurs. Since a pattern may consist of whole structures of classes, the abstraction takes places at a higher level than is normally possible using generalization alone. Fowler (1997) describes a number of patterns that recur in many business modelling situations, such as accounting, trading and organization structure.

Figure 8.11 shows Fowler's Accountability pattern as an illustration of an analysis pattern in practice. For the sake of simplicity, we will discuss only the class structure, although patterns are normally documented in more detail than this (see Chapter 15). An accountability structure may be of many kinds, such as management or contract supervision. In the case of Agate, this pattern could apply to several different relationships: that between a manager and a member of staff they supervise, that between a client and a client contact, or that between a client and a campaign manager. Since the

[2] The four authors of the book are known as the 'Gang of Four' (GOF) and the book is known as the GOF book.

[3] CORBA (Common Object Request Broker Architecture) is a general definition for a mechanism that uses Object Request Brokers (ORB) to provide interoperability between applications on different machines. We discuss its application in Chapters 17 and 19.

details of the relationship itself have been abstracted out as AccountabilityType, this one class structure is sufficiently general to be adapted to any of these relationships, given an appropriate set of attributes, operations and associations to other classes specific to the application model. The generalization of Person and Organization as Party similarly allows the pattern to represent relationships between individuals, organizations, or a mixture of the two.

Figure 8.12 shows a possible application of this analysis pattern to the Staff Contact relationship at Agate.

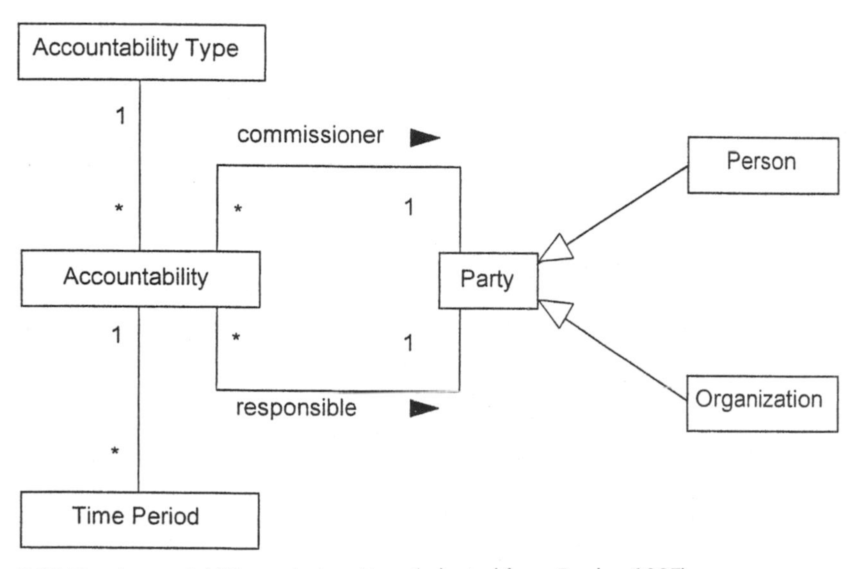

Figure 8.11 The Accountability analysis pattern (adapted from Fowler, 1997).

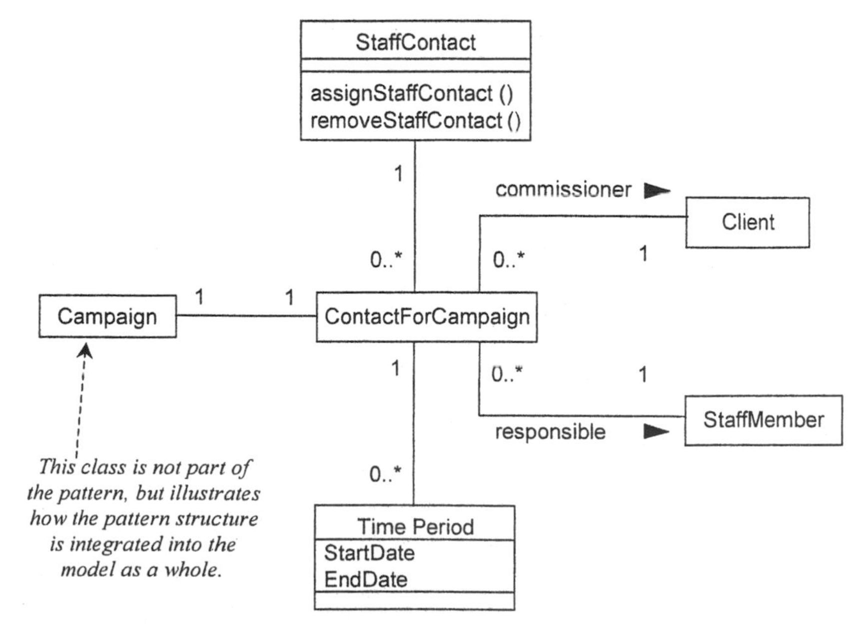

Figure 8.12 Accountability pattern applied to Agate's StaffContact relationsip.

Another pattern that might be useful in developing an analysis model for Agate is the Composite pattern (Gamma et al., 1995). However, since this pattern is useful from both analysis and design perspectives, we will defer its introduction until Chapter 15.

8.5 Summary

The primary purpose of refining the requirements model is to maximize opportunities for the reuse of components. These components may be of several kinds. Software components can reduce the effort involved in coding and implementation. Design components can reduce the time required to develop and test a design, and also act as a store of knowledge that embodies best known practice. Analysis components can reduce the time taken to develop a deep understanding of the system. In principle, whatever their form, opportunities for the use of components occur in only three ways. Existing components may be imported from other sources beyond the project boundaries. This requires careful evaluation of the degree of commonality between the requirements of the current project and the features of available components. Second, new reusable components may be developed for use on more than one part of the current project. For this to be possible, the requirements model must be at a sufficient level of abstraction for the common features between different aspects of the project to become evident. Finally, new reusable components may be developed for export to other projects. Here, too, it is necessary for the requirements modelling to identify those aspects of the project that can be generalized for use in other contexts.

The abstraction mechanisms of the object-oriented approach, particularly generalization and composition, can make a significant contribution to all three approaches to the reuse of components in systems development. So far, these benefits have been mainly apparent during the design and coding activities, but the starting point for finding reuse opportunities is during requirements analysis.

Patterns are a relatively new approach to documenting, sharing and reusing useful insights into many aspects of the software development activity. This chapter has considered how patterns can be used in software development, in particular from an analysis perspective. Later chapters will discuss in more detail the application of patterns to address issues of architecture (Chapter 13), design (Chapter 15) and organization (Chapter 21).

Review Questions

8.1 What are the advantages of components?

8.2 Explain why the NIH syndrome occurs.

8.3 What does object-orientation offer that helps to create reusable components?

8.4 Distinguish composition from aggregation.

8.5 Why are operations sometimes redefined in a subclass?

8.6 What is an abstract class?

8.7 Why is encapsulation important to creating reusable components?

8.8 Why is generalization important to creating reusable components?

8.9 When should you not use generalization in a model?

8.10 What does the term pattern mean in the context of software development?

8.11 How do patterns help the software developer?

8.12 What is an antipattern?

Case Study Work, Exercises and Projects

8.A Find out from your library about the coding system that is used for classifying books, videos, etc. Draw part of the structure in UML notation as a generalization hierarchy. Think up some attributes for 'classes' in your model to show how the lower levels are progressively more specialized.

8.B Choose an area of commercial activity (business, industry, government agency, etc.) with which you are familiar. Identify some ways in which its products show the use of generalization, and some ways that components used as inputs show the use of generalization.

8.C In Section 8.3.2, we established that generalization probably was an appropriate way of modelling the association between `Advert` and `NewspaperAdvert`. Identify the other possible subclasses for `Advert` from Section A1.3 and repeat the checks for each. Which of them pass (if any)? Do you think there are really only two levels to the hierarchy? Explain your reasoning. Redraw the Agate class diagram to include all the generalizations that you feel are justified.

8.D For each of your new Advert subclasses, suggest appropriate attributes, operations and aggregation or composition structures.

8.E Re-read the case study material for FoodCo and identify possible generalizations or compositions. Add these to the class diagram you drew for Exercises 7.C and 7.D.

8.F Consider your class diagram for FoodCo. Try to identify possible applications of the Transaction–Transaction Line Item pattern, or of the Accountability pattern and redraw your diagram as appropriate.

Further Reading

- All standard texts on object-orientation have one or more chapters that discuss the role of generalization and aggregation or composition in software and requirements reuse. Rumbaugh et al. (1991) give one of the clearest summaries. This particular book pre-dates the development of UML, but Rumbaugh was later one of the three amigos (along with Jacobson and Booch) who founded the UML standard. UML incorporates a great deal of OMT notation, particularly that used for the class diagram.
- Jacobson et al. (1999) give some very useful attention to the architectural and reuse issues involved in relation to composition, generalization and the identification of appropriate packages during requirements analysis.
- For a broad-ranging introduction to patterns, Gamma et al. (1995) and Buschmann et al. (1996) should be on the essential reading list of any software developer. The 'Pattern Languages of Program Design' (known as the PLOP) books (Coplien and Schmidt, 1995; Vlissides et al., 1996; Martin et al., 1998) catalogue a wide range of patterns for all aspects of software development.
- Fowler (1997) describes a number of analysis patterns that may be applied at this stage of modelling, although these are not presented in UML. Coad et al. (1997) also gives a number of analysis and design patterns; most are presented in Coad notation, but some are also shown in OMT notation and in Unified notation (an early version of UML).
- The patterns home page can be found at hillside.net/patterns/patterns.html. Further useful patterns are stored in the Portland Pattern Repository at www.c2.com/ppr.

CHAPTER 9

Object Interaction

OBJECTIVES

In this chapter you will learn

- how to develop object collaboration from use cases
- how to model object collaboration using an interaction sequence diagram
- how to model object collaboration using an interaction collaboration diagram
- how to cross-check between interaction diagrams and a class diagram.

9.1 Introduction

Communication and collaboration between objects is fundamental to object-orientation. This fact mirrors the world in which we live, where most human endeavour involves communication and collaboration between individuals. For example, each employee in a manufacturing organization has specialized tasks. Different employees collaborate with each other in order to satisfy a customer request. This involves communicating to request information, to share information and to request help from each other. In a similar way an object-oriented application comprises a set of autonomous objects, each responsible for a small part of the system's overall behaviour. These objects produce the required behaviour through collaboration, by exchanging messages that request information, that give information or that ask another object to perform some task. In Section 9.2 we explore the concept of collaboration in more detail.

We have already analysed (Chapter 7) use cases to determine the nature of the collaboration between the objects involved. This results in the identification of classes, their attributes and their associated responsibilities[1]. As we move to design these object interactions have to be specified more precisely. This involves deciding how to represent responsibities as operations. CRC cards were suggested in Section 7.6 as a

[1] This process is known as use case realization in the Unified Software Development Process.

supporting technique to aid the analysis and the resulting identification and allocation of responsibilities. CRC cards can also be used effectively when designing object interaction in more detail in terms of operations. UML itself provides two notations for modelling communication between objects: interaction sequence diagrams and interaction collaboration diagrams. Interaction sequence diagrams are introduced in Section 9.3 and collaboration diagrams are explored further in Section 9.4. In Section 9.5 we examine the issues concerned with interaction diagrams and model consistency.

9.2 Object Interaction and Collaboration

When an object sends a message to another object, an operation is invoked in the receiving object. For example, in the Agate case study there is a requirement to be able to determine the current cost of the advertisements for an advertising campaign. This responsibility is assigned to the `Campaign` class. For a particular campaign this might be achieved if the Campaign object sends a message to each of its `Advert` objects asking them for their current cost. In a programming language, sending the message `getCost()` to an `Advert` object, might use the following syntax.

```
advertCost = anAdvert.getCost()
```

Note that in this example the advert object is identified by the variable name `anAdvert`.

The cost of each advert returned by the operation `getCost()` is totalled up in the attribute `actualCost` in the sending object, `Campaign`, so in order to calculate the sum of the costs for all adverts in a campaign the above statement must be executed repeatedly. However, rather than think in terms of operation invocation we use the metaphor of message passing to describe object interaction, as this emphasizes that objects are encapsulated and essentially autonomous. Message passing can be represented on an object diagram, as in Figure 9.1 where the message is shown as an arrow. This notation is rather like that of the collaboration diagram (Section 9.4) and the similarity demonstrates that the latter is a natural extension of the class diagram.

It can be difficult to determine what messages should be sent by each object. In this case, it is clear that the `getCost()` operation should be located in the `Advert` class. This operation requires data that is stored in the `advertCost` attribute, and this has been placed in `Advert`. We can also readily see that an operation that calculates the cost of a `Campaign` must be able to find out the cost of each `Advert` involved. But this is a simple collaboration and the allocation of these operations is largely dictated by the presence of particular attributes in the classes. More complex requirements may involve the performance of complex tasks, such that an object receiving one message must itself send messages that initiate further collaboration with other objects.

It is an aim of object-oriented analysis and design to distribute system functionality appropriately among its classes. This does not mean that all classes have exactly equal levels of responsibility but rather that each class should have appropriate responsibilities. Where responsibilities are evenly distributed, each class tends not to be

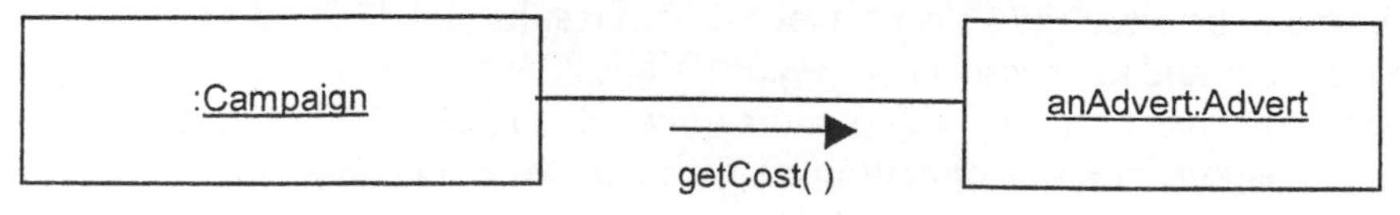

Figure 9.1 Object messaging.

unduly complex and, as a result is easier to develop, to test and to maintain. A class that is relatively small and self-contained has a much greater potential for reuse than one that is large, one that is complex or one that has responsibilities or functionality that are not clearly focused.

An appropriate distribution of responsibility among classes has the important side-effect of producing a system that is more resilient to changes in its requirements. When the users' requirements for a system change it is reasonable to expect that the application will need some modification, but ideally the change in the application should be of no greater magnitude than the change in the requirements. An application that is resilient in this sense costs less to maintain and to extend than one that is not. Figure 9.2 illustrates this.

The focus of a model of object interaction is to determine the most appropriate scheme of messaging between objects in order to support a particular user requirement. As we saw in Chapter 6, user requirements are first documented by use cases. Each use case can be seen as a dialogue between an actor and the system that results in objects performing tasks so that the system can respond in the way that is required by the actor. For this reason many interaction diagrams explicitly include objects to represent the user interface (boundary objects) and to manage the object communication (control objects). When such objects are not shown explicitly it can be assumed in most cases that they will need to be identified at a later stage. The identification and specification of boundary objects is in part an analysis activity and in part a design activity. During analysis our concern is to identify the nature of a dialogue in terms of the user's need for information and his or her access to the system's functionality.

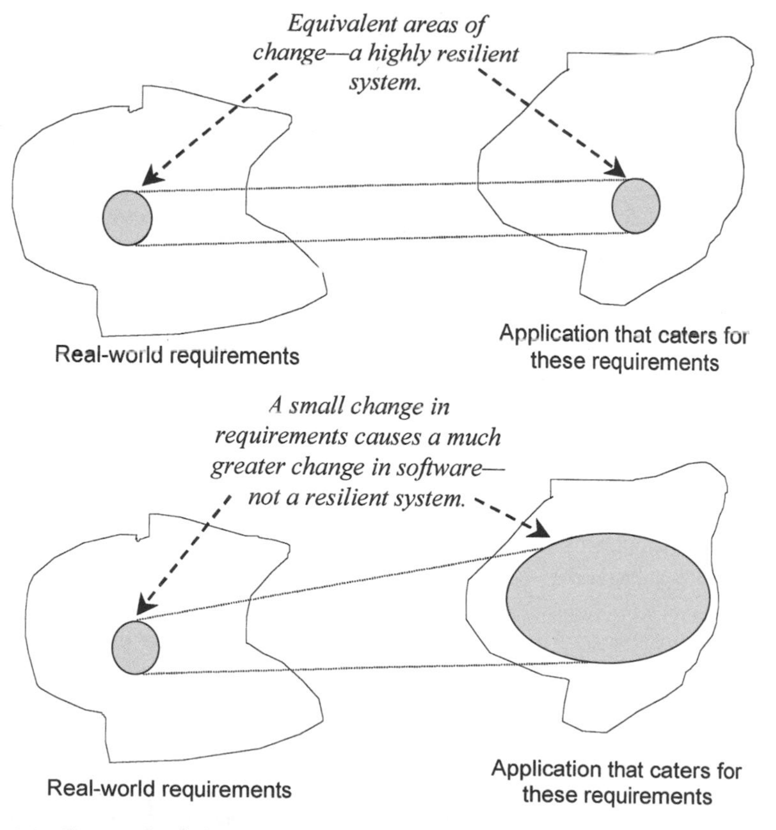

Figure 9.2 Resilience of a design.

The detailed specification of boundary objects that can manage the dialogue is primarily a design activity and is discussed in Chapters 16 and 17.

UML (OMG, 2001, page 3-100) discusses object interaction within the context of a collaboration and defines a collaboration as follows.

> 'The structure of Instances playing roles in a behavior and their relationships is called a *Collaboration*.'

The behaviour mentioned above can be that of an operation or a use case (or any other behavioural classifier in UML). A particular object instance may play different roles in different contexts or collaborations and may play more than one role in a given collaboration. Figure 9.1 illustrates a very simple collaboration showing the two participating objects and the link between them. Strictly speaking a collaboration does not show the messages that are being sent.

In order to illustrate the preparation of interaction diagrams we build on the CRC card analysis of the use case `Add a new advert to a campaign` that was discussed in Section 7.6. The use case description used in Chapter 7 is repeated here.

The Campaign Manager selects the required Campaign for the Client concerned and adds a new Advert to the existing list of adverts for that campaign. The details of the Advert are completed by the Campaign Manager.

The resulting CRC cards are shown in Figure 7.31. These form the basis for the interaction sequence diagrams that are developed in the next two sections.

9.3 Interaction Sequence Diagrams

An interaction sequence diagram (or simply a *sequence diagram*) is one of two kinds of UML interaction diagram. The other is the collaboration diagram, which we introduced in Chapter 7 and describe further in Section 9.4. In the UML Notation Guide (OMG, 2001, page 3-101) an interaction is defined as follows.

> *An Interaction* is defined in the context of Collaboration. It specifies the communication patterns between the roles in the Collaboration. More precisely, it contains a set of partially ordered[2] *Messages*, each specifying one communication; for example, what Signal to be sent or what Operation to be invoked, as well as the roles to be played, by the sender and receiver respectively.

This definition of an interaction is in terms of communicating roles and highlights the fact that these concepts can be applied in various contexts. Commonly, during analysis or design, object instances are modelled in terms of the roles they play and communicate by message passing.

A sequence diagram shows an interaction between objects arranged in a time sequence. Sequence diagrams can be drawn at different levels of detail and to meet different purposes at several stages in the development life cycle. The commonest application of a sequence diagram is to represent the detailed object interaction that occurs for one use case or for one operation. When a sequence diagram is used to model the dynamic behaviour of a use case it can be seen as a detailed specification of the use case. Those drawn during analysis differ from those drawn during design in two major respects. Analysis sequence diagrams normally do not include design objects nor do they usually specify message signatures in any detail.

[2] Partially ordered means that the messages may be placed in a time sequence and two or more messages may sent at the same time.

9.3.1 Basic concepts and notation

Figure 9.3 shows a sequence diagram for the use case Add a new advert to a campaign. The vertical dimension represents time and all objects[3] involved in the interaction are spread horizontally across the diagram. (The horizontal ordering of objects is arbitrary and has no modelling significance.) Time normally proceeds down the page. However, a sequence diagram may be drawn with a horizontal time axis if required, and in this case, time proceeds from left to right across the page. Each object is represented by a vertical dashed line, called a *lifeline*, with an object symbol at the top. A message is shown by a solid horizontal arrow from one lifeline to another and is labelled with the message name. Each message name may optionally be preceded by a sequence number that represents the sequence in which the messages are sent, but this is not usually necessary on a sequence diagram since the message sequence is already conveyed by their relative positions along the time axis.

UML uses the general term *stimulus* to describe an interaction between two objects that conveys information with an expectation of some action. Formally then, a message specifies the sender and receiver objects and the action of a stimulus. A message may correspond to calling an operation or raising a *signal*. In UML a signal is an asynchronous communication that may have parameters. When we discuss statecharts in Chapter 11 we will consider events. An event is the specification of an occurrence of significance and may for instance be the receipt of a message or a signal by an object. From a pragmatic modelling perspective we need to distinguish between synchronous and asynchonous messages.

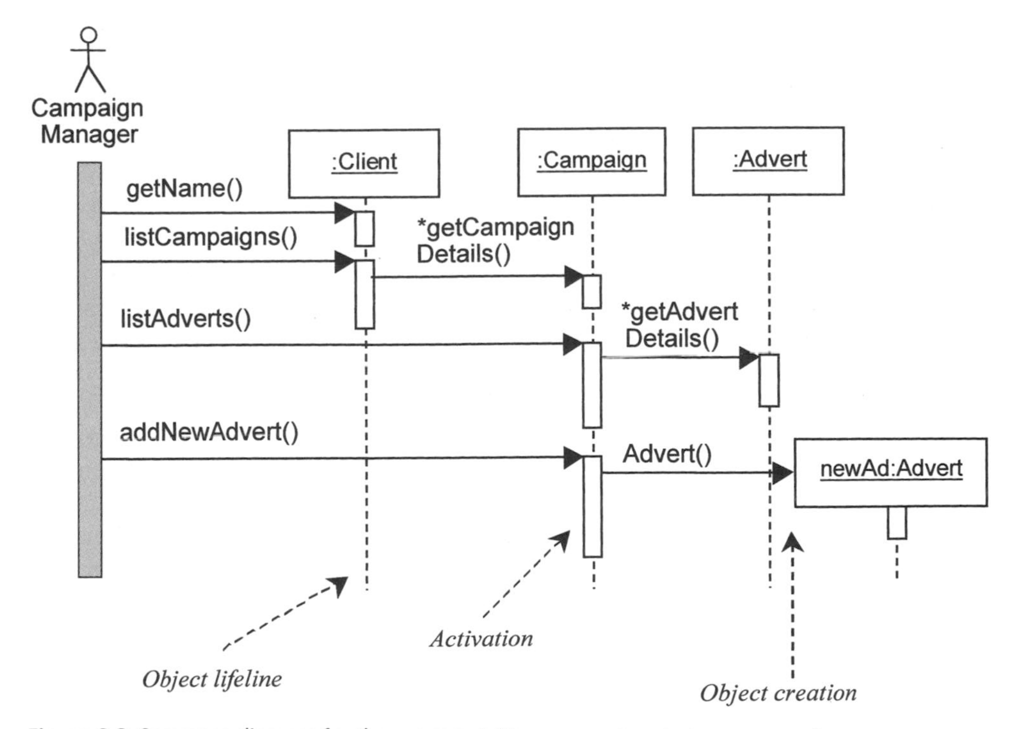

Figure 9.3 Sequence diagram for the use case `Add a new advert to a campaign`.

[3] More generally an interaction may involve *Classifier Roles*.

When a message is sent to an object, it invokes an operation[4] of that object. Once a message is received, the operation that has been invoked begins to execute. The period of time during which an operation executes is known as an *activation*, and is shown on the sequence diagram by a rectangular block laid along the lifeline. The activation period of an operation includes any delay while the operation waits for a response from another operation that it has itself invoked as part of its execution. The message name is usually the same as the particular operation that is being invoked. We will give examples of detailed specifications later.

Figure 9.3 shows a straightforward sequence diagram drawn without boundary or control objects. The `getName()` message is first message received by the `Client` and is intended to correspond to the `Campaign Manager` requesting the name of the selected `Client`. The `Client` object then receives a `listCampaigns()` message and a second period of operation activation begins. This is shown by the tall thin rectangle that begins at the message arrowhead. The `Client` object now sends a message `getCampaign Details()` to each `Campaign` object in turn in order to build up a list of campaigns. This repeated action is called iteration and is indicated by an asterisk (*) before the message. The conditions for continuing or ceasing an iteration may be shown beside the message name, although for reasons of space this is not shown in Figure 9.3.

This example of a continuation condition is written as follows.

```
[For all client's campaigns] *getCampaignDetails()
```

The Campaign Manager next sends a message to a particular `Campaign` object asking it to list its advertisements. The `Campaign` object delegates responsibility for getting the advertisement title to each `Advert` object although the `Campaign` object retains responsibility for the list as a whole (indicated by the continuation of the activation bar beyond the point where the message is sent).

When an advertisement is added to a campaign an `Advert` object is created. This is shown by the `Advert()` message arrow (this invokes the constructor[5] operation) drawn with its arrowhead pointing directly to the object symbol at the top of the lifeline. Where an object already exists prior to the interaction the first message to that object points to the lifeline below the object symbol. For example, this is the case for the `Campaign` object, which must exist before it can receive an `addNewAdvert()` message. The sequence diagram in Figure 9.3 corresponds directly to the interaction suggested by the CRC cards in Figure 7.31. In effect, the sequence diagram is a more formal representation of the same interaction, but with the messages and the sequence of interaction both made explicit.

Most use cases imply at least one boundary object that manages the dialogue between the actor and the system. In Chapter 17 we show a number of sequence diagrams that explicitly include interface classes, but these have been prepared from a design perspective. For example, Figure 17.13 shows a further revision to the sequence diagram for the use case `Check campaign budget`. If this is compared with the version shown in Figure 9.6, it can be seen that the management of the interaction between the user and the system is managed explicitly by the `CBWindow` class in Figure 17.13, whereas in Figure 9.7 the dialogue management is not shown at all. Some method-

[4] Though not explicitly specified in the UML specification, interaction sequence diagrams may be used to model interaction at the level of responsibilities as well.

[5] A constructor operation creates an object instance. In the executable system a constructor typically allocates memory for the new object and intializes attribute values. It is conventional in object-oriented programming languages to name constructors with the class name, hence the constructor operation name begins with a capital letter as does the class name.

ologies (for example USDP, which is described in Chapter 21) suggest that boundary classes should be identified during the analysis, while other approaches take the view that this is primarily a feature of design.

Larman (1998) approaches the question of interaction modelling from a very different perspective, and suggests that during the analysis activity, system interaction diagrams should be drawn that represent only the interface or boundary object and the messages that are exchanged between the user and the system as a whole.

Figure 9.4 shows an alternative sequence diagram for the use case `Add a New Advert to a Campaign` but drawn this time with boundary and control objects. Essentially this is in the style of the Unified Software Development Process. The boundary object, representing the user interface, is `:AddAdvertUI`. We have used the suffix UI to mean user interface. The control object is `:AddAdvert` and this manages the overall object communication. Although not shown in Figure 9.4, the interaction is initiated by the creation of the control object `:AddAdvert`. This then gets the client details before initiating the dialogue by creating the boundary class `:AddAdvertUI`.

Objects may be created or destroyed at different stages during an interaction[6]. On a sequence diagram the destruction of an object is indicated by a large X on the lifeline at the point in the interaction when the object is destroyed. An object may either be destroyed when it receives a message (this is shown in Figure 9.5) or it may self-destruct at the end of an activation if this is required by the particular operation that is being executed.

An object can send a message to itself. This is known as a *reflexive message* and is shown by a message arrow that starts and finishes at the same object lifeline. The use case `Check campaign budget` includes an example of this. For ease of reference the use case description is repeated below.

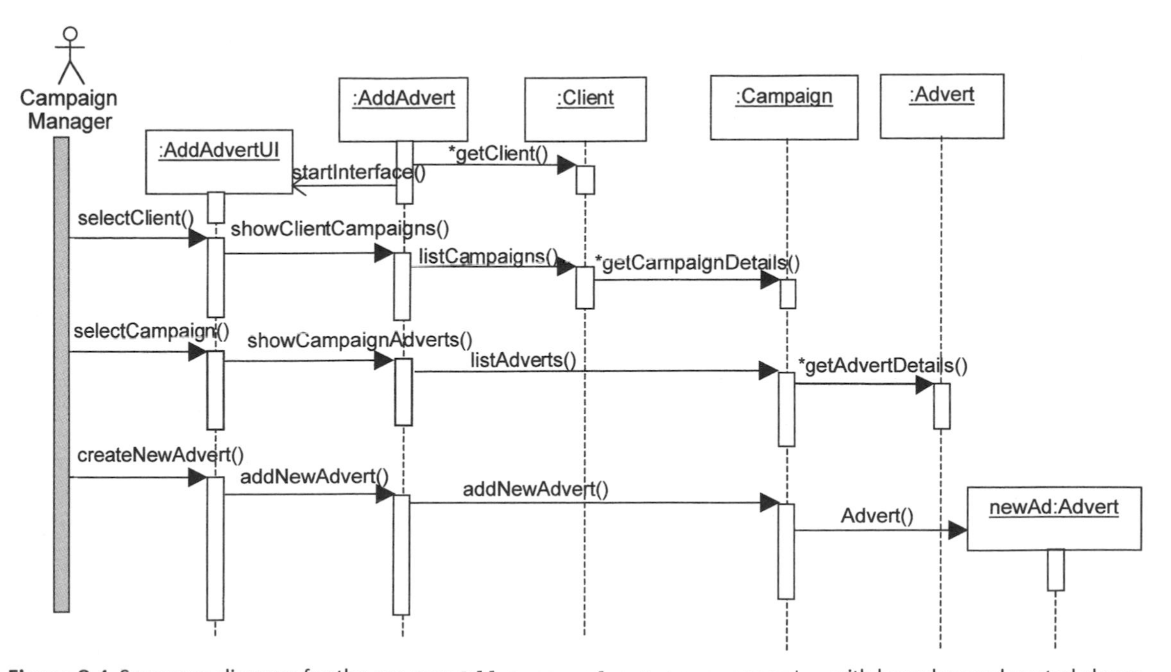

Figure 9.4 Sequence diagram for the use case `Add a new advert to a campaign` with boundary and control classes.

[6] Destructor operations are discussed in more detail in Chapter 14. In Figure 9.5 the `destroy()` message could simply be stereotyped «`destroy`».

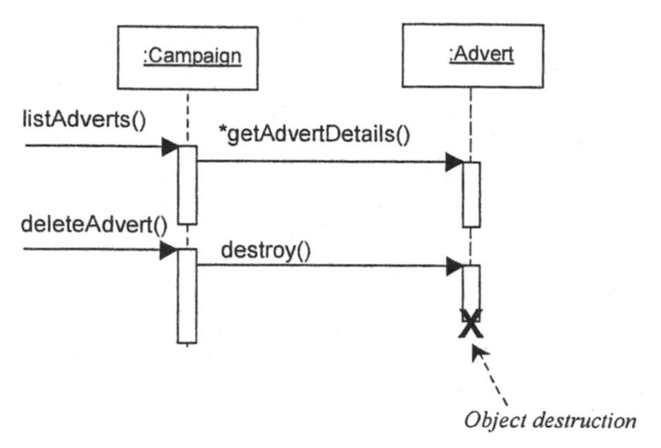

Figure 9.5 Object destruction.

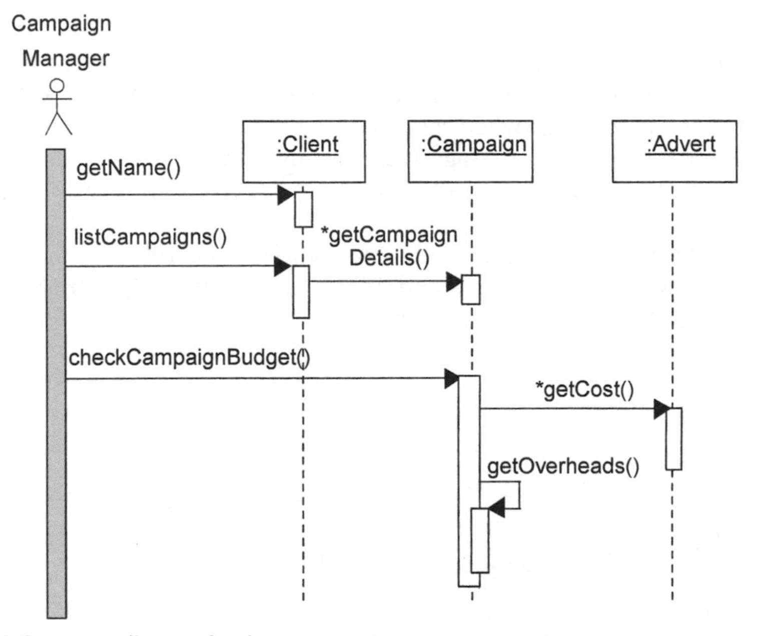

Figure 9.6 Sequence diagram for the use case `Check campaign budget`.

The campaign budget may be checked to ensure that it has not been exceeded. The current campaign cost is determined by the total cost of all the adverts and the campaign overheads.

The corresponding sequence diagram is shown in Figure 9.6 and this includes a reflexive message `getOverheads()` sent from a `Campaign` object to itself.

In this case the reflexive message invokes a different operation from the operation that sent the message and a new activation symbol is stacked on the original activation. (This is the shorter second rectangle shown offset against the first activation.) In certain circumstances an operation invokes itself on the same object; this is known as recursion and can be similarly represented but is not illustrated here[7].

Until this point our discussion has centred on simple use cases and correspondingly simple interactions. These are typical of many modelling situations, but more complex interactions also occur in many systems. It is sometimes also necessary to represent in

[7] If a limit has been set to the depth of recursion this can be recorded in a note on the diagram.

more detail the synchronization of messages. Figure 9.7 illustrates some variations in the UML notation.

The *focus of control* indicates times during an activation when processing is taking place within that object. Parts of an activation that are not within the focus of control represent periods when, for example, an operation is waiting for a return from another object. The focus of control may be shown by shading those parts of the activation rectangle that correspond to active processing by an operation. In Figure 9.7 the `Check campaign budget` use case is redrawn with foci of control shaded. The focus of control for the `checkCampaignBudget()` operation is initially with the `Campaign` object, but is then transferred to the `Advert` object and the activation rectangle in the `Campaign` object is now unshaded while the `Advert` object has the focus of control. The `checkCampaignBudget()` activation is also unshaded while the `getOverheads()` operation is activated by the reflexive message `getOverheads()`.

A *return* is a return of control to the object that originated the message that began the activation. This is not a new message, but is only the conclusion of the invocation of an operation. Returns are shown with a dashed arrow, but it is optional to show them at all since it can be assumed that control is returned to the originating object at the end of the activation in a destination object (asynchronous messages—see below—are exceptions). Returns are often omitted, as in Figure 9.4. Figure 9.7 explicitly shows all returns for the same interaction.

A *return-value* is the value that an operation returns to the object that invoked it. These are rarely shown on an analysis sequence diagram, and are discussed further in Chapter 10. For example in Figure 9.7 the operation invoked by the message `getName()` would have return-value of `clientName` and no parameters. In order to show the return-value the message could be shown as

```
clientName := getName()
```

where `clientName` is a variable of type `Name`.

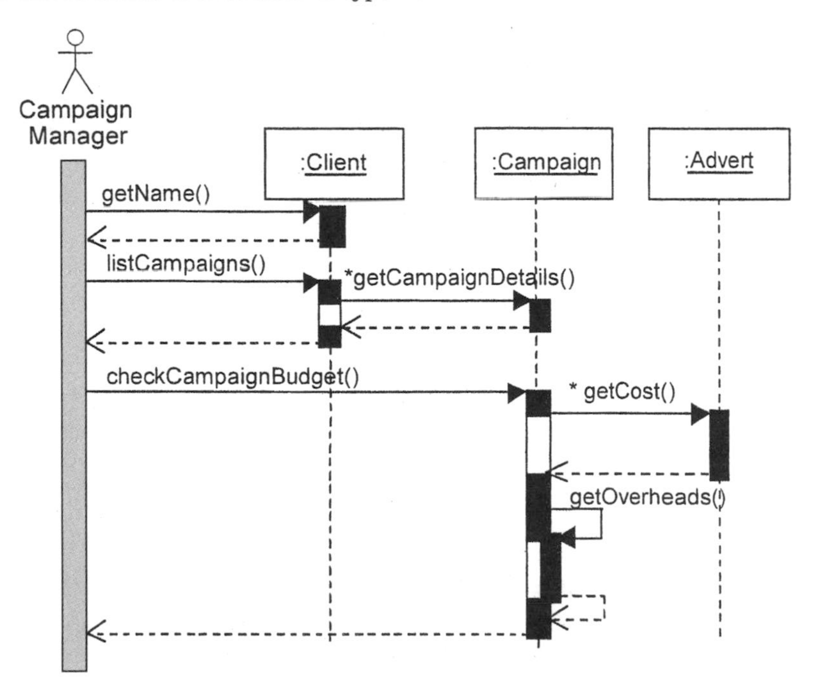

Figure 9.7 Sequence diagram for `Check campaign budget` showing foci of control and explicit returns.

A *synchronous message* or *procedural call* is shown with a full arrowhead (see Figure 9.8), and is one that causes the invoking operation to suspend execution until the focus of control has been returned to it. This is essentially a nested flow of control where the complete nested sequence of operations is completed before the calling operation resumes execution. This may be because the invoking operation requires data to be returned from the destination object before it can proceed. In Figure 9.7 the `Check campaign budget` use case is shown with procedural calls and explicit returns. Procedural calls are appropriate for the interaction since each operation that invokes another does so in order to obtain data and cannot continue until that data is supplied.

An *asynchronous message*, drawn with an open arrowhead as in Figure 9.8, does not cause the invoking operation to halt execution while it awaits a return. When an asynchronous message is sent operations in both objects may carry out processing at the same time. Asynchronous messages are frequently used in real-time systems where operations in different objects must execute concurrently, either for reasons of efficiency, or because the system simulates real-world activities that also take place concurrently. It may be necessary for an operation that has been invoked asynchronously to notify the object that invoked it when it has terminated. This is done by explicitly sending a message (known as a *callback*) to the originating object.

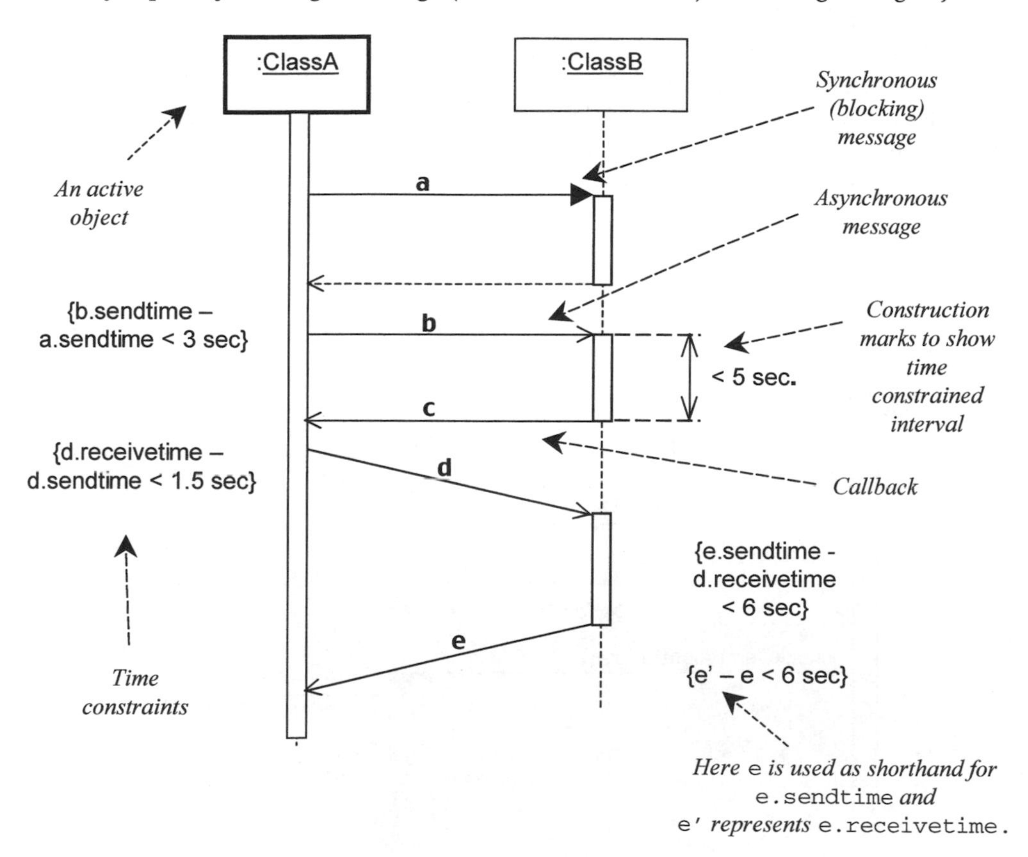

Figure 9.8 Sequence diagram showing different message types and time constraints.

9.3.2 Time constraints

A sequence diagram can be labelled to document time constraints in various ways. Labels may be included with, for example, descriptions of actions or any time constraints that apply to the execution of operations. In Figure 9.8 each of the messages is simply named with `a`, `b`, and so on. Time expressions may be associated with the name of the message so that time constraints can be specified for the execution of an operation or the transmission of a message. The standard functions `sendTime` (the time at which a message is sent by an instance) and `receiveTime` (the time an instance receives a message) give times when applied to message names. Thus `a.sendtime` gives the time that the message `a` is sent. Developers may invent other time functions (e.g. `elapsedTime`, `queuedTime`) if they are needed for special situations. Construction marks may also be used to show a time interval with a constraint. This is illustrated in Figure 9.8 to show the interval between the receipt of message `b` and sending message `c`. Time constraints are frequently used in modelling real-time systems where the application must respond within a certain time, typically for reasons of safety or efficiency. For most other information systems time constraints are not significant and only the sequence of the messages matters.

So far we have only considered message arrows that have been drawn horizontally across the sequence diagram and at right angles to the object lifelines. Drawing a message arrow in this fashion indicates that the time taken to send a message is not significant in comparison to the time taken for operation execution. There is consequently no need to model another activity during the period while a message is in transit. In some applications the length of time taken to send a message is itself significant. For example, in distributed systems messages are sent over a network from an object on one computer to another object on a different computer. If the transit time for a message is significant the message arrow is slanted downwards so that the arrowhead (the arrival of the message) is below (later than) the tail (the origination of the message). The asynchronous messages `d` and `e` shown in Figure 9.8 illustrates this. Where it does not cause confusion the `sendTime` for a message can be represented by the message name (label) only and the `receiveTime` can be represented by the name with the prime symbol (written as `e'`, for example). The time contraint for the transmission of the message `e` is shown using this style of notation.

9.3.3 Branching

The interactions that we have considered so far have only one execution path, although some have iterations during their execution. Some interactions have two or more alternative execution pathways. Each reflects a branch in the possible sequence of events for the use case it represents. The notation for branching is illustrated in Figure 9.9. This shows a sequence diagram for the use case `Add a new advert to a campaign if within budget` (an extension of the use case `Add a new advert to a campaign`). The use case description is as follows.

A new advertisement is added to a campaign by the campaign manager only if the campaign budget is not exceeded by adding the new advert. If adding the advertisement would cause the budget to be exceeded then a campaign budget extension request is generated. This will be recorded for later reference. The budget extension request is printed and sent to the client at the end of the day.

The first part of this sequence diagram is identical to that for `Check campaign budget` but only the `checkCampaignBudget()` message has been shown. Compare

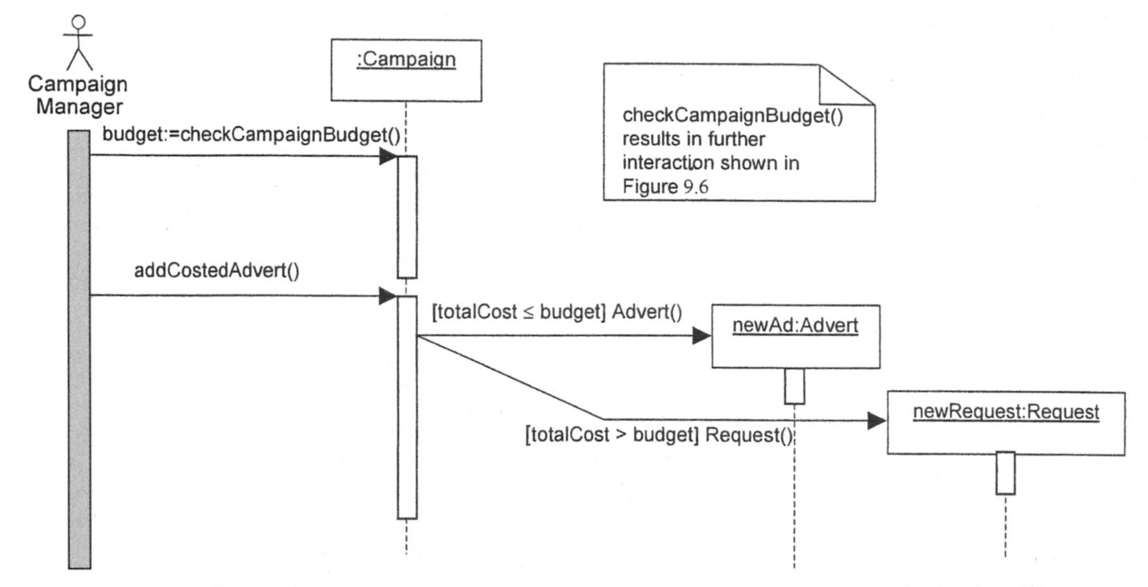

Figure 9.9 Sequence diagram for `Add new advert to a campaign if within budget` showing branching.

this diagram with Figure 9.6 where all the messages that result from the execution of the operation `checkCampaignBudget()` are shown. It is perfectly legitimate to show varying degrees of detail on sequence diagrams but the diagram should be annotated appropriately so that this is explicit. In this example we have annotated the diagram to this effect. This relationship will result in a change to the use case diagram in Figure A2.2, where `Add a new advert to a campaign if within budget` needs to be added and shown with an «`include`» relationship with `Check campaign budget`.

The branching is seen where two messages `Advert()` and `Request()` both start from the same point on the `Campaign` lifeline. Each branch is followed only if the branch condition is true; this is shown in square brackets before the message label. The fact that the `Advert()` message is above the `Request()` message does not imply a time sequence since the two branches are actually alternative execution pathways. Only one branch is followed during any one execution of the use case.

The branching notation can be used at a generic level to create a sequence diagram that represents all possible sequences of interaction for a use case. Such a generic diagram will typically show communication between anonymous objects or roles rather than particular instances. In general looping and branching constructs correspond respectively to iteration and decision points in the use case. When drawn at an instance level a sequence diagram shows a specific interaction between particular objects. The two kinds of sequence diagram (generic and instance level) are equivalent to one another if the interactions implied by the use case contain no looping or branching constructs.

9.3.4 Managing sequence diagrams

On occasions it is necessary to link two or more sequence diagrams together. It may be that a single sequence diagram is too complex and unwieldy to represent an interaction in an easily assimilable fashion. Perhaps the interaction involves too many lifelines to place on a single diagram or perhaps there is a sub-sequence that is common to several interactions. Another possibility is that part of the interaction involves complex messaging between members of a group of objects and that this part of the interaction is best shown separately.

One approach to dealing with the problem is to split a complex diagram into two or more smaller diagrams with the connections between the diagrams indicated by message arrows that are left in mid-air and do not end at a lifeline. For example the sequence diagram in Figure 9.4 is redrawn as two sequence diagrams in Figure 9.10 and Figure 9.11, each of which shows fewer lifelines that the original. This approach relies on clear annotation in each diagram to show how it is related to other diagrams.

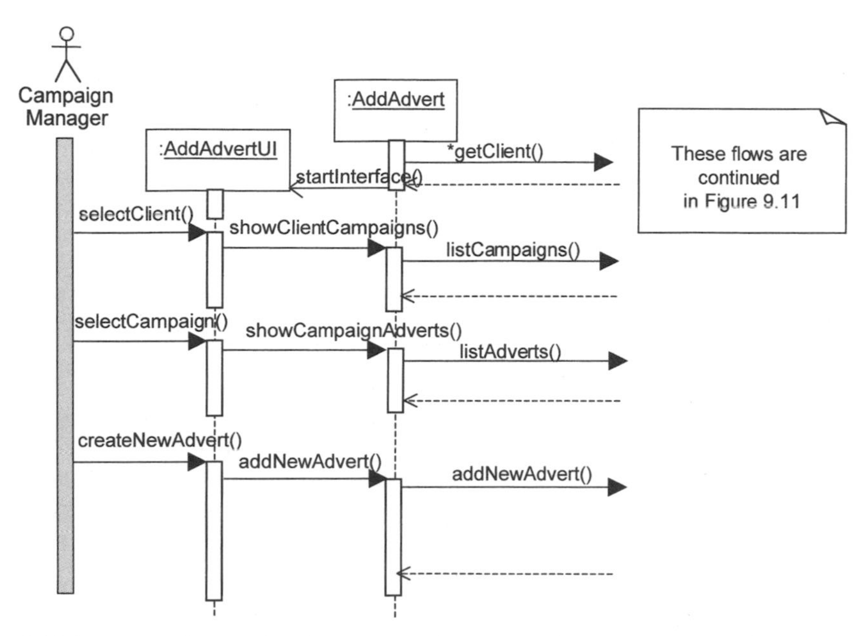

Figure 9.10 First part of interaction for use case `Add a new advert to a campaign`.

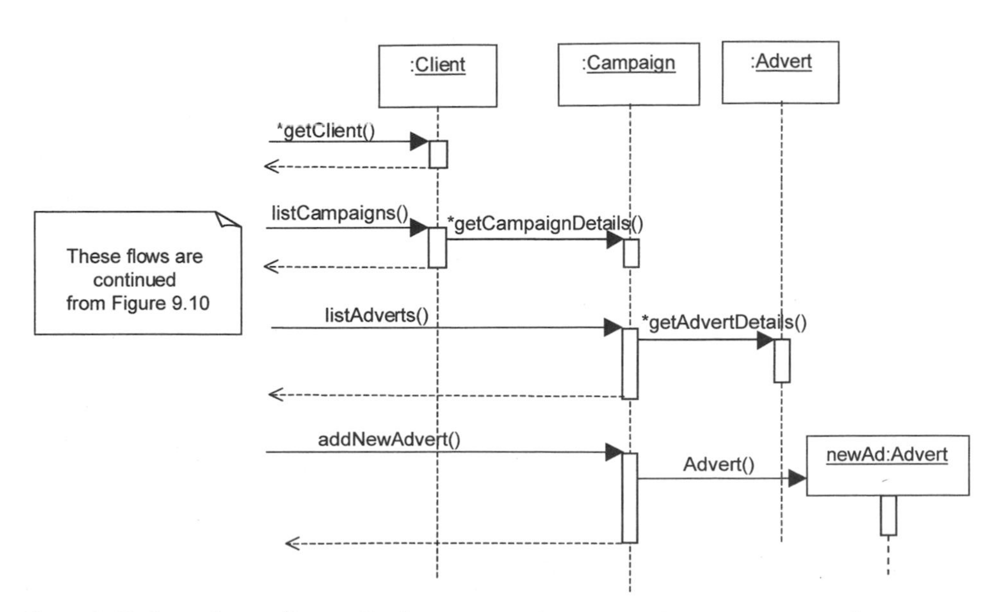

Figure 9.11 Second part of interaction for use case `Add a new advert to a campaign`.

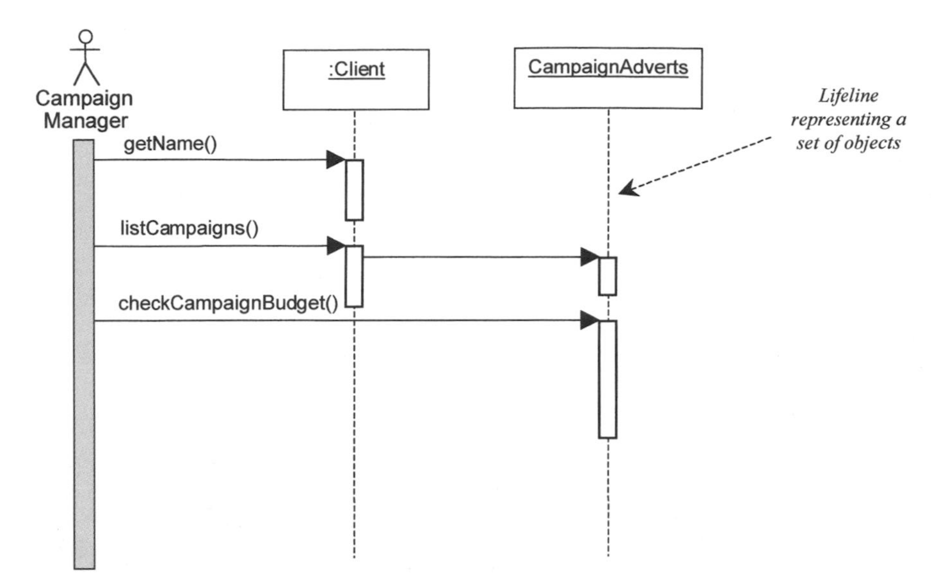

Figure 9.12 Alternative sequence diagram with object grouping for use case `Check campaign budget`.

Another approach is to represent a group of objects by a single lifeline. This is illustrated in Figure 9.12 where the lifeline for the object group `CampaignAdverts` represents the `Campaign` and `Advert` objects (see Figure 9.6 for this interaction without the object grouping). This approach is particularly useful when drawing an interaction at a high level and not showing the detailed interaction within a group of objects. Note that it is implicit in this notation that received messages are dealt with by interaction between members of the group or by a single object in the group. However, in Figure 9.9 the notation itself did not indicate that `checkCampaignBudget()` resulted in further interaction, this is only indicated by the accompanying note.

9.3.5 Modelling real-time systems and concurrency

Real-time systems are broadly characterized by the need to respond to external events within tight time constraints. Partly for this reason, they frequently exhibit concurrent behaviour in the form of simultaneous execution pathways or *threads of control*. An application that has concurrent execution always includes some objects that co-ordinate and initiate threads of control; these are known as *active objects* and instances of active classes. In addition, real-time applications usually include many other objects that work only within a thread of control; these are known as *passive objects* and belong to passive classes. Active objects or classes are shown with a heavy border on interaction diagrams and are frequently composites with embedded parts (the interface or boundary class as a composite is discussed in Chapter 17).

It is important for a sequence diagram of a concurrent system to show clearly which threads of control are active at any time. Control is represented by the synchronous and asynchronous messages introduced earlier in this section. In this usage, a full (synchronous) arrowhead represents the yielding of a thread of control and an open or stick (asynchronous) arrowhead represents a message that is sent without control being yielded.

9.4 Collaboration Diagrams

Collaboration diagrams are the second kind of interaction diagram in the UML notation set. They have already been introduced in Chapter 7 where they were used to represent the collaboration that realizes a use case. We examine the notation for collaboration diagrams in more detail here.

9.4.1 Basic concepts and notation

Collaboration diagrams have many similarities to sequence diagrams. They express the same information in a different format, and, like sequence diagrams, they can be drawn at various levels of detail and during different stages in the system development process. Due to their similar content, collaboration diagrams can be used for the automatic generation of sequence diagrams and vice versa. The most significant difference between the two types of interaction diagram is that a collaboration diagram explicitly shows the links between the objects that participate in a collaboration[8]. Unlike sequence diagrams, there is no explicit time dimension.

In a collaboration diagram the interaction is drawn on what is essentially a fragment of a class or object diagram, as can be seen in Figure 9.13. This example is drawn at quite a simple level of detail (but note that it includes a boundary object `:AddAdvertUI` and the control object `:AddAdvert`). This level of detail is often sufficient to capture the nature of a collaboration. Since the diagram has no time dimension the order in which messages are sent is represented by sequence numbers. In this diagram the sequence numbers are written in a nested style (for example, 3.1 and 3.1.1) to indicate the nesting of control within the interaction that is being modelled. Thus the operation `showCampaignAdverts()` passes control to the operation `listAdverts()`, which has one deeper level of nesting. A similar style of numbering is used to indicate branching constructs.

Typically there is more than one possible interaction for a particular use case and each of the alternative interactions will have different strengths and weaknesses. The alternatives arise because of the different possible allocations of responsibility. For example, although feasible, the interaction in Figure 9.13 may have some undesirable features. The message `getCampaignDetails` from `Client` to `Campaign` requires the Client object to return these details to the `AddAdvert` object. If the campaign details only include the campaign names then a relatively small amount of data is being passed from `Campaign` to `Client` and then onto `AddAdvert`. This may be acceptable. On the other hand, if the campaign details also include the start and finish dates for each campaign and the campaign budget, then much more data is being passed through `Client`. In these circumstances the `Client` object is now responsible for providing significant amounts of data about the campaigns, and arguably this should be the responsibility of the `Campaign` objects themselves. Instead data about campaigns can be passed directly from `Campaign` to `AddAdvert`. This is shown in Figure 9.14 where `AddAdvert` takes responsibility for getting the campaign details directly from the `Campaign` objects. In this interaction the `Client` object is only responsible for providing `AddAdvert` with a list of its campaigns[9]. This is an appropriate responsibility for `Client`.

[8] More generally it shows the *association roles* (normally links) that exist between the *classifier roles* (normally instances).

[9] This will be a list of object identifiers that is then used by the `AddAdvert` to navigate to each `Campaign` object in turn.

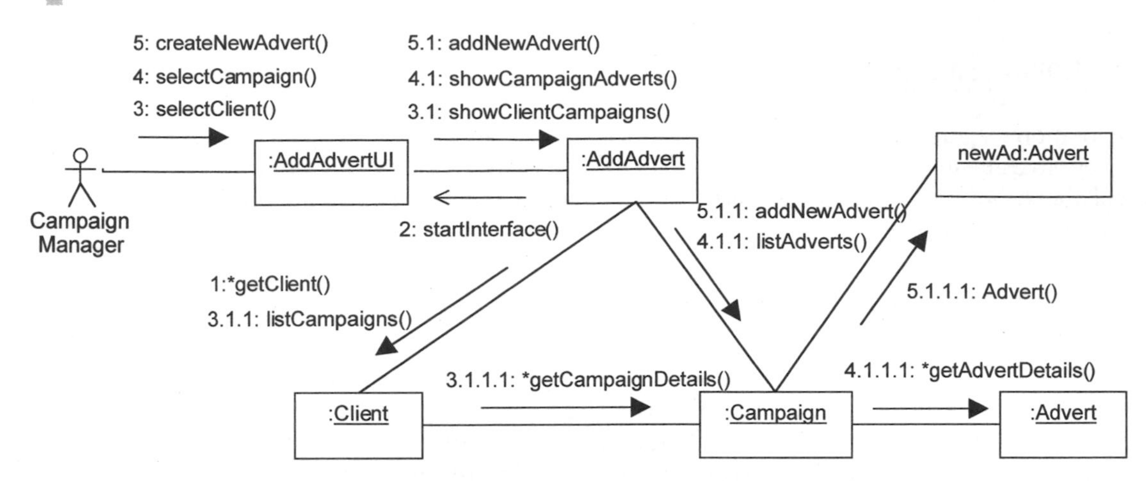

Figure 9.13 Collaboration diagram for the use case `Add a new advert to a campaign`.

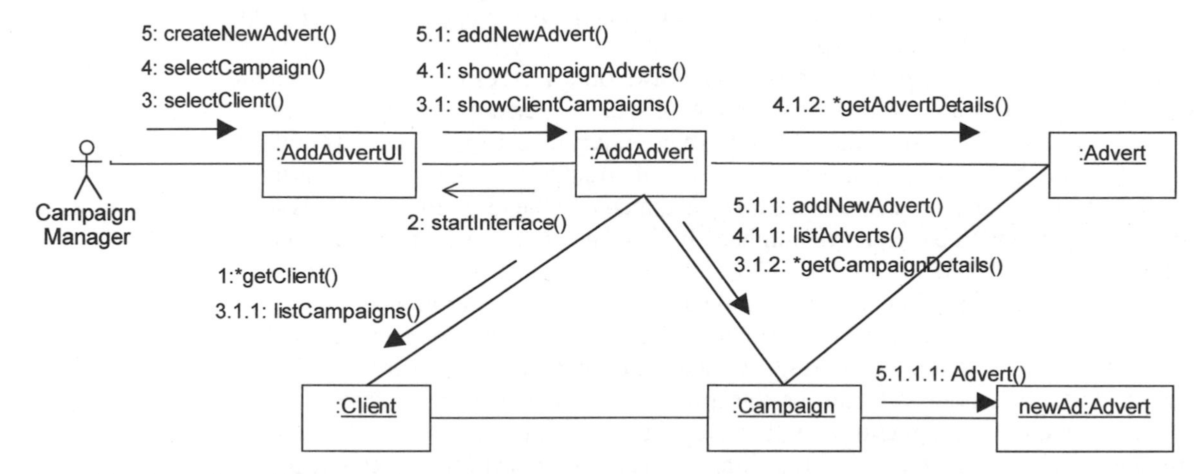

Figure 9.14 Alternative collaboration diagram for the use case `Add a new advert to a campaign`.

9.4.2 Message labels in collaboration diagrams

Messages on a collaboration diagram are represented by a set of symbols that are the same as those used in a sequence diagram, but with some additional elements to show sequencing and recurrence as these cannot be inferred from the structure of the diagram. Each message label includes the message signature and also a sequence number that reflects call nesting, iteration, branching, concurrency and synchronization within the interaction. The formal message label syntax is as follows:

```
[predecessor] [guard-condition] sequence-expression [return-value
':='] message-name'('[argument-list]')'
```

(Characters in single quotes such as `':='` are literal values. The other terms will be replaced in actual messages by relevant values. Terms in square brackets `[...]` are optional. Terms in braces `{...}` are repeated zero or more times.)

A *predecessor* is a list of sequence numbers of the messages that must occur before the current message can be enabled. This permits the detailed specification of branching pathways. The message with the immediately preceding sequence number is assumed to be the predecessor by default, so if an interaction has no alternative pathways the predecessor list may be omitted without any ambiguity. The syntax for a predecessor is as follows:

```
sequence-number { ',' sequence-number } '/'
```

The `'/'` at the end of this expression indicates the end of the list and is only included when an explicit predecessor is shown.

Guard conditions are written in Object Constraint Language (OCL) (see Chapter 10), and are only shown where the enabling of a message is subject to the defined condition. A guard condition may be used to represent the synchronization of different threads of control.

A *sequence-expression*[10] is a list of integers separated by dots `('.')` optionally followed by a *name* (a single letter), optionally followed by a *recurrence* term and terminated by a colon. A sequence-expression has the following syntax:

```
integer { '.' integer }[name] [recurrence] ':'
```

In this expression *integer* represents the sequential order of the message. This may be nested within a loop or a branch construct, so that, for example, message 5.1.3 occurs after message 5.1.2 and both are contained within the activation of message 5.1. In Figure 9.11 messages 4.1.1 and 4.1.2 are nested within the activation of message 4.1. The *name* of a sequence-expression is used to differentiate two concurrent messages since these are given the same sequence number. For example, messages 3.2.1a and 3.2.1b are concurrent within the activation of message 3.2. Recurrence reflects either iterative or conditional execution and its syntax is as follows:

Branching: `'['condition-clause']'`
Iteration: `'*''['iteration-clause']'`

Some further sample message labels are listed in Figure 9.15.

Figure 9.16 illustrates the use of a collaboration diagram to show the interaction for a single operation, in this case `checkCampaignBudget()`, which is one of the operations shown in the sequence diagrams in Figures 9.6 and 9.7.

Collaboration diagrams are preferred to sequence diagrams by some developers as they offer a view of object interaction that is easy to relate to a class diagram, due to the

Type of message	Syntax example
Simple message	`4: addNewAdvert()`
Nested call with return value *The return value is placed in the variable* `name`	`3.1.2: name:= getName()`
Conditional message *This message is only sent if the condition* `[balance > 0]` *is true*	`[balance > 0] 5: debit(amount)`
Synchronization with other threads *Message* `4: playVideo()` *is invoked only once the two concurrent messages* `3.1a and 3.1b` *are completed*	`3.1a, 3.1b/ 4: playVideo()`

Figure 9.15 Examples of the syntax for various types of message label.

[10] Sequence-expressions are commonly omitted from interaction sequence diagrams as the sequence is normally implied by the relative position of the messages one after the other. If because of branching the sequence is ambiguous, sequence-expressions should be included.

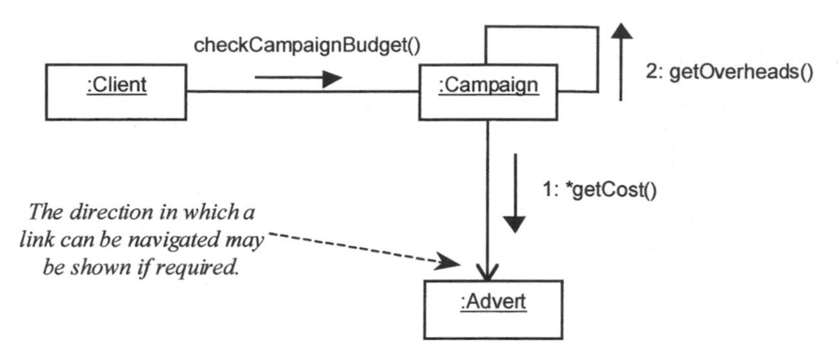

Figure 9.16 Collaboration diagram for the operation `checkCampaignBudget()`.

visibility of object links. Generally, collaboration diagrams are probably more useful during analysis activities while sequence diagrams are better at representing design detail. Collaboration diagrams are used to describe analysis use case realizations as typically the messages are not fully specified at this stage. However, when there are many messages between two objects in one interaction then a collaboration diagram is more difficult to read than the equivalent sequence diagram. In particular sequence diagrams also offer more explicit visual cues to the duration of each activation and show detailed design interactions more clearly. We believe that it is seldom useful to draw both kinds of interaction diagram for the same purpose as they are logically equivalent. Some developers use sequence diagrams for the generic model of the interaction and collaboration diagrams for specific scenarios. The choice of which is the more appropriate usually depends on the nature of the interaction and the purpose of the diagram. Often neither offers a clear advantage and it should be a matter of organizational policy as to which is normally prepared.

9.5 Model Consistency

The preparation of interaction diagrams (either sequence or collaboration diagrams) involves the allocation of operations to classes. These operations should be listed against the correct classes in the class diagram and if operation signatures have been specified in full these must be consistent. The interaction diagrams and the class diagrams should be mutually consistent. A good CASE tool enforces this consistency at a syntactic level, usually by prompting the developer with a list of operations currently allocated to a class when he or she adds a message that is sent to an object of that class. If a corresponding operation has not been defined in the destination class the appropriate operation is added automatically to the class definition.

But to ensure full consistency between a class diagram and a set of related interaction diagrams requires more than the simple syntactic check described above. Every sending object in an interaction diagram must have the ability to send its message to the destination object and this requires it to know the identity or *object reference* of the destination object. There are only two ways that a sending object can know the reference of a destination object. It may be already known by the sending object via a direct link, which really means that an association exists between the respective classes to which the objects belong. Alternatively, a sending object may obtain the reference it needs indirectly from another object (usually of a different class) that has a link with the destination object. The representation and placement of the object references that represent associations is a design issue that we discuss in detail in Chapter 14. At this stage it is sufficient to ensure that there is some possible pathway via object links (deduced from associations on the class diagram) that connects a sending object to a destination object. Any inconsistency between an interaction diagram and the corresponding class diagram indicates a flaw in one or the other. There may be a necessary association missing from the class diagram, for instance. Note that the existence of an association does not by itself guarantee the existence of any particular link. Where the minimum multiplicity of an association is zero there may not be any object that has a link for that association. If the multiplicity is one (or more), then each object must have at least one link. All message pathways should be analysed carefully.

When both sequence and collaboration diagrams are prepared for the same interaction the threads of execution specified in each must be exactly the same. Statechart diagrams (described in Chapter 11) document information about messages from the perspective of an object rather than an interaction, and it is also important to check for consistency between the statechart diagram for a class and all interaction diagrams that involve objects of that class.

9.6 Summary

Object interaction is a crucial feature of the object-oriented approach to systems development. When discussing object interaction we use the metaphor of message passing to describe the mode of collaboration between objects. Developing interaction diagrams requires a careful analysis of the use cases and may involve the use of CRC cards (see Chapter 7). UML uses two equivalent modelling techniques to describe interactions—sequence diagrams and collaboration diagrams—that provide a rich notation. Collaboration diagrams are more commonly used when analysing use cases to prepare analysis use case realizations. Sequence diagrams may be used effectively to represent detailed design specifications of interactions. When working with complex interactions it may be necessary to use several linked diagrams. UML provides various notational alternatives for this. It is common to have more than one possible interaction for a use case and it is a pragmatic judgement as to which is most appropriate. The design considerations that inform such judgements are discussed in detail in Chapter 14. An integral part of the process of developing interaction diagrams is ensuring that they and the class diagrams for an application are mutually consistent.

Review Questions

9.1 List two specific features of bad object-oriented modelling that are discouraged by the use of collaboration diagrams.

9.2 What are the benefits of keeping all classes reasonably small and self-contained?

9.3 What are the main differences between sequence diagrams and collaboration diagrams?

9.4 What are the essential parts of a message label (i) in a sequence diagram and (ii) in a collaboration diagram?

9.5 What is an object lifeline?

9.6 What is meant by the focus of control?

9.7 How do asynchronous messages differ from synchronous messages (i) in their behaviour and (ii) in their notation?

9.8 In what circumstances are sequence numbers in a collaboration diagram written in nested style (e.g. 3.2.1)?

9.9 What consistency checks should be applied to interaction diagrams?

9.10 Describe three ways in which complex interactions may be represented using UML.

Case Study Work, Exercises and Projects

Exercises 9.A-9.C are based on the use cases listed in Exercise 7.A and the use case realizations developed in Exercise 7.B.

9.A For each of the use cases prepare a sequence diagram.

9.B For the use case `Start line run` identify an alternative interaction and prepare a sequence diagram for this interaction.

9.C Critically compare the two interactions that you have identified for the use case `Start line run` and with suitable justification determine which is the most appropriate.

9.D Using a CASE tool with which you are familiar, enter several use case realizations including at least one collaboration and one sequence diagram (e.g. the FoodCo models).

9.E Critically evaluate the extent to which the CASE tool supports UML and the consistency check that is necessary between the different diagrams.

Further Reading

- Larman's (1998) use of sequence diagrams during analysis is both interesting and very different from the approach advocated in this book. Rumbaugh et al. (1991) and Booch (1994) discuss their variants of message modelling in detail. Buschmann et al. (1996) provides interesting examples of system sequence diagrams using a notation from which the UML notation for sequence diagrams has developed.

CHAPTER 10

Specifying Operations

OBJECTIVES

In this chapter you will learn

- why operations need to be specified
- the difference between algorithmic and non-algorithmic methods
- how to interpret different ways of specifying operations
- how to specify operations using one method.

10.1 Introduction

Operation specifications play a similar role in the project repository to that of other entries, such as attribute specifications. They support a graphical model by adding precision so that users can confirm the correctness of the model, and designers can use them as a basis for software development. But they are potentially the most complex of all entries in the repository, since they explain the detailed behaviour of the system.

We consider the need for specifying operations (Section 10.2), and introduce the 'contract' (Section 10.3) as a kind of black box specification. If the behaviour of an operation is simple, a contract that describes only its external interface may be all that is required, and if its behaviour is not yet understood in any detail, a black box specification may be all that is possible. Often there is also a need to describe an operation's logic, or internal behaviour.

The two general ways of doing this are respectively called 'algorithmic' (or 'procedural') and 'non-algorithmic' (or 'declarative'). A non-algorithmic approach is generally preferred in object-oriented development, but in some situations only an algorithmic approach is sufficiently expressive. We review some established techniques, in particular decision tables and pre- and post-condition pairs (Section 10.4.1) and Structured English (Section 10.4.2). None of these are specific either to UML, or to an object-oriented approach, but all can be used to specify operations in a UML model.

A full description is beyond the scope of this book, so we introduce them at an overview level.

UML does not require any specific techniques or notations for specifying operations, but activity diagrams (first introduced in Chapter 5) can be used to express the logic of an operation in a graphical form (Section 10.4.2). The UML has also a formal language known as the Object Constraint Language (OCL), which is intended mainly for specifying general constraints on a model. We introduce OCL (Section 10.5) and show how it can be used in specifying operations (Section 10.6).

10.2 The Role of Operation Specifications

Each operation specification is a small but necessary step on a path that begins with a user's idea of a business activity, and leads ultimately to a software system made up of collaborating objects with attributes and methods. From an analysis perspective, an operation specification is created at a point when the analyst's understanding of some aspect of an application domain can be fed back to users, ensuring that the proposals meet users' needs. From a design perspective, an operation specification is a framework for a more detailed design specification that later guides a programmer to a method that is an appropriate implementation of the operation in code. An operation specification can also be used to verify that the method does indeed meet its specification, which in turn describes what the users intended, thus checking that the requirements have been implemented.

Novice programmers often do not appreciate the need to design, still less specify, an operation before beginning to write it in program code. This is partly because beginners are given such simple tasks, e.g. to write a program that can calculate and display the area of a rectangle. More importantly, the student is shielded from the activity of requirements analysis. In effect, the teacher has already carried this out, and the student is presented with its results as a starting point: 'There is a need for a program to calculate the area of a rectangle'. Why? An answer given to a student will be put in educational terms, such as: 'This will help you to develop important basic skills in . . .'

Of course, the situation just described is quite artificial, and most students know this perfectly well. But it is only once the complexity or scale of a software system reaches a certain threshold that the production of code too early becomes extremely inefficient, and very possibly disastrous. To code a relatively small sub-task in a large system requires some understanding of the ways in which that sub-task will interact with other sub-tasks. If this understanding has not yet been achieved, assumptions must be made, and these may later turn out to be inappropriate, even disastrous, for the system as a whole.

Object-oriented programming is generally more immune to this kind of problem than other programming approaches, but it is still important to describe the logical operation of the planned software as early as possible. Modelling object interaction (see Chapter 9) is part of this description process, as it helps to determine the distribution of behaviour among the various classes. A detailed description of individual operations must also now be provided.

There are differences of opinion on how much specification should be done. On the one hand, Rumbaugh et al. (1991) suggest that only operations that are 'computationally interesting' or 'non-trivial' need be specified. 'Trivial' operations (e.g. those that create or destroy an instance, and those that get or set the value of an attribute) need not be specified at all. Further, operation specifications are kept simple in form, and consist only of the operation signature and a description of its 'transformation' (i.e. its logic). On the other hand, Allen and Frost (1998) recommend the specification of all operations, although the level of detail may vary according to the anticipated needs of the designer. We recommend the latter approach, mainly because of the problems that can arise later in a project if full documentation is not maintained. It is important to keep at least to a minimal documentation standard, even for operations that are very simple.

Each operation has a number of characteristics, which should be specified at the analysis stage. Users must confirm the logic, or rules, of the behaviour. The designer and the programmer responsible for the class will be the main users of the specification, as they need to know what an operation is intended to do: does it perform a calculation, or transform data, or answer a query? Designers and programmers of other parts of the system also need to know about its effects on other classes. For example, if it provides a service to other classes, they need to know its signature. If it calls operations in other classes or updates the values of their attributes, this may establish dependencies that guide how these classes should be packaged during design or implementation.

Defining operations should neither be begun too early, nor left too late. For Allen and Frost (1998), this task should be left until the class diagram has stabilized. In a project where the development activity has been broken down at an early stage to correspond to separate sub-systems, this may refer only to that part of the class diagram which relates to a particular sub-system. But for any given part of the model, it is important to create all operation specifications before moving into the object design activity.

10.3 Contracts

The term 'contract' is a deliberate echo of legal or commercial contracts between people or organizations. Signing (or becoming a party to) a contract involves making a commitment to deliver a defined service to an agreed quality standard. For example, a small ground-care company has a contract to mow the grass on the lawn in front of the Agate headquarters building. The contract stipulates how often the grass must be cut (every two weeks from April to October), the maximum height of the grass immediately after it is cut (no more than 3 centimetres) and how much Agate will pay for the service (£80 per cut). The contract does *not* spell out how the work will be done—for example, what type of mower should be used (electric or petrol, cylinder or rotary), how many staff or mowers should be involved, or in which direction the lawn should be cut.

In the language of system theory, a contract is an interface between two systems. In this example, Agate is a business system and the ground-care company is a system for mowing Agate's grass. The contract defines inputs and outputs, and treats the grass-mowing system to some extent as a black box, with its irrelevant details hidden. Which details are deemed irrelevant is always a matter of choice, and any contract can specify that some details of the implementation should be visible to other systems. For example, Agate's directors might not wish to permit the ground-care contractor to use toxic pesticides or weedkillers. This can be included as a constraint in the contract.

Meyer (1988, 1991) was one of the first to draw an analogy between commercial contracts and service relationships between objects. The use of the term is now widespread in object-oriented development since it stresses the encapsulation of classes and sub-systems in a model. Cook and Daniels (1994) use the concept extensively in the Syntropy™ methodology, and Allen and Frost (1998) apply it also in the SELECT Perspective methodology.

One of Meyer's principal arguments for using the analogy of a contract is that design-by-contract helps to achieve a software design that is correct in terms of its requirements. During requirements analysis, we do not yet need the full technical rigour that is required of a design specification, but there is still a clear advantage in adopting an approach that can later be extended seamlessly through design into implementation. Specification by contract means that operations are defined primarily in terms of the services they deliver, and the 'payment' they receive (usually just the operation signature).

Contracts can also be applied at a much higher level of abstraction than individual operations. Larman (1998) describes the use of contracts to define services provided by a system as a whole. Whether written for a single operation, for the behaviour of the system as a whole, or for some intermediate packaged component, the structure of a contract is very similar. A commercial contract usually identifies the parties, the scope (i.e. the context in which it applies), the agreed service, and any performance standards that apply. In just the same way, in object-oriented modelling we identify the nature of the service provided by the server object, and what must be provided by the client object in order to obtain the service. These various aspects can be summarized as follows.

- The intent or purpose of the operation.
- The operation signature including the return type (probably established during interaction modelling).
- An appropriate description of the logic (the following sections present some alternative ways of describing the logic of an operation).
- Other operations called, whether in the same object or in other objects.
- Events transmitted to other objects.
- Attributes set during the operation's execution.
- The response to exceptions (e.g. what should happen if a parameter is invalid).
- Any non-functional requirements that apply[1].

Most of this is self-explanatory, but the critical part of an operation specification is the logic description, and it is to this that we turn in the next section.

10.4 Describing Operation Logic

Rumbaugh et al. (1991) suggest an informal classification of operations that is a useful starting point in considering the various ways of describing their logic. First, there are operations that have side-effects. Possible side-effects include the creation or destruction of object instances, setting or returning attribute values, forming or breaking links with other objects, carrying out calculations, sending messages or events to other objects or any combination of these. A complex operation may do several of these things, and, where the task is at all complex, an operation may also require the collab-

[1] This list of features is adapted from Larman (1998) and Allen and Frost (1998).

oration of several other objects. It is partly for this reason that we identify the pattern of object collaboration before specifying operations in detail. Second, there are operations that do not have side-effects. These are pure queries; they request data but do not change anything within the system.

Like classes, operations may also have the property of being either `{abstract}` or `{concrete}` (although this decision is often the result of design considerations, and is therefore not always made when an operation is first specified). Abstract operations have a form that consists at least of a signature, sometimes a full specification, but they will not be given an implementation (i.e. they will not have a method). Typically, abstract operations are located in the abstract superclasses of an inheritance hierarchy. They are always overridden by concrete methods in concrete subclasses.

A specification may be restricted to defining only external and visible effects of an operation, and we may choose either an algorithmic or a non-algorithmic technique for this. A specification may also define internal details, but this is effectively a design activity.

10.4.1 Non-algorithmic approaches

A non-algorithmic approach concentrates on describing the logic of an operation as a black box. In an object-oriented system this is generally preferred for two reasons. First, classes are usually well-encapsulated, and thus only the designers and programmers responsible for a particular class need concern themselves with internal implementation details. Collaboration between different parts of the system is based on public interfaces between classes and sub-systems implemented as operation signatures (or message protocols). As long as the signatures are not changed, a change in the implementation of a class, including the way its operations work, has no effect on other parts of the system[2]. Second, the relatively even distribution of effort among the classes of an object-oriented system generally results in operations that are small and single-minded. Since the processing carried out by any one operation is simple, it does not require a complex specification.

Even in non object-oriented approaches, a declarative approach has long been recognized as particularly useful where, for example, a structured decision is made, and the conditions that determine the outcome are readily identified, but the actual sequence of steps in reaching the decision is unimportant. For situations like this, structured methods made use of non-algorithmic techniques such as decision tables and pre- and post-condition pairs (described in the following sections).

Decision tables

A decision table is a matrix that shows the *conditions* under which a decision is made, the *actions* that may result and how the two are related. They cater best for situations where there are multiple outcomes, or actions, each depending on a particular combination of input conditions. One common form shows conditions in the form of questions that can be answered with a simple yes or no. Actions are listed, and checkmarks are used to show how they correspond to the conditions. The following is an example of a possible application in the Agate case study. Figure 10.1 shows a corresponding decision table.

[2] In practice, the situation is more complex than this might suggest. Some object-oriented languages, e.g. C++, allow a lot of flexibility as to how much of the internal implementation of a class is visible to other classes.

Conditions and actions	Rule 1	Rule 2	Rule 3
Conditions			
Is budget likely to be overspent?	N	Y	Y
Is overspend likely to exceed 2%	–	N	Y
Actions			
No action	X		
Send letter		X	X
Set up meeting		X	

Figure 10.1 A decision table with two conditions and three actions, yielding three distinct rules.

When a campaign budget is overspent, this normally requires prior approval from the client, otherwise Agate is unlikely to be able to recover the excess costs. A set of rules has been established to guide Campaign Managers when they identify a possible problem. If the budget is expected to be exceeded by up to 2%, a letter is sent notifying the client of this. If the budget is expected to be exceeded by more than 2%, a letter is sent and the staff contact also telephones the client to invite a representative to a budget review meeting. If the campaign is not thought likely to exceed its budget, no action is taken.

The vertical columns with Y, N and X entries are known as *rules*. Each rule is read vertically downwards, and the arrangement of Ys and Ns indicates which conditions are true for that rule. An X indicates that an action should occur when the corresponding condition is true (i.e. has a Y answer). We can paraphrase the table into text as follows.

Rule 1. If the budget is not overspent (clearly in this case the scale of overspend is irrelevant, indicated by a dash against this condition), no action is required.

Rule 2. If the budget is overspent and the overspend is not likely to exceed 2%, a letter should be sent.

Rule 3. If the budget is overspent and the overspend is likely to exceed 2%, a letter should be sent and a meeting set up.

A single rule may have multiple outcomes that overlap with the outcomes of other rules. Decision tables are very useful for situations that require a non-algorithmic specification of logic, reflecting a range of alternative behaviours. But this is relatively unusual in an object-oriented system, where thorough analysis of object collaboration tends to minimize the complexity of single operations.

Pre- and post-conditions

As its name suggests, this technique concentrates on providing answers to the following questions.

- What conditions must be satisfied before an operation can take place?
- What are the conditions that can apply (i.e. what states may the system be in) after an operation is completed?

Let us consider an example from Agate. The operation `Advert.getCost()`[3] was first discussed in Section 8.2. Let us suppose that it has the following signature.

```
Advert.getCost():Money
```

[3] We have suffixed brackets to the names of operations to distinguish them from attributes. This does not necessarily mean that they are operations with no parameters, only that the parameters are not shown.

This operation has no pre-condition. (We may note that the object sending the message must know the identity of the object that contains the operation, but this is not in itself a pre-condition for the operation to execute correctly when invoked). The post-conditions should express the valid results of the operation upon completion. In this case, a money value is returned (for simplicity, we ignore the question of valid values for an advert cost, but we should note that in reality this attribute may be able to take only a limited range of values, depending on business constraints).

Pre-condition: none
Post-condition: a valid money value is returned

More complex examples can easily by constructed from the use case descriptions, or by consulting users if existing descriptions are not sufficiently detailed. Consider the use case `Assign staff to work on a campaign`. This involves calling the operation `Campaign.assignStaff()` for each member of staff assigned. Let us assume that the signature of this operation is as follows:

```
Campaign.assignStaff(creativeStaff)
```

This example has one pre-condition: a calling message must supply a valid `creative Staff object`. There is one post-condition: a link must be created between the two objects.

Pre-condition: `creativeStaffObject` is valid
Post-condition: a link is created between `campaignObject` and `creativeStaffObject`

Let us look at one more example from Agate, with more complex conditions. This is taken from the use case `Change the grade for a member of staff` (we assume that the use case is being invoked for a member of creative staff). This use case involves several operations including:

`CreativeStaff.changeGrade()`
`StaffGrade.setFinishDate()`
`StaffGrade()` *(the constructor operation that creates a new instance of this class)*

We examine only one of these in detail, `CreativeStaff.changeGrade()`, but our specification must still recognize calls made to other operations during execution. Let us assume that the operation signature is as follows:

```
CreativeStaff.changeGrade(gradeObject, gradeChangeDate)
```

The pre-conditions are straightforward, consisting only of a valid `GradeObject` and `gradeChangeDate`. The post-conditions are more involved, as once the operation is completed we should expect several effects to have taken place. A new instance of `StaffGrade` is created, and this is linked to the appropriate `creativeStaffObject` and `gradeObject` (by a `StaffOnGrade` link). The new `staffGradeObject` is also linked to the previous `staffGradeObject` (by a `PreviousGrade` link). Attribute values in the new `staffGradeObject` are set by its constructor operation (including `gradeStartDate`, which is set equal to the supplied parameter `gradeChangeDate`). The attribute `StaffGrade.gradeFinishDate` in the previous instance is also set, through a message to invoke the operation `StaffGrade.setFinishDate()`. A full logic description is thus as follows:

pre-conditions: `creativeStaffObject` is valid
`gradeObject` is valid
`gradeChangeDate` is a valid date

post-conditions: a new `staffGradeObject` exists
the new `staffGradeObject` is linked to the `creativeStaffObject`
the new `staffGradeObject` is linked to the previous one
the value of the previous `staffGradeObject.gradeFinishDate` is set equal to `gradeChangeDate`

For many operations in an object-oriented model, such a specification would be sufficiently detailed. It must meet the following two tests.

- A user should be able to check that it correctly expresses the business logic.
- A class designer should be able to produce a detailed design of the operation for a programmer to code.

However, while a declarative approach to operation specification usually meets all the needs of object-oriented development, there is still sometimes a case for using an algorithm. One example would be a requirement that involves carrying out a calculation where the sequence of steps is significant, and neither a designer nor a programmer could reasonably be expected to come up with a formula that produces the correct result.

10.4.2 Algorithmic approaches

An *algorithm* describes the internal logic of a process or decision by breaking it down into small steps (the word derives from al-Kwarazhmi, an Arab mathematician of the ninth century). The level of detail to which this is done varies greatly, depending on the information available at the time and on the reason for defining it. An algorithm also specifies the sequence in which the steps are performed. In the field of computing and information systems, algorithms are used either as a *description* of the way in which a programmable task is currently carried out (this is their purpose in operation specification), or as a *prescription* for a program to automate the task. This dual meaning again reflects the differing perspectives of analysis (understanding a problem and determining what must be done to achieve a solution) and design (the creative act of imagining a system to implement a solution). An algorithmic technique is almost always used during method design, because a designer is concerned with the efficient implementation of requirements, and must therefore select the best algorithm available for the purpose. But algorithms can also be used with an analysis intention. A major difference here is that there is no need for the analyst to worry about efficiency, since the algorithm need only illustrate accurately the results of the operation.

Control structures in algorithms. Algorithms are generally organized procedurally, which is to say that they use the fundamental programming control structures of sequence, selection and iteration. We can illustrate this in the Agate case study by considering the operation that calculates the total cost of a campaign. This operation is invoked during the use case `Check campaign budget`. For ease of reference, the use case description is repeated below.

> The campaign budget may be checked to ensure that it has not been exceeded. The current campaign cost is determined by the total cost of all the adverts and the campaign overhead costs.

Let us suppose that there is a precise (though simple) formula for this calculation, based on summing the individual total costs of each advert, and adding the campaign overhead costs. For further simplicity, let us assume that the overhead cost part of the calculation simply involves multiplying the total of all other costs by an overhead rate (this approximates to normal accounting practice). To convey an understanding of the calculation, we can begin by representing it as a mathematical formula.

```
total_campaign_cost = ( sum of all advert_costs ) * overhead_rate
```

This does not explicitly identify all the steps, but a sequence can be deduced. In fact, several possible sequences can be deduced, but any sequence that always produces a correct result will do. One possible sequence, at a very coarse level of detail would include the following steps:

1. add up all the individual advert costs;
2. multiply the total by the overhead rate;
3. the resulting sum is the total campaign cost.

For such a relatively simple calculation as this one, the formula itself would almost certainly serve better as a specification, but some are a lot more complex. When it is necessary to specify the sequence of calculation in more detail, we can use Structured English for this.

Structured English

This is a 'dialect' of written English that is about halfway between everyday non-technical language and a formal programming language. When it is necessary to specify an operation procedurally, this is the most useful and versatile technique. Its advantages include the possibility, with a little care, of retaining much of the readability and understandability of everyday English. It also allows the construction of a formal logical structure that is easy to translate into program code. Structured English is very easy to write iteratively, at successively greater levels of detail, and it is easily dismantled into components that can be reassembled in different structures without a lot of reworking. The logical structure is made explicit through the use of keywords and indentation, while the vocabulary is kept as close as possible to everyday usage in the business context. Above all, expressions and keywords that are specific to a particular programming language are avoided. The result ideally is something that a non-technical user is able to understand, alter or approve, as necessary, while it should also be useful to the designer. This means it must be capable of further development into a detailed program design without undue difficulty.

The main principles of Structured English are as follows. A specification is made up of a number of simple sentences, each consisting of a simple imperative statement or equation. Statements may only be combined in restricted ways that correspond to the sequence, selection and iteration control structures of structured programming. The very simplest specifications contain only sequences, and differ little from everyday English except in that they use a more restricted vocabulary and style (many organizations have their own Structured English house style). Here are some statements that illustrate a typical style of Structured English:

```
get client contact name
sale cost = item cost * (1 - discount rate )
calculate total bonus
description = new description
```

Selection structures show alternative courses of action, the choice between them depending on conditions that prevail at the time the selection is made. For example, an *if-then-else construct*, which has only two possible outcomes, is shown in the following fragment:

```
if client contact is 'Sushila'
  set discount rate to 5%
else
  set discount rate to 2%
end if
```

If the two alternatives are not really different actions, but are rather a choice between doing something and not doing it, the 'else' branch can be omitted. The following fragment shows this simpler form:

```
if client contact is 'Sushila'
  set discount rate to 5%
end if
```

Note that in each case the end of the structure is marked by `end if`. This important marker cannot be omitted. It allows the entire structure to be treated logically as an element, as if it were a single statement in a sequence.

Multiple outcomes are handled either by a *case* construct or by a *nested-if*. This fragment illustrates the case structure:

```
begin case
  case client contact is 'Sushila'
    set discount rate to 5%
  case client contact is 'Wu'
    set discount rate to 10%
  case client contact is 'Luis'
    set discount rate to 15%
  otherwise
    set discount rate to 2%
end case
```

The 'otherwise' branch of a case construct can be omitted if it is not required, although it is generally good practice to include a catch-all to ensure completeness. The next fragment shows the same selection specified using a nested-if construct.

```
if client contact is 'Sushila'
  set discount rate to 5%
else
  if client contact is 'Wu'
    set discount rate to 10%
  else
    if client contact is 'Luis'
      set discount rate to 15%
    else
      set discount rate to 2%
    end if
  end if
end if
```

This also illustrates how indentation can help the readability of a specification. For each corresponding set of control statements (lines beginning with 'if', 'else' and 'end if'), the indentation from the left margin is the same. This helps to show which sequence statements ('set discount rate to 10%', etc.) belong to each structure.

The third type of control structure is iteration. This is used when a statement, or group of statements, needs to be repeated. Typically this is a way of applying a single operation to a set of objects. Logically, once something has begun to occur repeatedly, there must be a condition for stopping the repetition (unless the repetition is to continue indefinitely). There are two main forms of control of iteration. These differ in whether the condition for ending the repetition is tested before or after the first loop. The next two examples show typical applications of each kind of structure. In the first, the test is applied before the loop is entered, so that if the list is empty no bonus is calculated.

```
do while there are more staff in the list
  calculate staff bonus
  store bonus amount
end do
```

In the second iteration example below, the test is applied after the loop is exited. This ensures that the action will be processed (or attempted) at least once. Note that the line at the end beginning `until` acts as an end-of-structure marker, just like the `end do` above.

```
repeat
  allocate member of staff to campaign
  increment count of allocated staff
until count of allocated staff = 10
```

Complex structures in Structured English. Different types of structure can be nested inside each other, as in the next fragment.

```
do while there are more staff in the list
  calculate bonus for this staff member
  begin case
    case bonus > £250
      add name to 'star of the month' list
    case bonus < £25
      create warning letter
  end case
  store bonus amount
end do
format bonus list
```

The operation mentioned near the beginning of this section (`Check campaign budget`) also illustrates the use of all three control structures, although in this case there is no nesting.

```
do while there are more adverts for campaign
  get next advert
  get cost for this advert
  add to cumulative cost for campaign
end do
set total advert cost = final cumulative cost
set total campaign cost = total advert cost × overhead rate
get campaign budget
if total campaign cost > campaign budget
  generate warning
endif
```

A Structured English specification can be made as complex as it needs to be, and it can also be written in an iterative, top-down manner. For example, an initial version of an algorithm is defined at a high level of granularity. Then, provided the overall structure is sound, more detail is easily added progressively. In refining the level of detail, structures can be nested within each other to any degree of complexity, although in practice it is unlikely that even the most complex operation would need more than two to three levels of nesting at most. It is in any case sensible to limit the complexity. One often-quoted guideline is that a Structured English specification should not be longer than one page of typed A4, or one screen if it is likely to be read within a CASE tool environment—although in practice the acceptable length of a section of text depends on the context.

The style in all the examples given above is based on that of Yourdon (1989), but this should not be taken as necessarily prescriptive. What passes for acceptable style varies widely from one organization to another, and in practice an analyst should follow the house style, whatever that happens to be.

Pseudo-code

Pseudo-code differs from Structured English in that it is closer to the vocabulary and syntax of a specific programming language. There are thus many different dialects of pseudo-code, each corresponding to a particular programming language. They differ from each other in vocabulary, in syntax and in style. Structured English avoids language specificity primarily to avoid reaching conclusions about design questions too early. Sometimes there seems no good reason to hold back, for example, because the final implementation language has been decided early in the project. This can be misleading, as it may be desirable at a later stage to redevelop the system in a different programming language. If the operations have been specified in a language-specific pseudo-code, it would then be necessary to rewrite them.

However language-specific it may be, pseudo-code remains only a skeleton of a program, intended only to illustrate its logical structure without including full design and implementation detail. In other words, it is not so much a fully developed program as an outline that can later be developed into program code. The following pseudo-code for `Check campaign budget` can be compared with the Structured English version above.

```
{
  { while more adverts:
    next advert;
    get advertcost;
    cumcost = cumcost + advertcost;
  endwhile;
  }
  { campaigncost = cumcost × ohrate;
  get campaignbudget;
  case campaigncost >= campaignbudget:
    return warningflag;
  endcase
  }
}
```

Note that, while this pseudo-code resembles C in its syntax, it is not actually written in C. Pseudo-code requires further work to turn it into program code.

Activity diagrams

Activity diagrams can be used to specify the logic of procedurally complex operations. The notation of activity diagrams was introduced in Chapter 5; in this section we illustrate their role in operation specification. When used for this purpose, activity states in the diagram usually represent steps in the logic of the operation. This can be done at any level of abstraction, so that, if appropriate, an initial high level view of the operation can later be decomposed to a lower level of detail. Figure 10.2 shows a simple example.

There is no complex behaviour to be shown at this level of abstraction. However, if we consider the operation logic at a more detailed level (as we might do when moving into the design activity) some selection logic may become apparent. Figure 10.3 illustrates one possibility.

This example contains a single selection, but is otherwise similar in structure to the version in Figure 10.2. Of course, an operation may include iteration as well as selection, and this, too, is straightforward to model using an activity diagram. Figure 10.4 shows the activity diagram for an operation `prepareBonusList()` (for comparison, see the earlier version of this modelled in Structured English). Figure 10.5 models the operation `CreativeStaff.calculateBonus()`, which also requires the use of selection and iteration.

Activity diagrams are inherently very flexible in their use, and therefore a little care should be exercised when they are employed in operation specification. A diagram may be drawn to represent a single operation on an object, but it may just as easily be drawn to represent a collaboration between several objects (for example one that realizes a use case). Figure 10.6 illustrates this for the use case `Check campaign budget` (compare with the corresponding sequence diagram later in Figure 10.8).

Finally, an activity diagram can be drawn for a more abstract collaboration between larger components of a system, or between entire systems. A single diagram does not necessarily translate into a single operation; whether or not it does is essentially a design decision.

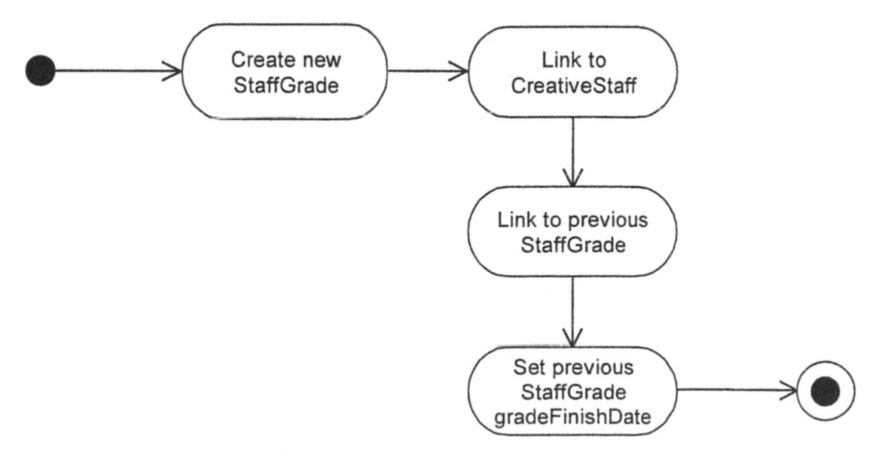

Figure 10.2 An activity diagram showing the main steps for the operation `CreativeStaff.changeGrade()`.

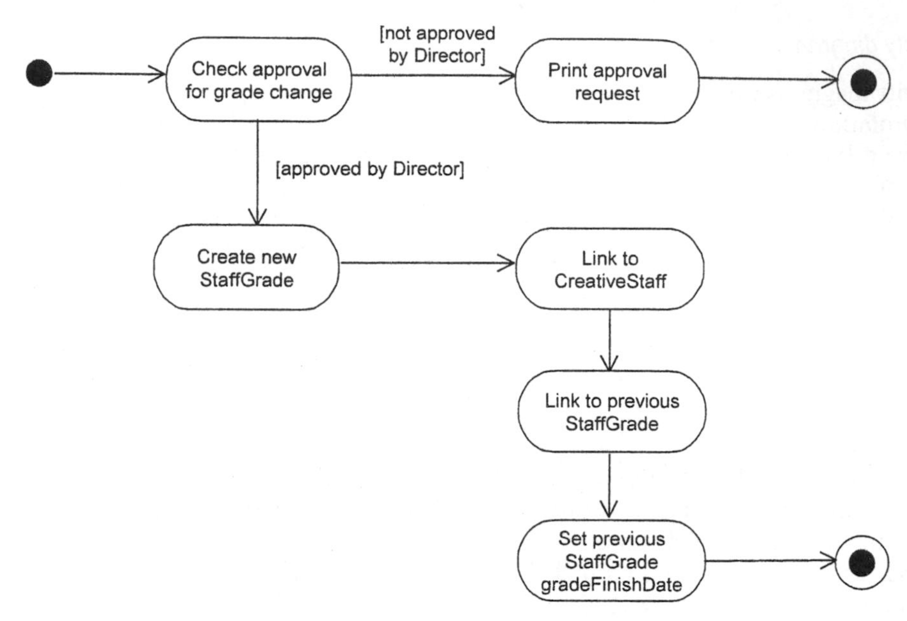

Figure 10.3 A more complex activity diagram for `CreativeStaff.changeGrade()` with an initial selection to check that approval has been given.

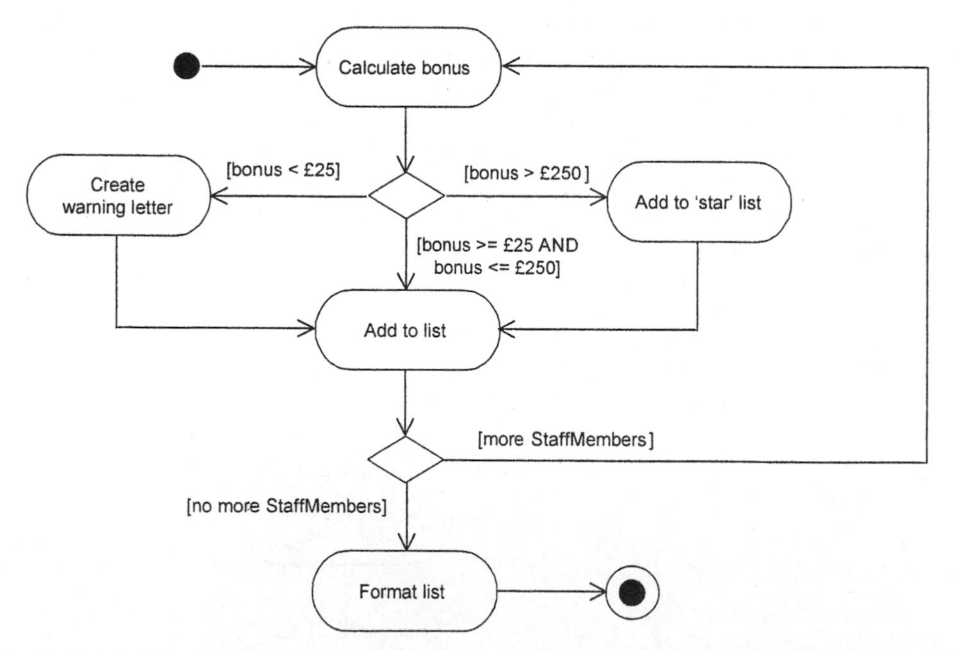

Figure 10.4 Activity diagram for `prepareBonusList()` showing selection and iteration structures. Compare with the Structured English example given earlier.

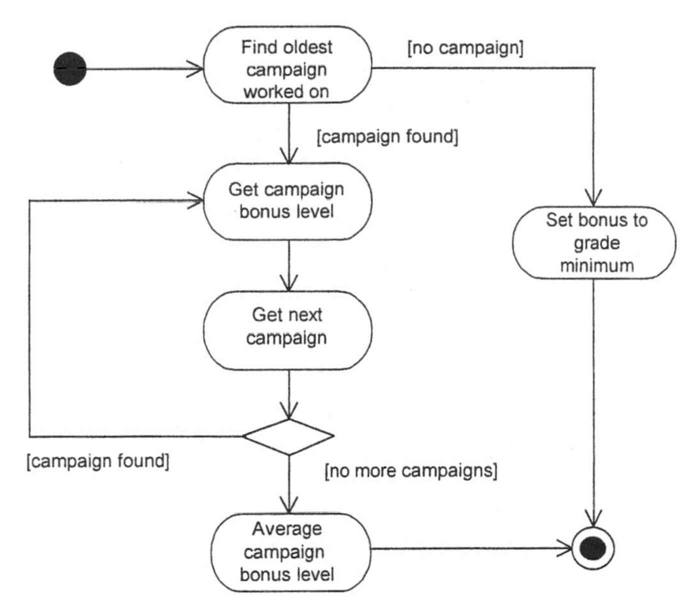

Figure 10.5 Activity diagram for `CreativeStaff.calculateBonus()`.

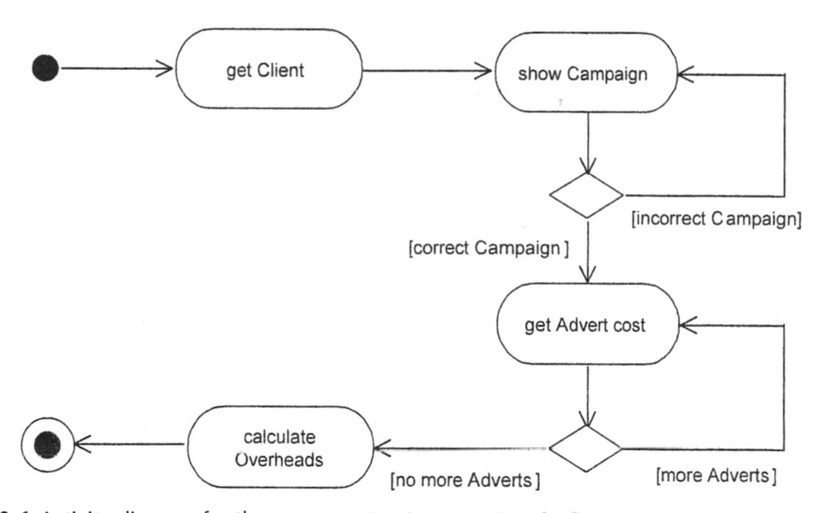

Figure 10.6 Activity diagram for the use case `Check campaign budget`.

10.5 Object Constraint Language

In drawing any class diagram, much of the time and effort is spent in working out what constraints apply. For example, the multiplicity of an association represents a constraint on how many objects of one class can be linked to any object of the other class. This particular example can be adequately expressed in the graphical language of the class diagram, but this is not equally so for all constraints. Among those for which it is not true are many of the constraints within operation specifications. For example, many pre- and post-conditions in a contract are constraints on the behaviour of objects that are party to the contract. Sometimes the definition of such constraints can be done in an informal manner (as in the examples in Section 10.4), but where greater precision is required, OCL provides a formal language.

OCL expressions are constructed from a collection of pre-defined elements and types, and the language has a precise grammar that enables the construction of unambiguous statements about the properties of model components and their relationships to each other. Most OCL statements consist of the following structural elements:

- A *context* that defines a domain within which the expression is valid. This is often an instance of a specific type, for example an object in a class diagram. A link (i.e. an instance of an association) may also be the context for an OCL expression.
- A *property* of that instance which is the context for the expression. Properties may include attributes, association-ends and query operations.
- An OCL *operation* that is applied to the property. Operations include (but are not restricted to) the arithmetical operators `*`, `+`, `-` and /, set operators such as `size`, `isEmpty` and `select` and type operators such as `oclIsTypeOf`.

OCL statements can also include OCL *keywords* that include the logical operators such as **and**, **or**, **implies**, **if**, **then**, **else** and **not** and the set operator **in**, printed in bold to distinguish them from other OCL terms and operations. Together with the non-keyword operations mentioned above, these can be used to define quite complex pre-and post-conditions for an operation.

Figure 10.7 gives some examples of expressions in OCL, mainly adapted from the OCL Specification, which is part of the UML Specification (OMG, 2001). All have an object of some class as their context. The figure shows examples of OCL syntax and an interpretation of the meaning of each.

OCL can specify many constraints that cannot be expressed directly in diagrammatic notation, and is thus useful as a precise language for pre- and post-conditions. The general syntax for operation specification is as follows:

```
Type::operation(parameter1:type,parameter2:type):return type
  pre: parameter1 operation
       parameter2 operation
  post:result = ...
```

OCL expression	Interpretation
`Person` `self.gender`	In the context of a specific person, the value of the property 'gender' of that person—i.e. a person's gender.
`Person` `self.savings >= 500`	The property 'savings' of the person under consideration must be greater than or equal to 500.

`Person`
`self.husband->notEmpty` **implies**
`self.husband.gender = male`
	If the set 'husband' associated with a person is not empty, then the value of the property 'gender' of the husband must be male. The boldface denotes an OCL keyword, but has no semantic import in itself.

`Company` `self.CEO->size <= 1`	The size of the set of the property 'CEO' of a company must be less than or equal to 1. That is, a company cannot have more than 1 Chief Executive Officer.
`Company` `self.employee->select (age < 60)`	The set of employees of a company whose age is less than 60.

Figure 10.7 Examples of some expressions in OCL.

Note that the contextual type is the type (for our purposes, normally a class) that owns the operation as a feature. The **pre:** expressions are functions of operation parameters, while **post:** expressions are functions of self, of operation parameters, or of both. OCL expressions can be written with an explicit **Context** declaration.

The following example illustrates this usage, together with an **inv:** label to denote an invariant (first introduced in Section 5.2.2).

```
Context Person inv:
        self.age >= 0
```

The invariant here is merely that a person's age must always be greater than or equal to zero—arguably, this should not need specification, but poorly specified computer systems often get the really obvious things wrong! For a complete list of keywords, see Section 6.4.8 of the UML Specification (OMG, 2001). The context of an OCL expression associated with a diagram (such as a class or collaboration diagram) is often obvious; when this is the case, the declaration can be omitted.

One particularly useful feature of OCL is its ability to define two values for a single property using the postfix `@pre`. As you might expect, this refers to the previous value of a property, and it can only be used in post-condition clauses. A typical use is to constrain the relationship between the values of an attribute before and after an operation has taken place. For example, the decision specified in Figure 10.1 defines different actions depending on changes in the estimated cost of a campaign in comparison with its budget. If the new estimated cost is greater than the old estimated cost, but exceeds the budget by no more than 2%, the value of this attribute is set to `true`, flagging a need to generate a warning letter to the client[4]. We can model this in a very simple way by adding an attribute `Campaign.clientLetterRequired`. We can write part of the logic in OCL as follows:

```
Context Campaign inv:
  post: if estimatedCost > estimatedCost@pre and
           estimatedCost > budget and
           estimatedCost <= budget * 1.02 then
           self.clientLetterRequired : Boolean = 'true'
        endif
```

This describes an invariant that can be included in operation specifications, and will help to define tests that check if the system displays the correct behaviour when a campaign budget has changed. (However, note this example is intended only to illustrate the notation. In practice, it is unlikely that this is really how we would model this requirement.)

Operation specifications frequently include invariants. When an invariant is associated with an operation specification it describes a condition that always remains true for an object, and which must therefore not be altered by an operation side-effect. Formal definition of invariants is valuable because they provide rigorous tests for execution of the software.

For example, the value of `Campaign.estimatedCost` should always equal the sum of all associated `Advert.estimatedCost` values multiplied by the current overhead rate. In OCL[5], this might be written as follows:

[4] If the budget is exceeded but the estimated cost has not increased, then we assume that a letter has already been sent, and so the flag does not need to be set.

[5] Note that this makes some assumptions about the way that the classes are designed, and is really intended only to illustrate the style of this kind of statement when written in OCL.

```
Campaign
  inv: self.estimatedCost = ohRate * self.adverts.estimatedCost->sum
```

In this example, the context is the Campaign class. To use an invariant within an operation specification, it can be written simply as an additional clause beginning **inv:**

```
ClassName::operation(parameter1:type,parameter2:type):return type
   pre:  parameter1...
         parameter2...
   post: result1...
         result2...
   inv:  invariant1...
         invariant2...
```

For an example from the Agate case study, we revisit the operation `creativeStaff.changeGrade()`, for which a logic specification was constructed in Section 10.4.1. To help make sense of this specification, it is also worth referring back to the analysis class diagram in Figure A3.14. In particular, note the recursive association from `StaffGrade` to itself. However, remember also that, as we saw in Chapter 8, `CreativeStaff` is a subclass of `Staff Member` and therefore inherits the same associations and roles (Chapter A4 includes a revised analysis class diagram that shows this specialization). Here is the main part of the operation specification rewritten in OCL:

```
CreativeStaff::changeGrade(grade:Grade, gradeChangeDate:Date)
  pre: grade oclIsTypeOf(Grade)
       gradeChangeDate >= today
  post:self.staffGrade[grade]->exists
       self.staffGrade[previous]->notEmpty
       self.staffGrade.gradeStartDate = gradeChangeDate
       self.staffGrade.previous.gradeFinishDate = gradeChangeDate
```

10.6 Creating an Operation Specification

Figure 10.8 shows the sequence diagram for the use case `Check campaign budget` first introduced in Chapter 9 (the use case description is repeated above in Section 10.4.2 and an activity diagram is shown in Figure 10.6). In this particular example the message `checkCampaignBudget` invokes the operation `Campaign.checkCampaignBudget()`.

A specification for `Campaign.checkCampaignBudget()` is given below. We have used different fonts to signpost the specification as follows. This font (Arial) labels the specification structure, `while this (Courier) highlights its content.` *Comments on the reasoning behind the specification are formatted like this.*

Operation specification: `checkCampaignBudget`

Operation intent: `return difference between campaign budget and actual costs.`

The invocation appears not to require any parameters, but does have a return type that we can expect it to contain a numerical value. Let us assume that there is a `Money` *type available. The signature is shown below, followed by the pre- and post-conditions.*

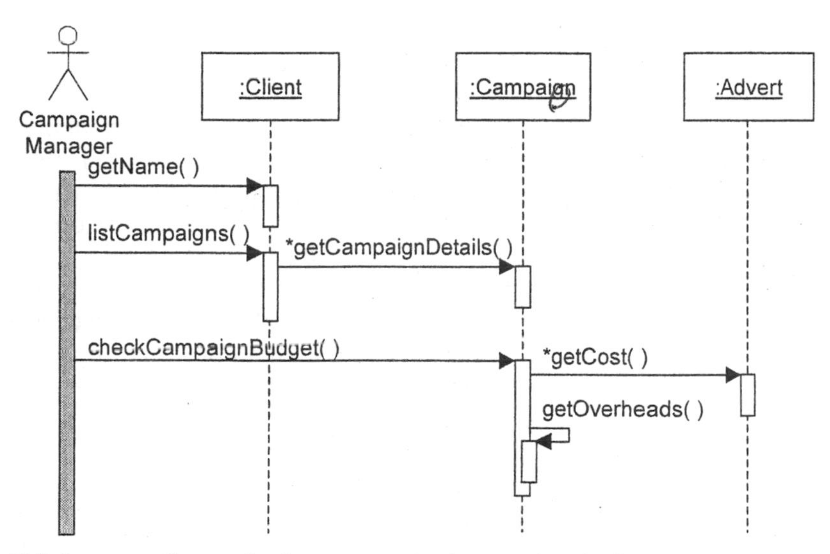

Figure 10.8 Sequence diagram for the use case `Check campaign budget`.

operation signature: `Campaign::checkCampaignBudget()`
`budgetCostDifference:Money`

logic description (pre- and post-conditions):

```
pre:self ->exists
post:campaignBudget - self.estimatedCost
     committedExpenditure = self.adverts.estimatedCost->sum
```

As can be seen from the sequence diagram, this operation calls two other operations and these must be listed. In a full specification, full signatures would be recorded, but we omit this detail here.

Other operations called: `Advert.getCost, self.getOverheads`
Events transmitted to other objects: `none`

The only messages are those required to call the operations just mentioned, whose return values are required by this operation. An 'event' is a message that starts another distinct thread of processing (see Chapter 11).

Attributes set: `none`

This is a query operation whose only purpose is to return data already stored within the system.

Response to exceptions: `none defined`

Here we could define how the operation should respond to error conditions, e.g. what kind of error message will be returned if a calling message uses an invalid signature.

Non-functional requirements: `none defined`

Several non-functional requirements may be associated with the operation, but these need to be determined through discussion with users. They may include, for instance, response time under live conditions (enquiries that are made frequently typically require a faster response) or the format of the output (e.g. if there is a house standard that overspent budgets are displayed in red). However, these are really design issues, and would be noted at this stage only if the information happens to be available at the time.

10.7 Summary

Operation specifications are the most detailed description of the behaviour of a system model. As such, they are also one of the more significant elements in the project repository. They provide an important link between the system's users, who typically possess a detailed understanding of the required system behaviour, and the designers and programmers who must implement this in software. Accurate specification of operations is essential if the software is to be coded correctly.

In this chapter we introduced the 'contract' as a framework for specifying operations, in terms of the service relationship between classes. Contracts are a particularly useful element of operation specification since they concentrate on the correctness of each object's behaviour.

We also described several techniques for describing operation logic. Non-algorithmic techniques, such as decision tables and pre- and post-condition pairs, take a black box approach and concentrate on specifying only the inputs to an operation (its pre-conditions) and the intended results of an operation (its post-conditions). In many cases, particularly where the operations themselves are simple, this is all the specification that a programmer needs to code the operation correctly.

Algorithmic techniques, such as Structured English, pseudo-code and activity diagrams, take a white box approach, and this means that they concentrate on defining the internal logic of operations. These techniques are particularly useful when an operation is computationally complex. They are also useful when we need to model some larger element of system behaviour, such as a use case, that has not yet been decomposed to the level of individual operations that can be assigned to specific classes.

Many elements of an operation specification can be written in OCL (UML's Object Constraint Language). OCL is intended for use as a formal language for specifying most types of constraints on an object model, and this includes operation pre- and post-conditions and invariants.

Review Questions

10.1 What are the two main purposes of an operation specification?

10.2 To what kinds of situation are decision tables particularly suited?

10.3 Why is it important to specify both pre- and post-conditions for an operation?

10.4 What are the main differences between algorithmic and non-algorithmic approaches to operation specification?

10.5 Why are non-algorithmic (or declarative) approaches generally preferred in object-oriented development?

10.6 Why are operation specifications in an object-oriented project likely to be small?

10.7 What are the three kinds of control structure in Structured English?

10.8 What is a sensible limit on the size of a Structured English specification?

10.9 What are the three components of most OCL expressions?

10.10 What is an invariant?

Case Study Work, Exercises and Projects

10.A Consider the first sequence diagram you drew for the use case `Start line run` from the FoodCo case study (in Exercise 9.A). Choose an operation in one of the classes involved in this ISD and write a contract for it. Make reasonable assumptions where necessary, and use a pre- and post-conditions approach for describing its logic.

10.B Consider the decision table in Figure 10.1. Suppose you have learned that an extra condition must be taken into account: the rules in the current table actually apply only to campaigns with a total budget of £5,000 or over, but for smaller campaigns the thresholds for each action are different. Thresholds for smaller campaigns are as follows. For an expected overspend of less than 10%, no action is taken. For expected overspends of 10%-19%, a letter is sent. For an expected overspend of 20% or more a letter is sent and a meeting is arranged. Draw a new version of the table that caters for small campaigns.

10.C Redraw the original decision table in Figure 10.1 as an activity diagram. Do the same for your new decision table from Exercise 10.B.

10.D Consider the decision table in Figure 10.1. Which of the three control structures are required to convert this into a Structured English specification? Re-write the decision table in Structured English format.

10.E Find out how a decision tree differs from a decision table (e.g. from one of the books listed in Further Reading). Produce a decision tree that corresponds to the decision table in Figure 10.1. What are the relative advantages and disadvantages of decision trees, decision tables and Structured English?

Further Reading

- Most techniques for operation specification were in regular use long before object-oriented analysis became widespread, and for this reason many of the best treatments are in textbooks that deal with structured approaches to systems analysis.
- Yourdon (1989) gives a thorough presentation of Structured English and pre- and post-conditions, while Senn (1989) gives good coverage of decision tables.
- Larman (1998) describes a contract-based approach to object-oriented analysis and design, with examples taken through to Java code.
- Meyer (1997) is perhaps the definitive text on design-by-contract in object-oriented software engineering. This book is very comprehensive, but note that it is aimed at technically-oriented readers.
- Further examples of operation specifications can be found in Case Study Chapter A4.

11

CHAPTER

Specifying Control

OBJECTIVES

In this chapter you will learn

- how to identify requirements for control in an application
- how to model object life cycles using statecharts
- how to develop statechart diagrams from interaction diagrams
- how to model concurrent behaviour in an object
- how to ensure consistency with other UML models.

11.1 Introduction

The various types of UML notation that have been introduced so far enable us to model the static structure of an application (class diagrams) and the way in which objects interact (sequence and collaboration diagrams). Another important aspect of an application that must be modelled is the way that its response to events can vary depending upon the passage of time and the events that have occurred already. For an application such as a real-time system it is easy to understand that the response of the system to an event depends upon its state. For example, an aircraft flight control system should respond differently to events (for example, engine failure) when the aircraft is in flight and when the aircraft is taxiing along a runway. A more mundane example is that of a vending machine, which does not normally dispense goods until an appropriate amount of money has been inserted. This variation in behaviour is determined by the state of the machine—which depends on whether or not sufficient money has been inserted to pay for the item selected. In reality, of course, the situation is more complicated than this. For example, even when the correct amount of money has been inserted, the machine cannot dispense an item that is not in stock. It is important to model state dependent variations in behaviour such as these since they represent constraints on the way that a system should behave.

Objects can have similar variations in their behaviour dependent upon their state. The concepts of state and event are discussed in Section 11.2. These variations in behaviour represent important constraints on the way that an object behaves and are dictated by the requirements for the system. UML uses statecharts to model states and state dependent behaviour for objects and for interactions. The notation used in UML is based upon work by Harel (1987) and was adopted by OMT (Rumbaugh et al., 1991) and also in the second version of the Booch approach (Booch, 1994). We introduce the UML notation in Sections 11.3 and 11.4. The procedures for preparing statecharts are explained in Section 11.5.

There is an important link between interaction diagrams and statecharts. A model of state behaviour in a statechart captures all the possible responses of a single object to all the use cases in which it is involved. By contrast, a sequence or a collaboration diagram captures the responses of all the objects that are involved in a single use case. A statechart can be seen as a description of all the possible life cycles that an object of a class may follow. It can also be seen as a more detailed view of a class. Some CASE tools handle statecharts as a child diagram to a class on a class diagram. We consider issues regarding consistency between models in Section 11.6 and examine quality guidelines in Section 11.7.

It is not only object-oriented approaches that have recognized the importance of modelling the lifetime behaviour of entities. For example, SSADM (Skidmore, Mills & Farmer, 1994) uses entity life histories for this purpose, and some approaches to the development of real-time systems, such as that of Yourdon, model the state behaviour of a system or sub-system.

The statechart is a versatile technique, and can be used within an object-oriented approach for other purposes than the modelling of object life cycles. In Chapter 17 we show how to use statecharts to build models of human–computer dialogues.

11.2 States and Events

All objects have a state. The current state of an object is a result of the events that have occurred to the object, and is determined by the current value of the object's attributes and the links that it has with other objects. Some attributes and links of an object are significant for the determination of its state while others are not. For example, in the Agate case study `staffName` and `staffNo` attributes of a `StaffMember` object have no impact upon its state, whereas the date that a staff member started his or her employment at Agate determines when the probationary period of employment ends (after six months, say). The `StaffMember` object is in the `Probationary` state for the first six months of employment. While in this state, a staff member has different employment rights and is not eligible for redundancy pay in the event that they are dismissed by the company.

The UML specification (OMG, 2001, page 3-138) defines a state as follows:

> A state is a condition during the life of an object or an interaction during which it satisfies some condition, performs some action or waits for some event . . .. Conceptually, an object remains in a state for an interval of time. However, the semantics allow for modelling 'flow-through' states which are instantaneous[1], as well as transitions that are not instantaneous.

The possible states that an object can occupy are limited by its class. Objects of some classes have only one possible state. For example, in the Agate case study a `Grade`

[1] Note that allowing instantaneous states is a recent modification to UML.

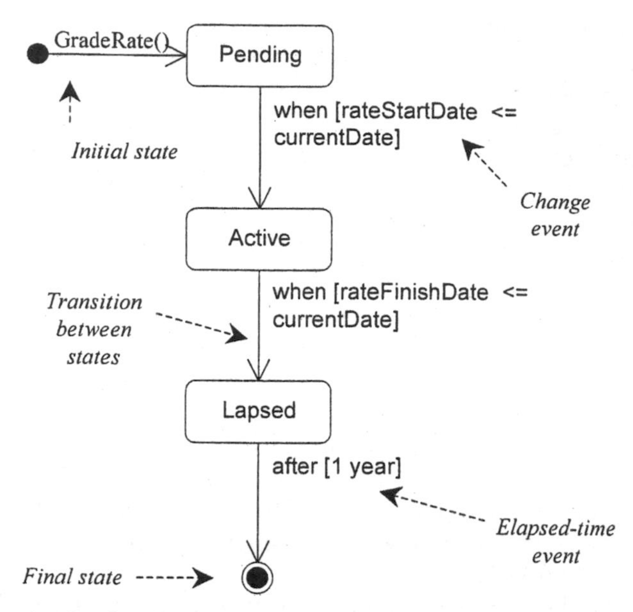

Figure 11.1 Statechart for the class *Grade Rate*.

object either exists or it does not. If it exists it is available to be used, and if it does not exist it is not available. Objects of this class have only one state, which we might name `Available`. Objects of other classes have more than one possible state. For example, an object of the class `GradeRate` may be in one of several states. It may be `Pending`, if the current date is earlier than its start date, `Active`, if the current date is equal to or later than the start date but earlier than the finish date (we assume that the finish date is later than the start date), or `Lapsed`, if the current date is later than the finish date for the grade. If the current date is at least a year later than the finish date then the object is removed from the system. The current state of a `GradeRate` object can be determined by examining the values of its two date attributes (alternatively, the `GradeRate` class might have a single attribute[2] (an enumerated type—that has an integer value for each possible state) with values that indicate the current state of an object). It is important to note that movement from one state to another for a `GradeRate` object is dependent upon events that occur with the passage of time. Figure 11.1 shows a statechart for `GradeRate`.

Movement from one state to another is called a *transition*, and is triggered by an *event*. When its triggering event occurs a transition is said to *fire*. A transition is shown as a solid arrow from the source state to the target state. An event is an occurrence of a stimulus that can trigger a state change and that is relevant to the object or to an application. For example, the cancellation of an advert at Agate is an event that will change the state of the `Advert` object being cancelled. Just as a set of objects is defined by the class of which they are all instances, events are defined by an *event type* of which each event is an instance. For example, the cancellation of an advert in the CheapClothes jeans campaign is one instance of an event, and the cancellation of an advert in the Soong Motor Co Granda campaign is another instance. Both are defined by the event type `cancellationOfAdvert()`. It is usually event types that are modelled, but we refer to them simply as events. An event can have parameters and a return value, and in an object-oriented system it is implemented by a message.

[2] An attribute that holds the value of the current state of an object is sometimes known as a state variable.

Events can be grouped into several general types. A *change event* occurs when a condition becomes true. This is usually described as a Boolean expression, which means that it can take only one of two values: true or false. Change events are annotated by the keyword *when* followed by the Boolean expression in parenthesis. This form of conditional event is different from a guard condition that is only evaluated at the moment that its associated event fires.

A *call event* occurs when an object receives a call for one of its operations either from another object or from itself. Call events correspond to the receipt of a call message and are annotated by the signature of the operation as the trigger for the transition.

A *signal event* occurs when an object receives a signal[3]. As with call events the event is annotated with the signature of the operation invoked. There is no syntactic difference between call events and signal events. It is assumed that a naming convention is used to distinguish between them.

An *elapsed-time event* is caused by the passage of a designated period of time after a specified event (frequently the entry to the current state). Elapsed-time events are shown by time expressions as triggers for the transitions. The time expression is placed in parentheses and should evaluate to a period of time. It is preceded by the keyword *after* and if no starting time is indicated it reflects the passage of time since the most recent entry to the current state.

The basic syntax for a call or signal event is:

```
event-name '(' parameter-list ')'
```

where the parameter-list contains parameters of the form:

```
parameter-name ':' type-expression
```

separated by commas. (Characters in single quotes, such as `'('`, are *literals* that appear as part of the event.)

11.3 Basic Notation

The *initial state* (in other words the starting point) of a life cycle is indicated by a small solid filled circle. The initial state is a notational convenience, and an object cannot remain in its initial state but must immediately move into another named state. In Figure 11.1 the `GradeRate` object enters the `Pending` state immediately on its creation. A transition from the initial state can optionally be labelled with the event that creates the object. The end point of a life cycle (in other words its final state) is shown by a bull's-eye symbol. This too is a notational convenience, and an object cannot leave its final state once it has been entered. All other states are shown as a rectangle with rounded corners and should be labelled with a meaningful name. In this example all transitions except the transition from the initial state are triggered by change events. The statechart for a `GradeRate` object is very simple, since it enters each state only once. Some classes have much more complex life cycles than this. For example, a `BookCopy` object in a library system may move many times between the states `OnLoan` and `Available`.

Figure 11.2 illustrates the basic notation for a statechart with two states for the class `Campaign` and one transition between them. A transition should be annotated with a *transition string* to indicate the event that triggers it.

[3] An asynchronous communication.

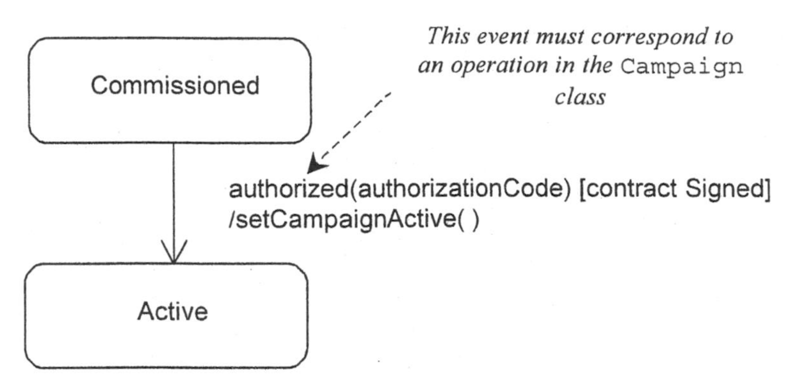

Figure 11.2 Fragment of the statechart for the class `Campaign`.

For call and signal events the format of the transition string is as follows:

```
event-signature '['guard-condition']' '/' action-expression
```

The event signature takes the following form:

```
event-name '('parameter-list')'
```

A *guard condition* is a Boolean expression that is evaluated at the time the event fires. The transition only takes place if the condition is true. A guard condition is a function that may involve parameters of the triggering event and also attributes and links of the object that owns the statechart. A guard condition is contained in square brackets—`'['...']'`. In Figure 11.2 the guard condition is a test on the `contractSigned` attribute in the class `Campaign` and since the attribute is Boolean it could be written as follows:

```
[contractSigned]
```

This expression evaluates to true only if `contractSigned` is true. A guard condition can also be used to test concurrent states of the current object or the state of some other reachable object. Concurrent states are explained later in Section 11.4.2.

An *action-expression* is executed when an event triggers the transition to fire. Like a guard condition, it may involve parameters of the triggering event and may also involve operations, attributes and links of the owning object. In Figure 11.2 the action-expression begins with the `'/'` delimiter character and is the execution of the `Campaign` object's operation `setCampaignActive()`.

An action-expression may comprise a sequence of actions and include actions that may generate events such as sending signals or invoking operations. Each action in an action string is separated from its preceding action with a semi-colon. An example of an action-expression with multiple actions is shown in the transition string below:

```
left-mouse-down(location) [validItemSelected] / menuChoice:=
pickMenuItem(location); menuChoice.highlight()
```

The sequence of actions in an action-expression is significant since it determines the order in which they are executed. In the example above if the actions were in the reverse order the value of `menuChoice` is likely to be different when the `highlight()` message is sent and the effect of the event would be different. Actions are considered to be atomic (that is, they cannot be subdivided) and cannot be interrupted once they have been started. Once initiated this action must execute fully before any other action

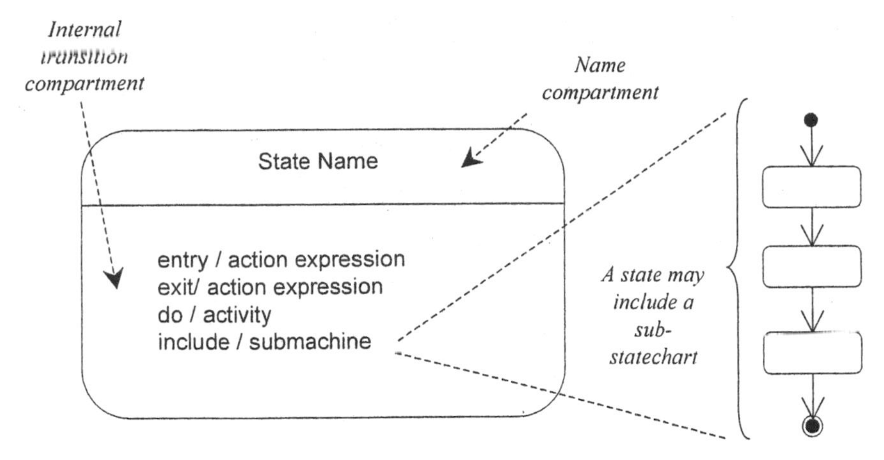

Figure 11.3 Internal actions and activities for a state.

is considered. This is know as *'run-to-completion'* semantics. An action will actually take time to execute in a live application but this is dependent only upon internal factors such as the complexity of the program code and the speed of the processor. Unlike the duration of an action, the duration of a state is normally dependent upon external events in the application environment.

So far we have considered only action-expressions that are associated with a transition. It can also be useful to model internal actions or activities associated with a state. These actions may be triggered by events that do not change the state, or by an event that causes the state to be entered or by an event that results in exiting the state. In Figure 11.3 the state symbol is shown with two compartments, a name compartment and an internal transitions compartment. Actions or activities that are associated with the state are listed in the latter. Note that an activity may persist for the duration of the state unlike actions that are transitory. It is thus possible to interrupt an activity.

Two kinds of internal event have a special notation. These are the *entry event* and the *exit event*, respectively indicated by the keywords *entry* and *exit*. These cannot have guard conditions as they are invoked implicitly on entry to the state and exit from the state respectively. Entry or exit *action-expressions* may also involve parameters of incoming transitions (provided that these appear on all incoming transitions) and attributes and links of the owning object. It is important to emphasize that any transition into a state causes the entry event to fire and all transitions out of a state cause the exit event to fire.

Activities are preceded by the keyword do and have the following syntax:

```
'do' '/' activity-name '(' parameter-list ')'
```

It is also possible to show that a state contains substates by using the keyword *include* followed by the name of the contained sub-statechart or submachine. Complex states may be represented by a statechart nested within the state. The nesting of one statechart within another allows the representation of highly complex behaviour. When an activity in a state ends (i.e. the nested statechart reaches its final state) the state is considered completed and the object makes a transition triggered by the completion of this activity. Alternatively an activity may persist as long as the object remains in the state, in which case it does not trigger a transition from the state. The activity will only end when some other specified event triggers a transition from the state.

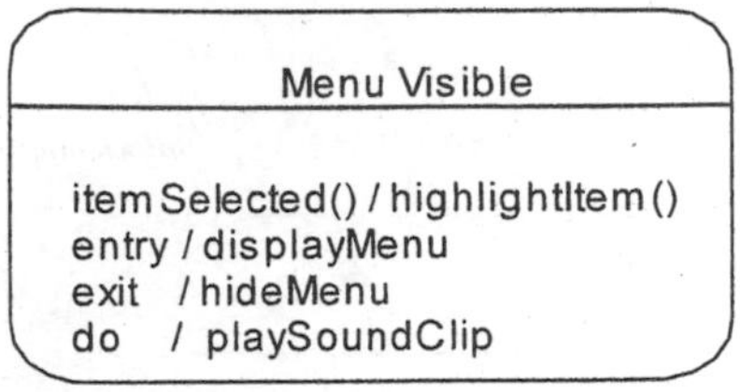

Menu Visible

itemSelected() / highlightItem()
entry / displayMenu
exit / hideMenu
do / playSoundClip

Figure 11.4 `Menu Visible` state for a `DropDownMenu` object.

In general an internal transition occurs in response to an event and results in an action but does not cause a change in state. The format for internal transitions other than those caused by entry to and exit from the state is as follows:

```
event-name '(' parameter-list ')' '['guard-condition']' '/'
action-expression
```

An event name may be listed more than once if the guard conditions are different. Figure 11.4 shows the `Menu Visible` state for a `DropDownMenu` object.

In this example, the entry action causes the menu to be displayed. While the object remains in the `Menu Visible` state, the activity causes a sound clip to be played and, if the event `itemSelected()` occurs, the action `highlightItem()` is invoked. It is important to note that when the event `itemSelected()` occurs the `Menu Visible` state is not exited and entered and as a result the exit and entry actions are not invoked. When the state is actually exited the menu is hidden.

Figure 11.5 shows a statechart for the class `Campaign`. The transition from the initial state to the `Commissioned` state has been labelled only with an action-expression that comprises the operations `assignManager()` and `assignStaff()`. Execution of these operations ensures that when a campaign is created a manager and member(s) of staff are assigned to it[4]. The operations are triggered by the event that creates a `Campaign` object. The transition from the `Completed` state to the `Paid` state has a guard condition that only allows the transition to fire if total amount due (`paymentDue`) for the `Campaign` has been completely paid (note that this guard condition allows a `Campaign` to enter the `Paid` state when the customer overpays).

The recursive transition from the `Completed` state models any payment event that does not reduce the amount due to zero or beyond. Only one of the two transitions from the `Completed` state (onc of which is recursive) can be triggered by the `paymentReceived` event since the guard conditions are mutually exclusive. It would be bad practice to construct a statechart where one event can trigger two different transitions from the same state. A life cycle is only unambiguous when all the transitions from each state are mutually exclusive.

If the user requirements were to change, so that an overpayment is now to result in the automatic generation of a refund, the statechart can be changed as follows. Since the action that results from an overpayment is different from the action that results from a payment that reduces `paymentDue` to zero, a new transition is needed from the `Completed` state to the `Paid` state. The guard conditions from the `Completed` state must also be modified. Figure 11.6 shows a statechart that captures this requirement. It is important to appreciate that the statecharts in Figures 11.5 and 11.6 are not equivalent to each other, but capture different versions of the users' requirements.

[4] Unless the specifications for these operations permit a null option.

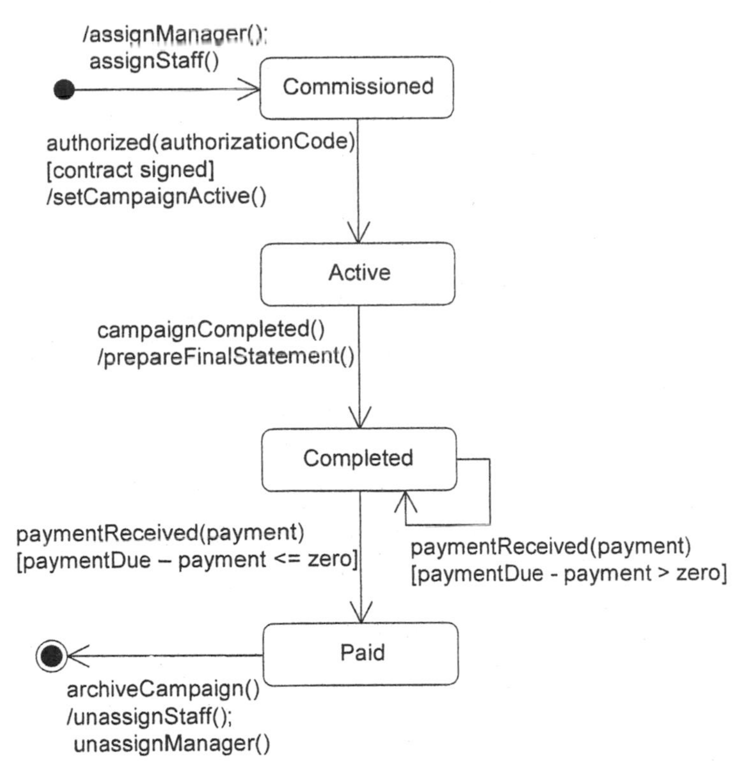

Figure 11.5 Statechart for the class `Campaign`.

11.4 Further Notation

The statechart notation can be used to describe highly complex time-dependent behaviour. Hierarchies of states can be nested and concurrent behaviour can also be represented.

11.4.1 Nested states

When the state behaviour for an object or an interaction is complex it may be necessary to represent it at different levels of detail and to reflect any hierarchy of states that is present in the application. For example, in the statechart for `Campaign` the state `Active` encompasses several *substates*. These are shown in Figure 11.7 where the `Active` state is seen to comprise three disjoint substates: `Advert Preparation`, `Scheduling` and `Running Adverts`. This diagram now shows a single state which contains within it a nested state diagram. In the nested statechart within the `Active` state, there is an initial state symbol with a transition to the first substate that a `Campaign` object enters when it becomes active. The transition from the initial pseudostate symbol to the first substate (`Advert Preparation`) should not be labelled with an event but it may be labelled with an action, though it is not required in this example. It is implicitly fired by any transition to the `Active` state. A final pseudostate symbol may also be shown on a nested state diagram. A transition to the final pseudostate symbol represents the completion of the activity in the enclosing state (i.e. `Active`) and a transition out of this state triggered by the

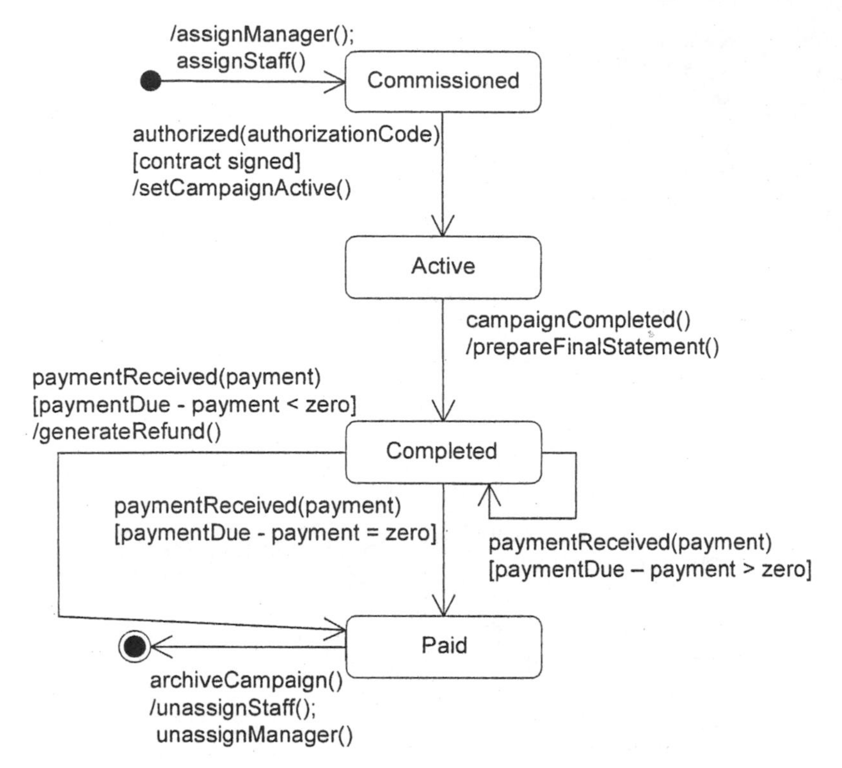

Figure 11.6 A revised statechart for the class `Campaign`.

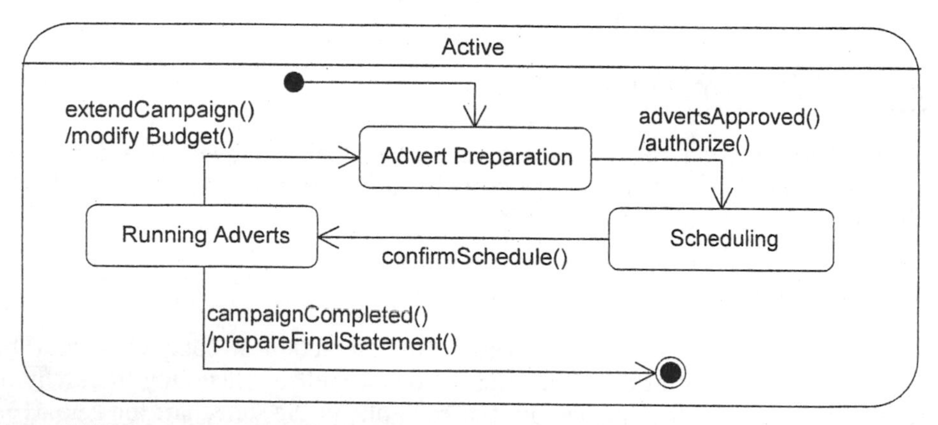

Figure 11.7 The `Active` state of `Campaign` showing nested substates.

completion event. This transition may be unlabelled (as long as this does not cause any ambiguity) since the event that triggers it is implied by the completion event.

When a campaign enters the `Active` state in Figure 11.7 it first enters the `Advert Preparation` substate, then if the adverts are approved it enters the `Scheduling` substate and finally enters the `Running Adverts` substate when the schedule is approved. If the campaign is deemed completed the object leaves the `Running Adverts` substate and also leaves the `Active` enclosing state, moving now to the `Completed` state (see Figure 11.5). If the campaign is extended while in the Running Adverts substate the `Advert Preparation` substate is re-entered (Figure 11.7). A high level statechart for the class `Campaign` can be drawn to include within the main diagram the detail that is shown in the nested statechart for the `Active` state if so desired. If the detail of the submachine is not required on the higher level statechart or is just too much to show on one diagram the higher level statechart can be annotated with the hidden decomposition indicator icon (two small state symbols linked together) as shown in Figure 11.8. The submachine `Running` is referenced using the include statement.

11.4.2 Concurrent states

Objects can have concurrent states. This means that the behaviour of the object can best be explained by regarding it as a product of two distinct sets of substates, each state of which can be entered and exited independently of substates in the other set. Figure 11.9 illustrates this form.

Suppose that further investigation reveals that at Agate a campaign is surveyed and evaluated while it is also active. A campaign may occupy either the `Survey` substate or the `Evaluation` substate when it is in the `Active` state. Transitions between these two states are not affected by the campaign's current state in relation to the preparing and running of adverts. We model this by splitting the `Active` state into two concurrent nested statecharts, `Running` and `Monitoring`, each in a separate sub-region of the `Active` statechart. This is shown by dividing the state icon with a dashed line. These concurrent substates for the `Active` state of the `Campaign` class are shown in Figure 11.9.

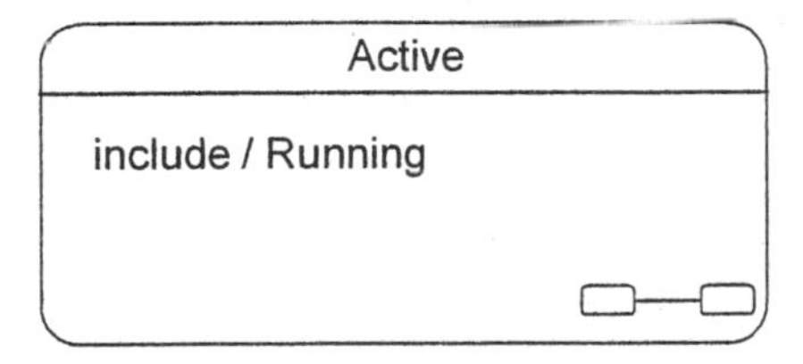

Figure 11.8 The `Active` state of `Campaign` with detail hidden.

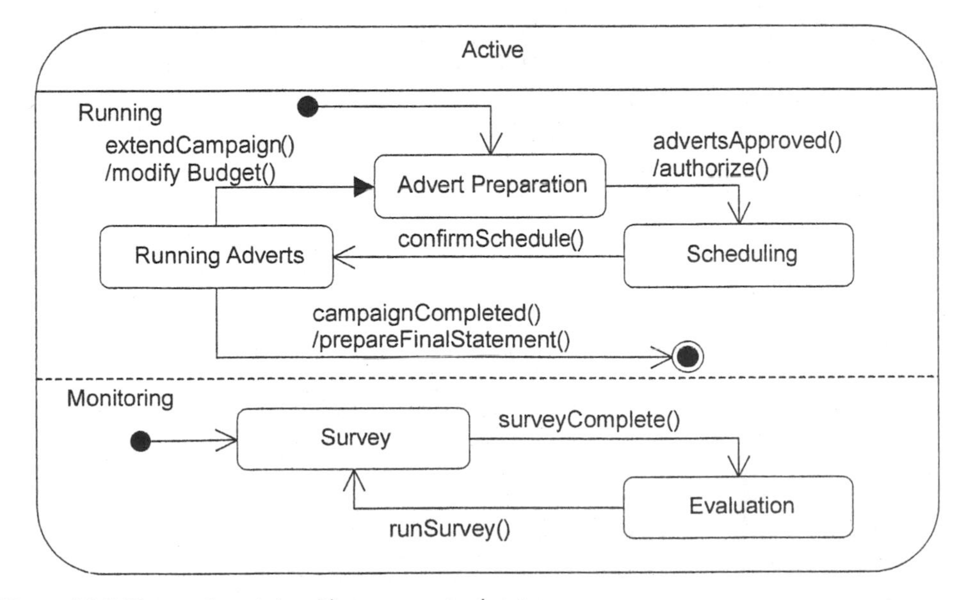

Figure 11.9 The `Active` state with concurrent substates.

A transition to a complex state such as this one is equivalent to a simultaneous transition to the initial states of each concurrent statechart. An initial state must be specified in both nested statecharts in order to avoid ambiguity about which substate should first be entered in each concurrent region. A transition to the `Active` state means that the `Campaign` object simultaneously enters the `Advert Preparation` and `Survey` states. A transition may now occur within either concurrent region without having any effect on the state in the other concurrent region. However, a transition out of the `Active` state applies to all its substates (no matter how deeply nested). In a sense, we can say that the substates inherit the `campaignCompleted()` transition from the `Active` state (shown in Figure 11.6) since it applies implicitly to them all. This is equivalent to saying that an event that triggers a transition out of the `Active` state also triggers a transition out of any substates that are currently occupied. The nested statechart `Monitoring` does not have a final state and when the `Active` state is exited one of the two states `Survey` or `Evaluation` will be occupied. The model in Figure 11.9 gives no indication as to which of these states will be occupied though UML offers synchronization constructs (Figure 11.10) if we need to use them. Inherited transitions can be masked if a transition with the same trigger is present in one of the nested statecharts (as is the case for the `campaignCompleted()` transition from the `Running Adverts` state in Figure 11.9).

Figure 11.10 shows the use of synchronization bars to show explicitly how an event triggering a transition to a state with nested concurrent states causes specific concurrent substates to be entered and also shows that the super-state is not exited until both concurrent nested statecharts are exited.

11.5 Preparing a Statechart

Statecharts can be prepared from various perspectives. The statechart for a class can be seen as a description of the ways that use cases can affect objects of that class. Use cases give rise to interaction diagrams (sequence diagrams or collaboration diagrams) and these can be used as a starting point for the preparation of a statechart.

Interaction diagrams show the messages that an object receives during the execution of a use case. The receipt of a message by an object does not necessarily correspond to an event that causes a state change. For example, simple 'get' messages (e.g. `getTitle()`) and query messages (e.g. `listAdverts()`) are not events in this sense. This is because they do not change the values of any of the object's attributes, nor do they alter any of its links with other objects. Some messages change attribute values without changing the state of an object. For example, a message `receivePayment()` to a `Campaign` object will only cause a change of state to `Paid` if it represents payment at least of the full amount due.

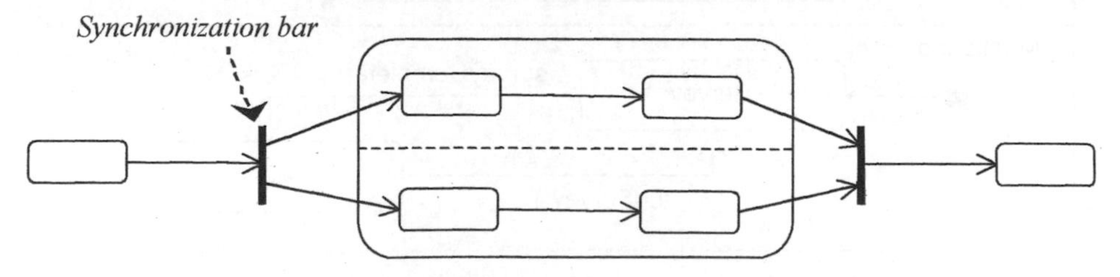

Figure 11.10 Synchronized concurrent threads.

11.5.1 A behavioural approach

Figure 11.11 shows a sequence diagram for the use case `Record completion of a campaign`. The receipt of the message `campaignCompleted()` by a `Campaign` object is an event from the perspective of the Campaign object. In this example this event is a call event and causes the `campaignCompleted()` operation to invoked triggering a transition from the `Active` state to the `Completed` state. Incoming messages to an object generally correspond to an event and trigger a state change. Allen and Frost (1998) describe the use of interaction diagrams to develop a statechart as a behavioural approach.

The preparation of a statechart from a set of interaction diagrams using this behavioural approach has the following sequence of steps.

1. Examine all interaction diagrams that involve each class that has heavy messaging.
2. Identify the incoming messages on each interaction diagram that may correspond to events. Also identify the possible resulting states.
3. Document these events and states on a statechart.
4. Elaborate the statechart as necessary to cater for additional interactions as these become evident, and add any exceptions.
5. Develop any nested statecharts (unless this has already been done in an earlier step).
6. Review the statechart to ensure consistency with use cases. In particular, check that any constraints that are implied by the statechart are appropriate.
7. Iterate steps 4, 5 and 6 until the statechart captures the necessary level of detail.
8. Check the consistency of the statechart with the class diagram, with interaction diagrams and with any other statecharts.[5]

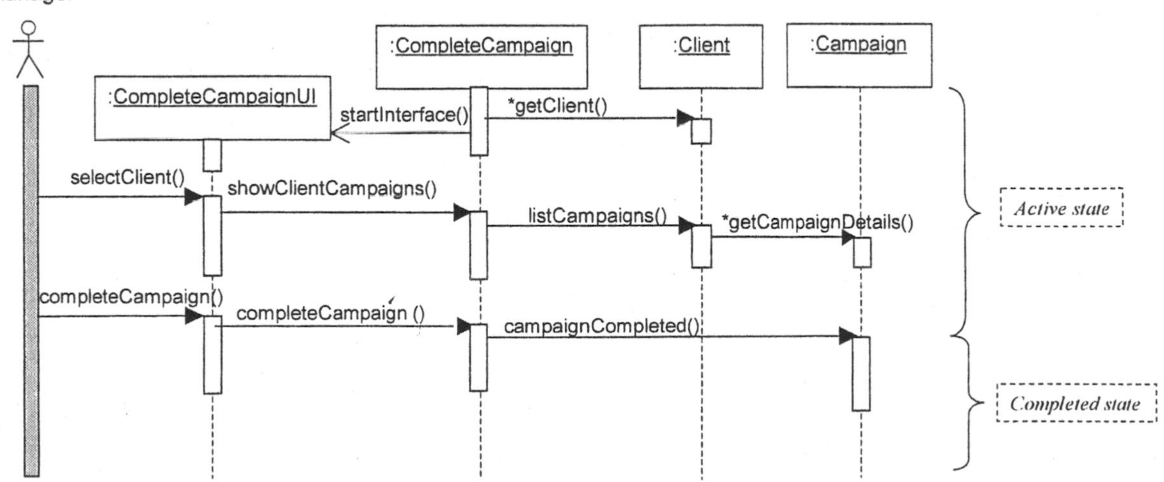

Figure 11.11 Sequence diagram for use case Record completion of a campaign.

[5] This step should also include checking of consistency with any class diagram constraints and constraints defined in operation specifications. In operation specifications this typically involves consideration of pre- and post-conditions and invariants.

The sequence diagram in Figure 11.11 has been annotated to indicate the state change that is triggered by the event `campaignCompleted()`. In order to identify all incoming messages that may trigger a state change for an object, all interaction diagrams that affect the object should be examined (sequence diagrams are probably easier to use for this purpose than collaboration diagrams, but this is a matter of personal preference). Analysis of the interaction diagrams produces a first-cut list of all events (caused by incoming messages) that trigger state changes, and also a first-cut list of states that the object may enter as a result of these events. If only major interactions have been modelled then the lists will not be complete but they can still provide an effective starting point.

The next step is to prepare a draft statechart for the class. Figure 11.12 illustrates the level of detail that might be shown in a first-cut statechart for the `Campaign` class. This would need to be expanded in order to reflect any events that have not been identified from the interaction diagrams, and also to include any exceptions. Complex nested states can be refined at this stage. A review of the statechart in Figure 11.12 results in

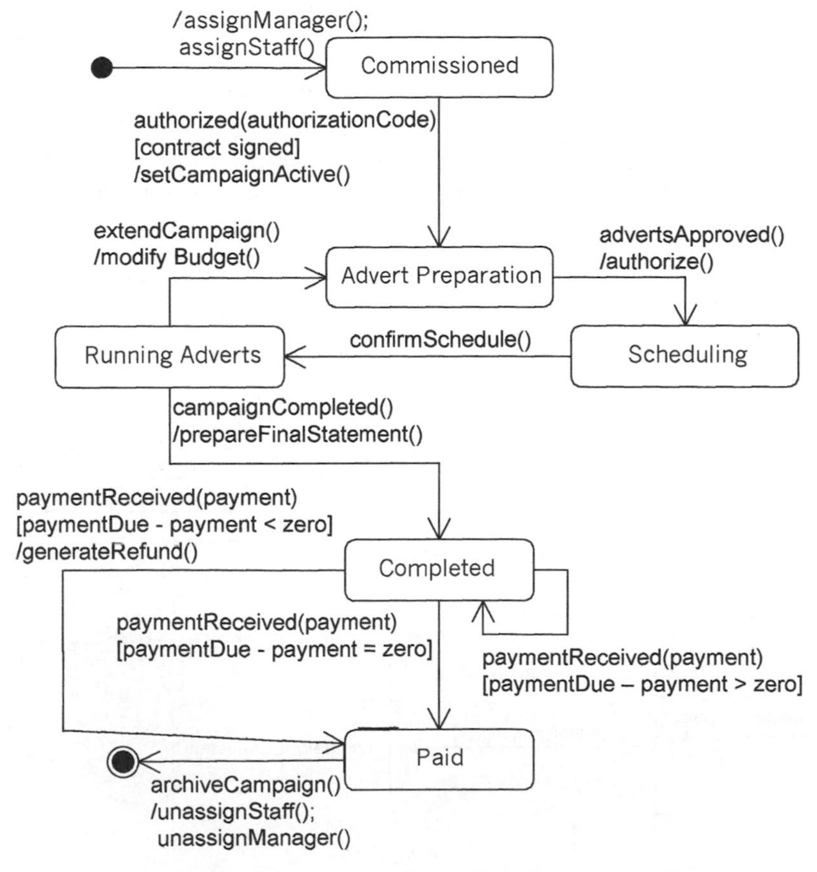

Figure 11.12 Initial statechart for the `Campaign` class—a behavioural approach.

the addition of the `Active` state to encompass the states `Advert Preparation`, `Scheduling` and `Running Adverts` (shown in the revised statechart in Figure 11.13).

The statechart is then compared to use cases in order to check that the constraints on class behaviour shown in the statechart satisfy the requirements documented in the use case. In this example the states `Surveying` and `Evaluating` have not yet been included. These might be identified in a final sweep up to check that the statechart is complete, and could then be added as concurrent states within the `Active` state.

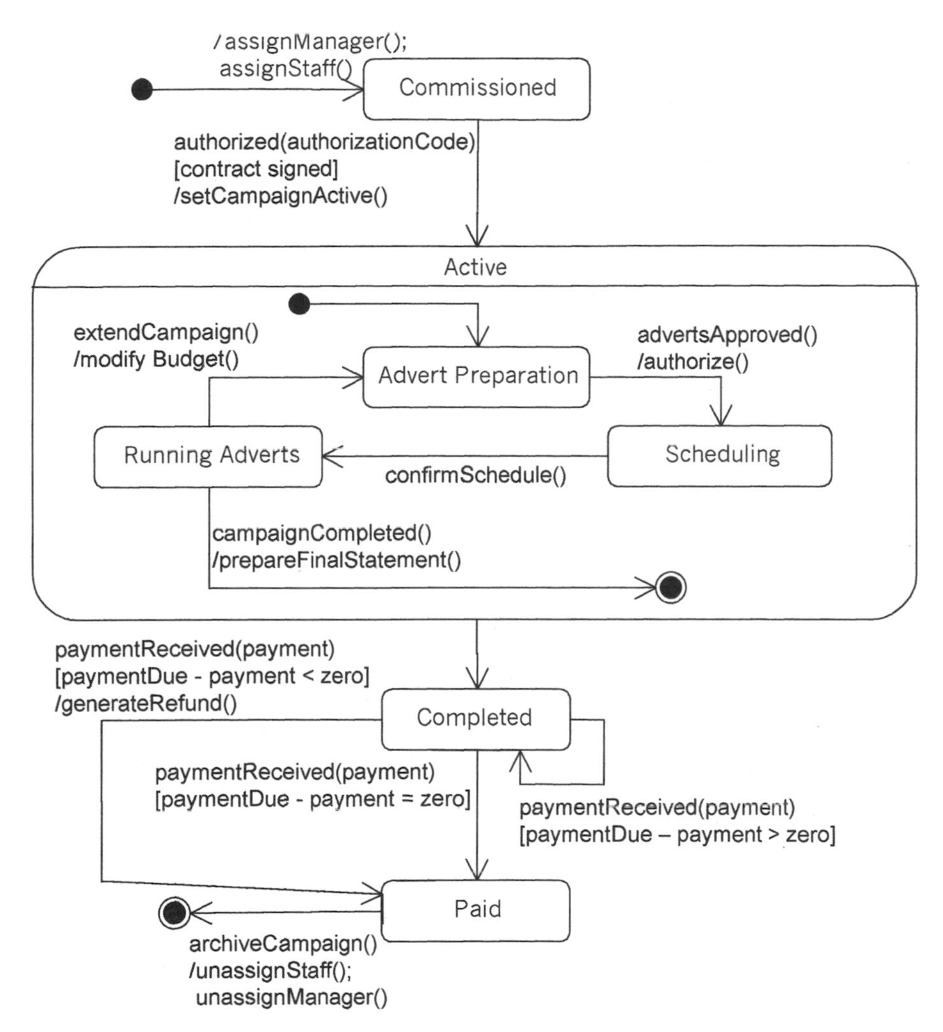

Figure 11.13 Revised statechart for the `Campaign` class.

Let us suppose that further investigation of behaviour that can affect a campaign reveals that in some circumstances a campaign can be cancelled. This is not permitted after a campaign has been completed but it can be cancelled while it is in the `Commissioned` or in the `Active` state. In either case cancellation costs are calculated for billing to the client. If the campaign is active then advertisement schedules are also cancelled. A final statechart that includes this additional requirement is shown in Figure 11.14. In this version the transition `campaignCompleted()` is shown explicitly from the nested concurrent state `Running Adverts` to the state `Completed`. When a transition like this fires any exit actions for the other concurrent states that are occupied are performed.

11.5.2 A life cycle approach

An alternative approach to the preparation of statecharts is based on the consideration of life cycles for objects of each class. This approach does not use interaction diagrams as an initial source of possible events and states. Instead, they are identified directly from use cases and from any other requirements documentation that happens to be available. First, the main system events are listed (at Agate '*A client commissions a new campaign*' might be one of the first to consider). Each event is then examined in order to determine which objects are likely to have a state dependent response to it.

The steps involved in the life cycle approach to state modelling are as follows:

1. Identify major system events.
2. Identify each class that is likely to have a state dependent response to these events.
3. For each of these classes produce a first-cut statechart by considering the typical life cycle of an instance of the class.
4. Examine the statechart and elaborate to encompass more detailed event behaviour.
5. Enhance the statechart to include alternative scenarios.
6. Review the statechart to ensure that is consistent with the use cases. In particular, check that the constraints that the statechart implies are appropriate.
7. Iterate through steps 4, 5 and 6 until the statechart captures the necessary level of detail.
8. Ensure consistency with class diagram and interaction diagrams and other statecharts.

The life cycle approach is less formal than the behavioural approach in its initial identification of events and relevant classes. It is often helpful to use a combination of the two, since each provides checks on the other. A life cycle approach might produce Figure 11.5 as an initial first-cut statechart for the `Campaign` class but further elaboration should still result in the statechart shown in Figure 11.14.

11.6 Consistency Checking

The need for consistency between different models was discussed in Chapter 9 in relation to interaction diagrams. Statecharts must also be consistent with other models.

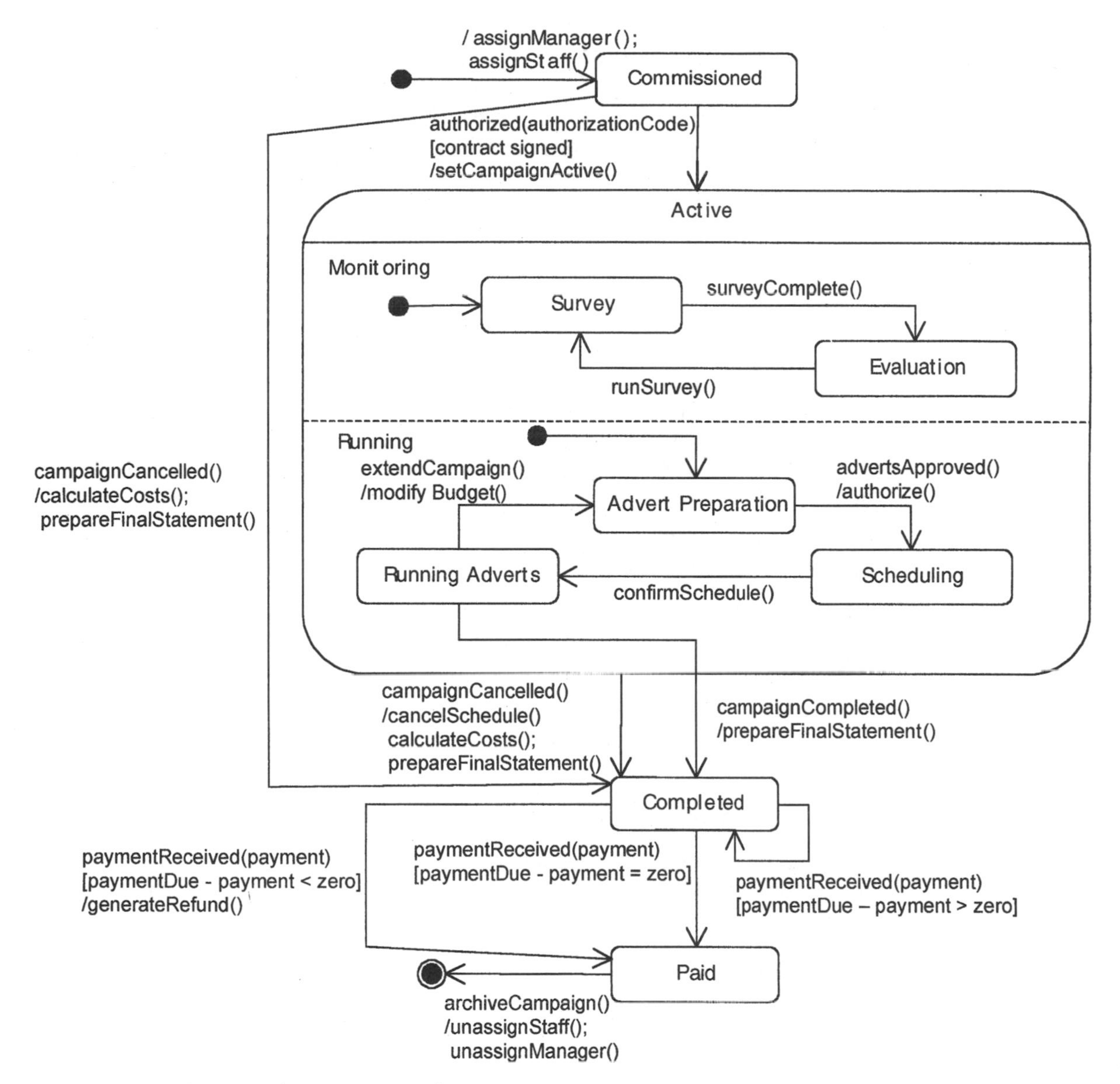

Figure 11.14 Final version of `Campaign` statechart.

- Every event should appear as an incoming message for the appropriate object on an interaction diagram.
- Every action should correspond to the execution of an operation on the appropriate class, and perhaps also to the despatch of a message to another object.
- Every event should correspond to an operation on the appropriate class (but note that not all operations correspond to events).
- Every outgoing message sent from a statechart must correspond to an operation on another class.

Consistency checks are an important task in the preparation of a complete set of models. This process highlights omissions and errors, and encourages the clarification of any ambiguity or incompleteness in the requirements.

11.7 Quality Guidelines

Preparing statecharts is an iterative process that involves refining the model until it both captures the semantics of the object life cycle and is effective for communication among team members. Listed below are a series of general guidelines that aid the production of good quality statecharts.

- Name each state uniquely to reflect what is happening for the duration of the state or what the state is waiting for.
- Do not use composite states unless the state behaviour is genuinely complex.
- Do not show too much complexity on a single statechart. If there are more than seven states consider using substates. Even with a small number of states a statechart may be too complex if there are a large number of transitions between them. Arguably the statechart in Figure 11.14 would be better represented on three diagrams, one for the high-level statechart with the detail of the `Active` state hidden and one diagram for each of the two submachines, `Running` and `Monitoring`.
- Use guard conditions carefully to ensure that the statechart describes possible behaviour unambiguously.

Statecharts are not flowcharts. Activity diagrams (see Chapters 6 and 10) are used to model procedural behaviour. Typical symptoms of statecharts that are too much like flowcharts include:

- Most transitions fired by state completion.
- Many messages sent to 'self' reflecting code re-use rather than actions triggered by events.
- States do not capture state dependent behaviour associated with the class.

Of course, a model that was intended to be statechart but turns out to be an activity diagram describing procedural flow may be a valuable model, it just is not a statechart.

11.8 Summary

The specification of the control aspects of an application is an important aspect of both analysis and design. The control aspects of an application are described in part by interaction diagrams but these focus only on a use case or an operation. In order to capture fully the control constraints for each class it is necessary to model the impact of events on that class and to model the resulting state changes with their attendant limitations on behaviour. It is only necessary to prepare statecharts for classes that have state dependent variations in behaviour. UML's statechart notation permits the construction of detailed models that may include the nesting of states and the use of concurrent states to capture complex behaviour.

Statecharts must be checked for consistency with their associated class and interaction diagrams and this may highlight the need to make modifications to these other models.

The notations provided by UML are very detailed and should be used with some care. There is no advantage in producing a statechart or an activity diagram that utilizes every UML feature unless this is really necessary for the application that is being modelled. Ideally, state models should be kept as simple as possible but should have sufficient detail to make them unambiguous and informative. The use of multiple nested states does not aid clarity unless the behaviour being described is itself complex.

Review Questions

11.1 Define event, state and transition.

11.2 What is the effect of a guard condition?

11.3 Why should all the guard conditions from a state be mutually exclusive?

11.4 What does it mean to say that an object can be in concurrent states?

11.5 How do nested states differ from concurrent states?

11.6 What is the difference between an action and an activity?

11.7 What UML modelling element has its behaviour partly described by a statechart?

11.8 What are the indications that a statechart has not been drawn to model state changes?

11.9 Against which other UML diagrams should a statechart be cross-checked?

11.10 What cross checks should be carried out?

Case Study Work, Exercises and Projects

11.A Using the interaction sequence diagrams that you prepared for Exercises 9.A–9.C, list events that affect a `ProductionLine` object and identify appropriate states for this class.

11.B Prepare a statechart for the class `ProductionLine`.

11.C List any changes that may have to be made to the class diagram for the FoodCo case study in the light of preparing this statechart.

Further Reading

- Statecharts have been used widely to model complex control behaviour. Various non-object-oriented approaches have used statecharts very effectively. In particular, the texts by Ward and Mellor (1985, 1986) and Hatley and Pirbhai (1987) provide detailed descriptions of their application in real-time applications.
- From an object-oriented perspective both Rumbaugh et al. (1991) and Booch (1994) provide useful descriptions of the Harel (1987) notation used in UML. Object-oriented real-time development approaches are well discussed by Douglass (1999) and also by Selic et al. (1994). The latter text is based on the ROOM (Real-time Object-Oriented Modeling) approach.
- Cook and Daniels (1994) give an interesting alternative perspective on the modelling of events and states.
- Useful advice on preparing statecharts can also be found in the Rational Unified Process (Rational, 2000). For a more recent view of statecharts from Harel, see Harel and Politi (1998), which presents the STATEMATE approach.

CHAPTER

Agate Ltd Case Study Further Analysis

Agate Ltd

A4.1 Introduction

In this chapter we show how the Analysis Model presented in Chapter A3 has been refined in a further iteration. The refinement has been carried out with two particular aims in mind.

First, we aim to improve our understanding of the domain and thereby increase the general usefulness of the model in a wider context. This essentially means identifying opportunities for reuse through the elaboration of generalization, composition and aggregation structures in the class model, as described in Chapter 8.

Second, we aim to improve the level of detail of the model and also the accuracy with which it reflects user requirements. This aim is addressed partly through appropriate allocation of behaviour to classes, derived from the analysis of class interaction using sequence diagrams and statecharts. In association with this activity, we also seek to specify the behavioural aspects of the model in more detail through the specification of operations. The related techniques are described in Chapters 9, 10 and 11.

Once these activities have been undertaken the analysis class diagram is revised to reflect our greater understanding of the domain and of the requirements.

The following sections include:

- samples of the sequence diagrams and statecharts that help us to understand the behavioural aspects of the model;
- specifications for some operations that capture this behaviour and communicate it to the designers;
- a revised analysis class diagram that shows the effects of further analysis on the static structure of the model.

Together, the class diagram and operation specifications comprise an analysis class model.

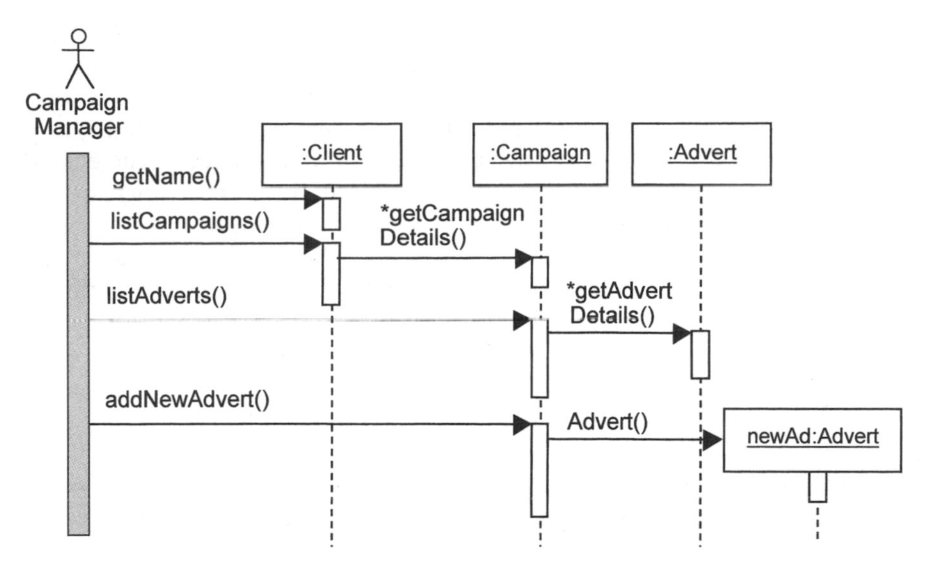

Figure A4.1 Sequence diagram for `Add a new advert to a campaign`.

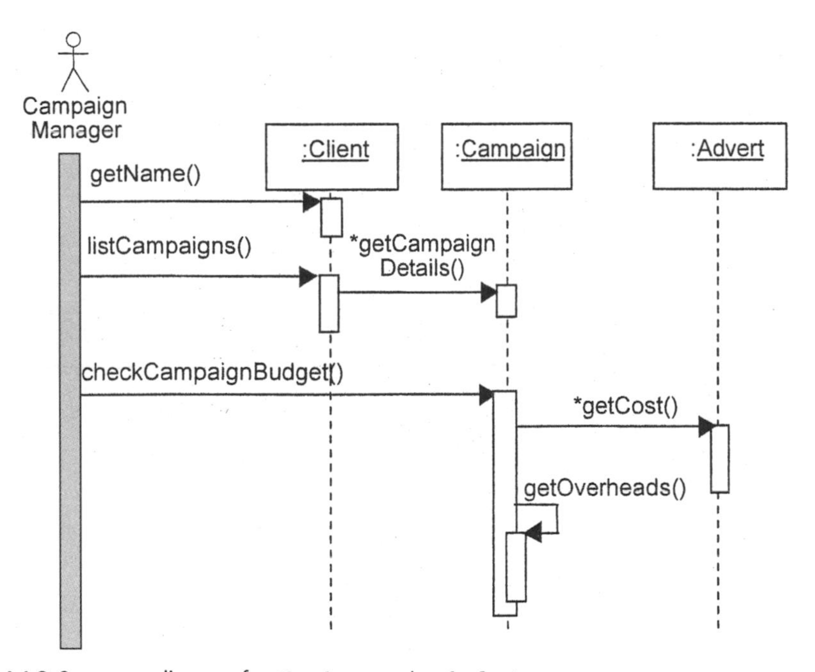

Figure A4.2 Sequence diagram for `Check campaign budget`.

A4.2 Sequence Diagrams

The first sequence diagram shown here in Figure A4.1 is for the use case `Add a new advert to a campaign`. The second sequence diagram, shown in Figure A4.2, is for the use case `Check campaign budget`. Both these sequence diagrams are discussed in some detail in Chapter 9; note that for simplicity we show here the version of `Add a new advert to a campaign` that does not include boundary and control classes.

Sequence diagrams help the requirements analyst to identify at a detailed level the operations that are necessary to implement the functionality of a use case. It is worth mentioning that, although at this point we are still primarily engaged in analysis—in other words, an attempt to understand the demands that this information system will fulfil—there is already a significant element of design in our models. There is no one correct sequence diagram for a given use case. Instead, there are a variety of possible sequence diagrams, each of which is relatively more or less satisfactory in terms of how well it meets the needs of the use case. The sequence diagrams illustrated here are the product of experimentation, judgement and several iterations of modelling carried out by analysts and users together.

A4.3 Statecharts

In this section we present the final statechart for `Campaign` (Figure A4.3), which has already been discussed at some length in Chapter 11, and an initial statechart for `StaffMember` (Figure A4.4), which is presented here for the first time. These represent

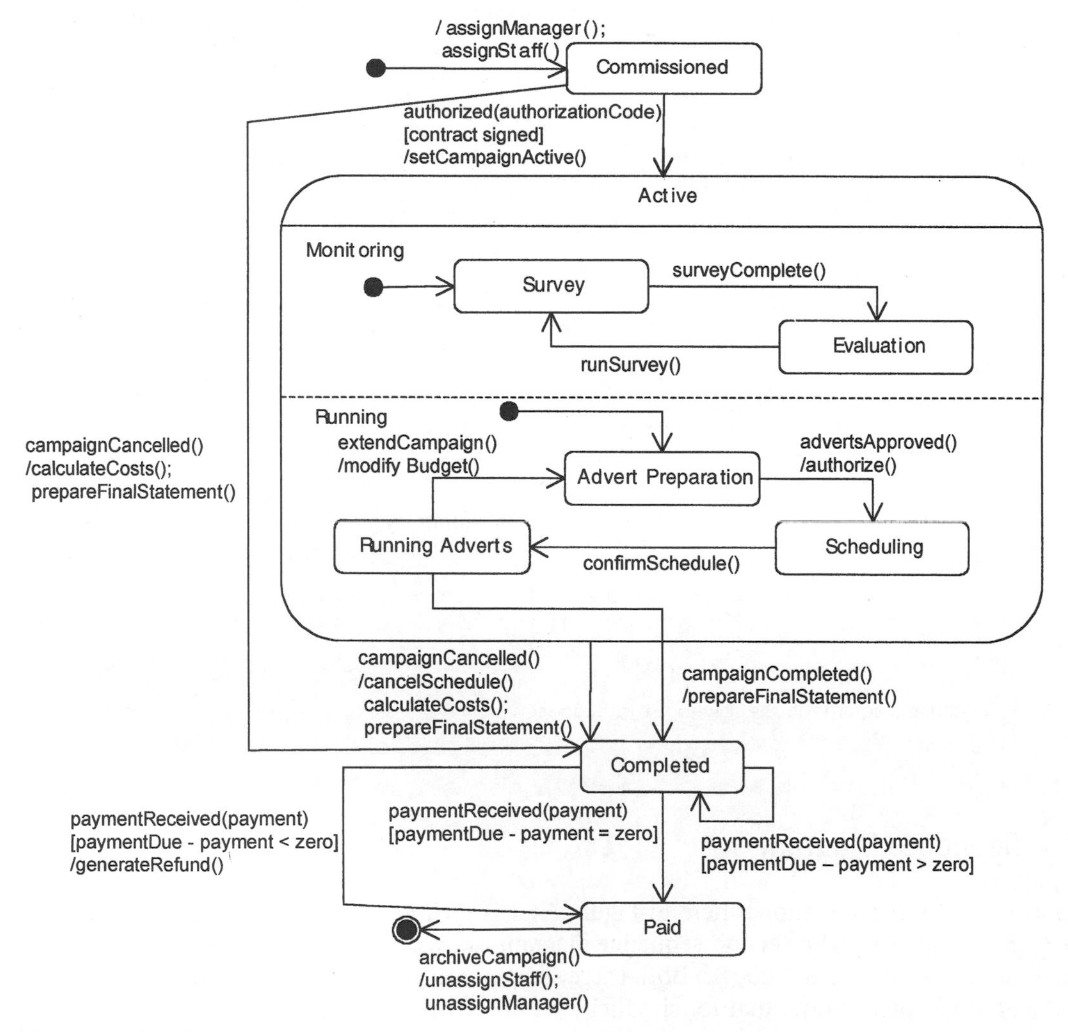

Figure A4.3 Statechart for `Campaign`.

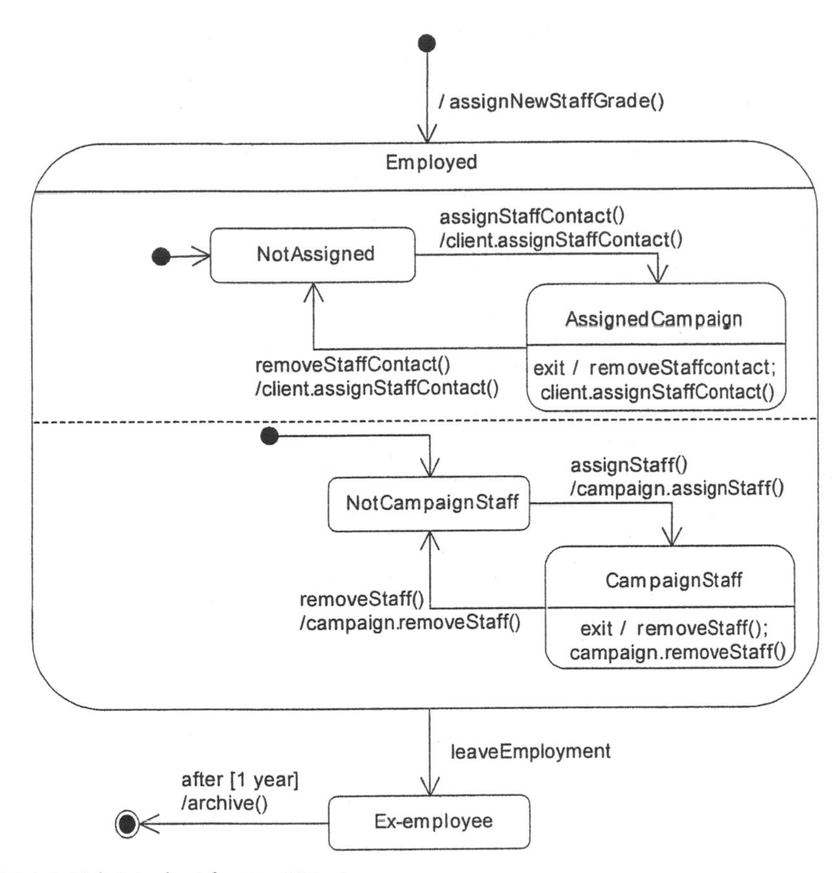

Figure A4.4 Initial statechart for `StaffMember`.

the behaviour of objects of significant classes in the `Campaign Management` and `Staff Management` analysis packages, respectively.

In conjunction with sequence diagrams, statecharts help to identify the operations that are required and to allocate those operations to appropriate classes. All operations shown on sequence diagrams and statecharts are added to the relevant class definitions. Each operation must also in due course be specified, and it is to this that we turn in the next section.

A4.4 Operation Specifications

The operation specifications given below define all operations identified for the sequence diagram `Check campaign budget`, which is shown above in Figure A4.2.

Note that in all cases the logic of the operation is very simple; for some it consists of little more than returning the value of an attribute. Each operation, and, indeed, each object, has responsibility for only a small part of the processing required to realize the use case.

By reading the operation specifications in conjunction with the sequence diagram, it is easy to see how the `Client`, `Campaign` and `Advert` objects collaborate to realize this use case.

Note that this view of collaborating objects is simplified to some extent, in that it does not include control and boundary objects and their operations. However, operations in these objects are no more complex than those shown below, since their primary role is simply to call and co-ordinate operations on the entity objects.

Context: `Campaign`

Operation specification: `checkCampaignBudget`

Operation intent: `return campaign budget and actual costs. It is assumed that the difference between can be calculated by an operation in the control class that coordinates the use case.`

operation signature: `Campaign::checkCampaignBudget()`
`campaignBudget:Money, committedExpenditure:Money`

logic description (pre- and post-conditions):

```
pre:  self->exists
post: campaignBudget = self.estimatedCost
      committedExpenditure = self.adverts.estimatedAdvertCost->sum
      result = campaignBudget, committedExpenditure
```

Other operations called: `Advert.getCost, self.getOverheads`

Events transmitted to other objects: `none`

Attributes set: `none`

Response to exceptions: `none defined`

Non-functional requirements: `none defined`

Operation specification: `getCampaignDetails`

Operation intent: `return the title and budget of a campaign.`

operation signature: `Campaign::getCampaignDetails()`
`title:String, campaignBudget:Money`

logic description (pre- and post-conditions):

```
pre:  self->exists
post: result = self.title, self.estimatedCost
```

Other operations called: `none`

Events transmitted to other objects: `none`

Attributes set: `none`

Response to exceptions: `none defined`

Non-functional requirements: `none defined`

Operation specification: `getOverheads`

Operation intent: `calculate the total overhead cost for a campaign.`

operation signature: `Campaign::getOverheads() campaignOverheads:Money`

logic description (pre- and post-conditions):

```
pre:  self->exists
post: result = self.campaignOverheads
```

Other operations called: `none`

Events transmitted to other objects: `none`

Attributes set: `none`

Response to exceptions: `none defined`

Non-functional requirements: `none defined`

Context: `Client`

Operation specification: `getName`

Operation intent: `return the client name.`

operation signature: `Client::getName()`
`self.name`

logic description (pre- and post-conditions):

```
pre:  self->exists
post: result = self.name
```

Other operations called: `none`

Events transmitted to other objects: `none`

Attributes set: none
Response to exceptions: none defined
Non-functional requirements: none defined

Operation specification: listCampaigns
Operation intent: return a list of campaigns for a client.
operation signature: Client::listCampaigns()titles:String[]
logic description (pre- and post-conditions):
pre: self->exists
post:result = self.Campaign->collect(title)
Other operations called: Campaign.getCampaignDetails
Events transmitted to other objects: none
Attributes set: none
Response to exceptions: none defined
Non-functional requirements: none defined

Context: Advert
Operation specification: getCost
Operation intent: return the actual cost for an advert.
operation signature: Advert::getCost()actualAdvertCost:Money
logic description (pre- and post-conditions):
pre: self->exists
post:result = self.actualAdvertCost
Other operations called: none
Events transmitted to other objects: none
Attributes set: none
Response to exceptions: none defined
Non-functional requirements: none defined

A4.5 Further Refinement of the Class Diagram

Figure A4.5 shows the revised analysis class diagram, after inheritance and aggregation structures have been added. For reasons of space, all attributes and operations have been suppressed from this view.

Figure A4.6 shows an excerpt from the analysis class diagram, detailing the generalization and aggregation structure for Advert with attributes and operations visible. This partial diagram reflects a further iteration of investigation and requirements modelling, which revealed that there is a requirement to keep track of the various elements used to create an advertisement. This is because photographs, music clips, and so on can often be used for more than one advertisement in a campaign, and it has been a problem to identify and retrieve these elements when they are needed.

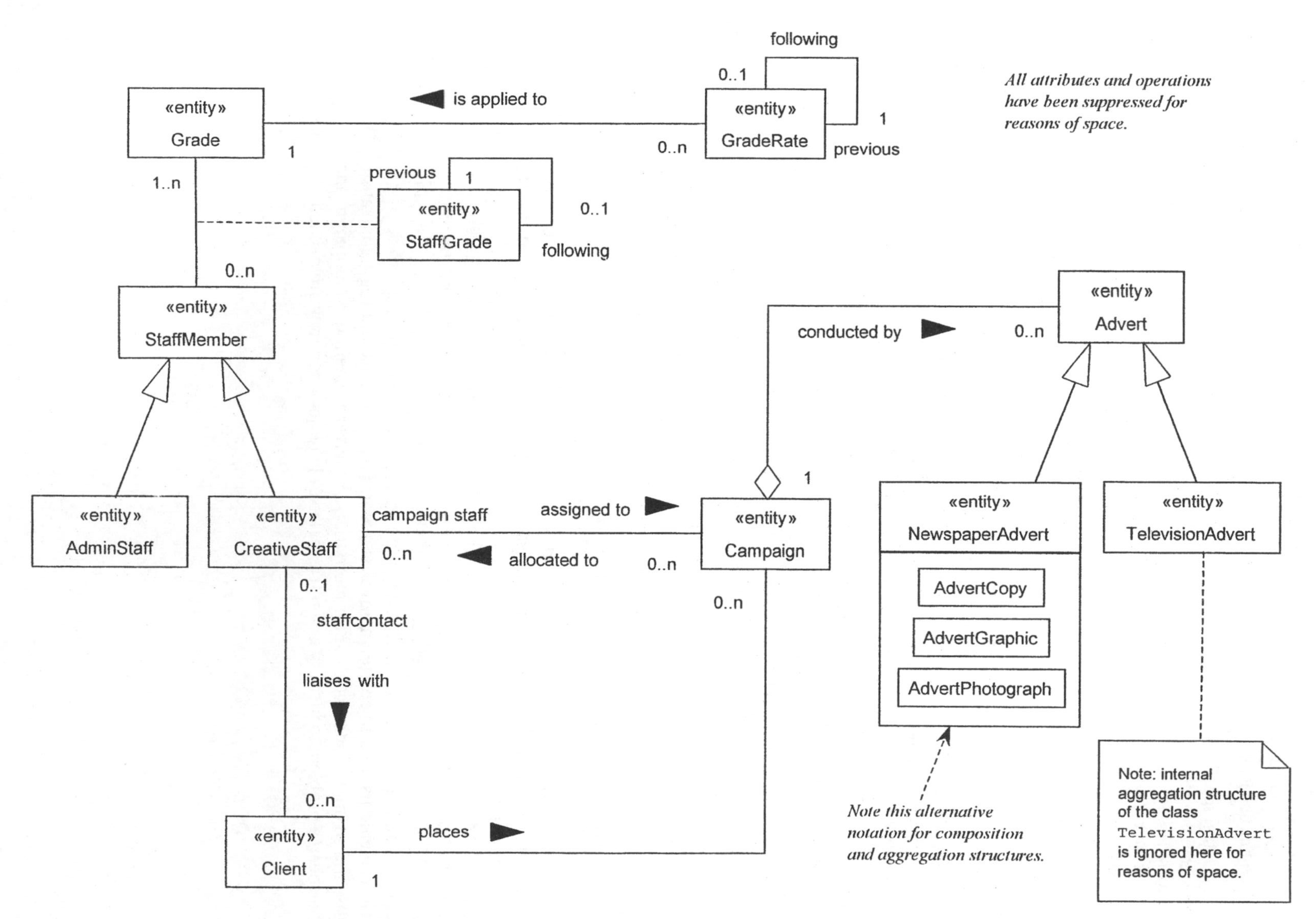

Figure A4.5 Revised analysis class diagram with generalization and aggregation structures.

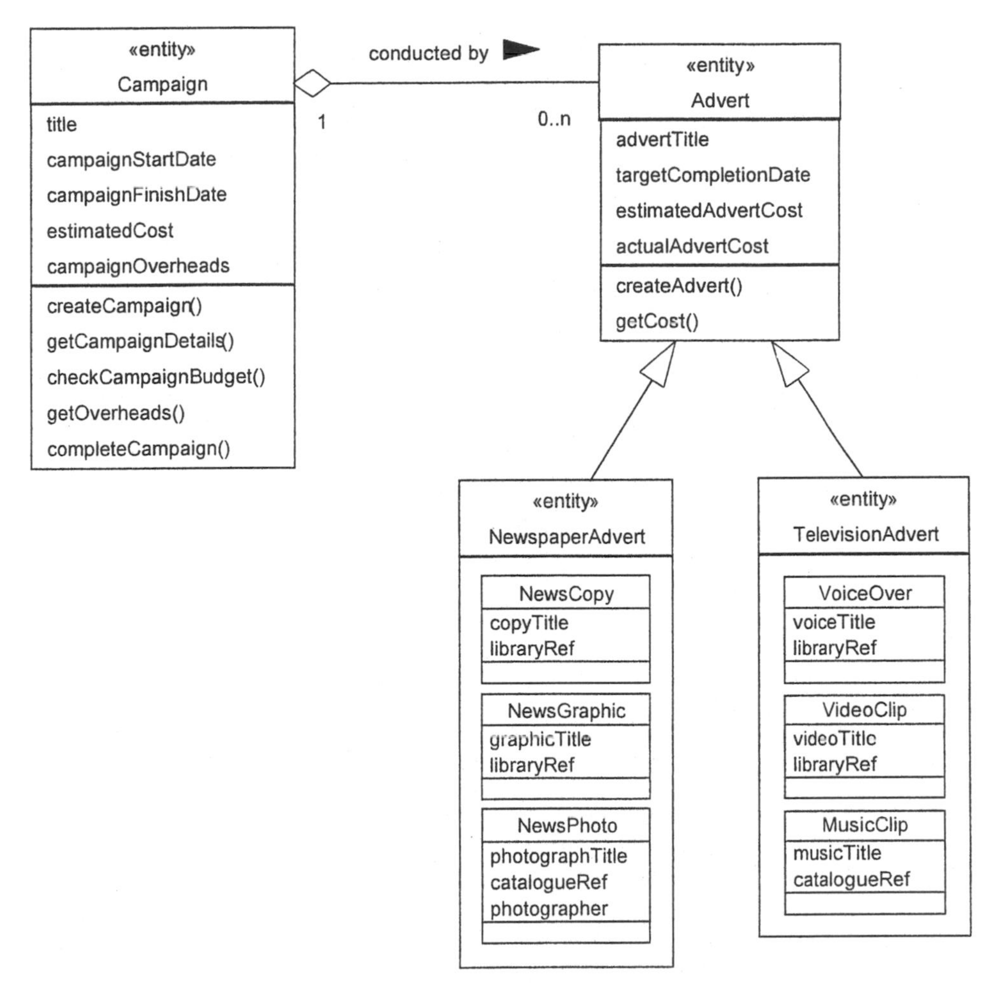

Figure A4.6 Generalization and aggegation structure for `Advert`.

A4.6 Further Activities of Requirements Analysis

Figure A4.7 shows an activity diagram that illustrates the relationship between the products of the Analysis Model before and after this iteration of analysis. Some details are worth highlighting.

- The analysis class diagram is now termed a model, since it includes some detailed class definition. In particular, all operations should now be specified at least in outline.
- Some parts of the analysis model may be substantially unchanged during this iteration, for example, the collaboration diagrams and the glossary. Although this is not necessarily the case, we have shown these as unaffected in Figure A4.7.
- As a result of the operation specification activity, many attributes may also have been specified in more detail. Some, particularly those that are required to provide parameters to operations in other classes, will certainly now be typed. We have not shown this yet, since the typing of attributes is essentially a design activity. But in practice, some design decisions are made in parallel with the more detailed analysis that we describe in this chapter.

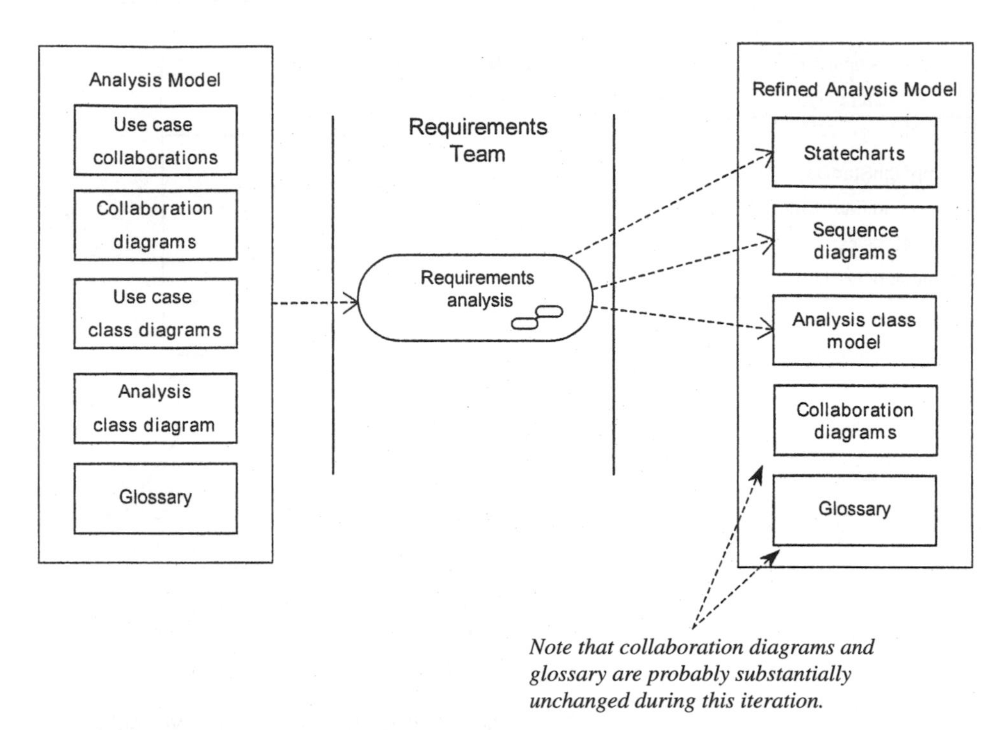

Figure A4.7 High level activity diagram showing how elements of the analysis model are created or updated during this iteration of analysis.

Figure A4.8 shows a more detailed view of the activities that are carried out and the products directly used or affected during this iteration. In this diagram, we have tried to suggest a sensible outline sequence for carrying out the various activities. However, it should be noted that this is no more than a guide, and is certainly not meant to be prescriptive. An iterative approach should always be followed that is sensitive to the needs of the project, to the skill of the developers and to the often haphazard manner in which understanding grows during the modelling and analysis of requirements.

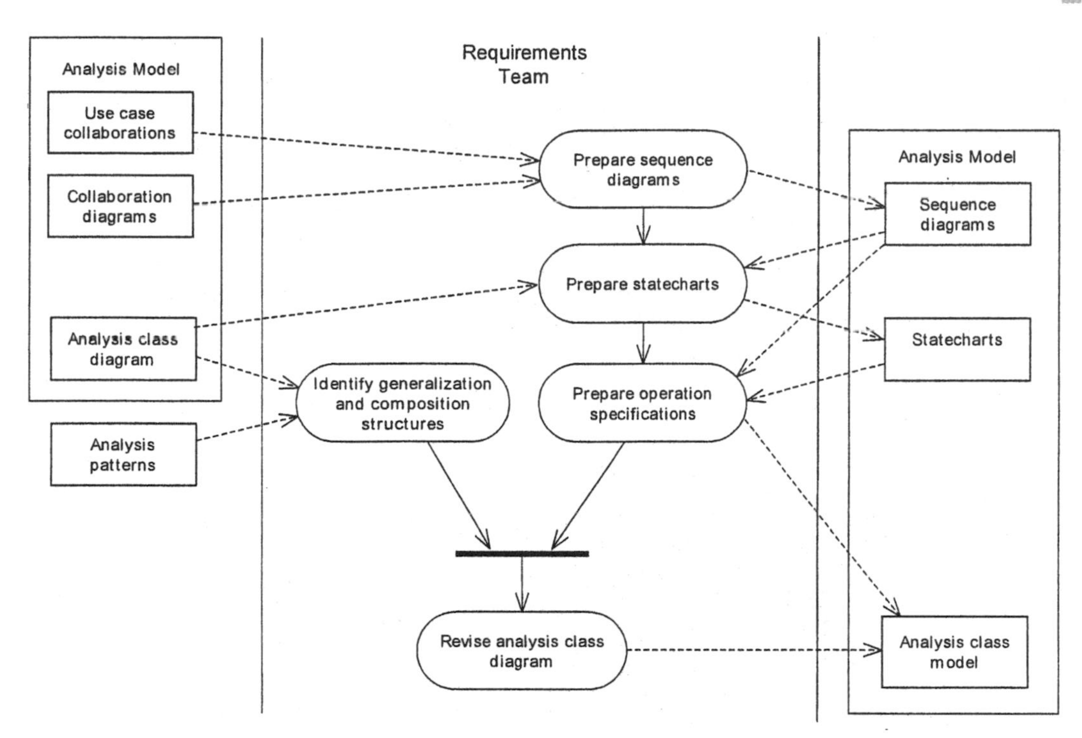

For clarity, we have detailed only those activities and products that are mostly directly involved in this iteration. Note also that the flow of activities is indicative and is not intended to be prescriptive.

Figure A4.8 The activities that are carried out and the products directly used or affected during this iteration of analysis.

12 CHAPTER

Moving into Design

OBJECTIVES

In this chapter you will learn

- the difference between analysis and design
- the difference between logical and physical design
- the difference between system and detailed design
- the characteristics of a good design
- the need to make trade-offs in design.

12.1 Introduction

Two questions that may be asked by those who are new to systems analysis and design are 'What is the difference between analysis and design?' and 'Why treat analysis and design as separate activities?' In the development of information systems, as in the development of many kinds of system, the process of analysis is distinguished from the process of design (Section 12.2). Analysis is often said to be about the 'What?' of a system, and design is described as being about the 'How?'. Design can start before or after the decision has been made about the hardware and software to be used in implementing the system. Implementation-independent or logical design is distinguished from implementation-dependent or physical design (Section 12.3). Design also takes place at two main levels: system design, which addresses architectural aspects that affect the overall system, and detailed design, which addresses the design of classes and the detailed working of the system (Section 12.4). In producing a design for a system, a designer will be working within a framework of general quality criteria and will also be trying to achieve measurable objectives for the design that are specific to the particular system (Section 12.5). Some of these objectives may conflict with one another, and constraints on the design may result in the need for trade-offs to be made.

12.2 How Is Design Different from Analysis?

Design has been described by Rumbaugh (1997) as stating 'how the system will be constructed without actually building it'. The models that are produced by design activities show how the various parts of the system will work together; the models produced by analysis activities show what is in the system and how those parts are related to one another.

The word *analysis* comes from a Greek word meaning to break down into component parts. When we analyse an organization and its need for a new system, the analysis activity is characterized as asking *what* happens in the current system and *what* is required in the new system. It is a process of seeking to understand the organization, investigating its requirements and modelling them. The result of this analysis activity is a specification of what the proposed system will do based on the requirements.

Design is about producing a solution that meets the requirements that have been analysed. The design activity is concerned with specifying *how* the new system will meet the requirements. There may be many possible design solutions, but the intention is to produce the best possible solution in the circumstances. Those circumstances may reflect constraints such as limits on how much can be spent on the new system or the need for the new system to work with an existing system. Jacobson et al. (1992) regard design as part of the construction process (together with implementation). The systems designer has his or her attention focused on the implementation of the new system, while the systems analyst is focused on the way the business is organized and a possible better organization; the foci of these two activities are very different.

A simple example of this can be seen in the Agate case study. Analysis identifies the fact that each `Campaign` has a `title` attribute, and this fact is documented in the class model. Design determines how this will be entered into the system, displayed on screen and stored in some kind of database together with all the other attributes of `Campaign` and other classes.

Design can be seen either as a stage in the systems development life cycle or as an activity that takes place within the development of a system. In projects that follow the waterfall life cycle model (Figure 3.2), the Analysis stage will be complete before the Design stage begins. However, in projects that follow an iterative life cycle, design is not such a clear-cut stage, but is rather an activity that will be carried out on the evolving model of the system. Rumbaugh (1997) distinguishes between the idea of design as a stage in the waterfall life cycle and design as a process that different parts of the model of the system will go through at different times.

In the Unified Process (Jacobson et al., 1999), design is organized as a workflow—a series of activities with inputs and outputs—that is independent of the project phase. In the Rational Unified Process (Kruchten, 1999), analysis and design are combined into a single workflow—the analysis activities produce an overview model, if it is required, but the emphasis is on design—and the workflow is similarly independent of the project phase. We have adopted a similar approach to the Unified Process in the process outlined in Chapter 5. A project consists of major phases (Inception, Elaboration, Construction and Transition); each phase requires one or more iterations, and within the iterations, the amount of effort dedicated to the activities in each workflow gradually increases and then declines as the project progresses. The difference between this kind of approach and the traditional waterfall model is that in the traditional approach Analysis, Design, Construction and other stages in the waterfall are both activities and stages: during the Analysis stage, for example, all the analysis activity is meant to take place. Real projects are not like this: during the early part of the project, maybe called Analysis, some design activity may take place; during the later part of the project, maybe called Design, some analysis activity may take place. Process models such as the Unified Process recognize this and give the phases different names to decouple them from the activities. As long as less and less analysis and more and more design take place as the project develops, the project is making progress.

Despite this, many projects still treat analysis and design as separate stages rather than activities that gradually elaborate the model as the project progresses. In the rest of this section, we explain some of the reasons why this approach is taken.

12.2.1 Design in the traditional life cycle

In large-scale projects that follow a traditional systems development life cycle there are a number of advantages to making a clear break between analysis and design. These are concerned with:

- project management,
- staff skills and experience,
- client decisions, and
- choice of development environment.

Project management. The project manager will have an overall budget in terms of money and staff time within which the system must be developed. Some proportion of those resources will have been allocated to analysis and some to design. In order to manage and control the project effectively, the project manager will want to have a clear idea of how much time is spent on each of these activities. If the two activities are allowed to merge, then the management of the project becomes more difficult. If all the time is being spent on analysis, then the project will fall behind schedule, whereas if all the time is being spent on design, then it is likely that the requirements have not been properly understood.

Staff skills and experience. Analysis and design may be carried out by staff with different skills and experience. Staff with job titles such as business analyst and systems analyst will have the skills and expertise to carry out the analysis, while systems architects and systems designers will have an understanding of the technology available to deliver the solution and will carry out the design.

Client decisions. The clients will want to know what they are paying for. The end of analysis is often a decision point in a project. The clients will be provided with a specification of the system that can be traced back to their requirements and they will have to agree to this or *sign it off* before work progresses to design. In some projects, the client may be presented with a number of alternative specifications that differ in scope (what parts of the system will be computerized). In this case, the client must choose which of these alternative systems to take forward into design.

Choice of development environment. In many projects the hardware and software that will be used to develop and deliver the finished system will not be known at the time of the analysis stage. There may be good reasons for delaying the choice of hardware and software until the requirements for the system have been determined by the analysis, especially in the rapidly changing world of information technology in which some new technology always seems to be due for release. Because the choice of hardware, development language and database will affect the design, it may be necessary to make a break between analysis and design so that decisions about the development environment can be made. This point is discussed in more detail in the next section.

12.2.2 Design in the iterative life cycle

There are also advantages to be gained from using an iterative life cycle such as the Unified Software Development Process. These are concerned with:

- risk mitigation,
- change management,
- team learning and
- improved quality.

Risk mitigation. An iterative process enables the identification of potential risks and problems earlier in the life of a project. The early emphasis on architecture and the fact that construction, test and deployment activities are begun early on, make it possible to identify technological problems and take action to reduce them. Integration of subsystems is begun earlier and is less likely to throw up unpleasant surprises at the last minute.

Change management. Users' requirements do change during the course of a project, often because of the time that projects take, and often because until they see some results they may not be sure what they want. In a waterfall life cycle, changing requirements are a problem, in an iterative life cycle there is an expectation that some requirements activities will still be going on late in the project, and it is easier to cope with changes. It is also possible to revise decisions about technology during the project, as the hardware and software available to do the job will almost certainly change during the project.

Team learning. Members of the team, including those concerned with testing and deploying, are involved in the project from the start, and it is easier for them to learn about and understand the requirements and the solution from early on. They are not then suddenly presented with a new and unfamiliar system. It is also possible to identify training needs and provide the training while people are still working on an aspect of the system.

Improved quality. Testing of deliverables begins early and continues throughout the project. This helps to prevent the situation where all testing is done in a final 'big bang' and there is little time to resolve the bugs that are found.

The use of object-oriented techniques helps to take advantage of an iterative life cycle. Before object-oriented approaches were developed, structured analysis and design was the dominant approach to analysis and design. In structured approaches, a clear distinction between analysis and design is made in terms of the types of diagram that are used. During analysis data flow diagrams are used to model requirements, whereas structure charts or structure diagrams are used to model the design of the system and the programs in it.

One of the arguments put forward for the use of object-oriented approaches is that the same model (the class diagram or object model) is used right through the life of the project. Analysis identifies classes, those classes are refined in design, and the eventual programs will be written in terms of classes. While this so-called *seamlessness* of object-oriented methods may seem like an argument for weakening the distinction between analysis and design, when we move into design different information is added to the class diagram, and other different diagrams are used to support the class diagram.

Rumbaugh (1997) distinguishes between analysis and design in terms of the amount of detail that is included in the model. On a continuum, the analysis stage provides an abstract model of 'what to do' while the design stage documents 'exactly how to do it'. As the project moves from one end of this continuum to the other, additional detail is added to the model until a clear specification of 'how to do it' is provided. This additional detail is added in the form of diagrams such as collaboration diagrams, state diagrams and deployment diagrams that supplement the information in the class diagram. The class diagram is also enhanced during design by the addition of detail about attributes and operations and additional classes to handle the implementation of the user interface, communication between layers and data storage.

12.3 Logical and Physical Design

At some point in the life of a system development project a decision must be made about the hardware and software that are to be used to develop and deliver the system—the hardware and software platform. In some projects this is known right from the start. Many companies have an existing investment in hardware and software, and any new project must use existing system software (such as programming languages and database management systems) and will be expected to run on the same hardware. This is more often the case in large companies with mainframe computers. In such companies the choice of configuration has been limited in the past to the use of terminals connected to the mainframe. However, client-server architectures (see next section) and open system standards, that allow for different hardware and software to operate together, have meant that even for such companies, the choice of platform is more open. For many new projects the choice of platform is relatively unconstrained, and so at some point in the life of the project a decision must be made about the platform to be used.

Some aspects of the design of systems are dependent on the choice of platform. These will affect the system architecture, the design of objects and the interfaces with various components of the system. Examples include the following.

- The decision to create a distributed system with elements of the system running on different machines will require the use of some *middleware* such as is provided by CORBA to allow objects to communicate with one another across the network. This will affect the design of objects.
- The decision to write programs in Java and to use a relational database that supports ODBC (Object Data Base Connectivity) will require the use of JDBC (Java Data Base Connectivity) and the creation of classes to map between the objects and the relational database.
- The choice of Java as a software development language will mean that the developer has the choice of using the standard Java AWT (Abstract Windowing Toolkit), the Java Swing classes or proprietary interface classes for designing the interface.
- Java does not support multiple inheritance; other object-oriented languages such as C++ do. If the system being developed appears to require multiple inheritance then in Java this will have to be implemented using Java's interface mechanism.
- If the system has to interface with special hardware, for example bar-code scanners, then it may be necessary to design the interface so that it can be written in C as a *native method* and encapsulated in a Java class, as Java cannot directly access low-level features of hardware.

Java has been used here as an example. The same kinds of issues will arise whatever platform is chosen.

It is also the case that there are many design decisions that can be made without knowledge of the hardware and software platform.

- The interaction between objects to provide the functionality of particular use cases can be designed using interaction diagrams or collaboration diagrams.
- The layout of data entry screens can be designed in terms of the fields that will be required to provide the data for the objects that are to be created or updated, and the order in which they will appear on the screen can be determined. However, the exact nature of a textbox and whether it is a Borland C++ `TEdit`, a Delphi `Edit`, a Java `TextField`, a C# `TextBox` or something else can be left until later.
- The nature of commands and data to be sent to and received from special hardware or other systems can be determined without needing to design the exact format of messages.

Because of this, design is sometimes divided into two stages. The first is *implementation-independent* or *logical* design and the second is *implementation-dependent* or *physical* design. Logical design is concerned with those aspects of the system that can be designed without knowledge of the implementation platform; physical design deals with those aspects of the system that are dependent on the implementation platform that will be used.

Having an implementation-independent design may be useful if you expect a system to have to be re-implemented with little change to the overall design but on a different platform, for example a Windows program that is to be ported to MacOS and Linux, or a program that must run on different types of handheld using Windows CE, EPOC and PalmOS.

In many projects, design begins after hardware and software decisions have been made. However, if this is not the case, then the project manager must ensure that the plan of work for the project takes account of this and that logical design activities are tackled first. In an iterative project life cycle, logical design may take place in the early design iterations, or if the system is partitioned into sub-systems, the logical design of each sub-system will take place before its physical design.

12.4 System Design and Detailed Design

Design of systems takes place at two levels: system design and detailed design. System design is concerned with the overall architecture of the system and the setting of standards, for example for the design of the human–computer interface; detailed design is concerned with designing individual components to fit this architecture and to conform to the standards. In an object-oriented system, the detailed design is mainly concerned with the design of objects. These two levels of design are described briefly in the next sections and in more detail in Chapters 13 and 14 (where detailed design is dealt with as class design).

12.4.1 System design

During system design the designers make decisions that will affect the system as a whole. The most important aspect of this is the overall architecture of the system. Many modern systems use a client–server architecture in which the work of the system is divided between the clients (typically PCs on the users' desks) and a server (usually a Unix or Windows NT machine that provides services to a number of users). This raises questions about how processes and objects will be distributed on different machines, and it is the role of the system designer or system architect to decide on this. The design will have to be broken down into sub-systems and these sub-systems may be allocated to different processors. This introduces a requirement for communication between processors, and the systems designer will need to determine the mechanisms used to provide for this communication. Distributing systems over multiple processors also makes it possible for different sub-systems to be active simultaneously or concurrently. This concurrency needs to be designed into the system explicitly rather than left to chance.

Many organizations have existing standards for their systems. These may involve interface design issues such as screen layouts, report layouts or how on-line help is provided. Decisions about the standards to be applied across the whole system are part of system design, whereas the design of individual screens and documents (to comply with these standards) is part of detailed design.

When a new system is introduced into an organization, it will have an impact on people and their existing working practices. Job design is often included in system design and addresses concerns about how people's work will change, how their interest and motivation can be maintained, and what training they will require in order to carry out their new jobs. How people use particular use cases will be included in the detailed design of the human–computer interface.

12.4.2 What happens in traditional detailed design?

In the 1960s and 1970s, detailed design was seen as consisting of four main activities:

- designing inputs,
- designing outputs,
- designing proccsses and
- designing files.

Designing inputs meant designing the layout of menus and data entry screens; designing outputs was concerned with the layout of enquiry screens, reports and printed documents; designing processes dealt with the choice of algorithms and ensuring that processes correctly reflected the decisions that the software needed to make; and designing files dealt with the structure of files and records, the file organization and the access methods used to update and retrieve data from the files.

The development of structured design methods made two major changes to the way in which design was carried out. First, the work of Jackson (1975) provided a method for designers to design programs by using a technique to match the structure of the inputs and outputs with the structure of data to be read from or written to files.

The second major contribution came from Yourdon and Constantine (1979) who defined a series of criteria that could be used in breaking systems and programs down into modules to ensure that they are easy to develop and maintain. These criteria concern two issues: *cohesion* and *coupling*. Criteria to maximize desirable types of cohesion have as their aim the production of modules—sections of program code in whatever language is used—that carry out a clearly defined process or a group of processes that are functionally related to one another. This means that all the elements of the module contribute to the performance of a single function. Poor cohesion is found when processes are grouped together in modules for other reasons. Examples of poor types of cohesion include:

- when processes are grouped together for no obvious reason (*coincidental cohesion*),
- because they handle logically similar processes such as inputs (*logical cohesion*),
- because they happen at the same time—for example when the system initializes—(*temporal cohesion*) and
- because the outputs of one process are used as inputs by the next (*sequential cohesion*).

By aiming to produce modules that are functionally cohesive, the designer should produce modules that are straightforward to develop, easy to maintain and have the maximum potential to be reused in different parts of the system. This will be assisted if coupling between modules is also reduced to the minimum.

Criteria to minimize the coupling between modules have as their aim the production of modules that are independent of one another and that can be amended without resulting in knock-on effects to other parts of the system. Good coupling is achieved if a module can perform its function using only the data that is passed to it by another module and using the minimum necessary amount of data. Poor coupling is found in the following circumstances:

- modules that rely on data in global variables or data in common blocks (used in languages such as COBOL and FORTRAN) that other modules may change,
- modules that are designed to need large amounts of data to be passed to them as parameters and
- modules that are not cohesive because they perform several functions and therefore require control information as well as data to be passed as parameters so that a decision can be made within the module about which function is required.

Modules with low coupling and high cohesion are the aim of structured design and programming techniques. In Chapter 14 we shall explain how these issues still apply to the design of object-oriented systems. Objects can be viewed in terms of cohesion—each class in the class diagram should encapsulate data and functionality that clearly belongs together. It is no less possible to produce classes that reflect poor cohesion, for example temporal cohesion, as it is to produce modules with poor cohesion. Classes with poor cohesion will reflect the structure of the particular application and will not be reusable in other circumstances. The assignment of responsibilities to classes, that is determining which operations belong in which classes, contributes to the cohesiveness and reusability of classes. Cohesion and coupling also apply to the operations of classes. Larman (1998) discusses the assignment of responsibilities to objects in order to achieve high cohesion and low coupling, and these issues are discussed in more detail in Chapter 14.

12.4.3 What do we add in O-O detailed design?

Traditionally, detailed design has been about designing inputs, outputs, processes and file or database structures; these same aspects of the system also have to be designed in an object-oriented system, but they will be organized in terms of classes. During the analysis phase of a project, concepts in the business will have been identified and elaborated in terms of classes, and use cases will have been identified and described. The classes that have been included in the class diagram will reflect the business requirements but they will only include a very simplistic view of the classes to handle the interface with the user, the interface with other systems, the storage of data and the overall co-ordination of the other classes into programs. These classes will be added in design with greater or lesser degrees of detail depending on the hardware and software platform that is being used for the new system.

Different authors describe these additional aspects of the system in different ways. Coad and Yourdon (1991) call the business classes the problem domain component and develop three further components in the design phase:

- human interface component,
- data management component and
- task management component.

Coad et al. (1997) propose a slightly different set of components:

- human interface component,
- data management component and
- system interaction component.

In their designs, windows play the co-ordinating role played by the classes in the task management component. Larman (1998) proposes an architecture based on three layers:

- presentation layer (windows and reports),
- application logic layer and
- storage layer.

However, his application logic layer includes both domain concepts—equivalent to the problem domain component—and services, which may include interfaces to the database and to other systems. Figure 12.1 shows these layers as UML packages.

What should be apparent, whatever the terminology, is that it is necessary to decide on the architecture that will be used and that this will involve the designers in designing classes to handle these aspects of the system. The amount of work involved may vary from system to system and component to component, depending on the hardware and software platform to be used. In particular, the design of these components offers opportunities to reuse existing classes.

12.4.4 Aspects requiring attention in O-O detailed design

Certain aspects of the detailed design require special attention in the development of object-oriented systems. These include reuse and assignment of responsibilities to classes.

One of the arguments for the use of object-oriented languages is that they promote reuse through encapsulation of functionality and data together in classes and through the use of inheritance. This is not just a programming issue, but one that also affects

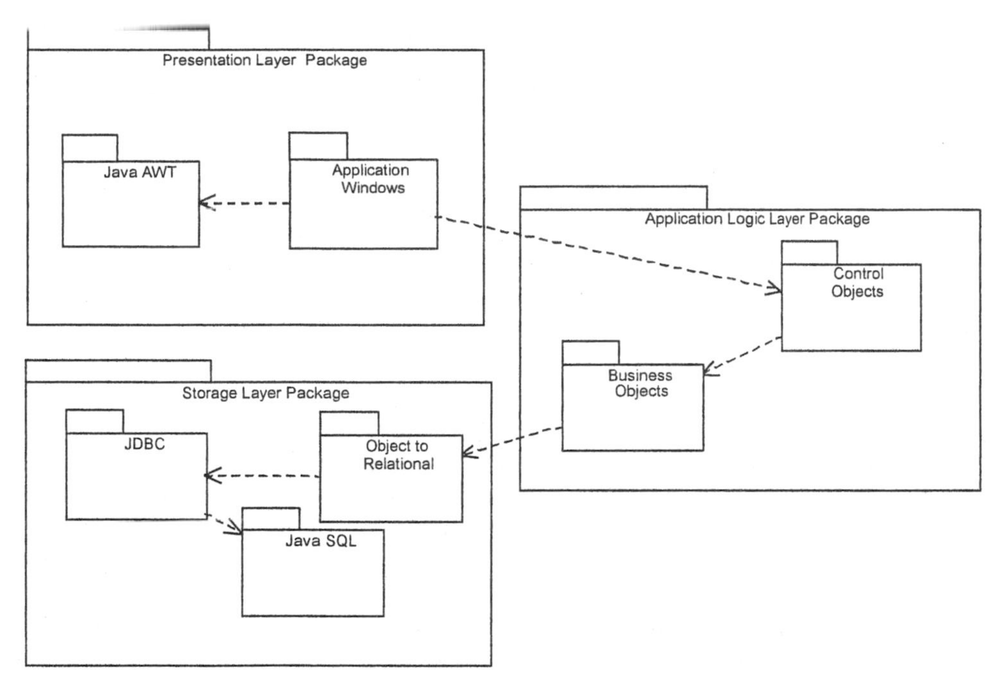

Figure 12.1 UML packages representing layers in the three-tier architecture.

analysis and design. There is a growing recognition of the need to reuse analysis results in object-oriented systems development. Design reuse already takes place at two levels: first through the use of design patterns, which are discussed in detail in Chapter 15; and second by recognizing during design that business classes that have been identified during analysis may be provided by reusing classes that have already been designed within the organization, or even bought in from outside vendors. There is a move in the software industry towards the use of *components* that provide this kind of functionality and that can be bought from vendors (Allen and Frost, 1998). CASE tool suppliers such as SELECT Software Tools have added component management software to their range of products, and IBM, through its San Francisco project, is making general-purpose business classes available as part of component packages. This issue is addressed in more detail in Chapter 20.

The assignment of responsibilities to classes is an issue that is related to reuse. Larman (1998) highlights this activity as the main task in design. In an object-oriented system, it is important to assign responsibility for operations to the right classes, and there is often a choice. In the FoodCo system, there will be a need to produce invoices for customers that include the calculation of Value Added Tax (VAT). (Value Added Tax is a tax used throughout Europe that is applied at each stage of the supply chain and not just as a purchase tax paid by the final end-user or consumer.) The calculation of VAT could be carried out by one of a number of classes in the model (Figure 12.2).

- Invoice—which organizes the total information for the whole sale.
- InvoiceLine—which contains the detail of each item sold and to which the tax applies.
- Product—to which different VAT rates may apply.
- TaxRate—which carries the details of the percentage that applies for each valid rate.

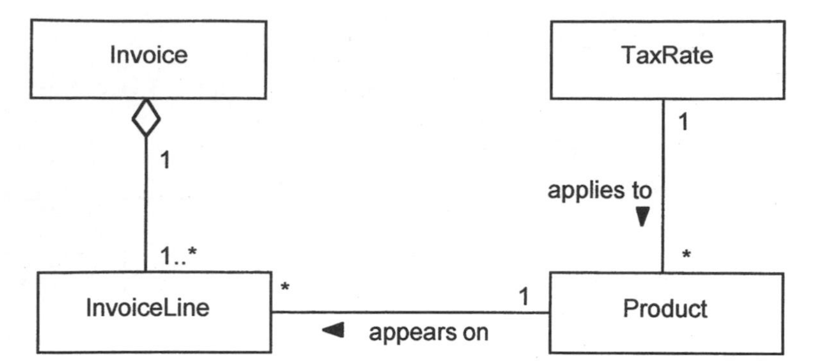

Figure 12.2 Partial class diagram for FoodCo.

If the designer makes the wrong decision, the resulting class will be less reusable and may constrain the design of other classes. If the responsibility for tax calculation is allocated to `Invoice` or `InvoiceLine`, then this has implications for `CreditNote` and `CreditNoteLine`, which may also need to calculate tax. If it is assigned to `Product`, then it cannot be reused in the Agate project where VAT applies to services as well as products. Clearly it needs to be assigned to `TaxRate` in order to maximize the reuse that can be made of the classes in this design.

12.5 Qualities and Objectives of Analysis and Design

At the end of the previous section, reusability was highlighted as something that designers of object-oriented systems are trying to achieve in their design. Reusability is not the only objective of design. There are a number of other criteria for a good design. Perhaps the most obvious measure of design quality is whether the finished application is of high quality. This assumes that the analysis that preceded the design work was itself of high quality. However, this is a rather vague and circular way of assessing quality: in this section we look briefly at aspects of quality in analysis and in more detail at quality in design. Design quality cannot be measured only against general criteria, and in Section 12.6 we consider more specific measures that can be applied to particular projects.

12.5.1 What makes for good analysis?

The cost of fixing faults in a system increases as the system progresses through the system's development life cycle. If an error occurs in the analysis of a system, it is cheaper to fix it during the analysis phase than it is later when that error may have propagated through numerous aspects of the design and implementation. It is most expensive to fix it after the system has been deployed and the error may be reflected in many different parts of the system. The quality of the design is, therefore, dependent to a large extent on the quality of the analysis.

Some methodologies have explicit quality criteria that can be applied to the products of every stage of the life cycle, but these quality criteria typically check syntactic aspects of the products, that is whether the notation is correct in diagrams, rather than semantic aspects, that is whether the diagrams correctly represent the organization's requirements. To provide a sound foundation for design, analysis should meet the following four criteria:

- correct scope,
- completeness,
- correct content and
- consistency.

These are described in more detail below.

Correct scope. The scope of a system determines what is included in that system and what is excluded. It is important, first that the required scope of the system is clearly understood, documented and agreed with the clients, and second that everything that is in the analysis models *does* fall within the scope of the system. In the case of the Agate system, it is not a requirement to replace the existing accounting system that is used to invoice clients. It is, however, a requirement that the new system should interface with the accounting system to provide for the transfer of data relating to financial aspects of advertising campaigns. The scope of the system therefore excludes use cases for accounting but should include use cases both to handle the entry of data that will be transferred to the accounting system and to handle the transfer itself. Coad et al. (1997) include a *not this time* component with their other four components (problem domain, human interface, data management and system interaction). The not this time component is used to document classes and business services that emerge during the analysis but are not part of the requirements this time. This is a useful way of forcing consideration of the scope of the system.

Clearly determining the scope of a system before design starts is a strong argument for making a distinction and a break between the analysis and design phases of a project. If the scope of the Agate system is agreed before design begins, then the project manager can plan more accurately for the work that will have to take place during the design phase. If the specification is allowed to evolve iteratively, then the management of Agate could introduce new requirements well into the life of the project that could have a severe impact on the ability to complete the project on time. An example could be the requirement to store not just data about adverts but digital multimedia versions of the components of the adverts themselves so that mock-ups of the adverts can be edited together and shown to clients on screen. An additional requirement of this nature would transform the project from a relatively straightforward business information system to a complex multimedia system.

Completeness. Just as there is a requirement that everything that is in the analysis models is within the scope of the system, so everything that is within the scope of the system should be documented in the analysis models. Everything that is known about the system from the requirements capture should be documented and included in appropriate diagrams. Often the completeness of the analysis is dependent on the skills and experience of the analyst. Knowing what questions to ask in order to elicit requirements comes with time and experience. However, analysis patterns and strategies, as proposed by Coad et al. (1997) and Fowler (1997), can help the less experienced analyst to identify likely issues. (The use of patterns, which draw on past experience, can be a good way of ensuring that the analysis is effective.)

Non-functional requirements should be documented even though they may not affect the analysis models directly. Rumbaugh (1997) suggests that some of the requirements found during analysis are not analysis requirements but design requirements. These should be documented, but the development team may only have to consider them once the design phase has begun. An example in the Agate system is the requirement that the system should be usable in different offices around the world and should handle multiple currencies. This would be noted during analysis; during design, it will mean that the system must be designed to support localization (adaptation to local needs) and to display different national currency symbols (perhaps using the Unicode standard).

Correct content. The analysis documentation should be correct and accurate in what it describes. This applies to textual information, diagrams and also to quantitative features of the non-functional requirements. Examples include correct descriptions of attributes and any operations that are known at this stage, correct representation of associations between classes, particularly the multiplicity of associations, and accurate information about volumes of data. Accuracy should not be confused with precision. FoodCo owns 1,500 acres of land (to the nearest 100 acres). To state that the company owns 1,700 is inaccurate. To state that it owns 1,523 is more precise. To state that it owns 1,253 is still inaccurate, although the precision gives a spurious impression of accuracy.

Consistency. Where the analysis documentation includes different models that refer to the same things (use cases, classes, attributes or operations) the same name should be used consistently for the same thing. Errors of consistency can result in errors being made by designers, for example, creating two attributes with different names that are used in different parts of the system but should be the same attribute. If the designers spot the inconsistency, they may try to resolve it themselves, but may get it wrong because the information they have about the system is all dependent on what they have received in the specification of requirements from the analysts.

Errors of scope or completeness will typically be reflected in the finished product not doing what the users require; the product will either include features that are not required or lack features that are. Errors of correctness and consistency will typically be reflected in the finished product performing incorrectly. Errors of completeness and consistency will most often result in difficulties for the designers; in the face of incomplete or inconsistent specifications, they will have to try to decide what is required or refer back to the analysts.

One general way of ensuring that the analysis models reflect the requirements is to use walkthroughs. Walkthroughs are described by Yourdon (1985) (and also in an appendix to Yourdon, 1989). They provide a structured review with other analysts and interested parties of the products of the analysis phase.

12.5.2 What makes for good design?

The quality of the design will clearly be reflected in the quality of the finished system that is delivered to the clients. Moreover, in the same way as the quality of analysis affects the work of designers, the quality of the design has an impact on the work of the programmers who will write the programme code in order to implement the system based on the design. Some of the criteria given below for a good design will bring benefits to the developers, while some will provide benefits for the eventual users of the system.

12.5.3 Objectives and constraints

The designers of a system seek to achieve many objectives that have been identified as the characteristics of a good design since the early days of information systems development. Yourdon and Constantine (1979) cite efficiency, flexibility, generality, maintainability and reliability; DeMarco (1979) proposes efficiency, maintainability and buildability; and Page-Jones (1988) suggests that a good design is efficient, flexible, maintainable, manageable, satisfying and productive. These latter two points highlight issues concerned with human–computer interaction and remind us of the need for the design to produce a usable system. Other characteristics of a good design are that it should be functional, portable, secure and economical; in the context of object-oriented systems, reusability is a priority objective.

Functional. When we use a computer system, we expect it to perform correctly and completely those functions that it is claimed to perform; when an information system is developed for an organization, the staff of that organization will expect it to meet their documented requirements fully and according to specification. So, for example, the staff of Agate will expect their system to provide them with the functionality required to document advertising campaigns, record notes about campaigns and store information about the advertisements to be used in those campaigns. If it does not perform these functions, it is not fully functional. Referring back to Rumbaugh's definition of design as 'how the system will be constructed without actually building it', a functional design should show how every element of the required system will work.

Efficient. It is not enough that a system performs the required functionality; it should also do so efficiently, in terms both of time and resources. Those resources can include disk storage, processor time and network capacity. This is why design is not just about producing any solution, but about producing the best solution. This objective may apply to Agate's requirement to store textual notes about ideas for campaigns and advertisements. A sample two lines of text containing twenty words takes up 97 bytes in text format, but stored in the format of a well-known word-processor takes up 12,800 bytes (this is without the use of any special fonts or styles). A poor design might use object linking and embedding (OLE) to handle the word-processing of the notes but would pay a severe penalty in increased storage requirements!

Economical. Linked to efficiency is the idea that a design should be economical. This applies not only to the fixed costs of the hardware and software that will be required to run it, but also to the running costs of the system. The cost of memory and disk storage is very low compared to 20 years ago, and most small businesses using Microsoft Windows probably now require more disk space for their programs than they do for their data. However, the growth of multimedia systems for business purposes may make it once more important to calculate the storage requirements of a system carefully.

Reliable. The system must be reliable in two ways: first, it should not be prone to either hardware or software failure; second it should reliably maintain the integrity of the data in the system. Hardware reliability can be paid for: some manufacturers provide systems with redundant components that run in parallel or that step in when an equivalent component fails; RAID (redundant arrays of inexpensive disks) technology can provide users with disk storage that is capable of recovering from failure of one drive in an array. The designers must design software reliability into the system. In physical design, detailed knowledge of the development environment is likely to help ensure reliability.

Reliability depends to some extent on the ability of the system to be tested thoroughly. A well analysed and designed system will specify the valid and invalid combinations of data that can be handled. It will also show clearly the structure of the system and which elements of the system are dependent on others so that testing can work up through classes, groups of classes, sub-systems and eventually the whole system.

Secure. Systems should be designed to be secure against malicious attack by outsiders and against unauthorized use by insiders. System design should include considerations of how people are authorized to use the system and policies on passwords.

It should also cover protection of the system from outsiders, including firewalls in either hardware or software to protect the system from access via public networks, such as the Internet. In European countries that are members of the European Union, there are data protection laws that are designed to protect the interests of individuals about whom data is held in information systems, such as the Data Protection Acts (1984 and 1998) and the Computer Misuse Act (1990) in the UK. Where such legislation exists, the designer should ensure that the design of the system will comply with its requirements.

Flexible. Some authors treat flexibility as the ability of the system to adapt to changing business requirements as time passes. Yourdon and Constantine call this feature *modifiability*. By flexibility they mean the ability to configure the system to handle different circumstances based on control values that are not compiled into the system but are available for the user to set at run-time. In the Agate system, this could be reflected in the choice of ODBC as the means to access the database. This provides a standard mechanism for accessing databases, and changing the ODBC driver used would allow the system to access a local or a remote version of the database, or for the system to be migrated to a different database engine at a later date. Another possibility would be to ensure that all the prompts and error messages used by the system are held in an external data file that can be loaded up when the program runs or in response to a menu option. This allows for the creation of multiple files of messages and would enable users to set the language that they wish to use. Java's use of Unicode, which provides a character set that includes ideographic characters (for example, Chinese, Japanese and Korean) as well as all the world's alphabets, would enable a system to be developed for Agate that could be localized for each location in which it is used.

General. Generality describes the extent to which a system is general-purpose. It is more applicable to utility programs than to large information systems. However, it includes the issue of *portability*, which applies to the Agate system that is to be developed in Java so that it can run on different hardware (PCs and Apple Macs). The system may also exhibit generality from the point of view of the developers who may wish to use the same system for other clients in the advertising industry. Reuse is discussed below.

Buildable. From the perspective of the programmer who has to write the program code to build the system, it is important that the design is clear and not unnecessarily complex. In particular the physical design should relate closely to the features that are available in the development language. Not all object-oriented languages offer the same features, for example in the visibility of attributes and operations (public, private, protected, friend etc.), in the ability to handle multiple inheritance, or in the availability of utility classes such as collections or linked lists in the base language. Designs that rely on features such as these will force the programmer to work around them if a different language is being used from the one that the designer had in mind.

Manageable. A good design should allow the project manager to estimate the amount of work involved in implementing the various sub-systems. It should also provide for sub-systems that are relatively self-contained and can be marked off as completed and passed on for testing without fear that changes to other parts of the system still in development will have unforeseen consequences on them.

Maintainable. Maintenance is cited as taking up as much as 60% of the data-processing budget of organizations. Maintenance activities include fixing bugs, modifying reports and screen layouts, enhancing programs to deal with new business requirements, migrating systems to new hardware and fixing the new bugs that are introduced by all of the above. A well-designed and documented system is easier to maintain than one that is poorly designed and documented. If maintenance is easy then it is less costly. It is particularly important that there is a close match between the developed program code and the design. This makes it easier for the maintenance programmer to understand the intention of the designer and to ensure that it is not subverted by the introduction of new code.

Usable. Usability includes a range of aspects including the idea, mentioned above, that a system should be both satisfying and productive. It may seem odd to suggest that people should enjoy using their computer systems and find it a satisfying experience. However, if you think about the times that you have used a computer system and have found it a source of dissatisfaction, then you can perhaps imagine a satisfying system as one with an absence of dissatisfying features. Many of the features that contribute to user satisfaction are characteristic of good human–computer interface (HCI) design. For example, the concept of *affordance* (meaning that objects on the interface suggest their function) can reduce the number of errors made by users. Reducing error rates and ensuring that if users do make an error it is clear both where they went wrong and how to recover from the error can contribute to the satisfaction of users. Productivity can be enhanced by ensuring that the tasks that users wish to carry out using the system are straightforward to carry out and do not introduce an overhead of keystrokes or mouse-clicks to achieve. If usability requirements have been captured (see Section 6.2.2), then the design should take these into account. Usability is considered in more detail in Chapter 16.

Reusable. Reusability is the Holy Grail of object-oriented development. Many of the features of object-oriented systems are geared to improve the possibility of reuse. Reuse affects the designer in three ways. First, he or she will consider how economies can be made by designing reuse into the system through the use of inheritance; second, he or she will look for opportunities to use design patterns, which provide templates for the design of reusable elements; and third, he or she will seek to reuse existing classes either directly or by sub-classing them. Design patterns are described in detail in Chapter 15. Existing classes could be classes that have been developed for other projects, classes in class libraries that are associated with the development language (such as the Java AWT) or classes that are bought in from outside vendors. To date, object-oriented development has not achieved the levels of reuse that were expected. In order to reuse a software class, a designer must be aware of the existence of the class, and be able to determine both that its interface matches the interface for the class that he or she requires and that the methods of the class match those required. It is arguable that in order to determine whether an available class matches requirements, the required class must already have been designed. The economies from reuse thus appear during the construction of the software and require a change to a culture of project management that supports reuse; this means that project managers must be able to recognize the effort that is saved by **not** writing and testing lines of code (because a class is being reused). The development of strategies to parcel up classes as components and the provision of component management software are an attempt to develop support for reuse (Allen and Frost, 1998). Chapter 20 covers some of these issues.

There is clearly some overlap between the categories that have been listed here. Aspects of maintainability overlap with flexibility, generality with reuse, efficiency with economy. What is often the case however is that some design objectives will conflict with one another. This more often happens at the level of specific objectives rather than general ones such as those described above. However, it should be possible to see that functionality, reliability and security could all conflict with economy. Many of the conflicts result from constraints that are imposed on the system by the users' non-functional requirements.

Constraints arise from the context of the project as well as from the users' requirements. The clients' budget for the project, the timescale within which they expect the system to be delivered, the skills of staff working on the project, the need to integrate the new system with existing hardware or systems, and standards set as part of the overall systems design process can all constrain what can be achieved. Resolving conflicts between requirements and constraints results in the need for compromises or trade-offs in design. A couple of examples should illustrate how these can occur.

- If the users of Agate's new system require the ability to change fonts in the notes that they write about campaigns and adverts, then they will want to be able to edit notes with the same kind of functionality that would be found in a word-processor. As pointed out above, this will seriously impact the storage requirements for notes. It will also have an effect on network traffic, as larger volumes of data will need to be transferred across the network when users browse through the notes. The designers will have to consider the impact of this requirement. It may be that the users will have to accept reduced functionality or the management of Agate will have to recognize that their system will have higher costs for storage than first envisaged. Compromise solutions may involve only transferring the text of a note (without the overhead of all the formatting information) when users are browsing a note and transferring the full file only when it needs to be viewed or edited. However, this will increase the processing load on the server. Another compromise solution might be to use a different file format such as RTF (rich text format) rather than the word-processor format. For the short text file discussed above this reduces the byte count to 1,770 while retaining formatting information.
- Agate would like the system to be configurable so that prompts, help and error messages are displayed in the language of the user. This means that each prompt and error message must be read into the programs from data files or the database. While this is good software design practice and makes the system more flexible, it will increase the workload of the designers when they design elements of the interface. Without this requirement, it is enough for each designer to specify that messages such as 'Campaign' or 'Not on file' appear on screen; there is a minimal need for liaison between designers. With this requirement, the designers will need to draw up a list of prompts, labels and error messages that can be referred to by number or by a key so that the same message is used consistently wherever it is applicable. This means that the programmers will not hard code messages into the system, but will refer to them as elements in an array of messages, for example. While this increases the flexibility and to some extent the maintainability of the system, it is likely to increase the cost of the design phase.

It is important that these design decisions are clearly documented and the reasoning behind compromises and trade-offs is recorded.

12.6 Measurable Objectives in Design

In the previous section, we discussed some of the general objectives of the designers in a system development project. Some objectives are specific to a particular project, and it is important to be able to assess whether these objectives have been achieved. One way of doing this is to ensure that these objectives are expressed in measurable terms so that they can be tested by simulation during the design phase, in prototypes that are built for this purpose or in the final system.

Measurable objectives often represent the requirements that we referred to as non-functional requirements in Chapter 6. They also reflect the fact that information systems are not built for their own sake, but are developed to meet the business needs of some organization. The system should contribute to the strategic aims of the business, and so should help to achieve aims such as:

- provide better response to customers, or
- increase market share.

However, such aims are vague and difficult to assess. If they are expressed in measurable terms, then it is possible to evaluate whether they can be achieved by the design, or whether they have been achieved by the finished system. Ideally they should be phrased in a way that shows how these objectives are attributable to the system. If a company expects to increase its market share as a result of introducing a new computer system but does not, it should be possible to tell whether this is a failure of the new system or the outcome of other factors outside the control of the system developers such as economic recession. The system may contribute to business objectives such as those above by providing better information or more efficient procedures, but for the objectives to be measurable they need to be phrased as operational objectives that can be quantified, such as:

- to reduce invoice errors by one-third within a year, or
- to process 50% more orders during peak periods.

These set clear targets for the designers and a way of checking whether these objectives can be achieved (within the constraints on the system) and whether they have been achieved once the system is up and running.

12.7 Planning for Design

In a waterfall life cycle project, the transition from the analysis phase to the design phase also gives the project manager the opportunity to plan for the activities that must be undertaken during design.

- If the hardware and software platform has not been decided upon, then design will begin as logical design, but the project manager must plan for the time when the platform is known and physical design can begin.
- The system architecture must be agreed and system standards must be set that will affect the design of individual sub-systems.
- If the designers are not familiar with all aspects of the platform, time must be allowed for training, or additional staff must be brought in, perhaps contractors with expertise in particular aspects of the hardware or software.
- Design objectives must be set and procedures put in place for testing those that can be tested by simulation during the design process.
- Procedures must also be put in place for resolving conflicts between constraints and requirements and for documenting and agreeing any trade-offs that are made.
- The amount of time to be spent on design of different aspects of the system (described in Chapters 13, 14, 16, 17 and 18) must be agreed.

In a project that uses an iterative life cycle, there is not the same transition. Nonetheless, design activities require a different mindset from analysis activities. Even if the same staff are playing the different analysis and design roles, they need to be clear about the roles that they are playing. Iteration planning requires project managers to think about the balance of activities from the different workflows that will be required in each iteration.

12.8 Summary

While analysis looks to the business in order to establish requirements, design looks to the technology that will be used to implement those requirements. An effective design will meet general objectives that will make the system easier to build and maintain, and more usable and functional for the end-users. The design of a system should also meet specific objectives relating to the business needs of the users, and these specific objectives should be phrased in quantifiable, operational terms that allow them to be tested. This process of design takes place in the context of constraints that are imposed by the users, their budget and existing systems, the available technology and the skills and knowledge of the design and development team.

In Chapters 13 and 14 we describe system design and detailed design of the classes in the required system. Chapter 15 explains how patterns can be used to assist the design process. Chapters 16 and 17 look specifically at the design of the human-computer interface, and Chapter 18 discusses the design of data storage.

Review Questions

12.1 What are the advantages of separating the analysis and design phases of a project?

12.2 What are the advantages of an iterative life cycle?

12.3 Users at Agate require a report of unpaid campaigns. Which of the following aspects of the report represents analysis, logical design and physical design?

The size of the paper and the position of each field in the report.

The fact that the user wants a report of completed campaigns that have not yet been paid for by the client.

The selection of the business objects and their attributes used by the report.

12.4 Which of the following sentences describing an element of the FoodCo system represents analysis, logical design and physical design?

The reason for stopping a run will be selected from one of the values displayed in a listbox (Java Choice) in the Record Line Stop dialogue window.

When a production line stops during a run, the reason for stopping will be recorded.

The reason for stopping a run will be entered into the system by selecting from a list of valid reasons.

12.5 What is meant by *seamlessness* in object-oriented systems development?

12.6 What are the differences between system design and detailed design?

12.7 Explain the difference between cohesion and coupling.

12.8 What aspects of the system are added to the class diagram(s) in object-oriented detailed design?

12.9 List four quality criteria for good analysis.

12.10 List twelve quality criteria for good design.

12.11 Re-read the description of the FoodCo case study in Case Study B1. Identify any constraints that you think might be imposed on the design of the new system.

12.12 Based on the same information try to identify possible measurable objectives for the new FoodCo system.

12.13 Agate wants the new system to provide access to the same data from every office around the world. Maintaining a network that is constantly connected between all the offices is considered too expensive, and using a network that dials up remote offices as required would provide response times that are too slow. What kind of compromise solution can you come up with to this problem?

Case Study Work, Exercises and Projects

12.A FoodCo requires a data entry screen for entering details of staff holidays. Without knowing what software or hardware is going to be used to develop this data entry screen, list as many features of the design as you can that are not dependent on the implementation platform.

12.B Design applies to a wide range of artefacts, for example cars, buildings, books and packaging. Choose some artefact that you use and try to identify what makes for a good design in this context. Are there aspects that do not apply to systems design? Are there aspects of systems design that should perhaps apply to the design of artefacts that you use?

12.C Find out what laws (if any) exist in your country to protect computer systems against malicious attack from hackers. What implications does the law have for the design of systems?

12.D One aspect of system design is concerned with the setting of standards across the system. Choose a system that you use regularly as part of your work or study and try to identify standards that are applied to the design of the interface. (This could be a software package you use, the library system you use to access your library catalogue, or even Windows.)

12.E In Section 12.5.2 we pointed out that some criteria for good quality in design will bring benefits to the designers, while others will bring benefits to the eventual users of the system. Try to decide which of the characteristics discussed in Section 12.5.3 bring benefits to the designers as well as the end users.

Further Reading

- For those with an interest in the historical development of systems design, the classics of structured design are Jackson (1975) and Yourdon and Constantine (1979). DeMarco (1979) deals with structured analysis. Two more recent books that are likely to be more easily available are Yourdon (1989) and Page-Jones (1988).
- If you are interested in an approach to the analysis and design of requirements that is completely different from object-oriented approaches, SSADM (Structured Systems Analysis and Design Method) makes a very clear distinction between requirements analysis, logical design and physical design. A separate stage in SSADM is used to carry out the choice of development environment (Technical System Options). Skidmore, Mills and Farmer (1994), Goodland and Slater (1995) or any other book on SSADM explains the way in which these stages are handled.
- More detail is provided in Chapters 13 to 18 on specific aspects of the design task.

CHAPTER 13

System Design

OBJECTIVES

In this chapter you will learn

- the major concerns of system design
- the main aspects of system architecture, in particular what is meant by subdividing a system into layers and partitions
- how to apply the MVC architecture
- which architectures are most suitable for distributed systems
- how design standards are specified.

13.1 Introduction

In Chapter 12 we saw how the move from analysis to design requires the consideration of many issues that are not directly related to the functional requirements of the information system. These are the non-functional requirements, and they reflect the way that the system is to be constructed and used, rather than what it does. Design is concerned with establishing how to deliver the functionality that was specified in analysis while, at the same time, meeting non-functional requirements that may sometimes conflict with each other.

System design is focused on making high-level decisions concerning the overall structure of the system. The main activities in system design are introduced in Section 13.2, and possible software architectures are discussed in Section 13.3. The design of those parts of the system that have to operate concurrently is addressed in Section 13.4. Other aspects of system design include the allocation of different parts of a system to run on different computers (Section 13.5) and the selection of suitable data management strategies (Section 13.6). System design also involves establishing protocols and standards for the design activities and the products that are produced. For example, a standard naming format should be used throughout the project, and program code must be written in an agreed style (Section 13.7). Finally, we consider the importance of prioritizing design trade-offs (Section 13.8), and how implementation issues impact on design (Section 13.9). The technology available for the construction of the application has a significant impact upon the design process as a whole and upon the decisions that are made during system design. It is a truism, but none the less worth stating, that the design activity requires detailed knowledge of the technologies that are available for use. The decisions that are made during system design inform the rest of the design process.

13.2 The Major Elements of System Design

The system design activity specifies the context within which detailed design will occur. A major part of system design is defining the *system architecture*. The meaning and scope of the term architecture for computerized information systems is much debated but it is generally accepted that it is an important feature of the delivered system. Evidence of this shift can be found in the employment market in the increasing number of jobs advertised that have the title of System Architect. The architecture of a system is concerned with its overall structure, the relationships among its major components and their interactions. If the system being considered contains human, software and hardware elements then its architecture includes how these elements are structured and how they interact. On the other hand, if the system being considered comprises software and hardware, then its architecture only concerns these elements. It is important to consider the structure of the software elements of the system and this is termed the *software architecture*. The hardware architecture of a system is discussed in this book in terms of the computers and peripherals required for the system and how software is allocated to them.

The architecture of the information system is first considered early in the project during the requirements capture and analysis activities. This first view of the system architecture is driven significantly by the use cases and then informs the continuing requirements capture and analysis activities. This forms a useful basis from which to develop the design architecture. The detailed software architecture of a computerized information system develops as the design process continues into object design[1] but it is important to identify an overall system architecture within which the detail can be refined. High-level architectural decisions that are made during system design determine how successfully the system will meet its non-functional objectives (e.g. performance, extensibility) and thus its long-term utility for the client. Reuse is one of the much-vaunted benefits of object-orientation and poor software architecture usually reduces both the reusability of the components produced and the opportunity to reuse existing components.

During system design the following activities are undertaken.

- Sub-systems and major components are identified.
- Any inherent concurrency is identified.
- Sub-systems are allocated to processors.
- A data management strategy is selected.
- A strategy and standards for human–computer interaction are chosen.
- Code development standards are specified.

[1] Also referred to as class design.

- The control aspects of the application are planned.
- Test plans are produced.
- Priorities are set for design trade-offs.
- Implementation requirements are identified (for example, data conversion).

Rumbaugh et al. (1991) proposed a similar list of activities, and equivalent activities for real-time structured design are cited by Goldsmith (1993). The product of the system design process is not a detailed specification. It is rather a specification of the design context that can guide the developers as they make detailed design decisions regarding the design specification from which construction will proceed.

13.3 Software Architecture

Software architecture, like system architecture has no generally agreed definition and can be interpreted differently depending upon the context. For Booch (1994), the architecture of a system is its class structure. He suggests that part of the architecture is the way in which classes are grouped together. Rumbaugh et al. (1991) use the term system architecture to describe the overall organization of a system into sub-systems. Buschmann et al. (1996, p. 384) give the following definition of a software architecture.

> A *software architecture* is a description of the sub-systems and components of a software system and the relationships between them. Subsystems and components are typically specified in different views to show the relevant functional and non-functional properties of a software system. The software architecture of a system is an artefact. It is the result of the software design activity.

A contrasting view is offered by Soni et al. (1995), who identify the four different aspects of software architecture shown in Figure 13.1.

In terms of object-oriented development, the conceptual architecture is concerned with the structure of the static class model and the connections between the components of the model. The module architecture describes the way the system is divided into sub-systems or modules and how they communicate by exporting and importing data. The code architecture defines how the program code is organized into files and directories and grouped into libraries. The execution architecture focuses on the dynamic aspects of the system and the communication between components as tasks and operations execute.

There are alternative ways of delineating different aspects of architecture. For example, a logical architecture might comprise the class model, while a physical architecture is concerned with mapping the software onto the hardware components. However, in all cases these different aspects combine to define a software architecture for the system. There are various architectural styles, each of which has characteristics that make it more or less suitable for certain types of application. We will consider

Type of architecture	Examples of elements	Examples of relationships
Conceptual	Components	Connectors
Module	Sub-systems, modules	Exports, imports
Code	Files, directories, libraries	Includes, contains
Execution	Tasks, threads, object interactions	Uses, calls

Adapted from Weir and Daniels (1998)

Figure 13.1 Four aspects of software architecture according to Soni et al.

some of the major alternatives. It is worth noting that software architectures have been documented in the patterns form by Buschmann et al. (1996, 2000) amongst others.

13.3.1 Sub-systems

A sub-system typically groups together elements of the system that share some common properties. An object-oriented sub-system encapsulates a coherent set of responsibilities in order to ensure that it has integrity and can be maintained. For example, the elements of one sub-system might all deal with the human–computer interface, the elements of another might all deal with data management and the elements of a third may all focus on a particular functional requirement.

The sub-division of an information system into sub-systems has the following advantages.

- It produces smaller units of development.
- It helps to maximize reuse at the component level.
- It helps the developers to cope with complexity.
- It improves maintainability.
- It aids portability.

Each sub-system should have a clearly specified boundary and fully defined interfaces with other sub-systems. A specification for the interface of a sub-system defines the precise nature of the sub-system's interaction with the rest of the system but does not describe its internal structure (this is a high level use of contracts, which are described in Chapter 9). A sub-system can be designed and constructed independently of other sub-systems, simplifying the development process. Sub-systems may correspond to increments of development that can be delivered individually as part of an incremental life cycle (if the developers are using the spiral life cycle model discussed in Chapter 3).

Dividing a system into sub-systems is an effective strategy for handling complexity. Sometimes it is only feasible to model a large complex system piece by piece, with the sub-division forced on the developers by the nature of the application. Splitting a system into sub-systems can also aid reuse as each sub-system may correspond to a component that is suitable for reuse in other applications. A judicious choice of sub-systems during design can reduce the impact on the overall system of a change to its requirements. For example, consider an information system that contains a presentation sub-system which deals with the human–computer interface (HCI). A change to the data display format need not affect other sub-systems. Of course there may still be some changes to the requirements that affect more than one sub-system. The aim is to localize the consequences of change, so that a change in one sub-system does not trigger changes in other sub-systems (sometimes referred to as the ripple effect). Moving an application from one implementation platform to another can be much easier if the software architecture is appropriate. An example of this would be the conversion of a Windows 98 application so that it could run in a Unix environment. This would require changes to the software that implements the human–computer interface. If this were dealt with by specialized sub-systems then the overall software change is localized to these sub-systems. As a result, the system as a whole is easier to port to a different operating environment.

Each sub-system provides services for other sub-systems, and there are two different styles of communication that make this possible. These are known as *client–server* and *peer-to-peer* communication and are shown in Figure 13.2.

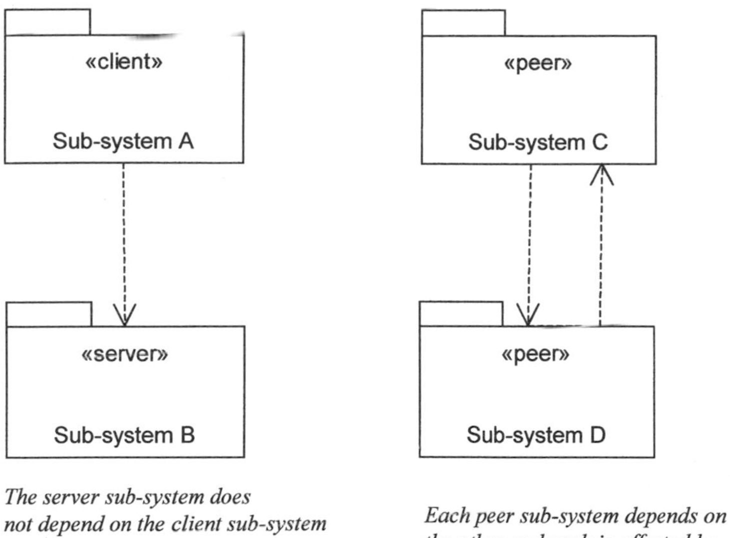
«client»
Sub-system A
«server»
Sub-system B
«peer»
Sub-system C
«peer»
Sub-system D
The server sub-system does not depend on the client sub-system and is not affected by changes to the client's interface.
Each peer sub-system depends on the other and each is affected by changes in the other's interface.

Figure 13.2 Styles of communication between sub-systems.

Client–server communication requires the client to know the interface of the server sub-system, but the communication is only in one direction. The client sub-system requests services from the server sub-system and not vice versa. Peer-to-peer communication requires each sub-system to know the interface of the other, thus coupling them more tightly. The communication is two way since either peer sub-system may request services from the other.

In general client–server communication is simpler to implement and to maintain, as the sub-systems are less tightly coupled than they are when peer-to-peer communication is used. In Figure 13.2 the sub-systems are represented using packages that have been stereotyped to indicate their role. Component and deployment diagrams can also be used to model the implementation of sub-systems (see Chapter 20).

13.3.2 Layering and partitioning

There are two general approaches to the division of a software system into sub-systems. These are known as *layering*—so called because the different sub-systems usually represent different levels of abstraction[2]—and *partitioning*, which usually means that each sub-system focuses on a different aspect of the functionality of the system as a whole. In practice both approaches are often used together on one system, so that some of its sub-systems are divided by layering, while others are divided by partitioning.

Layered sub-systems

Layered architectures are among the most frequently used high-level structures for a system. A schematic of the general structure is shown in Figure 13.3.

[2] Or layers of service.

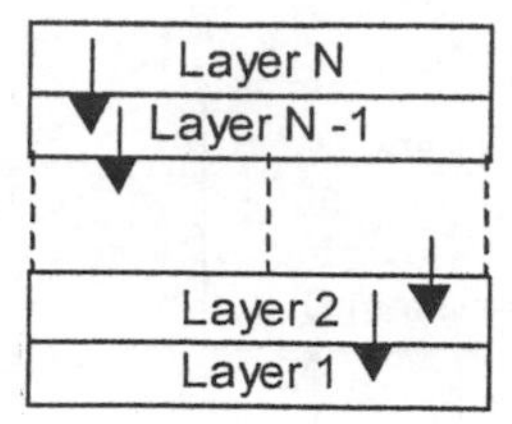

Closed architecture—messages may only be sent to the adjacent lower layer.

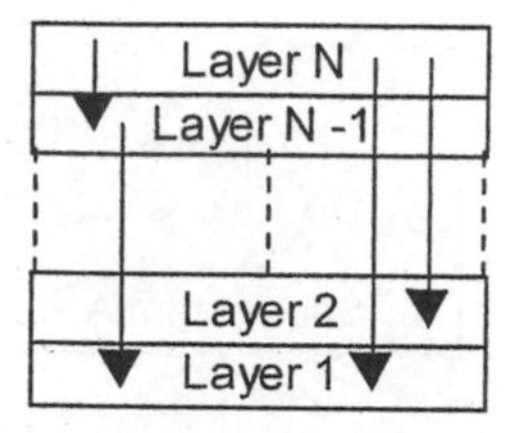

Open architecture—messages can be sent to any lower layer.

Figure 13.3 Schematic of a layered architecture.

Each layer corresponds to one or more sub-systems, which may be differentiated from each other by differing levels of abstraction or by a different focus of their functionality. It works like this: the top layer uses services provided by the layer immediately below it. This in turn may require the services of the next layer down. Layered architectures can be either open or closed, and each has its particular advantages. In a closed layered architecture a certain layer (say layer N) can only use the services of the layer immediately below it (layer N - 1). In an open layered architecture layer N may directly use the services of any of the layers that lie below it.

A closed architecture minimizes dependencies between the layers and reduces the impact of a change to the interface of any one layer. An open layered architecture produces more compact code since the services of all lower level layers can be accessed directly by any layer above them without the need for extra program code to pass messages through each intervening layer. However this breaks the encapsulation of the layers, increases the dependencies between layers and increases the difficulty caused when a layer needs to be changed.

Networking protocols provide some of the best known examples of layered architectures. A network protocol defines how computer programs executing on different computers communicate with each other. Protocols can be defined at various levels of abstraction and each level can be mapped onto a layer. The OSI (Open Systems Interconnection) 7 Layer Model was defined by the International Standardization Organization (ISO) as a standard architectural model for network protocols (Tanenbaum, 1992). The structure provides flexibility for change since a layer may be changed internally without affecting other layers, and it enables the reuse of layer components. The OSI 7 Layer Model is illustrated in Figure 13.4.

Buschmann et al. (1996) suggest that a series of issues need to be addressed when applying a layered architecture in an application. These include:

- maintaining the stability of the interfaces of each layer;
- the construction of other systems using some of the lower layers;
- variations in the appropriate level of granularity for sub-systems[3];
- the further sub-division of complex layers;
- performance reductions due to a closed layered architecture.

The OSI model has seven layers only because it covers every aspect of the communication between two applications, ranging from application-oriented processes to

[3] The context determines an appropriate size for the sub-systems.

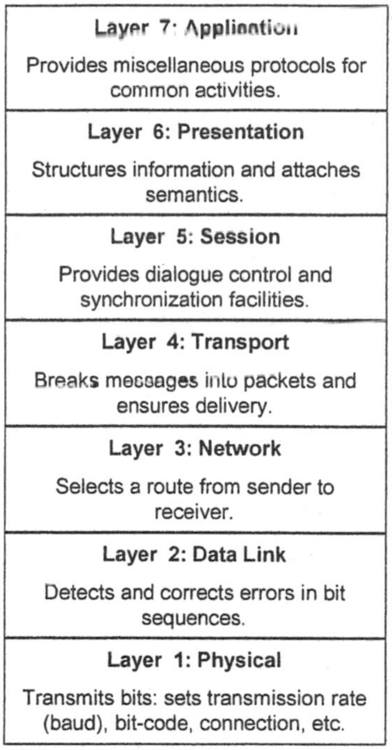

Layer 7: Application

Provides miscellaneous protocols for common activities.

Layer 6: Presentation

Structures information and attaches semantics.

Layer 5: Session

Provides dialogue control and synchronization facilities.

Layer 4: Transport

Breaks messages into packets and ensures delivery.

Layer 3: Network

Selects a route from sender to receiver.

Layer 2: Data Link

Detects and corrects errors in bit sequences.

Layer 1: Physical

Transmits bits: sets transmission rate (baud), bit-code, connection, etc.

Figure 13.4 OSI 7 Layer Model (adapted from Buschmann et al., 1996).

Application

Data formatting

Data management library classes

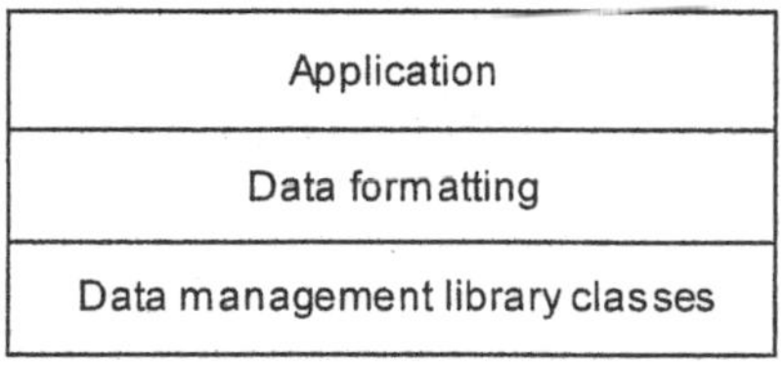

Figure 13.5 Simple layered architecture.

drivers and protocols that directly control network hardware devices. Many layered architectures are much simpler than this. Figure 13.5 shows a simple example of a three layer architecture.

The lowest layer of the architecture in Figure 13.5 consists of data management library classes. The layer immediately above this, the data formatting layer, uses services that are provided by the data management library classes in order to get data from a database management system. This data is formatted before it is passed upwards to the application layer. Supposing this system were to be modified to allow it to use a different database management system, the layered architecture limits major changes to the data management library class layer with some possible changes to the data formatting layer.

The following steps are adapted from Buschmann et al. (1996), and provide an outline process for the development of a layered architecture for an application. Note that this does not suggest that the specification of a system's architecture is a rule-based procedure. The steps offer guidelines on the issues that need to be addressed during the development of a layered architecture.

1. Define the criteria by which the application will be grouped into layers. A commonly used criterion is level of abstraction from the hardware. The lowest layer provides primitive services for direct access to the hardware while the layers above provide more complex services that are based upon these primitives. Higher layers in the architecture carry out tasks that are more complex and correspond to concepts that occur in the application domain.
2. Determine the number of layers. Too many layers will introduce unnecessary overheads while too few will result in a poor structure.
3. Name the layers and assign functionality to them. The top layer should be concerned with the main system functions as perceived by the user. The layers below should provide services and infrastructure that enable the delivery of the functional requirements.
4. Specify the services for each layer. In general it is better in the lower layers to have a small number of low-level services that are used by a larger number of services in higher layers.
5. Refine the layering by iterating through steps 1 to 4.
6. Specify interfaces for each layer.
7. Specify the structure of each layer. This may involve partitioning within the layer.
8. Specify the communication between adjacent layers (this assumes that a closed layer architecture is intended).
9. Reduce the coupling between adjacent layers. This effectively means that each layer should be strongly encapsulated. Where a client–server communication protocol will be used, each layer should have knowledge only of the layer immediately below it.

One of the simplest application architectures has only two layers—the application layer and a database layer. Tight coupling between the user interface and the data representation would make it more difficult to modify either independently, so a middle layer is often introduced in order to separate the conceptual structure of the problem domain. This gives the architecture shown in Figure 13.6, which is commonly used for business-oriented information systems. Note that it is essentially the same as the three-tier architecture suggested in Figure 12.1 (user interface, business objects and database tiers).

A common four layer architecture separates the business logic layer into application logic and domain layers, and this is illustrated in Figure 13.7. The approach that has been adopted during the analysis activity of use case realization results in the identification of boundary, control and entity classes. It is easy to see that it is possible to map the boundary classes onto a presentation layer, the control classes onto an application logic layer and the entity classes on a domain layer. Thus from an early stage in the

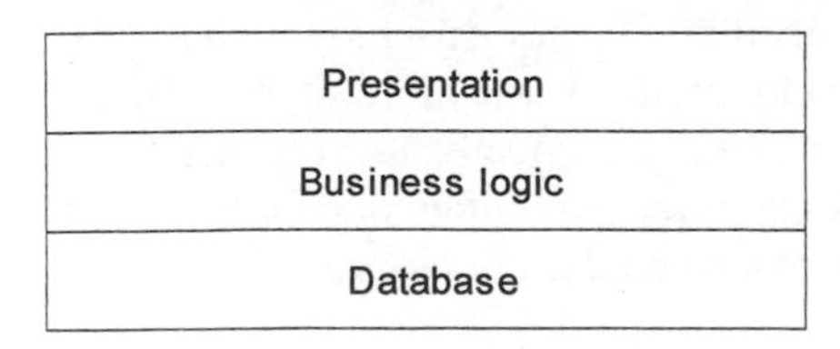

Figure 13.6 Three layer architecture.

development of an information system some of element of layering is being introduced into the software architecture. However, it is important to appreciate that as we move through design, the allocation of responsibility amongst these types of class may be adjusted to accommodate non-functional requirements.

Separating the application logic layer from the domain layer may be further justified because several applications share (or are likely to share) one domain layer, or because the complexity of the business objects forces a separation into two layers. It can also be used when the objects are physically distributed (see Chapter 19). However, it must be emphasized that there is no perfect solution to this kind of design problem. There are only solutions that have different characteristics (perhaps different levels of efficiency or maintainability). A good design solution is one that balances competing requirements effectively.

Layered architectures are used quite widely. J2EE™ (Sun Java Centre, 2001) adopts a multi-tiered[4] approach and an associated patterns catalogue has been developed. The architecture has five layers (client, presentation, business, integration and resource tiers) and the patterns catalogue addresses the presentation, business and integration tiers.

Partitioned sub-systems

As suggested earlier, some layers within a layered architecture may have to be decomposed because of their intrinsic complexity. Figure 13.8 shows a four layer architecture for part of Agate's campaign management system that also has some partitioning in the upper layers. In this example the application layer corresponds to the analysis class model for a single application, and is partitioned into a series of sub-systems. These sub-systems are loosely coupled and each should deliver a single service or coherent group of services. The `Campaign Database` layer provides access to a database that

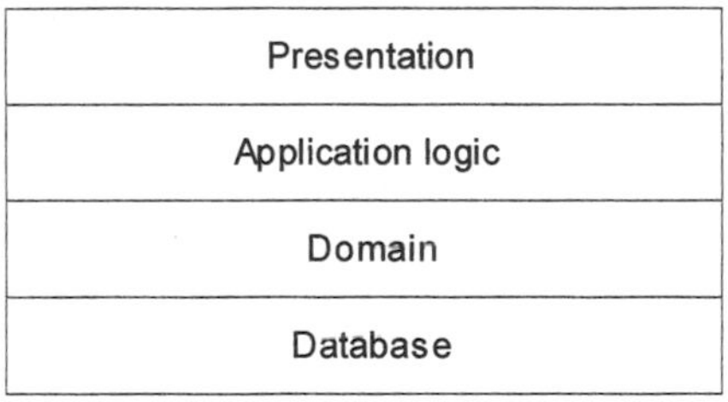

Presentation
Application logic
Domain
Database

Figure 13.7 Four layer architecture.

Presentation layer	Advert HCI Sub-system	Campaign Costs HCI Sub-system
Application layer	Advert Sub-system	Campaign Costs Sub-system
	Campaign Domain	
	Campaign Database	

A single domain layer supports two application sub-systems.

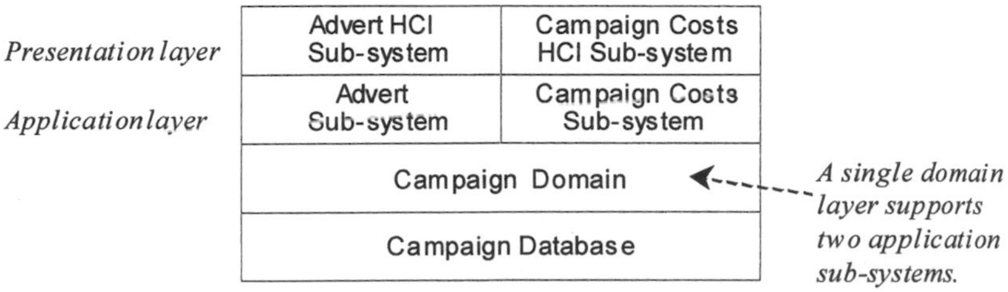

Figure 13.8 Four layer architecture applied to part of the Agate campaign management system.

[4] These use the term tier as broadly equivalent to layer.

contains all the details of the campaigns, their adverts and the campaign teams. The `Campaign Domain` layer uses the lower layer to retrieve and store data in the database and provides common domain functionality for the layers above. For example, the `Advert` sub-system might support individual advert costing while the `Campaign Costs` sub-system uses some of the same common domain functionality when costing a complete campaign. Each application sub-system has its own presentation layer to cater for the differing interface needs of different user roles[5].

A system may be split into sub-systems during analysis because of the system's size and complexity. However, the analysis sub-systems should be reviewed during design for coherence and compatibility with the overall system architecture.

The sub-systems that result from partitioning should have clearly defined boundaries and well specified interfaces, thus providing high levels of encapsulation so that the implementation of an individual sub-system may be varied without causing dependent changes in the other sub-systems. The process of identifying sub-systems within a particular layer can be detailed in much the same way as for sub-system layers.

13.3.3 Model–View–Controller

Many interactive systems use the Model–View–Controller (MVC) architecture. This structure was first used with Smalltalk but has since become widely used with many other object-oriented development environments. The MVC architecture separates an application into three major types of component: models that comprise the main functionality, views that present the user interface, and controllers that manage the updates to views. This structure is capable of supporting user requirements that are presented through differing interface styles, and it aids maintainability and portability.

It is common for the view of an information system that is required for each user to differ according to their role. This means that the data and functionality available to any user should be tailored to his or her needs. The needs of different types of user can also change at varying rates. For both these reasons it makes sense to give each user access to only the relevant part of the functionality of the system as a whole. For example, in the Agate case study many users need access to information about campaigns, but their perspectives vary. The campaign manager needs to know about the current progress of a campaign. She is concerned with the current state of each advertisement and how this impacts on the campaign as a whole—is it prepared and ready to run, or is it still in the preparation stage? If an advert is behind schedule does this affect other aspects of the campaign? The creative artist also needs access to adverts but he is likely to need access to the contents of the advert (its components and any notes that have been attached to it) as well as some scheduling information. A director may wish to know about the state of all live campaigns and their projected income over the next six months. This gives at least three different perspectives on campaigns and adverts, each of which might use different styles of display. The director may require charts and graphs that summarize the current position at quite a high level. The campaign manager may require lower level summaries that are both textual and graphical in form. The graphic designer may require detailed textual displays of notes with a capability to display graphical images of an advert's content. Ideally, if any information about a campaign or an advert is updated in one view then the changes should also be reflected immediately in all other views. Figure 13.9 shows a possible architecture, but some problems remain.

[5] This example is for illustrative purposes only. Our analysis class model for Agate is too small to justify this kind of partitioning in practice.

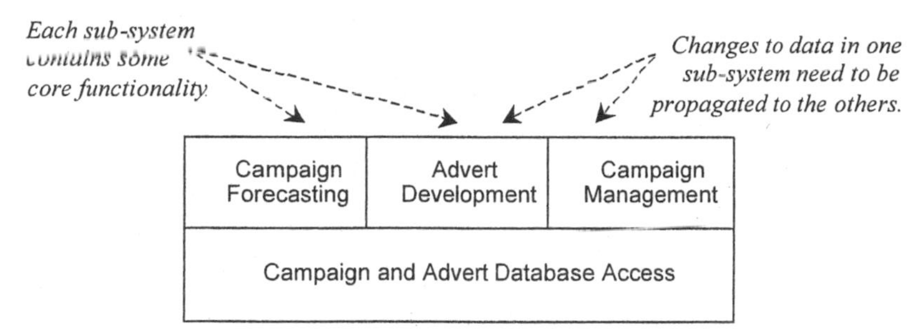

Figure 13.9 Multiple interfaces for the same core functionality.

The design of such varied and flexible user interfaces that still incorporate the same core functionality is likely to be expensive because elements of functionality may have duplicated for different interfaces. This makes the software more complex and thus also more error prone. There is an impact on maintainability too, since any change to core functionality will necessitate changes to each interface sub-system.

We repeat below some of the difficulties that need to be resolved for this type of application.

- The same information should be capable of presentation in different formats in different windows.
- Changes made within one view should be reflected immediately in the other views.
- Changes in the user interface should be easy to make.
- Core functionality should be independent of the interface to enable multiple interface styles to co-exist.

While the four layer architecture in Figure 13.8 resolves some of these problems it does not handle the need to ensure that all view components are kept up to date. The MVC architecture solves this through its separation of core functionality (model) from the interface and through its incorporation of a mechanism for propagating updates to other views. The interface itself is split into two elements: the output presentation (view) and the input controller (controller). Figure 13.10 shows the basic structure of the MVC architecture. The responsibilities of the components of an MVC architecture are listed below.

Model. The model provides the central functionality of the application and is aware of each of its dependent view and controller components.

View. Each view corresponds to a particular style and format of presentation of information to the user. The view retrieves data from the model and updates its presentations when data has been changed in one of the other views. The view creates its associated controller.

Controller. The controller accepts user input in the form of events that trigger the execution of operations within the model. These may cause changes to the information and in turn trigger updates in all the views ensuring that they are all up to date.

Propagation Mechanism. This enables the model to inform each view that the model data has changed and as a result the view must update itself. It is also often called the dependency mechanism.

Figure 13.11 represents the capabilities offered by the different MVC components as they might be applied to part of the campaign management system at Agate.

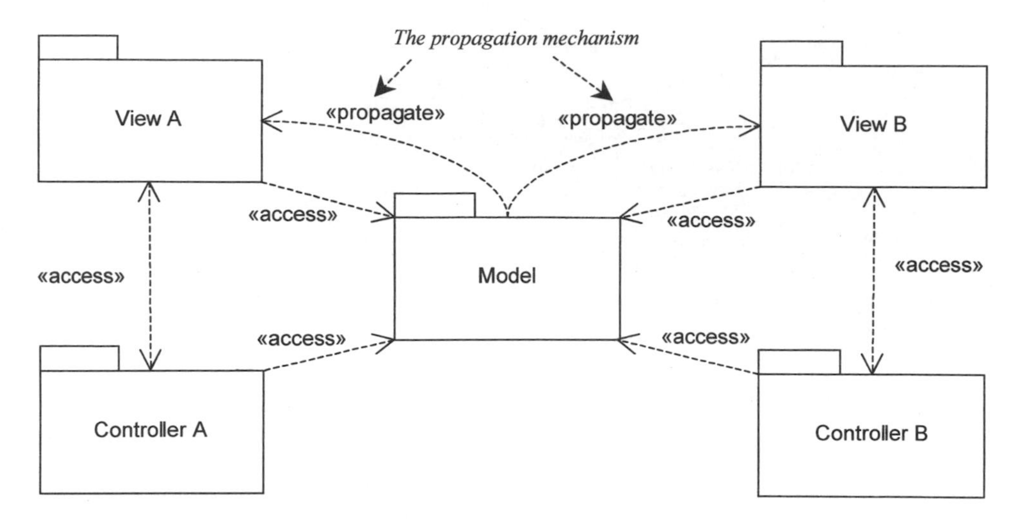

Figure 13.10 General structure of Model–View–Controller (adapted from Hopkins and Horan, 1995).

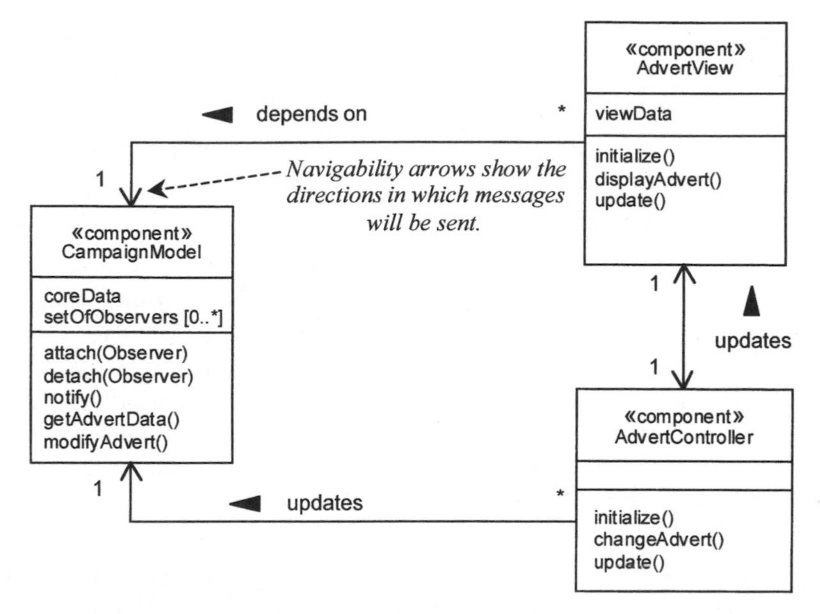

Figure 13.11 Responsibilities of MVC components, as applied to Agate.

The operation `update()` in the `AdvertView` and `AdvertController` components triggers these components to request data from the `CampaignModel` component[6]. This model component has no knowledge of the way that each view and controller component will use its services. It need only know that all view and controller components must be informed whenever there is a change of state (a modification either of object attributes or of their links).

The `attach()` and `detach()` services in the `CampaignModel` component enable views and controllers to be added to the `setOfObservers`. This contains a list of all components that must be informed of any change to the model core data. In practice there would be separate views, each with its own controller, to support the requirements of the campaign manager and the director.

[6] In this example the `CampaignModel` will hold details of campaigns and their adverts.

The interaction sequence diagram in Figure 13.12 illustrates the communication that is involved in the operation of an MVC architecture. (The choice of message type—synchronous or asynchronous—shown in this diagram is only one of the possibilities that could be appropriate, the features of the implementation environment would influence the actual design decision.) An AdvertController component receives the interface event `changeAdvert()`. In response to this event the controller invokes the `modifyAdvert()` operation in the `CampaignModel` object. The execution of this operation causes a change to the model.

For example, the target completion date for an advertisement is altered. This change of state must now be propagated to all controllers and views that are currently registered with the model as active. To do this the `modifyAdvert()` operation invokes the `notify()` operation in the model that sends an update() message to the view. The view responds to the `update()` message by executing the `displayAdvert()` operation which requests the appropriate data from the model via the `getAdvert-Data()` operation. The model also sends an `update()` message to the `Advert-Controller`, which then requests the data it needs from the model.

One of the most important aspects of the MVC architecture is that each model knows only which views and controllers are registered with it, but not what they do. The `notify()` operation causes an update message to all the views and controllers (for clarity, only one view and one controller are shown in the diagram, but interaction with the others would be similar). The `update()` message from the model is in effect saying to the views and controllers: 'I have been updated and you must now ensure that your data is consistent.' Thus the model, which should be the most stable part of the application, is unaffected by changes in the presentation requirements of any view or

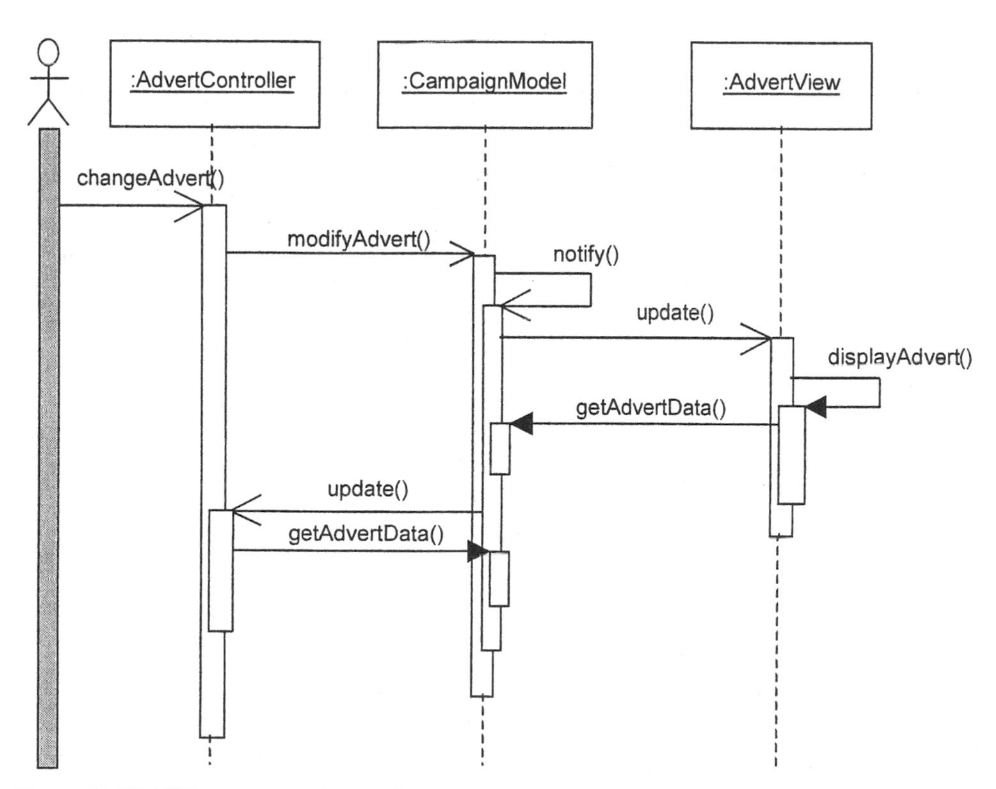

Figure 13.12 MVC component interaction.

controller. The change propagation mechanism can be structured so that further views and controllers can be added without causing a change to the model. Each of these may support different interface requirements but require the same model functionality. However, since views and controllers need to know how to access the model in order to get the information they require, some changes in the model will inevitably still cause changes in other components.

Other kinds of communication may take place between the MVC components during the operation of the application. The controller may receive events from the interface that require a change in the way that some data is presented to the user but do not cause a change of state. The controller's response to such an event would be to send an appropriate message to the view. There would be no need for any communication with the model.

13.3.4 Architectures for distributed systems

Distributed information systems are becoming more common as communications technology improves and becomes more reliable. An information system may be distributed over computers at the same location or different locations. Since Agate has offices around the world, it may need information systems that use data that is distributed among different locations. If Agate grows, it may also open new offices and require new features from its information systems. An architecture that is suitable for distributed information systems needs also to be flexible so that it can cope with change. A distributed information system may be supported by software products such as distributed database management systems or object request brokers (these are discussed in Chapter 19).

A general *broker* architecture for distributed systems is described by Buschmann et al. (1996). A simplified version of the broker architecture is shown in Figure 13.13.

A broker component increases the flexibility of the system by decoupling the client and server components. Each client sends its requests to the broker rather than communicating directly with the server component. The broker then forwards the service request to an appropriate server. A broker may offer the services of many servers and part of its task is to identify the relevant server to which a service request should be forwarded. The advantage offered by a broker architecture is that a client need not know where the server is located, and it may therefore be stored on either a local or a remote computer. Only the broker needs to know the location of the servers that it handles.

Figure 13.14 shows a sequence diagram for client-server communication using the broker architecture. The diagram is drawn with asynchronous message types but the actual implementation may involve both synchronous and asynchronous message

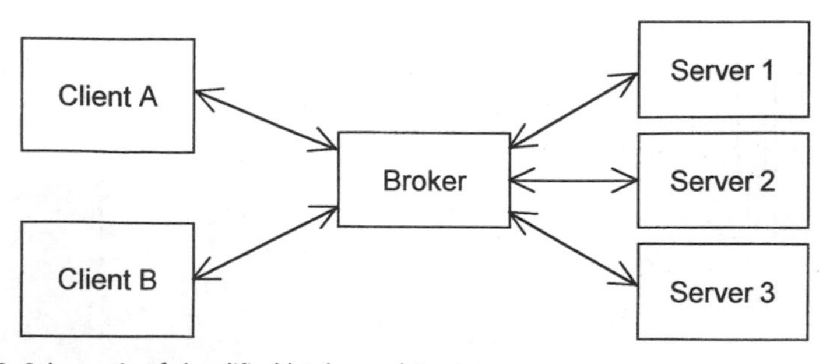

Figure 13.13 Schematic of simplified broker architecture.

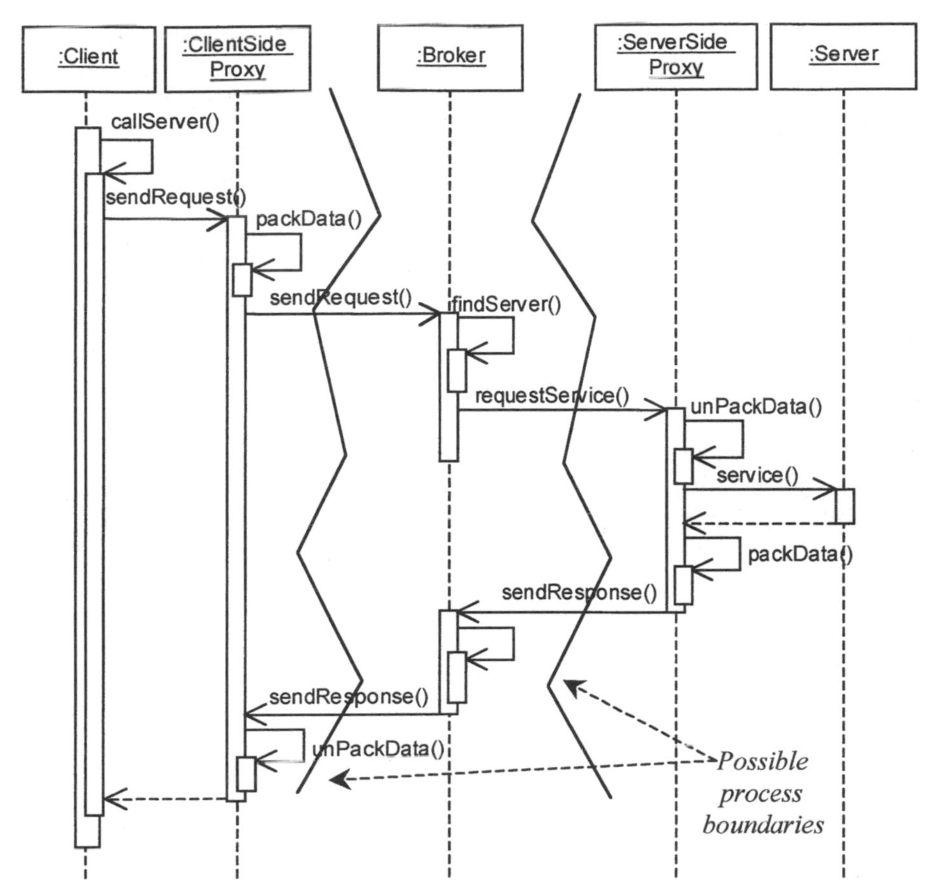

Figure 13.14 Broker architecture for local server (adapted from Buschmann et al., 1996).

types[7]. In this example the server sub-system is on a local computer. In addition to the broker itself, two additional *proxy* components have been introduced to insulate the client and server from direct access with the broker. On the client side a `ClientSide Proxy` receives the initial request from the client and packs the data in a format suitable for transmission. The request is then forwarded to the `Broker` which finds an appropriate server and invokes the required service via the `ServerSideProxy`.

The `ServerSideProxy` then unpacks the data and issues the service request sending the `service()` message to the `Server` object. The `service()` operation then executes and on completion control returns to the `ServerSideProxy` (this is shown explicitly by the dashed message return arrow). The response is then sent to the `Broker` which forwards it to the originating `ClientSideProxy`. Note that these are both new messages and not returns. The reason for this is that a broker does not wait for each response before handling another request. Once its `sendRequest` activation has been completed the broker will in all probability deal with many other requests, and thus requires a new message from the `ServerSideProxy` object that causes it to enter a new activation. Unlike the broker, the `ClientSideProxy` has remained active; this then unpacks the message and the response becomes available to the `Client` as control returns (again shown explicitly for clarity).

[7] In the versions of UML up to release 1.4 the open arrow-head signified an uncommitted message, that is one whose type has not been determined. This was a useful notational feature that is no longer part of UML. An asynchronous message used to be represented by a half arrow-head.

Figure 13.15 shows a schematic broker architecture that uses *bridge* components to communicate between two remote processors. Each bridge converts service requests into a network specific protocol so that the message can be transmitted.

13.3.5 Organization structures for architecture and development

Dividing a system into sub-systems has benefits for project management. Each sub-system can be allocated to a single development team that can operate independently of other teams, provided that they adhere to the interface requirements for their sub-system. Where a sub-system must be split between two development teams there is a heavy communications overhead that is incurred in ensuring that the different parts of the sub-system are constructed to consistent standards. In such cases the structure of either the organization or of the software tends to change so that they become more closely aligned with each other; this helps to minimize the communications overhead and is sometimes known as Conway's Law[8] (Coplien, 1995). If a sub-system that is being developed by more than one team is cohesive, and the way it is split between teams has no apparent functional basis, then the teams may coalesce in practice and operate as one. Teams that are working on the same sub-system are sometimes inhibited from merging—say, because they are located on different continents. The sub-system should then be treated as if it were two separate sub-systems. An interface between these two new sub-systems can be defined and the teams can then operate autonomously. Where the allocation of one sub-system to two teams is such that one team deals with one set of requirements, and another deals with a different set of requirements, the sub-system can also be treated as if it were actually two sub-systems, with a defined interface between them.

13.4 Concurrency

In most systems there are many objects that do not need to operate concurrently but some may need to be active simultaneously. Object-oriented modelling captures any inherent concurrency in the application principally through the development of interaction diagrams and statecharts. The examination of use cases also helps with the identification of concurrency. There are several ways of using these models to identify

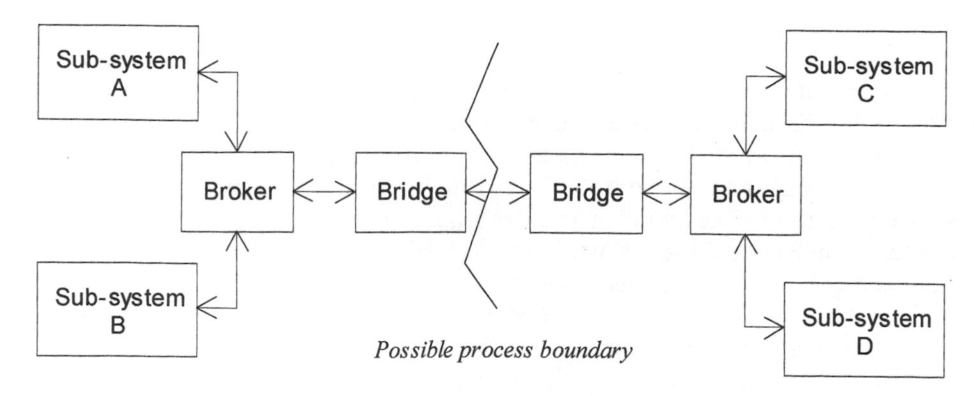

Figure 13.15 Schematic of broker architecture using bridge components.

[8] This is an example of an organizational or process pattern.

circumstances where concurrent processing may be necessary. First, a use case may indicate a requirement that the system should be able to respond simultaneously to different events, each of which triggers a different thread of control. Second, if a statechart reveals that a class has complex nested states which themselves have concurrent substates, then the design must be able to handle this concurrency. The statechart for the class `Campaign` has nested concurrent states within the `Active state` (see Figure 11.14) and there may be the possibility of concurrent activity. In this particular example, the concurrent activity that occurs in the real world need not necessarily be represented as concurrent processing in the computerized information system.

In cases where an object is required to exhibit concurrent behaviour it is sometimes necessary to split the object into separate objects in order to avoid the need for concurrent activity within any one object. Concurrent processing may also be indicated if interaction diagrams reveal that a single thread of control requires that operations in two different objects should execute simultaneously, perhaps because of asynchronous invocation. This essentially means that one thread of control is split into two or more active threads. An example of this is shown in Figure 13.16.

Different objects that are not active at the same time can be implemented on the same logical processor (and thus also on the same physical processor—this distinction is explained below). Objects that must operate concurrently must be implemented on different logical processors (though perhaps still on the same physical processor).

The distinction between logical and physical concurrency is as follows. There are a number of ways of simulating the existence of multiple processors using only a single physical processor. For example, some operating systems (Unix and Windows NT) allow more one than one task to appear to execute at the same time, and are thus called multi-tasking operating systems. In fact, only one task really takes place at any one time, but the operating system shares the processor between different tasks so quickly that the tasks appear to execute simultaneously. Where there are no tight time constraints a multi-tasking operating system can provide a satisfactory implementation of concurrency. But it is important to ensure that the hardware configuration of the computer can cope with the demands of multi-tasking.

When there are tight time constraints a scheduler sub-system can be introduced that ensures that each thread of control operates within the constraints on its response time.

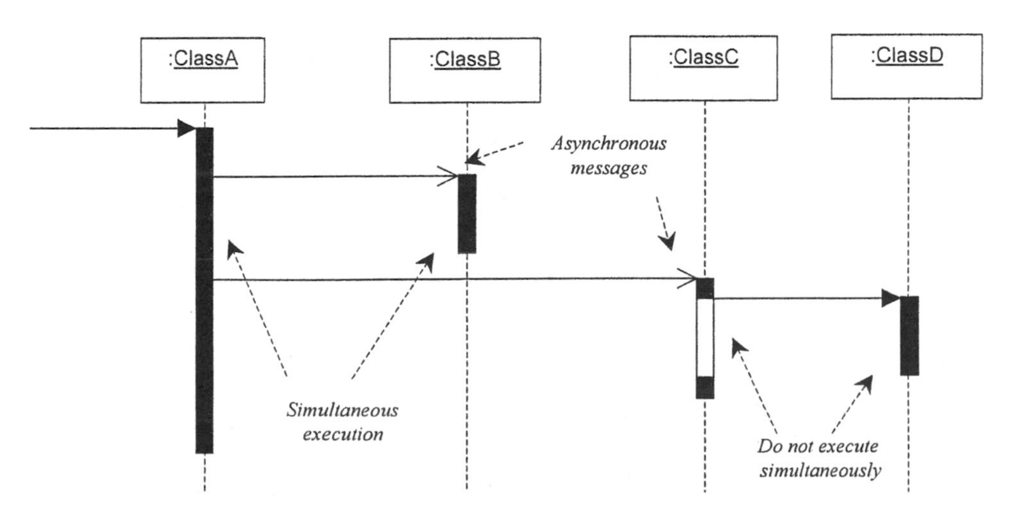

Figure 13.16 Concurrent activity in an interaction diagram.

Figure 13.17 illustrates a possible relationship between a scheduler and the other parts of a system. Events that are detected by the I/O (input/output) sub-systems generate interrupts in the scheduler. The scheduler then invokes the appropriate thread of control. Further interrupts may invoke other threads of control and the scheduler allocates a share of physical processor time to each thread.

Another way of implementing concurrency is to use a multi-threaded programming language (such as Java). These permit the direct implementation of concurrency within a single processor task. Finally, a multi-processor environment allows each concurrent task to be implemented on a separate processor.

Most concurrent activity in a business information system can be supported by a multi-user environment. These are designed to allow many users to perform tasks simultaneously. Multi-user concurrent access to data is normally handled by a separate database management system (DBMS)—these are introduced briefly in Section 13.6 and are discussed in more detail in Chapter 17.

13.5 Processor Allocation

In the case of a simple, single-user system it is almost always appropriate for the complete system to operate on a single computer. The software for a multi-user information system (all or part of it) may be installed on many computers that all use a common file-server. More complex applications sometimes require the use of more than one type of computer, where each provides a specialized kind of processing capability for a specific sub-system. An information system may also be partitioned over several processors either because sub-systems must operate concurrently, or because some parts of the application need to operate in different locations (in other words, it is a distributed system). Information systems that use the Internet or company intranets for their communications are being built more frequently. Such distributed information systems operate on diverse computers and operating systems.

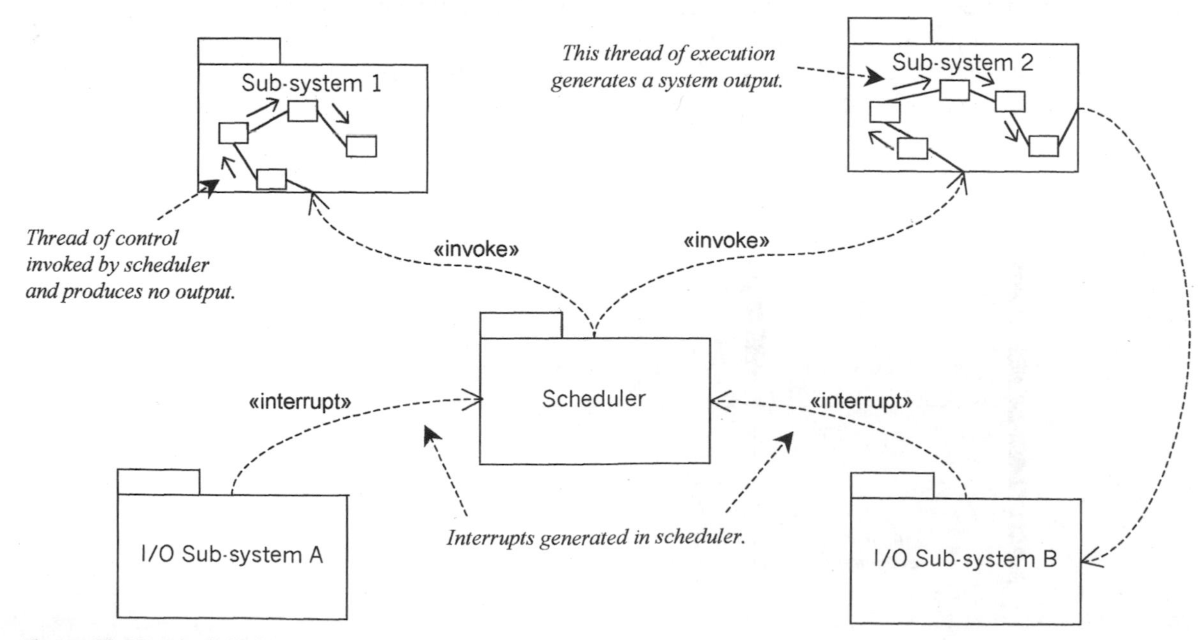

Figure 13.17 Scheduler handling concurrency.

The allocation of a system to multiple processors on different platforms involves the following steps.

- The application should be divided into sub-systems.
- Processing requirements for each sub-system should be estimated.
- Access criteria and location requirements should be determined.
- Concurrency requirements for the sub-systems should be identified.
- Each sub-system should be allocated to an operating platform—either general purpose (PC or workstation) or specialized (embedded micro-controller or specialist server).
- Communication requirements between sub-systems should be determined.
- The communications infrastructure should be specified.

The estimation of processing requirements requires careful consideration of such factors as event response times, the data throughput that is needed, the nature of the I/O that is required and any special algorithmic requirements. Access and location factors include the difficulties that may arise when a computer will be installed in a harsh operating environment such as a factory shop floor.

13.6 Data Management Issues

Suitable data management approaches for an information system can vary from simple file storage and retrieval to sophisticated database management systems of various types. In some applications where data has to be accessed very rapidly, the data may be kept in main memory while the system executes. However, most data management is concerned with storing data, often large volumes, so that it may be accessed at a later stage either by the same system or by another. A file-based approach to data management is suitable for simple data management, but the data storage and retrieval must be coded explicitly within the application. If the data is to be accessed by multiple users simultaneously, the management of the data storage and retrieval directly in the application becomes complex. However, a requirement for fast access to data may justify the use of a file-based approach and its resulting code complexity.

Database management systems (DBMS) provide various facilities that are useful in many applications. A DBMS typically offers support for:

- different views of the data by different users,
- control of multi-user access,
- distribution of the data over different platforms,
- security,
- enforcement of integrity constraints,
- access to data by various applications,
- data recovery,
- portability across platforms,
- data access via query languages and
- query optimization.

These capabilities make a DBMS the obvious choice for many applications. However, a DBMS does have a significant performance overhead and the standard data access mechanisms may be inappropriate for specialized systems. Once a decision has been made to use a DBMS the most appropriate type must be selected. A relational DBMS is likely to be appropriate if there are large volumes of data with varying (perhaps ad hoc) access requirements. An object-oriented DBMS is more likely to be suitable if specific transactions require fast access or if there is a need to store complex data structures and there is not a need to support a wide range of transaction types. A third type of DBMS is emerging—the object-relational DBMS—that is similar to an object-oriented DBMS in its support for complex data structures, but that also provides effective querying facilities. In some systems there may be different data management requirements for different sub-systems and it may be best then to use a mix of DBMS types. These issues and the detailed design consequences of the choice of DBMS are explored further in Chapter 18.

The DBMS used by the application will typically be accessed from programs using specialized class libraries to provide the database access functionality. It is rarely the case today that a developer must construct low-level primitives in order to access a DBMS since suitable commercial class libraries are widely available.

13.7 Development Standards

All information systems development projects should operate with clearly defined guidelines within which all members of the development team work. Many organizations will have corporate style guides that govern the production of software development artefacts, including the delivered system. In some organizations these corporate guides may be adapted for particular development projects. Modelling standards have been discussed in Chapter 5. From the design perspective it is important to specify guidelines for the development of I/O sub-systems and their interfaces and standards for the development of code.

13.7.1 HCI guidelines

Standards for the human–computer interface are an important aspect of the design activity, since it is with the interface that users actually interact. Some characteristics of good dialogues and the subject of style guides for HCI are discussed in Chapter 16.

13.7.2 Input/output device guidelines

Where an application interacts with mechanical or electronic devices such as temperature and pressure sensors or actuators that control heaters or motors, it is equally important to develop guidelines. The objective is to use a standard form of interface with the devices so that hardware may be changed or updated without occasioning any changes to the core system functionality. Sensor and controller devices usually have fully specified communications protocols and standardization is probably best achieved by encapsulating all direct access with each device in a single I/O device object. An I/O device class can be subclassed so that for each particular device involved with the application there is a class that deals with its communications protocol. Since all the I/O device classes would then be subclasses of one inheritance hierarchy they can easily be constructed to provide consistent interfaces to the rest of

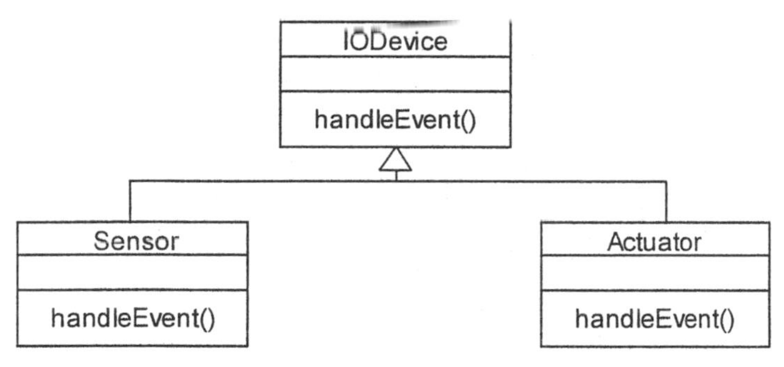

Figure 13.18 I/O device hierarchy.

the application. One way of achieving this is by the use of polymorphism, as shown in Figure 13.18. In this diagram, each subclass overrides the abstract operation `handleEvent()`, so that each can respond to a `handleEvent()` message in its own appropriate way, but other objects in the system do not necessarily need to know which particular type of object will receive the message.

13.7.3 Construction guidelines

Construction guidelines might not appear relevant at this stage in a systems development project. However, they are pertinent to system design because there is a growing tendency for developers to use CASE environments that have code generation capabilities. It is also the case that when a rapid development approach or an iterative approach is followed, the demarcation between design and construction activities can become blurred.

Construction guidelines will normally include advice on the naming of classes, of operations and of attributes, and where this is the case these guidelines are also applicable during the analysis activity. Wherever possible, consistent naming conventions should be enforced throughout the project since this makes it easier to trace an analysis class directly through to its implementation. Other guidelines for construction might relate to the use of particular software features (for example, using only standard language constructs in order to aid portability) and the layout of the code. These issues are addressed in more detail in Chapter 19.

13.8 Prioritizing Design Trade-offs

Design frequently involves choosing the most appropriate compromise. The designer is often faced with design objectives that are mutually incompatible and he or she must then decide which objective is the more important. The requirements model may indicate the relative priorities of the different objectives but, if it does not, then it is useful to prepare general guidelines. These guidelines must be agreed with clients since they determine the nature of the system and what functionality will be delivered. Guidelines for design trade-offs ensure consistency between the decisions that are made at different stages of development. They also ensure consistency between different sub-systems. However, no guidelines can legislate for every case. Design experience and further discussions with the client will remain necessary to resolve those situations that cannot be anticipated—at least some of these occur on almost every project.

13.9 Design for Implementation

The introduction of an information system within an organization frequently requires some form of data conversion. System design must consider how best to deal with this and other system initialization issues. Special programs may be designed and written that convert data from one format to another and procedures may have to be specified for the manual entry of data that configures the final system. A new system may need a batch input capability so that existing data can be loaded, even where the day-to-day use of the system is intended to be fully interactive with on-line update of all data.

The design and construction of the data conversion or initialization sub-systems may represent a significant part of development and can be complex. Logistical considerations such as the overall volumes of data and the time available for conversion and initialization will have a significant impact upon the approach that is taken. The data of an enterprise represents one of its most valuable resources and every care must be taken not to corrupt it or to lose it.

13.10 Summary

System design focuses on determining a suitable architectural structure for the system and defines the context within which the remaining design activity is performed. We have identified the following activities as germane to system design.

- Identify sub-systems and major components. Splitting a system into sub-systems provides many benefits, both technical and managerial. Three major styles of software architectures were discussed, layering and partitioning, the MVC and the broker architecture.
- Identify any inherent concurrency. UML provides models of the system that help to identify concurrent behaviour. Different design strategies have to be adopted depending upon the implementation environment and the response requirements for the system.
- Allocate sub-systems to processors. Computerized information systems commonly operate on multiple computers. In some cases it is important to partition a system across a series of computers (specialized by capability or location) to ensure that user requirements are addressed.
- Select data management strategies. The alternatives range from a simple file storage and retrieval mechanism to relational, object-oriented or object-relational DBMS.
- Specify development standards. Style guides are applicable to almost all systems development products. They ensure a uniformity of presentation and help ensure quality.
- Set priorities for design trade-offs. A clearly defined set of priorities is advisable to ensure that appropriate design decisions are made.
- Identify implementation requirements (e.g. data conversion). The requirements for the installation of the system are often ignored during system design.

System design determines the degree to which a system satisfies its non-functional objectives and consequently the degree to which it can effectively support the user requirements both now and in the future.

Review Questions

13.1 Why is an open layered architecture more difficult to maintain?

13.2 What are the disadvantages of the closed layer architecture?

13.3 What advantages would there be if the Advert HCI sub-system in Figure 13.8 were designed to have direct access to the Campaign Database layer?

13.4 What are the main differences between the MVC architecture and the layered and partitioned architecture?

13.5 Explain how the update propagation mechanism works in the MVC architecture.

13.6 In what sense does a broker decouple two sub-systems that need to communicate with each other? How does this work?

13.7 Why is it sometimes necessary to design information systems that have explicitly concurrent behaviour?

13.8 What ways are there of simulating concurrency in the execution of a system?

13.9 How should you go about allocating system tasks to processors?

13.10 What facilities are typically offered by a DBMS?

Case Study Work, Exercises and Projects

13.A Develop a series of steps for the identification of partitioned sub-systems within a layer in a layered architecture. Use the process for the identification of layers described in Section 13.3.2 as a starting point. Highlight any significant differences that you feel exist between the two processes.

13.B Suggest a suitable layered architecture with any necessary partitioning for the FoodCo case study by following the procedures defined above.

13.C Identify a suitable hardware architecture or architectures for the FoodCo case study and justify your decisions.

13.D Identify a suitable data management approach for the FoodCo case study and, if appropriate, specify the type of DBMS to be used.

Further Reading

- Buschmann et al. (1996 and 2000) provide further details of the architectures discussed in this chapter and describe other interesting alternatives.
- Both Rumbaugh et al. (1991) and Booch (1994) offer useful advice on system design issues.
- Goldsmith (1993) provides a non-object-oriented perspective on system design and it is interesting to consider the many issues that are the same in both approaches.
- Sommerville (1992) is an excellent text that covers a broad range of software engineering issues.
- Shlaer and Mellor (1988 and 1992) describe an alternative approach to object-oriented system design. Rather than the elaboration of the analysis models that we describe they suggest an approach called Recursive Design to produce the design model.

14 CHAPTER

Object Design

OBJECTIVES

In this chapter you will learn

- how to apply criteria for good design
- how to design associations
- the impact of integrity constraints on design
- how to design operations
- the role of normalization in object design.

14.1 Introduction

Object design is concerned with the detailed design of the objects and their interactions. It is completed within the overall architecture defined during system design and according to agreed design guidelines and protocols. Object design is particularly concerned with the specification of the attribute types, how operations function and how objects are linked to other objects. There are many sources of information that guide the object design process (Figure 14.1). During the design process the analysis models undergo some degree of transformation. There is a commonly accepted view that changes made to analysis artefacts to produce the design model should be kept to a minimum, as the analysis model is a coherent and consistent description of the requirements (Ward and Mellor, 1985, Goldsmith, 1993). Object design produces a detailed specification of the classes and uses the UML notation described in Section 14.2 for attributes and operation signatures. An important aspect of every class is what attributes (if any) and operations are generally accessible, and this is considered in Section 14.3. In Section 14.4 we consider a series of criteria, including coupling and cohesion, that help to produce good object-oriented designs.

Much design activity focuses on adding detail to the analysis specification and makes no change to the structures identified during analysis. However, there may be a need to

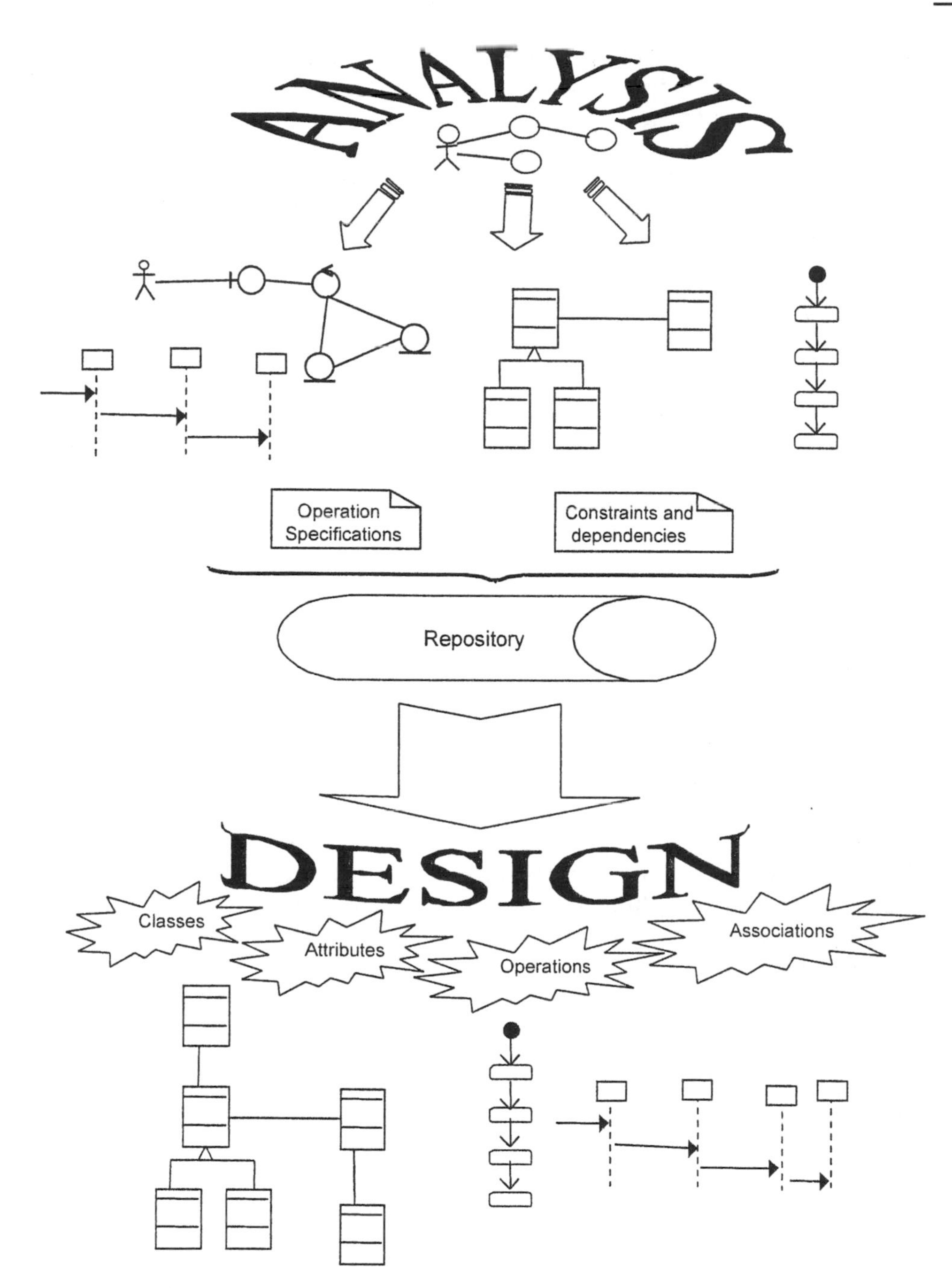

Figure 14.1 Information sources for object design.

modify the analysis structures for various reasons. The associations modelled in analysis have to be designed to minimize coupling. A series of common designs for associations are discussed in Section 14.5. Analysis will have identified integrity constraints that have to be adhered to by the information system if it is to maintain its consistency and integrity. Various strategies for the design of integrity constraints are examined in Section 14.6. The functionality of the information system is defined in the operation specifications. These have now to be translated into suitable designs (Section 14.7). If the information system is going to utilize relational database technology it may be appropriate to normalize the data structures (Section 14.8). This technique may also be useful for decomposing complex classes. Concurrent activities may be placed in separate sub-systems and may require an analysis class to be split into two design classes. As new classes are added to the design model to support data management and the operation of the interface, analysis classes have to be modified to build associations and communication pathways to these new sub-systems. As we shall see in Chapter 15 the use of design patterns offers structures and modes of interaction that support non-functional requirements.

14.2 Class Specification

14.2.1 Attributes and operation signatures

Attributes

During analysis we have not considered in detail the data types of the attributes, although on occasions it may be useful to record data type information in analysis. For example, an attribute `temperature` might be a floating-point data type if it holds the temperature in Centigrade or it might be an enumerated data type if it holds one of the values, 'hot' or 'cold'. The attribute has a different meaning and would be manipulated differently for each of these data types and it is important to determine during analysis which meaning is appropriate.

Common primitive data types include Boolean (true or false), Character (any alphanumeric or special character), Integer (whole numbers) and Floating-Point (decimal numbers).[1] In most object-oriented languages more complex data types, such as Money, String, Date, or Name can be constructed from the primitive data types or may be available in standard libraries. An attribute's data type is declared in UML using the following syntax:

```
name ':' type-expression '=' initial-value '{'property-string'}'
```

The name is the attribute name, the `type-expression` is its data type, the `initial-value` is the value the attribute is set to when the object is first created and the `property-string` describes a property of the attribute, such as constant or fixed. The characters in single quotes are literals. The attribute name is the only feature of its declaration that is compulsory.

Figure 14.2 shows the class `BankAccount` with attribute data types declared. The attribute `balance` in a `BankAccount` class might be declared with an initial value of zero using the syntax:

```
balance:Money = 0.00
```

[1] A list of Java primitive data types can be found in Deitel and Deitel (1997).

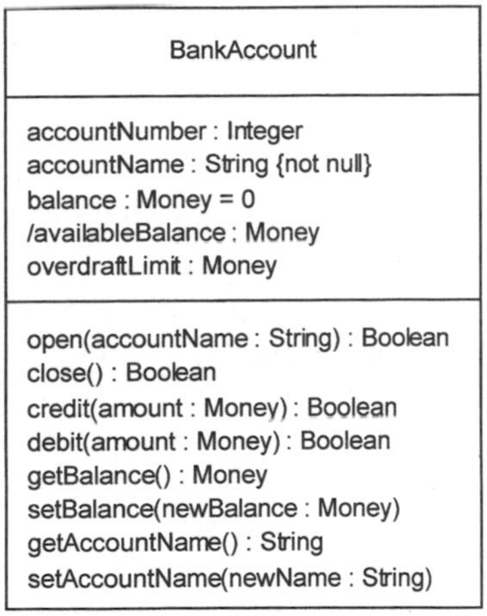

```
BankAccount
----------------------------------------
accountNumber : Integer
accountName : String {not null}
balance : Money = 0
/availableBalance : Money
overdraftLimit : Money
----------------------------------------
open(accountName : String) : Boolean
close() : Boolean
credit(amount : Money) : Boolean
debit(amount : Money) : Boolean
getBalance() : Money
setBalance(newBalance : Money)
getAccountName() : String
setAccountName(newName : String)
```

Figure 14.2 `BankAccount` class.

The attribute `accountName` might be declared with the property string indicating that it must have a value and may not be null using the syntax:

```
accountName : String {not null}
```

Attribute declarations can also include arrays. For example, an `Employee` class might include an attribute to hold a list of qualifications that would be declared using the syntax:

```
qualification[0..10]: String
```

This declaration states that the attribute qualification may hold from zero to 10 qualifications.

Operations

Each operation also has to be specified in terms of the parameters that it passes and returns. The syntax used for an operation is:

```
operation name '('parameter-list ')'':' return-type-expression
```

An operation's *signature* is determined by the operation's name, the number and type of its parameters and the type of the return value if any. The `BankAccount` class might have a `credit()` operation that passes the amount being credited to the receiving object and has a Boolean return value. The operation would be defined using the syntax:

```
credit(amount : Money): Boolean
```

A `credit()` message sent to a `BankAccount` object could have the format :

```
creditOK = accObject.credit(500.00)
```

where `creditOK` holds the Boolean return value that is available to the sending object when the `credit()` operation has completed executing. This Boolean value may be tested to determine whether the `credit()` operation performed successfully. In an object-oriented language like Java this would typically be done as follows:

```
try{
   accObject.credit(500.00);
} catch (UpdateException) {
   //some error handling;
}
```

The UML is a modelling language. It does not determine what operations should be shown in a class diagram. It provides the notation to use and suggestions on presentation, but does not tell the analyst or designer what to include and what not to include.

Some authors of books on object-oriented analysis and design give specific guidelines about operations. Coad and Yourdon (1991) talk about services: an object provides services to other objects. They talk about services as being like responsibilities at the system level, but then go on to specify services as what we call the operations of classes. Yourdon (1994) talks about some services being implicit services. These are services to create instances of objects, to modify attributes of instances, to select instances based on some kind of key or identifier, and to delete instances. They say that these usually need not be shown on diagrams, as they clutter up the diagrams and make them difficult to read. (However, all operations have to be specified somewhere, and it is important to recognize that a class may have several different constructors for instances of that class.) They also point out that sometimes it is important to be able to see these services. However, this is an issue about the functionality offered by CASE tools rather than methodologies. Ideally, it should be possible to switch off the display of any operation that the analyst or designer does not wish to have displayed in the operations compartment in a class.

An alternative approach would be to follow the way that some CORBA tools work. The IDL2JAVA tool will generate two Java operations for every attribute: one to set the values of the attribute and one to get the value of the attribute. The following example shows the operations generated to set and get the value of the attribute `smallUnit` in a `Currency` class.

Note that the naming convention of the operations differentiates between get (reader) and set (writer) operations by the number of parameters and the return type.

```
short smallUnit;
...
public void smallUnit(short smallUnit) {
  // implement attribute writer...
  smallUnit = smallUnit;
}
public short smallUnit() {
  // implement attribute reader...
  return smallUnit;
}
```

The framework for this fragment of code (together with the rest of the Java implementation of this CORBA interface) was generated automatically from the following IDL interface definition.

```
interface Money {
    attribute long largeUnit;
    attribute short smallUnit;
    attribute string format;
    Money Money( in long large, in short small);
    string toString();
};
```

So, if you have a CASE tool that can generate set and get operations for every attribute, you do not need to include them in your class diagram.

We have shown primary operations on some class diagrams for pedagogic reasons to emphasize their presence and hence have taken a pragmatic approach. In most parts of the book we have shown those operations that were useful to show. We have usually included the operations that are not primary operations (where primary operations are the create, destroy, get and set operations). Typically, these are the operations that Coad and Yourdon describe as algorithmically complex. We have included some primary operations, usually because they are referred to in the text or some related diagram.

One commonly held approach is normally not to show primary operations on analysis class diagrams as it can be assumed that such functionality is available. During analysis issues such as the visibility of operations or the precise data types of attributes may not have been finally decided. However, when completing a design class diagram it may be important to indicate that certain primary operations have public or protected visibility and as such these may justifiably be shown on the diagram. Those that are private may be omitted as they do not constitute part of the class' public interface.

Exceptionally primary operations may usefully be included on analysis class diagrams either if they reflect particular functionality that has to be publicly visible or if it is important to indicate that more than one constructor, for example, is required. A class may need more than one constructor if objects could be instantiated in one of several initial states that require different input parameters. Each constructor would have a different signature.

There are clearly alternative approaches and it is important that appropriate documentation standards are clearly defined at the outset of a project so that the absence of primary operations on a class diagram is not misinterpreted.

14.2.2 Object visibility

The concept of encapsulation was discussed in Chapter 4 and is one of the fundamental principles of object-orientation. During analysis various assumptions have been made regarding the encapsulation boundary for an object and the way that objects interact with each other. For example, it is assumed that the attributes (or more precisely the values of the attributes) of an object cannot be accessed directly by other objects but only via 'get' and 'set' operations (primary operations) that are assumed to be available for each attribute. Moving to design involves making decisions regarding which operations (and possibly attributes) are publicly accessible. In other words we must define the encapsulation boundary.

Figure 14.2 shows the class `BankAccount` with the types of the attributes specified and the operation parameters defined. The class has the attribute `balance`, which we might assume during analysis, can be accessed directly by the simple primary operations `getBalance()` and `setBalance()`. However, the balance should be updated through the operations `credit()` and `debit()` that contain special processing to check whether these transactions should be permitted and to ensure that the transactions are logged in an audit trail[2]. In these circumstances, it is important that changes to the value of the `balance` attribute can only occur through the `debit()` and `credit()` operations. The operation `setBalance()` should not be publicly available for use by other classes. Note also that the attribute `availableBalance` is a derivable

[2] In general, an audit trail records the details of transactions so that they can be checked to ensure that, for example, fraud has not occurred.

attribute indicated in UML by the symbol '/'. A derivable attribute is one whose value can be calculated or determined from the value of other attributes.

Meyer (1997) introduces the term 'secret' to describe those features that are not available in the public interface. Programming languages designate the non-public parts of a class, which may include attributes and operations, in various ways. The four commonly accepted terms[3] used to describe *visibility* are listed in Figure 14.3. Visibility may also be shown as a property string.

```
Balance : Money {visibility = private}
```

To enforce encapsulation the attributes of a class are normally designated private (Figure 14.4). The operation `setBalance()` is also designated private to ensure that objects from other classes cannot access it directly and make changes that are not recorded in the audit trail. Private operations can, of course, be invoked from operations in the same class. Commonly, complex operations are simplified by factoring out procedures into private operations.

In Figure 14.4 the operation `getBalance()` is assigned protected visibility so that subclasses of `BankAccount` can examine the value of the `balance` attribute. For

Visibility symbol	Visibility	Meaning
+	Public	The feature (an operation or an attribute) is directly accessible by an instance of any class.
–	Private	The feature may only be used by an instance of the class that includes it.
#	Protected	The feature may be used either by instances of the class that includes it or of a subclass or descendant of that class.
~	Package	The feature is directly accessible only by instances of a class in the same package.

Figure 14.3 Visibility.

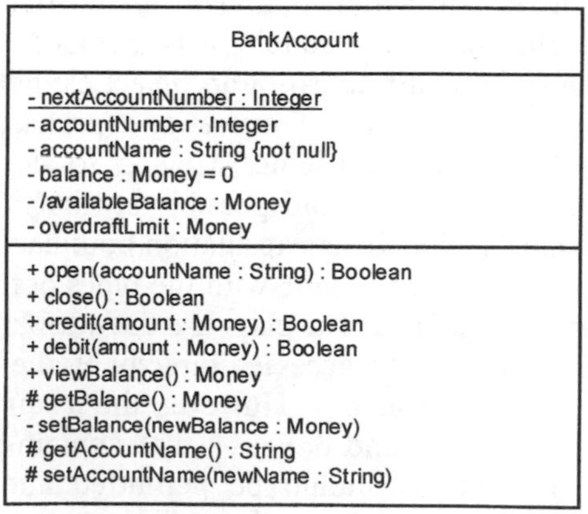

```
BankAccount
------------------------------------------
- nextAccountNumber : Integer
- accountNumber : Integer
- accountName : String {not null}
- balance : Money = 0
- /availableBalance : Money
- overdraftLimit : Money
------------------------------------------
+ open(accountName : String) : Boolean
+ close() : Boolean
+ credit(amount : Money) : Boolean
+ debit(amount : Money) : Boolean
+ viewBalance() : Money
# getBalance() : Money
- setBalance(newBalance : Money)
# getAccountName() : String
# setAccountName(newName : String)
```

Figure 14.4 `BankAccount` class with visibility specified.

[3] The precise meaning of the non-public categories of visibility depends on the programming language being used. When a designer is determining the visibility of parts of a model, he or she must be aware of the visibility (or scoping) offered by the implementation environment. Package visibility is a recent addition to UML.

example, the debit() operation might be redefined polymorphically in a JuniorBankAccount subclass. The redefined operation would use getBalance() to access the balance and check that a debit would not result in a negative balance.

The attribute nextAccountNumber in Figure 14.4 is an example of a class-scope attribute (indicated by underlining). A class-scope attribute occurs only once and is attached to the class not to any individual object. In this example nextAccountNumber holds the account number for the next new BankAccount object created. When a new BankAccount is created nextAccountNumber is incremented by one. The attribute accountNumber is an example of an instance-scope attribute (hence no underlining). Each BankAccount object has an instance-scope accountNumber attribute which holds its unique account number.

14.3 Interfaces

On occasions a class (or some other component) may present more than one external interface to other classes or the same interface may be required from more than one class. An interface in UML is a group of externally visible (i.e. public) operations. The interface contains no internal structure, it has no attributes, no associations and the implementation of the operations is not defined. Formally, an interface is equivalent to an abstract class that has no attributes, no associations and only abstract operations. Figure 14.5 illustrates two alternative notations for an interface. The simpler of the two UML interface notations is a circle. This is attached by a solid line to the classes that support the interface. For example, in Figure 14.5 the Advert class supports two interfaces, Manageable and Viewable, that is, it provides all of the operations specified by the interface (and maybe more). The circle notation does not include a list of the operations provided by the interface type, though they should be listed in the repository. The dashed arrow from the CreativeStaff class to the Manageable interface circle icon indicates that it uses or needs, at most, the operations provided by the interface.

The alternative notation[4] uses a stereotyped class icon. As an interface only specifies the operations and has no internal structure, the attributes compartment is omitted. This notation lists the operations on the diagram. The *realize* relationship, represented by the dashed line with a triangular arrowhead, indicates that the client class (e.g. Advert) supports at least the operations listed in the interface (e.g. Manageable or Viewable).[5] Again the dashed arrow from CreativeStaff means that the class needs or uses no more than the operations listed in the interface. The notation used for the *realize* relationship (the triangular arrowhead) is deliberately reminiscent of the notation for inheritance, as in a sense Advert inherits the Manageable interface. This concept can be implemented using the 'interface' programming language construct in Java.

[4] Normally only one of these notations would be used in a diagram.

[5] The realize relationship can be used between two classes (rather than between a class and an interface) to mean that the client (at the tail of the arrow) provides at least the operations defined in the supplier class (at the arrowhead), but does not necessarily support any of the structure (i.e. the attributes of the class).

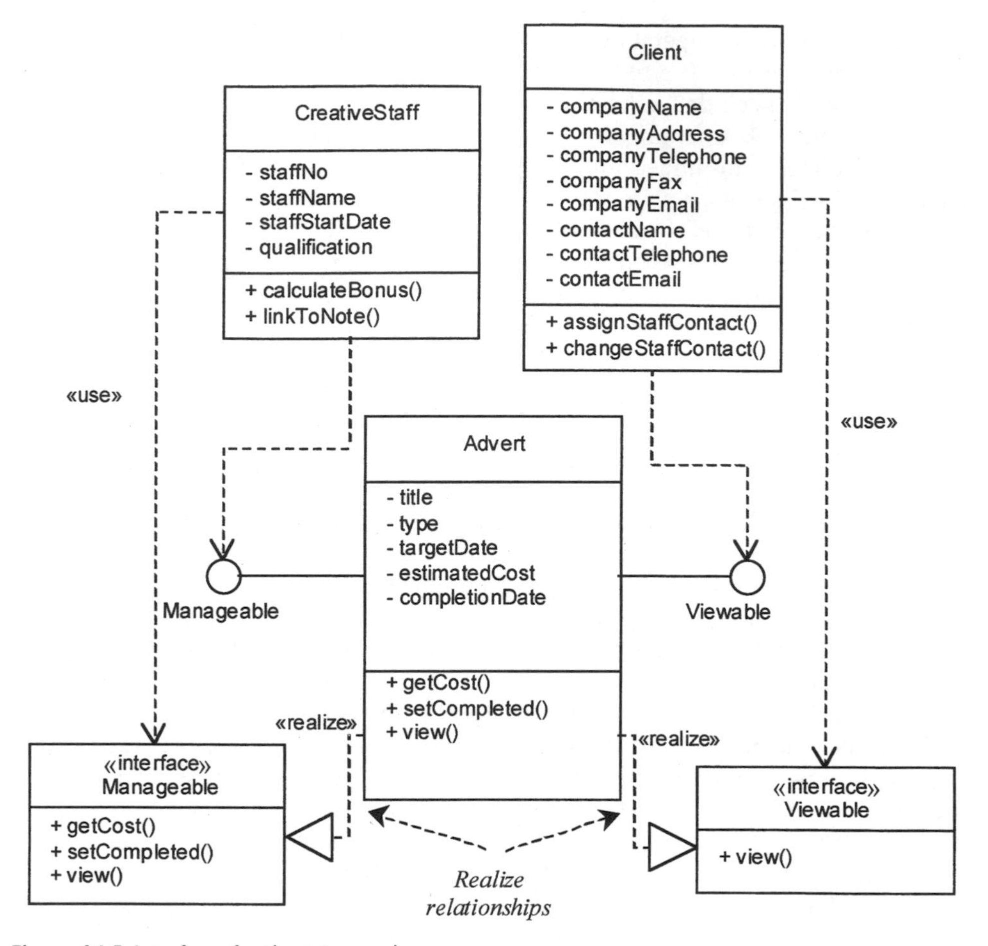

Figure 14.5 Interfaces for the `Advert` class.

14.4 Criteria for Good Design

14.4.1 Coupling and cohesion

In Chapter 12 we stated that coupling and cohesion are among the criteria for good design. In fact, it is good practice in analysis to prepare models that, as far as possible, satisfy these criteria. Coupling describes the degree of interconnectedness between design components and is reflected by the number of links an object has and by the degree of interaction the object has with other objects. Cohesion is a measure of the degree to which an element contributes to a single purpose. The concepts of coupling and cohesion are not mutually exclusive but actually support each other. Coad and Yourdon (1991) suggested several ways in which coupling and cohesion can be applied within an object-oriented approach. Larman (1998) also considers the application of these criteria. The criteria can be used within object-orientation as described below (adapted from Coad and Yourdon, 1991).

Interaction Coupling is a measure of the number of message types an object sends to other objects and the number of parameters passed with these message types. Interaction coupling should be kept to a minimum to reduce the possibility of changes rippling through the interfaces and to make reuse easier. When an object is reused in another application it will still need to send these messages (unless the object is modified before it is reused) and hence needs objects in the new application that provide these services. This complicates the reuse process as it requires groups of classes to be reused rather than individual classes. (In Chapter 8 we introduced the idea of the *component* as the unit of reuse and discuss it further in Chapter 20. Components are groups of objects that together provide a clearly defined service.)

Inheritance Coupling describes the degree to which a subclass actually needs the features it inherits from its base class. For example, in Figure 14.6 the inheritance hierarchy exhibits low inheritance coupling and is poorly designed. The subclass `LandVehicle` needs neither the attributes `maximumAltitude` and `takeOffSpeed` nor the operations `checkAltitude()` and `takeOff()`. They have been inherited unnecessarily. In this example it would appear the base class, `Vehicle`, would perhaps be better named `FlyingVehicle` and the inheritance relationship is somewhat suspect. A land vehicle is not a kind of flying vehicle (not normally anyway). However, many systems developers view designs with a small degree of unnecessary inheritance as being acceptable if the hierarchy is providing valuable reuse and is meaningful. It can be argued that if attributes and operations are inherited unnecessarily it is merely a matter of not using these features in the subclass. However, a subclass with unnecessary attributes or operations is more complex than it needs to be and objects of the subclass may take more memory than they actually need. The real problems may come when the system needs maintenance. The system's maintainer may not realize that some of the inherited attributes and operations are unused and may modify the system incorrectly as a result. Alternatively the system's maintainer may use these unneeded features to provide a fix for a new user requirement, making the system even more difficult to maintain in the future. For these reasons, unnecessary inheritance should be kept as low as possible.

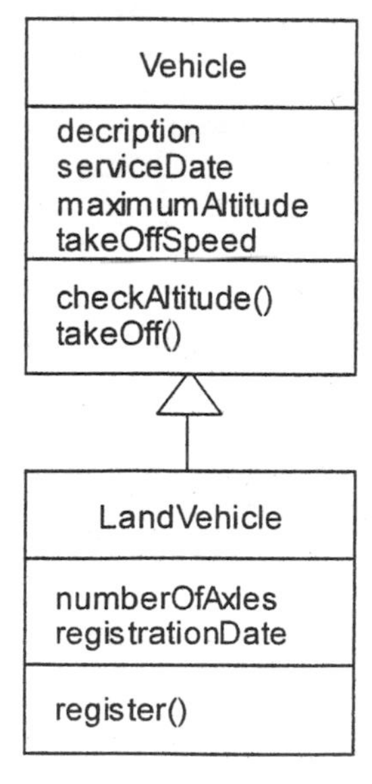

Figure 14.6 Inheritance coupling.

Operation Cohesion measures the degree to which an operation focuses on a single functional requirement. Good design produces highly cohesive operations, each of which deals with a single functional requirement. For example in Figure 14.7 the operation `calculateRoomSpace()` is highly cohesive.

Class Cohesion reflects the degree to which a class is focused on a single requirement. The class `Lecturer` in Figure 14.7 exhibits low levels of cohesion as it has three attributes (`roomNumber`, `roomLength` and `roomWidth`) and one operation (`calculate RoomSpace()`) that would be more appropriate in a class `Room`. The class `Lecturer` should only have attributes that describe a `Lecturer` object (e.g. `lecturerName` and `lecturerAddress`) and operations that use them.

Specialization Cohesion addresses the semantic cohesion of inheritance hierarchies. For example in Figure 14.8 all the attributes and operations of the `Address` base class are used by the derived[6] classes: the hierarchy has high inheritance coupling. However, it is neither true that a person is a kind of address nor that a company is a kind of address. The example is only using inheritance as a syntactic structure for sharing attributes and operations. This structure has low specialization cohesion and is poor design. It does not reflect meaningful inheritance in the problem domain. A better design is shown in Figure 14.9, in which a common class `Address` is being used by both the `Person` and `Company` classes. All the design criteria explained above may be applied at the same time to good effect.

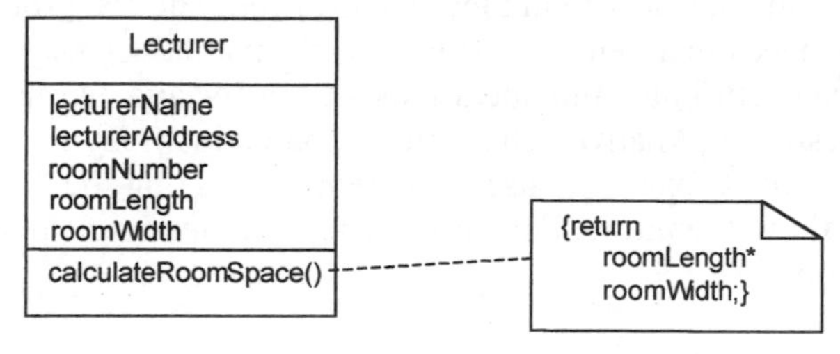

Figure 14.7 Good operation cohesion but poor class cohesion.

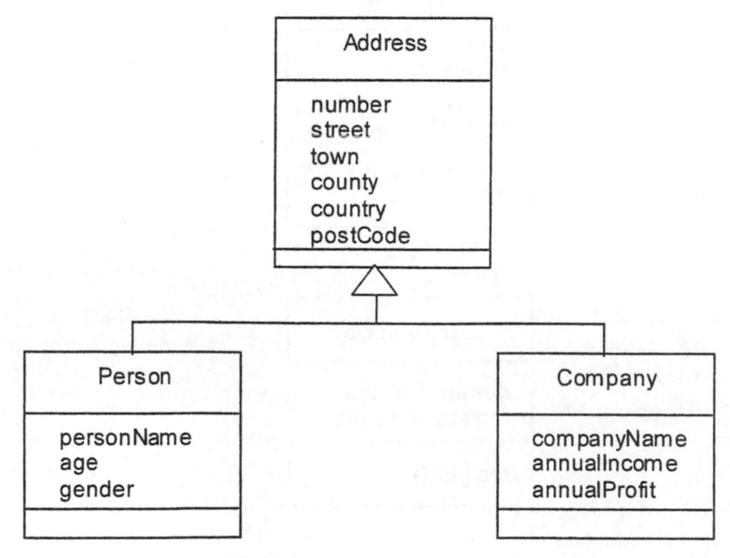

Figure 14.8 Poor specialization cohesion.

[6] A derived class is another term for subclass, the superclass is known as the base class.

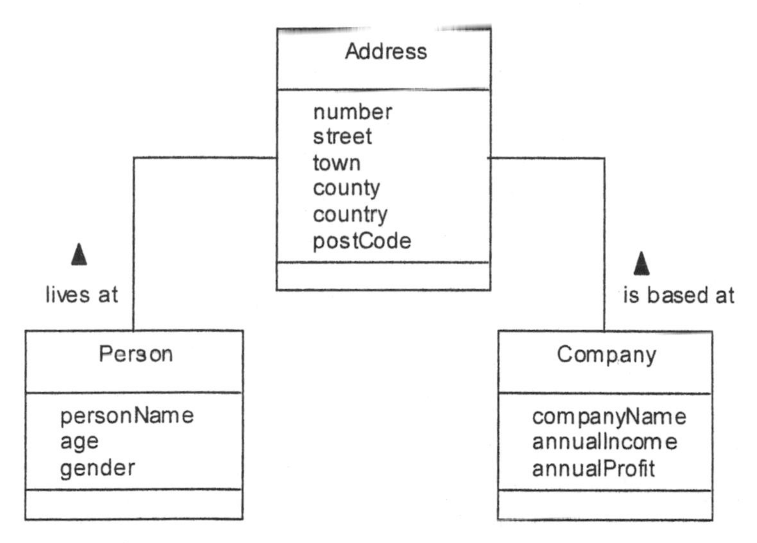

Figure 14.9 Improved structure using `Address` class.

14.4.2 Liskov Substitution Principle

The *Liskov Substitution Principle (LSP)* is another design criterion that is applicable to inheritance hierarchies. Essentially the principle states that, in object interactions, it should be possible to treat a derived object as if it were a base object. If the principle is not applied then it may be possible to violate the integrity of the derived object. In Figure 14.10 objects of the class `MortgageAccount` cannot be treated as if they are objects of the class `ChequeAccount` because `MortgageAccount` objects do not have a debit operation whereas `ChequeAccount` objects do. The `debit` operation is declared private in `MortgageAccount` and hence cannot be used by any other object. Figure 14.10 shows an alternative structure that satisfies LSP. Interestingly, this inheritance hierarchy has maximal inheritance coupling, and enforcing the LSP normally produces structures with high inheritance coupling.

14.4.3 Further design guidelines

Coad and Yourdon (1991) and Yourdon (1994) suggest further design guidelines that are included in the list below.

Design Clarity. A design should be made as easy to understand as possible. This reinforces the need to use design standards or protocols that have been specified.

Don't Over-Design. Developers are on occasions tempted to produce designs that may not only satisfy current requirements but may also be capable of supporting a wide range of future requirements. Designing flexibility (or any other non-functional requirement) into a system has a cost, the system may take longer to design and construct but this may be offset in the future by easier and less expensive modification. However, it is not feasible to design for every eventuality. Systems that are over-designed in first instance are more difficult to extend if the modifications are not sympathetic to the existing structure.

Control Inheritance Hierarchies. Inheritance hierarchies should be neither too deep nor too shallow. If a hierarchy is too deep it is difficult for the developer to understand easily what features are inherited. There is a tendency for developers new to O-O to produce over-specialized hierarchies, thus adding complexity rather than

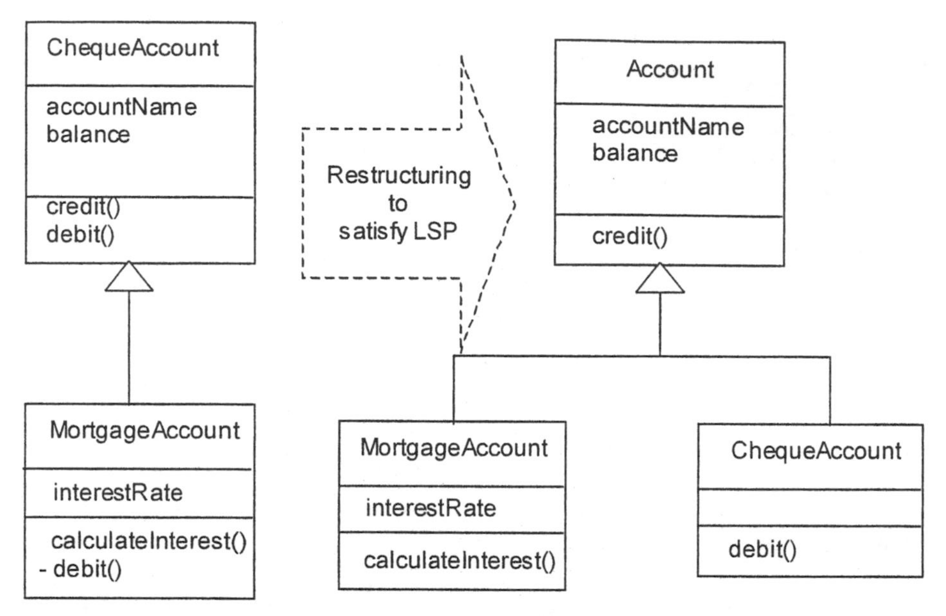

Figure 14.10 Application of the Liskov Substitution Principle.

reducing it. Yourdon (1994) suggests that hierarchies of up to nine levels are manageable, whereas Rumbaugh et al. (1991) suggest that more than about four or five levels is too many.

Keep Messages and Operations Simple. In general it is better to limit the number of parameters passed in a message to no more than three (of course, a single parameter might be a complete object). Ideally an operation should be capable of specification in no more than one page.

Design Volatility. A good design will be stable in response to changes in requirements. It is reasonable to expect some change in the design if the requirements are changed. However, any change in the design should be commensurate with the change in requirements. Enforcing encapsulation is a key factor in producing stable systems.

Evaluate by Scenario. An effective way of testing the suitability of a design is to role play it against the use cases using CRC cards if appropriate.

Design by Delegation. A complex object should be decomposed (if possible) into component objects forming a composition or aggregation. Behaviour can then be delegated to the component objects producing a group of objects that are easier to construct and maintain. This approach also improves reusability.

Keep Classes Separate. In general, it is better not to place one class inside another. The internal class is encapsulated by the other class and cannot be accessed independently. This reduces the flexibility of the system.

14.5 Designing Associations

An association between two classes indicates the possibility that links will exist between instances of the classes. The links provide the connections necessary for message passing to occur. When deciding how to implement an association it is important to analyse the message passing between the objects tied by the link.

14.5.1 One-to-one associations

In Figure 14.11 objects of the class Owner need to send messages to objects of the class `Car` but not vice versa. This particular association may be implemented by placing an attribute to hold the object identifier (some authors prefer to use the term object reference) for the `Car` class in the Owner class. Thus `Owner` objects have the `Car` object identifier and hence can send messages to the linked `Car` object. As a `Car` object does not have the object identifier for the `Owner` object it cannot send messages to the Owner object. The owns association is an example of a one-way association: the arrowhead on the association line shows the direction along which it may be navigated.

So before an association can be designed it is important to decide in which direction or directions messages may be sent. (If messages are not sent in either direction along an association, then the need for its existence should be questioned.) Essentially we are determining the navigability of the association.

In general an association between two classes A and B should be considered with the questions:

1. Do objects of class A have to send messages to objects of class B?
2. Does an A object have to provide some other object with B object identifiers?

If either of these questions is answered 'yes' then A objects need B object identifiers. However if A objects get the required B object identifiers as parameters in incoming messages, A objects need not remember the B object identifiers. Essentially, if an object needs to send a message to a destination object it must have the destination object's identifier either passed as a parameter in an incoming message just when it is required, or the destination object's identifier must be stored in the sending object. An association that has to support message passing in both directions is a two-way association. A two-way association is indicated with arrowheads at both ends[7]. As discussed earlier, it is important to minimize the coupling between objects. Minimizing the number of two-way associations keeps the coupling between objects as low as possible.

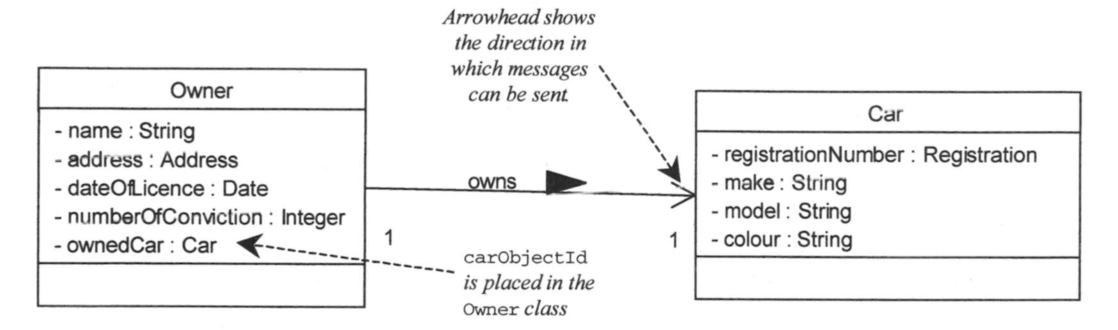

Figure 14.11 One-way one-to-one association.

[7] In UML a two-way association may be represented by drawing the association without the navigability arrowheads. However, an association without arrowheads may also represent an uncommitted association, that is, an association for which navigability is not yet decided. We suggest that arrowheads are always used to represent navigability.

14.5.2 One-to-many associations

In Figure 14.12, objects of the class `Campaign` need to send messages to objects of the class `Advert` but not vice versa. If the association between the classes was one-to-one, the association could be implemented by placing an attribute to hold the object identifier for the `Advert` class in the `Campaign` class. However, the association is in fact one-to-many and many `Advert` object identifiers need to be tied to a single `Campaign` object. The object identifiers could be held as a simple one-dimensional array in the `Campaign` object but program code would have to be written to manipulate the array.

Another way of handling the group of `Advert` object identifiers that is more amenable to reuse, is to place them in a separate object, a collection object that has operations to manage the object identifiers and that behaves rather like an index of adverts for the `Campaign` object. This is shown in the class diagram fragment in Figure 14.13. There will be many instances of the collection class, as each `Campaign` object has its own collection of `Advert` object identifiers. Notice that the `Advert Collection` class has operations that are specifically concerned with the management of the collection. The `findFirst()` operation returns the first object identifier in the list and the `getNext()` gets the next object identifier in the list.

When a `Campaign` object wants to send a message to each of its `Advert` objects the `Campaign` object first sends a `findFirst()` message to the class to get the first object identifier. The `Campaign` object can now send a message to its first `Advert` object. The `Campaign` then uses `getNext()` to get the next object identifier from the collection class and sends the message to the next `Advert` object. The `Campaign` object can

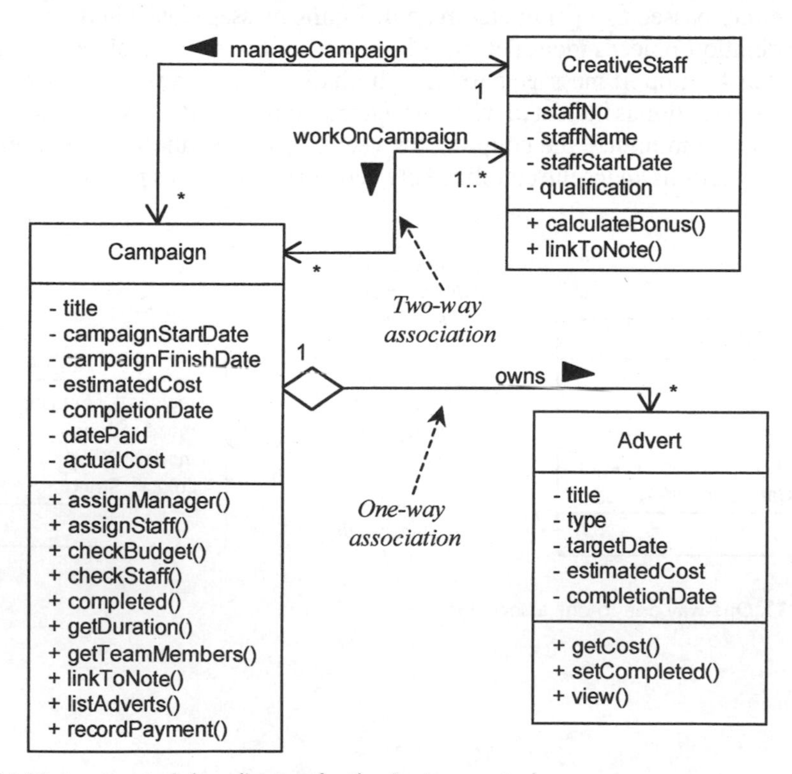

Figure 14.12 Fragment of class diagram for the Agate case study.

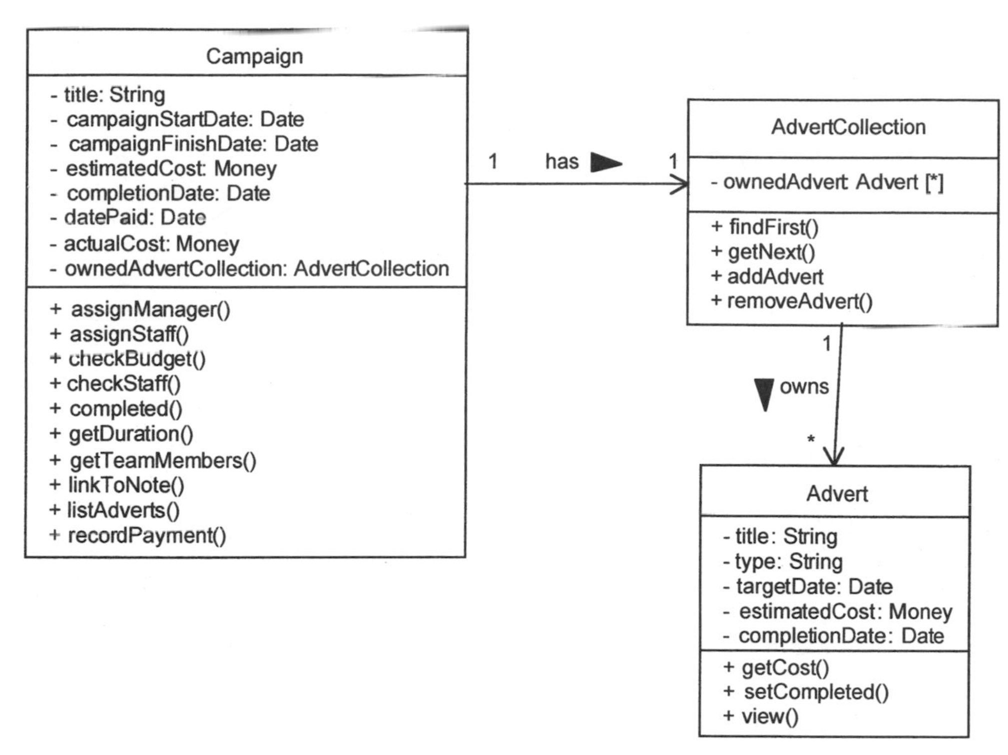

Figure 14.13 One-to-many association using a collection class.

then iterate through the collection of object identifiers and send the message to each of the `Advert` objects in turn[8].

Figure 14.14 shows the sequence diagram for the interaction that would enable the `Campaign` object to prepare a list of its adverts with their titles. (The labels on the left hand side of the diagram define the condition on the iteration and the scope of the iteration.) The `Campaign` object holds the object identifier of the collection class so that it can send messages to it. As an `Advert` object does not have the object identifier for the `Campaign` object to which it belongs, it cannot send messages to the `Campaign` object.

14.5.3 Many-to-many associations

The design of the many-to-many association `workOnCampaign` between `Creative Staff` and `Campaign` (see Figure 14.12) follows the principles described above. Assuming this is a two-way association, each `Campaign` object will need a collection of `CreativeStaff` object identifiers and each CreativeStaff object will need a collection of Campaign object identifiers. The designed association with the collection classes is shown in Figure 14.15. Both the `CreativeStaff` and `Campaign` classes contain an attribute to hold the object identifiers of their respective collection classes.

Collection classes can be designed to provide additional support for object navigation. For example, if there is a requirement to find out if an employee works on a campaign with a particular title, a message may be sent from the `CreativeStaff` object to each `Campaign` object the employee works on to get its title until either a

[8] O–O languages and class libraries provide various features to handle collection classes. For example in Java an Iterator could be applied to a collection class instance to provide the ability to iterate through the elements of the collection.

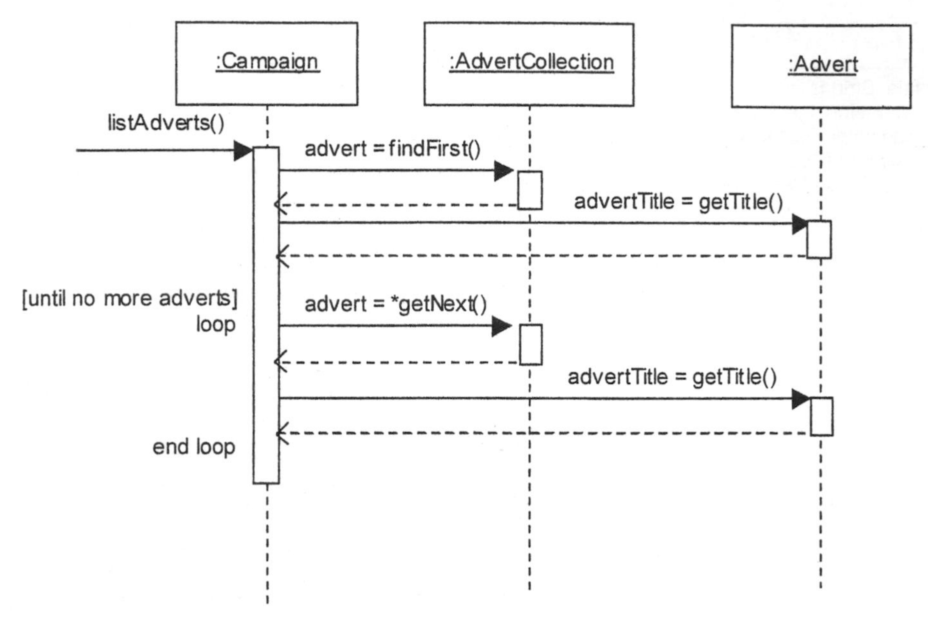

Figure 14.14 Sequence diagram for `listAdverts()`.

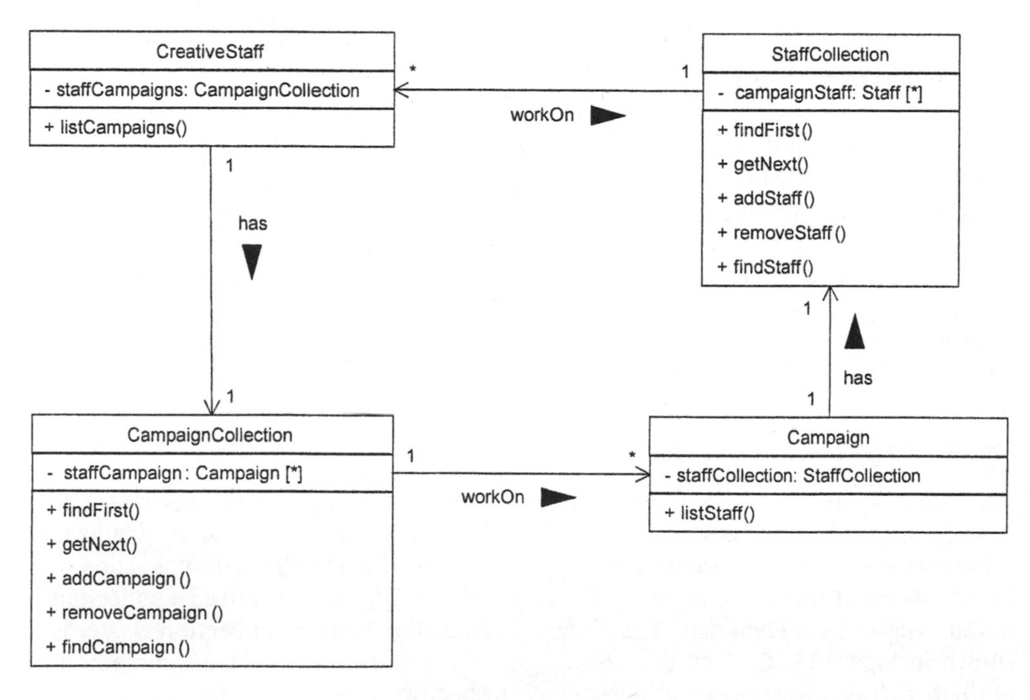

Figure 14.15 Two-way many-to-many association.

match is found or the end of the collection is reached. Two separate messages are required to access each `Campaign` object. So if an employee works on four campaigns a maximum of eight messages must be sent to find the campaign. In general, if the employee works on N campaigns a maximum of 2N messages must be sent.

An alternative search approach uses a `findCampaign()` operation in the collection class. This operation may be designed either to access an index in the collection object itself or to take responsibility to cycle through the `Campaign` objects searching for a title match. In the first case only the `findCampaign()` message is needed to find the campaign as the collection class indexes the campaigns by title. In the second case the collection object may have to send a maximum of N messages (one for each campaign on which the employee works). So in either case the inclusion of the `findCampaign()` operation in the collection class reduces the message passing required.

The management of object identifiers using collection classes may appear to increase appreciably the development effort required. In fact, the opposite is generally the case, as object-oriented languages normally provide collection classes of various types with standard collection management operations. The standard collection classes may offer various forms of indexing. They may also be subclassed to add additional application functionality. For example, Java provides, among others, standard List, Stack, Hashtable and Dictionary collection classes that may be subclassed to add application specific behaviour (Deitel and Deitel, 1999).

14.5.4 Keeping classes to a minimum

The association between `Campaign` and `AdvertCollection` is one-to-one (commonly the case with collection classes), and this suggests that one implementation strategy is to place the collection class inside the `Campaign` object. As discussed earlier in Section 14.4.3, this approach generally produces more complex classes and limits extensibility. However, in this case it is likely that collection class behaviour will largely be provided by a library class or by a feature of the development language being used, so the issue of increased complexity may not be so significant. The problem of any reduction in extensibility is also less significant in this case: as only a `Campaign` object would want to know which `Advert` objects are tied to it, and any request to access a `Campaign's Adverts` would be directed to the `Campaign` first. So, on balance, placing the collection class inside the `Campaign` class is a sensible design decision, and using library collection classes where possible maximizes reuse and reduces development effort. Clearly, if another class, apart from `Campaign`, needs to use the list independently of the `Campaign` class then it is more appropriate to keep the collection class separate.

14.6 Integrity Constraints

Systems analysis will have identified a series of integrity constraints that have to be enforced to ensure that the application holds data that is mutually consistent and manipulates it correctly. These integrity constraints come in various forms:

Referential Integrity that ensures that an object identifier in an object is actually referring to an object that exists.

Dependency Constraints that ensures that attribute dependencies, where one attribute may be calculated from other attributes, are maintained consistently.

Domain Integrity that ensures that attributes only hold permissible values.

14.6.1 Referential integrity

The concept of referential integrity as applied to a relational database management system (see Chapter 18) is discussed by Howe (2001). Essentially the same principles apply when considering references between objects. In Figure 14.12 the association `manageCampaign` between `CreativeStaff` and `Campaign` is two-way, and an object identifier called `campaignManagerId` that refers to the particular `CreativeStaff` object that represents the campaign manager is needed in `Campaign`. (`CreativeStaff` needs a collection of Campaign object identifiers to manage its end of the association.) To maintain referential integrity the system must ensure that the attribute `campaign ManagerId` either is null (not referencing any object) or contains the object identifier of a `CreativeStaff` object that exists. In this particular case the association states that a `Campaign` must have a `CreativeStaff` instance as its manager, and it is not correct to have a `Campaign` with a null `campaignManagerId` attribute. In order to enforce this constraint, the constructor for `Campaign` needs as one of its parameters the object identifier of the `CreativeStaff` object that represents the campaign manager, so that the `campaignManagerId` attribute can be instantiated with a valid object identifier.

Problems in maintaining the referential integrity of a `Campaign` may occur during its lifetime. For instance, the campaign manager, Nita Rosen, may leave the company to move to another job and Nita's `CreativeStaff` object will then be deleted[9]. Referential integrity is maintained by ensuring that the deletion of a `CreativeStaff` object that is a campaign manager always involves allocating a new campaign manager. The task of invoking the operation `assignManager()` is included in the `Creative Staff` destructor, and it will request the object identifier of the new campaign manager. Similarly, any attempt to remove the current campaign manager from a `Campaign` must always involve allocating the replacement.

The multiplicity of exactly one represents a strong integrity constraint for the system. In the example just discussed, it seems to be appropriate that a campaign should always have a manager, even when it has just been created. However, great care should be taken when assigning a multiplicity of exactly one (or in general a minimum of one) to an association, as the consequences in the implemented system can be quite dramatic. Let us imagine that the campaign manager, Nita Rosen, does leave Agate but that there is no replacement campaign manager available. The strict application of the integrity constraint implied by the `manageCampaign` association means that integrity can only be enforced if all the campaigns that Nita managed are deleted. Of course, because each `Advert` must be linked to a `Campaign`, all the `Advert` objects for each of the `Campaigns` must also be deleted in order to maintain referential integrity. This is an example of a cascade delete: deleting one object results in the deletion of many objects as referential integrity is applied. In the case of Agate, deleting the information about Nita's campaigns and their adverts would be disastrous. There are two solutions: either the constraint on the association is weakened by changing the cardinality to zero or one, or when Nita leaves a dummy `CreativeStaff` object is created and allocated as campaign manager to Nita's campaigns. Although the second solution is a fix, it has the advantage of providing an obvious place-holder, highlighting the problem of unmanaged campaigns but maintaining the integrity constraint. Of course, the minimum of one multiplicity was assigned to the association to reflect company policy that a campaign must always have a manager but this may not be a viable solution from a business perspective.

[9] This is a somewhat simplistic view. When an employee leaves the company the `CreativeStaff` object would be set to the state `Ex-employee`. In this case although the object still exists it is not appropriate for it to be referenced as a campaign manager.

14.6.2 Dependency constraints

Attributes are dependent upon each other in various ways. These dependencies may have been identified during analysis and must now be dealt with during design. A common form of dependency occurs when the value of one attribute may be calculated from other attributes. For instance, a requirement to display the total advertising cost may be satisfied either by storing the value in the attribute `totalAdvertCost` in the `Campaign` class or by calculating the value every time it is required. The attribute `totalAdvertCost` is a derived attribute and its value is calculated by summing the individual advert costs. Placing the derived attribute in the class reduces the processing required to display the total advertising cost as it does not require calculation. On the other hand, whenever the cost of an advert changes, or an advert is either added to or removed from the campaign, then the attribute `totalAdvertCost` has to be adjusted so that it remains consistent with the attributes upon which it depends. An example of the UML symbol ('/') used to indicate that a modelling element (attribute or association) is derived is shown in Figure 14.4.

In order to maintain the consistency between the attributes, any operation that changes the value of an `Advert`'s cost must trigger an appropriate change in the value of `totalAdvertCost` by sending the message `adjustCost()` to the `Campaign` object. The operation `adjustCost()` is an example of a *synchronizing operation*. The operations that have to maintain the consistency are `setAdvertCost()` and the `Advert` destructor. When a new advert is created the constructor would use `setAdvertCost()` to set the advert cost. This would invoke `adjustCost()` and hence ensure that the `totalAdvertCost` is adjusted. So any change to an `Advert`'s cost takes more processing if the derived attribute `totalAdvertCost` is used. Thus one part of the system executes more quickly while another part executes more slowly. Generally it is easier to construct systems without derived attributes, as this obviates the need for complex synchronizing operations. Derived attributes should only be introduced if performance constraints cannot be satisfied without them. If performance is an issue then one of the skills needed in design is how to optimize the critical parts of the system without making the other parts of the system inoperable.

Another form of dependency occurs where the value of one attribute is constrained by the values of other attributes. For example, let us assume that the sum of the total advertising cost, the staff costs, the management costs and the ancillary costs must not exceed the campaign's authorized budget. Any changes to these values must check that the authorized budget is not exceeded by the sum of the costs. If a change to any of these dependent values would cause this constraint to be broken, then some action should be taken. There are two possibilities. Either the system prohibits any change that violates the constraint, and an exception is raised, or it permits the change, and an exception is raised[10]. It is most likely that the violation of the constraint would occur as a result of an attempt interactively to change one of the constrained values and the exception raised would be a warning message to the user. If it is considered permissible for the constraint to be broken then all access to or reporting about these values should produce a warning message to the user.

Dependency constraints can also exist between or among associations. One of the simplest cases is shown in Figure 14.16 where the `chairs` association is a subset of the `isAMemberOf` association.

This constraint is stating that the chair of a committee must be a member of the committee, and it can be enforced by placing a check in the `assignChair()` operation

[10] An exception is a way of handling errors in a programming language.

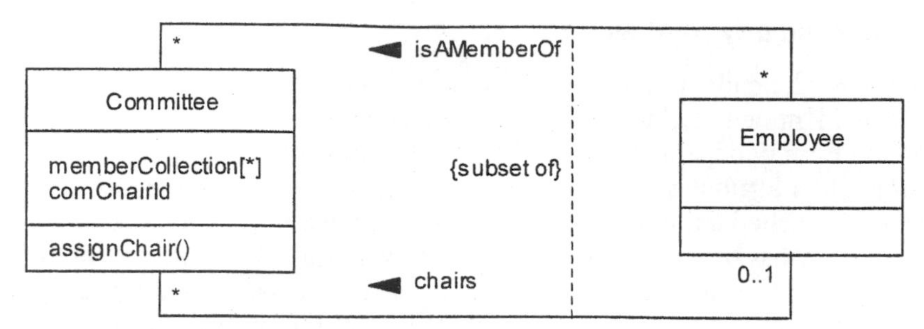

Figure 14.16 Constraints between associations.

in `Committee` to confirm that the `Employee` object identifier passed as a parameter is already in the collection class of committee members. More complex constraints may also exist that require several associations. Derived associations may also be introduced to improve performance if absolutely necessary and, as in the case of derived attributes, synchronizing operations are needed to ensure that the derived links are consistent with the links on which they depend.

14.6.3 Domain integrity

Domain integrity is concerned with ensuring that the values an attribute takes are from the appropriate underlying domain. For instance, the attributes from the `Cost` domain might reasonably be non-negative decimal values with two decimal places. These constraints may be viewed as an extended form of those implied by data types. The necessary integrity checking code is normally placed in the 'set' operations or in any interactive interface that permits the entry of values.

14.7 Designing Operations

The design of operations involves determining the best algorithm to perform the required function. In the simplest case, primary operations require little design apart from the inclusion of code to enforce integrity checks. For more complex operations, algorithm design can be an involved process. Various factors constrain algorithm design including:

- the cost of implementation,
- performance constraints,
- requirements for accuracy and
- the capabilities of the implementation platform.

Generally it is best to choose the simplest algorithm that satisfies these constraints, as this makes the operation easier to implement and easier to maintain. Rumbaugh et al. (1991) suggest that the following factors should be considered when choosing among alternative algorithm designs.

- *Computational complexity.* This is concerned with the performance characteristics of the algorithm as it operates on increasing numbers of input values. For example, the bubble sort algorithm has an execution time that is proportional to N × N where N is the number of items being sorted.
- *Ease of implementation and understandability.* It is generally better to sacrifice some performance to simplify implementation.

- *Flexibility.* Most software systems are subject to change and an algorithm should be designed with this in mind.
- *Fine-tuning the object model.* Some adjustment to the object model may simplify the algorithm and should be considered.

Designing the main operations in a class is likely to highlight the need for lower-level private operations to decompose complex operations. This process is much the same as traditional program design. Techniques such as step-wise refinement (Budgen, 1994) or structure charts (Yourdon and Constantine, 1979) may well be used to good effect. UML offers activity diagrams (see Chapters 5 and 10) as a technique both to document and to design operations. In circumstances where high levels of formality are required in operation design, formal specification techniques such as Z or VDM may be used[11].

Responsibilities identified during analysis may map onto one or more operations. The new operations that are identified need to be assigned to classes. In general, if an operation operates on some attribute value then it should be placed in the same class as the attribute. On occasions a particular operation may modify attributes in more than one class and could sensibly be placed in one of several classes. In choosing where to locate the operation, one view is that minimizing the amount of object interaction should be a major criterion, while another significant criterion is simplicity. However, in some cases it is not a clear-cut decision.

During analysis use case realization control classes are introduced to control the execution of use cases. Typically these control classes may be the best place for operations that are particular to the use case or that have no obvious owning entity class. Some designers may choose to allocate control class responsibility to boundary or entity classes during design to achieve performance or other implementation requirements. However, this results in boundary or entity classes that have less well focused functionality (lower class cohesion) and can make maintenance more difficult. This yet again reflects the trade-offs that have to be made during design.

14.8 Normalization

One form of dependency not discussed in Section 14.6.2 is functional dependency. For two attributes A and B, A is functionally dependent on B if for every value of B there is precisely one value of A associated with it at any given time. This is shown notationally as:

$$\mathbf{B} \rightarrow \mathbf{A}$$

Attributes may be grouped around functional dependencies according to several rules of normalization (see Section 18.3.1) to produce normalized data structures that are largely redundancy free (Howe, 2001). Normalization may be useful when using a relational database management system as part of the implementation platform or as a guide to decomposing a large, complex (and probably not very cohesive) object. Most object-oriented approaches to software development do not view normalization as essential, and structures that are not normalized are considered acceptable. In general however, object-oriented approaches, if applied with suitable quality constraints, will produce structures that are largely redundancy free.

[11] Z and VDM are formal languages that can be used to specify a system using mathematical entities such as sets, relations and sequences.

14.9 Summary

Object design is concerned with the detailed design of the system and is conducted within the architectural framework and design guidelines specified during system design. The detailed design process involves determining the data types of the attributes and defining the operation signatures. Interfaces may be specified. Associations have to be designed to support the message passing requirements of the operations. This involves determining how best to place object references in the classes. The application of integrity constraints is included in the design of operations. Operations have to be designed to enforce these integrity constraints. If derivable attributes are included in any of the classes then synchronizing operations are required to maintain their consistency. The detailed design process is guided by a series of criteria that incorporate the fundamental principles of coupling and cohesion. Normalization may be applied to complex classes as a means of decomposing them to produce redundancy-free structures.

Review Questions

14.1 What levels of visibility may be assigned to an attribute or an operation?

14.2 Why should attributes be private?

14.3 How does the application of the concepts of coupling and cohesion help to produce good object-oriented designs?

14.4 What are the advantages and disadvantages of applying the Liskov Substitution Principle?

14.5 How can collection classes be used when designing associations?

14.6 Under what circumstances should a collection of object references be included in a class?

14.7 How can referential integrity be enforced in an object-oriented system?

14.8 Under what circumstances is normalization useful during object-oriented design?

Case Study Work, Exercises and Projects

14.A Specify the attribute types and the operation signatures for the class `ProductionLine` in the FoodCo case study.

14.B For each association in which `ProductionLine` participates, allocate object identifiers to design the association.

14.C Show how referential integrity can be enforced for the associations designed in Exercise 14.B.

14.D For an object-oriented programming language of your choice investigate the language features available to support the use of collection classes.

Further Reading

- Rumbaugh et al. (1991) and Booch (1994) provide good advice on object design. Budgen (1994) offers descriptions of various design techniques.
- Meyer (1997) contains a comprehensive discussion of object-oriented software design and provides many interesting insights.
- There is an excellent treatment of normalization in Howe (2001) and detailed discussion of integrity constraints from a database perspective in Date (1995).

15

CHAPTER

Design Patterns

OBJECTIVES

In this chapter you will learn

- what types of patterns have been identified in software development
- how to apply design patterns during software development
- the benefits and difficulties that may arise when using patterns.

15.1 Introduction

Successful software development relies on the knowledge and expertise of the developer, among other factors. These are built up and refined during the developer's working life. A systems analyst or software engineer applies potential solutions to development problems, monitors their success or failure and produces more effective solutions on the next occasion. It is in the nature of software development that the same problems tend to recur. Different developers may expend a great deal of development time and effort on solving these problems from first principles each time they occur, and the solution that each produces may not be the most appropriate that could be achieved. This can result in information systems that are inflexible, difficult to maintain or have some other disadvantage such as inefficiency. The perpetuation of this cycle of reinventing the wheel has been encouraged partly because there were no effective mechanisms for communicating successful solutions to recurring problems.

Patterns provide a means for capturing knowledge about problems and successful solutions in software development. Experience that has been gained in the past can be reused in similar situations, thus reducing the effort required to produce systems that are more resilient, more effective and more flexible.

Patterns have been introduced in Section 8.4 where we considered analysis patterns in particular. In Chapter 14 we discussed several architectural patterns. In Section 15.2 we consider some more detailed aspects of patterns in general. Patterns are documented in various formats and these are considered in Section 15.3. Although patterns have been applied to many aspects of software development we focus our attention on some better-known design patterns (Section 15.4) to illustrate their application. Guidelines for using patterns are discussed in Section 15.5. Patterns have many benefits but there are also potential disadvantages associated with their use. These advantages and disadvantages are discussed in the context of the examples used in Section 15.6.

15.2 Software Development Patterns

15.2.1 Frameworks

There can be confusion between patterns and frameworks but it is important to distinguish between them. Frameworks are partially completed software systems that may be targeted at a specified type of application, for example sales order processing. An application system tailored to a particular organization may be developed from the framework by completing the unfinished elements and adding application specific elements. This may involve the specialization of classes and the implementation of some operations. Essentially the framework is a reusable mini-architecture that provides structure and behaviour common to all applications of this type.

The major differences between patterns and frameworks can be summarized as follows.

- Patterns are more abstract and general than frameworks. A pattern is a description of the way that a type of problem can be solved, but the pattern is not itself a solution.
- Unlike a framework, a pattern cannot be directly implemented in a particular software environment. A successful implementation is only an example of a design pattern.
- Patterns are more primitive than frameworks. A framework can employ several patterns but a pattern cannot incorporate a framework.

15.2.2 Pattern catalogues and languages

Patterns are grouped into catalogues and languages. A *pattern catalogue* is a group of patterns that are related to some extent and may be used together or independently of each other. The patterns in a *pattern language* are more closely related, and work together to solve problems in a specific domain. For example, Cunningham (1995) documented the 'Check Pattern Language of Information Integrity', which consists of eleven patterns that address issues of data validation. All were developed from his experience of developing interactive financial systems in Smalltalk.

One of these patterns, Echo, describes how data input should be echoed back to the user after it has been modified and validated by the information system (since Cunningham uses the Model–View–Controller structure he talks about this in terms of changes made by the model). Typically users enter small batches of values and then look at the screen to check that they have been correctly entered. The sequence in which a user can enter data into fields may not be fixed and so validation feedback should be given one field at a time. For example, a user enters a value as 5.236. This might be echoed back by the system as 5.24 (correctly rounded to two decimal places). The user receives direct visual feedback that the value has been accepted and how it has been modified.

15.2.3 Software development principles and patterns

Patterns are intended to embody good design practice and hence are based upon sound software development principles, many of which have been identified since the early days of software development and applied within other development approaches than object-oriented ones. Buschmann et al. (1996) suggest that the following are the key principles that underlie patterns:

- abstraction,
- encapsulation,
- information hiding,
- modularization,
- separation of concerns,
- coupling and cohesion,
- sufficiency, completeness and primitiveness,
- separation of policy and implementation,
- separation of interface and implementation,
- single point of reference and
- divide and conquer (this means breaking a complex problem into smaller, more manageable ones).

Most of these principles should already be familiar to the reader since they have been discussed in earlier chapters.

15.2.4 Patterns and non-functional requirements

Patterns can address the issues that are raised by non-functional requirements (Chapters 6 and 12). Buschmann et al. (1996) identify these as the important non-functional properties of a software architecture:

- changeability,
- interoperability,
- efficiency,
- reliability,
- testability and
- reusability.

These properties may be required of a complete system or a part of a system. For example, a particular set of functional requirements may be seen as volatile and subject to change. It is important to develop a structure for these requirements that can cope with change. Another requirement may be that a particular aspect of an application must be highly reliable. Again this requirement must be met by the design.

15.3 Documenting Patterns—Pattern Templates

15.3.1 Template contents

Patterns may be documented using one of several alternative templates. The *pattern template* determines the style and structure of the pattern description, and these vary in the emphasis they place on different aspects of patterns. The differences between pattern templates may mirror variations in the problem domain but there is no consensus as to the most appropriate template even within a particular problem domain. Nonetheless it is generally agreed that a pattern description should include the following elements (at least implicitly).

Name. A pattern should be given a meaningful name that reflects the knowledge embodied by the pattern. This may be a single word or a short phrase. These names become the vocabulary for discussing conceptual constructs in the domain of expertise. For instance, the names of three of the Gamma design patterns, Bridge, Mediator and Flyweight, give a strong indication of how they are intended to work.

Problem. This is a description of the problem that the pattern addresses (the intent of the pattern). It should identify and describe the objectives to be achieved, within a specified context and constraining forces. For example, one problem might be concerned with producing a flexible design, another with the validation of data. The problem can frequently be written as a question, for example 'How can a class be constructed that has only one instance and can be accessed globally within the application?' This question expresses the problem addressed by the Singleton pattern (discussed in Section 15.4.2).

Context. The context of the pattern represents the circumstances or preconditions under which it can occur. The context should provide sufficient detail to allow the applicability of the pattern to be determined.

Forces. The forces embodied in a pattern are the constraints or issues that must be addressed by the solution. These forces may interact with and conflict with each other, and possibly also with the objectives described in the problem. They reflect the intricacies of the pattern.

Solution. The solution is a description of the static and dynamic relationships among the components of the pattern. The structure, the participants and their collaborations are all described. A solution should resolve all the forces in the given context. A solution that does not resolve all the forces fails.

15.3.2 Other aspects of templates

A pattern template may be more extensive than the elements described above. Some other features that have figured in pattern templates are given below:

- an example of the use of a pattern that serves as a guide to its application;
- the context that results from the use of the pattern;
- the rationale that justifies the chosen solution;
- related patterns;
- known uses of the pattern that validate it (some authors suggest that until the problem and its solution have been used successfully at least three times—the *rule of three*—they should not be considered as a pattern);
- a list of aliases for the pattern ('also known as' or AKA);

- sample program code and implementation details (commonly used languages include C++, Java and Smalltalk).

Gamma et al. (1995) use a template that differs from that described above. Although very detailed, this does not explicitly identify the forces. Cunningham's (1995) *Checks Pattern Language of Information Integrity* is described in the Portland Form[1]. Variations in template style and structure make it difficult to compare patterns that are documented in different templates, and this limits their reusability since it is more difficult to use a pattern that is documented in an unfamiliar template.

15.4 Design Patterns

15.4.1 Types of Design Pattern

In the catalogue of 23 design patterns presented by Gamma et al. (1995) patterns are classified according to their scope and purpose. The three main categories of purpose that a pattern can have are *creational*, *structural* or *behavioural* (these are described in the following sections). The scope of a pattern may be primarily at either the class level or at the object level. Patterns that are principally concerned with objects describe relationships that may change at run-time and hence are more dynamic. Patterns that relate primarily to classes tend to be static and identify relationships between classes and their subclasses that are defined at compile-time. The Gamma patterns are generally concerned with increasing the ease with which an application can be changed, through reducing the coupling among its elements and maximizing their cohesion. The patterns are based on principles of good design that include the maximizing of encapsulation and the substitution of composition for inheritance wherever possible. Using composition as a design tactic produces composite objects whose component parts can be changed, perhaps dynamically under program control hence resulting in a highly flexible system. Nonetheless patterns will frequently use both inheritance and composition to achieve the desired result.

Changeability involves several different aspects (Buschmann et al., 1996)—maintainability, extensibility, restructuring and portability. Definitions of these terms vary but we use the following.

Maintainability is concerned with the ease with which errors in the information system can be corrected.

Extensibility addresses the inclusion of new features and the replacement of existing components with new improved versions. It also involves the removal of unwanted features.

Restructuring focuses on the reorganization of software components and their relationships to provide increased flexibility.

Portability deals with modifying the system so that it may execute in different operating environments, such as different operating systems or different hardware.

15.4.2 Creational patterns

A creational design pattern is concerned with the construction of object instances. In general, creational patterns separate the operation of an application from how its objects are created. This decoupling of object creation from the operation of the application gives the designer considerable flexibility in configuring all aspects of object

[1] Named after Portland, Oregon where it originated, this is basically free format text.

creation. This configuration may be dynamic (at run-time) or static (at compile-time). For example, when dynamic configuration is appropriate, an object-oriented system may use composition to make a complex object by aggregating simpler component objects. Depending upon circumstances different components may be used to construct the composite and, irrespective of its components, the composite will fulfil the same purpose in the application. A simple analogy illustrates this. A systems development department in an organization will vary in its composition from time to time. When permanent staff are on holiday, contract staff may be employed to perform their roles. This enables the department to offer the same service to the organization.

Creating composite objects is not simply a matter of creating a single entity but also involves creating all the component objects. The separation of the creation of a composite object from its use within the application provides design flexibility. By changing the method of construction of a composite object, alternative implementations may be introduced without affecting the current use.

Singleton Pattern

As an example we consider the creational pattern, Singleton, which can be used to ensure that only one instance of a class is created. In order to understand the use of the pattern we need to consider the circumstances under which a single instance may be required.

The Agate campaign management system needs to hold information regarding the company. For example, its name, its head office address and the company registration details need to be stored so that they can be displayed in all application interfaces and printed on reports. This information should be held in only one place within the application but will be used by many different objects. One design approach would be to create a global data area that can be accessed by all objects, but this violates the principle of encapsulation. Any change to the structure of the elements of global data would require a change to all objects that access them. The creation of a `Company` class overcomes this problem by encapsulating the company attributes (Figure 15.1). These are then accessible to other objects through the operations of the `Company` object. But there is still a problem with this proposal. An object that wants to use the `Company` object needs to know the `Company` object's identifier so that it can send messages to it. This suggests that the `Company` object identifier should be globally accessible—but again this is undesirable since it violates encapsulation.

Some object-oriented programming languages (including Java and C++) provide a mechanism that enables certain types of operations to be accessed without reference to a specified object. These are called *class* (or *static*) *operations*. This offers a solution to the problem of providing global access without the need to globally define the

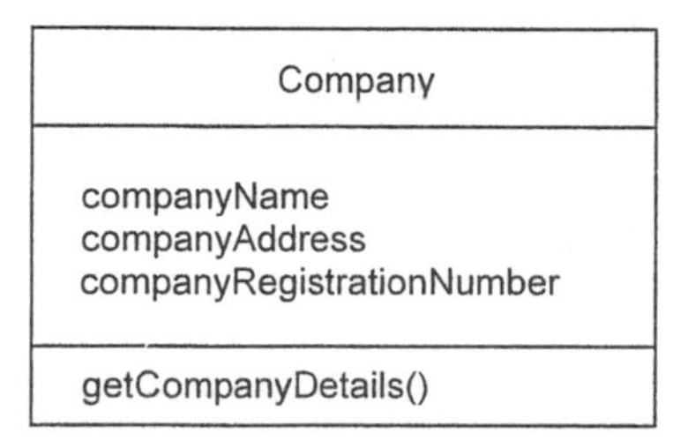

Figure 15.1 Company class for the Agate case study.

object identifier. For example, a static operation `getCompanyInstance()` can be defined in such a way that it will provide any client object with the identifier for the `Company` instance. This operation can be invoked by referencing the class name as shown below.

```
Company.getCompanyInstance()
```

When a client object needs to access the `Company` object it can send this message to the `Company` class and receive the object identifier in reply. The client object can now send a `getCompanyDetails()` message to the `Company` object.

There is one further aspect to this design problem. It is important that there should only be one instance of this object. To ensure system integrity the application should be constructed so that it is impossible to create more than one. This aspect of the problem can be solved by giving the `Company` class sole responsibility for creating a `Company` object. This is achieved by making the class constructor private so that it is not accessible by another object. The next issue that needs to be addressed is the choice of an event that causes the creation of the company object. Perhaps the simplest approach is to create the Company object at the moment when it is first needed. When the `Company` class first receives the message `getCompanyInstance()` this can invoke the `Company` class constructor. Once the Company object has been created, the object identifier is stored in the class (or static) attribute `companyInstance` so that it can be passed to any future client objects.

So far we have produced a design for the `Company` class (Figure 15.2) that provides a single global point of access via the class operation `getCompanyInstance()` and that also ensures that only one instance is created.

A simple version of the logic for the `getCompanyInstance()` operation is shown below.

```
If    (companyInstance == null)
      {
      companyInstance = new Company()
      } return companyInstance
```

The design may need to accommodate further requirements. Since Agate operates as a separate company in each country (each owned by the same multinational), variations in company law from country to country may necessitate different company registration details to be recorded for each country. This suggests a requirement for different types of `Company` class each with its own variation of the registration details. The creation of a separate subclass for each style of company registration details is a

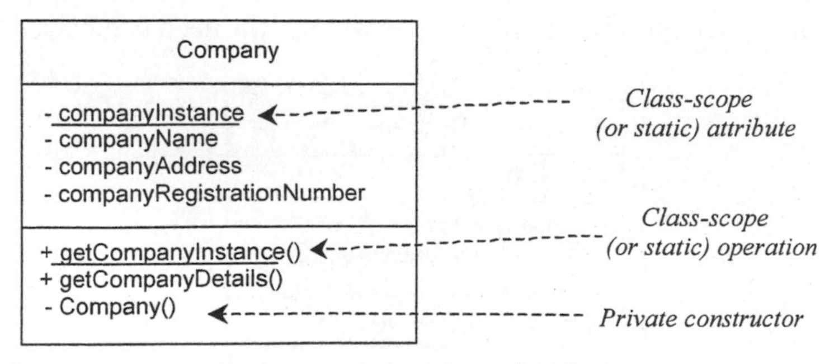

Figure 15.2 `Company` class with class-scope operation and attribute.

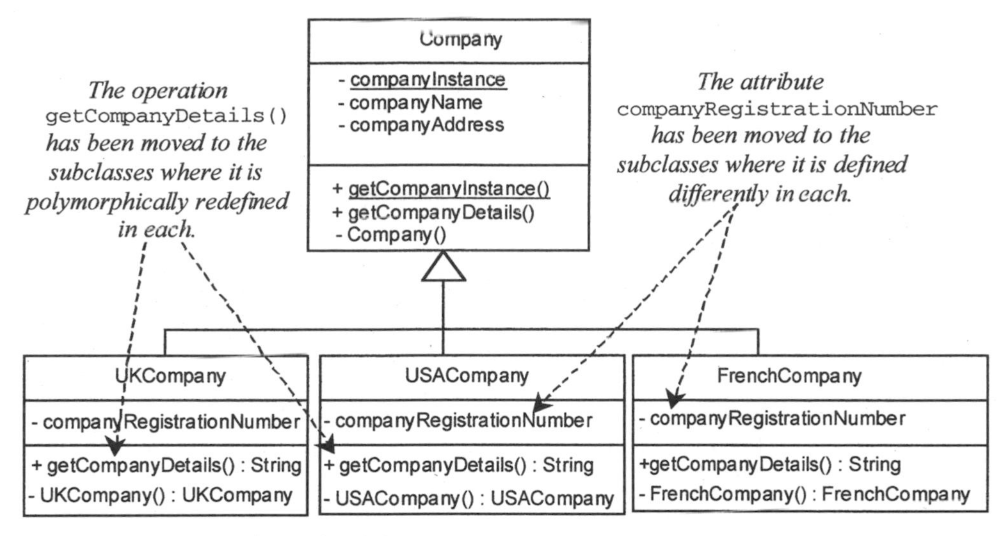

Figure 15.3 `Company` class with subclasses.

solution to this aspect of the problem (Figure 15.3). When the `getCompanyInstance()` operation is first called the appropriate subclass is instantiated. If the `Company` object has not yet been instantiated, its constructor operation can access details of the country (say, held in a `CurrentCountry` object) to determine which subclass should be instantiated.

This part of the design for the campaign management system now has:

- a class `Company` that is only instantiated once;
- an instance of this class that is globally accessible;
- different subclasses of `Company` that are instantiated as needed, depending on run-time circumstances.

This is an application of the Singleton pattern. The pattern is described below in more general language.

Name. Singleton.

Problem. How can a class be constructed that should have only one instance and that can be accessed globally within the application?

Context. In some applications it is important that a class have exactly one instance. A sales order processing application may be dealing with sales for one company. It is necessary to have a `Company` object that holds details of the company's name, address, taxation reference number and so on. Clearly there should only be one such object. Alternative forms of a singleton object may be required depending upon different initial circumstances.

Forces. One approach to making an object globally accessible is to make it a global variable but in general this is not a good design solution as it violates encapsulation. Another approach is not to create an object instance at all but to use class operations and attributes (called 'static' in C++ and Java). However, this limits the extensibility of the model since polymorphic redefinition of class operations is not possible in all development environments (for example C++).

Solution. Create a class with a class operation `getInstance()`, which, when the class is first accessed, creates the relevant object instance and returns the object identity

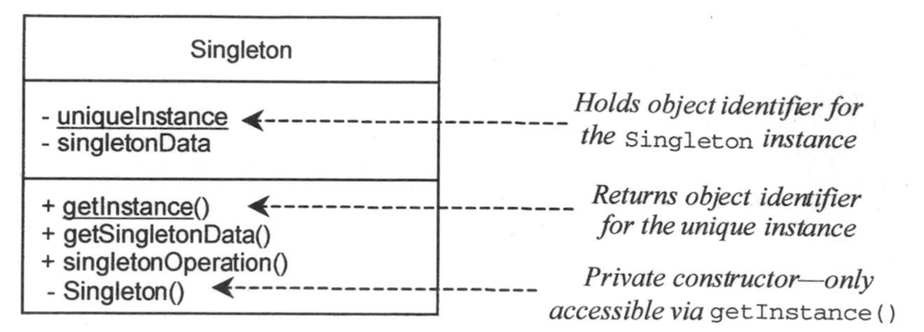

Figure 15.4 Creational Patterns—Singleton.

to the client. On subsequent accesses of the `getInstance()` operation no additional instance is created but the object identity of the existing object is returned. A class diagram fragment for the singleton pattern is shown in Figure 15.4.

The singleton pattern offers several advantages but also has some disadvantages.

+ It provides controlled access to the sole object instance as the Singleton class encapsulates the instance.
+ The namespace is not unnecessarily extended with global variables.
+ The Singleton class may be subclassed. At system start-up user-selected options may determine which of the subclasses is instantiated when the Singleton class is first accessed.
+ A variation of this pattern can be used to create a specified number of instances if required.
- Using the pattern introduces some additional message passing. To access the singleton instance the class scope method has to be accessed first rather than accessing the instance directly.
- The pattern limits the flexibility of the application. If requirements change and as a result the singleton class may have many instances then accommodating this new requirement necessitates significant modification to the system.
- The singleton pattern is quite well known and developers are tempted to use it in circumstances that are inappropriate. Patterns must be used with care.

15.4.3 Structural patterns

Structural patterns address issues concerned with the way in which classes and objects are organized. Structural patterns offer effective ways of using object-oriented constructs such as inheritance, aggregation and composition to satisfy particular requirements. For instance, there may be a requirement for a particular aspect of the application to be extensible. In order to achieve this, the application should be designed with constructs that minimize the side-effects of future change. Alternatively it may be necessary to provide the same interface for a series of objects of different classes.

Composite Pattern

It may be appropriate to apply the Composite structural pattern in a design for the Agate case study. In the following example we assume that further work is required to design a multimedia application that can store and play components of an advert.

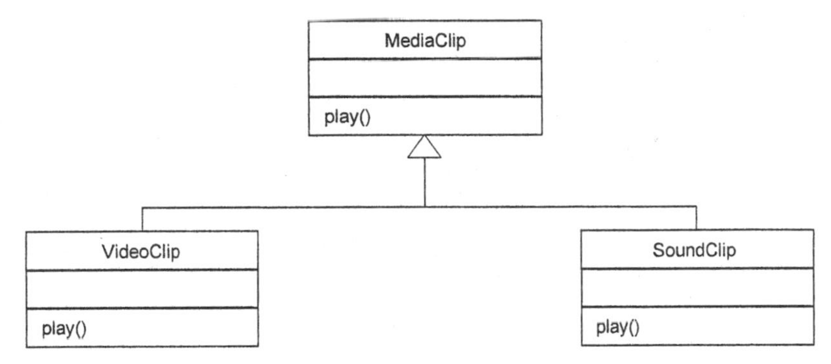

Figure 15.5 `MediaClip` inheritance hierarchy.

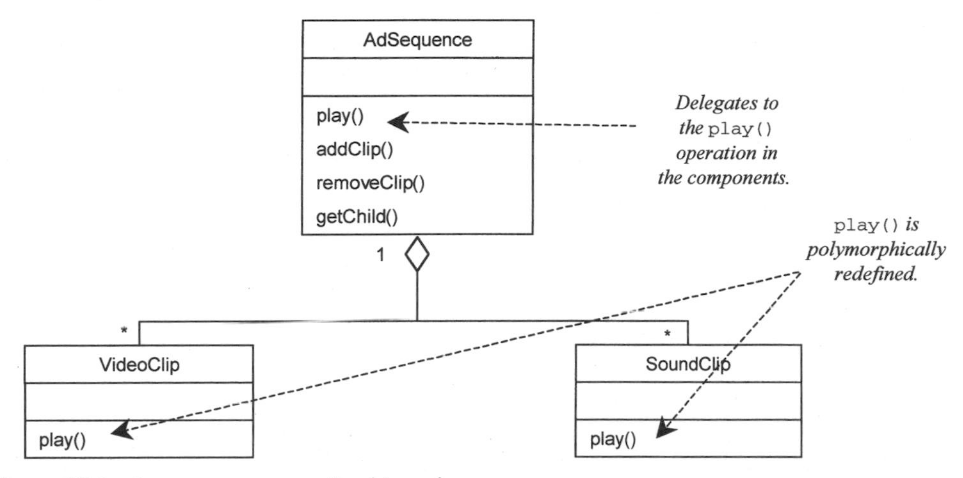

Figure 15.6 `AdSequence` aggregation hierarchy.

Here an advert is made up of sound clips and video clips each of which may be played individually or as part of an advert. The classes `SoundClip` and `VideoClip` have attributes and operations in common and it is appropriate that these classes are subclassed from `MediaClip` (Figure 15.5). But not all advert clips are primitive (that is, made up of only a single `MediaClip`). Some consist of one or more sequences of clips, such that each sequence is in turn an aggregation of `SoundClip` and `Video Clip` objects (Figure 15.6).

These two orthogonal[2] hierarchies can be integrated by treating `AdSequence` both as a subclass of `MediaClip` and also as an aggregation of `MediaClip` objects (see Figure 15.7).

All the subclasses have the polymorphically redefined operation `play()`. For the subclasses `VideoClip` and `SoundClip` this operation actually plays the object. But for an `AdSequence` object, an invocation of the `play()` operation results in it sending a `play()` message to each of its components in turn. This structure is a straightforward application of the Composite pattern, which is described more generally below.

Name. Composite.

Problem. There is a requirement to represent whole–part hierarchies so that both whole and part objects offer the same interface to client objects.

[2] The term literally means 'at right angles to each other'. It is more loosely used here to describe hierarchies that cannot be directly mapped onto each other.

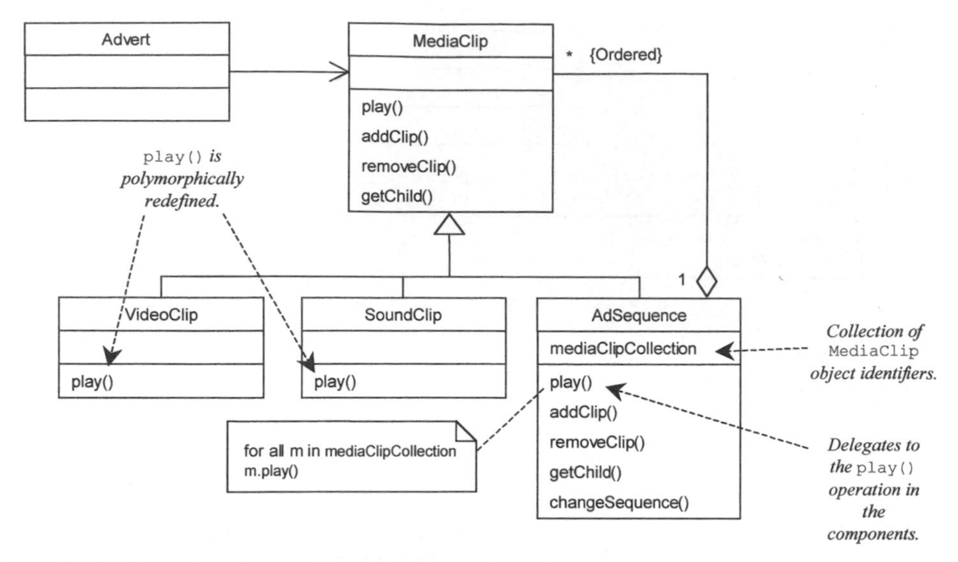

Figure 15.7 Integrating the two hierarchies for `MediaClip`.

Context. In an application both composite and component objects exist that are required to offer the same behaviour. Client objects should be able to treat composite or component objects in the same way. A commonly used example for the composite pattern is a graphical drawing package. Using this software package a user can create (from the perspective of the software package) atomic objects like circle or square and can also group a series of atomic objects or composite objects together to make a new composite object. It should be possible to move or copy this composite object in exactly the same way as it is possible to move or copy an individual square or a circle.

Forces. The requirement that the objects, whether composite or component, offer the same interface suggests that they belong to the same inheritance hierarchy. This enables operations to be inherited and to be polymorphically redefined with the same signature. The need to represent whole–part hierarchies indicates the need for an aggregation structure.

Solution. The solution resolves the issues by combining inheritance and aggregation hierarchies. Both subclasses, `Leaf` and `Composite`, have a polymorphically redefined operation `anOperation()`. In the `Composite` this redefined operation invokes the relevant operation from its components using a simple loop construct (Figure 15.8). The `Composite` subclass also has additional operations to manage the aggregation hierarchy so that components may be added or removed.

Further requirements may need to be considered for the Agate example. Perhaps `VideoClip` and `SoundClip` objects must be played in a particular sequence. This can be handled if the aggregate `AdSequence` maintains an ordered list of its components. This is shown in Figure 15.7 by the `{Ordered}` property on the aggregation association. Each component object can be given a sequence number, and two components that have the same sequence number are played simultaneously. The operation `changeSequence()` allows a component `MediaClip` object to be moved up or down within the sequence of clips in the advertisement (Figure 15.7).

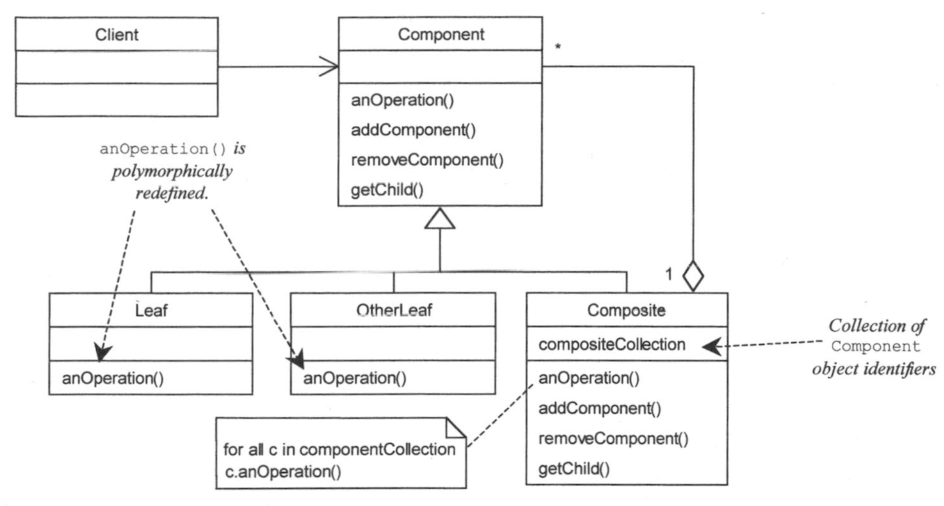

Figure 15.8 Structural patterns—Composite.

15.4.4 Behavioural Patterns

Behavioural patterns address the problems that arise when assigning responsibilities to classes and when designing algorithms. Behavioural patterns not only suggest particular static relationships between objects and classes but also describe how the objects communicate. Behavioural patterns may use inheritance structures to spread behaviour across the subclasses or they may use aggregation and composition to build complex behaviour from simpler components. The State pattern, which is considered below, uses both of these techniques.

State Pattern

Let us examine the Agate case study to determine whether it has features that may justify the application of the state pattern. First, are there any objects with significant state dependent behaviour? `Campaign` objects have behaviour that varies according to state; a `Campaign` object may be in one of four main states, as shown in Figure 11.13 (for simplicity we ignore the nested concurrent states of the `Active` state). Clearly a `Campaign` object's state changes dynamically as the campaign progresses, thus necessitating changes in the behaviour of the object.

For example, when a campaign is planned a `Campaign` object is created in the `Commissioned` state. It remains in this state until a campaign budget has been agreed and only then does it become possible to run advertisements, although some preparatory work may be done for the campaign in the meantime. Once a `Campaign` object enters the `Active` state all advert preparation and any other work that is done is subject to an agreed billing schedule. Several operations, for example `addAdvert()` and `calcCosts()`, will behave differently depending upon the state of the `Campaign` object. It would be possible to construct a working version of the software using the design for the `Campaign` class that is shown in Figure 15.9. However, this would be a complex class that is further complicated by state dependent operations such as `calcCosts()`, which would need to be specified with a series of case or if-then-else

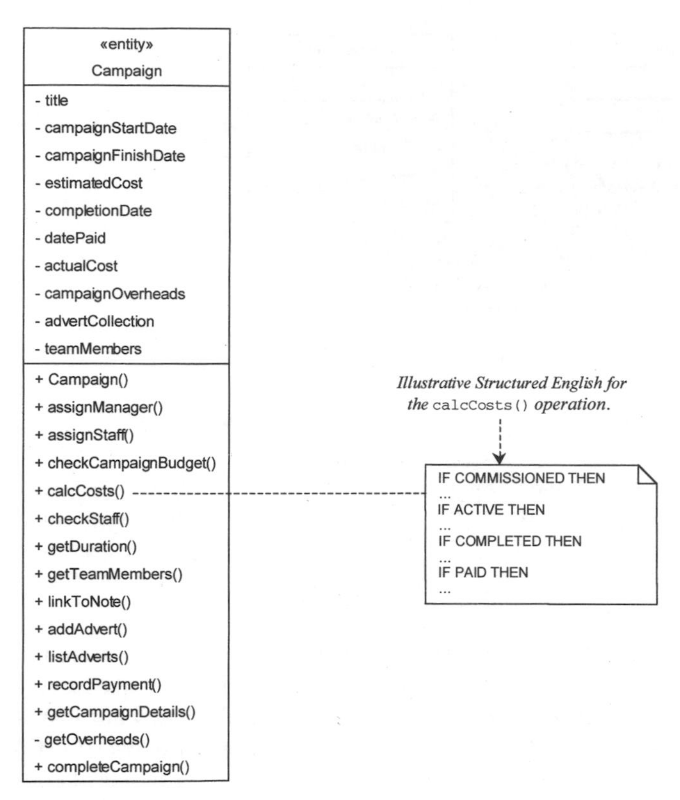

Figure 15.9 The `Campaign` class.

statements to test the state of the object. It would be simpler to subdivide the operations that have state dependent behaviour, which in this case would result in four separate `calcCosts()` operations, one for each state. The inclusion of `calcCosts Commissioned()`, `calcCostsActive()` and so on within `Campaign` would simplify the operation `calcCosts()`, but the class as a whole would become even more complex.

Another possibility is to create additional classes, one for each state so that each holds a state specific version of the operations, and this is how the State pattern works. A class diagram fragment illustrating this application of the State pattern is shown in Figure 15.10. Since the subclasses of `CampaignState` have no attributes specific to a particular `Campaign` object it is possible to have only one instance of each in the system. Thus there will be a maximum of four `CampaignState` objects, one for each state, and the additional overhead of manipulating the objects is unlikely to be significant (Figure 15.11). A variation on the Singleton pattern ensures that only one instance of each subclass can ever exist (this differs from its more usual application of ensuring that there is at most one object for a whole hierarchy).

The State pattern is described more generally below.

Name. State.

Problem. An object exhibits different behaviour when its internal state changes making the object appear to change class at run-time.

Context. In some applications an object may have complex behaviour that is dependent upon its state. In other words the response to a particular message varies according to the object's state. One example is the `calcCosts()` operation in the `Campaign` class.

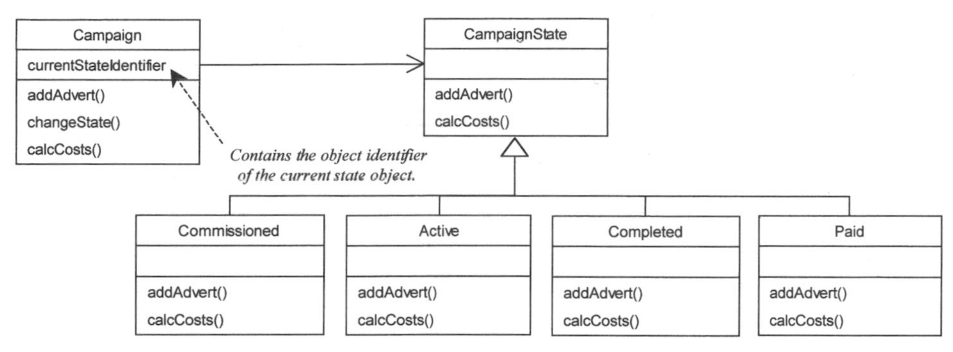

Figure 15.10 State pattern for Agate showing simplified version of `Campaign` class.

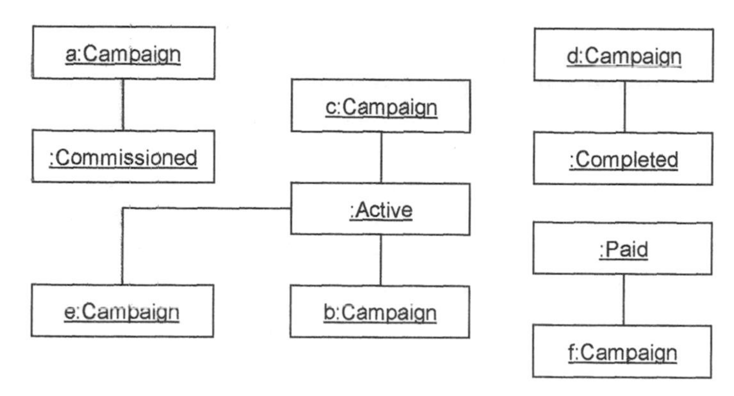

Figure 15.11 Some state pattern objects for Agate.

Forces. The object has complex behaviour that should be factored into less complex elements. One or more operations have behaviour that varies according to the state of the object. Typically the operation would have large, multi-part conditional statements depending on the state. One approach is to have separate public operations for each state but client objects would need to know the state of the object so that they could invoke the appropriate operation. For example four operations `calcCostsCommissioned()`, `calcCostsActive()`, `calcCostsCompleted()` and `calcCostsPaid()` would be required for the `Campaign` object. The client object would need to know the state of the `Campaign` object in order to invoke the relevant `calcCosts()` operation. This would result in undesirably tight coupling between the client object and the `Campaign` object. An alternative approach is to have a single public `calcCosts()` operation that invokes the relevant private operation (`calcCostsCommissioned()` would be private). However, the inclusion of a separate private operation for each state may result in a large complex object that is difficult to construct, test and maintain.

Solution: The state pattern separates the state dependent behaviour from the original object and allocates this behaviour to a series of other objects, one for each state. These state objects then have sole responsibility for that state's behaviour. The original object, shown as `Context` in Figure 15.12, delegates responsibility to the appropriate state object. The original object becomes an aggregate of its states, only one of which is active at one time. The state objects form an inheritance hierarchy.

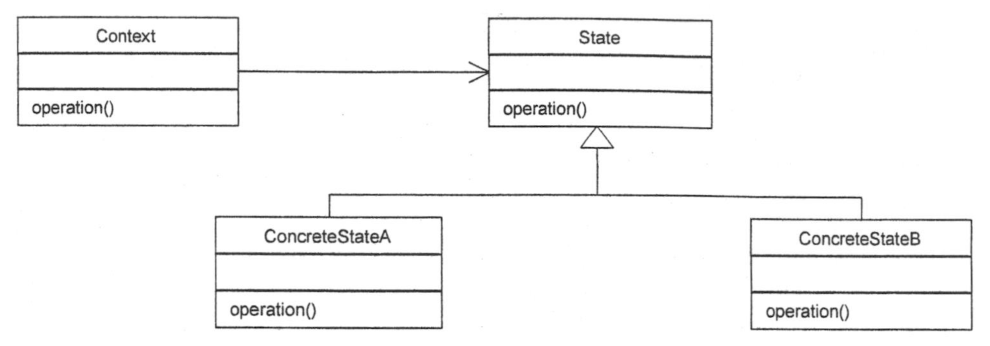

Figure 15.12 Behavioural patterns—State.

The responsibility for transitions from one state to another may either be given to the `Context` object or it may be shared among the State subclasses. If the rules for state changes are volatile and subject to change it may be better for the current State object to be responsible for the next transition. In this way the current state object always knows all the states into which the object may move next. However, this has the disadvantage of producing dependencies between state subclasses.

Use of the State pattern has both advantages and disadvantages, the latter particularly in terms of its possible side effects on system performance.

+ State behaviour is localized and the behaviour for different states is separated. This eases any enhancement of the state behaviour, in particular the addition of extra states.
+ State transitions are made explicit. The state object that is currently active indicates the current state of the Context object.
+ Where a state object has no attributes relevant to a specific Context object it may be shared among the Context objects. This State object is a Singleton!
- If the State objects cannot be shared among the Context objects each Context object will have to have its own State object thus increasing the number of objects and the storage requirements for the system.
- State objects may have to be created and deleted as the Context object changes state, thus introducing a processing overhead.
- Use of the State pattern introduces at least one extra message, the message from the Context class to the State class, thus adding a further processing overhead.

15.5 How to Use Design Patterns

The use of a pattern requires careful analysis of the problem that is to be addressed and the context in which it occurs. Before contemplating the application of patterns within a software development environment it is important to ensure that all members of the team receive appropriate training.

When a developer identifies a part of the application that may be subject to high coupling, a large, complex class or any other undesirable feature, there may be a pattern that addresses the difficulty. The following issues should be considered before employing a pattern to resolve the problem.

- Is there a pattern that addresses a similar problem?
- Does the pattern trigger an alternative solution that may be more acceptable?
- Is there a simpler solution? Patterns should not be used just for the sake of it.
- Is the context of the pattern consistent with that of the problem?
- Are the consequences of using the pattern acceptable?
- Are constraints imposed by the software environment that would conflict with the use of the pattern?

Gamma et al. (1995) suggest a seven-part procedure that should be followed after an appropriate pattern has been selected in order to apply it successfully.

1. Read the pattern to get a complete overview.
2. Study the Structure, Participants and Collaborations of the pattern in detail.
3. Examine the Sample Code to see an example of the pattern in use.
4. Choose names for the pattern's participants (i.e. classes) that are meaningful to the application.
5. Define the classes.
6. Choose application specific names for the operations.
7. Implement operations that perform the responsibilities and collaborations in the pattern.

A pattern should not be viewed as a prescriptive solution but rather as guidance on how to find a suitable solution. It is quite likely (in fact almost certainly the case) that a pattern will be used differently in each particular set of circumstances. At a simple level the classes involved will have attributes and operations that are determined by application requirements. Often a pattern is modified to accommodate contextual differences. For example, the `changeSequence()` operation shown in Figure 15.7 represents a variation on the State pattern and may also be a variation on the Singleton pattern. Alternatively a pattern may suggest some other solution to the developer.

It is critical to the use of patterns that pattern catalogues and languages should be made readily available to the developer. Many patterns are documented in hypertext on the Internet or on company Intranets. CASE tool support for patterns is developing and is provided by some vendors. An on-going area of research addresses the provision of suitable browsing capabilities for patterns. It is important to consider the way a pattern is documented so that it is easy for the developer to determine its applicability. We would contend that the minimum information needed is that described earlier in Section 15.3.

Software developers using patterns may wish to capture their experience in the pattern's format thus building up their own pattern catalogue or language. The process of identifying patterns is known as pattern mining and requires careful validation and management to ensure that the patterns that are captured are valuable. Again the pattern elements described earlier provide a checklist against which any candidate pattern can be compared.

If a pattern satisfies these criteria then its quality should be assured via a walkthrough. The most commonly used form of walkthrough for a pattern is known as a pattern writer's workshop. This involves a small group of pattern authors who constructively comment upon each other's patterns. The focus of a workshop helps the participants describe useful patterns effectively.

15.6 Benefits and Dangers of Using Patterns

One of the most sought after benefits of object-orientation is reuse. Reuse at the object and class level has proved more elusive than was initially expected. Patterns provide a mechanism for the reuse of generic solutions for object-oriented and other approaches. They embody a strong reuse culture. Within the design context, patterns suggest reusable elements of design and, most significantly, reusable elements of demonstrably successful designs. This reuse permits the transfer of expertise to less experienced developers so that a pattern can be applied again and again.

Another benefit gained from patterns is that they offer a vocabulary for discussing the problem domain (whether it be analysis, design or some other aspect of information systems development) at a higher level of abstraction than the class and object making it easier to consider micro-architectural issues. Pattern catalogues and pattern languages offer a rich source of experience that can be explored and provide patterns that can be used together to generate effective systems.

Some people believe that the use of patterns can limit creativity. Since a pattern provides a standard solution the developer may be tempted not to spend time on considering alternatives. The use of patterns in an uncontrolled manner may lead to over-design. Developers may be tempted to use many patterns irrespective of their benefits, thus rendering the software system more difficult to develop, maintain and enhance. When a pattern is used in an inappropriate context the side effects may even be disastrous. For example, the use of the State pattern may significantly increase the number of objects in the application with a consequent reduction in performance. In extreme cases, especially where the architecture is geographically distributed (as at Agate), this might prevent any successful use of the system. The introduction of any new approach to software development has costs for the organization. Developers need to spend time understanding the relevant pattern catalogues, they need to be provided with easy access to the relevant catalogues and they need to be trained in the use of patterns. Another aspect of the introduction of patterns is the necessary cultural change. Patterns can only be used effectively in the context of an organizational culture of reuse. Ironically, the introduction of a patterns approach may arouse less opposition to the encouragement of a reuse culture than an attempt to introduce procedures for code reuse. Developers need to think about how to apply a pattern to their current context, and thus there are greater opportunities for individual creativity.

These dangers emphasize that the use of patterns in software development requires care and planning. In this respect patterns are no different from any other form of problem solving: they must be used with intelligence. It is also important to appreciate that patterns only address some of the issues that occur during systems development. In no way should patterns be viewed as a 'silver bullet' that conquers all problems in systems development.

15.7 Summary

This chapter has considered how patterns can be used in software development. Patterns have been identified in many different application domains and are applicable at many different stages of the software development process. The most signifcant aspect of the growth of interest in patterns is the increased awareness of design issues that follows as a consequence. Patterns represent a significant change in the reuse culture in software development. Reuse need no longer be focused solely on elements of code, whether these are individual classes or complex frameworks, but can also realistically include the reuse of analysis or design ideas as described by patterns.

Related patterns are grouped together in catalogues. A pattern language is a group of patterns focused on a particular aspect of a problem domain so that when used together they provide solutions to the problems that arise.

Review Questions

15.1 What is the difference between a pattern and a framework and how is each used?

15.2 What are the main aspects of changeability?

15.3 Why is the class constructor private in the Singleton pattern?

15.4 What are the advantages of using the Singleton pattern?

15.5 What are the disadvantages of using the Singleton pattern?

15.6 What implementation problems may occur when using the State pattern?

15.7 What are the differences between a pattern language and a pattern catalogue?

15.8 List two general dangers and two general benefits of the use of patterns.

15.9 What seven steps are suggested by Gamma et al. for the effective use of patterns?

Case Study Work, Exercises and Projects

15.A Read the design patterns Bridge and Façade in Gamma et al. (1995) and rewrite their description using the structure of the template given in Section 15.3.

15.B In the FoodCo case study the `ProductionLine` class might be a candidate for design using the State pattern. Show how a variation of the State pattern could handle this requirement. What benefits and disadvantages are there in applying this solution?

15.C Where and how could the Singleton Pattern be used in the FoodCo case study? Prepare a design class specification for a suitable Singleton class.

Further Reading

- Gamma et al. (1995) and Buschmann et al. (1996) are two excellent texts that give important advice concerning software construction and should be on the essential reading list of any software developer. Buschmann et al. (2000) provide a further set architectural patterns. Even where the patterns they discuss are not directly relevant there is much to learn from their approach to solving design problems.
- Both Fowler (1997) and Coad et al. (1997) are useful sources of information from an analysis perspective.
- The *Pattern Languages of Program Design* (known as the PLOP) books (Coplien and Schmidt, 1995; Vlissides et al., 1996; Martin et al., 1998) catalogue a wide range of patterns for all aspects of software development.
- The patterns home page has links to a wealth of patterns and is well worth browsing. This can be found at http://hillside.net/patterns/patterns.html.
- Further useful patterns are available in the Portland Pattern Repository at http://c2.com/cgi-bin/wiki?PortlandPatternRepository. Organizational patterns can be found at http://i44pc48.info.uni-karlsruhe.de/cgi-bin/OrgPatterns.
- A series of Java related patterns (some architectural and some design) are available at http://developer.java.sun.com/developer/technicalArticles/J2EE/patterns/
- Some argue that the architectural approach adopted by Shlaer and Mellor (1988 and 1992) is pattern-based; they certainly have interesting ideas.

CHAPTER 16

Human–Computer Interaction

OBJECTIVES

In this chapter you will learn

- the importance of good user interface design
- what is meant by metaphors in human–computer interaction
- about different approaches to human–computer interaction
- how to apply the techniques of scenario-based design
- how standards and the law affect interface design.

16.1 Introduction

Designing the user interface can be critical in the development of an information system. The interface is what the users see. To them it is the system. Their attitude towards the entire system can be coloured by their experience of the user interface. Effective design of the interaction between people and the information systems they use is a discipline in its own right—human–computer interaction (HCI)—that combines the techniques of psychology and ergonomics with those of computer science. Before we move on to the design of the classes that make up the user interface in Chapter 17, we will address some of the HCI issues that influence the design of the user interface.

In Chapter 12 we raised the issue of designing the inputs and outputs of a system. This chapter is about the human factors aspects of designing the inputs and outputs. The inputs and outputs can be in the conventional form of data entry and enquiry screens and printed reports, or they can take the form of speech recognition, scanners, touch screens and virtual reality environments. We shall be concentrating on the conventional inputs and outputs used in information systems, although we recognize that the growth in the use of multimedia systems means that even quite conventional business information systems may incorporate multiple media as inputs and outputs.

There are two metaphors that are widely used to represent the user interface: first, the idea that the user is conducting a dialogue with the system, and second the idea that the user is directly manipulating objects on screen (Section 16.2). Much HCI work in the past has concentrated on producing guidelines for dialogue design, and we include a section on the characteristics of a good dialogue.

In the second part of the chapter, we consider an informal approach to the HCI design of a system and describe some of the activities that may be carried out. We then introduce three different approaches to HCI that reflect different beliefs about the best way to design the user interface of a system (Section 16.3). Examples of some of the techniques used in these approaches are given.

Finally, we make reference to the international standards for the ergonomics of workstation design and, for European readers, to the legal obligations for HCI design that are imposed by the European Union's directive on health and safety requirements for work with display screens (Section 16.4).

16.2 The User Interface

16.2.1 What is the user interface?

Users of an information system need to interact with it in some way. Whether they are users of FoodCo's tele-sales system entering orders made over the telephone by customers, or members of the public using a touch screen system to find tourist information, they will need to carry out the following tasks:

- read and interpret information that instructs them how to use the system;
- issue commands to the system to indicate what they want to do;
- enter words and numbers into the system to provide it with data to work with;
- read and interpret the results that are produced by the system either on screen or as a printed report; and
- respond to and correct errors.

It is important to note that these are mostly secondary tasks: they are concerned with using the system, not with the users' primary objectives. In the examples above, the primary tasks are to take a customer order and to find tourist information. If the system has been designed well, the secondary, system-related tasks will be easy to carry out; if it has not been designed well, the secondary tasks will intrude into the process and will make it more difficult for the users to achieve their primary tasks.

16.2.2 The dialogue metaphor

In the design of many computer systems, interaction between the user and the system takes the form of a *dialogue*. The idea that the user is carrying on a dialogue with the system is a *metaphor*. (A metaphor is a term that is used figuratively to describe something but is not applied literally.) There is no real dialogue in the sense of a conversation between people going on between the user and the computer[1], but as in dialogues between people, messages are passed from one participant to the other. Figure 16.1 shows the human–computer dialogue in schematic form. Figure 16.2 describes what is meant by each of the types of message that can be found in this dialogue.

[1] Although speech recognition and text to speech systems now make voice input and output possible, even if it is not yet a true dialogue.

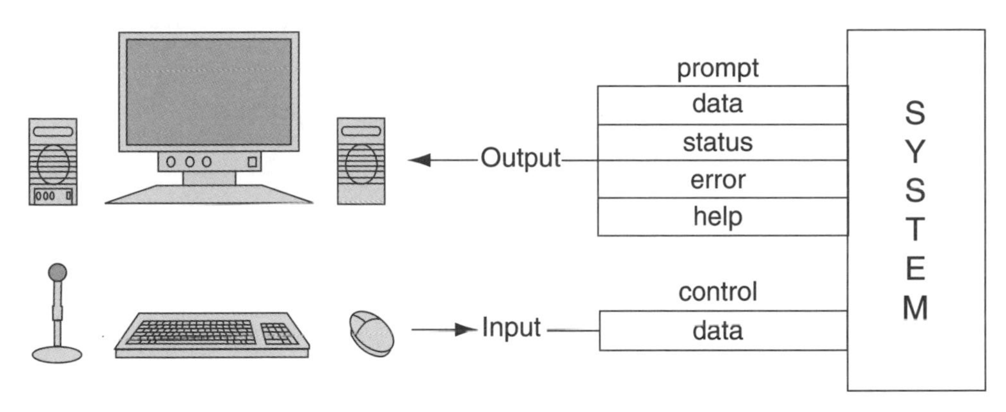

Figure 16.1 Schematic form of human–computer dialogue.

Output	prompt	request for user input
	data	data from application following user request
	status	acknowledgement that something has happened
	error	processing cannot continue
	help	additional information to user
Input	control	user directs which way dialogue will proceed
	data	data supplied by user

Figure 16.2 Types of messages in human–computer dialogue.

Figure 16.3 shows a sample screen layout from FoodCo's existing system which runs on a mini-computer with displays on dumb terminals. Although this only shows one screen, you can describe it in terms of the dialogue between the user and the system.

- The user may enter a command by selecting an option from a menu (not shown).
- The system responds with this data entry screen and automatically displays the Order Date and next Order No.
- The user enters a Customer Code.
- The system responds with the name and brief address of the customer as a confirmation that the correct number has been entered.

And so on . . .

The dialogue may not take exactly the same form each time that a user enters data into this screen. Sometimes, the user may not know the Customer Code and may have to use some kind of index look-up facility, perhaps entering the first few characters of the customer name in order to view a display of customers that start with those characters. Sometimes an order may consist of one line, usually it will consist of more, and if it consists of more than eight it will be necessary to clear those that have been entered from the screen and display space for a further eight lines to be entered. It also illustrates elements of the interface that support some of the message types listed in Figure 16.2. These are described in Figure 16.4.

```
CUSTORD1                 Customer Order Entry                  25/08/2001
--------------------------------------------------------------------------
Order Date  25/08/2001              Order No.   37291
Customer Code CE102_ Central Stores, Lytham St Annes
Customer Order Ref.  R20716___

   Prod.      Product                                    Unit     Line
   Code       Description                      Qty      Price    Price
01 12-75__    Sandwich spread 24x250g          __3      18.00    54.00
02 09-103_    Brown sauce 30x500g              _10      24.60   246.00
03 _______                                     ___
04 _______                                     ___
05 _______                                     ___
06 _______                                     ___
07 _______                                     ___
08 _______                                     ___
                                               Total            300.00
                                               Tax               52.50
                                               Order            ______
                                               Total            352.50
--------------------------------------------------------------------------
F1-Help    F2-Save    F3-Cancel  F4-New    F5-Cust.  F6-Prod.
F10-Exit                               Cust.     Lookup    Lookup
```

Figure 16.3 FoodCo customer order entry screen layout with sample data.

Output	prompt	request for user input and labels for automatically generated data, shown in bold, for example Customer Code
	data	automatic display of Order Date and next Order No., automatic calculation of totals and tax (shown in italics to distinguish it from input data)
	status	screen heading; could include display to confirm that a new order has been saved
	error	messages to warn of incorrect data entered, for example if a Customer Code is entered that does not exist or if a negative Quantity is entered
	help	additional information to user in response to the user pressing F1; may be general about the order entry screen or context-sensitive—specific to a particular type of data entry
Input	control	use of function keys to control dialogue
	data	numbers, codes and quantities typed in by user

Figure 16.4 Examples of types of messages in human–computer dialogue.

In the requirements model of the new system for FoodCo, there will be a use case for `Enter customer order`, as in Figure 16.5. This will be supported by a use case description, which may be quite brief early in the project. As the project progresses through further iterations, the use case description will be filled out in more detail. Not all the use cases will be for interactive dialogues: some will be for enquiries and some will be for printed reports. Figure 16.6 shows some of the use cases that the FoodCo sales clerks use. For each of these use cases there may be a sequence diagram to show the interaction between the collaborating objects. However, these sequence diagrams will not yet show all the details of the interaction between the user and the system at the interface. This will be covered in Section 17.5.

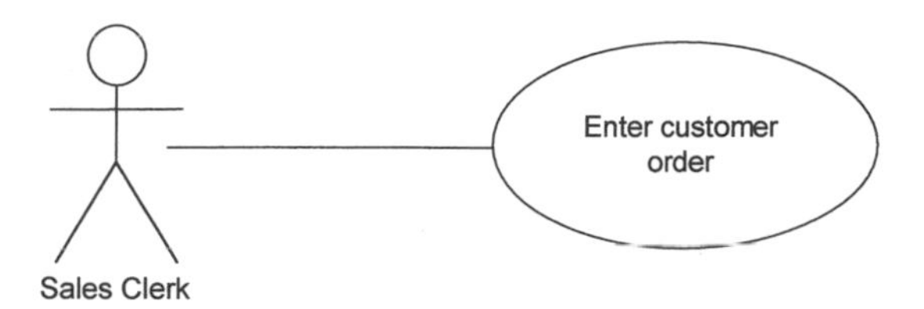

Figure 16.5 Use case diagram and description for `Enter customer order`.

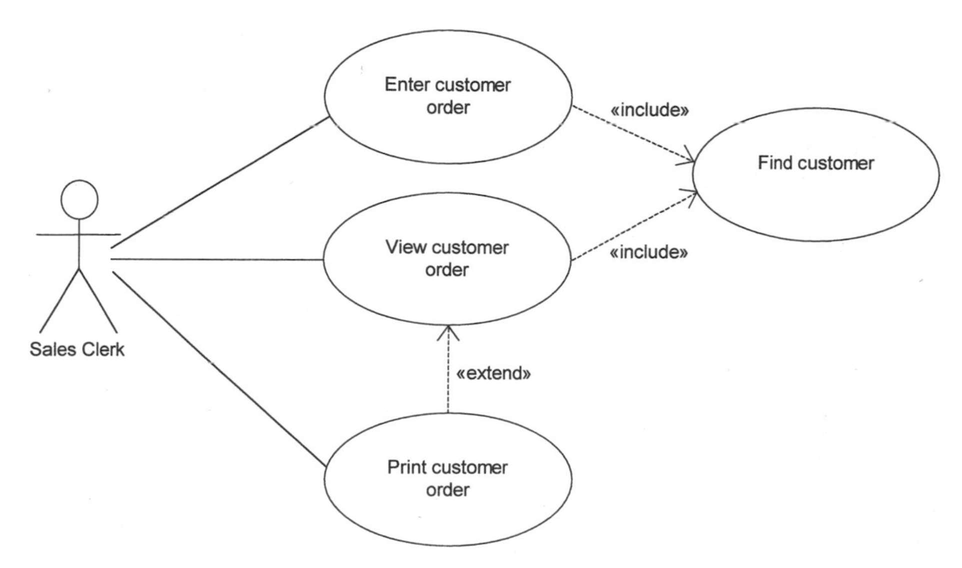

Figure 16.6 Use case diagram showing use cases used by sales clerks at FoodCo.

16.2.3 The direct manipulation metaphor

The other metaphor for the design of the user interface, which has become more widespread in the last few years, is the *direct manipulation* metaphor. Many people are now familiar with this through the use of GUIs. When you use a software package with this kind of interface you are given the impression that you are manipulating objects on the screen through the use of the mouse. This metaphor is reflected in the concrete nature of the terms that are used. You can:

- drag and drop an icon,
- shrink or expand a window,
- push a button and
- pull down a menu.

Such interfaces are *event-driven*. Graphical objects are displayed on the screen and the window management part of the operating system responds to events. Most such events are the results of the user's actions. The user can click on a button, type a character, press a function key, click on a menu item or hold down a mouse button and move the mouse. The design of user interfaces to support this kind of interaction is more complicated than for text-based interfaces using the dialogue metaphor. Figure 16.7 shows the interface of a Java program to implement the use case `Check campaign budget` for the Agate case study. In this use case, the user first selects the name of a client from a list box labelled Client. Having selected the client, a list of all active campaigns for that client is placed in the list box labelled Campaign. At this point, no campaign is selected, and the user can click on the arrow at the end of the list box to view the list and select a campaign. When a campaign has been selected, the user can click on the button labelled Check. The program then totals up the cost of adverts in that campaign, subtracts it from the budget and displays the balance as a money value (negative if the campaign is over budget). In this interface design, there is no point in the user selecting a campaign until a client has been selected or clicking the Check button until a client and a campaign have been selected. The designer may choose to disable the Campaign list box until the client has been selected, and disable the button until both client and campaign have been selected. Having checked one campaign, the user may choose a different client, in which case the contents of the Campaign list box have to be changed and the button disabled again until a different campaign has been selected. In Section 17.8 we shall use statechart diagrams to model the state of elements of a user interface like this in order to ensure that we have correctly specified the behaviour of the interface.

Windows like the one in the example above are usually called *dialogue boxes* in GUI environments. In terms of the metaphors that we have discussed, they combine elements of a dialogue with the user with direct manipulation of buttons and lists.

16.2.4 Characteristics of good dialogues

Many authors of books and reports on HCI have produced sets of guidelines to help designers to produce good designs for the user interface. Some such guidelines are specific to certain types of interface. Shneiderman (1997) proposes five high-level objectives for data entry dialogue design that date back to his original edition in 1986

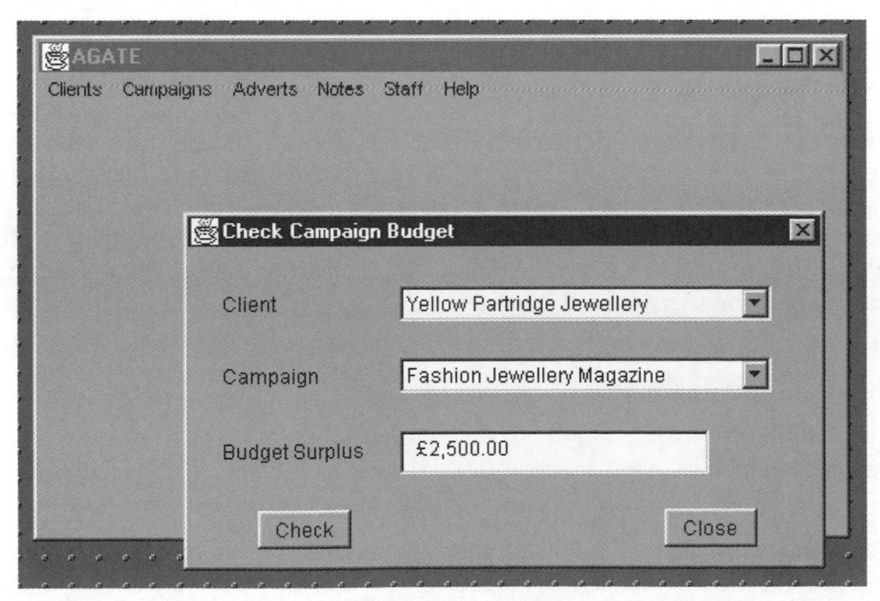

Figure 16.7 Interface for the use case `Check campaign budget` developed in Java.

when most interfaces were text-based. Other authors such as Gaines and Shaw (1983), also writing at a time when text-based interfaces were predominant, have proposed as many as seventeen.

Regardless of whether a system is being developed for a text-based environment or for a GUI environment, there are a number of important general characteristics of good dialogue design. These include:

- consistency,
- appropriate user support,
- adequate feedback from the system and
- minimal user input.

These are considered in turn below.

Consistency. A consistent user interface design helps users to learn an application and to apply what they know across different parts of that application. This applies to commands, the format for the entry of data such as dates, the layout of screens and the way that information is coded by the use of colour or highlighting. As an example of a command, if the user has to press function key F2 to save data in one part of the system, then they will expect the key to have the same effect elsewhere in the system. If it does not result in data being saved then they will have made an error. The outcome of this could be that pressing F2 does something else that the user did not expect, or that it does nothing, but the user thinks that they have saved the data and then exits without saving it. Whichever is the case, the user is likely to become annoyed or frustrated at the response of the system. Guidelines in corporate style guides or in those from Microsoft and Apple help to prevent this kind of user frustration. Style guides are discussed in Section 16.2.5.

Appropriate user support. When the user does not know what action to take or has made an error, it is important that the system provides appropriate support at the interface. This support can be informative and prevent errors by providing help messages, or it can assist the user in diagnosing what has gone wrong and in recovering from their error. Help messages should be context-sensitive. This means that the help system should be able to detect where the user has got to in a dialogue and provide relevant information. In a GUI environment, this means being able to detect which component of the interface is active (or has the *focus*) and providing help that is appropriate to that part of the interface. The help provided may be general, explaining the overall function of a particular screen or window, or it may be specific, explaining the purpose of a particular field or graphical component and listing the options available to the user. It may be necessary to provide a link between different levels of help so that the user can move between them to find the information they require. The hypertext style of help in Microsoft Windows provides this facility. Help information may be displayed in separate screens or windows, or it may be displayed simultaneously in a status line or using *tooltips* as the user moves through the dialogue or positions the cursor over an item. Many web page designers provide help about elements of their pages by displaying messages in the status line at the bottom of the browser window or by displaying a tooltip-style message in a box as the cursor moves over an item on the page.

Error messages serve a different purpose and require careful design to ensure that they inform rather than irritate the user. An error message that tells the user that he or she has just deleted an essential file and then expects the user to click on a button marked OK when it is anything but OK is likely to annoy the user. Error messages should explain what has gone wrong and they should also clearly explain what the user can or should do to recover the situation. This information should be in language that the user can understand. This may mean using jargon from the user's business that they will recognize and understand rather than using computer jargon. Figure 16.8 shows three different error message boxes for the same situation. Only one is really of any help to the user.

Warning messages can prevent the user from making serious errors by providing a warning or caution message before the system carries out a command from the user that is likely to result in an irreversible action. Warning messages should allow the user to cancel the action that is about to take place. Figure 16.9 shows an example of a warning message.

Adequate feedback from the system. Users expect the system to respond when they make some action. If they press a key during data entry, they expect to see the character appear on the screen (unless it is a control command or a function key); if they click on something with the mouse, they expect that item to be highlighted and some action from the system. Users who are uncertain whether the system has noticed their action keep on pressing keys or clicking with the mouse, with the result that these further keypresses and clicks are taken by the system to be the response to a later part of the dialogue, with unpredictable results. It is important that users know where they are in a dialogue or direct manipulation interface: in a text-based interface there should be a visible cursor in the current active field; in a GUI environment the active object in the interface should be highlighted. The Yes button in Figure 16.9 is highlighted in this way, and this means that it will respond to the user pressing `<Return>` on the keyboard.

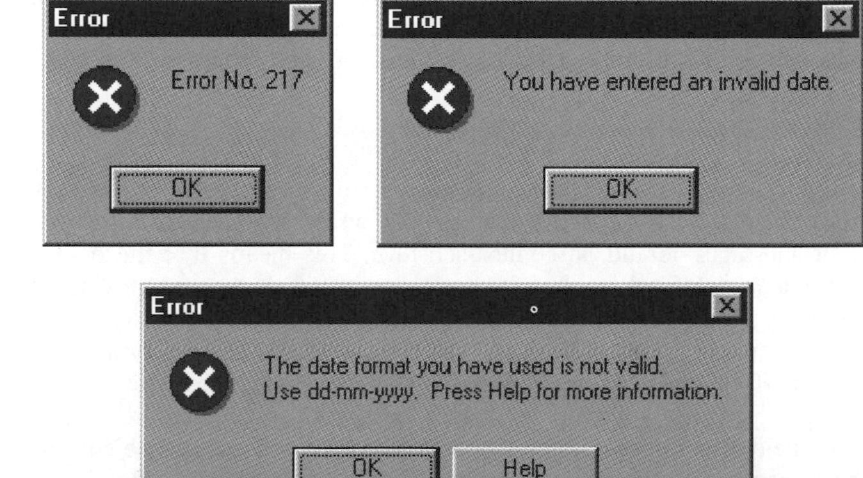

Figure 16.8 Example error messages for the same error.

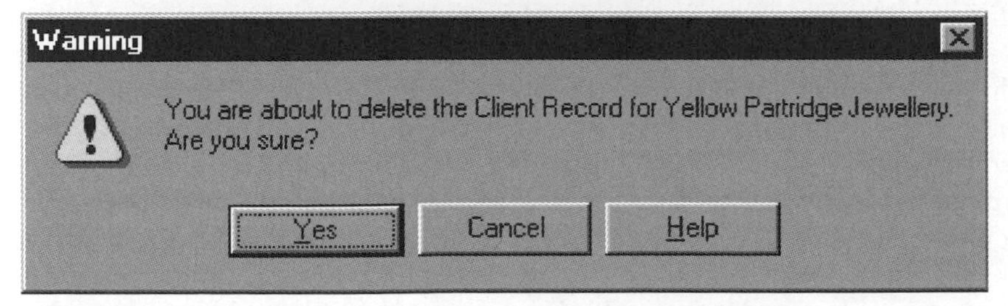

Figure 16.9 Example warning message.

The system's response time should be appropriate to the type of user action: responses to data entry keypresses should be instantaneous, while responses to commands in menus or by means of buttons may take longer. If a system response is going to take some time, the system should respond first with some kind of feedback indicating that it is busy. This can be changing the cursor to a different form, or it can be displaying a progress monitor that shows what proportion of the task has been completed. If possible, the user should be given the option of cancelling the command. The purpose of this kind of feedback is to reduce the user's uncertainty about whether the system has received the input and is doing something about it or is waiting for the next input.

Minimal user input. Users resent making what they see as unnecessary keypresses and mouse clicks. Reducing unnecessary input also reduces the risk of errors and speeds data entry. The interface should be designed to minimize the amount of input from the user. The user can be helped in this way by:

- using codes and abbreviations,
- selecting from a list rather than having to enter a value,
- editing incorrect values or commands rather than having to type them in again,
- not having to enter or re-enter information that can be derived automatically and
- using default values.

Some of these have a basis in the psychological aspects of the discipline of HCI. For example, being able to select values from a list rather than having to enter them from memory, allows the user to work by recognizing information rather than having to recall it.

It is also possible to reduce the amount of input as users become more familiar with a system by providing shortcuts or accelerators, key combinations that the user can use instead of selecting a command from a menu. However, these require the user to remember the key combinations and are less useful for new users who will find menus easier to use.

16.2.5 Style guides

In Section 16.2.4 consistency of the interface has been highlighted as one of the characteristics of good dialogue design. Some organizations provide standard guidelines for the design of user interfaces. One way in which standardization of user interface design has come about is through the domination of the PC market by Microsoft. Microsoft produces a book of guidelines *The Windows Interface Guidelines for Software Design* (Microsoft, 1997) that lays down the standards to which developers must adhere if they are to be granted Windows certification. Similar guidelines are available from Apple for the Apple Macintosh operating system—*Macintosh Human Interface Guidelines* (Apple, 1996). The effect of such guidelines is apparent in the similarity of many applications from different sources that make use of toolbars, status bars and dialogue boxes with buttons and other graphical components placed in similar positions. The benefit of this similarity for users is that different applications will look similar and behave in similar ways. This means that users can transfer their knowledge of existing applications to a new one and make inferences about the way that the new one will respond to certain types of user interaction.

Guidelines for user interface design are usually referred to as *style guides*, and large organizations with many different information systems produce their own style guides for the design of systems to ensure that all their applications, whether they are produced in-house or by outside software companies, conform to a standard set of rules that will enable users quickly to become familiar with a new application. Figure 16.3 reflects the use of an existing style guide in FoodCo. The layout of the screen with standard heading information at the top, the use of bold text to highlight prompts and labels, the position of the information about function keys and the use of specific function keys for particular commands are all standards within FoodCo for the design of text-based screens. This is important, as it means that a user can be confident that pressing function key F2 in any data entry screen will save the data on the screen.

The use of style guides and the characteristics of a good dialogue relate to dialogue and interface design in general. In the next part of this chapter, we consider how to ensure that the user interface is appropriate to the specific application for which it is being designed.

16.3 Approaches to User Interface Design

16.3.1 Informal and formal approaches

There are many different ways of designing and implementing the elements of the user interface that support the interaction with users. The choices that the designer makes will be influenced by a number of factors. These include:

- the nature of the task that the user is carrying out,
- the type of user,
- the amount of training that the user will have undertaken,
- the frequency of use and
- the hardware and software architecture of the system.

These factors may be very different for different systems. They are listed in Figure 16.10 for the FoodCo tele-sales system and a WAP (Wireless Access Protocol) tourist information system. Systems that are used by members of the public are very different from information systems used by staff. The WWW and WAP have made information systems available to people who are unlikely to receive training in using these systems, and who may have no experiencc of information systems in business settings.

This way of comparing the two systems and identifying factors that affect their design is very informal. More formal and methodical approaches to the analysis of usability requirements have been developed by researchers in the discipline of HCI. These approaches can be categorized under three headings:

- structured approaches,
- ethnographic approaches and
- scenario-based approaches.

These approaches are very different from one another. However, they all carry out three main steps in HCI design:

- requirements gathering,
- design of the interface and
- interface evaluation.

	Tele-Sales System	WAP Tourist Information System
The nature of the task that the user is carrying out	Routine task; closed solution; limited options.	Open-ended task; may be looking for information that is not available.
The type of user	Clerical user of the system; no discretion about use (must use it to do their job).	Could be anyone; discretion about use of system; novice in relation to this system.
The amount of training that the user will have undertaken	Training provided as part of job.	No training provided.
The frequency of use	Very frequent; taking an order every few minutes.	Very occasional; may never use it again.
The hardware and software architecture of the system	Mini-computer, dumb terminals with text screens, keyboard data entry. All software runs on the mini-computer. Structured programs with subroutines for data access and screen-painting.	Mobile telephone screen with keypad and scroll buttons to move through menus. WAP browser runs on mobile telephone, WAP gateway connects to server, which generates WML for WAP browsers and HTML for other browsers using XML and stylesheets.

Figure 16.10 User interface design factors for two systems.

Each of these approaches has similar objectives in each of these main steps. Typical objectives are shown in Figure 16.11. However, they differ in the ways that they set out to achieve these objectives. This is described below.

Structured approaches

Structured approaches to user interface design have been developed in response to the growth in the use of structured approaches to systems analysis and design during the 1980s. Structured analysis and design methodologies have a number of characteristics. They are based on a model of the systems development life cycle, which is broken down into stages, each of which is further broken down, for example into steps that are broken down into tasks. Specific analysis and design techniques are used, and the methodology specifies which techniques should be used in which step. Each step is described in terms of its inputs (from earlier steps), the techniques applied and the deliverables that are produced as outputs (diagrams and documentation). These approaches are more structured than the simple waterfall model of the life cycle, as they provide for activities being carried out in parallel where possible rather than being dependent on the completion of the previous step or stage. Typically such structured approaches use data flow diagrams to model processes in the system and take a view of the system that involves decomposing it in a top-down way. Structure charts or structure diagrams are used to design the programs that will implement the system.

Step	Objectives
Requirements gathering	Determine characteristics of the user population: types of user, frequency of use, discretion about use, experience of the task, level of training, experience of computer systems.
	Determine characteristics of the task: complexity of task, breakdown of task, context/environment of task.
	Determine constraints and objectives: choice of hardware and software, desired throughput, acceptable error rate.
Design of the interface	Allocate elements of task to user or system; determine communication requirements between users and system.
	Design elements of the interface to support the communication between users and system in the light of characteristics of the users, characteristics of the task and constraints on design.
Interface evaluation	Develop prototypes of interface designs.
	Test prototypes with users to determine if objectives are met.

Figure 16.11 Steps in HCI design and objectives in each step.

Proponents of structured approaches argue that they have a number of benefits.

- They make management of projects easier. The breakdown of the project into stages and steps makes planning and estimating easier, and thus assists management control of the project.
- They provide for standards in diagrams and documentation that improves understanding between the project staff in different roles (analyst, designer and programmer).
- They improve the quality of delivered systems. Because the specification of the system is comprehensive, it is more likely to lead to a system that functions correctly.

Advocates of structured approaches to HCI believe that similar benefits can be brought to HCI by adopting structured approaches. These approaches assume that a structured approach to analysis and design of a system is being used and that a structured approach to the HCI design can take place at the same time and be integrated to some extent into the project life cycle. Two examples of such approaches are discussed briefly below.

- STUDIO (STructured User-interface Design for Interface Optimization) developed with KPMG Management Consulting in the UK (Browne, 1994).
- The RESPECT User Requirements Framework developed for the European Union Telematics Applications Programme by a consortium of Usability Support Centres (Maguire, 1997).

Structured approaches make use of diagrams to show the structure of tasks and the allocation of tasks between users and the system. They also make extensive use of checklists in order to categorize the users, the tasks and the task environments. Evaluation is typically carried out by assessing the performance of the users against measurable usability criteria. STUDIO is used here as an example of a structured approach.

STUDIO is divided into Stages, and each Stage is broken down into Steps. The activities undertaken in each of the Stages are shown in Figure 16.12. STUDIO uses a number of techniques such as:

- task hierarchy diagrams,
- knowledge representation grammars,
- task allocation charts, and
- statecharts.

It is not possible to provide examples of all of these here. Statecharts are similar to those used in UML and based on the work of Harel (1988). Examples of statecharts applied to user interface design are included in Section 17.7. A sample task hierarchy diagram for `Take an Order` is shown in Figure 16.13. This diagram applies to the order entry screen of Figure 16.3. The diagram is read from top to bottom and left to right. In it, the boxes with a small circle in the top right-hand corner are selections, only one of which will take place each time an order is taken; the box with an asterisk in the top right-hand corner is an iteration, which will take place usually more than once.

Structured approaches may involve evaluation of the user interface designs in a laboratory situation. This reflects the need to have operational measures of usability that can be tested and used to assess the effectiveness of the design. These operational measures are derived from objectives that were gathered during the requirements analysis phase of the project described in Chapter 6. Examples include measures of how quickly users learn to use the system, error rates and time taken to complete tasks.

The National Physical Laboratory in the UK, one of the partners in the consortium that produced the RESPECT methodology, has a usability laboratory in which the interface designs are tested with users under laboratory conditions. One criticism of this approach is that people do not use systems under laboratory conditions, they use them in busy offices, noisy factories or crowded public spaces. Usability tests conducted in laboratories lack *ecological validity*. This means that they do not reflect the real conditions in the environment in which the systems will be used. As an alternative to laboratory-based usability studies, ethnographic approaches to the establishment of usability requirements, and to the testing of those requirements, have been proposed.

Stage	Summary of activities
Project Proposal and Planning	Decide whether user interface design expenditure can be justified. Produce quality plan.
User Requirements Analysis	Similar to systems analysis, with focus on gathering information relating to user interface design rather than general functionality.
Task Synthesis	Synthesize results of requirements analysis to produce initial user interface design. Produce user support documentation.
Usability Engineering	Prototyping combined with impact analysis to provide an approach to iterative development that is easy to manage.
User Interface Development	Handover of the user interface specification to developers to ensure that usability requirements are understood.

Figure 16.12 Summary of activities in each Stage of STUDIO.

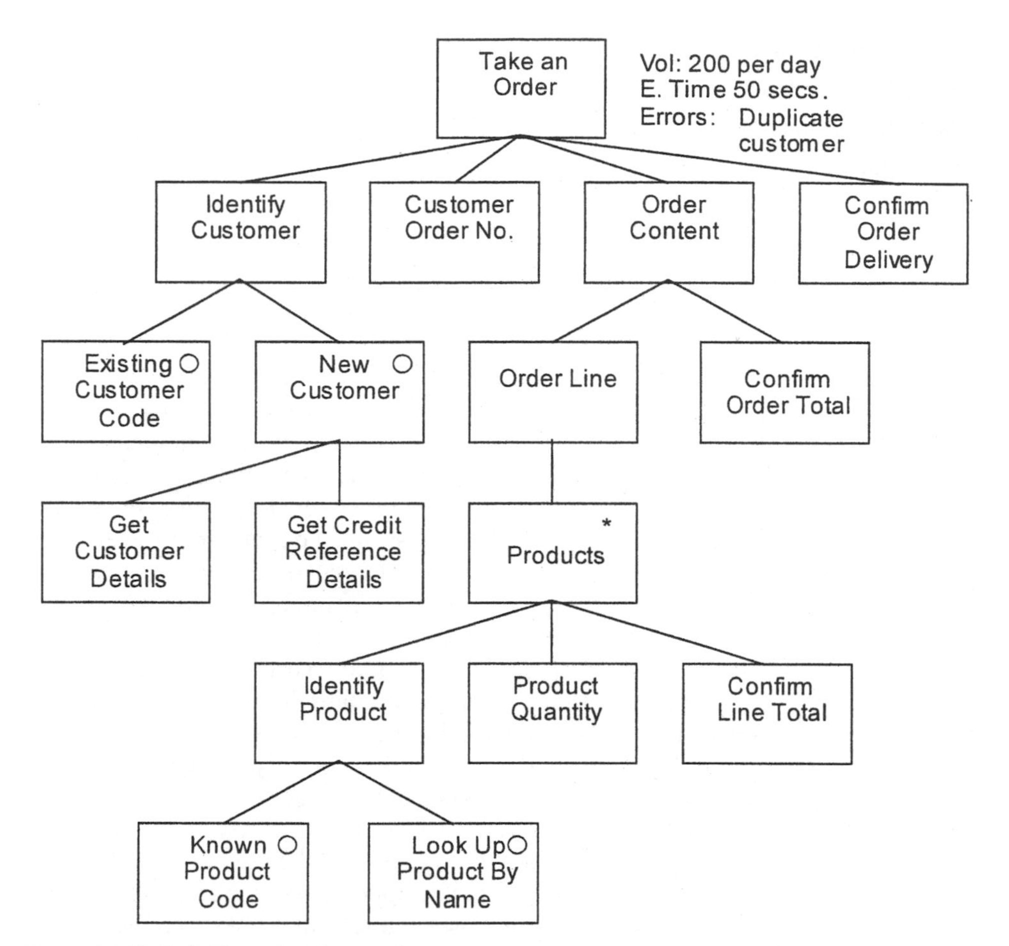

Figure 16.13 Task Hierarchy Diagram for `Take an Order`.

Ethnographic approaches

The term ethnography is applied to a range of techniques used in sociology and anthropology and reflects a particular philosophy about how scientific enquiry should be carried out in these disciplines. Researchers who employ an ethnographic method seek to involve themselves in whatever situation they are studying. They believe that only by doing this and becoming part of that situation can they truly understand and interpret what is happening. Ethnographic methods belong to a wide range of *qualitative* research methods. (Qualitative means 'concerned with the quality of something' and is usually taken to be the opposite of *quantitative* which means 'concerned about the quantity of something'. Quantitative methods typically rely on statistics to draw conclusions.) Hammersley and Atkinson (1995) provide a definition of ethnography.

> In its most characteristic form it involves the ethnographer participating, overtly or covertly, in people's daily lives for an extended period of time, watching what happens, listening to what is said, asking questions—in fact, collecting whatever data are available to throw light on the issues that are the focus of the research.

In HCI this means that the professional charged with carrying out the user interface design spends time with the users immersed in their everyday working life. Only by spending time in this way can the real requirements of the users be understood and documented. Ethnographic methods also emphasize that different users interpret their experience of using systems subjectively, and it is this subjective interpretation that the HCI professional must understand rather than assuming that the system can be assessed objectively.

Some HCI methods are criticized for failing to capture information about the context in which people are using systems, by focusing on the user and their tasks. Ethnographic approaches attempt to answer this criticism by the professional experiencing the daily working life of the people who will be the users of the system so that it can be better designed to meet their needs. Some structured approaches have also attempted to respond to the criticism about failing to take context into account, typically by adding some kind of contextual analysis questionnaire to the battery of checklists that they use.

There is no one ethnographic approach that covers the three steps in user interface development: requirements gathering, design of the interface, and interface evaluation. There are a number of approaches that can be classed as ethnographic, although their originators may use particular terms to describe their approaches.

Contextual enquiry is an approach developed by John Whiteside and others at Digital Equipment Corporation (Whiteside et al., 1988). It is used to carry out evaluation of the usability of a system in the users' normal working environment. The aim of contextual enquiry is to get as close to the users as possible and to encourage them to develop their interpretation of the system.

Participative or co-operative design and evaluation involve users actively in the design and evaluation of the user interface of systems (Greenbaum and Kyng, 1991). The social and political issues in the workplace that affect the use of systems are part of the material that is captured by these approaches.

Ethnographic approaches use a range of techniques to capture data: interviews, discussions, prototyping sessions and videos of users at work or using new systems. These data are analysed from different perspectives to gain insights into the behaviour of the users. Video is also used in other approaches, particularly in laboratory-based usability studies. Analysis of video can be particularly time-consuming.

Scenario-based approaches

Scenario-based design has been developed by John Carroll and others (Carroll, 1995). It is less formal than the structured approaches but more clearly defined than most ethnographic approaches. Scenarios are step-by-step descriptions of a user's actions that can be used as a tool in requirements gathering, interface design and evaluation. Use cases are similar to scenarios, and Carroll's book includes chapters by Ivar Jacobson and Rebecca Wirfs-Brock who have developed use cases and responsibility-based approaches to modelling interaction in object-oriented analysis and design. Of the three approaches discussed here, scenario-based design fits best with use case modelling.

Scenarios can be textual narrative describing a user's actions or they can be in the form of storyboards (a series of pictures that depict those actions), video mock-ups or even prototypes. Figure 16.14 shows a scenario that describes the actions of Peter Bywater from Agate when he demonstrates how he creates notes following the interview in Chapter 6.

Scenarios can be used like this in requirements gathering to document the actions that a user carries out in their current system. They can also be used to document ideas about how the user would see themselves using the new system. This is called envisioning the design. Alternative scenarios describing different approaches to the

Pete starts up the word-processor.

He types in a title for the note and changes its style to *Title*.

He types in two paragraphs describing his idea for an advertisement for the Yellow Partridge campaign to be used in fashion magazines in Europe during the summer of 2002.

He types his initials and the date and time.

He uses the short-cut keys to save the file.

The save-as dialogue box appears and, using the mouse, he changes to the *Summer 2002 Campaign* folder in the *Yellow Partridge* folder on the server.

He scrolls to the bottom of the list of files already in the folder and reads the title of the last note to be added, *Note 17*, he calls the new note *Note 18* and clicks on Save.

He exits from the word-processor.

Figure 16.14 Scenario describing Pete Bywater of Agate adding a new note.

design can be compared by the designers and the users. Figure 16.15 shows a scenario describing how a staff member at Agate might use the new system to create a new note about an advert.

For evaluation of the system more detailed scenarios are prepared so that the actual system can be compared against the expectations that the designer has of how the user will interact with it. Carroll (1995) claims that scenarios can be used in more than just these three ways. He lists the following roles for scenarios:

- Requirements analysis,
- User–designer communication,
- Design rationale,
- Envisionment,
- Software design,
- Implementation,
- Documentation and training,
- Evaluation,
- Abstraction and
- Team building.

Two of these are worth further comment: user–designer communication and design rationale.

In Chapter 6 we pointed out that the diagrams used by systems analysts and designers are used to communicate ideas, among other things. Information systems professionals need to communicate with the end-users of the systems that they are developing. Scenarios provide a means of communication that can be used by professionals and end-users to communicate about the design of the users' interaction with the system. They are simple enough that users can produce them without the need for the kind of training that they would need to understand class diagrams, for example. Scenarios can be used with use cases. The use cases can provide a description of the typical interaction;

The user selects Add a Note from the menu. A new window appears.
From the list box at the top of the window she selects the name of the client.
A list of campaigns appears in the list box below, and she selects a particular campaign.
A list of adverts appears in the next list box, and she selects a specific advert.
She types a few paragraphs into a text box to describe her idea for the advert. She fills the space on screen and a vertical scrollbar appears and the text in the text box scrolls up.
She enters her initials into a text box, and the system checks that she is allocated to work on that campaign.
The date and time are displayed by the system, and the Save button is enabled.
She clicks on the Save button and the word Saved appears in the status bar.
The text box, the text field for initials and the date and time are cleared.

Figure 16.15 Scenario describing how a user might add a note in the new system.

scenarios can be used to document different versions of the use case, for example, to document what happens when a user is adding a new note but is not authorized to work on the project they try to add it to. Use cases are concerned with the functionality offered by the system, while scenarios focus on the interaction between the user and the system.

Scenarios can be supported by additional documentation to justify design decisions that have been taken. Carroll (1995) calls these design justifications *claims*. The designer can document the reasoning behind alternative designs, explaining the advantages and disadvantages of each. Figure 16.16 shows some claims for the scenario in Figure 16.15. These usability claims from design can be checked during evaluation of the software or of prototypes.

Scenario-based design can result in large volumes of textual information that must be organized and managed so that it is easily accessible. There is a document management task to be undertaken that requires a rigorous approach to control different versions of scenarios and to cross-reference them to claims and feedback from users. Developers run the risk of delaying implementation while they work through and document alternative scenarios for different parts of the system. Rosson and Carroll (1995) present one way to try to prevent this happening. They use a computer-based tool to develop and document their scenarios and to develop working models of the scenarios in Smalltalk as they go along. This allows them to document software implementation decisions at the same time, and they propose that there are benefits to recording design decisions and software implementation decisions together in this way.

These three types of approach have been presented as though they were very separate. However, there are elements that they have in common. Some structured approaches have attempted to take on board the criticisms that they fail to address the context in which people work and use computer systems. Ethnographic methods may use the same data gathering techniques as other approaches, and may be used to provide information that can be used as the basis for drawing up scenarios. What they all share is a concern to enhance the usability of information systems and a recognition that usability issues must be integral to the design of computerized information systems.

The Save button is disabled until the user has selected a client and a campaign, entered some text and entered his or her initials. This prevents the user attempting to save the note before all data has been entered and getting an error message.

The initials of the user could be entered automatically from their network login, but observation shows that the creative staff often work together as a group and different people will come up with ideas that they record as notes. It would be inconvenient for them to be logging in and out of the system each time a different person wants to enter a new note. For this reason, they are required to enter their initials.

The initials, date, time and text fields are cleared after a note is saved, but the client, campaign and advert list boxes are left untouched so that the user can enter another note for the same advert or campaign without having to reselect these items.

Figure 16.16 Claims for the design scenario in Figure 16.15.

16.3.2 Achieving usability

People often talk about how *user-friendly* a piece of software is, but it is often very difficult to tell what it is they mean by this. As a concept it is very vague. Usability may seem like a similar concept, but the HCI community has developed definitions of usability that can be used to test a piece of software. Shackel (1990) produced definitions of four criteria that were originally developed in the 1980s.

- Learnability—how much time and effort is needed to achieve a particular level of performance.
- Throughput—the speed with which experienced users can accomplish tasks and the number of errors made.
- Flexibility—the ability of the system to handle changes to the tasks that users carry out and the environment in which they operate.
- Attitude—how positive an attitude is produced in users of the system.

In Chapter 6 we mentioned the International Standards Organization (ISO) definition of usability as 'the degree to which specific users can achieve specific goals within a particular environment; effectively, efficiently, comfortably and in an acceptable manner'. These criteria can be used in conjunction with the users' acceptance criteria documented during requirements gathering to assess how easy a software product is to use. Some of these can be quantified, for example, we can count the number of errors made by staff at FoodCo using the new system and compare that with the number of errors made with the old system and the objectives that they have set for the new system.

Sometimes conflicts will exist between different criteria and between usability criteria and other design objectives, and the designers will have to make compromises or trade-offs between different objectives. In particular, increasing flexibility is likely to conflict with the objective of developing the system at a reasonable cost.

16.4 Standards and Legal Requirements

In Section 16.2.5 we discussed style guides, which set standards for the design of user interfaces. Style guides like these determine the use of standard layouts, colour and function keys and the overall appearance of the system. The International Standards Organization (ISO) has produced standards that have a broader impact on the use of computer systems. ISO 9241 is an international standard for the ergonomic requirements for work with Visual Display Terminals, including both hardware and software. The standard covers physical aspects of the user's workstation (including positioning of equipment and furniture), the design of the computer equipment and the design of the software systems. ISO 14915 is a further standard, entitled *Multimedia User Interface Design—Ergonomic Requirements for human-centred multimedia interfaces*, and is a work in progress at the time of writing. These standards are intended to ensure the quality of systems and to prevent local standards becoming barriers to free trade.

In the European Union (EU), this has been taken one step further and the EU Council issued a directive on 29 May 1990 that has the force of law for member states. In the United Kingdom, for example, this directive has been implemented in the Health and Safety (Display Screen Equipment) Regulations 1992. Under these regulations all workstations must now comply with certain minimum requirements, and employers have a duty in law to ensure the health and safety of employees using display screen equipment.

The regulations provide a number of definitions:

- display screen equipment—any alphanumeric or graphic display screen;
- user—an employee who habitually uses display screen equipment as a significant part of his or her normal work (see table of criteria in the Health and Safety Executive guidance document);
- operator—self-employed person as above;
- workstation—display screen equipment, software providing the interface, keyboard, optional accessories, disk drive, telephone, modem, printer, document holder, work chair, work desk, work surface or other peripheral item and the immediate work environment around the display screen equipment.

The definition of display screen equipment excludes certain types of equipment, such as equipment in the cab of a vehicle, cash registers and some portable equipment.

As well as covering the physical equipment that is used by the user, the regulations cover environmental factors such as position of equipment, lighting, noise, heat and humidity in the workplace. Employers are required to:

- analyse workstations to assess and reduce risks,
- take action to reduce risks identified,
- ensure workstations meet the requirements of the regulations by the necessary dates,
- plan the work activities of users to provide breaks,
- provide eyesight tests for users,

- provide corrective appliances for eyes if required,
- provide training relevant to health and safety issues and workstations and
- provide information to employees about health and safety risks and measures taken to reduce them.

The analysis of workstations in order to reduce risks includes analysis of the software, and the guidelines published to assist employers to meet their responsibilities state the following requirements.

Employers must take into account the following principles in designing, choosing, commissioning and modifying software and in designing tasks for people that require them to use display screen equipment.

- The software that is used must be suitable for the task.
- The software must be easy to use and able to be adapted to the level of knowledge or experience of the operator or user.
- The employer is not allowed to use any kind of quantitative or qualitative checking facility without the knowledge of the operators or users.
- Systems must give feedback to operators or users about the performance of the systems that they are using.
- Systems must display information for users both in a format and at a pace that are adapted to the operators or users.
- The principles of software ergonomics must be applied, particularly to the way that people process data.

Clearly the effect of this is to require employers, and so also software developers, to demonstrate that they are applying good HCI practice in the way that they design software.

Many countries in the world have regulations in place to promote good practice in workstation use. The United States is a significant exception to this. The Occupational Safety and Health Administration (OSHA) has proposed rules designed to prevent musculo-skeletal disorders caused by poor work design, bad posture and repetitive activities and covering workstation design and layout. These were rejected by Congress, backed by industry lobbyists. However, this is the exception rather than the rule in developed countries, and there is plenty of on-line material on the subject.

16.5 Summary

System designers must take account of the requirements of the people who will use their software if they are to reduce errors and maximize the satisfaction of the users with the system. The user interface can be viewed as part of a dialogue between the user and the system and there are a number of characteristics of good dialogue design that can be used to ensure that the user is supported by the interface and assisted in carrying out their primary task.

It is possible to apply an informal approach to determining characteristics of the users, the task and the situation that will affect the interface design, or to apply a more formal approach using either structured, ethnographic or scenario-based techniques or some combination of these. The main aim of this is to produce software that can be demonstrated to meet the usability requirements of the people who will use it. This may be done in order to ensure compliance with international standards or it may be to meet legal requirements in some countries.

Review Questions

16.1 Think of a computerized information system that you use regularly. This could be a library system, an automated teller machine (ATM) that you use to get cash, a database that you use in your work or any other system that you are familiar with. Write down which elements of the interface support the five tasks listed at the start of Section 16.2.1.

16.2 For each of the elements of the interface that you have listed in Question 16.1, write down your ideas about how they could be improved.

16.3 What is the difference between the dialogue and direct manipulation metaphors?

16.4 Make a list of direct manipulation metaphors that are used in a GUI that you are familiar with. Are there any metaphors that do not work as you might expect?

16.5 What are the four characteristics of good dialogues described in Section 16.2.4

16.6 Figure 16.9 shows the **Yes** button in a dialogue highlighted. What do you think is the risk associated with making this the active button by default?

16.7 For the system that you wrote about in Question 16.1, note down information relevant to the design factors in Figure 16.10.

16.8 List as many differences as you can think of between structured, ethnographic and scenario-based approaches.

16.9 Make your own list of what you think the advantages and disadvantages could be of structured, ethnographic and scenario-based approaches.

Case Study Work, Exercises and Projects

16.A Using a GUI that you are familiar with as an example, try to identify features that you think might be part of the style guidelines for that GUI.

16.B Using the four criteria for good dialogues discussed in Section 16.2.4 evaluate an application that you use regularly. Identify the ways in which it meets these criteria and the ways in which it does not meet the criteria. Suggest ways in which it could be improved.

16.C Write a scenario to describe what is done when Rik Sharma of FoodCo starts to plan staff allocation, based on the interview transcript in Exercise 6.B (Chapter 6). (Make sure that you concentrate on what he does and not on what is done by other staff at other times.)

16.D For the system that you wrote about in Question 16.1, identify measurable objectives that could be used to measure how usable that system is. (You may like to start by thinking about how long it takes you to use it and how many errors you make.)

16.E Find out whether there are any legal requirements on software designers to comply with legislation that covers ergonomics or HCI in your country. Write a short report to summarize these requirements as though you were an analyst reporting to your manager on this legislation.

Further Reading

- Many computer science and information systems courses now include HCI as a subject. If you have not come across HCI before and want to find out more, there are a number of suitable text-books, such as Booth (1989), Preece et al. (1994) and Dix et al. (1998). A classic text in this area is Shneiderman (1997), which has been updated to cover graphical user interfaces in more detail since its first publication.
- Style guidelines are available from some of the largest companies in the industry. Links are available to the on-line versions of these in the book's web site.
- For a structured method for user interface design Dermot Browne's book (Browne, 1994) provides a step-by-step approach to user requirements analysis, task analysis, usability and interface design. Browne also uses Harel statecharts to model the behaviour of the interface in a more thorough way than many authors. Ian Horrocks (1999) provides another view of how to use statecharts in interface design. John Carroll's book (1995) provides a good coverage of scenario-based methods, and is very practical in approach.
- In the UK, Her Majesty's Stationery Office (HMSO) publishes a booklet that explains the requirements of the Display Screen Regulations. This book's website includes links to other resources on health and safety and workstation ergonomics. Australian and Canadian government organizations provide a good starting point for investigating standards and legislation. The US OSHA also has a good website, despite the fact that its proposed ergonomics regulations were rejected by Congress.

CHAPTER 17

Designing Boundary Classes

OBJECTIVES

In this chapter you will learn

- what we mean by the presentation layer
- how prototyping can be applied to user interface design
- how to add boundary classes to the class model
- how to model boundary classes in sequence diagrams
- how design patterns can be applied to the design of the user interface
- how to model the control of the user interface using statechart diagrams.

17.1 Introduction

In Chapters 12 and 13 we introduced the three-tier system architecture. The presentation layer in this architecture contains the boundary classes that handle the interface with the user—usually windows and reports—or with other systems. We shall be concentrating here on interaction with the human user. We can use the techniques and diagrams of UML to model the user interface by adding detail to the boundary classes in the class model. These classes handle the user interface and allow us to design business (or entity) classes that do not contain details of how they will be presented. This enables the reuse of the business classes.

The reasons for adopting a layered architecture are revisited in Section 17.2. Different authors use different terms for the user interface or presentation layer, and these terms are briefly introduced.

Practical example window layouts are used to show how the interaction between the user and the system is designed. There is rarely one right solution to the problem of designing the user interface for a particular application, and prototyping can be used to try out different interface designs (Section 17.3). The UML notation for packages and for package dependency can be used to show how class diagrams can reference classes from reusable class libraries (Section 17.4) and how boundary classes can be placed in separate packages. The window layouts can be modelled as classes if required, but this is not always necessary unless the behaviour of the interface or of graphical objects in the user interface is the subject of the application being developed.

The sequence diagrams developed in Chapter 9 can be extended to include the detail of interaction with the boundary classes (Section 17.5), and the model of the boundary classes is developed iteratively as we increase our understanding of the interaction (Section 17.6). This can include the development of an inheritance hierarchy of boundary classes in order to model common features of dialogues and reports that occur in many use cases.

Patterns can be used to provide generic models for the way that the interaction will work. Many systems written in Smalltalk use the Model–View–Controller (MVC) architecture, which separates the model (domain or business classes) from classes that handle the interaction between user and system. In Section 17.7 we examine the MVC architecture in terms of patterns.

The dynamic behaviour of the user interface is modelled with statechart diagrams (Section 17.8). UML statecharts were used in Chapter 11 to model the response of objects to events that take place during their lifetimes and to show how they change state as time passes. The same notation can be used to show the state of the user interface and how it responds to events such as mouse clicks on buttons or the entry of text into data entry screens. A statechart for the control class that manages the user interface for one use case is developed.

17.2 The Architecture of the Presentation Layer

In Chapter 7 the idea of boundary classes was introduced, and in Chapters 12 and 13 a layered model of the system was presented. The three-tier architecture is a common way to separate out user interface classes from the business and application logic classes and from mechanisms for data storage. There are a number of reasons for doing this, and these are shown in Figure 17.1.

This is not to say that classes should contain no means of displaying their contents to the outside world. It is common practice to include in each class a `print()` method that can be used to test the classes before the presentation layer has been developed. Such methods typically take an output stream (a file or a terminal window) as a parameter and produce a string representation of their attributes on that stream. This enables the programmer to check the results of operations carried out by classes without needing the full system in place. (See Chapter 19 for more on testing.)

The three-tier architecture was discussed in Sections 12.1 and 13.3. Different approaches to object-oriented development use different names for the layers of the three-tier architecture. The Unified Process uses the terms boundary, control and entity classes for the three types of classes, and these are the terms that we have used. Coad and Yourdon (1991) call the presentation layer the Human Interaction Component and keep it separate from what they call the Problem Domain Component, which contains what we have called the entity classes. Developers using Smalltalk to implement systems have for many years adopted a similar approach using the Model–View–Controller (MVC) approach that was described in Chapter 13. In the (MVC) approach a system is divided into three components:

- Model—the classes that provide the functionality of the system;

Logical design	The project team may be producing analysis and design models that are independent of the hardware and software environment in which they are to be implemented. For this reason, the entity classes, which provide the functionality of the application, will not include details of how they will be displayed.
Interface Independence	Even if display methods could be added to classes in the application, it would not make sense to do so. Object instances of any one class will be used in many different use cases: sometimes their attributes will be displayed on screen, sometimes printed by a printer. There will not necessarily be any standard layout of the attributes that can be built into the class definition, so presentation of the attributes is usually handled by another class.
Reuse	One of the aims is to produce classes that can be reused in different applications. For this to be possible, the classes should not be tied to a particular implementation environment or to a particular way of displaying the attribute values of instances.

Figure 17.1 Reasons for separating business and user interface classes.

- View—the classes that provide the display to the user;
- Controller—the classes that handle the input from the user and send messages to the other two components to tell them what operations to carry out.

Whatever approach is chosen in a particular project, all these approaches share the objective of keeping the behaviour of the interface separate from the behaviour of the classes that provide the main functionality of the system. To use the anthropomorphic style of some authors about object-oriented systems, the entity classes 'know' nothing about how they will be displayed.

Taking a three-tier architectural approach does not necessarily mean that the different types of classes will end up running on different machines or even that they will be completely separate. It is useful to distinguish between the logical architecture of the system and the physical architecture. The physical architecture may combine layers of the logical architecture on a single physical platform or it may split logical layers across physical systems. If you are designing Java applets, some of the responsibilities for control will be located in the applet class itself, together with the responsibilities of the boundary class, while other control responsibilities may be located in classes on a server together with entity classes. In a distributed system, the entity classes might exist in different databases on different servers and the control classes would pull the data together from these different sources in order to deliver it to the boundary classes.

For the Agate system, we are going to keep the boundary, control and entity classes separate. The boundary classes will run on the users' machines, while the control classes will be located on servers, and the entity classes will initially be on local servers but may later be distributed in different offices. (We shall discuss mechanisms for achieving this in Chapters 18 and 19.)

In the next few sections, we shall develop the boundary classes by:

- prototyping the user interface,
- designing the classes,
- modelling the interaction involved in the interface and
- modelling the control of the interface using statecharts.

17.3 Prototyping the User Interface

Prototyping was discussed in Chapter 3 as an approach to the development life cycle and in Chapter 6 as a way of helping to establish what the requirements for a system are. In Chapter 6 we used it to produce models of the user interface. The UML models we have produced so far have been analysis and design diagrams; they are rather like an architect's drawings—they represent what the finished product will be like but they do not really show how it will look. A prototype is a model that looks, and to some extent behaves like the finished product, but is lacking in certain features; it is more like an architect's scale model of a new building.

There are different kinds of prototype that can be built. A prototype that only provides a model of the user interface is one example of a *horizontal prototype*. It is horizontal because it deals with one layer of the layered architecture of the system. A *vertical prototype* takes one sub-system of the whole system and develops it through each layer: user interface, business classes, application logic and data storage. A horizontal prototype need not only deal with the user interface; there may be circumstances where it is more important to prototype the middle tier in order to test whether an innovative aspect of the system functionality works correctly.

Another distinction is made between those prototypes that are developed further and eventually by an iterative process become part of the finished system, and those prototypes that are simply used to test out design ideas and then thrown away after they have served their purpose. This second kind of prototype is known as a *throwaway prototype*. A throwaway prototype can be built using any programming language that is suitable for the purpose. Figure 17.2 shows a prototype of the user interface for Agate that was created in a matter of minutes using Microsoft Visual Basic. Although this application will be developed in Java, a visual programming environment such as Visual Basic or Delphi can be used to produce prototypes of the user interface. In fact, the Java examples used in the book take no more time than the Visual Basic examples, as they are also produced in a visual programming environment—Visual Café for Java.

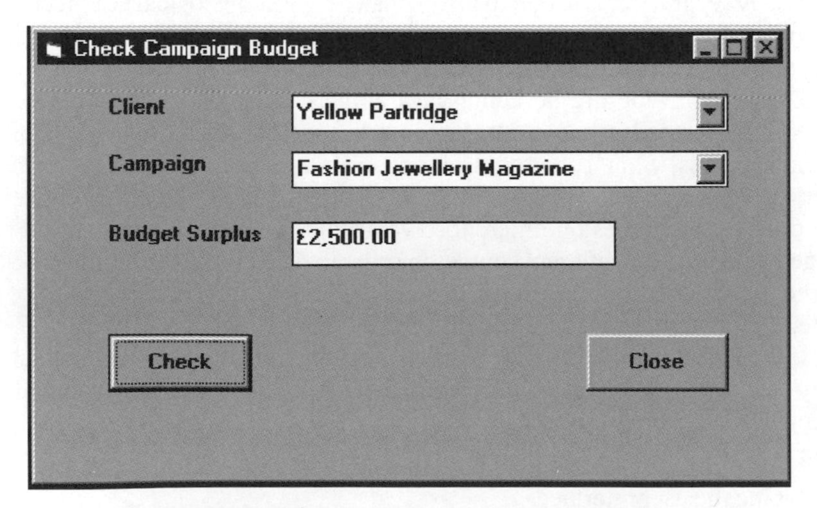

Figure 17.2 Visual Basic prototype of the `Check Campaign Budget` interface.

Visual programming environments can be used to develop prototypes of the user interface to applications. These can be shown to the users, and used, for example, in conjunction with techniques from scenario-based design (described in Chapter 16) to agree with users how they will interact with the user interface. In this way, prototypes can be used to design the interaction between the system and the user and establish a set of guidelines for how this interaction will take place.

Because visual programming environments are so easy to use, developers are often tempted to develop applications from the outside in: starting with the interface and linking the functionality to the visual components that appear on the screen. Without thorough analysis of the business requirements for the system, this can lead to a blurring of the distinction between the presentation layer and the business classes and application logic. Applications developed in this way often have a large amount of program code associated with interface objects such as buttons. This program code should be an operation of a control class or of one or more entity classes. If it is linked to a button it cannot be reused in the same way as if it is carefully encapsulated in a class. Typically, the programmer then needs to reuse the code in another window and copies and pastes the code to another button in the new window, then when a change is made to the code linked to one button it may not be copied to the code linked to the second button, and discrepancies creep into the system. This is not to say that it is not possible to develop good applications in visual environments; the important thing is that a thorough analysis should have been carried out first and the business objects and application logic should be kept separate from the visual components.

Prototyping can be used to try out alternative approaches to the same use case. In the example screen layout shown in Figure 17.2 we have assumed that the users will select first a client and then a campaign from dropdown lists. There are many possible alternatives to this. Three of these are:

- to use a separate look-up window for each class;
- to allow the user to enter part of a name (for example of a client) and for a list of close matches to be returned;
- to use a tree structure which shows the instances of clients and campaigns in a tree-like hierarchy.

Prototyping allows us to experiment with these approaches and build models which the users can try out for themselves. Figures 17.3 and 17.4 show screen shots of prototypes based on two of these different ways of handling the look-up process.

The choice of how the look-up in this use case is handled on screen will be determined by the style guidelines that were discussed in Chapter 16. It is important that style guidelines are agreed before implementation starts; prototyping can be used during design to try out various different interface styles and to get the users' agreement on which style will be followed.

Whichever style is adopted here will be adopted in other use cases in which the user needs to be able to look up clients from a list. The same approach will also be adopted in all use cases in which the user looks up any kind of class. For example, Figure 17.5 shows the same method as in Figure 17.4 being used to look up campaigns for a particular client in the use case for `Add a concept note`.

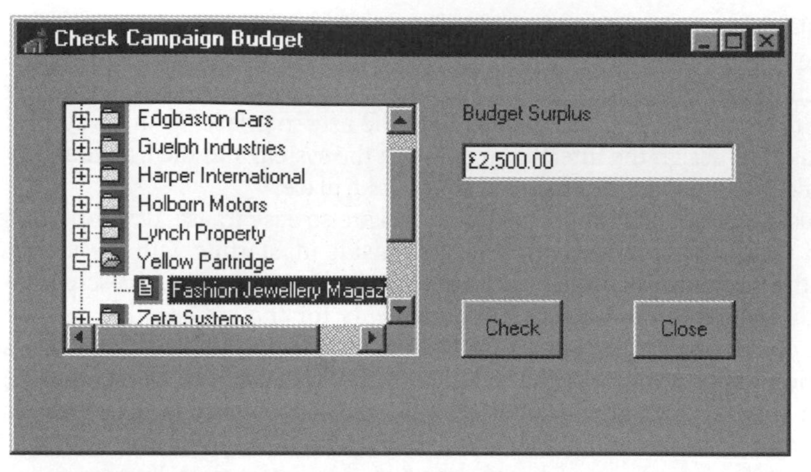

Figure 17.3 Prototype developed in Delphi using TreeView control.

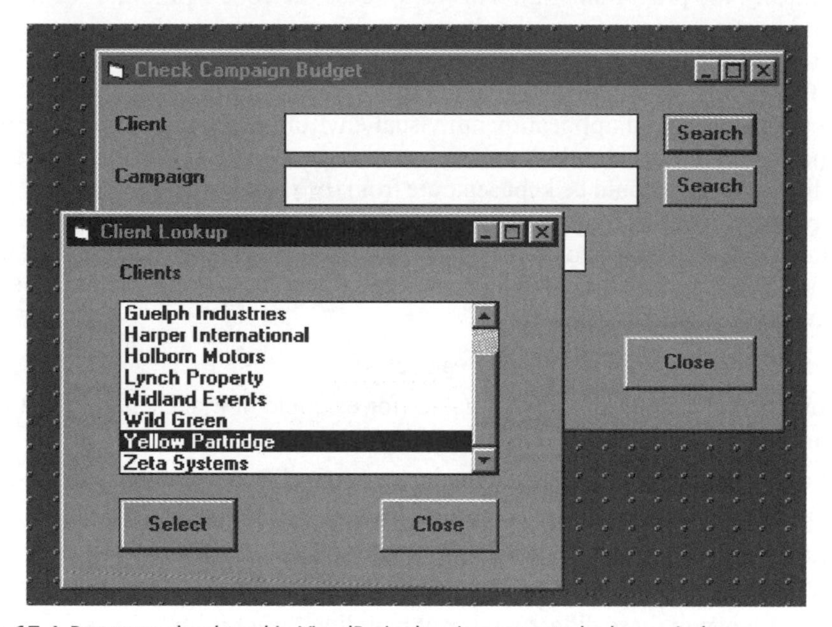

Figure 17.4 Prototype developed in VisualBasic showing separate look-up window.

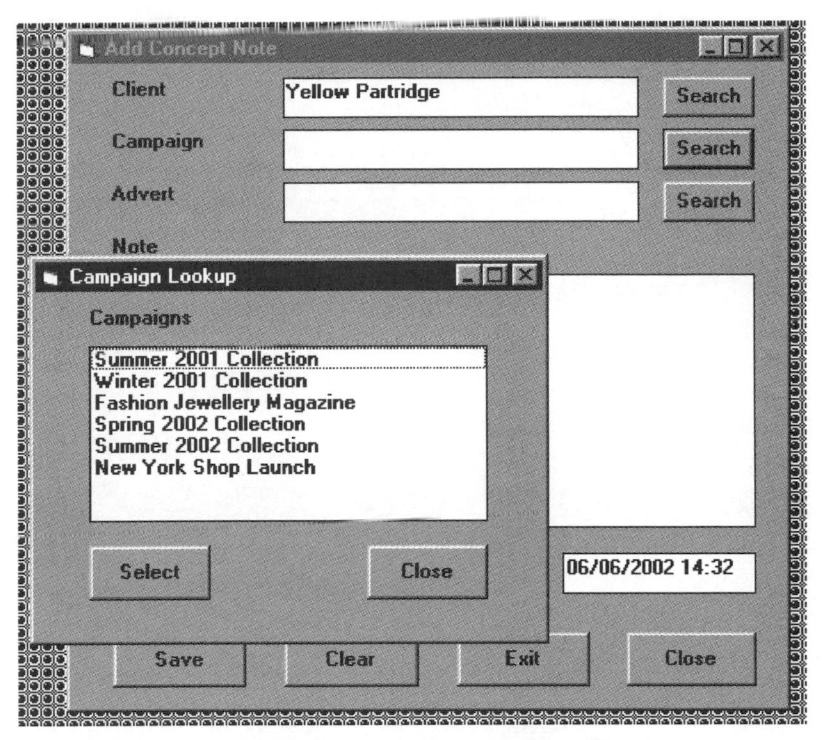

Figure 17.5 Use of the same style of look-up as in Figure 17.4 in a different use case.

17.4 Designing Classes

The next step is to design the classes that will provide the user interface. The use case for `Check campaign budget` is used as an example. At the simplest level, there will be an object that provides an interface onto the functionality of this use case. This could be a dialogue window like the one shown in Figure 17.6 (in the foreground). The analysis collaboration for this use case is shown in Figure 17.7.

In the simple analysis collaboration in Figure 17.7 we have not shown the class `Client` because it does not participate in the main functionality of the use case. In order to calculate what is left in the budget for a campaign we do not need the client object. However, in order to find the right campaign we do need the client: we home in on the right campaign in the context of the particular client. We need to be able to list all the clients and display them in the first dropdown. Once the client has been selected, we then need to list all the campaigns for that client in the second dropdown. In the approach that we have taken in this user interface design, using dropdowns rather than separate dialogue windows, we may want to add further control classes to the collaboration: one to list the clients the other to list the campaigns. This is shown in Figure 17.8.

If we had adopted the user interface style of Figures 17.4 and 17.5, then we should have separate user interface classes for each of these, and the collaboration would look like Figure 17.9.

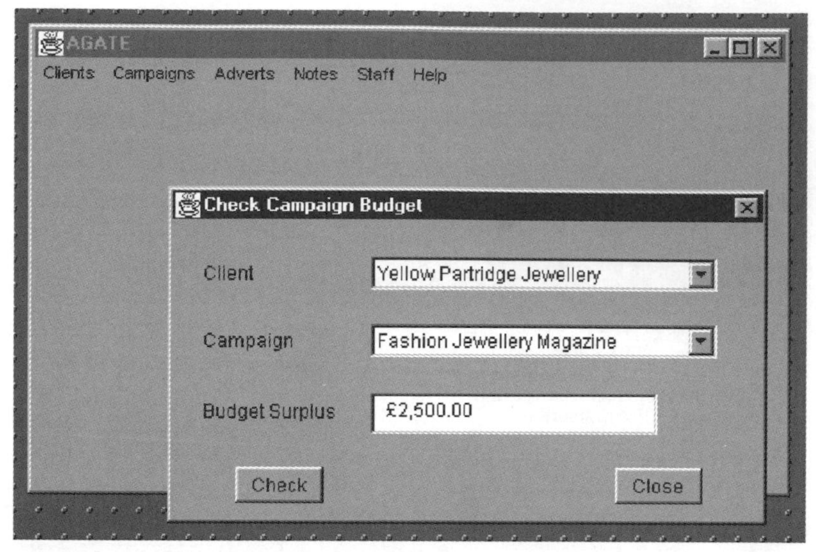

Figure 17.6 Dialogue window for the use case `Check campaign budget`.

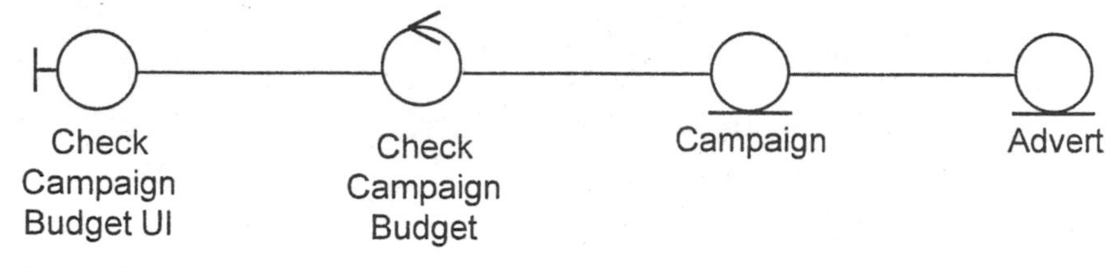

Figure 17.7 Collaboration for the use case `Check campaign budget`.

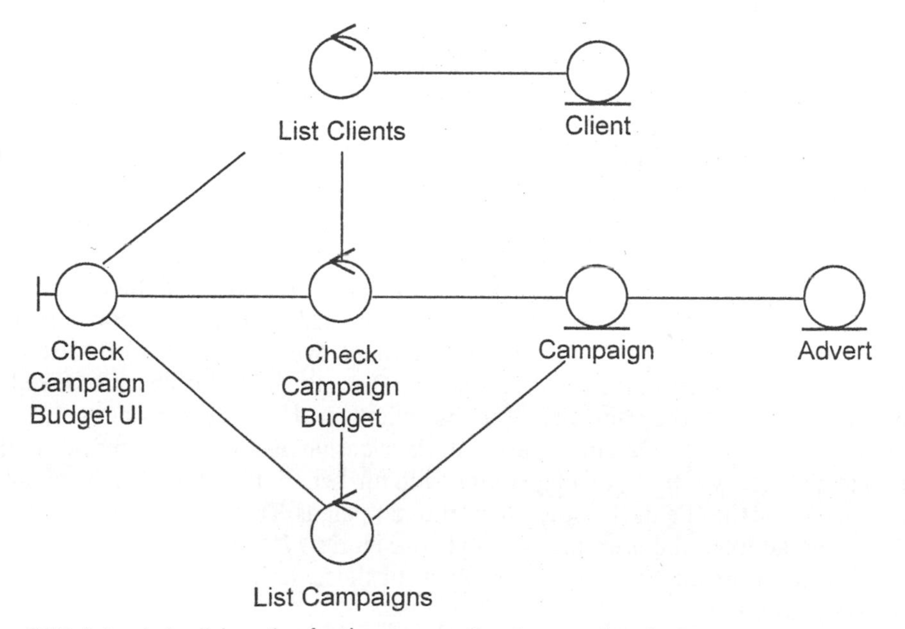

Figure 17.8 Extended collaboration for the use case `Check campaign budget`.

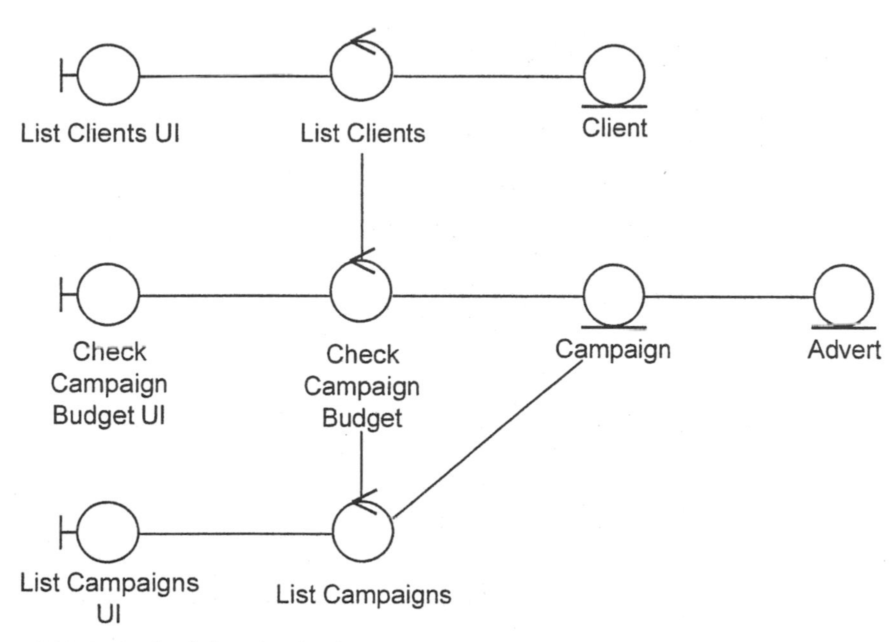

Figure 17.9 Revised collaboration for the use case `Check campaign budget.`

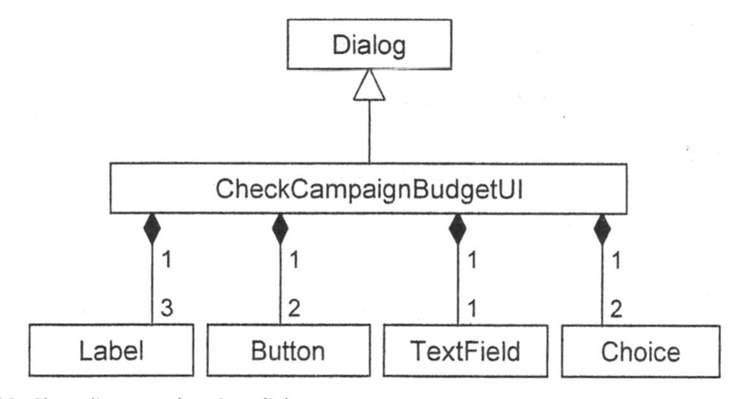

Figure 17.10 Class diagram showing dialogue components.

We will work with the collaboration of Figure 17.8. We shall treat instances of the boundary class `CheckCampaignBudgetUI` as single objects. We may not want to be concerned about the objects that make it up. In reality, this window may well be an instance of a subclass of a class such as `Dialog` that is available in a library of user interface classes, and it may contain a number of components: buttons, labels, dropdowns and a textbox. This can be shown in a class diagram, as in Figure 17.10 (`Choice` is the Java term for a dropdown list). The composition associations represent the fact that the `CheckCampaignBudgetUI` is made up of instances of the other classes. (Alternatively, this can be represented as a class with attributes for each of the components. This is shown in Figure 17.11, and makes it easier to draw a class diagram for the boundary classes.) The component classes that are used here all come from the Java Abstract Windowing Toolkit (AWT). The `CheckCampaignBudgetUI` class is dependent on the classes in the AWT, and this can be shown using packages in a class diagram, as in Figure 17.12.

CheckCampaignBudgetUI
- clientLabel : Label - campaignLabel : Label - budgetLabel : Label - checkButton : Button - closeButton : Button - budgetTextField : TextField - clientChoice : Choice - campaignChoice : Choice

Figure 17.11 Class for dialogue window showing dialogue components as attributes.

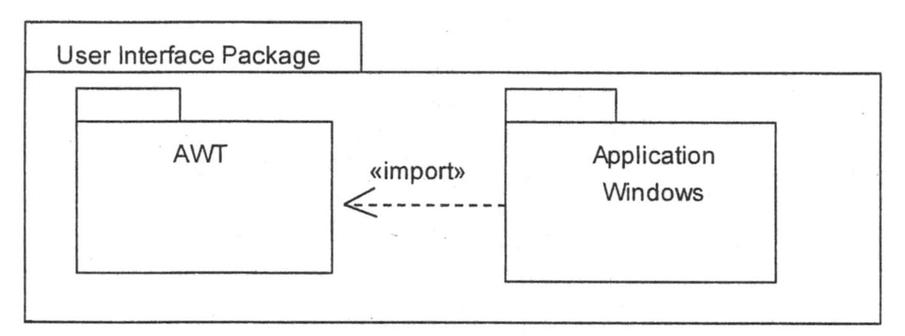

Figure 17.12 Class diagram to show dependency between classes in different packges.

The «import» stereotype on the dependency shows that it will be necessary to import the classes from this AWT package in order to make them available to be used with the classes in the Application Windows package. How this is done will depend on the language that is used for implementation. In Java it is done simply with a line of code:

```
import java.awt.*;
```

In Microsoft's new C# this is done with a using statement:

```
using System.WinForms;
```

We have used the «import» stereotype, as it is one of the standard UML elements listed in Appendix A of the UML Specification (OMG, 2001).

Figure 17.13 illustrates how classes from other packages can be shown in the class diagram by adding the pathname of the package, followed by two colons, to the name of the class. The Java AWT is only one example of this. Most object-oriented programming languages or development environments are provided with *class libraries* that contain many of the classes that are needed to build a working system. Microsoft, for example, provides the Microsoft Foundation Classes (MFC) which include all the classes such as buttons and text fields that are required to build a Windows interface to an application. These classes are grouped together and provided in what UML terms packages. This is an example of the reuse that is claimed as a benefit of object-oriented systems. These user interface classes, whether in Java or Visual C++ or another language, have been implemented by other developers and can be reused in many new applications.

If the application being designed is mainly concerned with the behaviour of the objects in the interface itself, for example a drawing package, a CASE Tool or an application with a strong visual element, then it may be advisable to model the user interface

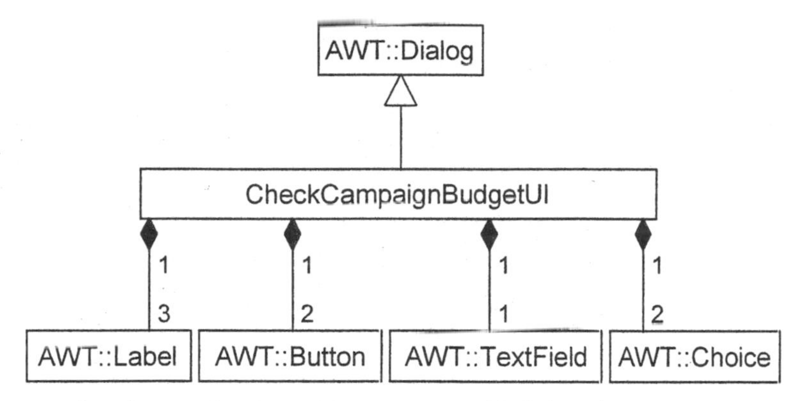

Figure 17.13 Class diagram showing AWT components with their package name.

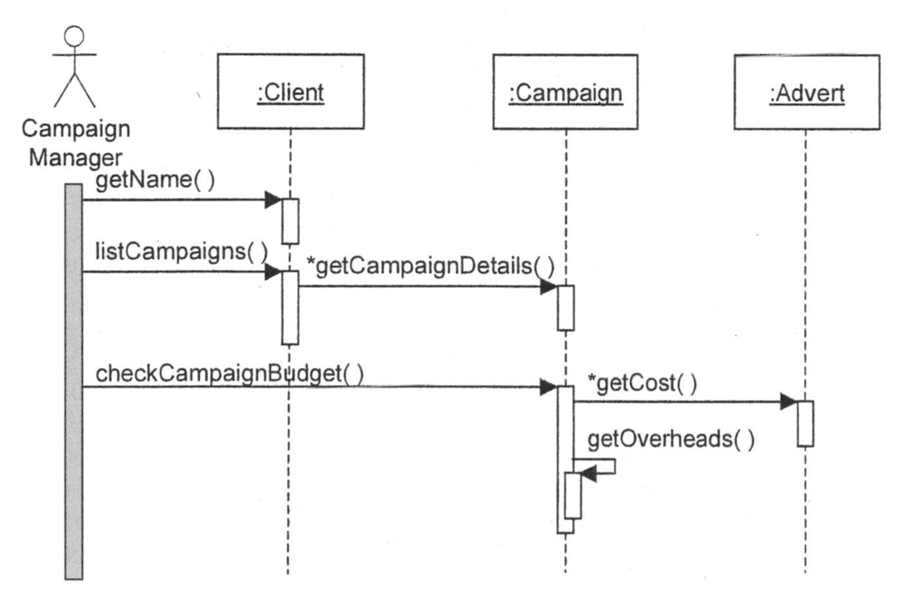

Figure 17.14 Sequence diagram for use case `Check campaign budget`.

using a class diagram as shown in Figure 17.13. In most applications in which the user interface will display text and numbers, it is not necessary to produce a model of the classes that make up the interface. It may be useful to show them in the style of Figure 17.11 in a class diagram in a separate package. Note that there are not normally associations among the classes in the interface package in the way that there are associations between entity classes in the domain model.

17.5 Designing Interaction with Sequence Diagrams

The sequence diagram for the use case `Check campaign budget` in Figure 10.8 did not show the boundary or control classes, but concentrated on the operations of the entity classes. We have shown it again in Figure 17.14. The collaboration diagram of Figure A3.8 showed the boundary and control classes, and we shall now elaborate the interaction in more detail.

We need to add the boundary and control classes of the collaboration in Figure 17.8 to this sequence diagram. (If we had adopted the design from Figure 17.9, then we would need to add three boundary classes rather than one.) Rather than trying to draw the entire sequence diagram in one go, we shall build up the interaction step by step. We are assuming here that it is an instance of the control class `CheckCampaignBudget` that is created first and that this creates a new instance of the `CheckCampaign BudgetUI` class to handle the user interface. As soon as it has created the boundary class, the control class needs to have the first dropdown populated with the names of all the clients, so it creates an instance of the control class `ListClients` and requests it to pass back the client names to the boundary class, passing it a reference to the boundary class in the message `(ccbUI)`. The instance of `ListClients` sends the message `addClientName(name)` repeatedly to the boundary class until it has finished. It then returns control to the `CheckCampaignBudget` instance and destroys itself. The main control class can now enable the boundary class, allowing the user to select a particular client. This is shown in Figure 17.15.

We have shown the names of return values and parameters in this diagram to illustrate what is happening. For example, the instance of `CheckCampaignBudget` needs to have a reference to the instance of `CheckCampaignBudgetUI (ccbUI)` so that it can pass it to the instance of `ListClients`. This enables `:ListClients` to send the `addClientName` message directly to `:CheckCampaignBudgetUI`. When `:Check CampaignBudgetUI` is created, it is passed a reference to the control class instance `:CheckCampaignBudget (this)`. In this way, it will be able to send messages back to the main control class when it needs to notify it of events.

Instances of `CheckCampaignBudgetUI` need to be able to respond to the message `addClientName(name)`. Many other boundary classes will need to allow the user to select a client from a dropdown, for example `AddConceptNoteUI`. We shall want to reuse `ListClients` in all the use cases where a list of clients has to be displayed in a boundary class, but cannot expect `ListClients` to know about all the different boundary classes to which it could send the message `addClientName(name)`. We can use the idea of an interface to specify the operations that all these boundary classes must respond to. (See Section 14.3 for an explanation of interfaces.) We could define

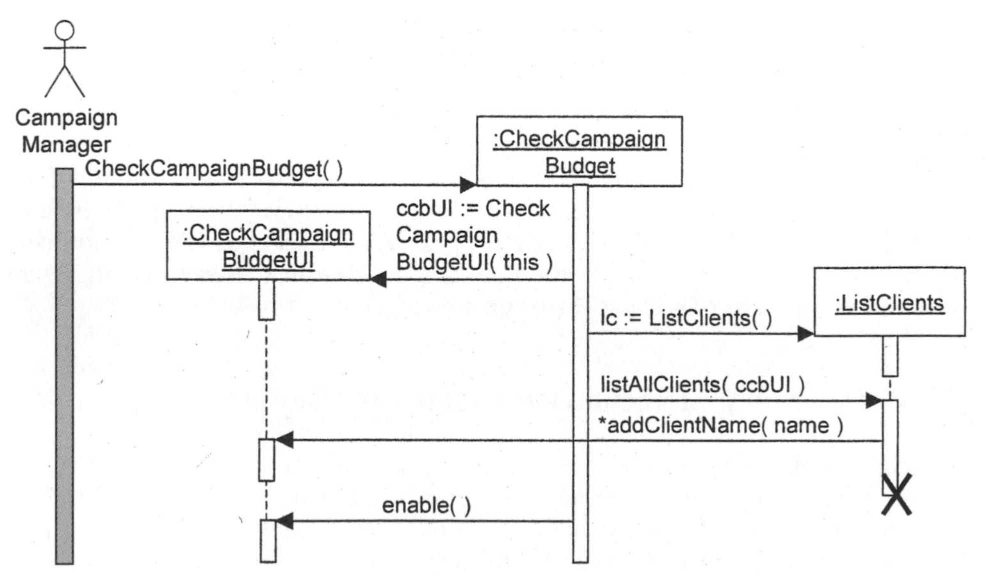

Figure 17.15 First part of detailed interaction for use case `Check campaign budget`.

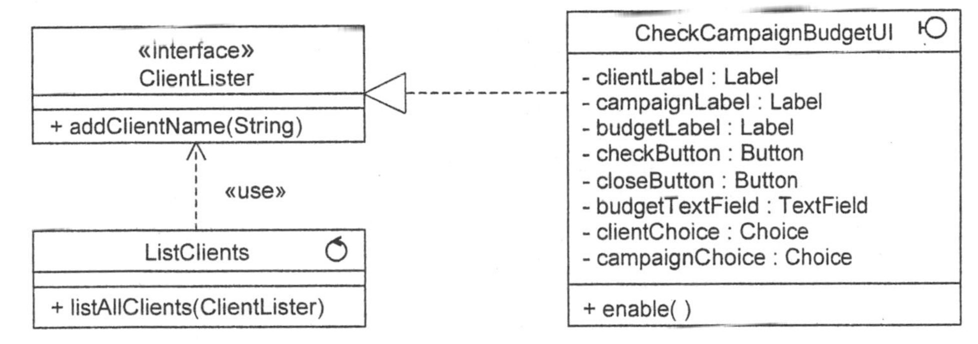

Figure 17.16 `ClientLister` interface.

```
import java.awt.*;
public class CheckCampaignBudgetUI extends Frame
implements ClientLister {
private Choice clientChoice;
...
  public void addClientName(String name) {
    clientChoice.addItem(name);
  }
...
}
```

Figure 17.17 Possible Java definition of `addClientName(name)`.

the interface `ClientLister`, and all the boundary classes that need to display a list of clients will realize it, as `CheckCampaignBudgetUI` does in Figure 17.16.

Design of other use cases may identify other operations that must be part of this interface, for example `clearAllClientNames()` or `removeClientName(name)`. The operation `addClientName(name)` must be implemented by program code in the class definition for `CheckCampaignBudgetUI`. For example, in Java it may be as in Figure 17.17 (with much code not shown).

One of the important features of this design is that any object that wants to manipulate the list of clients in the boundary object must do so through operations. We could change the design of the boundary class so that it displayed the list of clients in a scrolling list (a `List` in Java). We would then have to change the Java program that implements `CheckCampaignBudgetUI`, but we would not have to change the classes that use it, because the interface remains the same. (Other object-oriented languages use slightly different ways of achieving the same thing, but the principle is the same.)

Note that in Figure 17.15 we have not shown what the `ListClients` control class instance does to get the names of all the clients. This level of detail can be hidden in this diagram. There is no notation within UML to show that more detail is to be found in another sequence diagram, but we could include a note in the diagram. Figure 17.18 shows what happens when an instance of `ListClients` receives a `listAllClients` message. At this stage of the design we have not addressed how clients are to be stored in a database, so this is left vague. Design of the data storage is covered in Chapter 18. The rectangle with a condition below it in Figure 17.18 indicates that everything within it is repeated.

This kind of access to objects in a set of objects can be shown by the *multiobject* notation in a collaboration diagram. This is shown in Figure 17.19. The two superimposed rectangles represent the set of instances of `Client` that are being accessed in turn. The `getName()` message is then sent to each instance in turn.

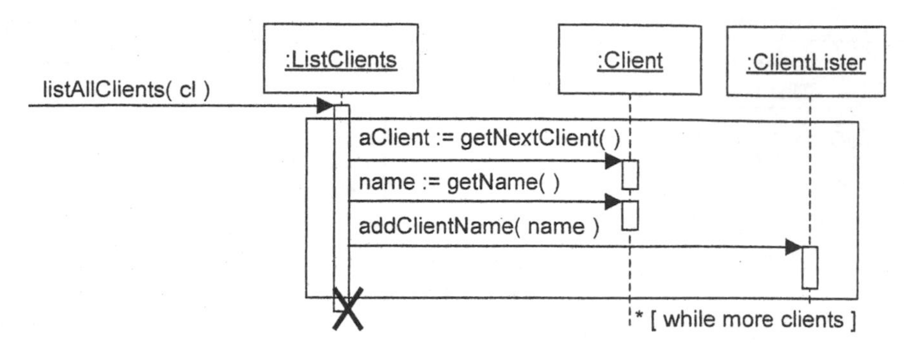

Figure 17.18 `listAllClients()` operation.

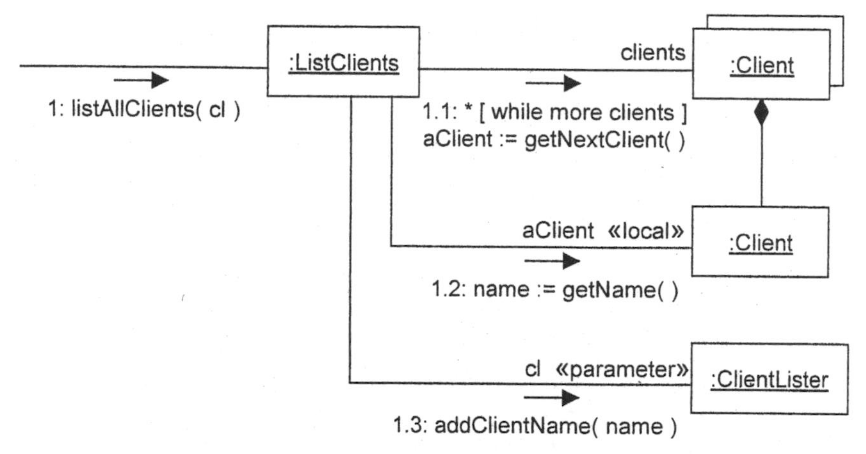

Figure 17.19 `listAllClients()` operation in a collaboration diagram.

Rosenberg and Scott (1999) use control classes as placeholders for operations that may later be assigned to other classes. This technique could be used in this example, where the `ListClients` control class may become part of some other class that handles the database access for the `Client` class. This approach can help in preventing an object-oriented system from ending up with a lot of classes like `ListClients` that are little more than wrappers for one or two operations, and more like programs than objects. On the other hand, this can break the architectural layering of the system, as the classes in the data storage layer will need to know about classes in the presentation layer.

Returning to the sequence diagram for this collaboration, the next event that will take place is that the user will select a client from the dropdown that has been populated with clients by the processes that we have just described. When a particular client has been selected, then the list of campaigns in the boundary class must be populated with only those campaigns in the database that belong to that client. This is shown in Figure 17.20.

Part of this sequence diagram is similar to the one in Figure 17.15. We would apply the same design principles to this, with a `CampaignLister` interface, which must also be implemented by `CheckCampaignBudgetUI`. The interaction is slightly different, as we need to pass the client to `ListCampaigns` so that it knows which client's campaigns to display. We can model the interaction within the boundary class in more detail if we wish.

When the instance of `CheckCampaignBudgetUI` is created it will add the instance of `Choice`, `clientChoice`, to itself and register an interest in events from the user that

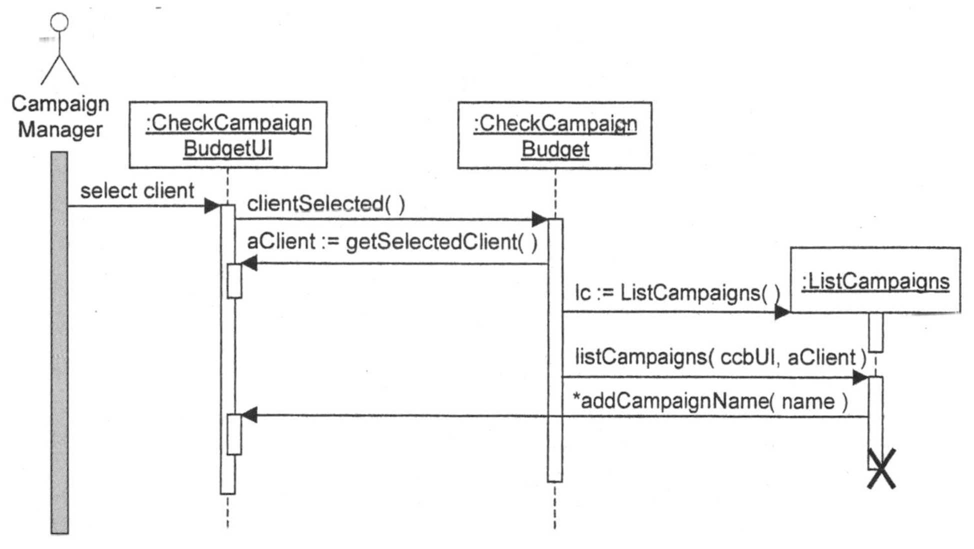

Figure 17.20 Second part of interaction for use case `Check campaign budget`.

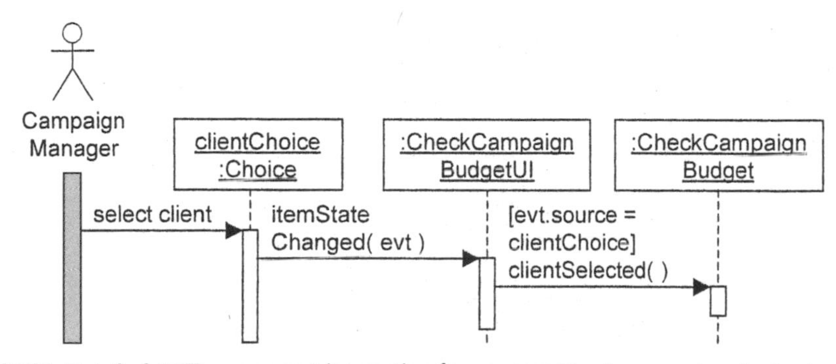

Figure 17.21 Detail of AWT component interaction for use case `Check campaign budget`.

affect `clientChoice`. (A `Choice` is a dropdown list.) In Java, for example, this means that `CheckCampaignBudgetUI` must implement the `ItemListener` interface. When an event takes place, like the user selecting a client in the dropdown, `:CheckCampaign BudgetUI` will be sent a message `itemStateChanged(evt)`, with the data associated with the event passed in the parameter `evt`. If the source of the event is `clientChoice`, then it should notify the control class by sending it the `clientSelected` message. Figure 17.21 shows this interaction.

At this point in the interaction, the user could either select a campaign from the list of campaigns or could select another client from the client list. In the latter case, then the interaction of Figure 17.20 could take place again. With the design as it is at present, this would result in the campaigns for the newly selected client being added onto the list of campaigns already in the dropdown. This is clearly incorrect, and this is where the interface needs an operation to clear the list: `clearAllClientNames()` was suggested for the `ClientLister` interface. The equivalent `clearAllCampaign Names()` is shown in Figure 17.22.

If the user does select a campaign, then the button, `checkButton`, should be enabled, and if the user clicks that, then the budget for the selected campaign should be calculated and displayed in the textfield. This is shown in Figure 17.23. We have not shown all the parameters and return values in this diagram, as it is very easy to clutter up such a diagram with additional text.

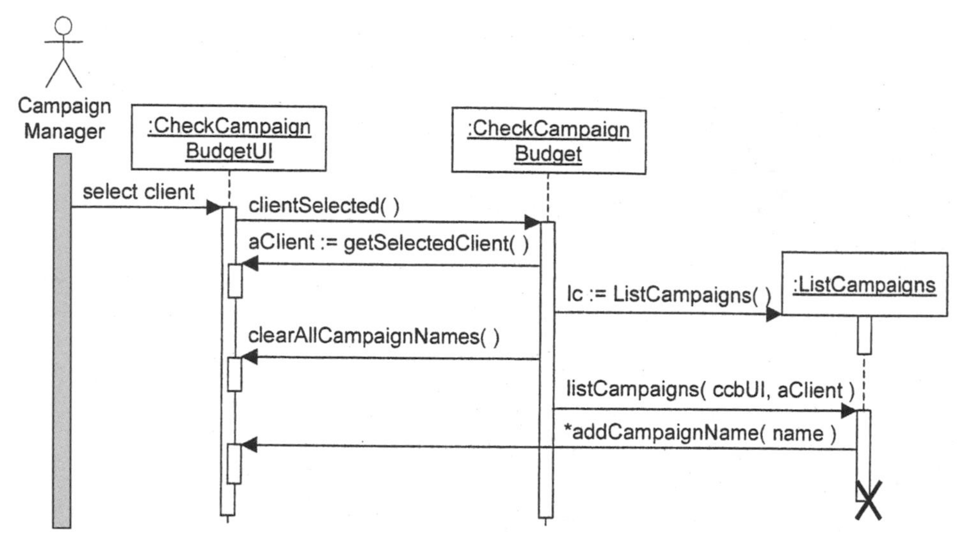

Figure 17.22 Revised second part of interaction for use case `Check campaign budget`.

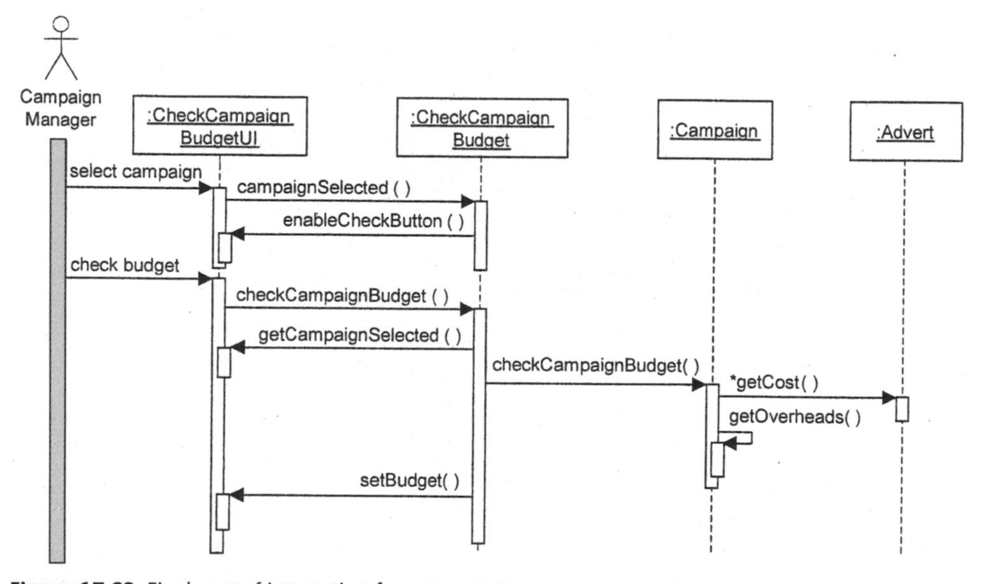

Figure 17.23 Final part of interaction for use case `Check campaign budget`.

In the same way as we did not show the detail of the interaction within the boundary class on the earlier diagrams, we have left the detail of how `:CheckCampaignBudgetUI` will get the values from the dropdowns and set the value of the textfield out of this diagram. This provides us with a clean interface, and `:CheckCampaignBudget` need know nothing of the internal workings of the boundary class. The display of the budget amount could be changed from a textfield to a label, or even digital speech, but the control class only needs to know that the boundary class will respond appropriately to the `setBudget` message.

There may be many places in the system where the same patterns of interaction as in Figures 17.18 and 17.21 take place. Rather than producing separate interaction sequence diagrams for all of these, we may choose to produce some generic sequence diagrams that show the pattern of interaction that is expected to take place when objects are listed in a dropdown or when an item is selected from a list.

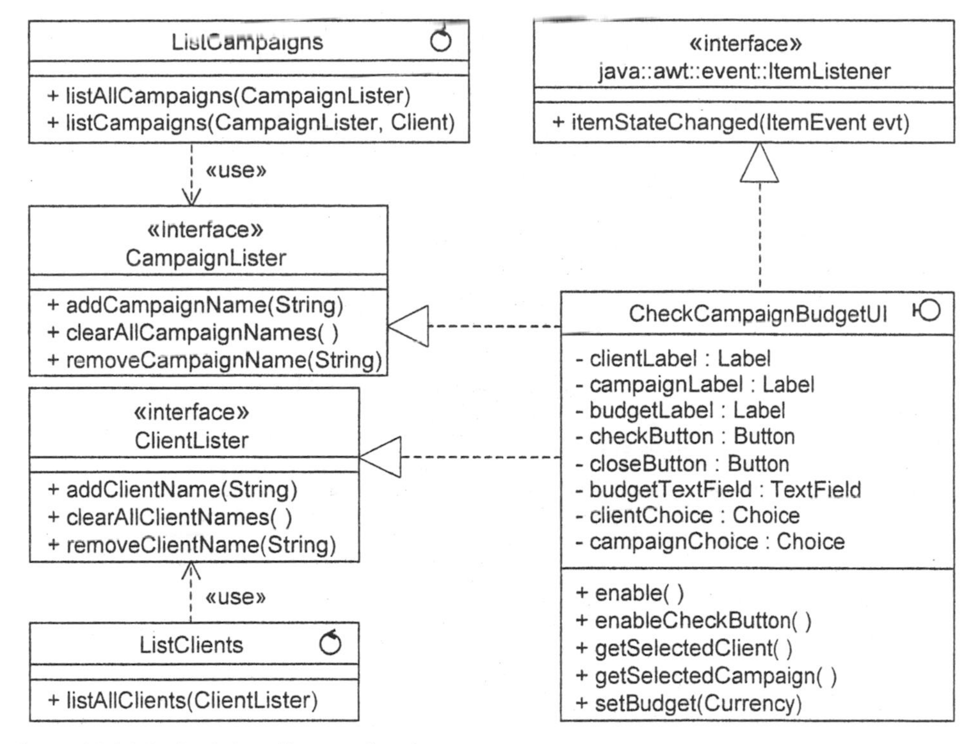

Figure 17.24 Revised class diagram showing `CheckCampaignBudgetUI`.

We can further extend the class diagram with the operations that have been identified for our boundary class. Figure 17.24 shows this for the `CheckCampaignBudgetUI` class. The code that manages the interface is the implementation of an operation of `CheckCampaignBudgetUI` that is invoked by the control class. It also shows the interfaces that it must realize, both some that are specific to this application and one that is part of the Java AWT event handling model.

Boundary classes include reports as well as screen displays, and reports can be shown in interaction diagrams as well. The simplest form of report is produced by opening an output stream to a device such as a printer. In this case each object that is to be printed out can be sent the message to print itself with a reference to the output stream as a parameter. This is shown in Figure 17.25 for a simple report of all clients, where a control class co-ordinates the printing of the report. (We have used collaboration diagrams for the following diagrams to show how they can be used for this kind of design as well.) If the user is required to enter parameters, for example selecting a client in order to print a report of all campaigns for that client, then a dialogue of some sort will also be required.

In the simple solution of Figure 17.25, the instances of `Client` are responsible for formatting themselves for output to the printed report. An alternative solution is to design a boundary class to handle the formatting of the attributes of one or more instances of one or more classes. This solution is shown in Figure 17.26.

The decision as to which of these approaches is best will depend on the development environment being used.

Control classes to manage the user interface like this are not always necessary. In Java applets, the Applet class can handle both the presentation and the control. In

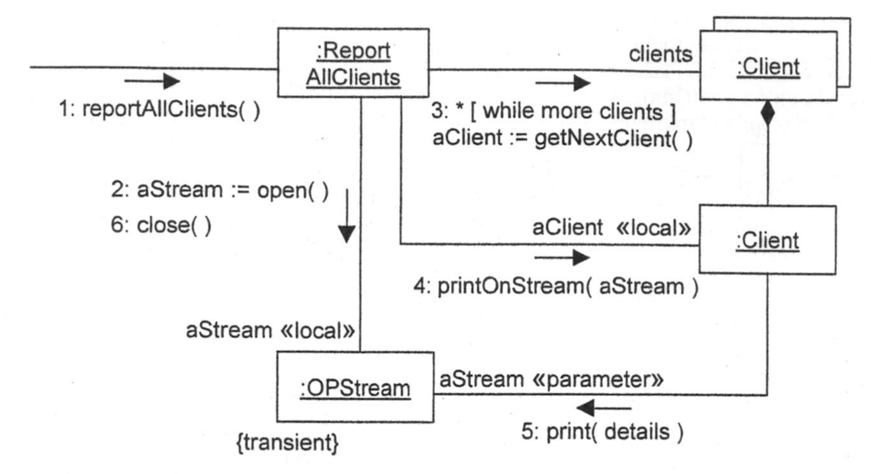

Figure 17.25 Report design in which the `Client` formats its own data.

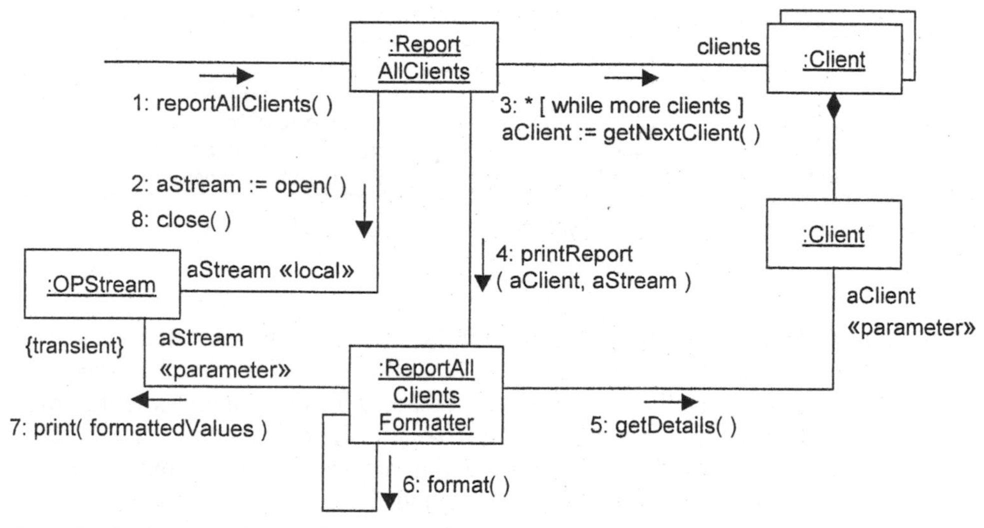

Figure 17.26 Report design in which a boundary class formats the data.

some applications, for example real-time applications where the attributes of objects are changing and those changes need to be reflected in different views of the data, it may be better to use a design based on the Smalltalk Model–View–Controller style of developing applications, which is discussed in Section 17.7.

17.6 The Class Diagram Revisited

The boundary classes can be added to the class diagram. They can be shown in a single diagram for the Application Windows package (see Figure 17.12) or in separate diagrams, grouped by type or subsystem. The buttons and other classes that are used to make these interface classes need not be shown in the diagrams. However, all the boundary classes in the application will have a package dependency to the package where the buttons and other classes are held.

There may be some commonality among boundary classes that can be abstracted out into an inheritance hierarchy. For example, all dialogue boxes that are concerned with printed reports may have a standard set of buttons, Print, Cancel and Select Printer, and radio buttons for Portrait and Landscape. These buttons and the associated event-handling mechanisms could be placed in a generic `PrintDialog` superclass, from which all other report dialogue boxes could inherit this functionality. If we had chosen the design for the lists of clients and campaigns that required a separate dialogue box for each, as in Figures 17.4, 17.5 and 17.9, then we would need to add a `LookupDialog` class which would be the parent of all look-up dialogues in the system. This is in turn a subclass of the Dialog class from the Java AWT package. The `Dialog` class is also the superclass of `PrintDialog`, the parent of all dialogues used to run reports. Figure 17.27 shows the beginnings of a possible inheritance hierarchy based on this approach. Note that LookupDialog and PrintDialog are both abstract classes: there will never be instances of either.

17.7 User Interface Design Patterns

We have suggested that in an application of this sort generic patterns of interaction can be documented for the design of the user interface and the boundary classes. It is also possible to use standard design patterns as we discussed in Chapter 15.

The Model–View–Controller (MVC) architecture is the classic object-oriented pattern for designing the user interface. It has been mentioned more than once in this chapter, although we have not used it here, and was described in Chapter 13. The Model class is the business object, the View is its presentation to the user and the Controller defines how the user interface responds to the user's actions.

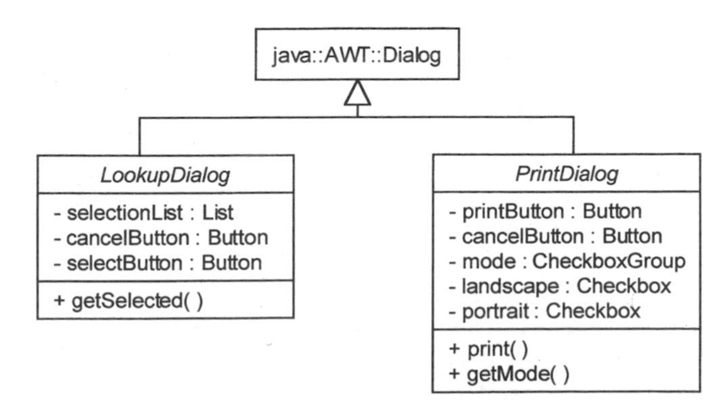

Figure 17.27 Beginnings of an inheritance hierarchy for possible dialog classes.

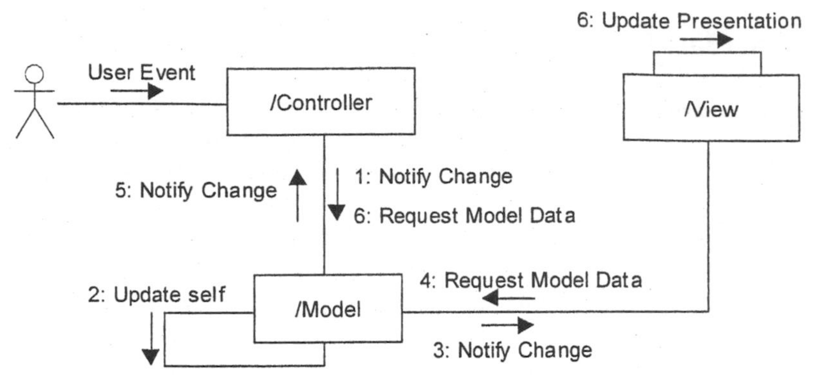

Figure 17.28 Model–View–Controller response to an external event.

This pattern is extensively used in Smalltalk systems, and the Java Swing libraries can also use MVC. In Smalltalk, Model, View and Controller are all available as classes from which developers can subclass the classes they need for their applications. These subclasses then inherit methods that provide for communication between the three components. When the controller detects interaction from the user it sends a message that is received by the relevant model class and tells it to modify itself; when a model class changes, the change is notified to the controller and the relevant view classes; the views then check with the model class to find out what has changed and update the data that is presented to the user. This sequence of interaction is shown in Figure 17.28 as a collaboration diagram using roles for the three elements. (Compare this with Figure 13.12.)

In the Smalltalk implementation of MVC, each Model class has an attribute that refers to a Controller. For each instance of a Model, this attribute can refer to an instance of a Controller, and that Controller can be any Controller subclass designed by the developers of an application. When an instance of a Model is created, it can be given a reference to the Controller that it must notify whenever any of its data changes. Similarly the Model is told which View or Views it must notify when it changes. Once the mechanism by which this operates is understood, and the links between the instances of the different classes are set up, this is relatively transparent in operation.

Java provides two similar mechanisms: first the Observer interface and Observable class, and second the EventListener subinterfaces and EventObject subclasses. The latter mechanism is used in the user interface and there are a number of subclasses of EventObject for different types of event, such as `MouseEvent` for mouse events and `ItemEvent` for checkboxes and lists, and different subinterfaces of EventListener to handle them (such as `MouseListener`, `ItemListener` and `ActionListener`). We used this mechanism to handle the event when the user selects a client in Figure 17.21. Because it involves a dropdown list (a `Choice`), we use `ItemEvent` and `ItemListener`.

Any class that implements the `ItemListener` interface must implement the method `itemStateChanged()`; this means that it must include code so that its instances can respond appropriately to the message `itemStateChanged(ItemEvent)` whenever they receive it. Any class that implements `ItemListener` can register its interest in the events that affect an instance of certain user interface components, such as the `Choice` and the `Menu` classes. This is done by sending the message: `AddItem Listener(ItemListener)` to an instance of one of these classes. It adds the new instance of `ItemListener` to a list it keeps of all the `ItemListeners` it must notify of any events. Then when the Menu is selected or the `Choice` is changed, it sends the

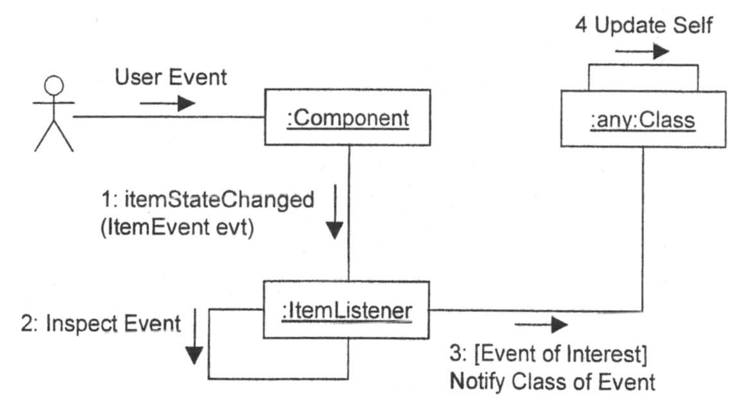

Figure 17.29 Java `ItemListener` response to an external event.

message `itemStateChanged()` to each of the objects in its list and passes them an `ItemEvent` object as a parameter. Each of the objects that receives the message can then inspect the `ItemEvent` object and decide whether it is interested in that particular event and whether it needs to take some action in response. The action could be to notify other classes of the event, particularly if the class that is implementing the `ItemListener` interface is a control class. This is shown in Figure 17.29. The use of the interface mechanism in Java means that the user interface component does not need to know the actual class of the class that implements `ItemListener`; as long as it implements the interface it will know what to do when it is sent the `itemStateChanged()` message.

Gamma et al. (1995) describe the MVC architecture in terms of three patterns: the Observer, the Composite and the Strategy design patterns.

- The Observer pattern provides a mechanism for decoupling business objects from their views. It is based on a publish–subscribe model. There are two main types of objects: Observers which are the View objects and Subjects which are the business objects. Observers can subscribe to a subject and ask to be notified of any changes to the subject. When a change takes place in the subject, it publishes information about the changes to all the observers that have subscribed to it. This is the basis of the Observer–Observable approach adopted in Java 1.1 (and later versions of Java). The Observer pattern is the core of the MVC architecture, but two other patterns also apply.
- The Composite pattern provides a means to structure objects into whole-part hierarchies. Most windowing interfaces use this approach to form composite views made up of smaller components. The class diagram of Figure 17.10 shows this kind of composite structure. The Composite pattern is a way of structuring views to include nested views. For example, a graphical display may represent the same model in two different views, a graph and a table of numbers that may be nested within another view.
- The Strategy pattern offers a way of defining a family of algorithms, encapsulating each one, and making them interchangeable. This allows the strategy to vary independently of client objects that use it. In MVC one controller can be replaced by another. For example, an on-screen calculator could be designed and built initially to respond to mouse clicks on the buttons on the screen. The controller for the calculator detects mouse clicks and notifies the internal application objects and the view objects about these mouse events. Later the calculator could be modified to

respond instead to keypresses on the computer numeric keypad, or to both mouse clicks and keypresses. The controller object can be replaced without having any impact on the application objects.

Design patterns provide a way of reusing design experience and best practice. Separating the concerns of the user interface and the control of interaction from the business application objects is a proven technique to produce good designs and maximize the opportunities for reuse of the classes that are designed and developed.

17.8 Modelling the Interface Using Statecharts

In Figure 17.10 we showed the classes that make up the user interface for the use case `Check campaign budget` in the Agate project. That diagram shows the static structure of the interface, but it provides no information about how it will behave in response to the user. The sequence diagrams developed in Section 17.6 and the prototype in Figure 17.6 provide additional information, but they tell us nothing about the permitted states of the interface. The sequence diagrams show the sequential view of the user working through the fields on the screen from top to bottom, but it is in the nature of GUI interfaces that the user can click on interface objects out of sequence. So what happens if the user clicks on the Check button before a client and a campaign have been selected? The user may choose to check more than one budget. What happens if they select a different client—how does that affect the other fields where data has already been selected? All these issues can be modelled using a statechart diagram. Statecharts were introduced in Chapter 11, and were used there to model the way that events affect instances of a class over its lifetime. They can also be used to model the short-term effects of events in the user interface. Browne (1994) uses statecharts in this way to model the user interface as part of the STUDIO methodology. Horrocks (1999) uses statecharts in a more rigorous way than Browne in his user interface–control–model (UCM) architecture and relates the use of statecharts to coding and testing of the user interface. Browne's approach leads to a bottom-up design of the interface, assembling statecharts for components into a complete model of an interface; Horrocks develops his statecharts in a top-down way, successively introducing nested substates where they are necessary. We are using Horrocks' approach in what follows.

For the example that follows, we are using the original design for the user interface with dropdowns for `Client` and `Campaign`, as in the prototype of Figure 17.6.

As a design principle in our user interfaces, we want to prevent users from making errors wherever possible rather than having to carry out a lot of validation of data entry in order to pick up errors that have been made. One way of doing this is to constrain what users can do when they are interacting with the interface. For example, in the `Check campaign budget` user interface it makes no sense to click the check button until both a client and a campaign have been selected. Rather than check whether a client and campaign have been selected every time the button is clicked, we can choose only to enable the button when we know that both have been selected. To do this we need to model the state of the user interface and it is this that we model using statecharts. This process involves five tasks.

- Describe the high-level requirements and main user tasks.
- Describe the user interface behaviour.

- Define user interface rules.
- Draw the statechart (and successively refine it).
- Prepare an event–action table.

We have simplified Horrocks' approach here. His book (Horrocks, 1999) provides a full and clear exposition of this approach.

Describe the high-level requirements and main user tasks

The requirement here is that the users must be able to check whether the budget for an advertising campaign has been exceeded or not. This is calculated by summing the cost of all the adverts in a campaign, adding a percentage for overheads and subtracting the result from the planned budget. A negative value indicates that the budget has been overspent. This information is used by a campaign manager.

Describe the user interface behaviour

There are five elements of the user interface: the client dropdown, the campaign dropdown, the budget textfield, the check button and the close button. These are shown in Figure 17.6.

The **client dropdown** displays a list of clients. When a client is selected, their campaigns will be displayed in the campaign dropdown.

The **campaign dropdown** displays a list of campaigns belonging to the client selected in the client dropdown. When a campaign is selected the check button is enabled.

The **budget textfield** displays the result of the calculation to check the budget.

The **check button** causes the calculation of the budget balance to take place.

The **close button** closes the window and exits the use case.

Define user interface rules

User interface objects with constant behaviour

- The client dropdown has constant behaviour. Whenever a client is selected, a list of campaigns is loaded into the campaign dropdown.
- The budget textfield is initially empty. It is cleared whenever a new client is selected or a new campaign is selected. It is not editable.
- The close button may be pressed at any time to close the window.

User interface objects with varying behaviour

- The campaign dropdown is initially disabled. No campaign can be selected until a client has been selected. Once it has been loaded with a list of campaigns it is enabled.
- The check button is initially disabled. It is enabled when a campaign is selected. It is disabled whenever a new client is selected.

Entry and exit events

- The window is entered from the main window when the `Check campaign budget` menu item is selected.

- When the close button is clicked, an alert dialogue is displayed. This asks 'Close window? Are you sure?' and displays two buttons labelled 'OK' and 'Cancel'. If 'OK' is clicked the window is exited; if 'Cancel' is clicked then it carries on in the state it was in before the close button was clicked.

Draw the statechart

At the top level, there are three states the application can be in. It can be in the `Main Window` (and we are assuming that this is modelled in detail elsewhere), in the `Check Budget Window` or in the `Alert Dialogue`. Figure 17.30 shows these top-level states.

Horrocks uses the convention of names of buttons in single quotes to represent button press events. We have used that notation here for the values that will be returned from the alert dialogue, but have used operation signatures for the other events, as we want to be able to check them against the sequence diagrams.

Within the `Check Budget Window` state, there are different substates of the user interface that must be modelled. Initially, no client is selected, then the user can select a client. Figure 17.31 shows the resulting two states. Because there will be actions associated with the user selecting a different client, we have shown the `clientSelected()` event returning to the `Client Selected` state if it occurs again.

Similarly, once the interface is in the Client Selected state, it can either be in the substate where a campaign has not yet been selected or the substate where a campaign has been selected. These two states are shown in Figure 17.32.

If the user interface is in the `Campaign Selected` state, then if the check button is pressed, the result will be displayed in the textfield, which will initially be blank. This is shown in Figure 17.33.

These various statecharts can be combined and nested within the top-level state for the `Check Budget Window`. This is shown in Figure 17.34.

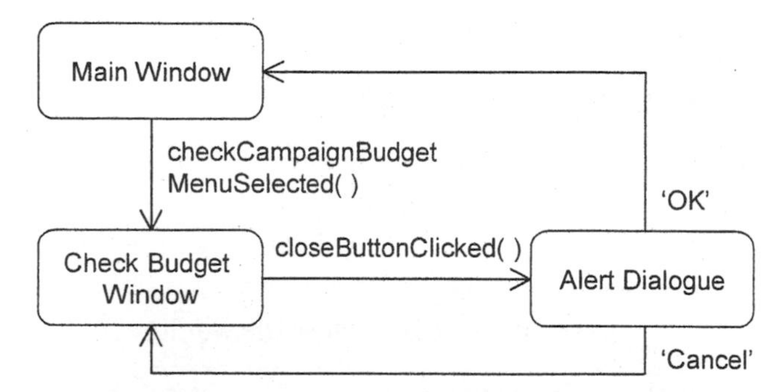

Figure 17.30 Top-level states.

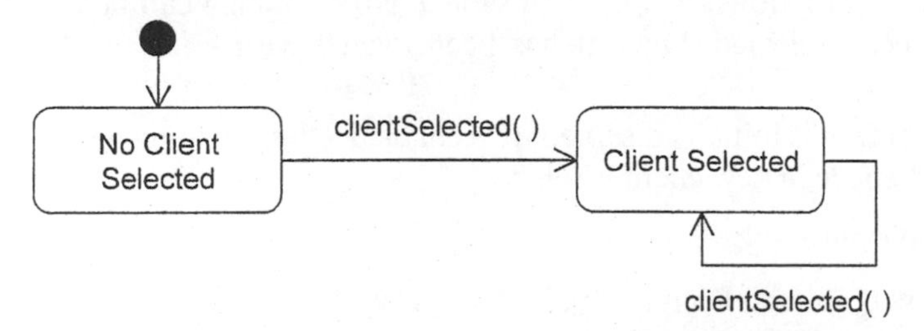

Figure 17.31 Client selection states within the state `Check Budget Window`.

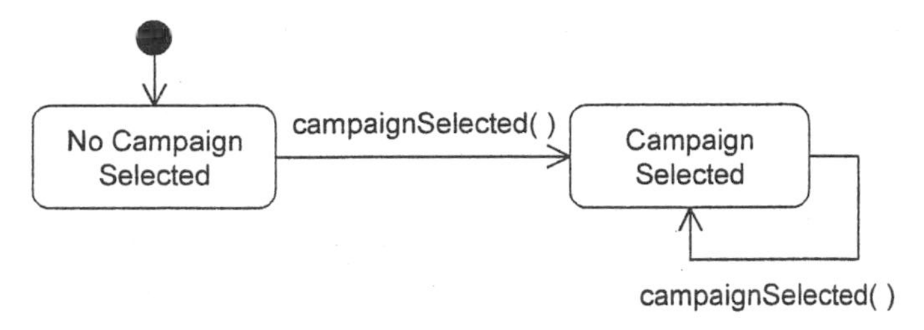

Figure 17.32 Campaign selection states within the state `Client Selected`.

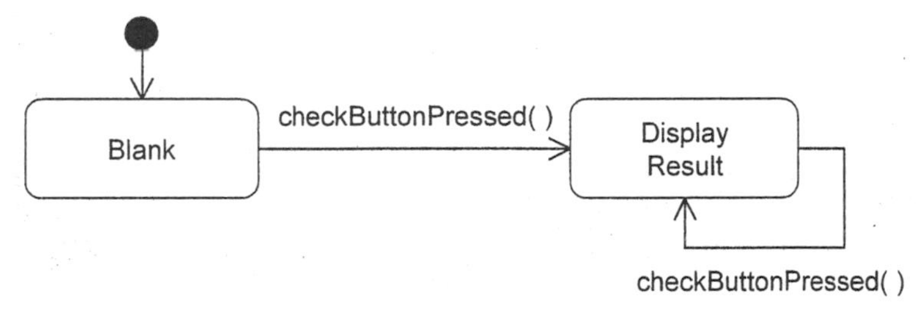

Figure 17.33 Display of result states within the state `Campaign Selected`.

Note the use of the *deep history indicator* where the `'Cancel'` event returns control from the `Alert Dialogue` to the `Check Budget Window`. The `H*` in a circle shows that when that transition takes place, it will return to the exact same state that it was in before the transition to the `Alert Dialogue` state, however far down in the nested hierarchy of states that was. This works like a memory. The state of the user interface before the `closeButtonClicked()` event is recorded, and the `'Cancel'` event returns it back to that recorded state.

Horrocks' approach has some notational differences from the UML standard. He numbers his states (as in Figure 17.34), as the numbers of the states can be stored as the values of the state variables that hold the information about the current state of the system. He also explicitly names the state variables in square brackets. Figure 17.35 shows an example of this notation. Using explicit state variables and numbers for states will help in coding the implementation of this design, and the numbered states make the production of the event–action table simpler. Figure 17.34 is slightly more complicated than it needs to be. States 2 and 4 have no real meaning; they can be treated as no more than a grouping of the enclosed substates to keep the number of states and transitions down. Figure 17.36 is a simplification. The simplified statechart has been used to prepare the event–action table in Figure 17.37.

Prepare an event–action table

UML statechart notation allows you to label transitions and states with actions. On a transition the action can be an action of the object itself or it can involve a message being sent to another object. Within states, *entry* and *exit actions* can be documented, as well as *do actions* that are carried out continuously while the object is in that state and *event actions* that are carried out if a particular event occurs while the object is in that state.

The use of these actions on statechart diagrams can make them very cluttered and difficult to read, especially if there are also guard conditions on the transitions as well as actions. UML allows you to put actions both on transitions and on states, although some authors on the subject suggest that you use either actions on transitions or actions on states, but not both.

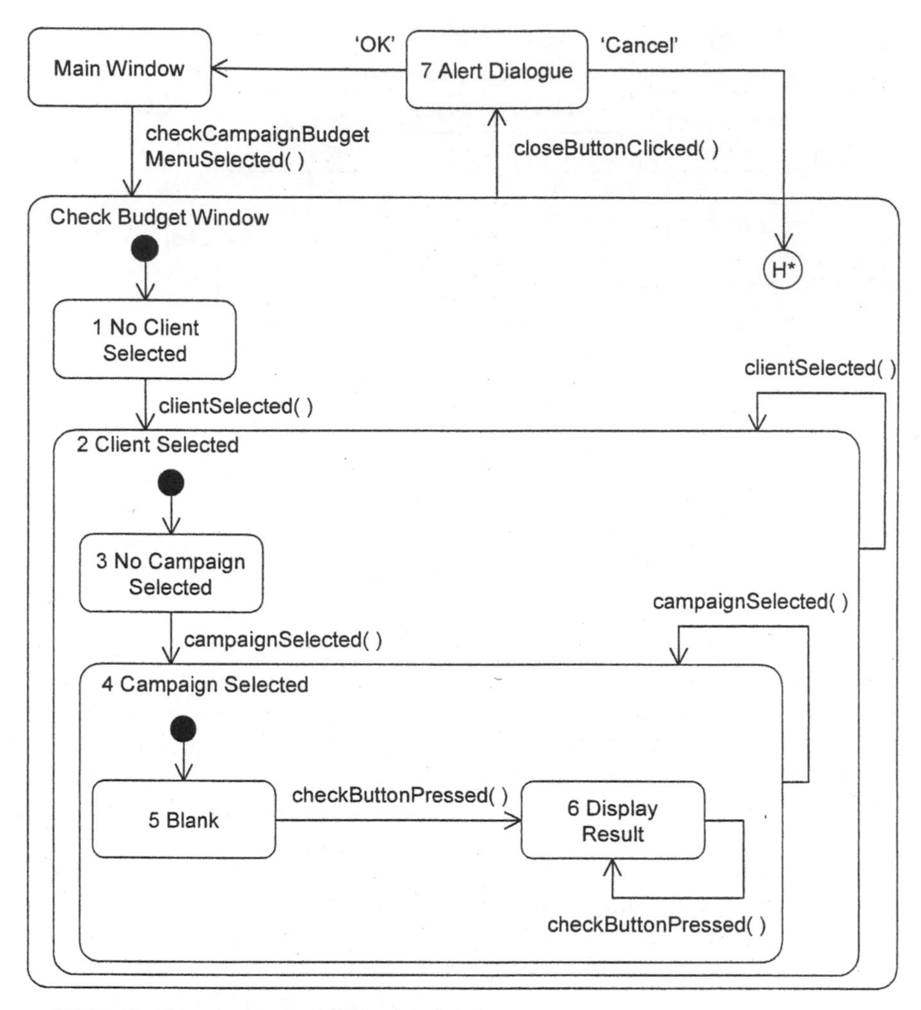

Figure 17.34 Combined statechart with nested states.

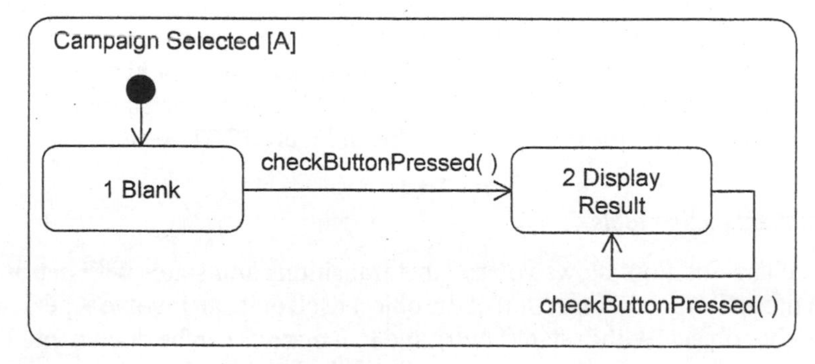

Figure 17.35 State variable and numbered states.

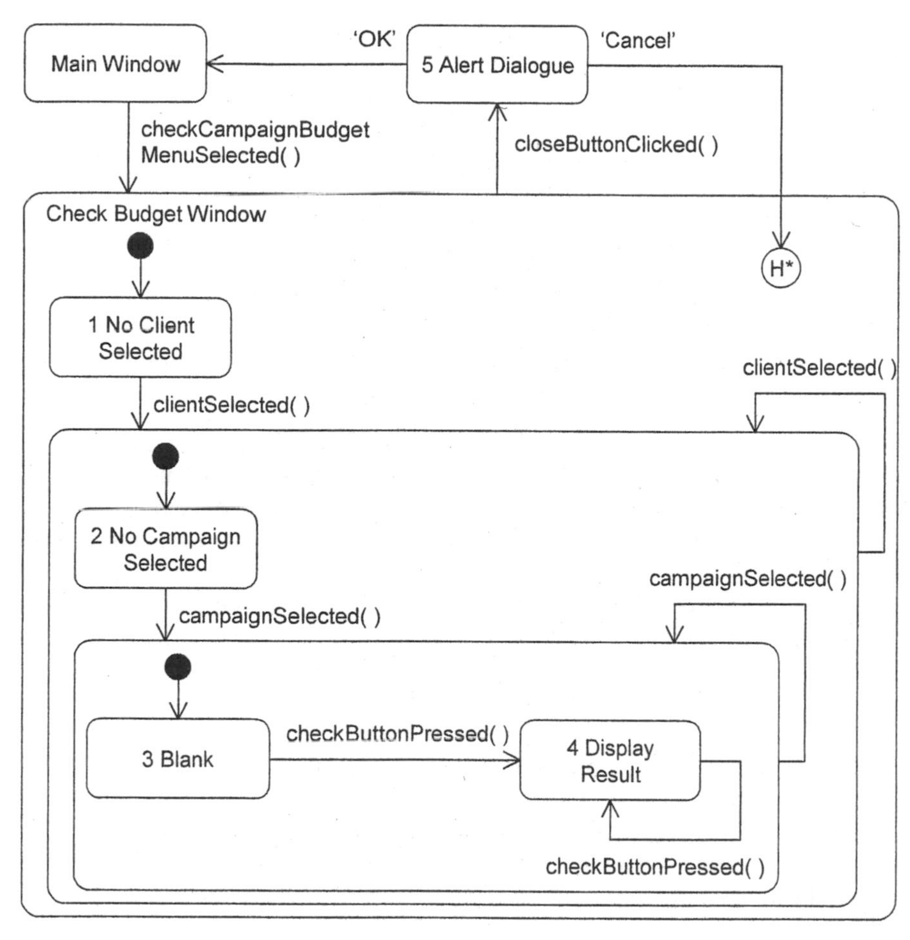

Figure 17.36 Simplified version of statechart in Figure 17.34.

Current State	Event	Action	Next State
–	Check Campaign Budget menu item selected.	Display CheckCampaignBudgetUI. Load client dropdown. Disable campaign dropdown. Disable check button. Enable window.	1
1	Client selected.	Clear campaign dropdown. Load campaign dropdown. Enable campaign dropdown.	2
2, 3, 4	Client selected.	Clear campaign dropdown. Load campaign dropdown. Clear budget textfield. Disable check button.	2
2	Campaign selected.	Clear budget textfield. Enable check button.	3
3	Check button pressed.	Calculate budget. Display result.	4
3, 4	Campaign selected.	Clear budget textfield.	3
4	Check button pressed.	Calculate budget. Display result.	4
1, 2, 3, 4	Close button clicked.	Display alert dialogue.	5
5	OK button clicked.	Close alert dialogue. Close window.	–
5	Cancel button clicked.	Close alert dialogue.	H*

Figure 17.37 Event–action table for Figure 17.36.

For complex statecharts, rather than displaying the actions in the statechart as in Chapter 11, an alternative is to list the actions in a table. This is an event–action table. From the point of view of the programmer, this table will be easier to use than a statechart labelled with actions. It should also make it easier to validate the statechart and to test the code once it has been implemented.

The event–action table lists the following values in columns.

- The current state of the object being modelled.
- The event that can take place.
- The actions associated with the combination of state and event.
- The next state of the object after the event has taken place. (If more than one state variable is used, these are shown in separate columns.)

Figure 17.37 shows an event–action table for the statechart of Figure 17.36.

We can now use this information to revisit the sequence diagrams. Indeed, if we know what the names of the messages or operations in the sequence diagrams are, we can use them in the event–action table instead of the natural language descriptions of the actions.

If we examine the first sequence diagram from Figure 17.15, we can see that we need some additional operations to be shown in the sequence diagram. The sequence diagram shows the boundary class being created, the client dropdown being loaded and the window being enabled, but we have not explicitly disabled the campaign dropdown and the check button. Figure 17.38 shows these additional operations.

We can apply the same approach to the sequence diagram of Figure 17.20, which shows what happens when the client is selected, and thus the transition from State `1` to State `2`. This is shown in Figure 17.39.

If the names of the events and operations have already been decided, because the sequence diagrams have been produced and the operations of classes have been

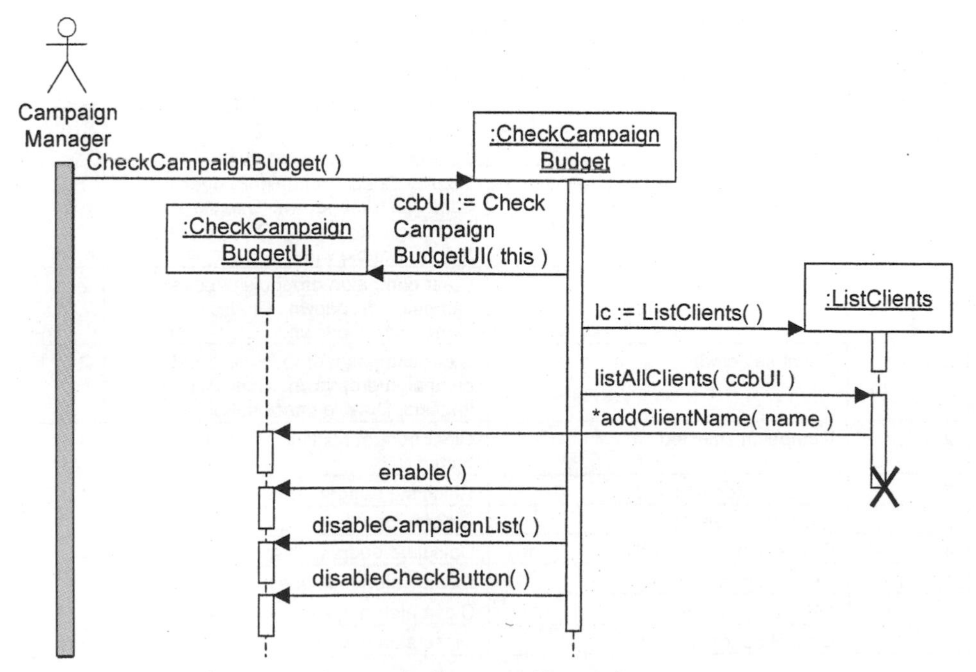

Figure 17.38 Revised sequence diagram for first part of interaction.

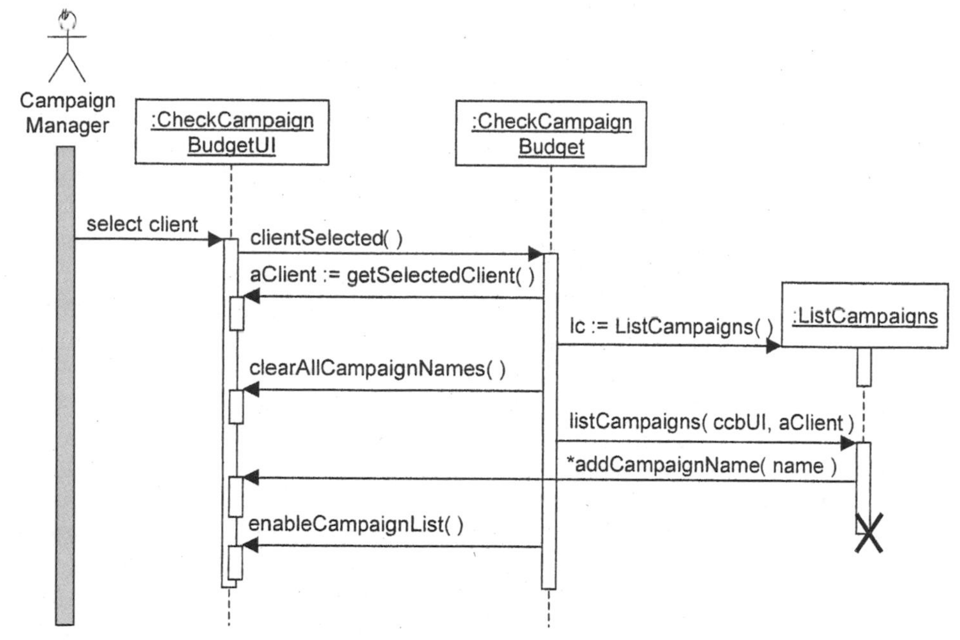

Figure 17.39 Revised sequence diagram for second part of interaction.

designed, then the event–action table can list them using these names. For example, in the transition from State `1` to State `2` in Figure 17.37, the event would be called `clientSelected()`, and the actions would be:

```
ListCampaigns();
CheckCampaignBudgetUI.clearAllCampaignNames();
ListCampaigns.listCampaigns();
CheckCampaignBudgetUI.enableCampaignList();
```

Working through the design of the interface in this way will lead to the addition of more operations to the class diagram in Figure 17.24. For example, the operation `disableCheckButton()` needs to be added to `CheckCampaignBudgetUI`, and the operations `enableCampaignList()` and `disableCampaignList()` need to be added either to the interface `CampaignLister` or to `CheckCampaignBudgetUI`.

17.9 Summary

Designing the interface objects requires us to model three important aspects of the interface. First, we need to determine the classes of the objects that will participate in the interaction with the user and decide on how we will reuse interface classes that are available in class libraries. The choice of interface objects will depend on the style guidelines that have been adopted for the system. These classes are shown in class and package diagrams. Second, we need to model the interaction with the user in sequence or collaboration diagrams. The way that the interaction is modelled will depend on the architecture that has been chosen for the system. The three-tier architecture and the Model–View–Controller architecture separate the boundary objects from the entity and control objects using well-defined methods. Third, we need to model the state of the interface to ensure that we understand how the interface will respond to events and what sequences of events are permitted. We use statechart diagrams to do this.

While carrying out these three modelling tasks, we can draw on design patterns to inform the way in which we select the interface classes and design the interaction between them and the business classes. Prototyping can be used to build models of the interface and test how it will work. Users should be involved in this process to ensure that the interface meets their requirements and to validate the analysts' understanding of their requirements for how tasks should be carried out.

Review Questions

17.1 Why should the user interface classes be kept separate from the business classes and application logic?

17.2 Explain the difference between vertical and horizontal prototyping.

17.3 What is meant by a throwaway prototype?

17.4 What does the «import» stereotype mean?

17.5 What role does each element of the Model–View–Controller architecture play?

17.6 What else do we use statechart diagrams for, apart from modelling the state of interface objects?

17.7 What are the five steps in preparing a statechart to model a user interface?

17.8 What information is held in an Event–Action table?

17.9 Convert the collaboration diagram of Figure 17.28 into a sequence diagram.

17.10 Convert the collaboration diagram of Figure 17.29 into a sequence diagram.

17.11 What are the differences between the MVC and Java `EventListener` approaches?

17.12 Convert the sequence diagram of Figure 17.38 into a collaboration diagram.

17.13 Convert the sequence diagram of Figure 17.39 into a collaboration diagram.

Case Study Work, Exercises and Projects

17.A Decide how you will handle the interaction between the user and the system for the use case `Record problem on line` for FoodCo. Draw a prototype user interface design.

17.B Draw a sequence diagram to include the interface objects that are needed for your prototype in Exercise 17.A.

17.C Draw a class diagram to show the classes that are used in the prototype from Exercise 17.A.

17.D Extend your class diagram from Exercise 17.C to show the superclasses of the interface classes.

17.E If you are familiar with a class library such as the Java AWT, Java Swing, Microsoft Foundation Classes or Borland Object Windows Library, then try to determine how your interface classes relate to classes in that class library.

17.F Produce a prototype for the use case `Record problem on line` using a language or visual programming environment with which you are familiar.

17.G Draw a statechart diagram for the interface to the use case `Record problem on line` to model the behaviour of your prototype developed in Exercise 17.F.

17.H Update your sequence diagram from Exercise 17.B to make sure that it reflects the statechart diagram of Exercise 17.G.

Further Reading

- The MVC architecture is explained in a number of books, particularly books which use Smalltalk as a programming language. Gamma et al. (1995) specifically explain it in terms of patterns.
- The Java EventListener event model was introduced into Java in Version 1.1. (Version 1.0 used a less efficient model.) There are a number of these EventListener interfaces for different kinds of events. Most introductory Java programming books explain them. The Swing classes can also use the MVC approach.
- Few books on object-oriented analysis and design provide much detail on statecharts, and those that do often provide simple models of telephone systems as examples rather than user interfaces. Browne (1994) is one of the few authors who seriously applies statechart diagrams to user interface design. Although he does not use the UML notation, his diagrams in Chapter 3 are similar enough to provide a clear idea of how they can be used to model the detail of user interface interaction. Horrocks (1999) applies a more rigorous software engineering approach to the use of statecharts to design interfaces. Browne's approach is bottom-up, while Horrocks' is top-down. Both use the statechart notation that was originally developed by Harel (1987). For a more recent view of statecharts from Harel, see Harel and Politi (1998), which presents the STATEMATE approach.

18 CHAPTER

Data Management Design

OBJECTIVES

In this chapter you will learn

- the different ways of storing persistent objects
- the differences between object and relational databases
- how to design data management objects
- how to extend sequence diagrams to include data management objects.

18.1 Introduction

Real information systems require persistent data: data that continues to exist even when the system is not active. In this chapter we discuss how persistent data can be modelled using UML. In Section 18.2, we explain what is meant by persistent data. Persistent data can be stored in files or in databases, and files are described in Section 18.3. Using a database management system (DBMS) offers a number of advantages over files. The use of a relational DBMS or an object DBMS will affect the design of the data management. Databases are introduced in Section 18.4.

Many object-oriented systems are constrained by organizations' existing investments in hardware and software and have to use relational database management systems (DBMS) to store data. *Normalization* can be used to design tables for a relational database. Alternatively there are rules of thumb that can be applied to convert a class diagram to a suitable set of tables. Both approaches are explained in Section 18.5.

Designing for an object DBMS will have a different impact on the design model, and this is discussed in Section 18.6. Depending on the object DBMS, making objects in the class diagram persistent can be straightforward. For many systems, there is a requirement to distribute data in different databases. Various mechanisms, such as CORBA (Common Object Request Broker Architecture), RMI (Remote Method Invocation) or EJB (Enterprise Java Beans) can be used to separate the persistent objects from the business logic in the application layer and interface objects. CORBA and EJB can provide the infrastructure for creating distributed databases. Section 18.7 briefly explains CORBA as an example of how this is done.

Database objects can be reflected in the UML models that are produced in design (Section 18.8). ODBC (Open Database Connectivity) and the specific Java implementation of this—JDBC (Java Database Connectivity)—can be used to link an object-oriented system to a relational database. Design patterns may be used to show how persistence can be designed into a system using the experience of many previous developers.

18.2 Persistence

18.2.1 The requirement for persistence

For some applications the data that is created or used while the application is running is not required after the application terminates. This applies mainly to simple applications: an example would be the on-screen calculators provided with GUI operating systems. Such data is called transient data.

Most applications, however, need to store data between one execution of the program and the next. In some cases, the data that is stored is secondary to the operation of the application. When you use a browser, one of the first things that happens as it loads is that it reads data from files that describe the user's preferences and record the last websites visited. The ability to store user settings and to configure applications in this way is an important factor in the application's usability, but it is not its primary purpose: a browser still works if the user's preferences and history are not available.

In the case of information systems, storing data is a primary requirement. Businesses and other organizations rely on their information systems to record data about other organizations, people, physical objects and business events and transactions. The data entered into such a system today will be required in the future; operations being carried out on data today rely on data that was stored in the past. Computerized information systems have replaced systems based on paper in files and ledgers, and must provide the same relatively permanent storage that is provided by paper-based systems. In most organizations, it is also important that data can be shared between different users. Data in the memory of a particular computer is not normally accessible to multiple users. It must be written away to some kind of shared data storage system so that other users can retrieve it when they require access to it.

This is what we mean by *persistent data*. It is data that must be stored in a secondary data storage system, not just in computer memory, that must be stored after the program that creates or amends it stops running, and that usually must be available to other users. Information systems also use transient data: for example, the results of calculations or lists of objects that are required for a particular purpose such as printing a report, but that are not required permanently and can be destroyed after they have been used.

In an object-oriented system, we are concerned with both *persistent objects* and *transient objects*. Persistent objects are those that must be stored using some kind of storage mechanism, while transient objects will be erased from memory after they have been used.

18.2.2 Overview of storage mechanisms

Ultimately all data in computer systems is stored in files of some sort. In Section 18.3, we explain the different kinds of file organizations and access methods that are available, and the purposes for which files are used. However, most information systems use a database management system of some sort in which to store their data. Database management systems provide a layer of abstraction that hides from the user the fact that the data is stored in files. If the database is a relational database, then the user of the database sees tables containing data. Each table may relate to part of a file, to a single file or to many files, but that is not important to the user of a relational database (who may be a designer or programmer). The way that the database stores tables in files is important to the database administrator, who has to be concerned with where the data is stored, taking backups and so on. If the database is an object database, then the user sees objects and links between them. Again these objects will be stored in files of some sort, but the designer or programmer does not need to know the details.

In an object-oriented system, a database of some sort is the most likely way of providing persistent storage for objects. However, it is possible to store objects in files. Most object-oriented languages provide mechanisms for converting objects into a form that can be written out to a file—*serializing* them—and for reading them back into memory from a file. This is unlikely to provide an efficient mechanism for a business information system. However, files can be used for many other storage purposes in object-oriented systems. They can be used to hold data that is transferred in from other systems, and in Section 19.6 we discuss the conversion of data from a system that is being replaced. Files can also hold configuration information, and in Section 18.3.5, we provide an example of how files can be used to localize the Agate system so that text items such as labels, button captions and menu entries are displayed in the language of the country where the application is being used.

The choice of database management system will have a significant impact on the work that is required of the system designers. If an object database is used, then there should be little work involved in designing the way that objects are stored. If a relational database is used, then more work is involved. In Section 18.5, we describe two approaches to converting classes to tables, and in Section 18.8, we describe some ways of designing persistence frameworks. In Section 18.8.5 we explain the tools and frameworks that are available to automate the process of mapping classes to tables.

18.2.3 Architecture for persistence

Part of the process of system architecture design is to determine how the requirements for the storage of persistent data will be met by the system. There may be trade-offs to be made between the requirements for a new system and the existing hardware and software that is available. Many organizations also have corporate standards for the database management systems that they use, and these will influence the architecture of a new system and the design of data storage.

Existing systems may have a different architecture from the one proposed for a new system, but there may be parts of the existing system that can be reused. This is often the case with databases, as organizations often have existing business systems and wish to use the data from those existing systems in new ways. If the new system is to be developed in an object-oriented language, it may be necessary to create a layer in the architecture that wraps the existing data so that it looks like objects even if it is stored in a relational database. This is shown in Figure 18.1. An example of an existing system

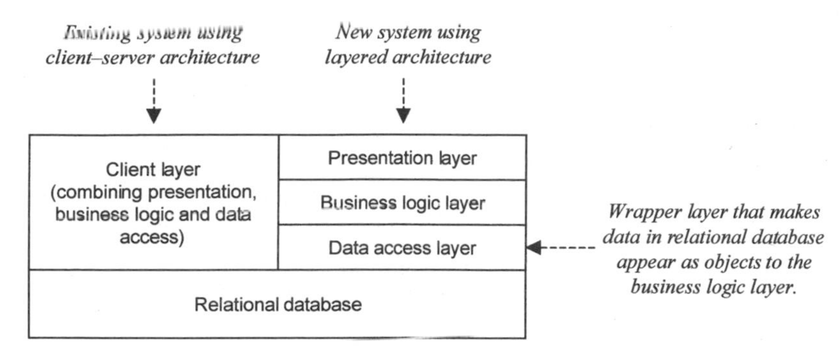

Figure 18.1 Layered architecture for existing and new systems sharing an existing relational database.

like this might use SQL-Server as its database and Visual Basic for the existing client programs. The Visual Basic client programs access the data in the database using ODBC (Open Database Connectivity) and SQL (Structured Query Language). The new system might use Java applications for the presentation layer, and Java components for the business logic layer, and for the data access layer. The Java components would provide a wrapper around the rows of data in the relational database, so that they appeared as objects to the business logic layer. The data access layer would probably still use ODBC—or in this case JDBC (Java Database Connectivity)—and SQL to connect to the database, but the objects used to access the data could be reused in other applications. If the data access layer is well designed and includes entity beans (a Java technique) that wrap up all the data in the database, it may be possible later to replace the old client–server application by adding to the presentation and business logic layers and reusing the data access layer.

For the architect designing the persistent storage of a system there are a number of questions to be answered.

- Are there parts of the system where storage in files is appropriate?
- Is the system truly an object-oriented system or a simple database system with a GUI user interface? For simple systems, it is possible to write programs in an object-oriented language such as C++ or Java, which provide a front-end to a database. The front-end could connect to the database using ODBC or JDBC and treat the data purely as rows of data in tables, without using any entity objects. We are not taking this approach here, as it is not appropriate for our case studies.
- Will the system use an existing DBMS or is there freedom to choose an appropriate DBMS? If an existing system is to be used, it will constrain the system design in some ways, but as we have shown above, it is possible to use a more flexible layered architecture to replace a client–server system while retaining the same DBMS.
- Will the system use a relational DBMS? If a relational DBMS is to be used then classes must be mapped to tables. This can be done using tools that automate this process, or by designing the tables and a suitable mechanism to fetch data from the database and assemble it into objects, and to save the objects back into the database when required.
- Will the system use an object DBMS? If an object DBMS is to be used, then work on designing the persistence mechanisms is likely to be much simpler.
- What is the logical layering of the system? A layered architecture is likely to be more flexible, separating the user interface, the business logic and the access to and storage of data.

- What is the physical layering of the system? More than one logical layer can reside on the same machine, for example the business logic and the data access can be on the same server. In large systems, there may be more than one machine providing the services of a single layer, for example several web-servers handling the presentation layer, which connect to two machines running the business logic, which connect to a single database server. If an application is delivered over the Internet or a company Intranet, then much of the presentation layer will reside on the web-server, where Active Server Pages (ASP), Java Server Pages (JSP) or some other related technology will be used to construct the web pages dynamically and deliver them to the client's browser.
- What is the distribution of the system? It is conventional for the presentation layer to be located on many client machines, but if the entity objects and the business logic are located on multiple machines, then the system must be designed to use mechanisms such as CORBA (Common Object Request Broker Architecture) or EJB (Enterprise Java Beans) to make it possible for clients to locate the objects they need to connect to in order to provide the functionality of the system.
- What protocols will be used to connect between layers of the system, particularly in a distributed architecture? Language- or operating system-specific protocols such as Java's Remote Method Invocation (RMI) or Microsoft's Distributed Component Object Model (DCOM) can be used but restrict the design to implementation on certain platforms. Open standards such as CORBA or SOAP (Simple Object Access Protocol), which uses XML (Extensible Markup Language), make it possible to build component-based systems that are not tied to particular platforms.

Building large, distributed systems is beyond the scope of this book, but in the rest of this chapter, we explain some of the mechanisms that can be used to design persistence into an object-oriented system, and we start with the simplest—files.

18.3 File Systems

The simplest means of persistent storage available for computerized information systems uses files. Most users of personal computers are familiar with the idea of files. Word-processors and spreadsheets store their data in files; browsers download them from websites where they are stored. At the simplest level, a file is a stream of bytes of data stored on some physical medium. In most cases, files are stored magnetically by adjusting the magnetic properties of the surface layer of a disk or tape. Files can also be stored optically on CD-ROMs and other optical storage systems, and electronically in special kinds of memory, such as the flash memory used in palmtops. However, the user is normally shielded from the physical implementation of the file system by the operating system of the computer or by a programming language, which provides high-level functions to create files, store data in them and retrieve that data.

18.3.1 File and record structures

Programming languages, and in some cases operating systems, also impose a structure on files. This structure breaks a file up into individual records, each of which groups together a number of fields representing the data that is to be held in the file. In the same way as each object contains a number of attributes, each one of which holds a particular kind of data about the object, each field in a record holds a particular kind of data about whatever it is that the record describes. Records in files can take different forms, described below.

- Fixed-length—Each record is made up of a number of fields, each of which has a fixed length in bytes. If the data in a particular field does not fill it, the field is padded out with special characters (often spaces). Each record is of the same, fixed length, and it is possible to skip from the beginning of one record to the beginning of the next by jumping that fixed number of bytes ahead.
- Variable length—Each record is made up of a number of fields, each of which may have a maximum length but has a minimum length of zero bytes. Fields are usually separated or delimited by a special character that would not appear in the data. Records may also be delimited by a special character. The length of each record may also be stored at the start of the record, making it possible to skip to the beginning of the next record by jumping that variable number of bytes ahead.
- Header and detail—Records may be of two types: each transaction recorded consists of a header record, followed by a variable number of detail records. This approach can be used with many business documents, such as orders, invoices and delivery notes that have a variable number of lines on them. Each record will contain a record type field. The number of detail records may be held in the header so that it is possible to tell where the next header record starts.
- Tagged data—The data may have a complex structure, as in object-oriented systems, and it may even be necessary to hold objects of different classes in the same file. Every object and attribute may be tagged with some kind of description that tells a program reading the file what the type of each item is. This approach is used for data in files that use Hypertext Markup Language (HTML) and Extensible Markup Language (XML).

Some systems store information about the structure of the file in a data dictionary, and this may be held in a separate file or at the start of the data file itself. This makes it possible to write programs which can read the data out of any file that uses this format: the program first reads the data dictionary information and configures itself to read the appropriate data structures from the rest of the file.

As well as having alternative ways in which the data can be structured within files, files can have different types of organization, can be accessed in different ways and can serve different purposes in a system.

18.3.2 File organization

There are three ways in which files can be organized: serial, sequential and random.

Serial organization. Each record in the file is written onto the end of the existing records in the file. If a record is to be deleted, the file must be copied from the start up to the deleted record, which is skipped, and then the rest of the file is copied back to the disk.

Sequential organization. In the basic form of sequential organization, each record is written to the file in some pre-determined order, usually based on the value of one of the fields in each record, such as an invoice number. Records must be deleted in the same way as for serial files. Each record must be added to the file in its appropriate place, and if it is necessary to insert a record into a file, the file is copied up to the point where the record is to be inserted, the new record is written to the file and then the rest of the file is copied.

Random organization. The term random is a poor way of describing the organization of random files, as the organization is really anything but random. The records are added to the file by means of precise algorithms that allow records to be written and read directly without having to read through the rest of the file. What this means is that if you choose any record *at random*, it should be possible to access it more or less straightaway without searching through the file. The algorithm usually converts a key field of each record into an address in the file that can be reached directly.

18.3.3 File access

Depending on the file organization chosen, different ways of accessing the data in the files are available to the designer. The main ones are serial, index-sequential and direct.

Serial access. Serial and basic sequential files can only be accessed serially. To find a particular record, it is necessary to read through the file, record by record, until the required record is located.

Index-sequential access. Access to sequential files can be improved by maintaining an index on the field that is used to order the data in the file (the key). Index-sequential files are used where it is necessary to read the file sequentially, record by record, and to be able to go straight to a particular record using its key. The indexing mechanism used for index-sequential files dates back to the time when mainframe operating systems made it possible to allocate the particular disks, cylinders and tracks where a file would be stored.

Records are stored sequentially within blocks (areas of the disk that have a defined size). Enough blocks are allocated to the file for the total anticipated number of records. Records are written into blocks in key order, but the blocks are not filled up from the start of the file, rather records are distributed evenly across the blocks, leaving space for new records in each block.

The index on the file can be dense or sparse. In a dense index, there is an entry for every key with a pointer to the first record in the file with that key (there may be more than one). In a sparse index, there is an entry for the last record in each block. To find a record, a program reads through the index until it finds a key value greater than the value of the key it is searching for. It then jumps to the block pointed to by that index entry and reads through it until it finds the required record.

To support large files, there may be two or more levels of index. For example, there may be a master index and a series of block indexes. The master index holds the value of the key field of the last record in each block index. Each block index holds the value of the key field of the last record of each block in a set of blocks. To find a record by its key, the master index is read until a key value is found that is greater than or equal to the key of the record being sought. This makes it possible to go to the block index for that record. The block index is then read until a key value is found that is greater than or equal to the key of the record being sought. This makes it possible to go to the block in which the record is held. The records in the block are then read sequentially until the desired record is found. A similar approach is taken in order to add records to an index-sequential file. The block in which the record is located is identified using the index, then the records in the block are read into memory and copied to the disk up to the point that the new record is to be inserted, the new record is written into the block, and then the rest of the records in the block are copied back to the block. Eventually some blocks will fill up, and it will be necessary to write some records into special overflow blocks. The addresses of the overflow blocks will be held in the blocks that have overflowed. Performance tuning of such files involves resizing them so that there are more blocks in the file and no records have to be stored in overflow blocks. Figure 18.2 shows the organization of data and indexes in an index-sequential file.

Index-sequential files have the advantage over sequential files that records can be read and written more quickly, although there is a storage overhead associated with maintaining the indexes. Compared to direct access, which is described next, there is also the overhead of the time taken to access the indexes before the data is reached.

Direct access. Direct access methods rely on the use of algorithms to convert the values of key fields in the records to addresses in the file. (The term random access is sometimes used.) The first and simplest of these is *relative addressing*. This access method requires the use of fixed length records and successive positive integers as keys. If each record is 200 bytes long, then record 1 will start at byte 1 of the file, record 2 at byte 201, record 3 at byte 401 and so on. It is possible to calculate the position of any record in the file by subtracting 1 from its key, multiplying the result by the size of the records and adding 1. Each record can be read directly by reading from that point in the file.

Hashed addressing is the second approach. This can use keys of any form. As with indexed sequential files a fixed number of blocks is initially allocated to the file. This is usually a prime number of blocks, as this helps to achieve a more even spread of records into blocks. The key is then hashed to determine to which block a particular record will be allocated. The hashing function is an algorithm that takes an ASCII string and converts it to an integer. There are many approaches. A simple approach is to take the characters in the string and convert them to their ASCII values (for example 'A' is 65). These ASCII values are summed together. The sum is divided by the number of blocks in the file and the remainder or modulo gives the number of the block in the file into which that record will be placed, starting at block 0. If a block fills up, then an additional block will be used as overflow, and its address will be held in the block that the record would have been stored in. Figure 18.3 shows the organization of a hashed direct access file and the calculation of a simplified version of the hashing algorithm based on just three characters of the key.

Master Index

Record key	Block address
Feng	1
Patel	2
Zarzycki	3

Block Index (Index block 2)

Record key	Block address
Finlayson	55
Gomez	56
Hanson	57
Jacobson	58
....	....
Patel	84

Data Records in blocks (for simplicity only the keys are shown)

Block address	Records				
55	Fern	Finch	Finlayson		
56	Finn	Firmin	Ford	Gangar	Gomez
57	Gordon	Govan	Hamer	Hanson	
58	Ho	Ibrahim	Jacobson		
....	....				

Figure 18.2 Schematic of indexes and data in an index-sequential file ordered by surname.

Records hashed on first three characters of key field

Khan → Kha
ASCII Values = 75, 104, 97
75 + 104 + 97 = 276
276 divided by 7 leaves a modulo of 3
So Khan will be added in Block 3.

Data records in blocks (for simplicity only the keys are shown)

Block no. in file	Records			
0	Hao			
1	Ford	Farmer		
2	Firmin	Firth		
3	Harris			
4	Hastings	Gomez	Jacobson	
5	Ibrahim	Finch	Fern	Gangar
6	Hanson			

Figure 18.3 Organization of a sample hashed direct access file.

Improving access. There are a number of ways of improving access to data in files. Files with a random organization can normally only be accessed directly or serially (for example to copy the file), so it is not possible to read through them sequentially in an order based on key fields. However, it is possible to add two extra fields to each record containing the key values of the next record and the previous record in sequence (a linked list). This makes it possible to read through the records sequentially, but adds a considerable overhead when a record is added or deleted.

A common way of improving access to a file is to add a secondary index. This is a similar approach to that used in adding an index to an index-sequential file. It is used when there is a requirement either to access records in a file based on the values in some field other than the key field, for example to find address records by postal code or zipcode, or to provide sequential access to a random file, by building an index of sequential keys. A separate file is created in which the keys are the values from the indexed field (for example the postal code) in all the records in the main file. Each record also contains either the keys or the block addresses of each of the records that contain that indexed field. This kind of index is known as an *inverted file*. There are various structures that can be used for indexes, such as B-trees, which have different benefits—in terms of speeding up retrieving records—and different disadvantages—in terms of adding an overhead to updates to the file.

18.3.4 File types

We have seen that as well as files that hold data, there can also be files that hold indexes to the data in the main files. Other types of files may be required in a file-based system.

Master files. Master files are the files that hold the essential, persistent data records for the system. In transaction-processing systems the master files are updated with details of transactions that are recorded in transaction files. Master files usually require some kind of direct access so that records can be updated quickly.

Transaction files. Transaction files record business transactions or events that are of interest to the organization and that are used to update records in master files. In a banking system, transactions that take place when customers withdraw cash from an automatic teller machine (ATM) may be recorded in a transaction file. At the end of the day, the transaction file is processed in order to update all the accounts of customers who have withdrawn cash.

Index files. Indexes are used to speed up access to records in master files, as described above. There are many index file structures that can be used, and the choice of index structure will depend on the nature of the data and the type of access required.

Temporary files or work files. During the course of processing data, it may be necessary to create a temporary file that is used to hold the results of one process before it is used in another process. For example, it may be necessary to sort records into a particular order so that a report can be produced, a work file containing the records in the correct order would be produced and then deleted once the report had been created. If you use a PC that prints your work in the background (while you get on with another task), then it is using spool files to hold the printed output that is sent to the printer. These spool files are deleted when they have been printed.

Backup files. Backup files may be direct copies of master files or transaction files that are held on the system and that allow the data to be restored if the originals are destroyed or corrupted. Alternatively, they may be files with a special structure that can be used to reconstruct the data in the system.

Parameter files. Many programs need to store information about items of data of which there is only one record. These are typically system settings, such as the name and address of the company using the software, and configuration information, such as the currency format to be used to print out money values or the format to be used for printing the date. Parameter files hold this kind of information.

In a project that uses files to store data, part of the role of the designer is to choose the appropriate file organization and access method for the storage of all the objects in the system. In most cases, objects will need to be stored and retrieved using some kind of object identifier, and direct access files will be required. However, there may be requirements for all the objects of a particular type to be retrieved from the file in sequence, which may indicate the need for an organization that supports sequential access, or for the addition of an index. Some object-oriented languages provide mechanisms for streaming objects out to a serial file. This is the case with both Smalltalk and Java. Java, for example, contains two classes called `ObjectOutputStream` and `ObjectInputStream` that can be used to write objects together with information about their class to a disk file and read them back again.

Many organizations use database management systems to hold data. In their systems, there will be no need for designers to make decisions about the physical file structures used to provide persistent storage for objects. However, there are still situations where files are the appropriate mechanism for storing data in an application. In the next section we present one example.

18.3.5 Example of using files

In the Agate system, one of the non-functional requirements is that the application can be customized for use in different countries with different languages. This means that all the prompts that are displayed in windows, labels on buttons and menus, and error and warning messages cannot be written into the classes in the presentation layer as string literals. For example, if a user interface class was implemented in Java using string literals, the line of code to create a cancel button would look like this.

```
Button cancelButton = new Button ("Cancel");
```

To use the program in French, someone would have to go through finding all the strings like this and translating them. Then there would be two versions of the program, and any changes would have to be made in both. This is unmanageable with just two languages, let alone several. In Java, it is possible to use the class `java.util.Locale` to hold information about the current locale in which an application is running. This includes information about the language, the country and a variant value. For example "fr" for French, "FR" for France and "EURO" to indicate that the country uses the Euro as its unit of currency from January 2002. For an application running in French in Canada, only the language code "fr" and the country code "CA" would be required.

Another Java class, `java.util.ResourceBundle`, uses the locale information to hold objects in memory, each of which is associated with a key value. It can load these objects from a file, and the name of the file is made up of the name of the resource and the language, country and variant codes. So for a resource called `UIResources`, designed to hold the values for all the prompts and labels, there could be different versions called `UIResources_fr_FR_EURO`, `UIResources_en_UK`, and `UIResources_en_US`, for France, England and the United States respectively.

When the user interface class is instantiated, it needs to find out its locale and then load the correct resources into memory with a line of Java like this.

```
resources = ResourceBundle.getBundle("UIResources",currentLocale);
```

Then the code to set up the cancel button becomes the following.

```
Button cancelButton = new Button(resources.getString("Cancel"));
```

The resource file is made up of records, each of which is on a separate line, with an equals sign to separate the key from the associated string, for example:

```
Cancel = Annuler
OK = OK
File = Fichier
```

for the French version. When the application is deployed, either the installation routine must install the correct resource files on the machines that will be running the user interface classes, or a full set of files must be deployed, and the appropriate one will be chosen at runtime. (See Chapter 19 for more detail about implementation and deployment.)

18.4 Database Management Systems

18.4.1 Files and databases

Files are appropriate for simple programs and for storing data that does not need to be shared and updated by many users. During the 1960s systems were built using files to store data; since the 1970s most large commercial systems have used databases of some sort to hold their data, and more importantly database management systems to organize and manage the tasks associated with storing and providing effective access to large volumes of data.

Using files to store data can result in a number of problems.

- As the number of applications grows, the number of different files grows. Some of these files may hold the same data for different applications in different formats. So, data is duplicated, taking up unnecessary storage space. This is known as redundancy.
- There is the risk that the updates to data in different applications will not be synchronized. For example, a customer address may be changed in one file but not in another, leaving the data inconsistent.
- Each application must contain its own mechanisms for storing the data in its set of files, if the data changes or the way that it is stored has to be changed, then each program within an application that accesses that data must be amended. This makes it difficult to add new programs to an application that uses some of the same data but also need to store additional data.
- As business requirements change, users may want to access the data in new ways, for example to produce a report combining data from different applications. This cannot be implemented without considerable programming effort.

The first step towards resolving these problems is to analyse the storage requirements of different applications across the organization and to build a corporate database that contains all the data from different applications. Each application then uses a subset of this database for its own requirements. The second step is to use a database management system to organize and manage the data and to decouple the storage mechanisms from the application programs. The aim of using a DBMS is to separate the details of the way that data is stored physically from the way that it is used in application programs. This is achieved by producing a logical model of the data that is independent of the demands of applications and that could be stored in different ways in the database. This is known as the Three-schema Architecture. Figure 18.4 shows the Three-schema Architecture. The external schema represents the ways in which data is used in application programs. The conceptual schema is a logical model of the data and is independent both of the external schema and of the details of how the data will be stored. The physical organization of the data is to be found in the internal schema, which defines the files used to store the data. The aim of this approach is to isolate the application programs from the details of how any particular item of data is stored. This is central to the way that relational DBMS work. Design for relational DBMS is described in more detail in Section 18.5.

DBMS provide more than just a means of storing data that can be shared across many applications. They provide tools and features that can be used to manage the data.

- Data definition language (DDL). The DDL is used to specify the data that is held in a database management system and the structures that are used to hold it.
- Data manipulation language (DML). The DML is used to specify updates and retrievals of the data in the DBMS.

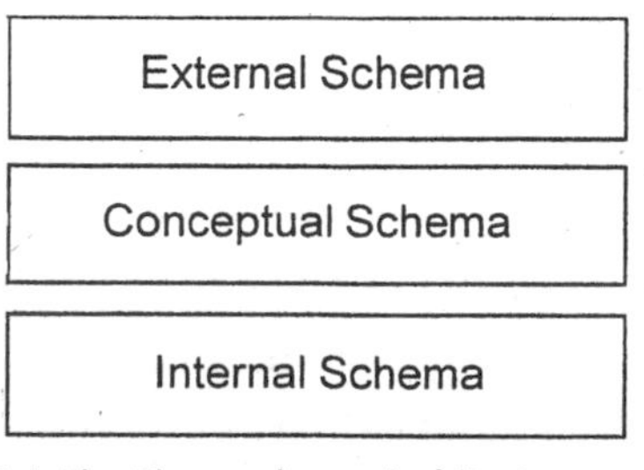

The view on data used by application programs.

The logical model of data that is separate from how it is used.

The physical storage of data in files and indexes.

Figure 18.4 The Three-schema Architecture.

- Integrity constraints. Constraints can be specified to ensure that the integrity of the data is maintained.
- Transaction management. Updates to the database can be specified as transactions in which all of the component updates must succeed, otherwise the entire transaction is *rolled back*, which means that it is not committed to the database.
- Concurrency. Many users can simultaneously use the database and update its contents.
- Security. Access to the data in the database can be controlled, and permissions granted to different users for different levels of access (for example in SQL, select, insert, update and delete).
- Tuning of storage. Tools can be used to monitor the way that data is accessed and to improve the structures in the internal schema in order to make the access more efficient. These changes can then be made without having any impact on the application programs.

As mentioned above, these structures in the internal schema will be files. An important feature of a DBMS is that the kind of file used, the access methods and the indexes that are held on the file are hidden from users of the DBMS (typically application programmers) and can be changed without affecting the programs that use that data.

These changes to the database do not happen automatically. For large systems a Database Administrator (DBA) must be employed to manage the database and to ensure that it is running efficiently. The DBA will be responsible for controlling the data dictionary that defines the conceptual schema of the database, for controlling access to data and for tuning the performance of the database. For smaller systems a DBA will not be needed, but someone will need to be responsible for managing the database.

In summary, the use of a DBMS based on the Three-schema Architecture has a number of advantages over the use of files to store data for an information system.

- The use of a conceptual schema can eliminate unnecessary duplication of data.
- Data integrity can be ensured by the use of integrity constraints and transaction management techniques.
- Changes to the conceptual schema, the logical model, should not affect the application programs, provided the external schema used by the application programs does not have to be changed.
- Changes to the internal schema, the physical storage of the data, have no impact on the conceptual schema and should not affect the application programs, except perhaps positively in enabling them to access data more efficiently. Compromises may have to be made between the needs of different application programs.
- Tools are available for tuning the performance of the database, for the back-up and recovery of data and to control security and access to data by multiple simultaneous users.

However the use of DBMS may also have disadvantages for organizations that decide to go down this route.

- There is a cost associated with investing in a large DBMS package.
- There is a running cost involved in employing staff to manage the DBMS.
- There will be a processing overhead in converting data from the database to the format required by the application programs.

The most widely used type of DBMS is the relational DBMS. For object-oriented systems one might hope to be able to use an object DBMS. However, in many situations, organizations have an existing relational DBMS and new object-oriented applications must share that corporate database. A relational database with C++ or Java as the application development language is still the most common platform in use.

18.4.2 Types of DBMS

The three main types of database that we are concerned with here are relational, object and hybrid object-relational.

Relational databases. The idea of relational databases was first suggested by Codd (1970). His proposal was followed by considerable research effort that led to the development of commercial relational database management systems (RDBMS) during the 1970s. However, it was not until 1986 that the American National Standards Institute published the first SQL standard based on this work (ANSI, 1986). SQL (Structured Query Language) is now the standard language for relational databases and provides both DDL and DML capabilities.

Relational databases have a theoretical foundation in set theory, and their operations are defined in terms of the *relational algebra*, a mathematical specification of the operations that can be carried out on *relations*. The essence of the relational model is to eliminate redundancy from data and to provide the simplest possible logical representation of that data. This is achieved by means of a series of steps that can be applied in analysing the data and that result in normalized data. This normalized data is held in relations or tables. This process simplifies a complex data structure until it can be held in a series of tables. Each table is made up of *rows* of data. Each row contains attribute values that are organized in *columns*. Each column contains data values of the same attribute type. The data in each row must be distinct and can be uniquely identified by some combination of attribute values in that row. Each attribute value in the table must be atomic, that is, it may not contain multiple values or be capable of being broken down further. Figure 18.5 shows the conventional form for representing tables on paper. In existing RDBMS, all data structures must be decomposed into this kind of two-dimensional table[1].

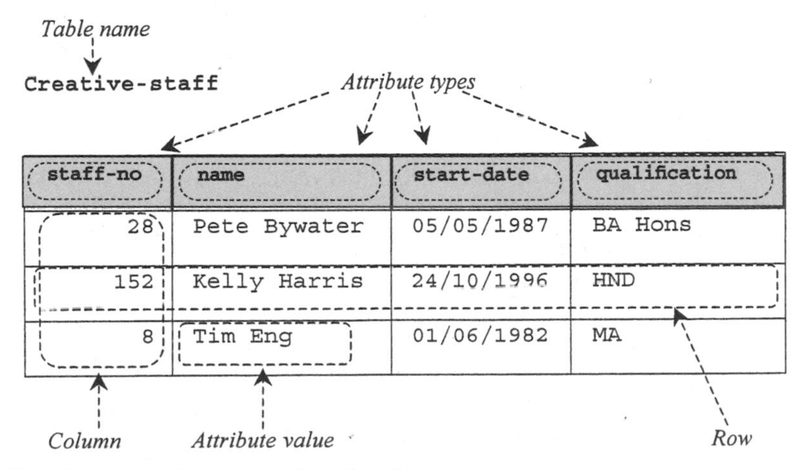

Figure 18.5 Conventional representation of a table.

[1] Date and Darwen (1998) argue that the relational model can handle complex data types, but it is the way that relational DBMS have been implemented that leads to their inability to handle these data types.

The weakness of current implementations of RDBMS lies in the fact that objects in object-oriented systems do not fit easily into this model. They can be broken down into tables, as is shown in Section 18.5, but there is a processing overhead associated with breaking them down and reconstructing them. References to other objects (represented by associations in the class diagram) must also be maintained when an object is stored in a relational database, and restored when it is retrieved, and even if the associated object is not itself in memory some mechanism must be created to allow it to be referenced and to be sent messages. Tables in RDBMS are linked to one another by common attribute values (foreign keys), whereas objects are linked to one another by references or pointers. Data in RDBMS is processed in sets, while data in ODBMS must be navigated through, following links from object to object.

Relational DBMS are currently the most widely used type of DBMS. They are based on a sound mathematical theory, they have been developed over a number of years, they are robust and they are efficient and flexible for the kind of data that they are designed to handle. The best known is probably Access; others include Oracle, SQL-Server, DB2, Informix, Ingres, Progress and Sybase. The growth of Linux has also led to the availability of free and open source databases such as MySQL.

Object databases. In contrast, objects in an object-oriented system are not flat, two-dimensional structures. Each object may contain other objects nested within it. For example, a `SalesOrder` object could contain an `Address` object that contains its own attributes, and a `Collection` of `OrderLine` objects, each of which is made up of two attributes. An example of this is shown in Figure 18.6 with a `SalesOrder` object with its class definition in UML notation. Object database management systems (ODBMS) have been developed to handle complex objects of this sort. Part of the motivation for the development of ODBMS has been the growth in the number of applications that use complex data structures. These include multimedia applications in which objects such as sounds, images and video clips are not easily represented in tables, and applications such as computer-aided design packages in which the designer may want to deal with different levels of abstraction, for example treating a sub-assembly in terms of its behaviour as a sub-assembly, in terms of the individual chips or in terms of the components such as logic gates that make up those chips. ODBMS provide services that make it possible to store complex objects of this type. Examples of ODBMS include Jasmine, ObjectStore and Ontos.

Object-relational databases. Object-relational databases combine the simplicity and efficiency of relational databases with the ability of object databases to store complex objects and to navigate through the associations between classes. The SQL standard is being updated to allow the relational model to incorporate many features of object-oriented systems such as user definable abstract data types, inheritance and operations. The open source product PostgreSQL is probably the most well-known hybrid DBMS. Oracle now includes some hybrid features. In what follows we shall focus on relational and object databases.

18.5 Designing for Relational Database Management Systems

18.5.1 Relational databases

RDBMS have been in use since the 1970s. They use mature technology and are robust. It is common for an object-oriented system to be built to use a relational DBMS. Relational databases hold data in flat two-dimensional tables whereas classes may have complex nested structures with objects embedded within other objects. If it is necessary

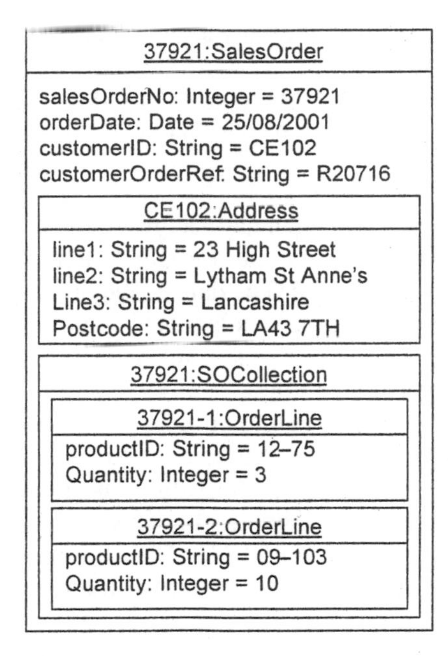

(a) *Complex object.*

SalesOrder

salesOrderNo: Integer
orderDate: Date
customerID: String
customerOrderRef: String
deliveryAddress: Address
orderLineList: SOCollection

(b) *Class with classes of nested objects as attributes.*

Figure 18.6 Composite `SalesOrder` object with equivalent UML class.

to use an RDBMS to provide the storage for a system built using an object-oriented programming language, then it will be necessary to 'flatten' the classes into tables in order to design the storage structures. When the system requires an instance of a class from the database, it will have to retrieve the data from all the tables that hold parts of that object instance and reconstruct it into an object. When a complex object instance has to be stored, it will have to be taken apart and parts of it will be stored in different tables. The designer of such a system has to decide on the structure of the tables to use to represent classes in the database. It should be emphasized that it is only the attribute values of object instances that are stored in an RDBMS; operations are implemented in the programming language used.

There are two ways in which classes can be converted to tables in a relational database. The first, normalization, is suitable for decomposing complex objects into tables. It is used in systems that are not object-oriented to design the structure of tables in databases. It can also be used during object design to simplify large complex objects that are not cohesive (see Chapter 14). The second approach is based on a series of rules of thumb that can be applied to classes in a class diagram to produce a set of table structures. In this section, we describe these two approaches, and in Section 18.8 we discuss the impact that this will have on the design of classes in the system.

18.5.2 Data modelling and normalization

In order to store the objects from an object-oriented system in a relational database, the objects must be flattened out. *Normalization* is an approach that is also used to convert the complex structures in business documents into tables to be stored in a relational database. A typical example of its use would be to design a set of tables to hold the data in a sales order like the FoodCo example in Figure 16.3. How then do we apply normalization? Normalization is based on the idea of *functional dependency*.

This kind of dependency was mentioned in Section 14.8. It was not discussed in detail there and is explained here.

For two attributes, A and B, A is functionally dependent on B if for every value of B there is exactly one value of A associated with it at any given time. This is shown as:

$$\mathbf{B} \rightarrow \mathbf{A}$$

Attributes may be grouped around functional dependencies according to the rules of normalization to produce normalized data structures that are largely free of redundancy. There are five *normal forms* of normalized data. The data is free of redundancy in *fifth normal form*. For practical purposes it is usually adequate to normalize data into *third normal form*. Normalization is carried out in a number of steps, and we shall apply these to an example from the Agate case study.

Analysis activity during a further iteration has identified a class called `InternationalCampaign`, and the attribute values of two instances of this class are shown in Figure 18.7. As stated earlier, we may wish to decompose this class into simpler classes because it is not cohesive (as suggested in Chapter 14) or we may need to decompose it into table structures for storage using an RDBMS. The same approach is used in both cases. Here we are applying normalization as part of the design for a relational database.

The first step is to remove any calculated values (derived attributes). There are no examples of this in the `InternationalCampaign` instances.

We now create a relation that is said to be in unnormalized form. Each `InternationalCampaign` is uniquely identified in this system by its `campaignCode`. This is the *primary key* attribute. Figure 18.8 shows the data from these instances in a table. Each `InternationalCampaign` is represented in a single row. Note that there are multiple values in some of the columns in each row. These relate to the locations where the campaign will run and the manager in each location.

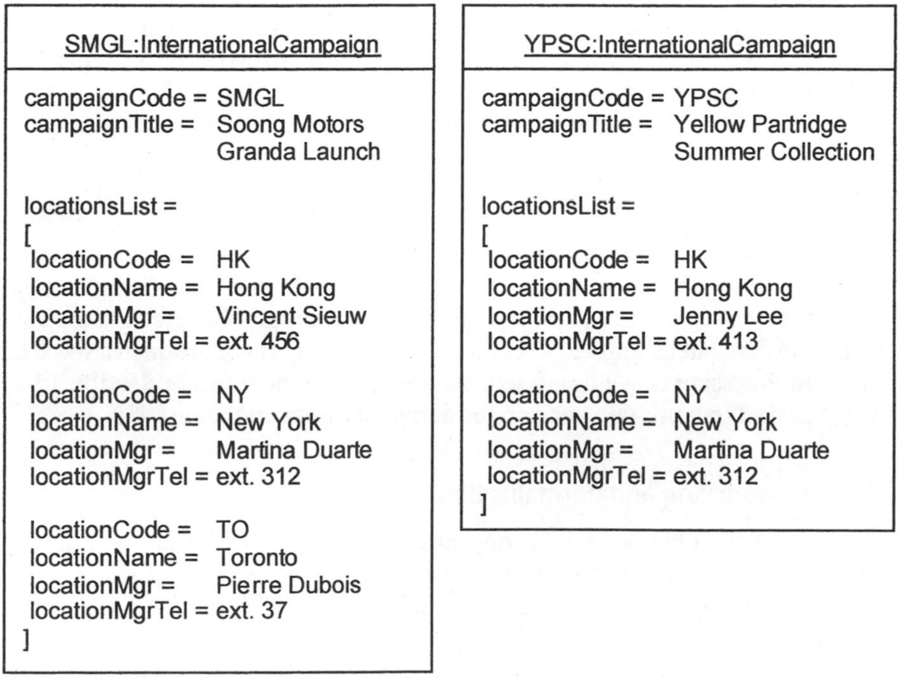

```
SMGL:InternationalCampaign

campaignCode = SMGL
campaignTitle = Soong Motors
                Granda Launch

locationsList =
[
 locationCode = HK
 locationName = Hong Kong
 locationMgr = Vincent Sieuw
 locationMgrTel = ext. 456

 locationCode = NY
 locationName = New York
 locationMgr = Martina Duarte
 locationMgrTel = ext. 312

 locationCode = TO
 locationName = Toronto
 locationMgr = Pierre Dubois
 locationMgrTel = ext. 37
]
```

```
YPSC:InternationalCampaign

campaignCode = YPSC
campaignTitle = Yellow Partridge
                Summer Collection

locationsList =
[
 locationCode = HK
 locationName = Hong Kong
 locationMgr = Jenny Lee
 locationMgrTel = ext. 413

 locationCode = NY
 locationName = New York
 locationMgr = Martina Duarte
 locationMgrTel = ext. 312
]
```

Figure 18.7 Example InternationalCampaign objects.

InternationalCampaign

campaign Code	campaign Title	location Code	location Name	locationMgr	location MgrTel
SMGL	Soong Motors Granda Launch	HK NY TO	Hong Kong New York Toronto	Vincent Sieuw Martina Duarte Pierre Dubois	456 312 37
YPSC	Yellow Partridge Summer Collection	HK NY	Hong Kong New York	Jenny Lee Martina Duarte	413 312

Figure 18.8 Table for sample International Campaigns.

InternationalCampaign-1

campaign Code	campaign Title	location Code	location Name	locationMgr	location MgrTel
SMGL	Soong Motors Granda Launch	HK	Hong Kong	Vincent Sieuw	456
SMGL	Soong Motors Granda Launch	NY	New York	Martina Duarte	312
SMGL	Soong Motors Granda Launch	TO	Toronto	Pierre Dubois	37
YPSC	Yellow Partridge Summer Collection	HK	Hong Kong	Jenny Lee	413
YPSC	Yellow Partridge Summer Collection	NY	New York	Martina Duarte	312

Figure 18.9 Revised table for International Campaigns without repeating groups.

A table is in *first normal form* (1NF) if and only if all row/column intersections contain atomic values. The table in Figure 18.8 does not conform to this criterion and must be redrawn as in Figure 18.9. These multiple values are often known as *repeating groups*.

The `campaignCode` no longer uniquely identifies each row in the table. Each row is identified by a combination of `campaignCode` and `locationCode`. These attributes form a candidate *primary key* for the table.

The data values have been flattened out into a two-dimensional table and could now be stored in a relational database as they are. However, there is redundancy that we want to eliminate from the data. If redundant data is held in the database, there is the risk that values will not be updated correctly. For example, if Martina Duarte's telephone extension number changes, the system must ensure that it is correctly updated in every row in which it appears. This is inefficient and prone to error.

The next step is to convert these relations to *second normal form* (2NF). A relation is in 2NF if and only if it is in 1NF and every non-key attribute is fully dependent on the primary key. Here the attribute `campaignTitle` is only dependent on `campaignCode`, and `locationName` is only dependent on `locationCode`. (These are sometimes called part-key dependencies.) The other attributes are dependent on the whole primary key. (Remember A is dependent on B if for every value of B there is exactly one value of A associated with it at a given time.) Figure 18.10 shows the creation of two new relations `Campaign` and `Location`.

The next step is to convert the tables to *third normal form* (3NF). A relation is in 3NF if and only if it is in 2NF and every attribute is dependent on the primary

InternationalCampaign-2

campaign Code	location Code	locationMgr	location MgrTel
SMGL	HK	Vincent Sieuw	456
SMGL	NY	Martina Duarte	312
SMGL	TO	Pierre Dubois	37
YPSC	HK	Jenny Lee	413
YPSC	NY	Martina Duarte	312

Campaign

campaign Code	campaignTitle
SMGL	Soong Motors Granda Launch
YPSC	Yellow Partridge Summer Collection

Location

location Code	location Name
HK	Hong Kong
NY	New York
TO	Toronto

Figure 18.10 2NF tables.

InternationalCampaign-3

campaign Code	location Code	locationMgr
SMGL	HK	Vincent Sieuw
SMGL	NY	Martina Duarte
SMGL	TO	Pierre Dubois
YPSC	HK	Jenny Lee
YPSC	NY	Martina Duarte

LocationManager

locationMgr	location MgrTel
Vincent Sieuw	456
Martina Duarte	312
Pierre Dubois	37
Jenny Lee	413

Campaign

campaign Code	campaignTitle
SMGL	Soong Motors Granda Launch
YPSC	Yellow Partridge Summer Collection

Location

location Code	location Name
HK	Hong Kong
NY	New York
TO	Toronto

Figure 18.11 3NF tables.

key and not on another non-key attribute. `Campaign` and `Location` are in 3NF, but in `InternationalCampaign-2 locationMgrTel` is dependent on `locationMgr` and not on the primary key. Figure 18.11 shows the tables in 3NF with the addition of a new table called `LocationManager`.

These relations can be shown in a diagram using the notation of entity–relationship diagrams, which are often used to represent the logical structure (conceptual schema) of relational databases. Note that this is not part of the UML notation. Figure 18.12 shows the relations of Figure 18.11 as an entity–relationship diagram. Some UML CASE tools can also produce entity–relationship diagrams, and some can generate the SQL statements to create the tables.

If we examine the attributes of these relations, we may come to the conclusion that `InternationalCampaign` was not a very well analysed class in the first place. It should perhaps be a subclass of `Campaign`. `LocationManager` appears to be nothing more than `CreativeStaff` with an association to `InternationalCampaign`.

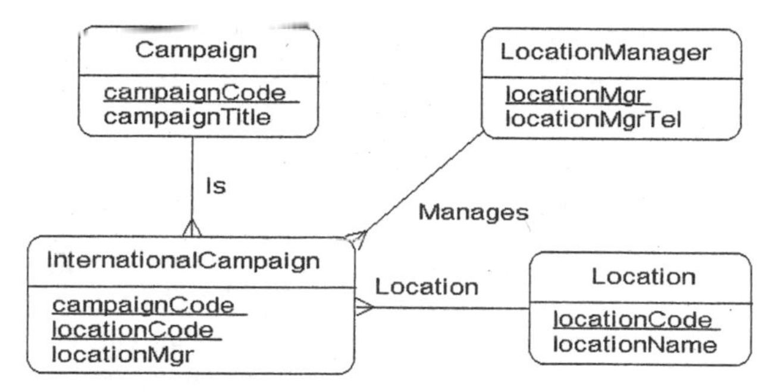

Figure 18.12 Entity–relationship diagram for tables of Figure 18.11.

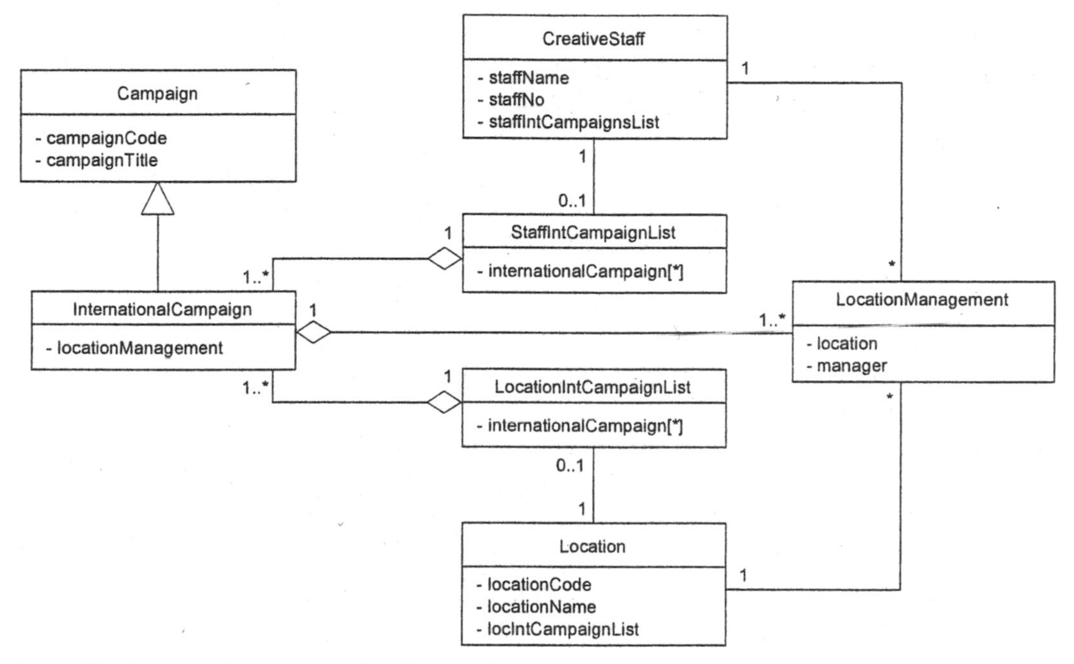

Figure 18.13 Excerpt from design class diagram for `InternationalCampaign`.

We also require a new class called `Location` with an association to `InternationalCampaign`. One possible design is shown in Figure 18.13. Note that in this class diagram we have included the collection classes that were introduced during object design (Chapter 14). There is presumably also an association between `StaffMember (the superclass of CreativeStaff)` and `Location`, but we should not rely on that to find out which Location an `InternationalCampaign` is running in, as the member of staff who is `LocationManager` could move offices.

The classes in Figure 18.13 raise a number of questions that should have been addressed during the analysis of requirements. Some of these are listed below.

- Does each campaign have a location? Is this the main location for an international campaign?
- Does each member of staff work in a location?
- Are there different versions of adverts for different locations?
- How are international campaigns costed? What currencies are used?

You can probably think of others. In an iterative life cycle, it is acceptable to be raising these issues at this stage. In a traditional waterfall life cycle, these issues should have been resolved during the analysis stage, but in reality may not have been.

18.5.3 Mapping classes to tables

An alternative approach to that provided by normalization is to follow a set of guidelines for how to map the classes and multiplicities in the class diagram to tables in a relational database design. A summary of the patterns that can be applied to this mapping can be found in Brown and Whitenack (1996). The following guidelines are derived from Rumbaugh et al. (1991) and Brown and Whitenack (1996).

- Classes with a simple data structure. These classes become tables.
- Object identifiers become primary keys. A unique identifier is generated for every object and can be used as a primary key in the relational table in which it is held. (Various schemes are available that guarantee a unique id for every object.)
- Classes that contain an instance of another class as an attribute. A separate table should be created for the embedded class. Objects of the embedded class should be allocated a unique object identifier. The object identifier should replace the embedded object in the table for the container class as a foreign key.
- Classes that contain collections. Allocate an object identifier to the class held in the collection. This class will be represented by a table. Create a separate table that contains two columns. The first holds the object identifiers of the objects that contain the collection; the second holds the object identifiers of the objects that are held in the collection.
- One-to-many associations can be treated like collections.
- Many-to-many associations become separate tables. Create a table that contains two columns. Each row contains a pair of object identifiers, one from each object participating in the association. (These are like two collections.)
- One-to-one associations are implemented as foreign key attributes. Each class gains an extra attribute in which to hold the object identifier of the associated object.

(A *foreign key* is used in relational databases to create the relationships between tables. `locationCode` in the `InternationalCampaign` table in Figure 18.11 is an example of a foreign key. Objects do not have keys, and this is why object identifiers are allocated to them. It may be possible to use an attribute that will have a unique value in each instance of a class as a foreign key.)

When a relational database is used, collection classes that exist only to provide access to a set of objects of the same class need not be part of the data that is stored in tables. If it is necessary to iterate through every instance of a particular class, this can be done by selecting every row from the table.

Inheritance poses more of a problem. There are three alternative ways of mapping an inheritance hierarchy to relational database tables.

- Only implement the superclass as a table. Attributes of subclasses become attributes of the superclass table and hold null values where they are not used. This approach is most appropriate where subclasses differ from their superclass more in behaviour than in attributes. A `type` attribute is required to indicate which subclass each row represents.

- Only implement the subclasses as tables. The attributes of the superclass are held in all the subclass tables. This only works if the superclass is abstract and there will be no instances of it.
- Implement all the classes (both superclass and subclasses) as separate tables. To retrieve the data for a subclass both its own table and the table of its superclass must be accessed. Again a `type` attribute may be required.

The solution that is chosen may depend on the requirements of the particular application or may be constrained by the use that will be made of the data in the database by other applications.

This brings us to a further aspect of relational databases: data is added to and retrieved from them using SQL statements. SQL provides both the DDL and DML for relational databases. Figure 18.14 shows the SQL statements necessary to create the tables of Figures 18.11 and 18.12 in Oracle generated by the CASE tool from the storage class diagram.

Figure 18.15 shows an SQL statement that finds all the international campaigns with the `locationName` 'Hong Kong'. There is a design decision to be made in deciding where to place the responsibility for this kind of requirement.

```
CREATE TABLE Campaign (
        campaignCode VARCHAR(6) NOT NULL,
        campaignTitle VARCHAR(50) NULL,
        PRIMARY KEY (campaignCode)
        );

CREATE TABLE InternationalCampaign (
        campaignCode VARCHAR(6) NOT NULL,
        locationCode VARCHAR(2) NOT NULL,
        locationMgr VARCHAR(30) NULL,
        PRIMARY KEY (campaignCode, locationCode)
        );

CREATE TABLE Location (
        locationCode VARCHAR(2) NOT NULL,
        locationName VARCHAR(20) NULL,
        PRIMARY KEY (locationCode)
        );

CREATE TABLE LocationManager (
        locationMgr VARCHAR(30) NOT NULL,
        locationMgrTel INT NULL,
        PRIMARY KEY (locationMgr)
        );
```

Figure 18.14 SQL statements to create tables of Figures 18.11 and 18.12.

```
SELECT campaignTitle FROM Campaign c, InternationalCampaign ic, Location l
        WHERE c.campaignCode = ic.campaignCode
        AND ic.locationCode = l.locationCode
        AND l.locationName = 'Hong Kong'
```

Figure 18.15 SQL statement to find campaigns running in Hong Kong.

- This SQL statement could be executed and only data for those objects that are required would be returned from the database. This replaces the interaction modelled in a sequence diagram with functionality provided by the DBMS.
- Alternatively, the data from all these tables could be retrieved from the database and used to instantiate the objects in the system. Each `InternationalCampaign` object could then be sent a message to check whether it includes the `Location` 'Hong Kong'. This will involve it sending a message to each associated `Location` object. This is more object-oriented, but will take a longer time to execute.
- The third alternative is to retrieve data from each table in turn, as though navigating through the structure of the class diagram, first the `InternationalCampaign` then each of the `Locations` for that `InternationalCampaign`. This approach requires use of indexes on the tables to make access possible in a reasonable time.

This kind of design decision trades off the pure object-oriented approach against the efficiency of the relational database. In order to retrieve this data into objects in a programming language such as Java, the SQL statements must be embedded in the program. During design, we have to decide which classes have the responsibility for accessing the database. In Section 18.6 we describe two different approaches to this design decision and show how they can be modelled in UML. However, before we address how we can model the database management responsibilities of the system, we need to consider object DBMS and what they have to offer the designer as an alternative to relational DBMS.

18.6 Designing for Object Database Management Systems

18.6.1 Object databases

Object DBMS differ from current relational DBMS in that they are capable of storing objects with all their complex structure. It is not necessary to transform the classes in the design model of the system in order to map them to storage objects. As you might expect, using an object database maintains the seamlessness that is claimed for object-oriented systems right through to the storage of objects in the database. Designing for an object database will have a minimal impact on the design of the system.

The standard for object databases has been set by the Object Data Management Group (ODMG) and is currently available in Version 2.0 (Cattell and Barry, 1997). The standard defines both the Object Definition Language (ODL) and the Object Manipulation Language (OML) for object databases. The ODL is similar to the DDL elements of SQL for relational databases but allows objects to maintain their complex structure: objects can contain other objects, including collections, as attributes. Figure 18.16 shows the ODL definition of the `InternationalCampaign`, `StaffMember` and `CreativeStaff` classes based on Figure 18.12 but with the `StaffIntCampaignList` as an embedded attribute of `CreativeStaff`. Note also that the client contact association with `Client` is shown as a one-to-one association to illustrate the syntax.

However, individual ODBMS do not necessarily conform to the standard. Eaglestone and Ridley (1998) describe both the ODMG standard and the syntax of O_2. Blaha and Premerlani (1998) describe how to map an object-oriented design to ObjectStore. Some object databases include a DML similar to SQL that can be used to query data in the database. Statements in this OML are usually embedded in the

```
interface     InternationalCampaign
(extent       international_campaigns
 key          (campaignCode, locationCode))
{
              attribute        String campaignCode;
              attribute        String locationCode;
}
interface     StaffMember
(extent       staff_members
 key          staffNo)
{
              attribute        Short staffNo;
              attribute        String staffName;
              attribute        Date staffStartDate;
}
interface     CreativeStaff: StaffMember
(extent       creative_staff
 key          staffNo)
{
              attribute        String qualification;
              attribute        List<InternationalCampaign>
                                              staffIntCampaignList
              relationship     Client isContact;
              inverse Client::hasContact;
}
```

Figure 18.16 ODL for `International Campaign`, `StaffMember` and `Creative Staff`.

```
IntCampaign * CreativeStaff::findIntCampaign ( string campaignCode )
{
        IntCampaign * intCampaignPointer;
        intCampaignPointer = staffIntCampaignList.getValue().query_pick(
                "IntCampaign*",
                "campaignCode == this->campaignCode",
                os_database::of(this));
        return intCampaignPointer;
}
```

Figure 18.17 Example query (highlighted in bold) in C++ for ObjectStore.

program code of an object-oriented language. Unlike SQL statements, which are usually built as strings and then executed by a call to a system function or passed to a database server, OML queries are extensions to the programming language that is being used. Figure 18.17 shows a C++ operation to query the collection of `InternationalCampaign` objects (abbreviated in the figure to `IntCampaign` to save space) associated with a particular `CreativeStaff` instance in order to find one that contains a particular `campaignCode`. The `campaignCode` would be passed as a parameter to the operation and not hard-coded as in the SQL of Figure 18.15.

Object databases provide much better facilities for navigating the structure of the classes in the class diagram. Figure 18.18 shows an excerpt from a class declaration and implementation in C++ that illustrates how the associations are created between classes using ObjectStore's macros. This excerpt declares an attribute of

```
os_relationship_1_m(CreativeStaff, isContact, Client, hasContact,
      os_Set<Client*>) isContact;

...

os_rel_1_m_body(CreativeStaff, isContact, Client, hasContact);
```

Figure 18.18 Example association declaration and implementation in C++ for ObjectStore.

`CreativeStaff` called `isContact`, the type of which is an `os_Set` of pointers to objects of type `Client`. The inverse association from `Client` is called `hasContact`. This technique makes it straightforward to implement the associations, and it is not necessary to manually insert object identifiers into the classes to handle the associations. This would make the collection classes of Figure 18.13 unnecessary.

Where an object contains a reference to another object as an attribute, it is possible to directly invoke its public methods or access its public attributes even if the object in question is not in memory. The great benefit of object databases is that they transparently materialize objects from the database into memory without the need for special action to be taken by the developer. Typically the only requirement is for some part of the application to make a connection to the database and for operations that read from or write to the database to be bracketed by commands to start and finish a transaction. This approach is used, for example, in ObjectStore for Java and C++.

The structure of the class diagram will require minimal changes to be used with an object database. Minor changes may be required. For example, if you use ObjectStore Persistent Storage Engine (PSE Pro) with Java you will find that some of the Java collection classes cannot be made persistent. However ObjectStore provides persistent equivalents for these classes. For example, wherever you have used a class such as a Hashtable and require it to be persistent, it must be replaced with an instance of the class OSHashtable.

Although object DBMSs make it easier to map an object-oriented design straight to a database, there is one area where the seamlessness breaks down. Apart from very simple operations, such as those to insert new values or to do simple arithmetic, ODL and hence object databases do not support the storage of operations in the database. Operations must still be implemented in an object-oriented programming language such as Java or C++.

18.7 Distributed Databases

Before we consider how to design the classes that handle data management for a database that is not object-oriented, it is worth considering the issue of design for distributed databases.

In a simple system, objects can be stored on the local machine in some kind of database, brought into the memory of the machine and sent messages to invoke their operations before being saved back to the database and removed from memory. This is not always possible. The objects that participate in a particular use case need not be on the same machine as the user interface, indeed they may be distributed on different machines. This is the essence of architectures based on multiple tiers or layers, which we discussed in Chapter 13 and used as the basis of the user interface design in Chapter 17.

There are several ways of communicating between layers on distributed machines, for example Remote Procedure Calls (RPC) in languages such as C or C++ and Remote

Method Invocation (RMI) in Java. The object-oriented industry standard, however, is CORBA (the Common Object Request Broker Architecture). CORBA Version 2.0 is defined as a standard by the Object Management Group (OMG, 1995), a different group from the ODMG. More recently, Simple Object Access Protocol (SOAP) has emerged as a contender in this marketplace. SOAP uses XML to encapsulate messages and data that can be sent from one process to another. We explain CORBA here as an example of this kind of communication protocol.

CORBA separates the interface of a class (the operations that it can carry out) from the implementation of that class. The interface can be compiled into a program running on one computer. An object instance can be created or accessed by name. To the client program it appears to be in memory on the same machine. However, it may actually be running on another computer. When the client program sends it a message to invoke one of its operations, the message and its parameters are converted into a format that can be sent over the network (known as marshalling); at the other end the server unmarshals the data back into a message and parameters and passes it to the implementation of the target object. This object then carries out the operation and, if it returns a value, that value is marshalled on the server, unmarshalled on the client, and finally provided as a return value to the client program. The marshalling and unmarshalling process makes it possible to pass objects with a complex structure over the network by flattening them out and reconstructing them at the other end.

CORBA achieves this by means of programs known as ORBs (Object Request Brokers) that run on each machine. The ORBs communicate with each other by means of an Inter-ORB Protocol (IOP). Over the Internet, the protocol used is IIOP (Internet IOP). To use this facility, the developer must specify the interface for each class in an Interface Definition Language (IDL). IDL defines the interface of the class in terms of its public attributes and operations. Figure 18.19 shows a sample IDL file for the `Location` class. The IDL file is then processed by a program that converts the interface to a series of files in the target language or languages. In the case of Java, this program is called IDL2JAVA and produces several files. These files include:

- a file that defines the interface in Java,
- a stub file that provides the link between the client program and the ORB,
- a file that provides a skeleton for the implementation on the server.

There are a number of other files that may be generated, but these are the important ones. The stub file implements the interface on the client and is compiled into the client program; the skeleton file also implements the interface (but on the host) and is amended by the developer to provide the implementation and compiled on the host.

Systems developed using CORBA can be set up so that the remote objects are located on a named machine and accessed by name. However, CORBA also provides services

```
module Agate
{  interface Location
   { attribute string locationCode;
     attribute string locationName;
     void addIntCampaign( in IntCampaign campaign );
     void removeIntCampaign( in string campaignCode );
     int numberOfCampaigns( );
   };
};
```

Figure 18.19 CORBA IDL for `Location` class.

for locating objects by name when it is not known where they are running. It also includes sophisticated services for locating objects that implement a certain interface and for interrogating an object to determine its interface (operations, parameter types and return types) in order to dynamically invoke its operations. However, for the vast majority of applications, these services are not required.

CORBA is known as *middleware*, as it acts as an intermediary between clients and servers. As such it enables the implementation of a three- or four-tier architecture that isolates the user interface and client programs from the implementation of classes on one or more servers. CORBA also provides interoperability between different languages: a Java client program can invoke operations on a C++ or Smalltalk object that exists on a separate machine. CORBA also makes it possible to encapsulate pre-existing programs written in non-object-oriented languages by wrapping them in an interface. To the client it looks like an object, but internally it may be implemented in a language such as COBOL.

18.8 Designing Data Management Classes

18.8.1 The layered architecture

In designing the boundary classes, one of our aims has been to keep them separate from the business logic and the entity objects.

In designing the presentation layer one of our aims was to isolate the entity classes in the system from the way that they are presented on screen and in reports and documents. We did this in order to maximize the reusability of the classes. Our aim is to do the same with the data storage layer. Classes are less reusable if they are tightly coupled to the mechanisms by which instances are stored in some kind of file system or database. We also want to decouple the entity classes from the business logic.

What are the options for locating the operations that handle the tasks of storing and retrieving objects?

1. We could add operations to each class to enable objects to save and store themselves.
   - This reduces reusability. Each class must now contain code to implement the operations that couple the class to the particular data storage mechanism used.
   - This breaches the idea of cohesion. Each business class should contain operations that are relevant to the behaviour of instances of that class, data storage methods belong in a data storage class.
   - If an object is not currently instantiated, how can we send it a message to invoke an operation to load itself?
2. We can get around this last problem by making the storage and retrieval operations class-scope methods rather than instance-scope methods (static methods in Java or static member functions in C++).
   - This still suffers from the first two problems listed above for option 1. The class is less reusable and lacks cohesion.
3. All persistent objects in the system could inherit methods for storage from an abstract superclass—`PersistentObject` for example rather than `Object` in Java.

- This has the effect of strongly coupling existing classes to the `PersistentObject` superclass, so all business classes end up inheriting from a utility class.

4. Where we have introduced collection classes into the design to manage collections of objects, we could make these collection classes responsible for storing and retrieving object instances.
   - This is closer to a solution. The collection classes are design artefacts—not part of the business classes in the system. However, we may wish to reuse the design, and we are coupling it to the storage mechanisms.
5. We could introduce separate classes into the system whose role is to deal with the storage and retrieval of other classes. This is the database broker approach.
   - This solution fits the layered architecture. These classes are part of the data storage layer.
   - The data storage classes are decoupled from the business classes. The business classes will contain nothing that indicates how they are to be stored. The same business classes can be reused unchanged with different storage mechanisms.
6. We could limit the number of new data storage classes to one. Different instances of this class would be created with attributes to hold the names of tables or files that are to be used to store and retrieve instances of their associated classes.
   - This parameterized version is more difficult to set up and more difficult to implement.
   - It requires some part of the system outside the database broker class to know what parameters to set for each instance that is created.

Option 5 is the approach that is favoured by most developers of object-oriented systems. It involves the use of a number of patterns. Larman (1998) describes it in some detail as a *persistence framework*, the main feature of which is the use of *database brokers*, which mediate between the business classes and the persistent storage and which are responsible for storing and retrieving objects. However, we shall first describe the use of option 3—inheritance from a persistent superclass—before looking at option 5.

18.8.2 PersistentObject superclass

A simple approach to the design of data storage classes is to design an abstract superclass `PersistentObject` that encapsulates the mechanisms for an object of any class to store itself in and retrieve itself from a database. Eriksson and Penker (1998) use this approach in order to keep their example case study application simple and independent of any vendor's DBMS. The `PersistentObject` superclass implements operations to get an object by object identifier, to store, delete and update objects and to iterate through a set of objects. These operations are implemented in terms of two abstract operations, to write and read objects, that must be implemented by each subclass that inherits from the `PersistentObject` superclass. This is shown in Figure 18.20 (adapted from Eriksson and Penker, 1998). This approach also uses an aspect of option 2, as the `getObject()` operation is a class-scope method rather than an instance-scope method (as are the others underlined in the class diagram of Figure 18.20).

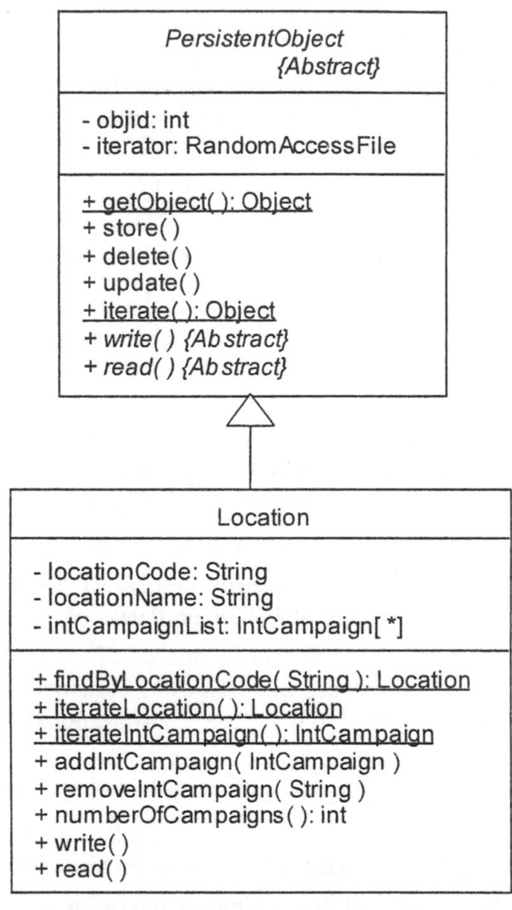

Figure 18.20 Excerpt from class diagram to show inheritance from `PersistentObject`.

The `PersistentObject` hides some of the detail of the implementation of the data storage from the business objects in the application. However, they must implement the `write()` and `read()` operations, and this will limit their reusability.

This approach does have the benefit of limiting the changes that will be made to sequence diagrams. Messages that have been shown being sent to object instances to select an instance or to iterate through a set of instances can be shown as being sent to the class rather than the instances. Figure 18.21 shows an example of this for the use case `Get number of campaigns for location`. We have shown the class `Location` using a constructor `Location()` to make the particular instance `:Location` available. While the object instance is being created in memory for this instance of the collaboration, strictly speaking, it already exists as an object and is just being *materialized* from the database.

However, the use of a persistent superclass is unlikely to be robust enough for business applications, and a more sophisticated approach, such as the database broker must be used.

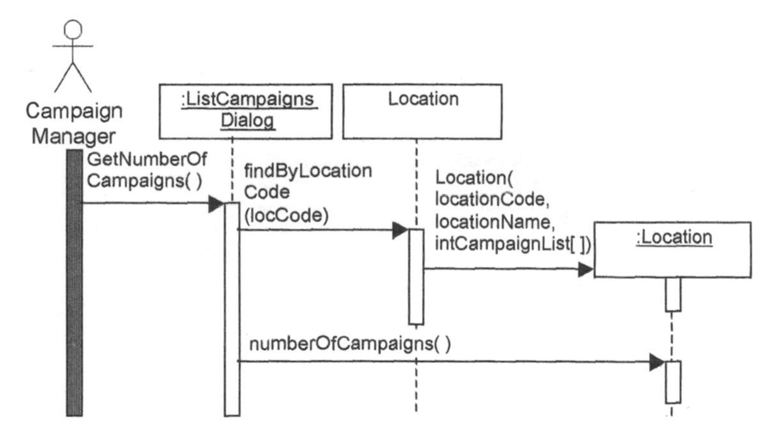

Figure 18.21 Sequence diagram for `Get number of campaigns for location` showing `Location` retrieving (or materializing) an instance.

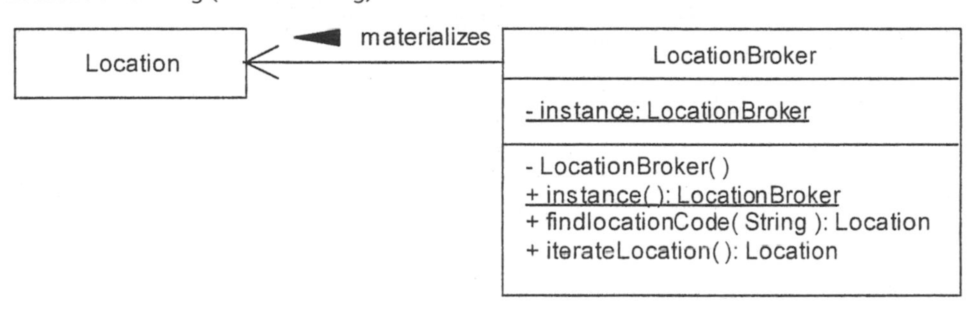

Figure 18.22 `Location` and `LocationBroker` classes.

18.8.3 Database broker framework

The database broker framework separates the business objects from the data storage implementation. The classes that provide the data storage services will be held in a separate package.

Our objective here is to separate the data storage mechanisms completely from the business classes. For each business class that needs to be persistent, there will be an associated database broker class that provides the mechanisms to materialize objects from the database and dematerialize them back to the database. A simple form of this is shown in Figure 18.22 for the `Location` class. The `LocationBroker` is responsible for the storage and retrieval of `Location` object instances. In order to ensure that there is only ever one `LocationBroker` instance, we can use the Singleton pattern (see Chapter 15). This means that we use a class-scope operation, but only to obtain an instance of the `LocationBroker` that can be used subsequently to access the database. The sequence diagram involving the `LocationBroker` is very similar to that of Figure 18.21, and is shown in Figure 18.23.

(Note that this diagram does not show the creation of the instance of `Location Broker`.)

Each persistent class in the system will require a broker class, so it makes sense to create a superclass that provides the services required by all these broker classes. Larman (1998) suggests two levels of generalization. At the top of his hierarchy is an abstract `Broker` class that provides the operation to materialize an object using its object identifier. This is then subclassed to produce different abstract classes of brokers for different kinds of storage, for example, one for a relational database and

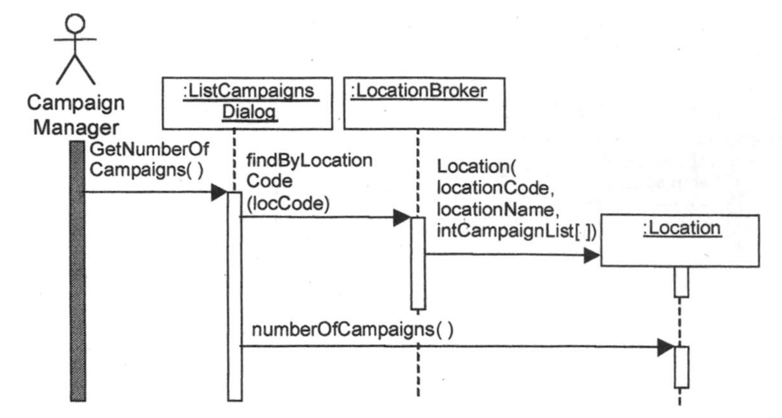

Figure 18.23 Sequence diagram for `Get Number of Campaigns for Location` showing `LocationBroker` retrieving an instance of `Location`.

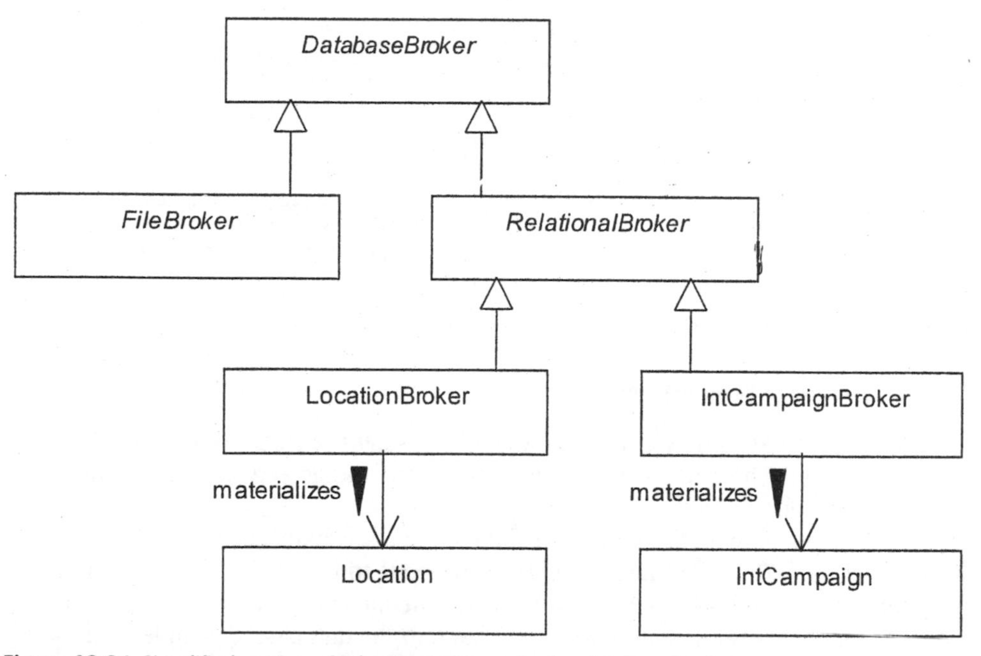

Figure 18.24 Simplified version of inheritance hierarchy for database brokers.

one for a file system. Finally, the appropriate broker is subclassed into the concrete classes for each persistent class in the system. A simplified version of this inheritance hierarchy is shown in Figure 18.24.

In the Agate case study, we could use JDBC to link the Java programs to a relational database. This will require the use of classes from the `java.sql` package, in particular `Connection`, which is used to make a connection to the database, `Statement`, which is used to execute SQL statements, and `ResultSet`, into which the results of SQL Select statements are placed (we can then iterate through the `ResultSet` retrieving each row in turn and extracting the values from each column). An appropriate driver will also be required. Figure 18.25 shows the associations between the `Relational Broker` abstract class and these other classes. The figure shows the Oracle JDBC driver; to access a database such as Access via a link from JDBC to ODBC, the appropriate driver would be `sun.jdbc.odbc::JdbcOdbcDriver`.

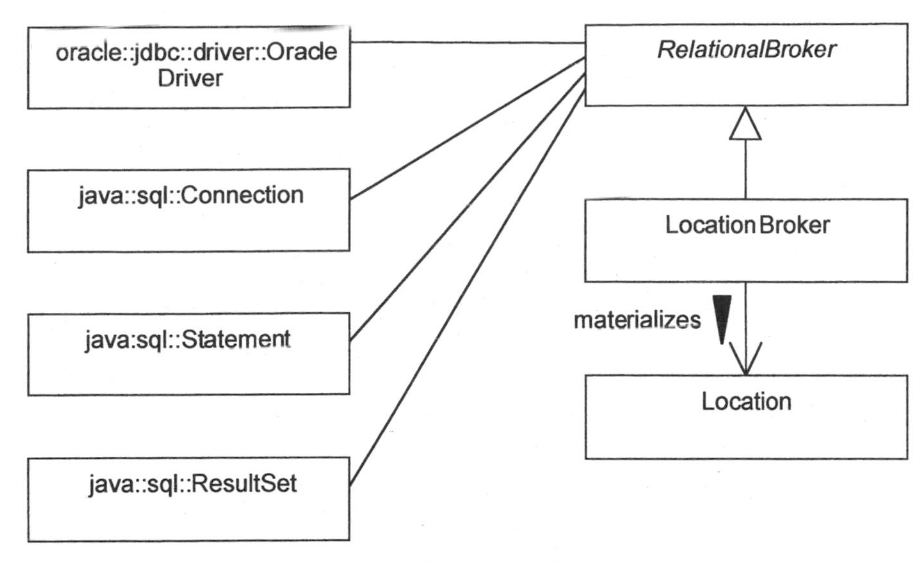

Figure 18.25 `RelationalBroker` class and classes from other packages.

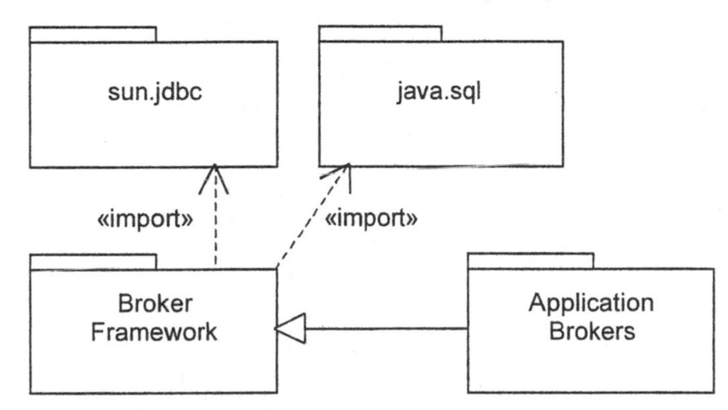

Figure 18.26 Class diagram showing packages for database brokers package.

The dependencies between the application classes and those in other packages can be shown using packages in a class diagram, as for the classes in the presentation layer in Chapter 17, Figure 17.12. This is shown in Figure 18.26.

This simple framework using database broker classes can be extended to deal with some of the problems that remain.

- The most important problem concerns the way in which persistent objects maintain references to other objects. If the `LocationBroker` retrieves an instance of `Location`, what happens when an operation of the Location requires it to send a message to one of its `IntCampaigns`. The `IntCampaign` will not necessarily have been retrieved from the database. The same applies to the many other operations that involve collaboration between objects.
- The second problem concerns the ability to manage transactions in which a number of objects are created, retrieved from the database, updated and deleted.

Two extensions to the database broker framework can be used to resolve these problems. The first uses the Proxy pattern to provide proxy objects for those objects that have not yet been retrieved from the database. The second uses caches to hold objects in memory and keep track of which have been created, updated or deleted.

The Proxy pattern (Gamma et al., 1995) provides a proxy object as a placeholder for another object until it is required. In this case, we can use proxies for each business class to link to, where there is an association with another object or objects. If no message is sent to the associated objects, then the proxy does nothing, if a message is sent then the proxy asks the relevant database broker to retrieve the object from the database, and once it has been materialized, the proxy can pass the message directly to it. Subsequently, messages can be sent directly to the object by the proxy, or the proxy can replace the reference to itself in the object that sent the message with a reference to the real object. For this to work, the proxy must hold the object identifier of the object that it is a placeholder for. When the object itself is retrieved from the database, the object identifier is effectively transformed into a reference to the object itself. The proxy class must also implement the same interface as the real class so that it appears to other objects as if it is the real thing.

Caches can be combined with this approach. The database broker can maintain one or more caches of objects that have been retrieved from the database. Each cache can be implemented as a hashtable, using the object identifier as the key. Either a single cache is maintained and some mechanism is used to keep track of the state of each object, or six caches can be maintained:

- new clean cache—newly created objects,
- new dirty cache—newly created objects that have been amended,
- new deleted objects—newly created objects that have been deleted,
- old clean cache—objects retrieved from the database,
- old dirty cache—retrieved objects that have been amended,
- old deleted objects—retrieved objects that have been deleted.

As objects are changed, the broker must be notified so that it can move them from one cache to the other. This can be achieved using the Observer–Observable pattern: the object implements Observable, and the broker inherits from Observer.

When the transaction is complete, the broker can be notified. If the transaction is to be committed, the broker can process each object according to which cache it is in:

- new clean cache—write to the database,
- new dirty cache—write to the databasc,
- new deleted objects—delete from the cache,
- old clean cache—delete from the cache,
- old dirty cache—write to the database,
- old deleted objects—delete from the database.

The cache or caches can be used by the proxy object to check whether an object is already available in memory. When it receives a message, the proxy can ask the broker for the object, if it is in a cache, the broker will return a reference to it directly, if it is not in the cache, the broker will retrieve it. Figure 18.27 shows the associations between the broker class, the caches and the proxy.

The collaboration between these classes can be seen in the collaboration diagram in Figure 18.28, which represents the following interaction.

A `Location` is in memory with an `IntCampaignProxy` as a placeholder for the real `IntCampaign` object that runs in that `Location`. In order to print a list of campaigns, the `title` of the `IntCampaign` is required.

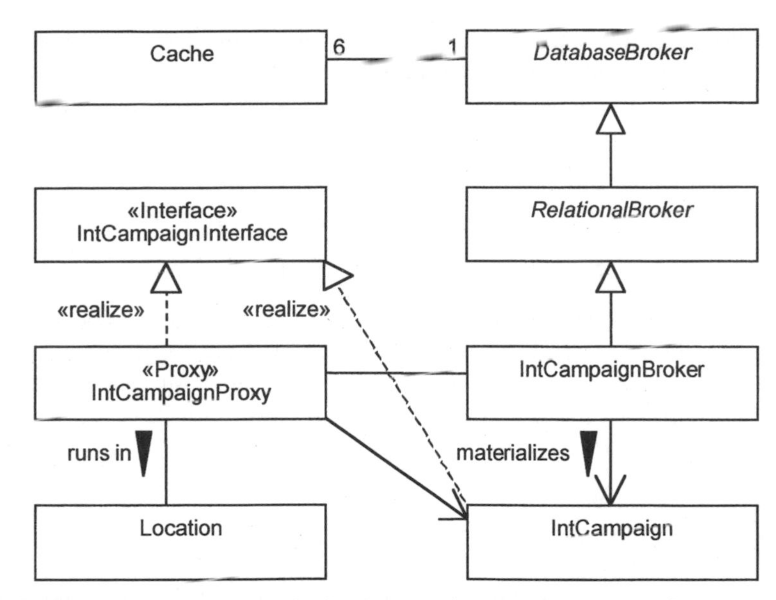

Figure 18.27 Extension of the database broker framework to include caches and proxies.

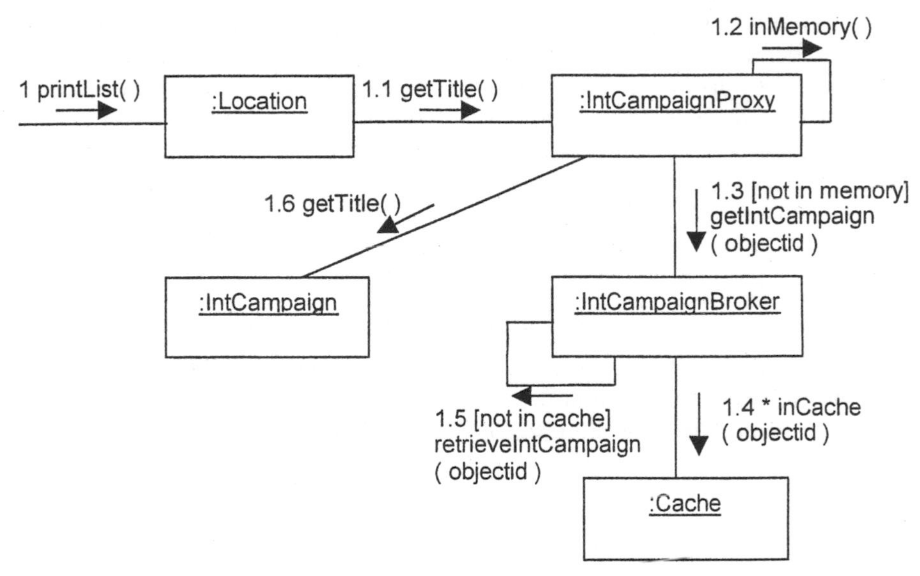

Figure 18.28 Collaboration diagram showing proxy, broker and cache objects collaborating to retrieve an object instance.

1.1 The `Location` sends the message `getTitle()` to the `IntCampaignProxy`.

1.2 The `IntCampaignProxy` checks whether the `IntCampaign` is in memory.

1.3 It is not, so it then requests the object by object identifier from the broker, `IntCampaignBroker`.

1.4 The `IntCampaignBroker` checks if the object is in a `Cache`.

1.5 It is not, so the `IntCampaignBroker` retrieves the object from the database and returns it to the `IntCampaignProxy`.

1.6 The `IntCampaignProxy` sends the `getTitle()` message to the `IntCampaign` object and returns the result to the `Location`.

This may appear to be overkill, but it enables us to maintain our objective of decoupling the business classes from the data storage mechanisms that are provided by the classes in the persistence framework. This should make it possible to migrate the data storage to a different platform without having an impact on the business classes or the application logic, simply by replacing the database broker class with the broker for a different database.

18.8.4 Using a data management product or framework

In the previous two sections we have considered ways of designing a persistence mechanism yourself. However, this is not necessary, as there are a number of products and frameworks available that provide a persistence mechanism for you. We shall consider two commercial products that will handle the mapping of objects to relational database tables, and the frameworks in the Java 2 Enterprise Edition (J2EE) that handle persistent objects.

Object–table mappings

Webgain Toplink Foundation Library for Java is a product that will take classes and map their attributes to columns in relational database tables. It can either map attributes to columns in existing tables, or it can generate the schema for the necessary tables from a class definition. It also generates Java classes to provide the persistence mechanism. There are versions of Toplink that work with application servers: BEA's Web Logic and IBM's WebSphere. Another product is CocoBase from Thought Inc. This is a single product that will either produce Java classes for what they call 'transparent persistence' or will generate code to work with the Enterprise Java Beans (EJB) 1.1 standard.

Application servers

In the last few years there has been a move from object and object-relational databases towards application servers. Java application servers provide an environment in which Enterprise Java Beans can run. EJB is part of the J2EE framework developed by Sun and other interested parties. J2EE provides a number of standards for building enterprise Java systems, primarily systems that will use the Web to deliver services. Various software providers have produced commercial and open source implementations of J2EE.

Application servers provide a mechanism for managing distributed objects or components. In the Java world, these objects and components are typically Enterprise Java Beans (EJBs). EJBs can be either session beans, components that encapsulate a piece of business logic, or entity beans, components that encapsulate one or more entity classes. Application servers provide a way for client applications to communicate with the business logic layer and for the business logic layer to communicate with entity objects. They can provide persistence services for the entity objects, storing them in a database and retrieving them when required. Some application servers come with their own database; others can be run in conjunction with a range of databases from other suppliers. As with CORBA, described in Section 18.7, there is work to be done in defining interfaces and writing classes that implement those interfaces, but it is likely to be significantly less work than developing your own persistence mechanism, and with the use of tools like TopLink and CocoBase, the amount of work can be reduced.

Using EJB and an application server also makes it possible to design systems in which the same entity objects support Web-based and conventional client applications.

18.9 Summary

The design of persistent storage for object-oriented systems is not straightforward. For simple systems it is possible to use files to store the objects. However, commercial systems require a more robust and sophisticated approach so that objects can be shared between applications and users. Database management systems provide the facilities to build robust, commercial strength information systems and offer a number of advantages. Object DBMS can be used and will have a less significant impact on the design of the classes in the system than if a relational DBMS is used. However, many organizations have an existing investment in a relational DBMS, and it may be necessary to build the system to use this database. In this case it is necessary to design tables, either by normalizing object instances or by following a set of guidelines for mapping classes and associations to tables. To decouple the business objects from the persistent storage mechanism, a persistence framework can be designed that can be extended to handle the resolution of object identifiers into references to real objects and that can use caches to manage transactions involving multiple objects.

The design of the persistent data storage mechanisms should ideally be carried out in conjunction with the object design activities of Chapter 14. If an object DBMS is being used, it may provide mechanisms by which to implement collection classes to handle associations (as discussed in Section 14.5), and class diagrams and interaction diagrams should reflect this. On the other hand, if a relational DBMS is being used, then the broker classes will play the role of collection classes and provide operations such as `findFirst()`, `getNext()` and `findObject()`, shown for example in Figure 14.15. If the object design is carried out before the data management design, then the collection classes that are produced as part of an implementation-independent design model can be converted to database broker classes at a later stage.

Benefits can be gained from using persistence frameworks and tools to develop the storage mechanisms in a system rather than designing your own.

Review Questions

18.1 Give one example each of a persistent and a transient object.

18.2 Explain the difference between different types of file organization and file access.

18.3 Of the different kinds of record type listed in Section 18.3.1 suggest which would be most appropriate for storing complex nested objects. Explain the reasons for your choice.

18.4 Outline the advantages and disadvantages of using a database management system over developing an application using files.

18.5 What is the key difference between a relational DBMS and an object DBMS?

18.6 List in your own words the three steps used in going from an unnormalized relation to a relation in third normal form.

18.7 What are the three ways of mapping the classes in an inheritance hierarchy to tables?

18.8 What is meant by OML and ODL?

18.9 What is the difference between ODBC and CORBA as ways of connecting to databases?

18.10 Explain what is meant by (i) a broker and (ii) a proxy.

Case Study Work, Exercises and Projects

18.A Find out what you can about localization mechanisms in a programming language or environment such as Java or Windows. What use do they make of files?

18.B Normalize the data in the Agate invoice in Figure 6.1. (Remember to remove the calculated values first.)

18.C Normalize the data in the FoodCo sales order entry screen of Figure 16.3. (Remember to remove the calculated values first.)

18.D Use the guidelines in Section 18.4.2 to decide on the tables necessary to hold the classes of Figure 14.12.

18.D Find information about a relational DBMS and an object DBMS and write a short report comparing the features they offer.

18.E Extend the sequence diagram of Figure 17.18 to show the use of a proxy class and database broker.

18.F Redraw your answer to 18.E as a collaboration diagram.

18.G Draw collaboration diagrams similar to Figure 18.28 to show what happens (i) when the `IntCampaign` is already in memory, and (ii) when it is in one of the caches.

Further Reading

- Codd's (1970) paper on relational databases was reprinted in the 25th Anniversary issue of the Communications of the ACM, which is more likely to be available in a library (1983, 26(1) pp. 64–69). This 25th Anniversary Issue is well worth looking at for other papers by some of the greats of computer science.
- Silberschatz et al. (1996) provide a good overview of database theory. Howe (2001) explains normalization in detail.
- Loomis (1995) deals with background to object databases and the functionality they offer, whereas Eaglestone and Ridley (1998) present the ODMG standard and provide a worked example case study using O_2.
- Java 2 Enterprise Edition (J2EE) provides a range of tools for developing scaleable distributed systems using Java. In particular, Enterprise Java Bean (EJB) entity beans are used to provide persistent storage for objects, and can provide a mapping between classes and tables in a relational database. Alur, Crupi and Malks (2001) explain how to use a number of design patterns with J2EE to get the best from the framework it provides.
- There are links to the web pages of the standards bodies—ANSI, ISO, ODMG and OMG—and to the web pages of various database providers in the website for this book.

CHAPTER A5

Agate Ltd Case Study—Design

Agate Ltd

A5.1 Introduction

In this chapter we show how part of the Analysis Model presented in Chapter A4 has been modified by the activities of design. The design activities have been concerned with finalizing the software architecture, designing the entity classes, their attributes, operations and associations, designing the boundary classes and the human–computer interaction, designing the mechanisms used for data storage, and designing the control classes. These activities have been explained in Chapters 13, 14, 15, 16, 17 and 18.

The following sections include:

- package diagrams to illustrate the overall software architecture;
- class diagrams to illustrate the classes in the design model;
- sequence diagrams to illustrate the interaction between instances of classes;
- a statechart diagram for the control of the user interface.

A5.2 Architecture

The architecture of the system has been designed to use Java Remote Method Invocation (RMI) for communication between the client machines and the server[1]. Control classes have been split into two layers. First there are the control classes that reside on the client machines (in the package `Agate Control Client`) and manage the interaction between users and the boundary classes. These control classes are essentially those that were designed in Chapter 17. Second, there are control classes that reside on the server. These control classes handle the interaction between the

[1] To meet the non-functional requirements relating to the distribution of the system, we will need a more complex architecture than this. The eventual solution will probably involve Java 2 Enterprise Edition (J2EE) and Enterprise Java Beans (EJB), and will require the use of application server software. For now we are presenting a design that is not so dependent on an expensive application server.

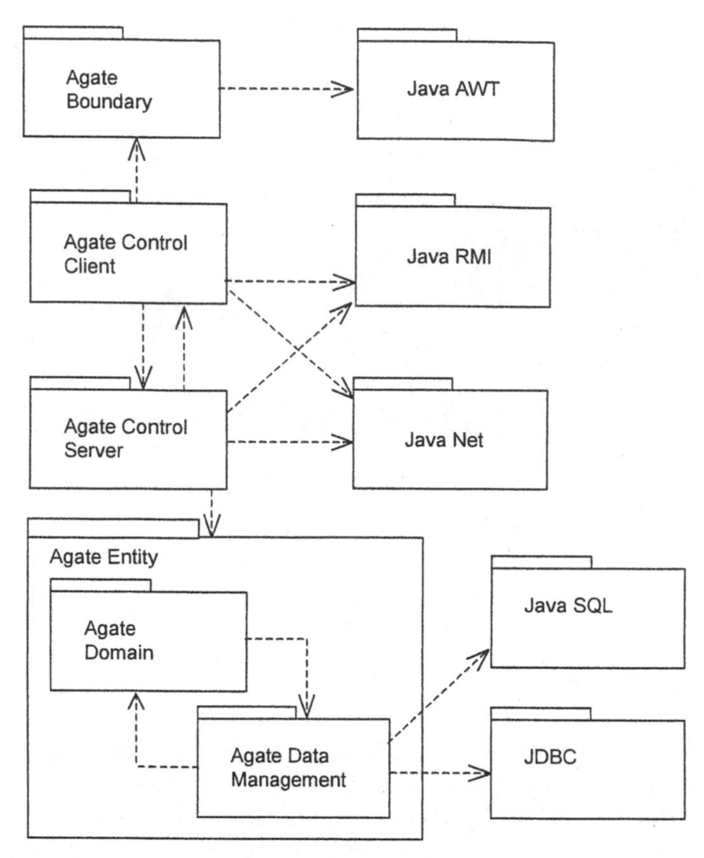

Figure A5.1 Package diagram for software architecture.

business logic of the application and the entity classes (and the associated data management classes). This helps to decouple the layers: the only communication between the clients and the server will be the communication between the client and server control classes, using RMI.

Not all control classes will have versions on both the clients and the server. For example, the `ListClients` and `ListCampaigns` classes in Figures 17.38 and 17.39 could just exist on the server, where they will have more immediate access to the entity and data mangement classes. One consequence of this will be visible in the sequence diagrams, where these two classes will no longer be passed references to the boundary class as a parameter, but will return their results to the control class on the client machine, which will set the values in the boundary class. This is shown in Figures A5.11 and A5.12.

On the server, we are using JDBC, and we will map the classes to relational database tables. A design based on the Broker pattern will be used to handle this.

A5.3 Sample Use Case

For the purpose of this case study chapter we are going to present the design of one use case, `Check campaign budget`, for which the boundary and control classes were designed in Chapter 17.

Figure A5.2 shows the design of the user interface for this use case. In the first iteration, we are not concerned with adding the extensions to the use case that handle printing of the campaign summary and campaign invoice.

A5.4 Class Diagrams

The packages on the architecture diagram have been named in a way that will allow us to use the Java package notation for classes. So, for example, the boundary classes will be in the package `Agate::Boundary`. This is the first package that we are illustrating here, and the classes we are concerned with are shown in Figure A5.3.

The boundary class `CheckCampaignBudgetUI` will implement the two interfaces `CampaignLister` and `ClientLister`. Note that some of the operations that were

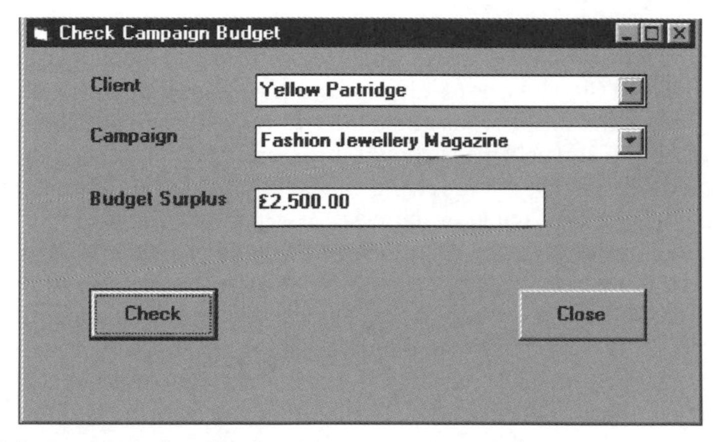

Figure A5.2 Prototype user interface for `Check campaign budget`.

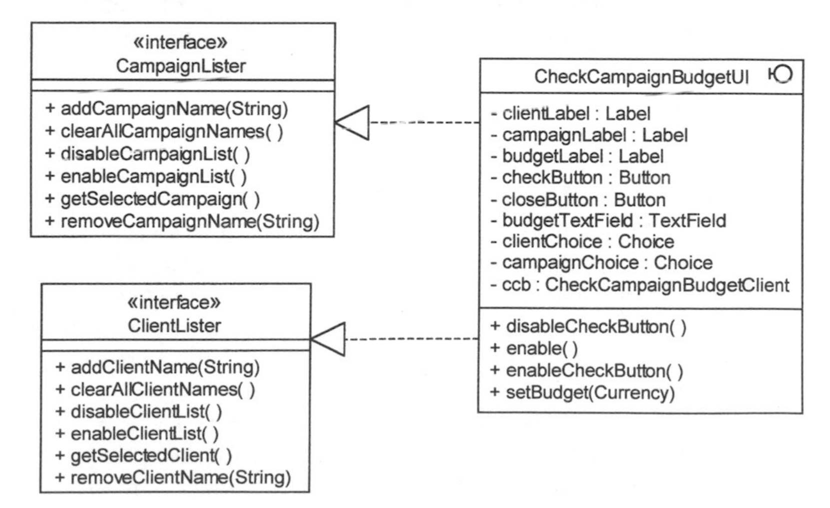

Figure A5.3 Relevant classes in the package `Agate::Boundary`.

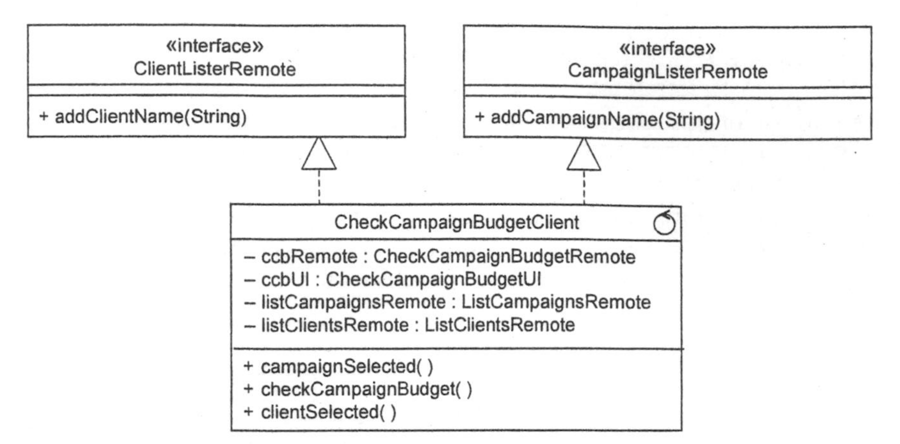

Figure A5.4 The class `Agate::Control::Client::CheckCampaignBudget`.

included in the class `CheckCampaignBudgetUI`, such as `getSelectedClient()` have been moved into the interfaces, as it is thought that they will apply to any class that implements these interfaces.

Because the control class `CheckCampaignBudget` will now be split, the version that resides on the client machines (now called `CheckCampaignBudgetClient`) must be able to respond to the messages `addCampaignName()` and `addClientname()`. We have used interfaces for this, because they have to be sent messages remotely by the control classes on the server. This is shown in Figure A5.4. Note also that this class will need to hold a reference to the version of itself that exists on the server. We have not shown the full package name in the class diagram, but the attribute `ccbRemote` will in fact be an instance of `Agate::Control::Server::CheckCampaignBudget Remote`. In fact there will be an instance of `Agate::Control::Server::Check CampaignBudgetServer` on the server, and for the object on the client to communicate with it via RMI it will have to implement the interface `Agate::Control:: Server::CheckCampaignBudgetRemote`. If `ListCampaigns` and `ListClients` only exist on the server, then they will also be in the same package and will implement the interfaces `ListCampaignsRemote` and `ListClientsRemote`.

All the classes that communicate via RMI will need to inherit from the Java RMI package. Rather than being subclasses of the default Java class `Object`, they will need to be subclasses of `java.rmi.server.UnicastRemoteObject`.

In Figure A5.5 we have shown the control classes that reside on the server and the remote interfaces which they must implement. Although we have not shown the full package names, the references to `ClientListerRemote` and `CampaignListerRemote` are to the interfaces in the package `Agate::Control::Client`, shown in Figure A5.4.

The entity classes that collaborate in this use case are `Client`, `Campaign` and `Advert`. They are shown in a first draft design in Figure A5.6. However, this design will only work for the kind of application where all the objects are in memory. We need to be able to deal with the process of materializing instances of these classes from the database, and when required, materializing their links with other object instances or collections of object instances. For example, when a particular `Client` is materialized, we do not necessarily want to re-establish its links with all its `Campaigns` and the instance of `StaffMember` that is its `staffContact`. The Broker pattern, which we discussed in Chapter 18, is a way of making it possible to materialize the objects that are linked to other objects only when they are required.

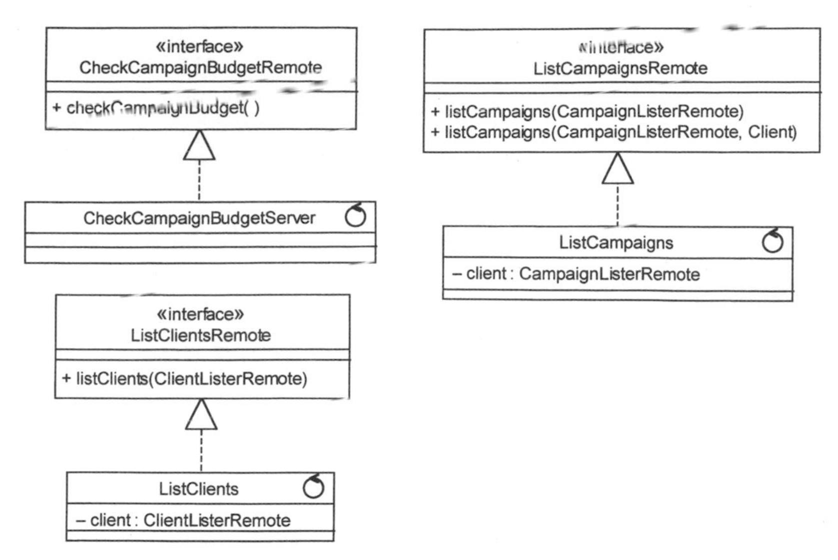

Figure A5.5 Relevant classes in the package `Agate::Control::Server`.

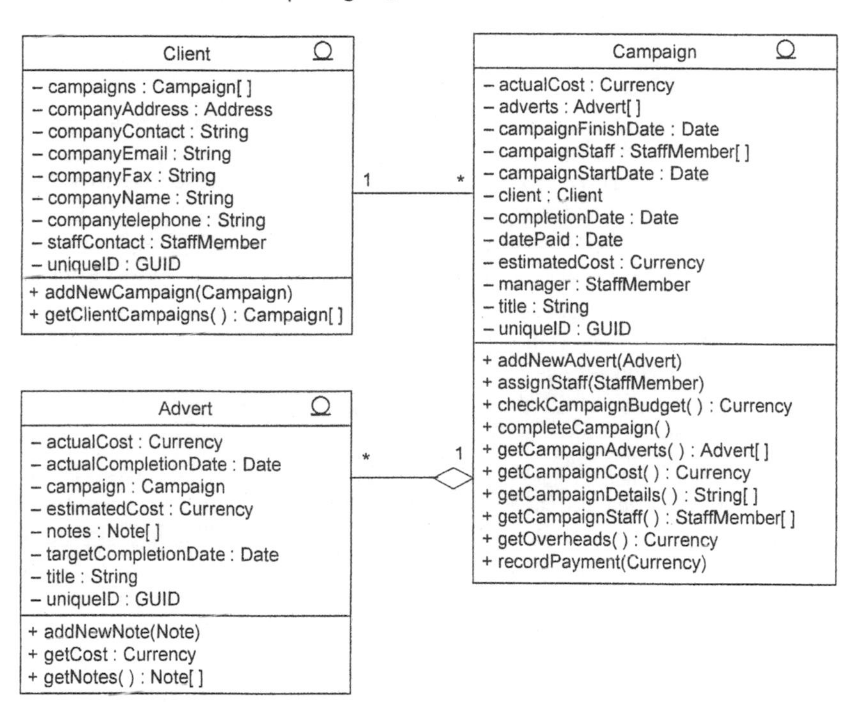

Figure A5.6 First draft design of some classes in the package `Agate::Entity::Domain`.

In order to achieve this, we can replace the references to the arrays of linked objects with references to the various subclasses of `Broker`, for example `ClientBroker`, `CampaignBroker` and `AdvertBroker`. Since these are still private attributes, they cannot be referred to directly by other objects, and their values can only be obtained by calling one of the operations of the object in which they are contained. The result of this is shown in Figure A5.7.

The broker subclasses could use the Singleton pattern (see Chapter 15). If this is done, then the design of the operations to return sets of whatever objects they are

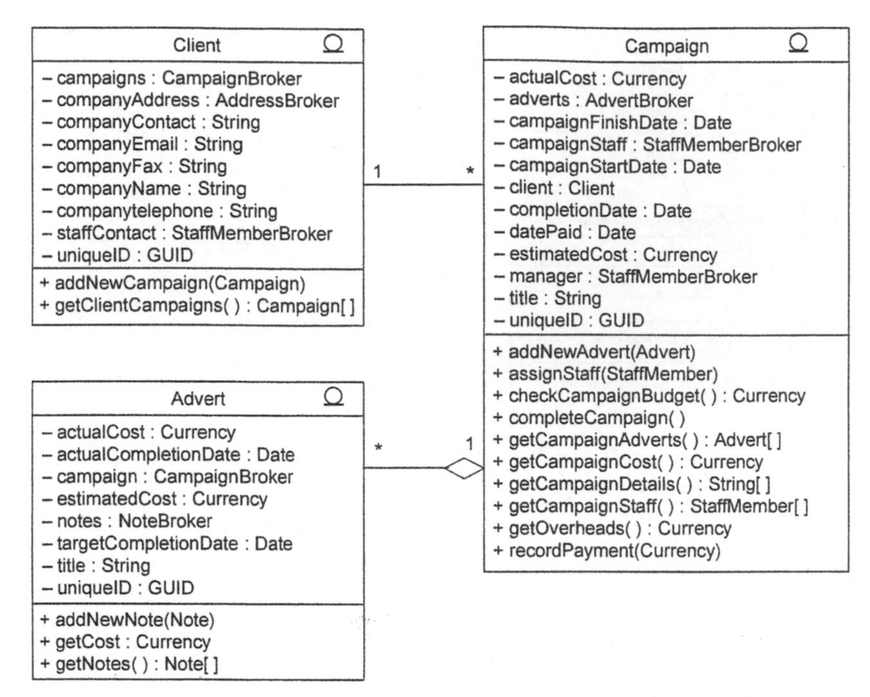

Figure A5.7 Second design of some classes in the package `Agate::Entity::Domain`.

acting as brokers for will have to be carefully designed to handle concurrent requests from different clients. Alternatively, there could be multiple instances of brokers, and they could be created and destroyed as required, or there could be a pool of brokers available in the server, and when an object needs a broker of a certain type, it would request one from the pool. Figure A5.8 shows the brokers that we are interested in for this use case. We have not used the Singleton pattern in this design.

These brokers will also be used directly by the control classes, for example, when they need to obtain a list of all the objects of a particular class in the system. The brokers have been shown with attributes in which to hold references to the objects necessary for connecting to the database and issuing queries. We have also assumed that having obtained a list of results, a broker may store it internally in a collection class and allow client objects to iterate through the list of results using an enumerator[2].

The brokers will be in the package `Agate::Entity::DataManagement`, together with any other necessary classes to handle the connection to the database. (In this design we are not using proxies or caches, in order to keep it relatively simple.)

The final piece of design necessary to enable the interaction of this use case realization to take place concerns how the control objects on the client machine will obtain references to control objects on the server. For this, we shall use the Factory pattern. A Factory class creates instances of other classes and returns a reference to the new instance to the object that requested it. This is shown in Figure A5.9.

So, an instance of the control class `CheckCampaignBudgetClient` on the client machine will request a Factory on the server to provide it with a reference to an instance of `CheckCampaignBudgetServer`. The Factory will create this instance and pass back a reference to it via the RMI connection with the client. From that point onwards, the client object can make direct requests to the control object on the server. When it is finished with it, it can destroy it, or ask the Factory to destroy it.

[2] A mechanism for working through a collection dealing with each object in turn.

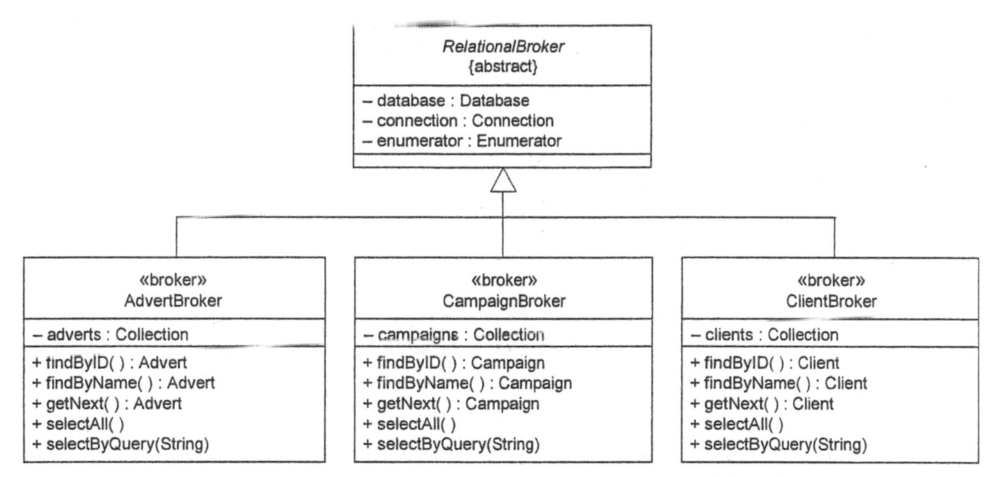

Figure A5.8 Broker classes in the package `Agate::Entity::DataManagement`.

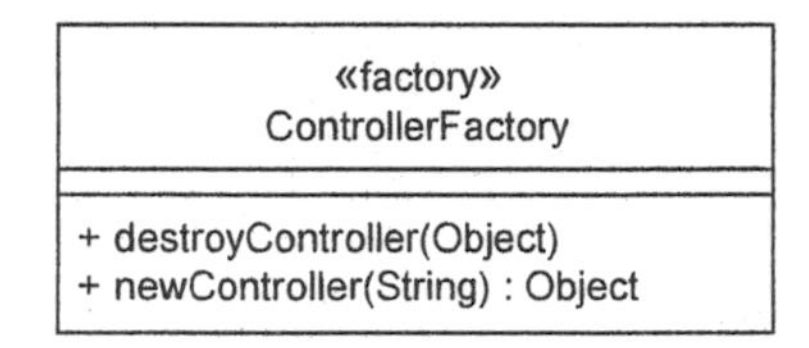

Figure A5.9 Factory class in the package `Agate::Control::Server`.

In a more sophisticated design, the Factory could hold a pool of already instantiated control classes ready for use. When a client requests an instance of a particular control class, the Factory will take one from the pool if it is available. When the client is finished with the instance, the Factory can put it back into the pool. We are not using pooling in this design, but it is an approach that is commonly used to improve the performance of servers to prevent delays while instances are created and destroyed on demand.

In this design the control class on the server only has one method. This class, `Agate::Control::Server::CheckCampaignBudgetServer` could be designed to hold the business logic for checking the budget of a campaign, but we have taken the decision to leave the responsibility for calculating whether or not the budget is overspent in the `Campaign` class. There is a case for giving this responsibility to the control class; then, if the business logic changes, it only has to be updated in the control class. However, this makes the entity objects little more than data stores.

Figure A5.10 shows the package diagram with the classes (but not the interfaces) from Figures A5.3 to A5.9.

A5.5 Sequence Diagrams

Figures A5.11 to A5.13 show the sequence diagrams from Chapter 17 revised to take account of the splitting of the control objects and the addition of the Factory class. The package names of objects are also shown.

Although we show the control class on the client as able to directly connect to the instance of `ControllerFactory` on the server, in reality it would have to request a reference to this object from a naming service or registry on the server, for example a running instance of the Java `rmiregistry`.

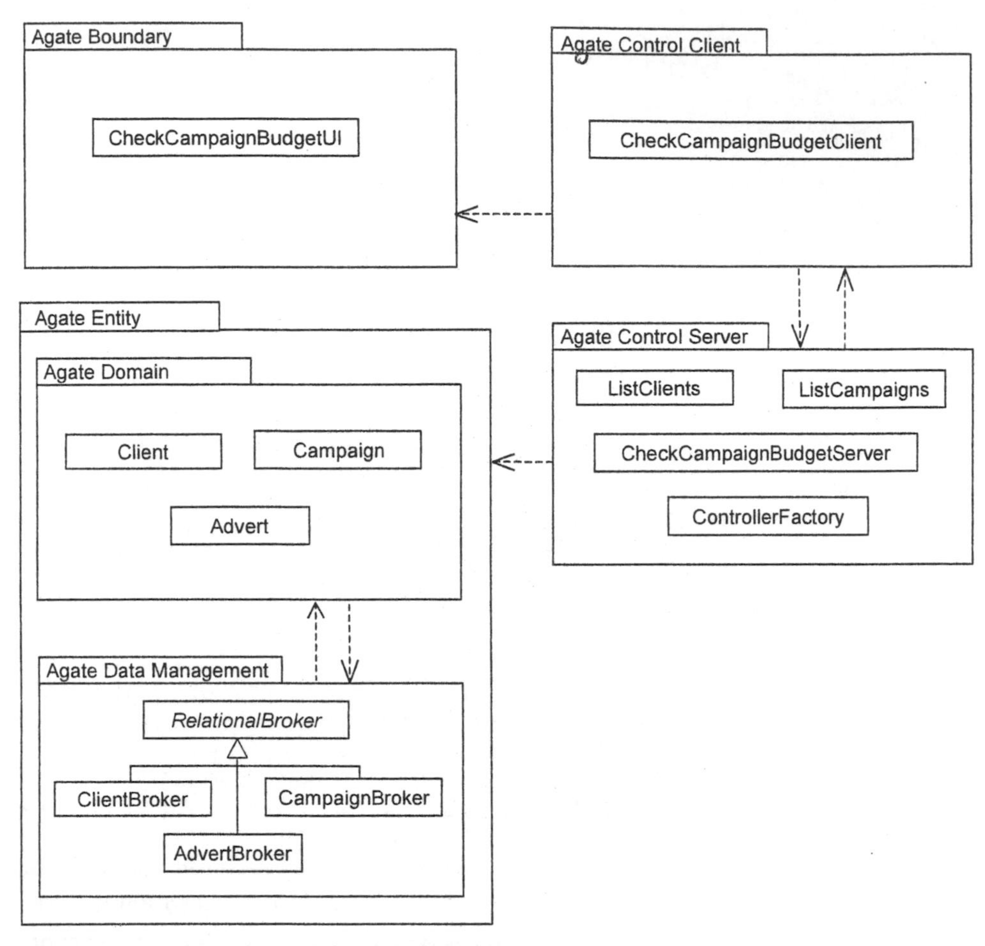

Figure A5.10 Package diagram showing classes.

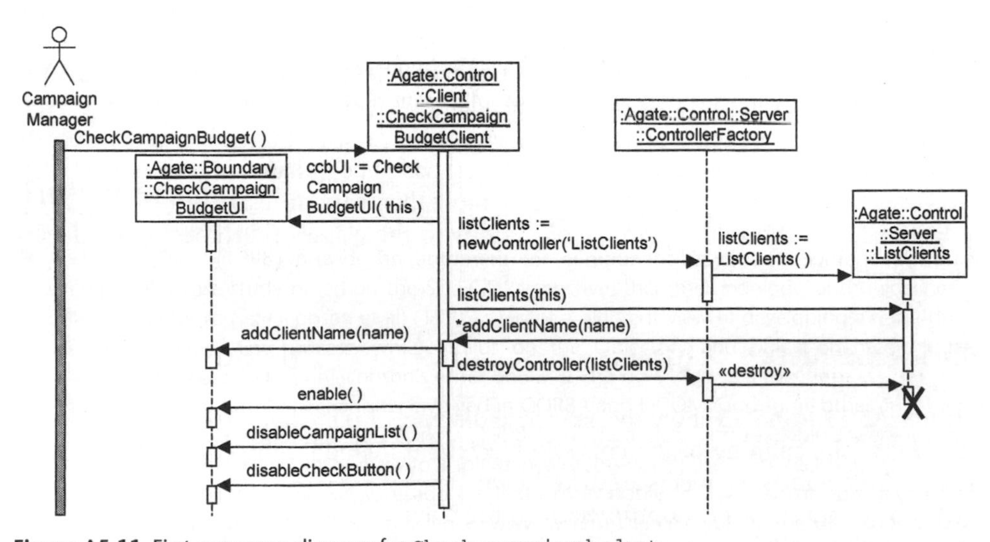

Figure A5.11 First sequence diagram for `Check campaign budget`.

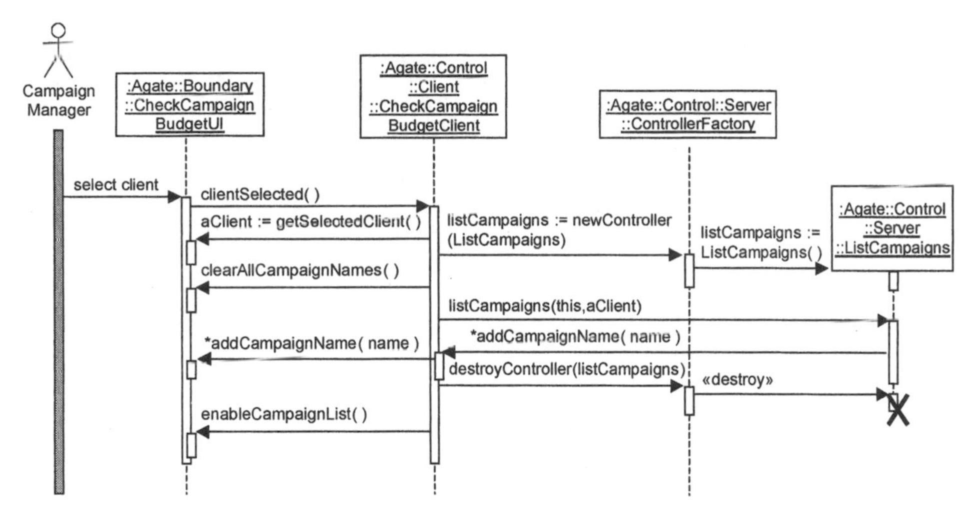

Figure A5.12 Second sequence diagram for `Check campaign budget`.

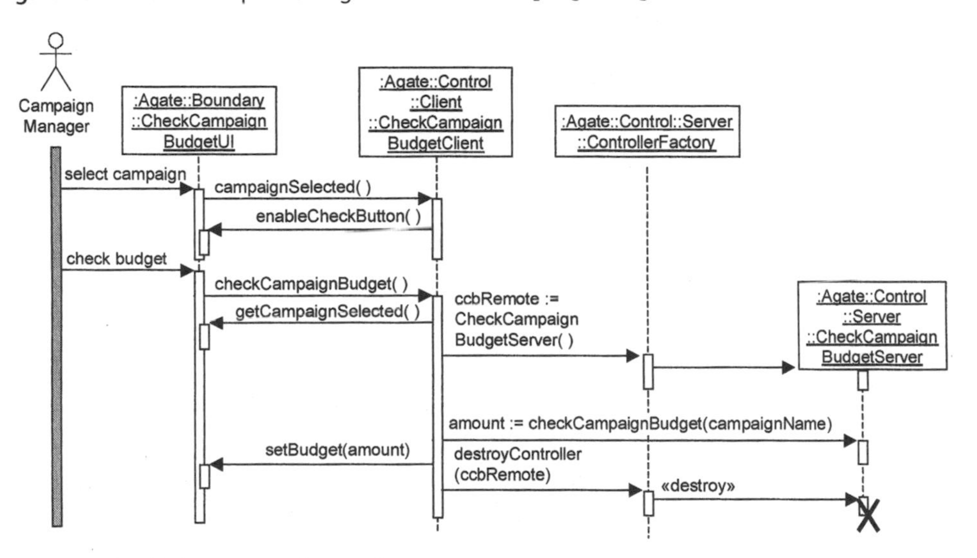

Figure A5.13 Third sequence diagram for `Check campaign budget`.

In Figure A5.14 we show the interaction between the control class, the brokers and the entity classes. Note how the broker classes perform the tasks involved in retrieving instances or sets of instances from the database.

We have used a simple approach for obtaining the adverts linked to a particular campaign, by having the broker return an array of `Adverts`. As mentioned above, this could return an enumerator so that the control class could iterate through the collection of `Adverts`.

A string named `query` has been passed to the `selectByQuery()` operation of the `AdvertBroker`. The exact format of this will depend on how the object-relational database mapping is set up. If the uniqueID attributes are used in the database as foreign keys, then the SQL statement will be something like:

```
SELECT * FROM adverts WHERE adverts.campaignID = '123456789';
```

and the ID of the particular campaign is added into the query string before it is passed to the broker.

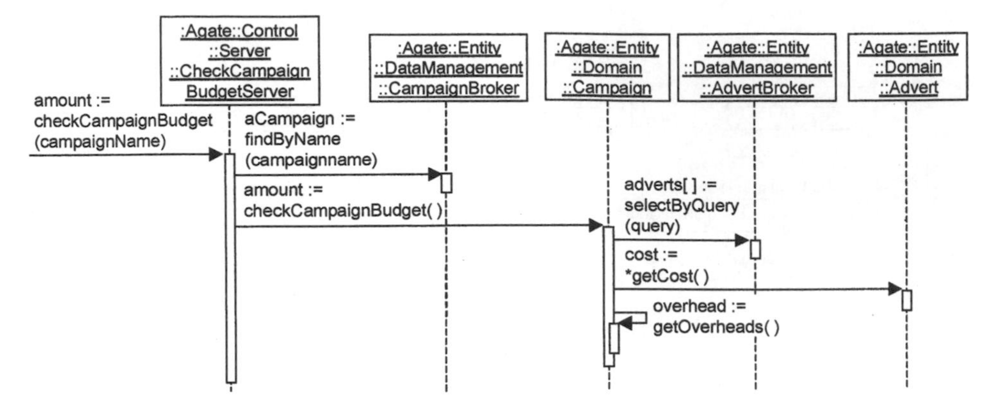

Figure A5.14 Sequence diagram for the operation `checkCampaignBudget()`.

A5.6 Database Design

Figure A5.15 shows the SQL to create the tables to map to the classes in Figure A5.7.

```
CREATE TABLE Clients
  (VARCHAR(30) uniqueID PRIMARY KEY NOT NULL,
  VARCHAR(30) companyAddress,
  VARCHAR(40) companyContact,
  VARCHAR(30) companyEmail
  VARCHAR(30) companyFax,
  VARCHAR(50) companyName NOT NULL,
  VARCHAR(30) companyTelephone,
  VARCHAR(30) staffContactID);
CREATE INDEX client_idx ON Clients (staffContactID, companyName);
CREATE TABLE Campaigns
  (VARCHAR(30) uniqueID PRIMARY KEY NOT NULL,
  FLOAT actualCost,
  DATE campaignFinishDate,
  DATE campaignStartDate,
  VARCHAR(30) clientID NOT NULL,
  DATE completionDate,
  DATE datePaid,
  FLOAT estimatedCost,
  VARCHAR(30) managerID,
  VARCHAR(50) title);
CREATE INDEX campaign_idx ON Campaigns (clientID, managerID, title);
CREATE TABLE Adverts
  (VARCHAR(30) uniqueID PRIMARY KEY NOT NULL,
  FLOAT actualCost,
  DATE actualCompletionDate,
  VARCHAR(30) campaignID NOT NULL,
  FLOAT estimatedCost,
  DATE targetCompletionDate,
  VARCHAR(50) title);
CREATE INDEX advert_idx on Adverts (campaignID, title);
```

Figure A5.15 SQL to create tables for the classes `Client`, `Campaign` and `Advert`.

The indexes are required to ensure that it is possible quickly to retrieve all the campaigns linked to a client, or all the adverts linked to a campaign. A character field has been used to hold the unique ID for each object. We are assuming that some mechanism will be used to generate these, but have not detailed it here. An alternative would be to use long integer values.

A5.7 Statecharts

Figure A5.16 shows the event–action table for the statechart of Figure A5.17. This statechart is the same as the one shown in Chapter 17.

Current State	Event	Action	Next State
–	Check Campaign Budget menu item selected.	Display CheckCampaignBudgetUI. Load client dropdown. Disable campaign dropdown. Disable check button. Enable window.	1
1	Client selected.	Clear campaign dropdown. Load campaign dropdown. Enable campaign dropdown.	2
2, 3, 4	Client selected.	Clear campaign dropdown. Load campaign dropdown. Clear budget textfield. Disable check button.	2
2	Campaign selected.	Clear budget textfield. Enable check button.	3
3	Check button pressed.	Calculate budget. Display result.	4
3, 4	Campaign selected.	Clear budget textfield.	3
4	Check button pressed.	Calculate budget. Display result.	4
1, 2, 3, 4	Close button clicked.	Display alert dialogue.	5
5	OK button clicked.	Close alert dialogue. Close window.	–
5	Cancel button clicked.	Close alert dialogue.	H*

Figure A5.16 Event–action table for Figure A5.17.

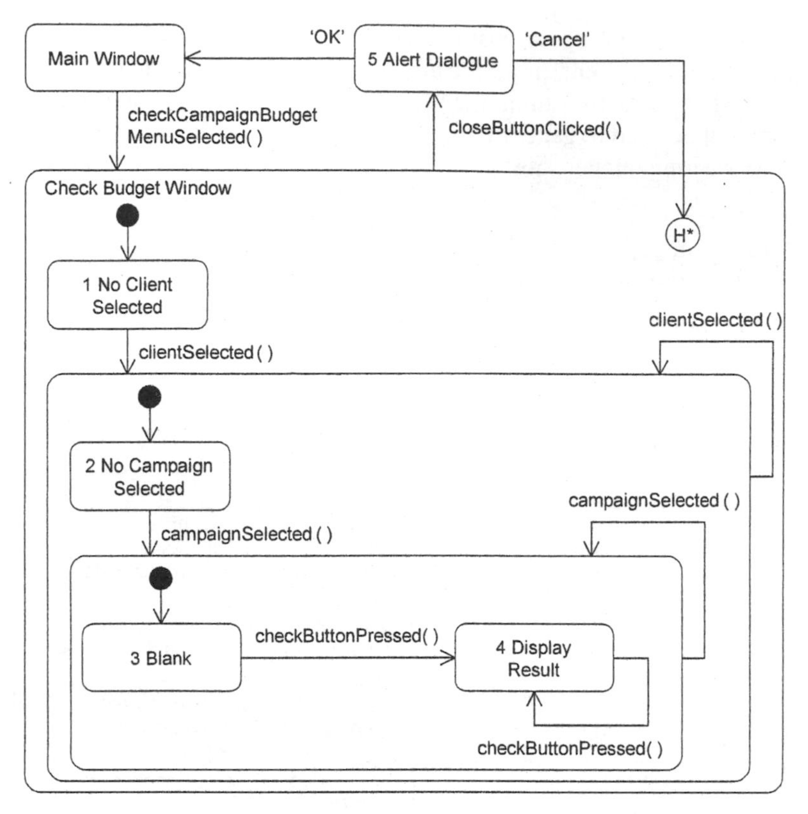

Figure A5.17 Statechart for control of the user interface in `Check campaign budget`.

A5.8 Activities of Design

The activities in the design workflow are shown in the activity diagrams of Figure A5.18 and A5.19.

In order to keep the diagram simple, we have shown the flow of activities in Figure A5.19 without dependencies on the products that are used and created. Although we have shown a flow through the activities from top to bottom, there will inevitably be some iteration through this workflow even within a major iteration.

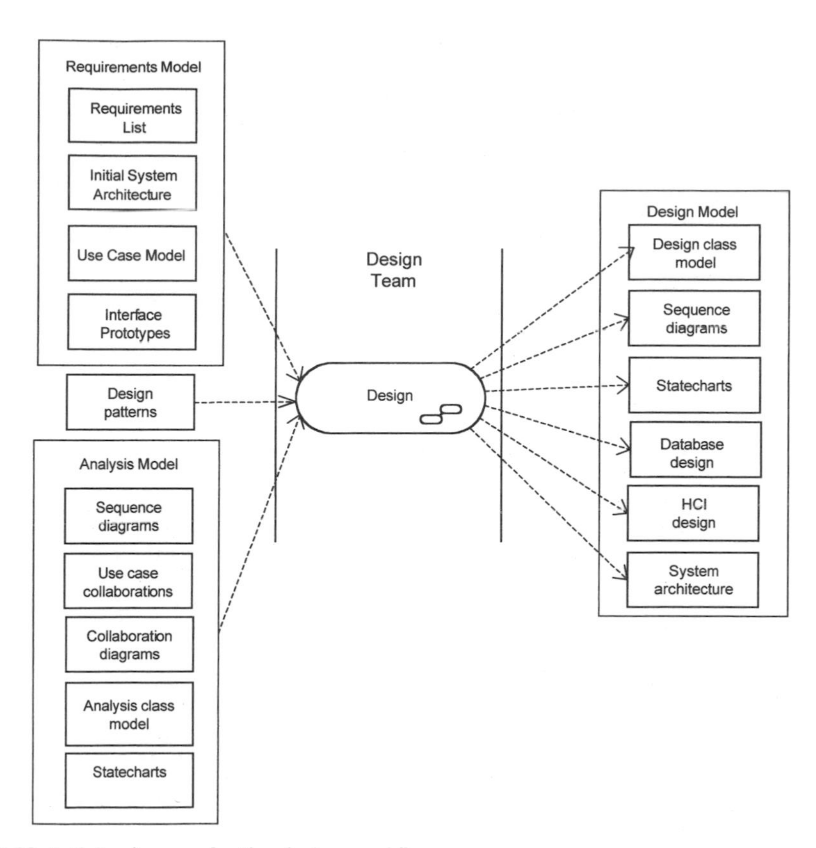

Figure A5.18 Activity diagram for the design workflow.

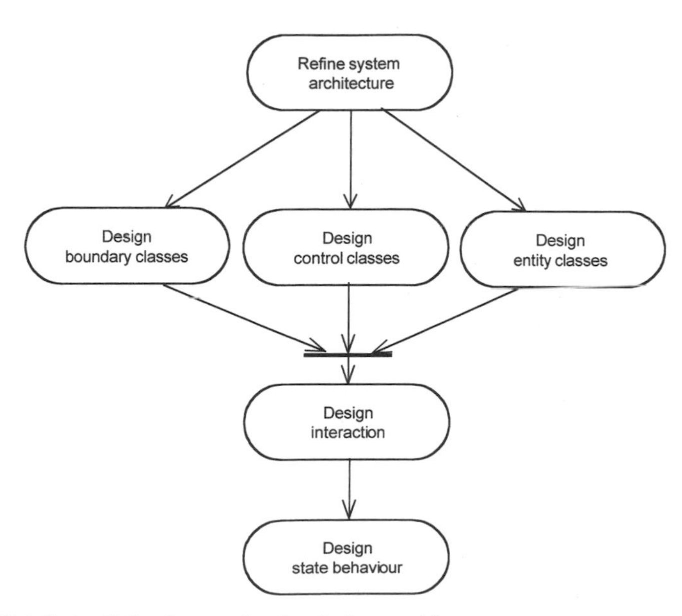

Figure A5.19 Detailed activity diagram for the design workflow.

19

CHAPTER

Implementation

OBJECTIVES

In this chapter you will learn

- how to draw component diagrams
- how to draw deployment diagrams
- the tasks that must be carried out in testing a new system
- how to plan for data to be converted from an existing system
- the alternative ways of introducing a new system into an organization
- the tasks that are undertaken as part of system maintenance and review.

19.1 Introduction

Implementation might be considered outside the scopc of analysis and design. However, in projects that use rapid application development techniques, the distinction between different roles tends to break down (see Chapter 21). Analysts in particular may have a role during implementation in dealing with system testing, data conversion and user training. In this chapter we address some of the issues concerned with the implementation of systems and the role of analysts and designers in this process. We focus on the various UML diagrams that can be used to plan and document the implementation of the software. We also cover more general aspects of system implementation that are independent of the development approach that is taken.

A range of different software packages are required for implementation, for example, languages and database management systems. It is important to keep track of the relationship between different elements of the system such as source code files, object code files and software libraries. It is also important to maintain standards in the software development process: classes, objects and variables should be named in ways that make their purpose clear to others and make it possible to trace from analysis through design to code; programs should be self-documenting and well-structured. (This is all covered in Section 19.2.)

UML provides two diagrams that can be used to document the implementation of a system. Component diagrams are used to document dependencies between the different elements of the system (Section 19.3). They can also be combined with deployment diagrams (Section 19.4) to show how the software components relate to the physical architecture of the system. For a large system these diagrams may be an unwieldy way of documenting the implementation, and it may be simpler to maintain tables of information using a spreadsheet.

The testing of a new system is an essential part of implementation and includes testing of individual components, sub-systems and the complete system (Section 19.5). A major task when a new system is introduced is to take data from an existing system or systems and transfer it into the new system (Section 19.6). The existing data may be held on paper or in a computerized information system that is being replaced. Temporary staff may have to be employed during the changeover to the new system. They and the existing staff will require training in how to use the system, and user documentation will have to be produced (Section 19.7). There are four different strategies for the introduction of a new system into an organization. These different approaches are appropriate to different circumstances, and each has its own advantages and disadvantages (Section 19.8). Finally, we address the maintenance issues that arise after a system has been implemented (Section 19.9).

19.2 Software Implementation

19.2.1 Software tools

The implementation of a system will require a range of tools. Ensuring that these are available in compatible versions and with sufficient licences for the number of developers who will be using them is part of the project management role. Many of these tools have been designed and developed to make the work of the system developer easier. In this section we describe each in turn.

CASE tools

Computer-aided software engineering tools allow the analysts and designers to produce the diagrams that make up their models of the system. CASE tools were discussed in detail in Section 3.6. There are now several CASE tools that support UML notation. If they have been implemented to use the UML XML Metadata Interchange format (XMI), it should be possible to exchange models between different vendors' tools. The repository for the project should also be maintained using the CASE tool to link the textual and structured descriptions of every class, attribute, operation, state and so on to its diagrammatic representation.

To ensure that the implementation accurately reflects the design diagrams, it may be possible to generate code in one or more programming languages from the documentation in the CASE tool. CASE tools exist that generate code for languages such as Visual Basic, C++ and Java. Some support the generation of SQL statements to create relational database tables to implement data storage, and the generation of the CORBA IDL for a distributed system. Some CASE tools provide support for reverse engineering from existing code to design models. When this is combined with code generation it is known as round-trip engineering.

Compilers, interpreters and run-time support

Whatever the language being used, some kind of compiler or interpreter will be required to translate the source code into executable code. C++ must be compiled into object code that can be run on the target machine. Smalltalk is interpreted; each command is translated as the program executes. Java is compiled into an intermediate bytecode format and requires a run-time program to enable it to execute. For applets, this run-time program is provided in the web browser, otherwise it is provided by the program called simply java or java.exe. C# can be compiled into bytecode in Microsoft Intermediate Language (MSIL) format for .NET applications.

Visual editors

Graphical user interfaces can be extremely difficult to program manually. Since the advent of Visual Basic, visual development environments have been produced for a wide range of languages. These enable the programmer to develop a user interface by dragging and dropping visual components onto forms and setting the parameters that control their appearance in a properties window. All the user interface examples in Chapters 6, 16 and 17 were produced in this way.

Integrated development environment

Large projects involve many files containing source code and other information such as the resource files for prompts in different human languages discussed in Chapter 18. Keeping track of all these files and the dependencies between them, and recompiling all those that have changed as a project is being built is a task best performed by software designed for that purpose. Integrated development environments (IDEs) incorporate a multi-window editor, mechanisms for managing the files that make up a project, links to the compiler so that code can be compiled from within the IDE, and a debugger to help the programmer step through the code to find errors. An IDE may also include a visual editor to help build the user interface and a version control system to keep track of different versions of the software.

Configuration management

Configuration management tools keep track of the dependencies between components and the versions of source code and resource files that are used to produce a particular release of a software package. Each time a file is to be changed, it must be checked out of a repository. When it has been changed it is checked in again as a new version. The tool keeps track of the versions and the changes from one version to the next. When a software release is built, the tool keeps track of the versions of all the files that were used in the build. To ensure that an identical version can be rebuilt, other tools such as compilers and linkers should also be under version control.

Some such tools are simple and easily available such as CVS and RCS, while others are for large scale distributed projects and require full-time administrators. Web interfaces are available for some, and these make it possible to check items in and out over the Internet for work on Open Source software.

There are standard protocols for version control software, which make it possible for users of editors, IDEs and CASE tools to check items out and in from within the tool.

Class browsers

In an object-oriented system, a browser provides a visual way of navigating the class hierarchy of the application and the supporting classes to find their attributes and operations. Smalltalk-80 was the first language to provide this kind of browsing capability. Some IDEs now provide it. The Java Application Programming Interface (API) is documented in HTML and can be browsed with a web browser.

Component managers

Chapter 20 discusses how software reuse can be achieved through the development of reusable components. Component managers provide the user with the ability to search for suitable components, to browse them and to maintain different versions of components.

DBMS

A large-scale database management system will consist of a considerable amount of software. If it supports a client–server mode of operation, there will be separate client and server components as well as all the tools discussed in Section 18.2.6. To use ODBC or JDBC will require ODBC software installed on the client. For any database, special class libraries or Java packages may be required on the client either during compilation or at run-time or both. ObjectStore PSE Pro includes a post-processor that is used to process Java class files to make them persistent.

CORBA

An ORB is required in order to use CORBA. It will include the IDL compiler that takes interface definitions in IDL and produces the interface, stub and skeleton files necessary to use CORBA. The pure-Java Caffeine software that runs with Visigenic's Visibroker for Java ORB will generate the necessary classes directly from Java code, even from compiled class files.

Testing tools

Automated testing tools are available for some environments. What is more likely is that programmers will develop their own tools to provide harnesses within which to test classes and sub-systems. Section 19.3 covers testing in more detail.

Installation tools

Anyone who has installed commercial software on a Windows PC or a Mac or used a package manager on Linux will have experienced one of these tools, which automate the creation of directories, the extraction of files from archives and the setting up of parameters or registry entries. To do this they maintain the kind of information that can be modelled using component and deployment diagrams (see Sections 19.3 and 19.4). In our experience, uninstallation tools do not work as well!

Conversion tools

In most cases data for the new system will have to be transferred from an existing system. Whereas once the existing system was usually a manual system, most projects nowadays replace an existing computerized system, and data will have to be extracted from files or a database in the existing system and reformatted so that it can be used to set up the database for the new system. Packages like Data Junction provide automated tools to extract data from a wide range of systems and format it for a new system.

Documentation generators

In the same way that code can be generated from the diagrams and documents in a CASE tool, it may be possible to generate technical and user documentation. In Windows there are packages such as Documentation Studio that can be used to produce files in Windows Help format. Java includes a program called javadoc that processes Java source files and builds HTML documentation in the style of the API documentation from special comments with embedded tags in the source code.

19.2.2 Coding and documentation standards

Even one person developing software on his or her own is likely to find at some point that they cannot remember the purpose of a class, an attribute or an operation in a program. On any project in which people collaborate to develop software, agreed standards for the naming of classes, attributes, operations and other elements of the system are essential if the project is not to descend into chaos. (See also Chapter 5 and Section 13.6.)

Naming standards should have been agreed before the analysis began. In this book we have tried to conform to a typical object-oriented standard.

- Classes are named with an initial capital letter. Words are concatenated together when the class name is longer than one word. Capital letters within the name show where these words have been joined together. For example `SalesOrderProxy`.
- Attributes are named with an initial lower case letter. The same approach is taken as for classes by concatenating words together. For example `customerOrderRef`.
- Operations are named in the same way as attributes. For example `getOrderTotal()`.

There are other standards. In C++ one convention is to use *Hungarian* notation: all member variable (attribute) names are prefixed by an abbreviation that indicates the type of the member variable, for example `b` for a Boolean, `i` for an integer, `f` for a float, `btn` for a button and `hWnd` for a handle to a window object. This can be particularly useful in languages that are not strongly typed, like Smalltalk, as it helps to enforce the consistent use of the same variable for the same purpose by different developers.

Consistent naming standards also make it easier to trace requirements from analysis through design to implementation. This is particularly important for class, attribute and operation names.

Not everything in a program can be deduced by reading the names of classes, attributes and operations. Beveridge (1996) in a book on Java programming gives five reasons for documenting code.

- Think of the next person. Someone else may be maintaining the code you have written.

- Your code can be an educational tool. Good code can help others, but without comments complicated code can be difficult to understand.
- No language is self-documenting. However good your naming conventions, you can always provide extra help to someone reading your code.
- You can comply with the Java coding standards. Your documentation will be in the same hypertext format as the Java API documentation.
- You can automate its production. The javadoc program is discussed below. It generates HTML from your comments.

(The comments about Java apply just as well to other languages.) Standards should be enforced for the way that comments are added to a program. This should include a block at the start of each class source file (or header file in C++) that describes the purpose of the class and includes details of its author and the date on which it was written. The amendment history of the source file can be included in this block. Every operation should begin with a comment that describes its purpose. Any obscure aspect of the code should be documented with a comment. If you are developing in Java, you can use the javadoc conventions to generate HTML documentation for your classes. You can embed HTML tags and javadoc tags in the comments, and javadoc will also use its own special tags to add information about the author and version to the HTML. The javadoc tags include:

```
@see classname—'See also' hypertext link to the specified class;
@version text—'Version' entry;
@author name—'Author' entry.
```

As well as technical documentation, there will be a need for user documentation that will be required for training users and for them to refer to once the system is in use. Standards need to be agreed with the users for the format of this documentation, which may be produced by analysts or by specialist technical authors.

19.3 Component Diagrams

In a large project there will be many files that make up the system. These files will have dependencies on one another. The nature of these dependencies will depend on the language or languages used for the development and may exist at compile-time, at link-time or at run-time. There are also dependencies between source code files and the executable files or bytecode files that are derived from them by compilation. Component diagrams are one of the two types of implementation diagram in UML. Component diagrams show these dependencies between software components in the system. Stereotypes can be used to show dependencies that are specific to particular languages. Figure 19.1 shows a component diagram that represents the dependency of a C++ source code file on the associated header file, the dependency of the object file on both and the dependency of an executable on the object file. Stereotypes can be used to show the types of different components.

An alternative representation of part of the diagram above is to use the UML interface notation to show the specification of a class (the header file in C++) as an interface and the body as the component, as in Figure 19.2.

This notation can be used in Java to show the dependency of classes on the interfaces of other classes. This is particularly appropriate for distributed systems using CORBA in which applications running on a client are dependent on the interfaces of

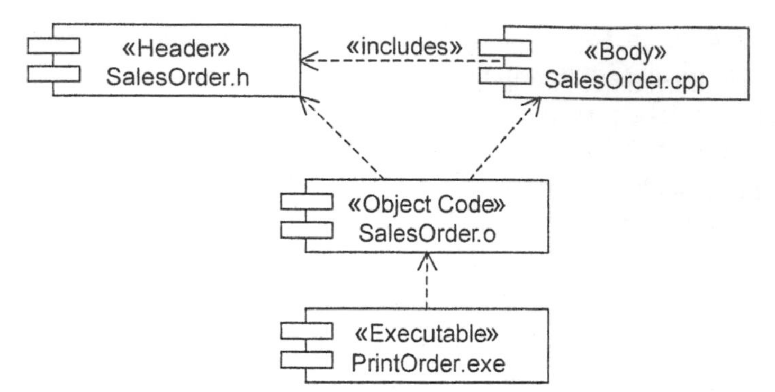

Figure 19.1 Component diagram showing dependencies in C++.

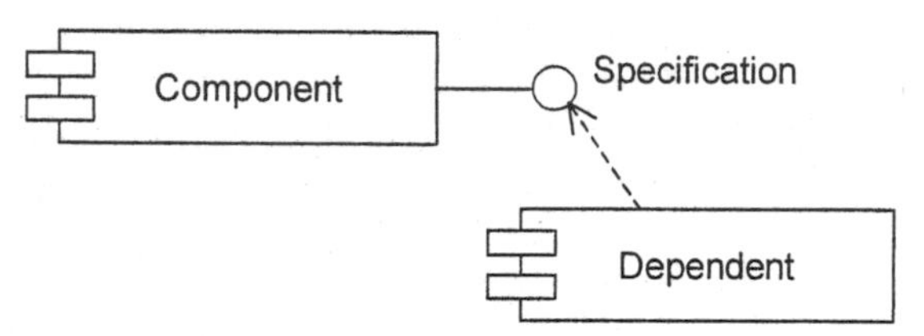

Figure 19.2 Dependency of a component on the interface of another component.

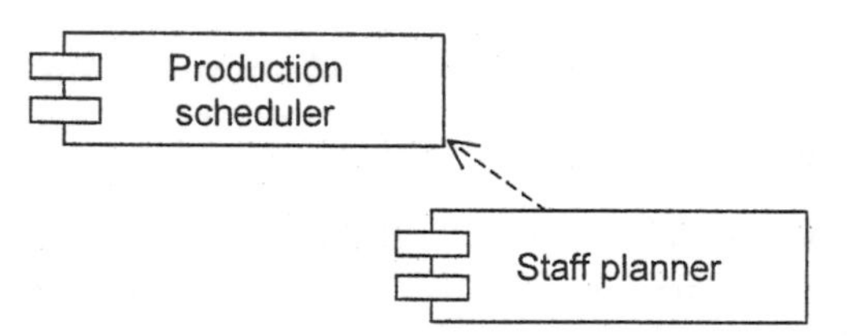

Figure 19.3 Example dependency between higher level components.

classes that are actually implemented on other machines on the network. Component diagrams need not be used at this low level, but can be used to show dependencies between large-scale components within a system as in Figure 19.3.

In Java, component diagrams can be used to show the dependency of classes on packages that contain the classes that they import. This is particularly important in a language such as Java where the availability of packages of classes at run-time is critical to the running of a program.

Active objects, typically processes running on a separate thread, can be shown in a component diagram. An example of this is shown in Figure 19.4.

Different authors use component diagrams in different ways. Fowler (1997) suggests that components correspond exactly to packages so that their use in deployment diagrams shows where each package is running on the system. Muller (1997) suggests that packages contain components as well as other packages. Strictly in UML packages should be used for model management: for organizing models into convenient parts that contain types of diagram or sub-systems. We would suggest the following distinction between them.

- Components in a component diagram should be the physical components of a system.
- During analysis and the early stages of design, package diagrams can be used to show the logical grouping of class diagrams (or of models that use other kinds of diagram) into packages relating to sub-systems.

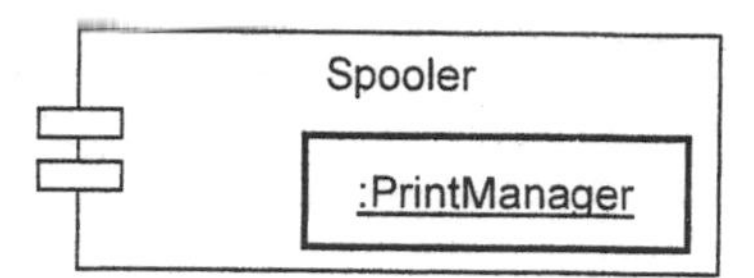

Figure 19.4 Active object within a component.

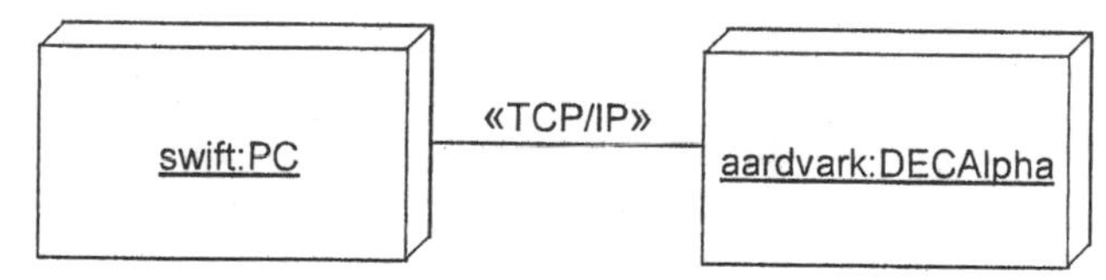

Figure 19.5 Simple deployment diagram.

- During implementation, package diagrams can be used to show the grouping of physical components into sub-systems.
- As we shall see in the next section, component diagrams can be combined with deployment diagrams to show the physical location of components of the system. The classes in one logical package may be distributed across physical locations in a physical system, and the component diagram and deployment diagram can be used to show this.

If you use component diagrams, it is advisable to keep separate sets of diagrams to show compile-time and run-time dependencies. However, this is likely to result in a large number of diagrams, and we suggest an alternative method of managing this information at the end of the next section.

Component diagrams show the components as types. If you wish to show instances of components you can use a deployment diagram.

19.4 Deployment Diagrams

The second type of implementation diagram provided by UML is the deployment diagram. Deployment diagrams are used to show the configuration of run-time processing elements and the software components and processes that are located on them. They are made up of nodes and communication associations. Nodes are typically used to show computers and the communication associations show the network and protocols that are used to communicate between nodes. Nodes can be used to show other processing resources such as people or mechanical resources. Nodes are drawn as 3D views of cubes or rectangular prisms, and the simplest deployment diagrams show just the nodes connected by communication associations as in Figure 19.5.

Deployment diagrams can show either types of machine or particular instances as in Figure 19.5 where `swift` is the name of a PC. Deployment diagrams can be shown with components and active objects within the nodes to indicate their location in the run-time environment. Figure 19.6 shows the location of the Sales database on the server and some components on client PCs.

If you try to show all the components of a system in deployment diagrams they are likely to become very large or difficult to read. They can serve the purpose of communicating information about the location of key components to other members of the team or to users. Indeed, most computer professionals will have drawn an informal diagram like this at some time in their working lives to show where different parts of a system are to be located. Deployment diagrams show the physical architecture of the system.

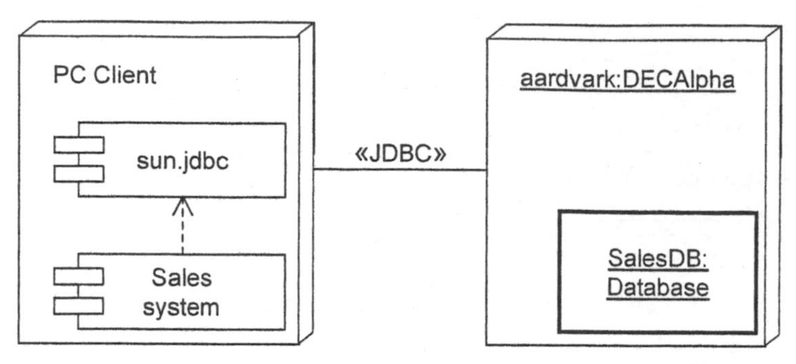

Figure 19.6 Deploment diagram with components on PC clients and an active object on the server.

Campaign database—compilation time dependencies				
	JDBC sun.jdbc.*	Campaign. .java	Campaign. Broker.java	Campaign. Proxy.java
JDBC sun.jdbc.*				
Campaign. java				
Campaign. Broker.java	✓	✓		
Campaign. Proxy.java		✓	✓	

Figure 19.7 Excerpt from example table to replace component diagram.

If you intend to use component and deployment diagrams to illustrate general principles about the way that the new system will be structured, then they are fine as a diagramming technique. However, if the aim of drawing implementation diagrams is to provide a complete specification of the dependencies between components at compile-time and run-time and the location of all software components in the implemented system, then this may be one of those cases where a picture is not worth a thousand words. Having implemented a system in Java that uses class files, classes from a visual editor, JDBC drivers, CORBA and ObjectStore Persistent Storage Engine on PCs, and Oracle on a workstation, we know that keeping track of all these dependencies and documenting which components have to be on which machines is not a trivial task. For most systems, this information may be easier to maintain in a tabular format, and a spreadsheet may be the best way of doing this.

Component diagrams can be replaced by a table that shows a list of all the software components down the rows and the same list across the top of the columns. It may be best to keep up to three tables (depending on the language used) for compile-time, link-time and run-time dependencies. For each case where a component is dependent on another, place a mark where the row of the dependent component intersects with the column of the component on which it is dependent. A simple example of this is shown in Figure 19.7.

In the same way, deployment diagrams can be replaced by a table that lists components down the rows and either types of machines or particular instances across the top of the columns. A mark is entered in the row–column intersection for every component that has to be on a particular machine or type of machine. If the exact location of components in a directory structure is important, then that location can be entered into the table. This is shown in Figure 19.8. Later this will form the basis of the information required for installing software onto users' machines for testing and eventual deployment.

19.5 Software Testing

19.5.1 Who carries out the testing?

One view of testing is that it is too important to be left to the programmers who have developed the software for the system. This is not meant as a criticism of programmers but reflects the fact that it is important that testing is carried out by someone whose assessment of the software will be objective and impartial. It is often difficult for programmers to see the faults in the program code that they have written. An alternative view is provided by Extreme Programming (XP) (Beck, 2000). XP is an approach to rapid application development in which programmers are expected to write test harnesses for their programs before they write any code. Every piece of code can then be tested against its expected behaviour, and if a change is made can easily be retested. XP is explained in Chapter 21.

Some organizations employ specialist software testers. The following paragraph is an excerpt from an advertisement in the British computer press for a post as a tester.

> A leading financial institution has an opportunity for a systems tester to work on a business critical project. Testing throughout the project life cycle, you will liaise closely with developers and team leaders to implement test cases and organize automated testing scripts. All testing is organized within a fully automated environment. With a background in testing you will have strong business acumen . . .

However, not all organizations can afford the luxury of specialist testers. Often the analysts who carried out the initial requirements analysis will be involved in testing the system as it is developed. The analysts will have an understanding of the business requirements for the system and will be able to measure the performance of the system against functional and non-functional requirements.

Campaign database—run-time locations		
	Client PC	**Database server**
JDBCsun.jdbc.*	c:\jdbc	
Campaign.class	c:\agate\campaign	
CampaignBroker.class	c:\agate\campaign	
CampaignProxy.class	c:\agate\campaign	
SQL*Net		✓
OCI Listener		✓

Figure 19.8 Excerpt from example table to replace deployment diagram.

The systems analysts will use their knowledge of the system to draw up a test plan. This will specify what is to be tested, how it is to be tested, the criteria by which it is possible to decide whether a particular test has been passed or failed, and the order in which tests are to take place. Based on their knowledge of the requirements, the analysts will also draw up sets of test data values that are to be used.

The other key players in the process of testing new software are the eventual users of the system or their representatives. Users may be involved in testing the system against its specification, and will almost certainly take part in final user acceptance tests before the system is signed off and accepted by the clients. If a use-case-driven approach to testing is used, the use cases are used to provide scenarios to form the basis of test scripts.

19.5.2 What is tested?

In testing any component of the system, the aim is to find out if its requirements have been met. One kind of testing seeks to answer the following questions.

> Does it do what it's meant to do?
> Does it do it as fast as it's meant to do it?

This is equivalent to asking 'Never mind how it works, what does it produce?' and is known as *black box* testing because the software is treated as a black box. Test data is put into it and it produces some output, but the testing does not investigate how the processing is carried out. Black box testing tests the quality of performance of the software. It is also necessary to check how well the software has been designed internally. This second kind of testing seeks to answer the following question.

> Is it not just a solution to the problem, but a *good* solution?

This is equivalent to asking 'Never mind what it's for, how well does it work?' and is known as *white box* testing because it tests the internal workings of the software and whether the software works as specified. White box testing tests the quality of construction of the software. In a project where reusable components are bought in, it may not be possible to apply white box testing to these components, as they may be provided as compiled object code. However, some suppliers will supply source code as well as compiled code, and there is a growing Open Source movement that makes this possible. As an aside, some organizations require, as part of their software contracts, that source code is placed in escrow. This means that a copy of the source code is lodged with a third party, usually a lawyer or a trade association, so that it is available to the client if the software company goes out of business. By this means the client ensures that they will be able to maintain and enhance the software even if its original developers are no longer able to.

Ideally, testers will use both white box and black box testing methods together to ensure:

- completeness (black box and white box),
- correctness (black box and white box),
- reliability (white box), and
- maintainability (white box).

However, the aim of any kind of testing is always to try to get the software to fail—to find errors—rather than to confirm that the software is correct. For this reason the test data should be designed to test the software at its limits, not merely to show that it copes acceptably with routine data.

Testing can take place at as many as five levels:

- unit testing,
- integration testing,
- sub-system testing,
- system testing and
- acceptance testing.

In an object-oriented system, the units are likely to be individual classes. Testing of classes should include an initial *desk check*, in which the tester manually walks through the source code of the class before compilation. The class should then be compiled, and the compilation should be clean with no errors or warnings. To test the running of a class the tester will require some kind of test program (the term *harness* is often used) that will create one or more instances of a class, populate them with data and invoke both instance methods and class methods. If pre-conditions and post-conditions have been specified for operations, as suggested in Chapter 10, then the methods that have been implemented will be tested to ensure that they comply with the pre-conditions and that the post-conditions are met when they have completed. Statechart diagrams can be used to check that classes are conforming to the behaviour in their specification.

It may be difficult to test classes in isolation. For the reasons that are discussed in Chapter 20 on reuse, most classes are coupled in some way to other classes in the system. Unit testing merges into integration testing when groups of classes are tested together. The obvious test unit at this point is the use case. The interaction between classes can be tested against the specification of the sequence diagrams and collaboration diagrams. User interface classes and data management classes will also have to be tested in conjunction with the classes in the application logic layer. If scenario-based design has been used (see Chapter 16), then the scenarios can form the basis for testing scenarios in which a use case can be tested against a typical business situation.

Use cases that share the same persistent data should be tested together. This is one form of sub-system testing in which the sub-systems are built around different business functions that make use of the same stored data.

If significant changes are made to a system, then some of the tests must be run again to ensure that the changes have not broken existing functionality. This is *regression testing*.

Testing is sometimes described as taking place at three levels.

Level 1

- Tests individual modules (e.g. classes).
- Then tests whole programs (e.g. use cases).
- Then tests whole suites of programs (e.g. the Agate application).

Level 2

- Also known as Alpha testing or verification.
- Executes programs in a simulated environment.
- Particularly tests inputs that are:

  - negative values when positive ones are expected (and vice versa),
  - out of range or close to range limits, or
  - invalid combinations.

Level 3

- Also known as Beta testing or validation.
- Tests programs in live user environment:
  - for response and execution times,
  - with large volumes of data, and
  - for recovery from error or failure.

A final stage of testing is *user acceptance* testing during which the system is evaluated by the users against the original requirements before the client signs the project off. Documentation produced during requirements capture and analysis will be used to check the finished product, in particular use case scenarios.

19.5.3 Test documentation

Thorough testing requires careful documentation of what is planned and what is achieved. This includes the expected outcomes for each test, the actual outcomes, and for any test that is failed, details of the retesting. Figure 19.9 shows part of a test plan for the Agate case study. It shows details of each test and its expected outcomes. The results of the actual tests will be documented in a separate, but similar format, with columns to show the actual result of each instance of each test and the date when each test was passed, and to document problems. Many organizations have standard forms for these documents or may use spreadsheets or databases to keep this information. The advantage of using a spreadsheet or database is the ability to produce reports that show what percentage of tests are complete. If requirements are held in a database, it is possible to link requirements to the tests that show whether they have been met and thus to provide a mechanism for tracing through from the original requirements to functionality in the finished system.

Testers should also watch out for unexpected results. Interaction between different operating systems can often cause unanticipated problems with different conventions for newline characters or case sensitivity of filenames. Problems such as these should be reported as bugs and recorded in a fault reporting package for action by the developers.

19.6 Data Conversion

Data from existing systems will have to be entered into a new system when it is introduced. The organization may have a mixture of existing manual and computerized systems that will be replaced by the new system. The data from these systems must be collated and converted into the necessary format for the new system. The timing of this will depend on the implementation strategy that is used (see next section), but it is likely to be a costly task, involving the use of staff time, the employment of temporary staff or the use of software to convert data from existing computer systems. These costs should have been identified in any cost benefit analysis that was carried out at the inception of the project.

Test no.	Test description	Test data	Expected result
234	Create a new Campaign.		Campaign Estimated Cost is set to £0.00
235	Add Advert 1 to Campaign.	Advert Estimated Cost = £500.00	Campaign Estimated Cost is set to £500.00
236	Add Advert 2 to Campaign.	Advert Estimated Cost = -£500.00	Negative value rejected. No change to Campaign Estimated Cost.
237	Add Advert 3 to Campaign.	Advert Estimated Cost = £300.00	Campaign Estimated Cost is set to £800.00
238	Set Advert 1 Completed.	Advert Actual Cost = £400.00	Campaign Estimated Cost is set to £700.00. Actual Cost is set to £400.00.

Figure 19.9 Excerpt from test plan for Agate.

If data is being collated from existing manual systems, it may be necessary to gather it from different sources. Data may be stored in different files, on index cards, in published documents, such as catalogues, or in other paper-based systems. If this data is going to be entered manually into the new system, by users keying it in, then the designers should draw up paper forms that can be used to collate the information so that it is all in one place when it is keyed in. Some data will only ever be entered when the system is started up, for example codes that are used in the system and will not be altered. Special data entry windows will be required for this kind of one-off activity.

Data from existing computer systems will have to be extracted from existing files and databases and reformatted to be usable with the new system. This provides an opportunity to clean up the data: removing out-of-date records and tidying up the values that are stored. Address and telephone number attributes of existing systems are likely to have been abused or misused by users. The work of converting the data may be done by using special programs written by the developers of the system, by employing consultants who specialize in this kind of work or by using commercial software that is capable of reading and writing data in a variety of formats, or even, as is the case with packages such as Data Junction, capable of working out the format of an unknown file.

The tasks involved in data conversion can be summarized as follows.

- Creating and validating the new files, tables or database.
- Checking for and correcting any format errors.
- Preparing the existing data for conversion:
  - verifying the existing data for correctness,
  - collating data in special forms for input, and
  - obtaining specially written programs to convert and enter the data.
- Importing or inputting the data.
- Verifying the data after it has been imported or input.

All the converted data may have to be ready for entry into the new system to meet a tight deadline, or it may be possible to enter it over a period of time. It is best to convert relatively static data such as product information and customer details first and leave dynamically changing files of information such as orders or other business transactions until last. It may be that only open orders should be taken over into the new system. The implementation strategy will determine the timescale for conversion.

Always carry out a trial data conversion exercise before doing it for real. One of the authors was involved in converting data from one stock and manufacturing system to another recently. In both packages (from the same supplier) the Part ID field was case-sensitive. However, in that same supplier's data conversion software all alphabetic characters in Part IDs were converted to upper case. Because upper and lower case letters were used to distinguish major and minor sub-assemblies of the same product, this caused problems that would have delayed the implementation if the error in the data conversion program had not been identified well before the planned implementation date.

19.7 User Documentation and Training

19.7.1 User manuals

As well as preparing the technical documentation for the system, analysts will be involved in producing manuals for end-users. The technical documentation will be required by the system manager and other staff responsible for running the system, and by staff who have to maintain the system. Ordinary users of the system, who will be using it to carry out their daily work tasks, require a different kind of documentation.

Users will require two kinds of manual. During training they will need training materials that are organized around the tasks that they have to carry out with the new system. On-line computer-based training materials can be developed so that users learn the tasks in a staged way. These may be in the form of self-study tutorials that users can work through independently of any formal training that is provided.

The users will also need a reference manual that they can refer to while they are using the system. The reference manual should be a complete description of the system in non-technical language. Many software companies employ technical authors to write manuals in language that users can understand. The manual should be organized for ease of use. This involves the author understanding how the user will carry out their tasks and the kind of problem that they will face. The manual should be organized around the users' tasks, and should be supplemented with a comprchensive index based on the terms that the users will be familiar with rather than the technical terms used by the system developers. Particular attention should be paid to exceptional circumstances and not just routine tasks.

The reference manual may be replicated in the on-line help so that the users can refer to it while they are using the system. However, it should also be available as a paper manual that the users can refer to if there is a problem with the system, or on a CD-ROM that can be loaded onto a separate machine.

19.7.2 User training

Temporary staff and existing staff will have to be trained in the tasks that they will carry out on the new system. Analysts are likely to be involved in the design of the training programme, the development of training materials, the planning of the training sessions and the delivery of the training itself.

Training programmes should be designed with clear learning objectives for the trainees. They will be using the system, and it is important that the training is practical and geared to the tasks that they will be performing. If it is too theoretical or technical, they will not find it useful. Training should be delivered 'just in time'—when the users need it—as they will forget much of what they are told within a short space of time, so training delivered a few weeks before it is required is likely to be wasted. On-line computer-based training using video and audio materials that users can refer to when they need it is likely to be of most use. If formal training sessions are used, then trainees should be given learning tasks to take away and carry out in their workplace. This implies that they will be allocated adequate time for training—skimping on training in order to save money is likely to be counter-productive. Staff will not get the best out of the system and are likely to become frustrated if they do not understand how to work the system. It is often worth following up after users have started using a new system to check that they are using it correctly.

19.8 Implementation Strategies

There are four main strategies for switching over to the new system:

- Direct changeover;
- Parallel running;
- Phased changeover;
- Pilot project.

Figure 19.10 shows three of these changeover strategies in diagram form. Each of them has its advantages and disadvantages.

Direct changeover means that on an agreed date users stop using the old system and start using the new system. Direct changeover is usually timed to happen over a weekend to allow some time for data conversion and implementation of the new system. This does not mean that everything happens in a couple of days, as preparatory work will have been carried out in advance. The advantages and disadvantages of this approach are:

+ the new system will bring immediate business benefits to the organization, so should start paying for itself straightaway;

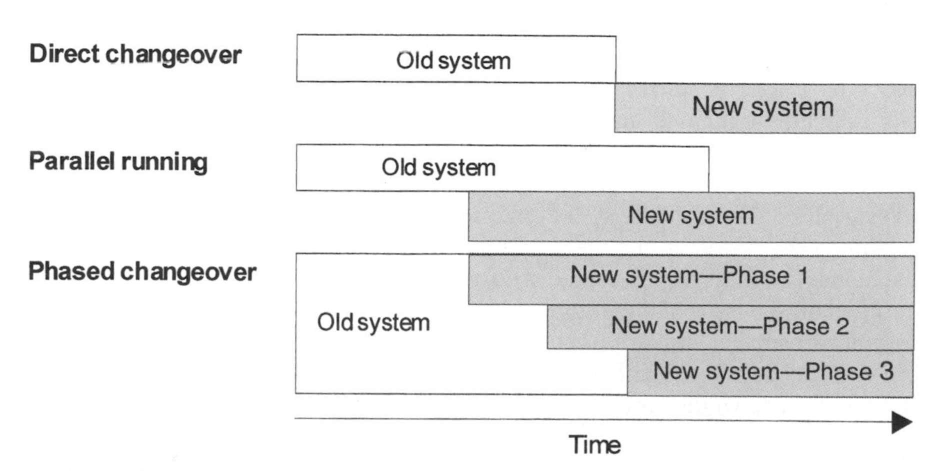

Figure 19.10 Changeover strategies.

+ it forces users to start working with the new system, so they will not be able to undermine it by using the old system;
+ it is simple to plan;
- there is no fallback if problems occur with the new system;
- contingency plans are required to cope with unexpected problems;
- the plan must work without difficulties for it to be a success.

Direct changeover is suitable for small-scale systems and other systems where there is a low risk of failure such as the implementation of established package software.

Parallel running allows the existing system to continue to run alongside the new system. The advantages and disadvantages of this approach are:

+ there is a fallback if there are problems with the new system;
+ the outputs of the old and new systems can be compared—so testing can continue;
- there is a high cost as the client must pay for two systems during the overlap period, and this includes the staffing necessary to maintain information in the old system as well as the new;
- there is a cost associated with comparing the outputs of the two systems;
- users may not be committed to the new system as it is easier to stick with the familiar system.

Parallel running should be used in situations where there is a high level of risk associated with the project and the system is central to the business operations of the organization.

In a *phased changeover*, the system is introduced in stages. The nature of the stages depends on the sub-systems within the software, but introduction into one department at a time may be appropriate. The advantages and disadvantages are:

+ attention can be paid to each individual sub-system as it is introduced;
+ if the right sub-systems can be chosen for the first stages then a fast return on investment can be obtained from those sub-systems;
+ thorough testing of each stage can be carried out as it is introduced;
- disaffection and rumour can spread through the organization ahead of the implementation if there are problems with the early phases;
- there can be a long wait before the business benefits of later stages are achieved.

Phased changeover is suitable for large systems in which the sub-systems are not heavily dependent on one another. The different phases can be introduced on a geographical basis or by department.

A variation on phased changeover is the use of a *pilot project* approach. This involves trialling the complete system in one department or on one site. The decision on extending the system to the rest of the organization depends on the success of the pilot scheme. The pilot project can be used as a learning experience, and the system can be modified on the basis of the practical experience in the pilot project. As such, pilots are suitable for smaller systems, or packaged software, as it is unlikely that a large-scale system will be developed and then introduced in a way that makes its full-scale implementation dependent on the success of a pilot.

19.9 Review and Maintenance

19.9.1 The next steps

The work of analysts, designers and programmers does not stop after a system has been implemented. There is a continuing requirement for staff to work on the new system. First, it is important that the organization reviews both the 'finished' product and the process which was undertaken to achieve it. This may be for contractual reasons, in order to check that the product meets requirements. However, there is a growing recognition of the need for organizations to learn from experience and to record and manage the organizational knowledge which results from this learning. If there were any problems during the lifetime of the project, then these should be reviewed and conclusions drawn about how they might be avoided in the future. The amount of time spent on different tasks during the project can be used as the basis for metrics to estimate the amount of time that will be required for similar tasks in future projects. Second, it is unlikely that the system will be working perfectly according to the users' requirements, and further work will have to be done. Third, in an object-oriented project, the design should be reviewed to identify candidate components for future reuse, although as we suggest later in this chapter, planning for reuse should begin in the early stages of a project. Identifying reusable software components is not an activity to be left until the completion of the project. This subject is covered in Chapter 20.

19.9.2 The review process and evaluation report

The review process will normally be carried out by the systems analysts who have been involved in the project from the start, although it is possible to involve outside consultants in the process for an impartial view. They will normally be supported by representatives of users and user management. The various stakeholders who have invested time, money and commitment in the project will all have an interest in the content of the evaluation report. The report can be very detailed or can provide an overview evaluation—like everything else in the project, there will be a cost associated with producing it. The report's authors should consider the following areas.

Cost benefit analysis. The evaluation should refer back to criteria that were set for the project at its inception. It may not be possible to determine whether all the benefits projected in the cost benefit analysis have been achieved, but most of the costs of development, installation, data conversion and training will have been incurred and can be compared with the projections.

Functional requirements. It is important to check that the functional requirements of the system have been met. Clearly, this is something that should have been taking place throughout the lifetime of the project, but a summary can now be produced. Any actions that were taken to reduce the functional requirements, perhaps to keep the project within budget or on schedule, should be documented for future action under the heading of maintenance. If large areas of functionality were removed to bring the project in on schedule or within budget, a new project should be considered. Major bugs should similarly be documented if they have emerged since the implementation of the new system.

Non-functional requirements. The system should be reviewed to ensure that it meets the targets for non-functional requirements that were documented during the requirements analysis stage. It is now possible to assess whether quantitative objectives for learnability, throughput, response times or reduction of errors have been achieved.

User satisfaction. Both quantitative and qualitative evaluations of the users' satisfaction with the new system can be undertaken, using questionnaires or interviews or both. The results should be treated carefully, as users can pick on quite minor problems or be influenced in their views of the finished product by experiences during the project.

Problems and issues. As stated above, this is an important part of the evaluation process. Problems that occurred during the project should be recorded. These problems may have been technical or political, and it is important to handle the political issues with tact. Including a section criticizing unco-operative users or obstructive user management will mean that some readers take no notice of the rest of the report. Solutions to problems should also be included in the report, as should an indication of who should be learning from this part of the process.

Positive experiences. It is all too easy to focus on the negative aspects of a completed project. It is worth recording what parts of the project went well and to give credit to those responsible.

Quantitative data for future planning. The evaluation report provides a place in which to record information about the amount of time spent on different tasks in the project, and this information can be used as the basis for drawing up future project plans. The quantitative data should be viewed in the light of the problems and issues that arose during the project, as the amount of time spent on a difficult task that was being tackled for the first time will not necessarily be an accurate predictor of how much time will be required for the same task in the future.

Candidate components for reuse. If these have not already been identified during the project itself, then they should be identified at this stage. There will be different issues to be addressed, depending on whether the project has been carried out by in-house development staff or external consultants. For in-house projects, the reuse of software components should be viewed as a process of recouping some of the investment made and being able to apply those reusable elements of the system in future projects. For projects undertaken by external consultants, it may highlight legal issues about who owns the finished software that should have been addressed in the contract at the start of the project.

Future developments. Any requirements for enhancements to the system or for bugs to be fixed should be documented. If possible, a cost should be associated with each item. Technical innovations that are likely to become mature technologies in the near future and that could be incorporated into the system in an upgrade should also be identified.

Actions. The report should include a summary list of any actions that need to be undertaken as a result of carrying out the review process, with an indication of who is responsible for carrying out each such action and proposed timescales.

The end of a project provides a good opportunity for managers of the development staff to review performance on the project as part of a separate review process, the results of which are not necessarily in the public domain. This may feed into staff appraisal procedures and can affect the payment of bonuses, promotion prospects and staff development issues. The latter can include the choice of the next project for a member of staff to work on, their training needs or opportunities to manage or act as a mentor to less experienced staff.

19.9.3 Maintenance activities

Very few systems are completely finished at the time that they are delivered and implemented, and there is a continuing role for staff in ensuring that the system meets the users' requirements. As well as maintenance of the system, there will be a need for support of users: providing initial and on-going training, particularly for new staff; improving documentation; solving simple problems; implementing simple reports that can be achieved using SQL or OQL without the need for changes to the system software; documenting bugs that are reported; and recording requests for enhancements that will be dealt with by maintenance staff. In large organizations these tasks are often handled by a helpdesk that supports all the organization's systems, and it may be appropriate for a member of the development team to join the helpdesk staff either temporarily or permanently. Whether or not this happens, helpdesk or support staff will need to be provided with training so that they can support the new system.

Maintenance involves more significant amendments to a system once it is up and running. Maintenance may be required for a number of reasons.

- There will almost certainly be bugs in the software that will require fixing. The use of object-oriented encapsulation should mean that it is easier to fix bugs without creating knock-on problems in the rest of the system. It is sometimes suggested that bug-fixing involves spending as much time fixing bugs that were introduced by the previous round of maintenance as it does in fixing bugs in the original system.
- In an iterative life cycle, parts of the system may be in use while further development is undertaken. Subsequent iterations may involve maintaining what has already been developed.
- Users request enhancements to systems virtually from day one after implementation. Some of these will be relatively simple, such as additional reports, and may be dealt with by support staff, while others will involve significant changes to the software and will require the involvement of a maintenance team. Often these user requests will reflect the fact that until the system is running and users have a chance to see what it can do, it is difficult for them to have a clear idea of what their requirements are.
- In some cases, changes in the way that the business operates or in its environment, for example new legislation, will result in the need for changes to the system.
- Similarly, changes in the technology that is available to implement a system may result in the need for changes in that system.
- Disasters such as fires that result in catastrophic system failure or loss of data may result in the need for maintenance staff to be involved in restoring the system from data back-ups. Procedures for handling disastrous system failure should be put in place before disasters take place.

In each of these cases, it is necessary to document the changes that are required. In the same way as it is necessary to have a system in place during a project for handling users' requests for changes to the requirements (a change control system), it is necessary to have a system for documenting requests for changes and the response of the maintenance team. This should include the following elements.

Bug reporting database. Bugs should be reported and stored in a database. The screen forms should encourage users to describe the bug in as much detail as possible. In particular, it is necessary to document the circumstances in which the bug occurs so that the maintenance team can try to replicate it in order to work out the cause.

Requests for enhancements. These should describe the new requirement in a similar amount of detail. Users should rate enhancements on a scale of priorities so that the maintenance team can decide how important they are.

Feedback to users. There should be a mechanism for the maintenance team to feed back to users on bug reports and requests for enhancements. Assuming that bugs affect the agreed functionality of the system, users will expect them to be fixed as part of the original contract or under an agreed maintenance contract. The maintenance team should provide an indication of how soon each bug will be fixed. Enhancements are a different matter. Depending on the contractual situation, enhancements may be carried out under a maintenance contract or they may be subject to some kind of costing procedure. Significant enhancements may cost large amounts of money to implement. They will require the same kind of assessment as the original requirements. They should not be left to maintenance programmers to implement as they see fit, but should involve analysts and designers to ensure that the changes fit into the existing system and do not have repercussions on performance or result in changes to sub-systems that affect others. This process itself may incur significant costs just in order to work out how much an enhancement will cost to implement. Significant enhancements should therefore be regarded as mini projects in their own right, and involve the professional skills of project managers, analysts and designers.

Implementation plans. The maintenance team will decide how best to implement changes to the system, and this should be carried out in a planned way. For example, significant additions to a class that affect what persistent data is stored in the database will require changes to the database structure, and may also require all existing instances of that class to be processed in order to put a value into the new attribute. This will probably have to take place when the system is not being used, for example over a weekend. Enhancements to the system may fall into one of four categories: those to be made at no cost; those that will be made at a cost to be agreed with the client; those that will be held over until a major upgrade to the software is made and that will be part of a future version; and those that cannot or will not be made in the foreseeable future.

Technical and user documentation. Amendments to a system must be documented in exactly the same way as the original system. Diagrams and repository entries must be updated to reflect the changes to the system. If this is not done, then there will be a growing divergence between the system and its technical documentation; this will make future amendments all the more difficult, as the documentation that maintenance analysts consult will not describe the actual system. Clearly, user documentation, training and help manuals as well as on-line help must all be updated.

In large organizations with many systems, staff in the information systems department may spend more time on maintenance of existing systems than they do on development of new systems. There is a growing movement for organizations to *outsource* their maintenance. This means handing over the responsibility for maintenance of a system to an external software development company under a contractual agreement that may also involve the provision of support. Some companies now specialize entirely in maintaining other people's software.

The other impact of the growth of the maintenance task is that the time will come when a system is such a burden in terms of maintenance that a decision must be made about its replacement. The review process that takes place at this stage may lead to the inception of a new project and to the systems development life cycle starting again.

19.10 Summary

The implementation of a new system involves a large number of different types of software package that are used to produce and to support the finished system. Component diagrams and deployment diagrams are the two UML implementation diagrams that can be used to document the software components and their location on different machines in the system. For large, complex installations, these diagrams may become unwieldy, and a table format in a spreadsheet or database may be easier to maintain.

Analysts and designers have a role during the implementation stage in maintaining system and user documentation and in providing user training. They also plan and carry out testing, plan for data conversion from existing systems and assist the project management in planning the appropriate implementation strategy for the system.

After a new system has been implemented, it is customary and advisable to carry out a post-implementation review. This will result in the production of an evaluation report to stakeholders that will measure the success of the project and identify issues and problems from which the organization should learn. Typically the evaluation report will include the following sections.

- Cost benefit analysis.
- Functional requirements.
- Non-functional requirements.
- User satisfaction.
- Problems and issues.
- Quantitative data for future planning.
- Opportunities for reuse.
- Future developments.
- Actions.

New systems rarely work exactly as expected, and maintenance must be carried out in order to ensure that the system is bug-free and meets users' requirements. Procedures must be put in place for the maintenance team (project manager, analysts, designers and programmers) to document the process and the changes that are made.

Review Questions

19.1 List the different categories of software that may be used in developing a system.

19.2 What packages have you used and which categories do they fall into?

19.3 What is the difference between a package diagram and a component diagram?

19.4 Draw a component diagram to show the run-time dependency between a Java class file, the java.exe run-time program and the Java classes in a zip file.

19.5 Draw a deployment diagram to show how a web browser and web server are located on different machines and the communication protocol they use.

19.6 List five tests that you would carry out on the FoodCo use case `Start Line Run`.

19.7 List the sections that you would include in a post-implementation evaluation report and explain the content of each section.

19.8 What is the difference between maintenance work carried out to fix bugs and work carried out to add requested enhancements to a system?

19.9 Why should decisions about enhancements not be left to maintenance programmers?

19.10 What tasks do maintenance staff undertake?

Case Study Work, Exercises and Projects

19.A Draw a component diagram for a program that you have written.

19.B Find out about the Library system in your school, college, university or local library. Draw a deployment diagram to show the physical hardware architecture.

19.C The FoodCo system will have a database on a central server and PCs in different offices that access that server. Draw a deployment diagram to show this architecture.

19.D There are commercial packages to automate the installation of software such as InstallShield for Windows and RPM for Red Hat Linux. Investigate how these packages maintain information about what components must be installed on a machine.

19.E Read the user manual or on-line help for a software package that you use. What examples can you find of good practice and bad practice in what has been written? How would you improve it?

19.F What would you include in the screen layouts for users to fill out for bug reports and enhancement requests? Produce a draft design for each of these two layouts.

Further Reading

- Implementation takes us on into areas that are outside the scope of this book. There are a number of excellent books on Java implementation, for example Deitel and Deitel (1997). Wutka's book on Java 2 Enterprise Edition (2001) gives a clear idea of the techniques involved in building J2EE systems. Orfali and Harkey (1998) explain CORBA and the alternatives.
- Many CASE tools now offer a wide range of functionality, including integration with configuration management tools, built-in IDEs for programming, code generation and reverse engineering, object-relational mapping tools, software metrics and deployment tools. Vendor websites are the best starting point for understanding what is on offer.

CHAPTER 20

Reusable Components

OBJECTIVES

In this chapter you will learn

- why reuse is regarded as important
- some of the problems with reuse
- how to plan systems development to achieve reuse of components
- the different techniques for obtaining and managing reuse
- how to design a reusable component.

20.1 Introduction

In the system life cycle, implementation is followed by maintenance, when the new system has any remaining bugs removed and enhancements made to it. Using object-oriented technology does not make the problems of removing bugs and enhancing systems go away, but it does add a possible further stage to the life cycle—the reuse stage. In this chapter we look at how object-oriented software can be reused. In particular we look at the idea of *componentware*—software packaged into components that can be reused as part of other systems.

In Section 20.2 we discuss approaches to developing component-based software: we raise issues concerning planning for reuse during the project and consider how to package software to make it reusable. We provide an example of industrial strength componentware and present an example from the case studies used in this book which demonstrates how software can be designed and packaged for reuse (Section 20.4). Chapter 21 addresses the project management issues associated with planning for reuse.

20.2 Why Reuse?

In Chapter 4 reusability was discussed as one of the reasons for adopting object-oriented development techniques and programming languages, in Chapter 8 inheritance and composition were discussed as two techniques that facilitate the development of reusable components, and in Chapter 12 we highlighted reusability as one of the characteristics of a good object-oriented design. Reusable software has been one of the objectives of developers for many years. Using top-down functional decomposition of designs in languages such as Fortran or C, the development of reusable libraries of functions has made it possible for programmers to save time and effort by reusing others' work. The growth of Visual Basic as a programming language was aided by the availability of controls that could be bought off the shelf and incorporated into applications to provide functionality that would be difficult for the less experienced programmer to develop—and in any case, why re-invent the wheel? Object-oriented languages have always been promoted as significantly enabling the reuse of software, and when Java was released in the mid 1990s, part of the hype surrounding the language was the ability to download and reuse services over the Internet. Why then is reuse regarded as so important?

The arguments for reuse are partly economic and partly concerned with quality.

- If some of the requirements of a project can be met by models or software components that have been developed on a previous project or are bought in from an outside supplier, then the time and money spent producing those models or code is saved. Although the saving will be partly offset by the cost of managing a catalogue of reusable models or code or of paying to buy them from elsewhere. Lim (1994), describing the situation at HP, cites improved productivity, faster time to market for products, fewer defects and a return on investment of 215% on one project and 410% on another.
- If a developer can reuse a design or a component that has been tested and proved to work in another application, then there is a saving in the time spent to test and quality assure the component. Jacobson et al. (1997) cite IBM as an example of a company that has invested in software reuse and that has reuse support centres that maintain a library of 500 zero-defect components in Ada, PL/X and C++.

Developers of object-oriented systems are often end-users of reusable components, when they use packages, libraries, classes, or controls in their chosen development environment. However, object-oriented systems have not achieved the level of reuse that was expected of them in terms of generating reusable components that can be applied again within the same organization. There are a number of reasons for this, some are technical and some are concerned with organizational culture.

- *Inappropriate choice of projects for reuse.* Not all organizations or projects within those organizations are necessarily suitable to take advantage of or act as to sources of reusable components.
- *Planning for reuse too late.* If reuse is appropriate, it is something that needs to be planned for even before a project starts, not an afterthought. By the time a project has been completed, it is likely that anything that might have been reusable will have been designed in such a way that it cannot easily be extracted from the rest of the system. To achieve reuse, the organization needs to be structured to support it, with the people and tools in place to make it possible.

- *The level of coupling between different classes in an object-oriented design.* Many people have thought of classes as the unit of reuse in object-oriented developments. However, when we come to design classes for different systems, it may be possible to identify similar classes that could be developed in a way that makes them of use in more than one system, but more often than not, the implementations of these classes will include attributes and associations that tie them in to other classes in the particular application of which they are a part.
- *The lack of standards for reusable components.* This has changed recently with developments in the technology of repositories in which to store components and with the introduction of standards such as the Object Management Group's CORBA Version 2.0 (OMG, 1995) and the W3C's SOAP (Simple Object Access Protocol).

In the following sections, we'll address each of these issues.

20.2.1 Choice of project

Not all projects are necessarily suitable for the development of reusable components. The two main factors that influence this are the nature of the business within which the software development is taking place and the maturity of the organization's object-oriented development.

Jacobson et al. (1997) identify four kinds of software business, which they suggest are suitable candidates for developing reusable components. In all of these they talk of the organization developing a Reuse-driven Software Engineering Business (RSEB).

- Organizations where creating an RSEB improves the business processes within the organization: large organizations with a considerable information systems infrastructure and a portfolio of projects to support business activities.
- Organizations producing hardware products that contain embedded software: they cite Hewlett-Packard and Ericsson as examples of this type of organization.
- Consultancy companies and software houses that develop software for external clients that have outsourced their information systems development: particularly those which target particular vertical markets (companies in the same kind of business).
- Developers of software products, such as Microsoft, where reusable components can be applied across a large product range and where end-users can also benefit from the interoperability of software through mechanisms such as DCOM (Distributed Component Object Model).

Small, one-off projects in small organizations are unlikely to bring significant benefits from building reuse into the software development lifecycle.

If we consider our two case study companies, we have to ask whether they would fall into the first of these categories. Although both of them are developing systems to support their business activities, are they large enough and with enough potential projects to justify taking a reuse-driven approach to these projects? Building the organizational structures to support reuse costs money, and that expense is only justified if it can be recouped by savings on other projects. In both cases, they do not have a developed information systems department, and we would probably have to say that they are not going to benefit from developing an RSEB.

If, however, an outside consultancy company is doing the development for both Agate and FoodCo, then there are a number of areas where reuse may be applicable. A software company looking at the two systems would identify areas such as managing information about staff that are common to both. If the software company specializes in a vertical market—media and advertising or food manufacturing—then there are going to be parts of the software systems that can be reused elsewhere. Indeed, if the software company has already introduced a reuse-driven approach to its business, the systems for Agate and FoodCo could be developed from a range of existing components tailored to the specific requirements of these companies.

20.2.2 Organizational structure

Jacobson et al. (1997), based on experience at Hewlett-Packard, describe organizations as typically going through six stages of development of a reuse culture. At each stage some benefit is to be gained, but it is unlikely that an organization can leap from a situation in which there is no reuse of design models or software taking place to one in which there is a complete organizational culture of reuse and the structures and tools are in place to support the consistent reuse of components in a way that brings the kind of business benefits that were mentioned earlier. The six stages are as follows.

- None. No code reuse takes place; everything is developed from scratch.
- Informal code reuse. Developers trust each other enough to begin to reuse each other's code in order to save time on development.
- Black-box code reuse. Particular pieces of code are engineered for reuse, and all developers are encouraged or required to use them to ensure a consistent approach and reduce maintenance costs.
- Managed workproduct reuse. An organizational structure is developed to manage reusable code, to maintain versions, to document functionality and to train developers.
- Architected reuse. In order to ensure that components work together, a common architecture is designed and applied to all development processes.
- Domain-specific reuse-driven organization. The organization's software development is geared to the production of reusable components for the business domain, and the culture and structure of the organization supports this approach.

Allen and Frost (1998) argue that despite moves to client–server and three-tier architectures, most software development organizations still have an application mindset, and software is developed for individual applications, even within the same organization, without regard for reuse. (Although many professional developers would counter this with the argument that they have always developed reusable libraries of code.) This may even be a step backwards from the developments of the 1970s and 1980s when the growth of corporate databases under the central control of database administrators meant that at least the data model was likely to be consistent across the whole organization.

To gain the benefits of an RSEB requires an incremental process of change within the organization, involving technical champions to argue the technical case and develop the software architecture, management champions who believe in the business benefit and will provide the support and investment to allow the change to take place, pilot projects to kick-start the process, education and training of developers to enable them to understand the process, and the development of support structures within the organization. Of these, the first is the most critical: to achieve effective reuse, the elements of the software architecture must be common across different systems.

One of the most significant requirements for support structures is that if developers are to use reusable components in their code they need some way of finding out what

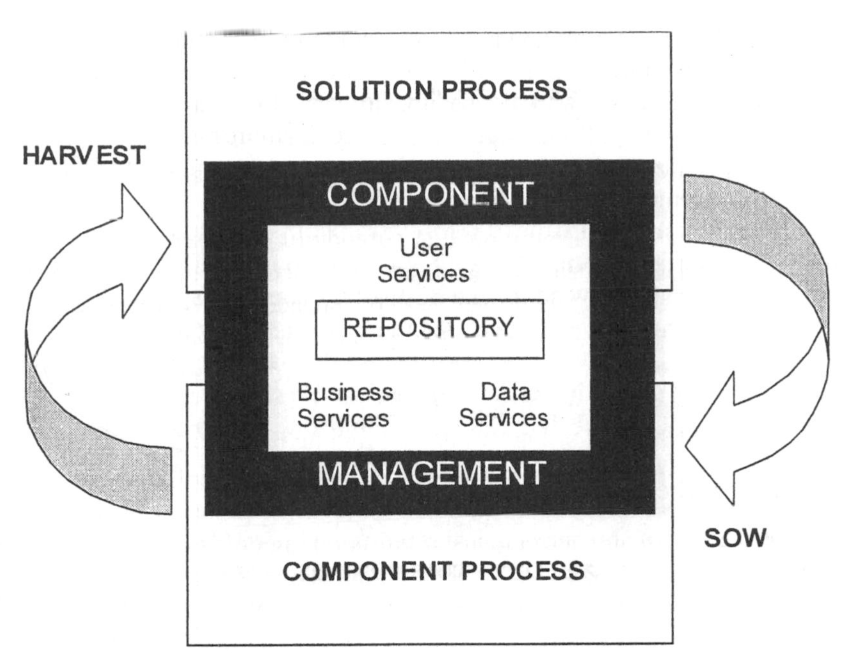

Figure 20.1 The SELECT Perspective service-based process (adapted from Allen and Frost, 1998).

components are available and what their specifications are. This requires software tools to manage a repository of components and staff to maintain the components in the repository and to document them.

Allen and Frost (1998) place a repository at the centre of their model of the development process for reusable components. Figure 20.1 shows this with the two complementary processes: sowing reusable components during development and harvesting reusable components for reuse in other projects. Their approach is described in more detail in Section 20.3.1.

20.2.3 Appropriate unit of reuse

In Chapter 8 we talked about components as the unit of rcuse, and we have used the term throughout this chapter so far. However, we have not yet defined what we mean by a component in this context.

If we again consider the case studies, there is a need for a `Client` or `Customer` class both in the Agate system and in the FoodCo system. During analysis, these two classes may look very similar, but as we move into design, the associations between these classes and others in their system will be resolved into specific attributes. The Agate `Client` class will have attributes to link it to `Campaigns` while the FoodCo `Customer` class will be linked to `SalesOrders`. If the development of both systems is being carried out by the same software company, then it requires a novel style of project management and organization to recognize this commonality in two different projects. If the commonality is recognized, then there is no guarantee of successful reuse unless a suitable architecture is developed that will support the reuse of the common elements of the `Client` class and allow it to be tailored to the requirements of the individual systems.

For example, either a `Client` class can be obtained from elsewhere and subclassed differently for each project, or a `Client` class can be written that is domain-neutral and then subclassed for each different project. This inheritance-based approach also helps to solve a problem that is common with software that is tailored to the needs of different customers of the software house: it clearly separates those parts of the class that are common to all users from those that have been tailored to specific needs. This helps with the installation of upgrades and prevents the changes made for one customer being implemented for all customers.

However, even if we can reuse the `Client` class in both applications by extending its functionality through inheritance, there are going to be other aspects of the `Client` class that we may or may not want to take through into another system. These include control classes and the business logic associated with the management of clients, related boundary classes and the mechanisms that manage the persistent storage of instances of `Client` in some kind of database. So can we reuse the class on its own?

Allen and Frost (1998) argue that the class is the wrong level of granularity at which to apply reuse. They argue that reuse should take place at the level of components rather than classes. They define a component as follows.

> A component is an executable unit of code that provides physical black-box encapsulation of related services. Its services can only be accessed through a consistent, published interface that includes an interaction standard. A component must be capable of being connected to other components (through a communications interface) to form a larger group.
>
> (Allen and Frost, 1998, p. 4)

However, this view of a component as executable code limits the types of reuse that can be made of a component. Jacobson et al. define a component as follows.

> A component is a type, class or any other workproduct that has been specifically engineered to be reusable.
>
> (Jacobson et al., 1997, p. 84)

This definition is more useful, as it does not limit the developer to only considering executable code for reuse. The intermediate products of the development life cycle—use case model, analysis model, design model, test model—can all be considered as candidates for reuse. There are two outcomes from this view.

- First, we may choose to reuse as components sub-systems that provide more functionality than just a single class. For example, again both Agate and FoodCo have requirements to manage information about employees. Rather than developing two separate systems that handle staff, grades, staff development and so on, we may aim to produce a single reusable sub-system that can be used in both companies' systems.
- Second, we may choose to reuse intermediate products. For example, if we have an existing sub-system to manage staff, we could reuse some of the use cases and related boundary, control and entity classes, but choose to leave out others. Tracing through from the analysis model to the design and test models of the system, we should also be able to reuse elements of these models.

Essentially, we are dealing here with the difference between black-box and white-box reuse. Allen and Frost are suggesting a black-box model in which the contents of the components are not visible to the reuser; Jacobson et al. are suggesting a white-box model in which the internals of the component are visible, giving more flexibility to the reuser about how they make use of the component.

Jacobson et al. also make the point that there are different mechanisms for reusing components. In Chapter 8 we discussed the use of inheritance and composition as mechanisms for reuse. However, using inheritance to subclass existing classes is not the only mechanism for reuse, and if the class is not the unit of reuse, then other mechanisms must be used. Jacobson et al. suggest the following:

- inheritance,
- the «include» relationship between use cases («uses» in version 1.0 of UML),
- extensions and extension points in use cases and classes,
- parameterization, including the use of template classes,
- building applications by configuring optional components into systems and
- generation of code from models and templates.

The last two are development processes rather than specific design structures, and make reuse easier to achieve.

20.2.4 Component standards

If we are talking about black-box reuse, then the potential for reuse depends on the software mechanisms for reusable components. If we want to consider white-box reuse, then the potential depends on the mechanisms for exchanging software models. In the latter case, UML is clearly a candidate for exchangeable, reusable software models, especially if CASE tool vendors implement the XMI (XML Metadata Interchange). In the former case, then we are dependent on the developers of programming languages and software development infrastructure to deliver appropriate tools to the development community to enable them to develop reusable components.

A number of programming languages and development environments provide mechanisms by which developers can package software into components. Figure 20.2 lists some of these. This table shows that the search for ways of promoting reuse through some kind of modular architecture is not new in the software development industry. Reuse has been an objective that has driven the design of programming languages and has informed the development of programming styles. However, the potential for developing reusable components has been increased recently by three factors.

- The development of CORBA as a standard for interoperability of components written in different languages and running on different platforms.
- The promotion of Java as an object-oriented language with relatively straightforward mechanisms for producing software in packages to deliver different services.
- The growth of the Internet and the World Wide Web, which has made it possible for people to make their software components easily available to a wide marketplace of potential reusers.

It is the platform independence of Java and CORBA that makes them different from the other languages and environments shown in Figure 20.2.

.NET, which is emerging at the time of writing, together with C#, the new language from Microsoft, may provide an alternative to Java. In particular, it provides a mechanism based on the SOAP Contract Language (SCL) to discover the services offered by Web Services, which describe themselves using XML. .NET also defines extensions to Microsoft's Portable Executable (PE) format so that metadata is stored with the bytecode in Microsoft Intermediate Language (MSIL) executables, allowing them to provide information about the services they offer in response to requests in the correct format. However, it is not clear whether or when Microsoft will provide

Language or development environment	Mechanism for component reuse
Borland Delphi	Object Pascal units compiled into .dll files—Dynamic Link Libraries
Microsoft Visual Basic	.vbx files—Visual Basic Extensions .ocx files
Microsoft Windows	.ole files—Object Linking and Embedding DDE—Dynamic Data Exchange .dll files—Dynamic Link Libraries COM—Common Object Model DCOM—Distributed Common Object Model
CORBA	.idl files—Interface Definition Language IOP—Inter-ORB Protocol
Java	.jar files—Java Archive packages JavaBeans
Microsoft.NET	MSIL—Microsoft Intermediate Language CLR—Common Language Runtime WSDL—Web Service Description Language

Figure 20.2 A sample of languages and development environments with mechanisms for reuse.

support for other operating system platforms, unlike Java and CORBA, which already offer interoperability on a range of platforms.

However, the existence of CORBA, Java and the WWW does not guarantee that effective reuse of components will take place. A strategy needs to be put in place within the organization to ensure that reuse is built into the systems development life cycle.

20.3 Planning a Strategy for Reuse

In some organizations, reuse may just be about making use of reusable components from elsewhere, using the kinds of mechanisms that are listed in Figure 20.2. In others, reuse will be about the kind of organizational change that we discussed in Section 20.2.2. In the rest of this section we describe two approaches to the introduction of a reuse strategy and then in Section 20.4 give an example of a commercially available reusable component package.

20.3.1 The SELECT Perspective

Allen and Frost (1998) describe the SELECT Perspective approach to the development of reusable components. At the level of practical techniques, this includes guidelines for the modelling of business-oriented components and for wrapping legacy software in component wrappers. They distinguish between reuse at the level of component packages, which consist of executable components grouped together, and service packages, which are abstractions of components that group together business services. The focus of this approach is to identify the services that belong together and the classes that implement them. Service classes in a single package should have a high level of internal interdependency and minimal coupling to classes in other packages.

In order to develop reusable components while achieving the development of a system to meet users' needs, the Perspective approach breaks the development process into two parts: the solution process and the component process. These two parts run in parallel and feed off each other.

- The solution process focuses on specific business needs and delivering services to meet the users' requirements. Its products have immediately definable business value. During the solution process, developers will draw on the component process in their search for reusable components that can be applied to the project.
- The component process focuses on developing reusable components in packages that group together families of classes to deliver generic business services. During the component process, the developers produce components that can be reused in the solution process. The component process also searches out opportunities to reuse services from existing legacy systems and legacy databases and from other packages of components.

Allen and Frost use the analogy of sowing and harvesting reusable services: the component process sows reuse and the solution process harvests services. Figure 20.1 shows the relationship between the two in diagrammatic form.

Central to Figure 20.1 is a *repository*. Software support is needed for effective component reuse to take place. This support takes the form of repository-based component management software. Components are placed in the repository as a means of publishing them and making them available to other users. The repository is made up of catalogues and the catalogues contain details of components, their specifications and their interfaces. Component management software tools provide the functionality for adding components to the repository and for browsing and searching for components. Component management software may be integrated with CASE tools to allow the storage of analysis and design models as well as source code and executables.

20.3.2 Reuse-driven Software Engineering Business

Jacobson et al. (1997) describe an approach to developing reusable software components that is rooted in Jacobson's OOSE and Objectory (Jacobson et al., 1992), and that uses the notation of UML Version 1.0. The approach is based on practical experience within Ericsson and Hewlett-Packard.

Unlike Allen and Frost, who consider components as executables or as packages of executables designed to deliver a particular service, Jacobson et al. consider reuse in terms of any of the work products of systems development. This means that models that are produced before the finished program code are candidates for reuse, and that artefacts other than classes, for example use cases, can be reused. However, the key point of this approach is that the design of systems to make use of reusable components requires an architectural process right from the start. And that means changing the way the business operates.

In order to transform the business into a reuse business, Jacobson et al. also draw on another book from the same stable. Jacobson et al. (1995) explains an approach to business process reengineering that is based on OOSE and Objectory. The task of developing a reuse business is a reengineering task that can be modelled using object-oriented business engineering, and that leads to the development of systems to support the RSEB. Jacobson et al. suggest that the end result is a business consisting of the following competence units: Requirements Capture Unit, Design Unit, Testing Unit, Component Engineering Unit, Architecture Unit, and Component Support Unit. These competence units are groupings of staff with particular skill-sets and the business data and documents for which they are responsible.

The emphasis in RSEB is to design an architecture for systems that support reuse from the start. This is done through three engineering processes: Application Family Engineering, Component System Engineering and Application System Engineering.

- Application Family Engineering (AFE) is an architectural process that captures the requirements for a family of systems and turns them into a layered architecture, consisting of an application system and a supporting component system.
- Component System Engineering (CSE) is the process of focusing on the requirements for the component system and developing the use cases, analysis models and design for reusable components to support application development.
- Application System Engineering (ASE) is the process of developing the requirements for applications and developing the use cases, analysis models and design to produce application software that makes use of the reusable component systems developed by CSE.

The life cycle for this kind of project is an iterative one. The engineering processes can run concurrently, with the emphasis changing as the project progresses. Model elements in the component systems are exposed to those in the application systems through façades. The Façade pattern is explained in Section 20.5. Jacobson et al. use the Façade pattern to organize components in packages and then expose those components to other packages. (We take a slightly different approach in Section 20.5.)

20.4 Commercially Available Componentware

Until recently, most commercially available components took the form of utilities or graphical user interface components. The best example of this is the wide variety of controls that are available for use with Microsoft Visual Basic. Originally these were supplied as add-ins in the form of .vbx files which could be included in the Visual Basic toolbar in the same way as the built-in controls, or in the form of OLE (Object Linking and Embedding) objects which allowed the functionality of other software packages such as word-processors to be embedded in applications. With the introduction of ActiveX, based on the Microsoft COM architecture, these add-ins are now available as .ocx files. If you look through the catalogue of a good software supplier that sells development tools, you will find pages of ActiveX controls that can be used in your applications and that provide the developer with the possibility of building a wide range of different functions into their software without having to reinvent the wheel. Examples include:

- serial communications,
- computer-aided design,
- project management including Gantt charts,
- spreadsheets,
- scientific charts and
- barcode reading and printing.

The use of standardized mechanisms to access the functionality of these controls has meant that other software companies can also write interfaces to them, and it is just as possible to use ActiveX controls in Borland Delphi or in Java (using Microsoft Visual J++) as it is in Visual Basic or Visual C++.

For applications written in Java, however, there is another mechanism that can be used: the JavaBean. JavaBeans (or just Beans) are components written in Java that typically combine a graphical element with the functionality to support it. Beans support the Component–Container model in the Java Abstract Windowing Toolkit (AWT), which means that Beans that encapsulate access to particular domain objects can be added into applications in the same way as other visual components can be added to the applications.

Most of these add-in controls provide generic capabilities rather than reusable components for particular types of business operations. Suppliers are now producing reusable components for business functions. Most notable among these developments is IBM's San Francisco project.

The San Francisco project provides distributed, server-based components for different types of business processes. San Francisco uses a layered architecture (shown in Figure 20.3). The Foundation layer provides a programming model to support distributed transactions, and uses a set of distributed object services and utilities written entirely in Java. It also provides an optional GUI framework written using JavaBeans. The Common Business Objects layer implements general purpose business objects together with the facts and rules required for any business application. This includes business objects such as company, business partner, address and calendar. Four components are provided in the Core Business Processes layer at present. These are:

- general ledger,
- accounts receivable and accounts payable,
- warehouse management and
- order management.

These have been built using design patterns, many of which were discovered as the project developed, and provide support for electronic commerce. IBM has integrated the San Francisco project into its WebSphere range of products for E-business development.

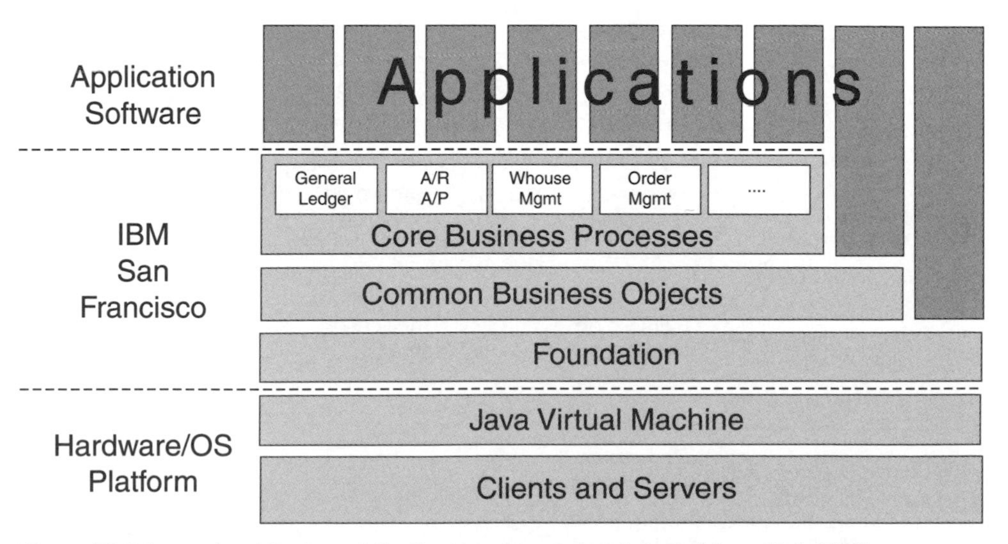

Figure 20.3 Layered architecture of the San Francisco project (adapted from IBM, 1998).

20.5 Case Study Example

A common feature of many applications is the need to control the access of members of staff to the different programs that make up a system. A non-functional requirement both for Agate and for FoodCo is to restrict access of staff to the use cases that they are permitted to use. This requirement can be summarized as follows.

Each program in the system will be represented by a use case in the use case diagram. One or more actors will be associated with each use case, and each actor may be associated with more than one use case. A member of staff will fill the role of one or more actors, and each actor will be represented by one or more members of staff. Each actor will be given access rights to use specific use cases (programs). A member of staff may only use those use cases (programs) for which one of the actor roles they fill has been given an access right.

This non-functional requirement in the context of the main systems can be viewed as the basis for functional requirements in a security sub-system. This sub-system is a potential candidate for the development of a reusable component. Figure 20.4 shows the use cases for this sub-system. It can be modelled in a class diagram in the same way as the business classes that meet the functional requirements of the system. Figure 20.5 shows the initial domain class diagram for this requirement. Two association classes, `ActorRole` and `AccessRight` have been included in the class diagram, as it was initially thought that there might be some data associated with the creation of links

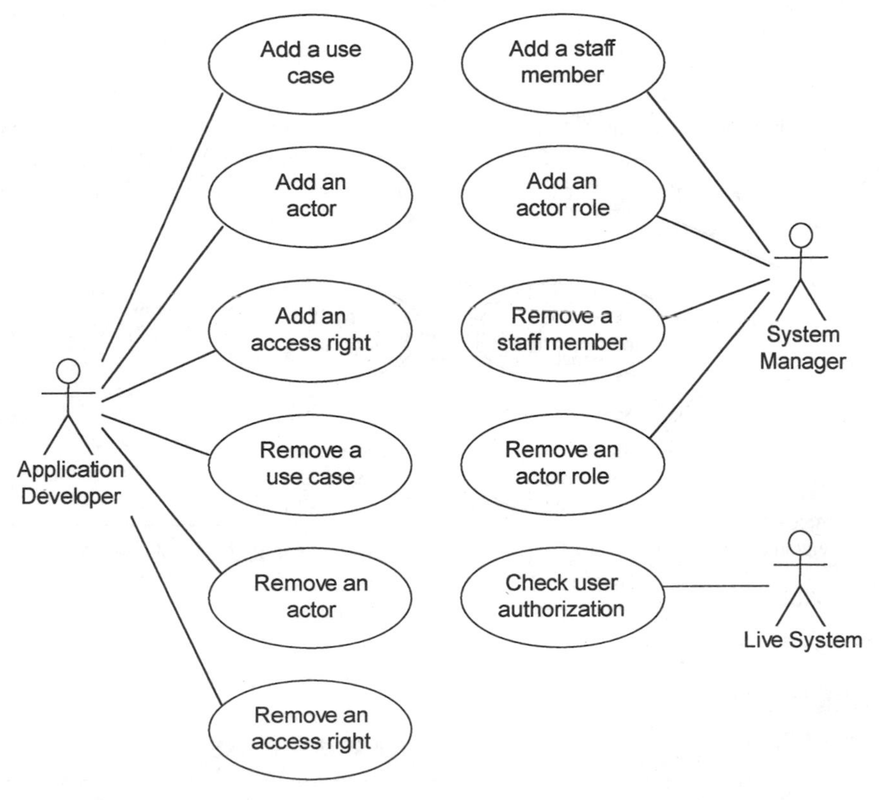

Figure 20.4 Use case diagram for security requirement.

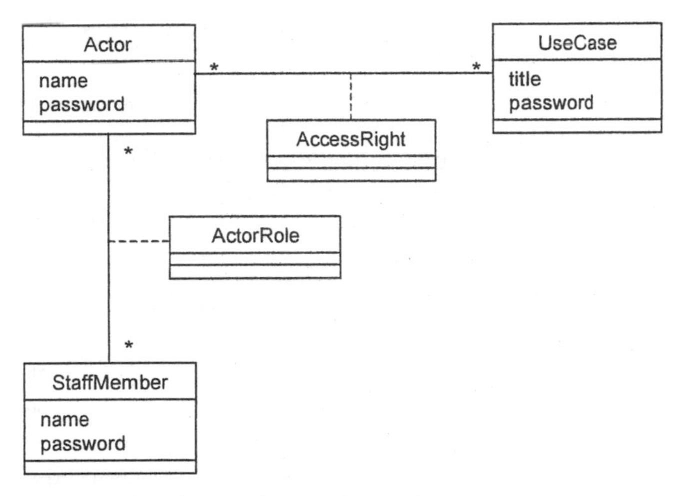

Figure 20.5 Initial domain class diagram for security requirement.

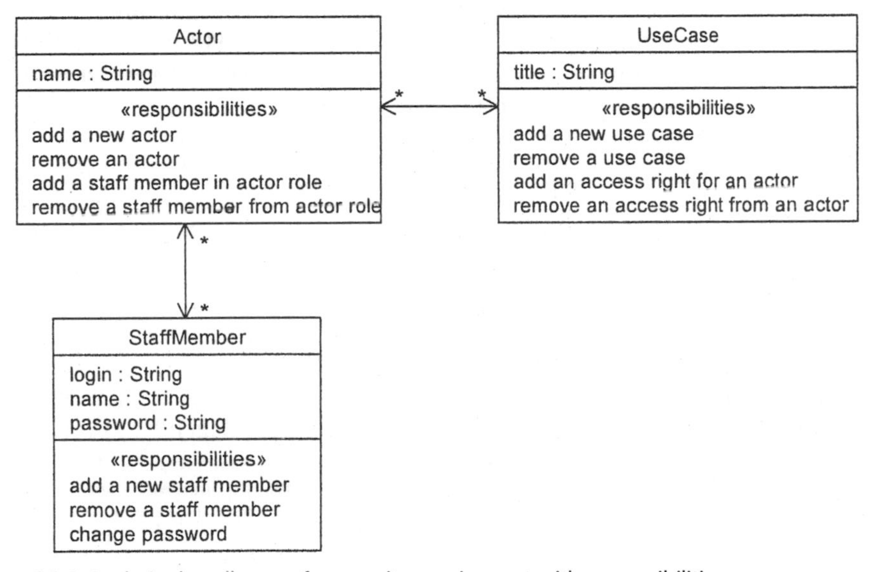

Figure 20.6 Analysis class diagram for security requirement with responsibilities.

between instances, for example the data that an access right was given, or the type of an access right (Read/Write/Update). However, further discussion with users and analysis of the requirements indicates that this is likely to make the sub-system more complicated than it needs to be, so they have been removed from Figure 20.6, which shows the analysis class diagram.

There are many design alternatives for this part of the system. The particular design alternative that we choose will affect the detailed design of this sub-system. If we look at it as though we are the software company developing software for both Agate and FoodCo, some of the alternatives are as follows.

- Do we design this sub-system with a set of boundary and control classes to support the use cases in Figure 20.4?
- Can we reuse the existing `StaffMember` class in the business domains of Agate and FoodCo (and any other companies with similar systems)? We do not really want to have to set up data about staff members in two places.

- What happens if we do reuse the `StaffMember` class in the business domain and then want to use this security sub-system to support a system that does not have `StaffMember` as an entity class?
- If this security sub-system is to be implemented for all the application software we develop, then we are going to have to make some classes in the software (presumably the control classes) aware of the interface to this sub-system. How do we extend these control classes: rewrite them, extend them or subclass them?
- How do we provide persistent data storage for this sub-system? Does it have its own storage mechanisms, or will it use the same storage mechanisms as whatever application it is supporting?
- What parts of this sub-system are we going to make visible to other applications? Are we going to make all the classes visible or are we going to provide a single interface?

We might choose to design the system so that when a user starts running the application, they are prompted for their name or user ID and a password. Alternatively, if they are required to log into a network anyway, the software could obtain the user's ID from the network operating system. Each time a user starts a new use case in the main application, the application will need to check with the security classes whether that user is authorized to use that particular use case.

The security requirement is not part of the business requirements of the domain applications, and we want to reuse the security software in other applications, so it makes sense to separate these classes from the rest of the software and put them in a package of their own. The security classes will require their own boundary classes, to allow the actors shown in Figure 20.4 to carry out the use cases. These will run on client computers and will be in a separate package within the overall security package. They will have dependencies on other packages that provide these services, such as the Java AWT. We have created two packages for control classes, one for classes that will run on the clients and control the boundary classes, and one for control classes that will run on the server. These control classes will have a dependency on the core security classes. Figure 20.7 shows these package dependencies. We have also shown a package to represent a business application that will be using the services of the security package to authenticate users. It is arguable whether this should have a dependency on the server control classes or on some kind of client package that hides the implementation. Whatever approach we take, we want to provide a clean interface to the functionality of the security sub-system for developers to use. It should be possible to design and implement a separate security client, which uses the interface to the security server control classes. Also, programmers should have a straightforward application programming interface (API) to the authentication service—the use case `Check user authorization`.

One way of doing this would be to replace the control classes in the `Security Server Control Classes` package with a single control class. This will make it easier for developers to reuse the package, and application programmers only need to know the API of this one class. However, that would lead to a single class with no attributes and a large number of operation implementations. An alternative approach is to leave the control classes as they are and to create a Façade class that has the operations from the control classes within the sub-system but does not contain their implementation.

This approach is based on a design pattern, called the Façade pattern. (See Chapter 15 for more on design patterns.) The Façade pattern is a Gang of Four structural pattern, and is described in Gamma et al. (1995) in the following terms.

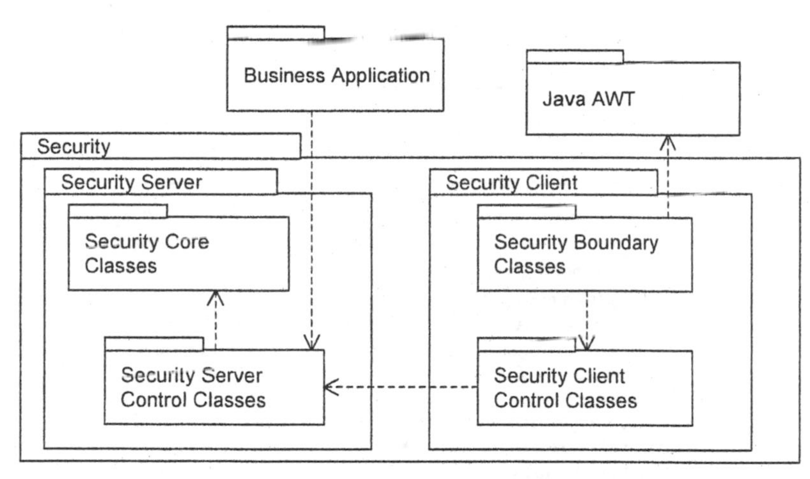

Figure 20.7 Package diagram showing security classes and dependencies.

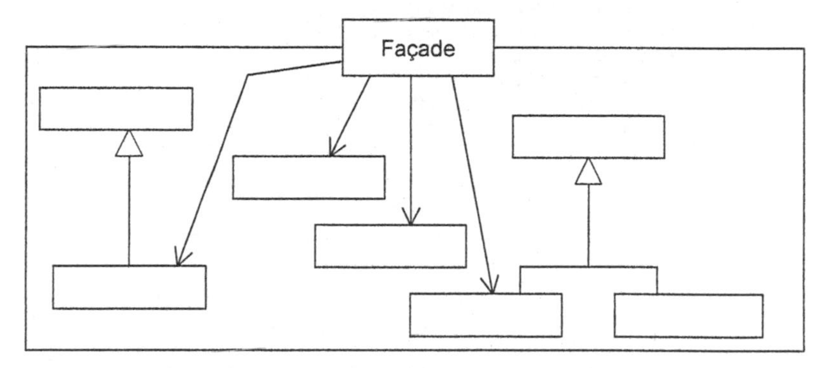

Figure 20.8 Structure of Façade pattern (adapted from Gamma et al., 1995, p. 187).

> **Intent**
> Provide a unified interface to a set of interfaces in a sub-system. Façade defines a higher-level interface that makes the sub-system easier to use.
> ....
>
> **Applicability**
> Use the Façade pattern when
> - you want to provide a simple interface to a complex sub-system . . .
> - there are many dependencies between clients and the implementation classes of an abstraction. Introduce a façade to decouple the sub-system from clients and other sub-systems, thereby promoting sub-system independence and portability.
> - you want to layer your sub-systems . . .
>
> (Gamma et al., 1995, pp. 185–186)

The structure of the Façade pattern is shown in Figure 20.8. We could use this structure to add a single class, called `SecurityManager` which provides the API to the functionality in the security package. Or we could add two separate Façade classes, one for the management of the security sub-system (adding staff members etc.), and one for the authentication service used by business applications. This is shown in the class diagram in Figure 20.9. We have added other operations that will be required in order to support the use cases for maintaining the information in the sub-system, for example to list all the actors for a particular use case.

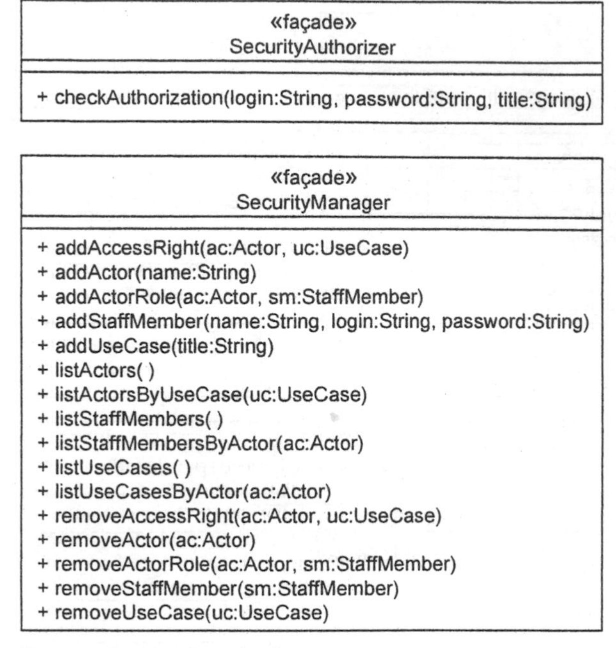

Figure 20.9 Class diagram showing Façade classes.

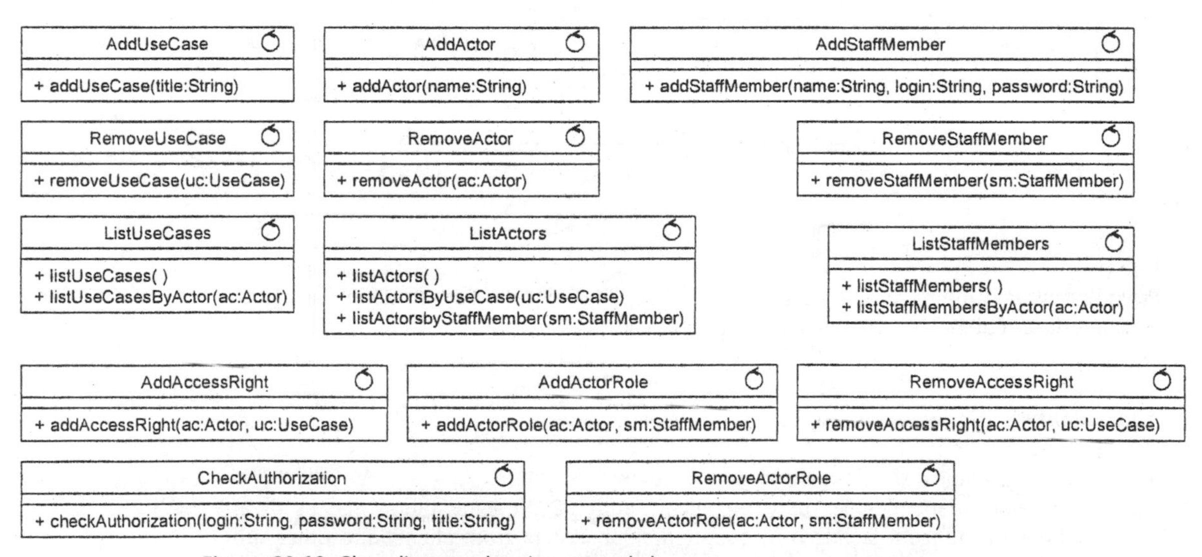

Figure 20.10 Class diagram showing control classes.

The control classes in the `Security Server Control Classes` package can probably be designed to be Singletons (see Section 15.4.2). Figure 20.10 shows the control classes in this package.

In order to make the security package as reusable as possible, it either needs to make use of whatever data storage mechanisms are used in the application with which it is supplied, or it needs to have its own mechanism for persistent storage. The simplest approach is to provide the security package with its own persistence mechanism. We can use an object-oriented database management system such as Object Store PSE Pro for Java to provide a persistence mechanism without having to worry about brokers and proxies. (Alternatively, given the relatively small volumes of data that are involved, it is possible for the persistence to be provided by using a system of files. If the data is to be stored in simple files, then it should be encrypted before storage. An encryption package could be added.)

In the design in Figure 20.11 we have added collection classes to provide entry points from the control classes to the lists of `Actors`, `UseCases` and `StaffMembers`. We have also added hashtables as collection classes to implement the associations between the classes. Adding a link between a `UseCase` and an `Actor` means adding the `Actor` to the `UseCase.actors` hashtable and adding the `UseCase` to the `Actor.useCases` hashtable.

The collection classes support operations that have been added to the façade class and the control classes. The intention of this is that it should be easy to display in a dialogue box a list of all the Actors that currently have AccessRights to a particular UseCase etc. They are also necessary to check the authorization of staff members for particular use cases by working through the list of actors for a particular staff member, and for each actor checking whether that actor has an access right to the use case

If we use ObjectStore then we shall need to base the collection classes on the persistent `OSHashtable` class that is provided with ObjectStore. Figure 20.12 shows

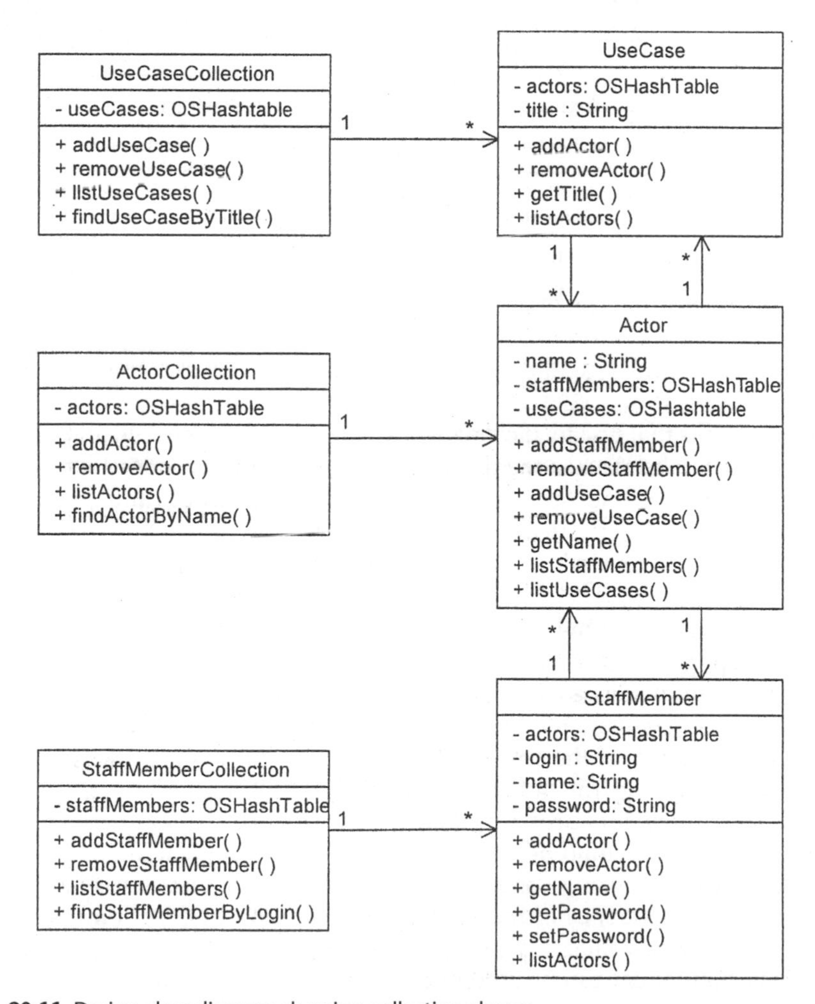

Figure 20.11 Design class diagram showing collection classes.

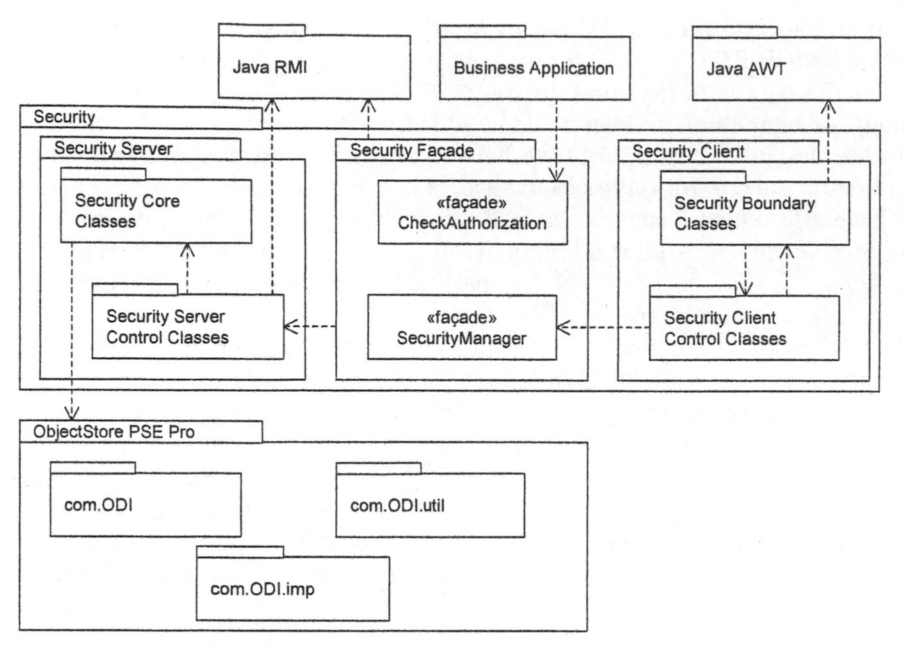

Figure 20.12 Package diagram showing dependencies of the `Security` package.

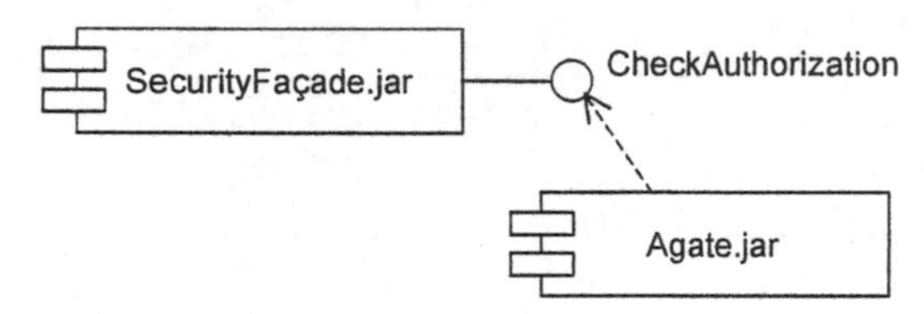

Figure 20.13 Component diagram showing dependency on the `SecurityFaçade` package.

the dependency between `Security` package and the package that contains the `OSHashTable` (`com.ODI.util`).

All the operations that update the database will require a reference to an ObjectStore database and must take place in the context of a transaction. The dependencies on the ObjectStore `Database` and `Transaction` classes (in `com.ODI` and `com.ODI.imp` respectively) are also reflected in the dependencies of Figure 20.12.

If we use Java Remote Method Invocation (RMI) to allow client packages to connect to the `Security` package, then we also require the dependency on the Java RMI package, which is also shown in Figure 20.12. We have also shown the façade packages in a separate `Security Façade` package.

Clearly, the security package itself should not be accessible to unauthorized users, so there will be a requirement for the security package to use its own services to restrict access to the developers of the system, who will set up the associations between use cases and actors, and to the system manager who will authorize staff within the company to use particular use cases by linking them to specific actor roles.

The security package could be implemented in Java, compiled and stored in Java Archive (.jar) files. This can be shown in a component diagram, as in Figure 20.13, which shows the dependency of the Agate application logic classes (also stored in a .jar file) on the `CheckAuthorization` interface that the `SecurityFaçade` package presents. This component can be used in both the Agate and FoodCo projects and in other future projects.

20.6 Summary

Adopting object-oriented software development is not on its own a sufficient cause for a business to benefit from the reusability that is claimed as one of the advantages of object-orientedness. Businesses thinking of taking advantage of reusability in object-oriented software need to take into account a number of factors.

- Is the organization the right kind to be able to develop reusable components as well as use them?
- How will it systematically move from a situation of little or no reuse to one in which reuse is built into the design of systems from the start?
- What organizational support systems and software tools are required to enable reuse?

Given a commitment to the introduction of a reuse business, an organization needs to reengineer its software development operations to provide the structures and support for a reuse culture and to train developers in this approach. For reuse to be effective, it should not be left until the maintenance stage but should be planned into the project from the outset. The architecture of the systems should be designed from the start to support reuse.

The SELECT Perspective is one methodology that focuses on techniques and project planning to achieve reuse. The RSEB of Jacobson et al. is another approach, and one which advocates an approach based on reengineering the business and designing architectures for reuse. Their approach is based on three processes, which produce a layered architecture, separating the component layer from the application layer, engineer the component layer and build the application layer on top of the component layer.

The introduction of CORBA, Java and the WWW make it easier for developers to produce and distribute reusable software components. Commercially available components include the products of IBM's San Francisco project, which provides business components for general ledger, for accounts receivable and payable and for warehouse and order management. The Façade pattern provides a means for hiding the complexity of components from application programmers by means of classes that handle the API to the classes in the component.

Review Questions

20.1 What are the benefits of reuse?

20.2 What are some of the obstacles to reuse?

20.3 Give Jacobson et al.'s definition of a 'component'.

20.4 Name three mechanisms for creating reusable components in different programming languages.

20.5 How does Allen and Frost's definition of a component differ from that of Jacobson et al.?

20.6 What are the three processes in Jacobson et al.'s approach to reuse, and what is meant by each?

20.7 What is the purpose of the Façade pattern?

Case Study Work, Exercises and Projects

20.A Describe the mechanisms that are available in an object-oriented language with which you are familiar for creating reusable components.

20.B A data encryption package is required to be added to the security package to provide services to encrypt different data types, such as the `Integer`, `String` and `Date` classes. Draw a class diagram (with packages and classes) to show new classes `CryptInteger`, `CryptString` and `CryptDate`. Access to `Encryption` and `Decryption` control classes is to be provided by a single Façade class called `EncryptionManager`. Include these in your diagram in suitable packages.

20.C Redraw the package diagram in Figure 20.12 to include the encryption package and add any new dependencies necessary.

20.D Draw a deployment diagram for FoodCo showing the security packages as components, some of which will be on client machines, and some of which will be on a server. (You may want to refer back to Chapter 19.)

20.E Complete the design of the security package and implement it in an object-oriented language such as Java, Smalltalk or C++. (This is suitable for a coursework assignment or small project.)

Further Reading

- Allen and Frost (1998) provide an argument for component-based development, a worked example of a case study based on the SELECT Perspective, their methodology, and guidelines on how to plan for reuse. Jacobson et al. (1997) present a different view of developing a reuse-driven software development process, which builds on the Objectory and object-oriented business reengineering approaches of Jacobson's other books. It also uses UML as a notation.
- For a comparison of the mechanisms involved in CORBA and DCOM (and some other approaches to distributed systems), Orfali and Harkey (1998) provide a clear and readable coverage which includes detailed instructions on how to implement systems which use these techniques.
- To find the wide range of software add-ins that are available as ActiveX controls, look at the website of a supplier such as System Science: www.systemscience.co.uk (in the UK). Look at IBM's website at www-4.ibm.com/software/ad/sanfrancisco/ for information about the IBM San Francisco project.

CHAPTER 21

Managing Object-Oriented Projects

OBJECTIVES

In this chapter you will learn

- how to manage iterative projects
- how to use project planning techniques
- the main features of DSDM and XP
- what metrics are available for object-oriented projects.

21.1 Introduction

Information systems development is a complex activity that requires careful management. We discussed the need for various models of the systems life cycle in Chapter 3 and the way in which they bring structure to a project. The discussion of the UML techniques in this book highlights the need to plan and manage the whole process. There are inter-dependencies between the artefacts of software development, and their production has to be co-ordinated if the process is to be efficient. A large software development project may involve many developers, some with specialized skills. The specialist in requirements capture is required early in the project, the expert in ODBMS implementation is needed during design and construction, and the installation and support teams become involved to some extent during design and construction and more fully when the information system is complete. The different activities may require different resources whose availability has to be planned. As suggested in Chapter 2, the timing of the installation may be critical for the success of the project. The management process is further complicated by the fact that the sequence of some activities may be significant. For example, testing of a system can only begin when at least some elements have been constructed, though of course test scripts and test harnesses may be prepared early in the project. Resource allocation and project planning are considered in Section 21.2.

Iterative development approaches present particular challenges. One critical issue is how to control the number of iterations. This was discussed briefly in Chapter 3 and we examine the question in more detail in Sections 21.3, and also in 21.4 where we introduce the Dynamic Systems Development Method (DSDM). Extreme Programming (XP) is another interesting approach that is highly iterative and this is discussed in Section 21.5.

A significant part of project planning is the determination of the resources required for each stage in the project. It is also important to establish the period for which each resource will be required. This can be based only upon an estimate of how long each task is likely to take. Estimates of task duration rely heavily upon experience but can also involve the measurement of various aspects of the proposed system. Many factors affect the length of time that it takes to complete a given activity and it is not possible to quantify them all. Some factors are measurable (at least to some extent): for example, project risk, system size and complexity. Less tangible aspects of the project such as staff expertise, team morale or the difficulty of using new technology are harder to measure. Section 21.6 introduces software metrics that can be used to estimate the influence that project characteristics have upon overall development time.

The patterns that were introduced in Chapter 8 related to analysis and those discussed in Chapter 15 are mainly useful in a design context. Other patterns have been developed that are applicable to the management of the development process. In Section 21.7 we briefly introduce the idea of development organization patterns. One of the difficulties of introducing object-oriented development to an organization is that, initially at least, some systems are based on an object-oriented approach while others are not. In Section 21.8 we outline some of the problems to which this gives rise.

The move towards object-orientation introduces a series of challenges for an organization. Staff must be trained, and existing software systems must be integrated with the new approach. Any major change, and a change in development method is only one example, requires careful planning in order to minimize the attendant risks (Section 21.9).

21.2 Resource Allocation and Planning

Given the complexity of project management there is a need for tools and techniques to support the process. Yourdon (1989) identifies three particular areas of the management of software development where modelling techniques can play a useful role:

- in the estimation of money, time and people required;
- in assisting the revision of these estimates as a project continues;
- in helping to track and manage the tasks and activities carried out by a team of software developers.

Many tools (e.g. CA SuperProject) have been developed to support the management of any type of project, not just those that are focused on software development.

21.2.1 Critical Path Analysis

The technique known as Critical Path Analysis (CPA) was developed in the late 1950s for use on major weapons development projects for the US Navy (Whitten, Bentley and Barlow, 1994). Originally known as Project (or Program) Evaluation and Review Technique (PERT)[1], it is also called Network Analysis and it has been widely used on many different types of project.

For the purposes of carrying out a critical path analysis, a project is viewed as a set of activities or tasks, each of which has an expected duration. Completion of an activity corresponds to a milestone or event for the project. Each milestone also represents the start of activities that are directly dependent on the completion of the predecessor or predecessors. CPA is based on an analysis of sequential dependencies among the activities, and uses the expected duration for each task to derive an estimate of the overall duration of the project. In particular, it identifies any inter-task dependencies that are critical to the project duration—collectively these are known as the *critical path*. The preparation of a CPA chart involves the following steps.

List all project activities and milestones

A sample list for the development of the Agate Campaign Management system is shown in Figure 21.1. Each activity is labelled with a letter and has a short description. The third column in the table contains a milestone number that represents the completion of that activity.

Determine the dependencies among the activities

Some activities cannot start until another (sometimes more than one) has been completed. The preceding activities are listed in column 4. For example, in Figure 21.1 the activity `Review use cases` must be completed before the activity `Identify classes` can begin.

Activity	Description	Milestone	Preceding activities	Expected duration	Staffing
A	Interview users	2	-	5	2
B	Prepare use cases	3	-	2	See A
C	Review use cases	4	A, B	2	3
D	Draft screen layouts	5	C	2	2
E	Review screens	6	D	2	2
F	Identify classes	7	C	2	3
G	CRC analysis	8	F	4	3
H	Prepare draft class diagram	9	F	5	3
I	Review class diagram	10	G, H	4	4

Figure 21.1 Project activity table for Agate.

[1] Strictly speaking PERT is more elaborate than CPA, using statistical measures in addition to critical path analysis, but the terms are generally used synonymously. CPA charts are sometimes known as activity diagrams (Skidmore et al., 1994) but this name risks confusion with UML activity diagrams.

Estimate the duration of each activity

There are several different approaches to this, due to the uncertainty involved in estimating task duration. One that is used widely is given by the following formula:

$$ED = \frac{MOT + (4 \times MLT) + MPT}{6}$$

where *ED* is the expected duration of a task, *MOT* is the most optimistic time, *MLT* is the most likely time and *MPT* is the most pessimistic time for its completion. The *ED* is thus a weighted average of the three estimates. Each *MOT* assumes that a task will not be delayed, even by likely events such as employee absence. The *MPT* assumes that most things that can go wrong will go wrong, and that completion of the activity will be delayed to the maximum plausible extent. Equipment will arrive late, technical problems will occur and some staff will be ill. (It would be unrealistic to take this to the extreme, for example, to assume that all development staff will be struck down by an influenza epidemic—estimates should be realistic.) The *ED*s are entered in the fifth column of the table. Note that in this example the staff requirements for activities A and B have been treated as one since the two activities are highly interdependent.

Draw the CPA chart

Two main styles of notation are used for CPA charts, known respectively as 'activity on the node' and 'activity on the arrow' diagrams. Both show the same information, but they look very different from each other. To avoid confusion we present only 'activity on the arrow' notation. In this style each milestone is represented by a circle divided into three compartments. One compartment is labelled with the milestone number, and the other two will hold the earliest start time (EST) and the latest start time (LST) (these terms are explained below) for all activities that begin at that milestone. Figure 21.2 illustrates the notation.

The first draft of a CPA chart shows dependencies between activities and their expected durations, since this information is known or can be estimated before the diagram is drawn. The partially completed CPA chart in Figure 21.3 represents graphically the activity precedences listed in table form in Figure 21.1.

Note the *dummy activity* between milestones 8 and 9 (there are others between 2 and 3 and between 6 and 10). The explanation for this is that activities H and G both depend on milestone 9 (the completion of activity F), and are also both predecessors to milestone 10 (where activity I begins). Since H and G may not necessarily finish at the same time, an extra milestone is needed for one of these events. In effect, milestone 8 represents the completion of G, regardless of whether or not H has finished (it is not

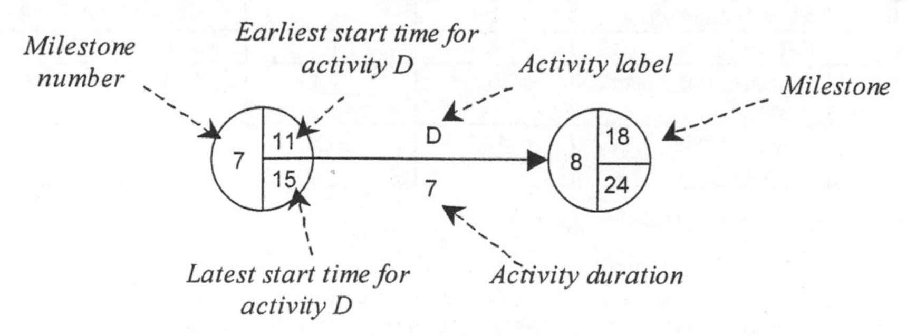

Figure 21.2 CPA notation.

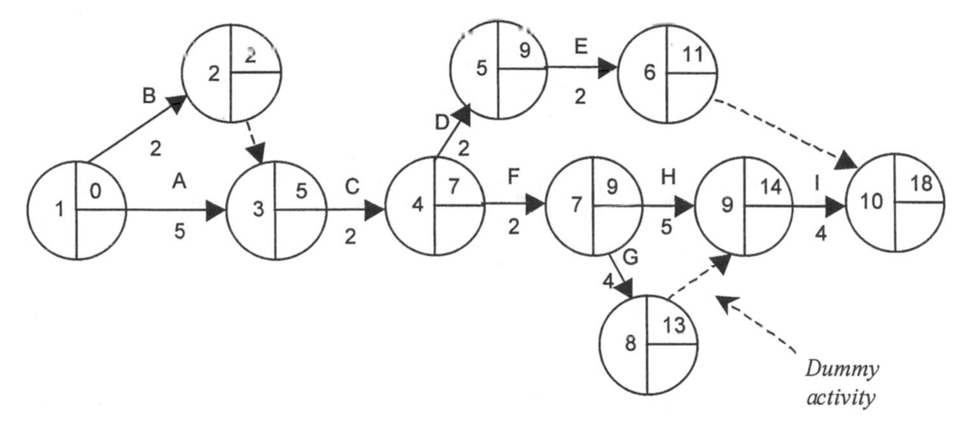

Figure 21.3 Partial CPA chart for the Agate advertising sub-system project.

significant which of the two activities is chosen to terminate at the extra milestone). A dummy activity (with ED of 0) then needs to be introduced in order to connect milestone 8 to milestone 9, thus preserving the sequence of dependencies.

The next step is to enter an earliest start time for each milestone. This is done by working through the diagram from the very first milestone to the very last (sometimes called a *forward pass*). It is a convention that the EST for the first milestone is set to 0 (elapsed time is usually measured in days, but any other unit of time can be used equally well). The EST for most other milestones is calculated simply by adding the EST of the immediately preceding activity to its duration. For example, activity C (the immediate predecessor for D) has an EST of 5 and an ED of 2, therefore the EST for activity D is 7. Where an activity has two or more predecessors, its EST is determined by the predecessor that has the latest completion time. For example, activity I is dependent upon the completion of both G and H, so its EST is set to the later of the two calculations. In general the EST for any milestone is set to the earliest time that *all* predecessor activities can be completed.

The next step requires completing the latest start time for each milestone. The latest start time is entered by working back from the last milestone (sometimes this is called a *backward pass*). The LST for the last milestone is set equal to its EST. Each preceding LST is then calculated as follows. The LST for most milestones equals the LST for its successor minus the ED of the intervening activity. For example, the LST for milestone 9 is 18 – 4 = 14 (the LST for milestone 10 minus the ED for activity I). Milestone 7 presents more of a problem as this has two successor activities, G and H. In such cases, two calculations are performed (or more, if there are more than two successors) and the earlier of the two answers is taken. For example, if the LST for milestone 7 were determined purely by activity G, this would give 14 – 4 = 10 (the LST for milestone 8 is 14). This would mean that activity G could afford to begin as late as time 10 without delaying any other activities. But a similar calculation for activity H gives 14 – 5 = 9. This means that if H begins later than time 9, its completion will be delayed beyond time 14, which in turn would delay milestones 9 and 10 and thus also activity I and the project completion. The LST for milestone 7 is therefore set to 9. In general, the LST for a milestone is set to the latest time that allows *every* activity that begins at that milestone to be completed by the LST for its succeeding milestone.

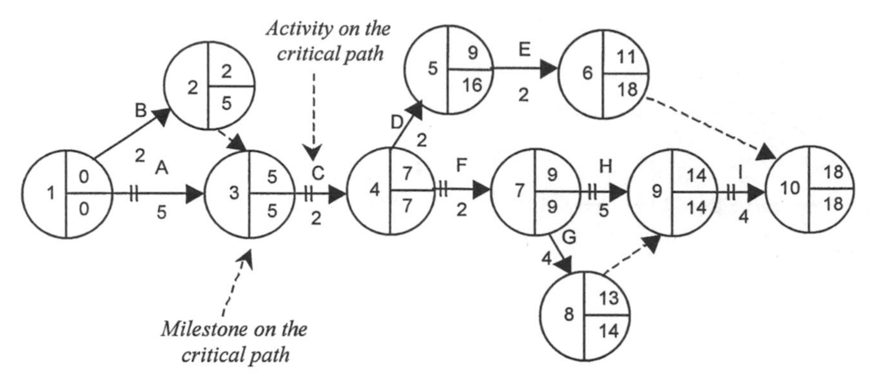

Figure 21.4 The critical path for the Agate advertising sub-system project.

Identify critical path

Once all LSTs have been entered onto the diagram, the *slack time* (or *float*) for each activity can be calculated. This is the difference between an activity's EST and its LST, and it represents the time by which that particular activity can be delayed without affecting the overall duration of the project. The path through all milestones that have a slack time of 0 is called the *critical path*. This is indicated by a double bar across activity arrows that connect the milestones. These milestones, and their intervening activities, are critical to the completion of the project on time. Milestones that have an EST that is different from their LST are not critical, in the sense that they have some scheduling flexibility. The completed diagram is shown in Figure 21.4.

A CPA chart is an effective tool for identifying those activities whose completion is critical to the completion of a project on time. If any activity that is on the critical path falls behind schedule then the project as a whole is behind schedule. However, while critical activities naturally receive the closest scrutiny, all project activities should be monitored. Delay even in a non-critical activity can, if it is sufficiently severe, alter the critical path.

CPA charts have limitations, chief among which is that they are not very useful for representing the extent of any overlap between activities. As a result, they are often used in conjunction with other techniques, particularly the Gantt chart (described in the following section).

21.2.2 Gantt charts

The Gantt chart (named for its inventor Henry Gantt) is a simple time-charting technique that uses horizontal bars to represent project activities. The horizontal axis represents time and is often labelled with dates or week numbers so that the completion of each activity can be monitored easily. Activities are listed vertically on the left of the chart, and the length of the bar for each activity corresponds to its ED.

A Gantt chart shows the overlap of activities clearly and this provides an effective way of considering alternative resource allocations. The Gantt chart can be drawn with either dashed lines or dashed boxes that show the slack time for non-critical activities.

A Gantt chart is also easy to convert into a stacked bar graph that can be used to show the way that the total resource allocation for a project changes over time. Figure 21.5 shows a Gantt chart and a staffing bar chart for the Agate advertising sub-system project. The staffing chart is derived as follows. The final column of Figure 21.1

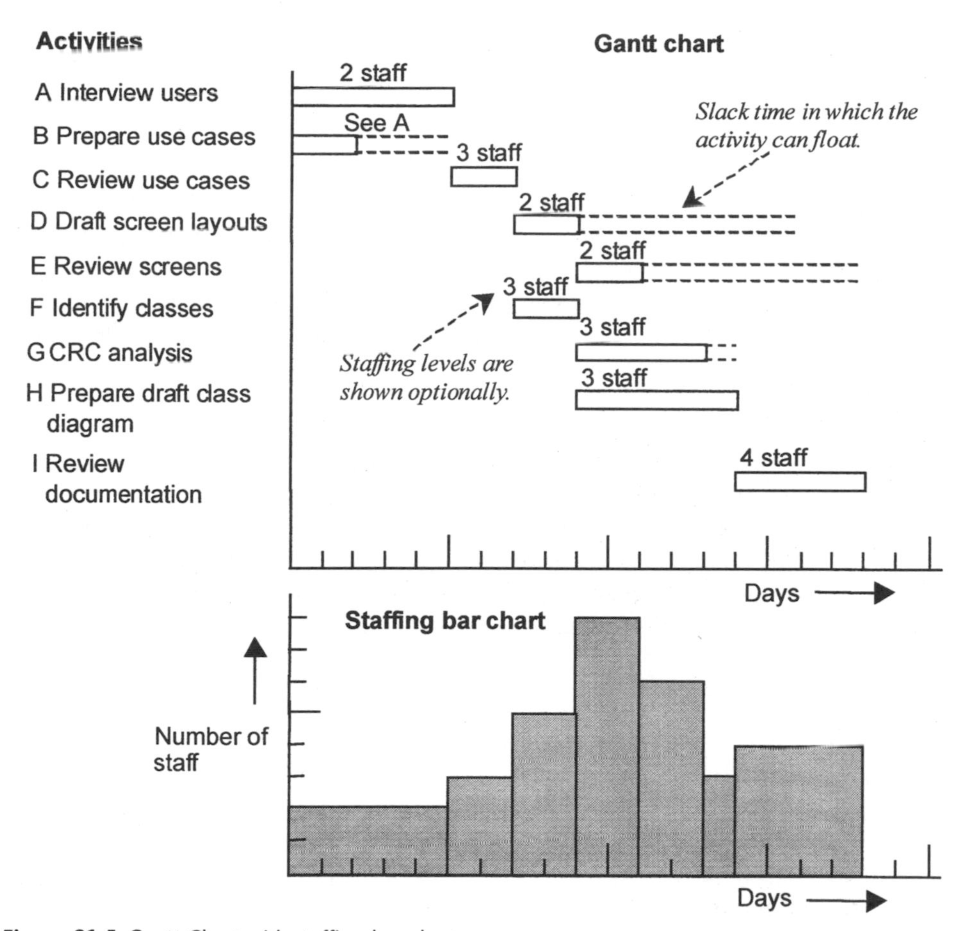

Figure 21.5 Gantt Chart with staffing bar chart.

gives a staff allocation for each activity. The Gantt chart is read vertically for each successive time interval to calculate the total number of staff required for all project activities combined. The result is shown as a vertical bar that indicates the total staffing required at that time.

Activities that are not on the critical path can have their start time adjusted. For example, activity E cannot begin until time 9 at the earliest, but it could start as late as time 16 without affecting project completion. A project manager can adjust the resource profile to accommodate staff availability, a process known as resource *smoothing*. For example, activities E, G and H can occur concurrently. H is on the critical path and cannot be moved, but the slack time for E allows it to begin at time 16 instead of time 9. The manager can minimize the overall resource requirement by rescheduling activity E to begin at time 14. This should be done with care, however, as it may move an activity onto the critical path. Figure 21.6 shows the smoothed resource profile.

The Gantt chart is a useful tool for resource monitoring. The progress of each activity can be shown independently and compared against planned progress. In Figure 21.6 the Gantt chart reflects the current state of a project. Activities A, B, C and F are complete. D has not been started and will shortly become critical. G is on schedule but H is behind schedule. Activities D and H need to be investigated by the project manager. Possible reasons for the delay include:

- an unexpected technical problem;
- staff absence;
- the complexity of the activity has been underestimated.

The project manager must decide how best to get the project back on schedule. Perhaps the allocation of more staff to activity H would resolve the difficulty, but throwing staff at an activity can be counter-productive for several reasons. First, if the additional staff are not familiar with the activity, they may require extensive briefing, thus reducing the time that is available to complete the work. Second, as team size increases so do the communication overheads. Third, some activities are limited in the maximum number of staff that can be involved. For example, a car repair may require 20 person-hours of work to complete. This does not mean that 20 mechanics working together could do the work in one hour. It is more likely that only, say, five mechanics could work productively on one car at any one time. The best possible repair time would therefore be four hours.

When a critical path activity is behind schedule it may not be possible to regain the lost time. A project manager then has really only two options, and ideally the choice should be discussed with the client. First, the project deadline can be moved to accommodate the delay. Second, the scope of the project or the quality of the product can be reduced to permit completion on time. The last approach requires an analysis of all

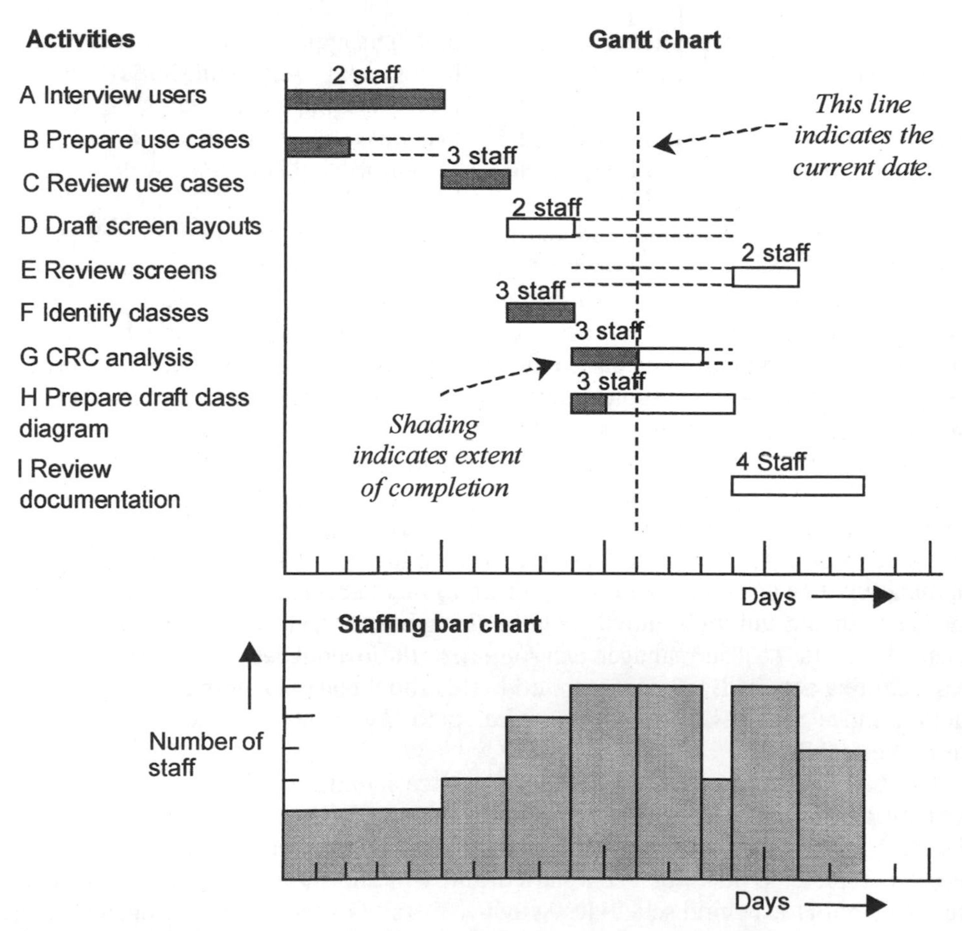

Figure 21.6 Gantt chart showing activity completion and smoothed staff profile.

uncompleted critical path activities in order to identify what can be omitted to reduce their EDs. The resultant changes may alter the critical path, and activities whose completion was not critical before may now become critical. Any attempt to reduce the scope of a project requires user involvement to ensure that only non-critical features are omitted, particularly during the first increment.

The process of planning a project is iterative. An alteration to the staff allocation for an activity will probably change its ED, which in turn may change the critical path. A project manager must operate within the constraints of staff and resource availability.

21.3 Managing Iteration

In Chapter 3 we considered prototyping as an iterative development activity and discussed the need to specify objectives for the prototyping activity so that they can provide criteria to control the number of iterations (Figure 3.6 shows an example of an iterative life cycle). At the end of an iteration the product, say a prototype, is evaluated against pre-defined objectives. In general terms this seems straightforward but in practice it can be difficult to determine whether the objectives of the iterative activity have been achieved.

Let us suppose that the interface for the Agate Campaign Management system is to be developed by prototyping, with the explicit objective of producing an interface with which the campaign staff are happy. However, although this objective may be worthwhile it is of little use for the management of the activity. Imagine that the users are never completely happy at the end of any iteration: they will continue to suggest further improvements without end. As the process continues, the nature of the modifications that are suggested at each iteration will change. Over time the improvements will become cosmetic and ultimately peripheral to the utility of the system. It would be sensible to end the iterative process before this point is reached. A more suitable objective for the exercise might be phrased as follows:

Continue the iterations until fewer than five cosmetic changes are requested on a single iteration.

It is still not clear how many iterations will be needed to satisfy this criterion. If the project has unlimited time and an unlimited budget this may not be a problem but this is unlikely to be the case. Additional criteria can be added to tighten up the objectives, such as the following.

The prototyping phase must be completed before the end of October and must not exceed 50 developer-hours.

21.4 Dynamic Systems Development Method

The Dynamic Systems Development Method (DSDM) is a management and control framework for rapid application development (RAD). The distinction between RAD and prototyping is sometimes blurred. A RAD approach aims to build a working system rapidly while a prototyping approach also builds rapidly, but usually only produces a partially complete system, typically to confirm some aspect of the requirement. Because both approaches aim to build software quickly, similar development environments are used and one approach to prototyping continues the development of a prototype incrementally until it becomes a working system. In effect this is a RAD development approach.

The traditional waterfall approach to systems development has deficiencies, particularly the time taken to deliver a working system and the inflexibility of the approach to requirements change. Iterative approaches to development can also be problematic. As suggested above it is sometimes difficult to cease the iterations when they become unproductive. In the early 1990s RAD became much more popular and was viewed as a way of matching systems development to the fast changing needs of business. However, there were until recently no commonly accepted structures for either the use or the management of RAD. The DSDM consortium was formed in 1994 to produce an industry standard definition of the RAD process and DSDM was subsequently defined. The DSDM framework defines structure and controls to be used in a RAD project but does not specify a development methodology. DSDM may be used with either an object-oriented or a structured methodology. DSDM takes a fundamentally different perspective on project control. Rather than viewing requirements as fixed and attempting to match resources to the project, DSDM fixes resources for the project, fixes the time available and then sets out to deliver only what can be achieved within these constraints.

DSDM is based upon nine underlying principles (Stapleton, 1997):

- Active user involvement is imperative. Many other approaches effectively restrict user involvement to requirements acquisition at the beginning of the project and acceptance testing at the end of the project. In DSDM users are members of the project team and include one known as an 'Ambassador' user.
- DSDM teams are empowered to make decisions. A team can make decisions that refine the requirements and possibly even change them without the direct involvement of higher management.
- The focus is on frequent product delivery. A team is geared to delivering products in an agreed time period and it selects the most appropriate approach to achieve this. The time periods are known as timeboxes and are normally kept short (2 to 6 weeks). This helps team members to decide in advance what is feasible. Products can include analysis and design artefacts as well as working systems.
- The essential criterion for acceptance of a deliverable is fitness for business purpose. DSDM is geared to delivering the essential functionality at the specified time.
- Iterative and incremental development is necessary to converge on an accurate business solution. Incremental development allows user feedback to inform the development of later increments. The delivery of partial solutions is considered acceptable if they satisfy an immediate and urgent user need. These solutions can be refined and further developed later.
- All changes during development are reversible. If the iterative development follows an inappropriate development path then it is necessary to return to the last point in the development cycle that was considered appropriate. Changes are limited within a particular increment.
- Requirements are initially agreed at a high level. Once requirements are fixed at a high level they provide the objectives for prototyping. The requirements can then be investigated in detail by the DSDM teams to determine the best way to achieve them. Normally the scope of the high level requirements is not changed significantly.
- Testing is integrated throughout the life cycle. Since a partially complete system may be delivered it must be tested during development, rather than after completion. Each software component is tested by the developers for technical compliance and by user team members for functional appropriateness.

- A collaborative and co-operative approach between all stakeholders is essential. The emphasis here is on the inclusion of all stakeholders in a collaborative development process. Stakeholders not only include team members, but others such as resource managers and the quality assurance team.

21.4.1 The DSDM life cycle

The DSDM life cycle has these phases:

- feasibility study,
- business study,
- functional model iteration,
- design and build iteration,
- implementation.

The relationships between the phases are shown graphically in Figure 21.7, and each is described in more detail below. Note that the last three are actually iterative processes, but for the sake of clarity this is not shown explicitly in Figure 21.7.

The *feasibility study* phase determines whether the project is suitable for a DSDM approach. It typically lasts only weeks, whereas the feasibility stage can last months on a traditionally run project. The study should also answer questions such as the following:

- Is the computerized information system technically possible?
- Will the benefit of the system be outweighed by its costs?
- Will the information system operate acceptably within the organization?

The *business study* phase identifies the overall scope of the project and results in agreed high level functional and non-functional requirements. Maintainability objectives are set at this stage and these determine the quality control activities for the remainder of the project. There are three levels of maintainability:

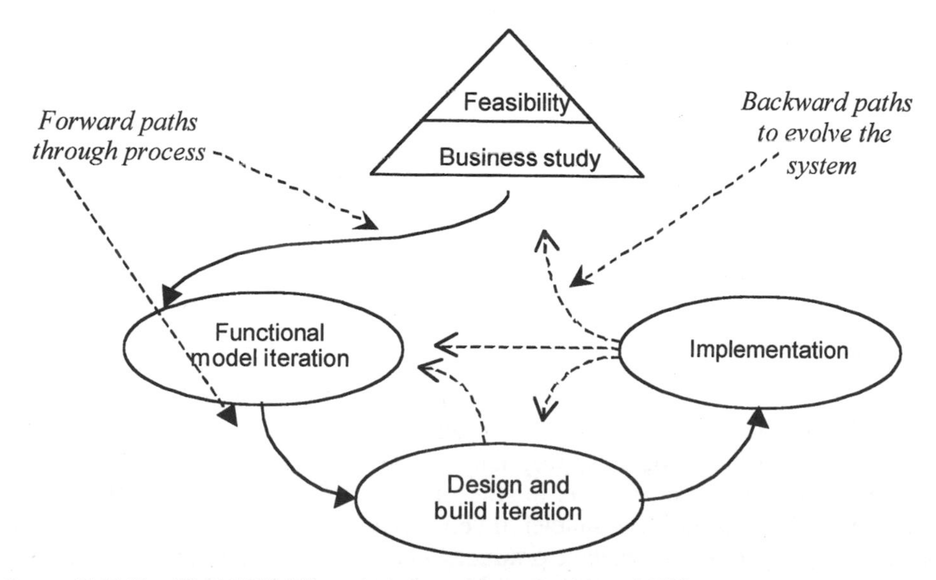

Figure 21.7 Simplified DSDM life cycle (adapted from Stapleton, 1997).

- maintainable from initial operation;
- not necessarily maintainable when first installed but this can be addressed later;
- short life-span system that will not be subject to maintenance.

Where the third level of maintenance is chosen care should be taken to ensure that the system is discontinued at the end of its allotted time, otherwise it may become subject to maintenance requests that are difficult to service.

The *functional model iteration* phase is concerned with the development of prototypes to elicit detailed requirements. The intention of DSDM is to develop prototypes that can ultimately be delivered as operational systems, so these must be built to be sufficiently robust for operational use and also to satisfy any non-functional requirements such as performance. When completed the functional model comprises high level analysis models and documentation together with prototypes that are concerned with detailed functionality and usability. During the functional model iteration the following activities are undertaken:

- the functional prototype is identified;
- a schedule is agreed;
- the functional prototype is created;
- the functional prototype is reviewed.

The *design and build iteration* phase is concerned with developing the prototypes to the point where they can be used operationally. The distinction between the functional model iteration and the design and build iteration is not clear-cut and both phases can run concurrently. The activities for the design and build iteration phase are very similar to those described above for the functional model iteration phase.

The *implementation* phase deals with the installation of the latest increment including user training. At this point it is important to review the extent to which the requirements have been met. If they have been fully satisfied the project is complete. If some non-functional requirements have yet to be addressed the project may return to the design and build iteration phase. If some element of functionality was omitted due to time constraints the project may return to the functional model iteration phase. If a new functional area is identified the project may return to the business study phase. The return flows of control are shown with dashed arrows in Figure 21.7. Implementation comprises the following iterative activities:

- producing user guidelines and gaining user approval;
- training users;
- implementing the system;
- reviewing the business requirements.

21.4.2 Timeboxing

Timeboxing is an approach for fixing the resource allocation for a project or a part of a project. It limits the time available for the refinement of requirements, design, construction and implementation as appropriate. A RAD project has a fixed completion date that defines an overall timebox for the project. A DSDM approach to project management will then identify smaller timeboxes within this, each with a set of prioritized objectives. Each timebox produces one or more deliverables that allow progress and quality to be assessed. Within a timebox the team have three major concerns. They must first carry out any investigation needed to determine the direction that should be taken for that part of the project. They must then develop and refine the specified deliverables. Finally they must consolidate their work prior to the final deadline.

It is sometimes difficult to prioritize the requirements that will be actioned during a timebox. One way of doing this is to apply the set of rules that are known as the *MoSCoW* rules (for Must . . . Should . . . Could . . . Want).

Must have requirements are crucial. If these are omitted the system will not operate. In DSDM the set of *Must have* requirements are known as the minimum usable subset.

Should have requirements are important but if necessary the system can operate usefully without them.

Could have requirements are less important and provide less benefit to the user.

Want to have but will not have this time around requirements can reasonably be left for development in a later increment.

All of these requirements are important for the final system but not to the same extent. If the full set cannot be addressed within a timebox, the MoSCoW categorization can be used to focus the requirements in an appropriate way.

21.5 Extreme Programming

Extreme Programming (XP) is a novel combination of elements of best practice in systems development. It was first publicized by Kent Beck (Beck, 2000) and incorporates a highly iterative approach to development. It has become well known in a relatively short period of time for its use of *pair programming* though it encompasses various other important ideas. Pair programming involves writing the program code in pairs and not individually. At first sight it would appear that this approach would significantly increase the staffing level and hence the cost of developing an information system but the advocates of XP claim otherwise.

Beck (2000) identifies the four underlying principles of XP as communication, simplicity, feedback and courage.

Communication. Poor communication is a significant factor in failing projects, XP highlights the importance of good communication among developers and between developers and users.

Simplicity. Software developers are sometimes tempted to use technology for technology's sake rather than seeking the simplest effective solution. Developers justify complex solutions as a way of meeting possible future requirements. XP focuses on the simplest solution for the immediate known requirements.

Feedback. Unjustified optimism is common in systems development. Developers tend to underestimate the time required to complete any particular programming task. This results in poor estimates of project completion, constant chasing of unrealistic deadlines, stressed developers and poor product quality. Feedback in XP is geared to giving the developers frequent and timely feedback from users and also in terms of test results. Work estimates are based on the work actually completed in the previous iteration.

Courage. The exhortation to be courageous urges the developer to throw away code that is not quite correct and start again rather than trying to fix the unfixable. Essentially the developer has to leave unproductive lines of development despite personal investment in the ideas.

XP also argues that embracing change is important and key to systems development and that development staff are motivated by producing quality work.

Requirements capture in XP is based on *user stories* that describe the requirements. These are written by the user and form the basis of project planning and the development of test harnesses. User stories are very similar to use cases though some proponents of XP suggest that there are key differences in granularity. A typical user story is about three sentences long and does not include any detail of technology. When the developers are ready to start writing the system they get detailed descriptions of requirements face to face with the customer. Beck (2000) talks about the systems development process as being driven by the user stories in much the same way that the USDP is use case driven. XP involves the following activities.

- *The planning game* involves quickly defining the scope of the next release from user priorities and technical estimates. The plan is updated regularly as the iteration progresses.
- The information system should be delivered in *small releases* that incrementally build up functionality through rapid iteration.
- A unifying *metaphor* or high level shared story focuses the development.
- The system should be based on as *simple design*.
- Programmers prepare unit *tests* in advance of software construction and customers define acceptance tests.
- The programme code should be restructured to remove duplication, simplify the code and improve flexibility—this is known as *refactoring*, and is discussed in Fowler (1999) in detail.
- Pair programming means that code is written by two programmers using one workstation.
- The code is owned collectively and anyone can change any code.
- The system is integrated and built frequently each day. This gives the opportunity for regular testing and feedback.
- Normally staff should work no more than forty hours a week.
- A user should be a full-time member of the team.
- All programmers should write code according to agreed standards that emphasize good communication through the code.

The XP approach is best suited to projects with a relatively small number of programmers—say no more than ten. In XP it is critical to maintain clear communicative code and to have rapid feedback. If a project precludes either of these then XP is not the most appropriate approach. One key feature of XP is that the code itself is the design documentation. This runs counter to some aspects of the approach suggested in this book. We have suggested that requirements are effectively analysed and suitable designs produced though the use of visual models using UML. Nonetheless XP does offer interesting insights into how software development projects can be organized and managed.

21.6 Software Metrics

Planning and managing a software development project requires the estimation of the resources required for each of its constituent activities. A resource estimate for an activity can be based upon subjective perceptions of the activity or it can be based upon measurements of size and complexity, either of the activity itself or of the artefact that is produced. DeMarco's (1982) frequently quoted aphorism sums it up:

> You can't control what you can't measure.

We are used to applying measures in many forms of human endeavour. For example, the productivity of a car factory can be measured quite accurately in terms of the time taken to construct a car. The use of metrics in software development is still rather limited. The term software engineering was coined in the 1960s as part of an attempt to introduce the greater degree of rigour that was seen in the management of other types of engineering. It is a feature of most engineering disciplines that careful measurement is used to assess the efficiency and effectiveness of artefacts and of processes. A *software metric* is a measure of some aspect of software development, either at project level—usually its cost or its duration—or at the level of the application—typically its size or its complexity.

Software metrics can be divided broadly into two categories: *process metrics* that measure some aspect of the development process and *product metrics* that measure some aspect of the software product. (We use the term software product broadly to include anything that is produced during a software development project, including for example analysis models and test plans as well as program code.) Two examples of process metrics are the project cost to date and the amount of time spent so far on the project. Product metrics relate to the information system that is under development. One of the simplest product metrics is the number of classes in an analysis class diagram.

Software metrics can also be categorized as *result* or *predictor* metrics, which are used respectively to measure outcomes and to quantify estimates. The current cost of a project is a result metric (even though this is only an interim outcome that will probably be modified tomorrow). A measure of class size (a crude measure might be a simple count of attributes and operations) would be a predictor metric, so called because it can be used as a basis for predicting the time that it will take to produce program code for that class. Result metrics are also known as control metrics (Sommerville, 1992) since they are used to determine how management control should be exercised. For example, a measurement of the current level of progress in the project is used to decide whether action is necessary to bring the project back onto schedule. The term 'predictor metric' is generally applied only to a measure of some aspect of a software product that is used to predict some other aspect of the product or of the project progress. Predictor metrics are not used solely for estimation. The results obtained from their application to a project may indicate, for example, that the system will be difficult to maintain or that it may offer very low levels of reuse. Since neither outcome is desirable, managers may attempt to change the design for the system or the process for its development in order to improve the system.

The validity of predictor metrics is based upon three assumptions, often made only tacitly by managers (Sommerville, 1992):

- there is some aspect of a software product that can be accurately measured;
- there is a relationship between the measurable aspect and some other relevant characteristics of the product;
- this relationship has been validated and can be expressed in a model or a formula.

The last of these assumptions suggests that a significant volume of historical data must be collected so that an appropriate statistical analysis can validate the relationship. In practice this is only feasible if the data collection is automated (for example, it is done by a CASE tool). Generally speaking, software developers do not regard the capture of metrics data as an important part of their role. Their attention is inevitably focused on the delivery of the product on time. Another factor that limits the uptake of metrics is the possibility of using them to monitor the performance of the developers themselves. This may give rise to concern among the developers about how data on their performance might be used. This is of course an ethical issue as well as a management one.

A number of metrics have been identified for use with structured analysis and design approaches. For example, De Marco (1982) developed a complexity metric known as the Bang Metric for use on structured analysis projects. Other metrics that focused on the degree of coupling and cohesion between program modules have also been suggested for use with a structured design approach. Most of these metrics were used little despite the introduction of automated CASE support.

A number of authors have identified metrics for object-oriented systems development (Lorenz, 1994; de Champeaux, 1997; Graham, 1995). For instance de Champeaux suggests a list of desirable features as part of a general description of a useful metric:

- either it is elementary and focuses on a single well-defined aspect, or it is an aggregation of elementary metrics;
- it is suitable for automated evaluation;
- gathering the metric data is not too costly;
- it is intuitive;
- application to a composite is equivalent to application to the components and summation of the individual results;
- the metric can be measured numerically and arithmetic operations are meaningful.

He lists a series of quality metrics, for example, a dependency metric that provides a measure of the stability of a system. When applied to a package or a sub-system this measures the degree of inter-package or inter-sub-system coupling. It is calculated by the formula:

$$I = \frac{CE}{(Ca + Ce)}$$

where I is the instability of the system, Ca is the level of afferent coupling (the number of classes outside the package that depend on classes within the package), and Ce is the level of efferent coupling (the number of classes outside the package upon which classes within the package depend) (adapted from de Champeaux, 1997).

When I is zero the package is maximally stable and has no dependencies on classes in other packages. When I is 1 the package is maximally unstable and has dependencies only on classes outside itself.

The ability of a package to absorb change is reflected (in part at least) by the ratio of abstract classes to all classes within the package. Where this is zero then the package consists solely of concrete classes and is difficult to change. A ratio of one indicates the presence of no concrete classes at all and it is easier to change.

Lorenz and Kidd (1994) suggest a wide range of metrics including metrics for application size and class size. Application size is essentially determined from the number of use cases and the number of domain classes[2] together with multiplying factors that reflect the complexity of the user interface. The size of a class is determined by the number of attributes and operations it has and the size of the operations. Size metrics such as this can be used to estimate the resource requirement for a project providing that appropriate historical data is available to derive and validate the relationship.

21.7 Process Patterns

We have already discussed analysis patterns in Chapter 8 and design patterns in Chapter 15. Coplien (1995) has defined a pattern language that is focused on the development process. The pattern language comprises both organization and process patterns. As with design patterns they capture elements of experience as problem–solution pairs. The patterns address issues such as team selection, organizational size, team structure, the roles of team members and so on. Conway's Law (Pattern 14) discusses how the architecture of the system comes to reflect the organizational structure or vice versa. Mercenary Analyst (Pattern 23) is concerned with producing project documentation successfully. This pattern suggests that it is frequently more effective to hire a technical writer who can focus solely on the documentation.

21.8 Legacy Systems

There are many definitions of the term legacy system. We take it to mean any computerized information system that has been in use for some time, that was built with older technologies (perhaps using a different development approach at different times) and, most importantly, that continues to deliver benefit to the organization. Most computerized information systems interact with other computerized information systems. They may share data, the output from one may be an input to another and so on. Any new information system is likely to need to interact with older legacy systems that have not been built using the same technologies. Redeveloping legacy systems so that they interact appropriately with new systems is likely to be prohibitively expensive and probably involves too much risk. These legacy systems may be critical to the operation of the organization. The problem is one of integrating new object-oriented systems with non-object-oriented systems.

One strategy that enables the interoperation of old and new is the use of an *object wrapper* (Graham, 1995). An object wrapper functions as an interface that surrounds a non-object-oriented system so that it presents an interface suitable for use with new object-oriented systems. Essentially the old system appears to be object-oriented. The form of the wrapper depends on the nature of the legacy system. Where the old system uses text or form-based screen interfaces the wrapper may involve program code that reads data from the screen and writes data to the screen (sometimes known as screen scrapers) using some form of virtual terminal.

When an organization embarks upon object-oriented software development there may be an intention to migrate all existing software to the new technologies. The cost

[2] Lorenz and Kidd use the terms 'scenario scripts' and 'key classes'.

and risk involved may constrain this but where it is feasible to migrate systems it is important to manage the process carefully. Wrappers may be used initially to provide an object-oriented interface to the system. Then, once the system has been wrapped, it can be redeveloped (perhaps one sub-system at a time) without affecting its interface to other systems. A further variation on this approach is to use the Façade pattern (see Chapter 20) to wrap sub-systems so that they can be migrated incrementally.

21.9 Introducing Object Technology

The introduction of any new way of working requires careful planning. Introducing object technology is a very significant change. Both the approach to systems development and the development technology are new. Staff must be trained in the principles of object-orientation, in new analysis and design techniques, in one or more object-oriented programming languages, in the use of new development environments and perhaps also in how to use a particular object-oriented database management system. Almost every aspect of the software development activity will change. This is not just a matter of training the developers; they also must gain experience in applying the new techniques and using the new technology. The safest way to achieve this is to apply object-orientation within a pilot project in the first instance (as suggested for FoodCo in Box B1.2). The move to object-orientation must be carefully planned and controlled and should ideally follow the steps shown below:

- identify a suitable pilot project;
- train the relevant staff;
- monitor the project carefully when it is under way;
- review the project implementation;
- identify the lessons learned and migrate this experience to other suitable projects.

Identification of a suitable pilot project. A pilot project should ideally not be subject to very tight timescales. The team is likely to require additional time to become familiar with the concepts and notation. The project should be amenable to an object-oriented approach in terms of its application domain and implementation technology.

Training. The move to object-orientation requires a different perspective on the problem domain. Training should be planned so that developers are trained in the new techniques immediately before they need to use them. It is important that a developer should understand the conceptual basis of object-orientation before embarking upon learning a new modelling notation (for example UML) or a new programming language (for example Java). Training and familiarization with the selected CASE toolset should be included. Users who will interact with the project should also be trained so that they can participate effectively in user reviews.

Monitoring the project. To gain maximum benefit from the move to object-orientation it is important to monitor its use within the organization. This provides feedback on the effectiveness of the training, the most appropriate way of applying the techniques and the project management requirements.

Review project implementation. A post-implementation review provides an opportunity to determine whether or not the expected benefits of the approach have been realized. It is important not to place unreasonable expectations upon a pilot project since unfulfilled aspirations may then over-shadow its successes. For instance, the potential benefits of reuse are unlikely to be achieved immediately after an organization has migrated to object-orientation. Even with careful management, significant levels of reuse may not appear for a further two or three years.

Migrate experience to other suitable projects. The experience of the pilot project will suggest adjustments that can be made to improve the way that object-orientation is to be used in the organization. Members of the initial project team may be used to seed other projects with their experience of the use of object-oriented development methods.

21.10 Summary

Systems development projects are similar to any other project in their need for sound management to ensure that they are completed within budget and on time. Some standard project management techniques used for this purpose include Critical Path Analysis and Gantt Charts, both of which can be used for project planning, to monitor resource utilization and to monitor project progress.

Object-oriented projects generally use a rapid development approach, and DSDM has emerged as the predominant framework for RAD project management. An object-oriented approach works particularly well with DSDM. XP provides an interesting alternative approach to systems development and provides some valuable insights.

The successful management of systems development relies on the quantification of various measures of effort, progress and complexity. A number of software metrics have been developed for this purpose including some that are intended specifically for use within object-oriented development.

The introduction of object-orientation to an organization should be carefully planned and managed. It is also important to ensure that existing computerized information systems can continue to operate alongside object-oriented systems, for example by the use of object wrappers.

The success of an information systems development project depends upon many factors, but in the final analysis the most important factor by far is the skill of the development staff. Effective project management enables developers to apply their knowledge and expertise in the most productive way.

When used by suitably skilled developers and managed by appropriately qualified managers, object-oriented development offers a means for capturing, modelling and building complex information systems that fully meet the needs of their users.

Review Questions

21.1 In CPA how is the slack time for an activity calculated?

21.2 In CPA how can you decide whether a milestone is on the critical path?

21.3 What are the advantages of a Gantt chart when compared to a CPA chart?

21.4 What are the advantages of a CPA chart when compared to a Gantt Chart?

21.5 What are the main life cycle stages for DSDM?

21.6 What are the MoSCoW rules and how do they help in the management of a timeboxed activity?

21.7 What are the underlying principles of XP?

21.8 What are the key elements of the XP approach?

21.9 How do object wrappers help integrate new object-oriented systems with existing non-object-oriented systems?

21.10 What factors should be considered when migrating to object technology?

Case Study Work, Exercises and Projects

21.A Prepare a CPA chart, a Gantt chart and a staffing profile for the project activities shown in the table at the foot of this page.

21.B FoodCo had never used object-oriented development methods until the start of the Production Costing System project. Identify other projects at FoodCo that would be alternative candidates for a pilot project, when considered as the first step in a company-wide migration process. Compare these in terms of their appropriateness.

21.C Critically evaluate a project management software package that is available to you. What changes to its functionality and non-functional characteristics would be required for it to be integrated with a CASE tool that supports UML?

Activity	Description	Mile-stone	Preceding activities	Expected duration	Staffing
A	Interview users	2		4	3
B	Prepare use cases	3		3	2
C	Review use cases	4	A, B	3	3
D	Draft screen layouts	5	C	4	2
E	Review screens	6	D	2	2
F	Identify classes	7	E	2	3
G	CRC analysis	8	F	3	2
H	Prepare draft class diagram	9	F	4	3
I	Review documentation	10	G, H	4	4

Further Reading

- De Marco (1982) provides a good introduction to the issues of managing software projects although the development techniques discussed are not object-oriented.
- Stapleton (1997) is a comprehensive and readable introduction to DSDM. Information is also available at www.dsdm.org. Newkirk and Martin (2001) provide a useful description of applying XP to a project. Further information about XP can also be found at www.xprogramming.com. The integration of XP and the Rational Unified Process is discussed in a white paper by Gary Pollice (2001) on the Rational website www.rational.com.
- Graham (1995) examines the migration to object technology and is well worth reading.
- Lorenz and Kidd (1994), Graham (1995) and de Champeaux (1997) address the use of object-oriented metrics. Melton (1995) is a collection of research articles on the subject, the first chapter being an interesting history of the development of software metrics.
- Coplien's pattern language (Coplien, 1995) describes many interesting patterns that will be familiar to project managers, albeit in many different guises.

22 CHAPTER

System Development Methodologies

OBJECTIVES

In this chapter you will learn

- what a systems development methodology is
- why methodologies are used
- the need for different methodologies
- the main features of one methodology.

22.1 Introduction

In this book our main focus has been on describing the techniques used during the analysis and design of an object-oriented information system. Underpinning this is an implicit assumption that projects move from one task to another, sometimes in sequence, sometimes looping back to repeat a task if it requires further work, sometimes choosing between alternative ways of proceeding, but always with a particular purpose in mind, and always under someone's control. In practice, the planning, organization and control of any large-scale project is itself a significant problem, and developers will need guidance on a wide range of questions that arise when the techniques are applied in a real-life situation (the project management issues are discussed in Chapter 21). The use of a set of modelling or documentation standards (such as that provided by UML) has an important part to play, but is not on its own enough. There is a need for a higher level view of the system development process, which promotes ways of organizing and managing the many different activities involved. This need is partly met by systems development methodologies. However, while the use of an appropriate methodology can help to guide the system development process and thus reduce the impact of many of the problems introduced in Chapter 2, this too cannot guarantee success.

In this chapter we describe what is meant by 'methodology', differentiating this from 'method' (Section 22.2). We discuss some reasons why methodologies are widely used in IS development, and also why they continue to evolve today (Section 22.3). Some object-oriented methodologies are briefly described (Section 22.4), and one, the USDP, is given a more detailed treatment (Section 22.5). We discuss some relative strengths and weaknesses of different methodologies. These factors are then considered in light of the need to select a methodology that is suited both to the organizational context and to the kind of projects to be carried out (Section 22.6). Finally, we touch on some of the competing intellectual traditions that gave rise to the many methodologies in use today (Section 22.7), in particular the distinction between the so-called soft and hard views of systems development.

22.2 'Method' and 'Methodology'

The techniques of system development must be organized into an appropriate developmental life cycle if they are to work together. For example, once an analyst has constructed collaboration diagrams for the main use cases, should the next steps be to convert these into sequence diagrams and write operation specifications, or should he or she now concentrate on preparing a class diagram and developing inheritance and composition structures? All of these tasks need to be completed at some point, but how is the analyst to know which is more appropriate at this specific point in the project? UML itself contains nothing that helps to make this decision.

The organization of tasks is not contained within the techniques themselves, and must be described at a higher level of abstraction. The *method* of a project is the term given to the particular way that the tasks in that project are organized. Sometimes this is called the *process* of software development, although process can also have a more all-embracing meaning, that includes what the tasks are, how they are carried out and how they are organized.

The words 'method' and 'methodology' are used interchangeably by many authors, but their meanings actually differ in a significant way. In order to plan and organize for the next project, project managers must be able to think at a still higher level. It is at this level that the term 'methodology' applies. A method is a step-by-step description of the steps involved in doing a job. Since no two projects are exactly alike, any method is specific to one project. A methodology is a set of general principles that guide a practitioner or manager to the choice of the particular method suited to a specific task or project. Or, to put it in familiar object-oriented terms, a methodology is a type while a method is its instantiation on one project. Figure 22.1 summarizes the different levels of abstraction involved.

An analogy might be the difference between the method followed by an amateur cook, who prepares a dish by following the recipe in a book, and a master chef, who has a deep understanding of ingredients, flavours, cookery tools and techniques. The amateur may use the recipe in a prescriptive way, while by contrast, a master chef may use the recipe as a guide, but will also vary the steps and ingredients in a creative way. This approach of selecting and adapting existing methods could reasonably be called a methodology. The difference lies in not always being bound slavishly to follow specific, pre-defined methods.

Increase level of abstraction	Example of application	Typical product
Task	Developing a first-cut class diagram for FoodCo.	A specific version of the FoodCo class diagram.
Technique	Description of how to carry out a technique, e.g. UML class modelling.	Any UML class diagram.
Method	Specific techniques used on a particular project (e.g. FoodCo uses cases, class model, collaboration diagrams, etc.) that lead to a specific product.	FoodCo's product costing system.
Methodology	General selection and sequence of techniques capable of producing a range of software products.	A range of object-oriented business applications.

Figure 22.1 Increasing level of abstraction in defining a project.

22.2.1 Methodology

A methodology in the domain of IS must cover a number of aspects of the project, although coverage varies from one to another. Avison and Fitzgerald (1988) describe a methodology as a collection of many components (not in the UML sense of the word). Typically, each methodology has procedures, techniques, tools and documentation aids that are intended to help the system developer in his or her efforts to develop an information system. There is usually also some kind of underlying philosophy that captures a particular view of the meaning and purpose of information systems development.

Checkland (1997), in a conference address that discussed the potential contribution of the wider systems movement to information systems development, gave a more general definition that captures well the notion of methodology as a guide to method. In his view, a methodology is a set of principles that in any particular situation has to be reduced to a method uniquely suited to that particular situation.

To give some examples of these aspects:

- The UML class diagram is a technique, and so is operation specification.
- Rational Rose is a tool.
- The activity represented by 'find classes by inspecting the use case descriptions' is an aspect of process. So is the advice that an analyst is usually the best person to write test plans.
- The advice that 'operation specifications should not be written until the class model is stable' is an aspect of structure, as it identifies a constraint on the sequence in which two steps should be performed.
- Analysis and design can (reasonably) be viewed as distinct stages.
- The statement 'object-oriented development promotes the construction of software which is robust and resilient to change' is an element of a systems development philosophy.

A package that contains enough information about each of these aspects of the overall development process is fit to be named a methodology. Many attempts have been made to capture the essence of methodology for software development, and the resulting methodologies are almost as varied as are the projects themselves. In practice, methodologies vary widely in philosophy, in completeness of definition or documentation, in coverage of the life cycle, and even in the type of application to which they are best suited.

Some authors even believe that the idea of a published methodology is misleading, if it is taken to mean that following a methodology leads inevitably to a good product. For example, Daniels (co-developer of the Syntropy methodology) argued in a conference presentation (1995) that an IS methodology is a means of *learning* the process of systems development, not a recipe for practising it. He compared a methodology to a ladder that, once used, can be thrown away (the metaphor is borrowed from Wittgenstein). We are safe to 'throw the ladder away', not because we do not need one, but because we now know how to build one of our own that is better suited to our organization and projects.

22.2.2 Logical views of a system

At a very abstract level of discussion, three complementary views of a real-world system must be understood in order to model it adequately for the purpose of conducting software development. These are:

- a data view, that describes the real-world system in terms of attributes and associations that must be stored within the software;
- a process view, that describes the operations that are (or need to be) carried out on that data;
- a temporal view, that captures the time sequence and time constraints on individual processes, and also the possible sequences of events that may impinge on the system.

In order to be useful, any IS methodology must include techniques for discovering, analysing and modelling the relevant content for each of these abstract views. These are packaged in different ways by the various modelling approaches. The history of information systems development represents an evolution of representations of the three system views, which have necessarily become much more sophisticated over time, as the systems being modelled have themselves grown larger and more complex.

22.3 Why Use a Methodology?

Over many decades, IS methodologies have been developed and introduced specifically to overcome those problems of software development projects that were perceived to be important at the time. However, to date, no methodology has been wholly successful in fulfilling its objectives, partly because computing is a highly dynamic field, and the nature of both projects and their problems is constantly changing. In a changing world, it is unlikely that yesterday's solution will ever completely solve today's problems.

Nevertheless, many advantages have been claimed for the use of a methodology, including the following.

- The use of a methodology helps to produce a better quality product, in terms of documentation standards, acceptability to the user, maintainability and consistency of software.
- A methodology can help to ensure that user requirements are met completely.
- Use of a methodology helps the project manager, by giving better control of project execution and a reduction in overall development costs.

- Methodologies promote communication between project participants, by defining essential participants and interactions, and by giving a structure to the whole process.
- Through the standardization of process and documentation, a methodology can even encourage the transmission of know-how throughout an organization.

If all these claims can be fulfilled in practice, the benefits would be clear. The evidence is mixed, partly because of the many dissimilarities between organizations, and also between the different types of project within any one organization. However, the picture in the UK is that approximately two-thirds of businesses use some form of IS methodology. Even where this actually means little more in practice than an in-house set of standards for documentation and procedures (as it certainly is in at least some cases), this still shows a consistent level of faith in the utility of a methodology of some kind.

22.4 A Brief Historical Review

Viewed historically, the earliest approach that might be thought of as a step towards an IS methodology is the traditional life cycle (TLC). Other life cycle models such as the spiral life cycle (described in Chapter 3) are further steps towards a methodological approach. However, as we saw in Section 22.2.1, a methodology is rather more than a life cycle.

22.4.1 Structured methodologies

Early structured methodologies, such as those authored by DeMarco (1979) and Gane and Sarson (1978), were introduced to overcome some of the problems encountered with the TLC. The central aim, suggested in the name, was to produce a 'structured' specification of the proposed software, with all functions, data storage, and interfaces between sub-systems clearly defined. To achieve this, it is necessary to define the techniques to be used and the deliverables at the end of each stage, as well as the stages of the life cycle. This all provides a more solid basis for project management, since the development process is more visible, and progress can be measured in terms of deliverables. A project manager using a structured methodology might be able to apply tests like this: 'If the data flow model and entity-relationship diagram have been approved, the analysis phase is complete.'

However, structured methodologies did not overcome all of the difficulties in systems development. For example, a manager still sometimes had no sound way of knowing whether a team was being optimistic in its progress reports. Anecdotes about project management in this period often take a rather ironic form: 'It's funny how a team can complete 90% of the data flow diagrams in 10% of the time, but then the other 10% of the work takes the other 90% of the time.' Secondly, most structured methodologies focus either on functional or data aspects of the system being modelled. For example, Yourdon's (1989) structured methodology is primarily process-oriented, while Information Engineering (Finkelstein, 1989) is primarily data-oriented. Whichever view holds sway, there is a danger of neglecting important aspects of the other. Even where both are given equal emphasis, there can be difficulty in co-ordinating the models of the two views.

Some structured methodologies attempted to cater for all views in a balanced way. One leading example was SSADM (Structured Systems Analysis and Design *Method*—

a good example of the method/methodology confusion!). For many years this has been a UK government standard for systems development and procurement (Skidmore et al., 1994). SSADM has separate but carefully correlated models of processes, data and temporal sequence. While SSADM has undoubtedly been very successful, any methodology that attempts this degree of co-ordination inevitably tends to grow large and unwieldy, partly due to the need for continual checking and cross-referencing between the separate models.

SSADM continues to evolve to this day, and a number of changes have been made in the most recent version, SSADM4+ (CCTA, 2000), to make the methodology more compatible with an object-oriented development approach. However, the fundamental models created in SSADM4+ are still organized around a logical separation of process and data. At heart, the methodology still appears structured, rather than object-oriented.

22.4.2 Object-oriented development

The main difficulty of structured methodologies in general is that most remain tied to a waterfall life cycle. This runs directly counter to the natural, iterative style of development for object-oriented software and makes it very difficult for a structured analysis and design approach to lead to an object-oriented implementation. In object-oriented development, as we saw in Chapter 3, an iterative cycle of development is both necessary and desirable. For this reason, throughout the 1990s, a great deal of research effort has resulted in a growing number of methodologies better suited to object-oriented software development.

We do not have the space to present a detailed introduction to the whole range of contemporary object-oriented methodologies. Instead, we concentrate on the USDP (Jacobson et al., 1999). This combines features of Objectory (Jacobson, 1992), Object Modelling Technique (Rumbaugh et al, 1991) and the Booch method (Booch, 1994), which were three of the leading object-oriented methodologies of the 1990s. The names of their leading authors are, of course, familiar as the three amigos who went on to develop UML and the USDP. Before turning our attention to USDP, however, we briefly consider the OPEN methodology, which, at the time of writing, is one of the few rivals to the USDP.

OPEN

OPEN (Object-oriented Process, Environment and Notation) is a methodology that, like the USDP, has a 'tailorable life cycle model' (Graham et al., 1998). Its authors define a methodology as consisting of the distinct elements shown in Figure 22.2. As a package, OPEN consists of a number of components, described in the following paragraphs.

Activities. An activity is a collection of tasks seen from a project manager's perspective. Activities are similar to stages and phases in other methodologies. Activities have both pre- and post-conditions, and are carried out within timeboxes (see Chapter 21). Some examples of activities are: *project initiation, requirements engineering, analysis and model refinement, project planning* and *build.*

Tasks. Within each activity, one or more tasks are carried out. These primarily represent the developer's view of the project, although management tasks are also included in a very comprehensive list (five pages of task names alone are given in *The OPEN Process Specification*). Some examples of tasks are: (draw) *rich pictures*

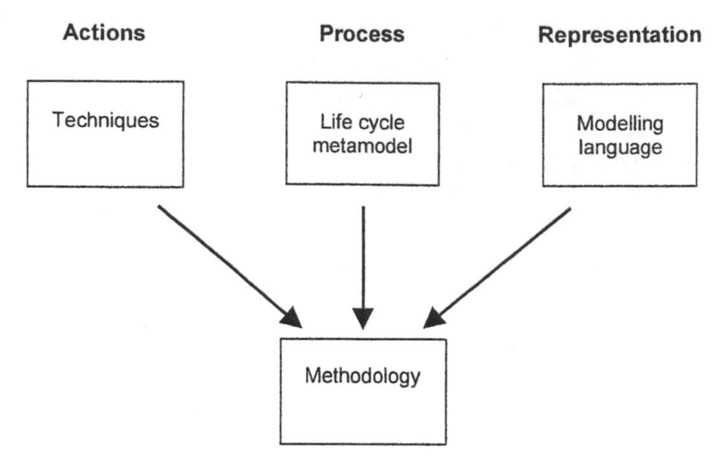

Figure 22.2 The OPEN view of a methodology.

(adopted from Soft Systems Methodology—see Section 22.7), *analyse user requirements*, *choose project team*, *design UI* (user interface) and *perform class testing*.

Techniques. Each technique describes how to carry out one or more tasks. The list of techniques is also comprehensive, and includes many that are not original to OPEN. For example, the *rich pictures* task (mentioned above) involves use of the *rich picture* technique, and the CRC technique is also used (see Chapter 8). Other examples of techniques are: *class internal design* and *object life cycle histories*.

Deliverables. Deliverables are the post-conditions for activities, and often also the pre-conditions for other activities. Some examples of deliverables are: *user requirements statement* (the main deliverable from the requirements engineering activity), *requirements specification* (this includes a physical prototype of the proposed system) and many diagrams such as *inheritance diagrams* and *deployment diagrams*.

Notation. For object modelling, OPEN is not tied to any particular notation. UML may be used (Henderson-Sellars and Unhelkar, 2000) but the original specification of the methodology used its own notation, called COMN (Common Object Modelling Notation). Further discussion of this topic is beyond the scope of this book, and so we give no example of COMN here.

22.5 The Unified Software Development Process

We have already briefly introduced the main principles that underlie the USDP in Chapter 5. In this section, we present a more detailed picture of the methodology.

Philosophy and principles. This part of the USDP should be familiar to anyone who has read much of the rest of this book, since the development process that we have followed is partly based upon, and broadly consistent with, the USDP. In particular, the USDP is a use-case driven, architecture-centric, iterative and incremental process. These terms can be explained briefly as follows.

The starting point for all modelling is some sort of interaction, called a use case, between a user and the software system under consideration. This interaction is the beginning of the modelling activity and also the fundamental unit from which later models are derived. Use cases are thus important in several different ways. Each use case is a thread that links a series of models from requirements to implementation; it is

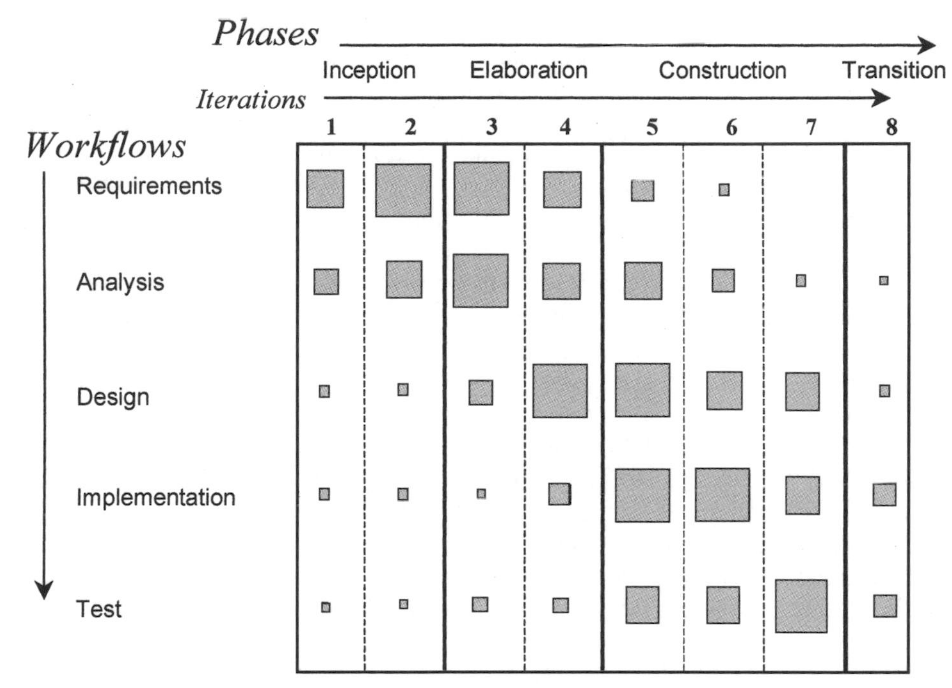

Figure 22.3 Phases, iterations and workflows in the Unified Software Development Process.

also a unit of delivery that has practical significance to users; it is a constant reminder to the systems developers that only the users' requirements really matter.

In the USDP, the resulting software architecture is an essential theme in modelling from the earliest stages of a project. This is reflected in the stereotyping of the classes that contribute to realizing a use case as boundary, control and entity classes.

Phases and workflows. Figure 22.5 repeats a diagram shown earlier in Chapter 5. This illustrates the relationship between the phases, iterations and workflows of the USDP. We do not need to dwell on workflows (requirements, analysis, design and so on) here, beyond noting that they are made up of activities, since these are the main subject of the greater part of this book. We explain the USDP view of activities a little later in this section. In the paragraphs that immediately follow, we explain the phases of USDP and relate these to the activities that are carried out within each phase.

While an activity is something that has particular meaning for the developers who carry it out, a *phase* is considered primarily from the perspective of the project manager. He or she must necessarily think in terms of milestones that mark the progress of the project along its way to completion.

Phases are sequential. A project passes through each phase in turn and then (usually) moves on to the next. The end of a phase is a decision point for the project management. When each phase is complete, those in charge must decide whether to begin the next phase or to halt development at that point. The focus of the manager's attention shifts as the project progresses from one phase to the next.

Within each phase, the activities are carried out in an iterative manner that can be summed up, albeit in a very simplistic way, as follows:

> Do some investigatation, model the requirements, analyse them, do some design, do some coding, test the code, then repeat the whole process.

There is no set rule that states how many iterations should be conducted within a phase; this is a matter for the project management team to judge, depending on the project characteristics and the available resources.

Within each phase, the workflows are essentially the same. All four phases include the full range of workflows from requirements to testing, but the emphasis that is given to each workflow changes between the phases. In the earlier phases, the emphasis lies more on the capture, modelling and analysis of requirements, while in the later phases the emphasis moves towards implementation and testing.

During the inception phase, the essential decision is that of assessing the potential risks of the project in comparison with its potential benefits. This judgement of project viability (or otherwise) during the inception phase resembles the feasibility stage of a waterfall life cycle. The decision will probably be based partly on a similar financial assessment (typically some sort of cost benefit analysis). One principal difference at this early stage is that the viability of a USDP project is much more likely to be judged partly also on the delivery of a small subset of the requirements as working software. During the inception phase, the main activities are thus requirements capture and analysis, followed by a small amount of design, implementation and testing. Another major difference is that, even at this early stage, there is the likelihood of iteration. That this is even possible, is due to the fact that the development approach is object-oriented.

During the elaboration phase, attention shifts to the reduction of cost uncertainties. This is done principally by producing a design for a suitable system that demonstrates how it can be built within an acceptable timescale and budget. As the emphasis shifts towards design, the proportion of time spent on design activities increases significantly. There is a further small increase in the time spent on implementation and testing, but this is still small in relation to the analysis and design activity.

The construction phase concentrates on building, through a series of iterations, a system that is capable of satisfactory operation within its target environment. Implementation and testing rapidly become core activities in this phase, with a move further away from design and towards testing as each iteration gives way to the next.

Finally, the transition phase concentrates on achieving the intended full capability of the system. This deals with any defects or problems that have emerged late in the project. It could also include system conversion, if an older system is being replaced (see Chapter 19).

Workers and activities. The USDP differentiates between the real people who are involved with any project, such as users, analysts, managers and customers, and the more abstract *worker*. This term denotes someone who plays a specified part in carrying out an activity. Some examples of workers are: use-case specifier, system architect, component engineer and integration tester. There need not be a direct one-to-one mapping between people and workers. An employee may play the part of different workers at different times, and, conversely, a group of people could represent a single worker engaged on an activity.

Most USDP activities can be partially defined in terms of the workers who carry them out, and the artefacts that either serve as inputs or are produced as outputs. Figure 22.4 illustrates this for the activity `Analyse a use case` (this can be compared with the process that we follow in Chapter 7).

As we mentioned above, a workflow can be seen as a flow of activities. Since each activity can be related to a worker who will carry it out, we can identify which workers will need to participate in the project. Figure 22.5 shows the Analysis workflow broken down into its constituent activities.

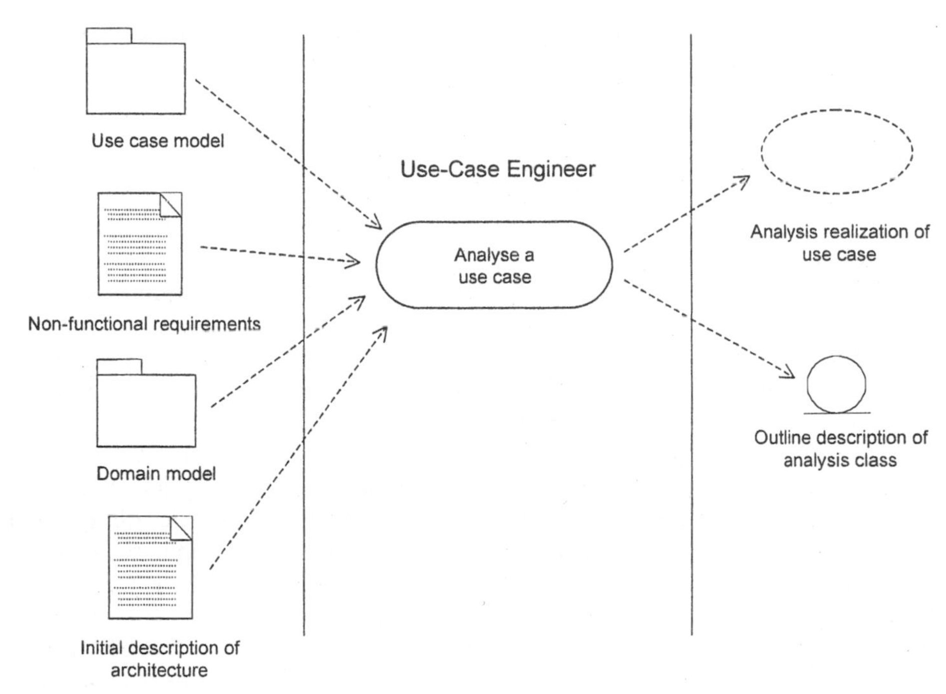

Figure 22.4 Inputs and outputs of the activity `Analyse a use case` (adapted from Jacobson et al., 1999).

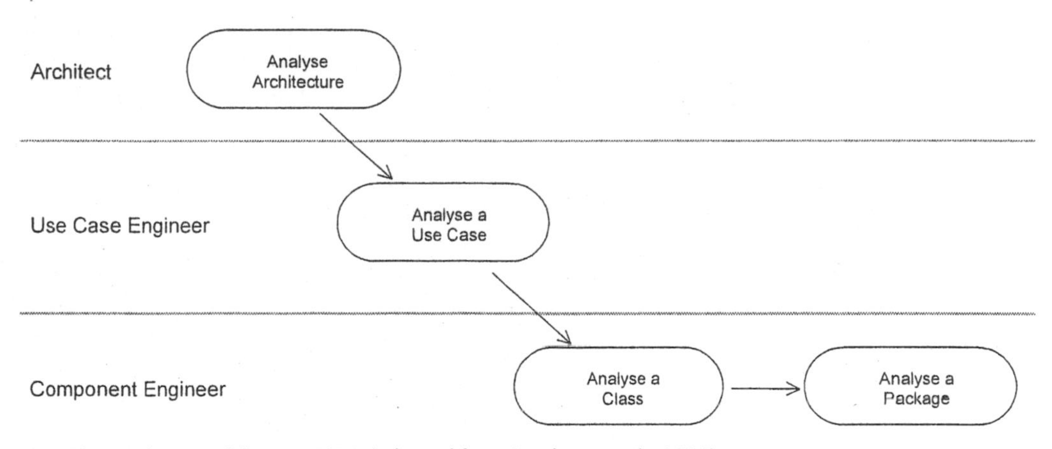

Figure 22.5 The analysis workflow in USDP (adapted from Jacobson et al., 1999).

Artefacts. By now, it should be reasonably clear what the main artefacts are in USDP. These will clearly include models, such as the Use Case Model or the Design Class Model, and products, such as an implementation package or sub-system. However, Jacobson et al. (1999) define an artefact very broadly as almost any kind of information that is created, used or amended by those who are involved in developing the system. This means that the term also covers small scale things, such as a class specification or an operation signature, and transient things such as notes and prototypes.

Summary. The USDP represents the most mature object-oriented methodology that has yet been released. In large part, this is due to its ancestry. A recent book by Rosenberg and Scott (1999) states that the Booch method provided good methods for detailed design and implementation, OMT had particular strengths in its tools for exploring the problem space and Objectory was particularly strong in its approach to defining the solution space, but that all three ingredients are necessary for overall success. The USDP strives to bring these together.

Many aspects of the USDP follow almost inevitably from its basis in object-orientation. For an appreciation of this, it is only necessary to make a high level comparison between USDP and OPEN. Both present a very similar structure, although names differ—for example, a USDP workflow is an OPEN activity and a USDP activity is an OPEN task. But both regard a methodology as essentially comprising a set of techniques, activities and products and both regard an expressive notation as essential.

However, in at least one respect USDP is distinct from most other object-oriented methodologies to date, and we regard this as probably its greatest weakness. As methodologies go, USDP is large and complex. There will inevitably be a significant learning curve involved wherever USDP is adopted within an organization.

Our experience suggests that complex methodologies (SSADM provides a historic example) tend to be fully adopted only when the organization has sufficient resources to provide thorough training and the culture to impose relatively strict discipline on development staff (SSADM was particularly successful in the UK public sector, where it was adopted as a Government standard). The complexity of USDP undoubtedly derives from a natural desire to retain the best features of the three contributing methodologies. But this is unlikely to encourage the adoption of the USDP in its complete form. In practice, many software development organizations may take from the methodology only what is easiest to implement. In some cases, this may amount to no more than the UML notation and a degree of lip service towards the methodology. Such a response would be both unfortunate and unnecessary. The development approach that we have advocated in this book provides one example of a subset of USDP, but many other approaches are possible that adhere to the spirit of the USDP without slavishly following its every detail.

22.6 Participative Design Approaches

Participatory Design (PD) is the name given to a collection of approaches to information systems development that share a guiding ethos more than they share any particular tools or techniques. Sometimes known as co-operative design, PD should perhaps be seen as a movement rather than as a methodology. The common ethos is based on the assumption that active involvement of users in the design and development activity is critical to the success of an information system. This is because a successful design for an information system is said to rely just as much on knowledge and understanding of the work that is to be supported as it does on knowledge of the possibilities and limitations of the available technology. In practice, most approaches to PD assume that users will be actively involved throughout the development life cycle, rather than in just the design activity. The approach has distinctly Scandinavian roots. Like the ETHICS methodology (See Section 22.8), participatory design can be related back to the workplace democracy movement of the 1960s. This had a particular impact on industrial relations, and subsequently also on information systems development theory and practice, in the Scandinavian countries.

We include the approach in this chapter because of its many points of contact with the development approach advocated in this book.

- Some approaches to PD, particularly that of Kyng and his collaborators (see, for example, Bødker et al., 1993) emphasize the use of *use scenarios*. In many respects these resemble the use cases of Objectory and USDP, although there are differences in granularity (the use scenario corresponds more closely to a set of related use cases). In Chapters 6 and 7 we introduced a use-case driven approach to requirements modelling and analysis.

- PD approaches usually emphasize prototypes and storyboards, often low-technology mock-ups drawn on paper or cardboard, for requirements capture. In Chapter 6 we presented storyboards and prototypes, both hand-drawn and computer-displayed, as important elements in establishing users' requirements and exploring alternative approaches to their realization.
- The typical PD lifecycle is experimental and iterative in nature, in recognition of the fact that design is always to some extent a learning experience for all participants. In Chapter 5, and in most subsequent chapters of this book, we have stressed the significance of an iterative approach to object-oriented development.
- The active participation of users in design and development is fundamental to all PD approaches, usually throughout the project life cycle. An iterative approach to object-oriented development naturally encourages the involvement of users at many stages of the life cycle. In Chapter 21 we introduced DSDM as a project management methodology that is thoroughly compatible with object-oriented development, and DSDM, like PD, insists on the active involvement of users in the project team.

Many styles of participative design have been proposed and practised, both in Western Europe and in North America. We try to give an understanding of the movement as a whole through a focus on only one example, the Co-operative Design approach developed by Morten Kyng at the University of Aarhus in Denmark. This is shown graphically in Figure 22.6.

The diagram illustrates a possible sequence of activities within a project and also shows whether the degree of responsibility for each activity that lies with users or with developers. Most activities are shared to some extent between the two, but we can see that some rely chiefly on the contributions of developers (for example, building the prototypes) while others rely primarily on the contribution of users (for example, evaluating the prototypes).

While Figure 22.6 might seem at first sight to suggest a waterfall life cycle, a more careful reading reveals that in fact the underlying life cycle is iterative. Two cycles of prototyping are shown followed by a final system build, but this can just as easily be interpreted as three cycles of iterative development. The approach lends itself equally well to any number of iterations and also to incremental approaches to delivery. The contribution of PD approaches to RAD, and to DSDM in particular, is quite clear.

Part of the background to PD is the belief among its proponents that the models built during the analysis and design of a proposed system fall into two general categories.

The first category includes representations of the work that is done by people who will become users of the system. In the Agate case study, the use case `Assign staff to work on a campaign` falls into this category. A problem can occur with models that fulfil this purpose (of representing work). Sometimes a user's lack of familiarity with the modelling language can cause misunderstanding about what a model is meant to convey. For this reason it is important that users themselves play a major role in constructing the models.

The second category includes representations of the system that is being designed. In the Agate case study, the sequence diagram for the same use case falls into this

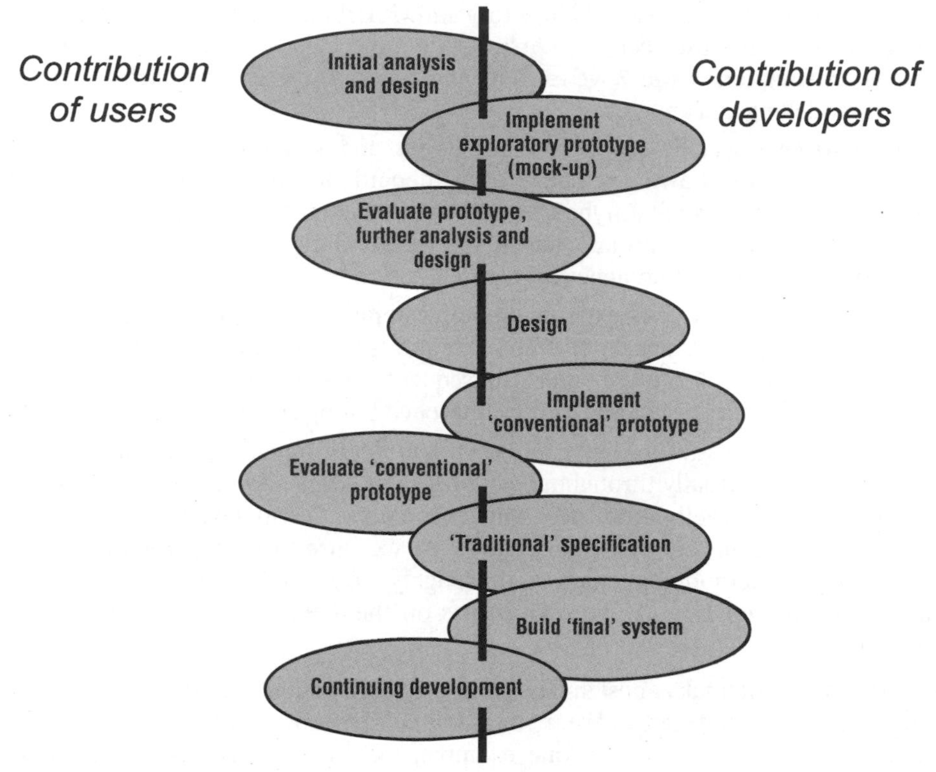

Figure 22.6 Illustrative lifecycle for Co-operative Design (adapted from Kyng, 1995).

category. A similar problem can occur with models that fulfil this purpose (of representing the system that is being designed). Only a user can really tell whether a given design fully takes into account all requirements together with the constraints and limitations that are imposed by work practices, working environment, technology and so on. Thus the activities of system design also require active user participation.

In practice, since it is important for both users and developers that their mutual knowledge and understanding should grow as the project progresses, it makes sense for both groups to share responsibility for the whole project from inception to completion.

22.7 Issues in Choosing a Methodology

The introduction of a methodology to an organization is not a trivial matter. There are many costs, some quite difficult to estimate. Staff must be trained in the techniques, structure and management of the new methodology, documentation must be purchased and software licences must be obtained for CASE tools that support the methodology. The indirect, hidden costs are often under-estimated. Productive time is lost during training, and for some time after the change there is also a reduction in productivity and an ongoing need for support from external consultants. This is true whether or not the organization already uses a methodology. Even with careful evaluation before a decision is made, followed by careful planning of the change, it is still wise to conduct a full-scale trial of a new methodology on a pilot project, which must also be chosen carefully. It would be unwise to risk the failure of a critical system, yet a pilot project must be sufficiently complex to put the new methodology to a thorough test.

The choice of the 'right' methodology is also fraught with difficulties, as there are now many hundreds to choose from, and these differ radically in their philosophies, their coverage of the life cycle and their suitability to particular application domains. Many factors that affect the appropriateness of a methodology, including type of project (large, small, routine or mission-critical), application domain (e.g. real-time, safety-critical, user-centred, highly interactive, distributed or batch-mode) and nature of the IS development organization.

An influential thinker on the management of software development is Humphrey (1989), whose 'process maturity model' suggests that organizations evolve through five successive levels of maturity. First comes an 'initial' level, where development activities are chaotic, and each developer uses ad hoc procedures that they have probably devised themselves. There are no common standards and no dissemination of good practice through the organization, so the success of any project depends solely on the skill and experience of the development team. Many organizations today are still at this level, where there is no point in introducing any methodology, since management have neither the skill nor the structures required to control it. Next comes the 'repeatable' level, where an organization has adopted some development standards and project management procedures. These allow successes to be repeated on later projects, and the organization can benefit from a methodology, since management procedures are capable of enforcing its application. However, while individual managers may repeat their successes, there is no clear understanding of which specific factors led to each success. It is unlikely that success can be generalized to different kinds of project or application, and the flexibility of the organization is still limited. A prescriptive methodology that defines all steps in some detail is more likely to be successful.

An organization at the 'defined' level has produced its own definition of the software process, and is able to standardize activities throughout the organization. A methodology can now be introduced more readily and is likely to produce greater benefits, since the organization already has a culture of working to defined procedures. But staff still adapt much more readily to a methodology that is in harmony with their current ways of working. The next step is typically to introduce a metrics programme (see Section 21.6), which, if successful, can lift the organization to the 'managed' level—but few organizations are yet at this level. Only a tiny handful have reached the final 'optimizing' level, where there is a capability for continuous improvement in all activities (corresponding to the general management approach called 'Total Quality Management').

22.8 Hard versus Soft Methodologies

This chapter would not be complete without some mention of the critical debate that turns on the distinction between hard and soft methodologies. The distinction has emerged principally from the broad systems movement, and, while there is no one precise definition of the difference, it is summarized in Figure 22.7.

To summarize, 'hard' is usually taken to mean objective, quantifiable or based on rational scientific and engineering principles. In contrast, 'soft' involves people issues and is ambiguous and subjective. The methodologies that we have discussed so far in this chapter all derive mainly from the hard tradition, although some influence of a soft approach can be discerned in Participative Design and also in the use case

Hard systems view	Soft systems view
The activity of IS development is all about building a technical system that is made only of software and hardware.	An IS also comprises the social context in which the technical system (software and hardware) will be used.
Human factors are chiefly important from the perspective of the software's usability and acceptability. Politics and group behaviour are only an issue for project managers.	A new IS impacts on interpersonal communication, social organization, working practices and much more, so human and social factors are paramount.
Organizations exist only to meet rational objectives, through the application of rational principles of business management. It is possible to be both rational and objective about the requirements for a new system.	Organizations are made up of individuals with distinct views and motivations, so any picture of requirements is subjective. It is not always possible even to reach a consensus. In practice, this means that the powerful decide, not the wise.
When requirements are uncertain or unclear, it is up to management to decide. Setting objectives is a principal role of management, and others should follow their lead.	If management has not accommodated the full range of views in the organization, encouraging managers to decide on the requirements may be completely counter-productive.

Figure 22.7 Some underlying assumptions of hard and soft systems approaches.

technique, since this aims at eliciting the practical, context-based requirements of individual users.

On the whole, those methodologies that might be characterized as principally soft in their orientation tend to focus more on making sure that the 'right' system is developed, than on how to actually develop the system. Their intellectual antecedents are diverse. For example, Mumford's Effective Technical and Human Implementation of Computer Systems (ETHICS) (Mumford, 1995) derives from the socio-technical school of social theory, and has a considerable degree of overlap with participative design in its concerns. Meanwhile Checkland's influential Soft Systems Methodology (SSM) (Checkland, 1981; Checkland and Scholes, 1990; Checkland and Holwell, 1998) is based on a set of philosophical ideas about the nature of systems at a conceptual level. However, in spite of their very different origins, both provide ways of exploring and agreeing the characteristics of the organization as a system, before any attempt is made to define a specific information system that will help users meet their goals. This approach is very different from that taken by most 'hard' methodologies, which tend to assume that the purpose and nature of the organization can, to a large extent, be taken for granted.

One way of looking at this is to say that soft and hard methodologies cover different parts of the life cycle. In this view, a soft methodology is more useful in the earlier stages of the life cycle, particularly when there is uncertainty about the goals or strategy of the organization as a whole. A hard approach will be more appropriate once any initial uncertainties and ambiguities have been resolved (insofar as this is possible), since the emphasis then shifts to a specific project with relatively clear goals and boundaries.

This has led to the suggestion that, in certain situations, hard and soft methodologies can complement each other, and can be used together to help overcome some of the perennial difficulties in systems development. Flynn (1998) proposes a 'contingency framework' shown in Figure 22.8, which aims at helping to select an appropriate methodology for a specific organizational context.

'Requirements uncertainty' is the extent to which requirements are unknown or subject to debate or disagreement, and also whether they are expected to change during development. For example, a new system intended primarily to automate an existing manual system may have relatively low requirements uncertainty. Agate's Campaign

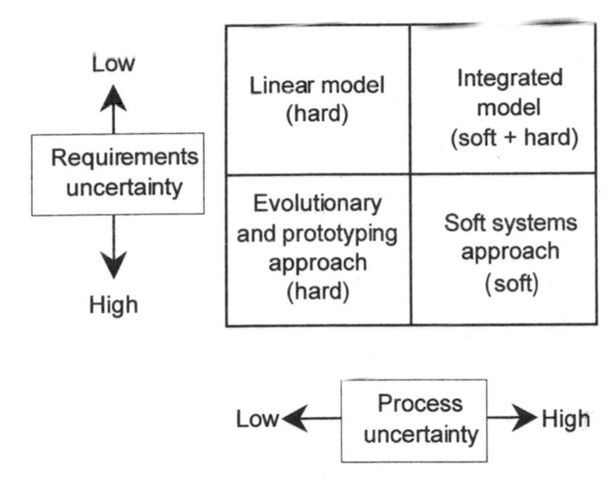

Figure 22.8 The contingency framework can be used to help select an appropriate methodology for a given organizational context (adapted from Flynn, 1998).

Management system might fall into this category. 'Process uncertainty' refers to the degree of doubt about the best way to build the proposed system. For example, it may be very difficult to decide on the best process if a new system will use untried technology, with unpredictable effects on the organization's employees. A project intended to introduce Electronic Commerce to an organization with no experience of it might fall in this category. The term 'Linear Model' refers to a sequential life cycle model like the waterfall model.

A project is rated along both dimensions, and this helps to indicate an appropriate development approach. For example, where both the requirements and process are clear from the outset, a linear model of development is recommended, which in practice might either mean using a traditional structured methodology, or procuring a ready-made solution. At the other extreme, a soft approach is recommended, so that the character of the problem is clarified before any further action is taken.

In seeking to merge together a soft and a hard methodology, the development team is really trying to devise a unique method suited to the project. This implicitly recognizes the complementary nature of their strengths and weaknesses.

22.9 Summary

We began this chapter by considering two definitions of 'methodology', and went on to discuss how the concept of methodology differs from method. This is an important distinction, since the development approach is an important factor in the success or failure of a project. Many methodologies have been developed over the years, stemming from very different traditions, and each in some way attempting to counter a perceived shortcoming in other contemporary methodologies. The 1990s were a prolific time for the spread of object-oriented methodologies, among which the USDP looks set to be the most popular. The last decade has also seen a great deal of research into the feasibility of merging hard and soft methodologies in the hope of being able to meet a wider range of demands and improve the overall success rate of IS development projects.

Review Questions

22.1 What is the difference between 'methodology' and 'method'?

22.2 Distinguish between 'task' and 'technique', and give some examples of each.

22.3 What are the three logical views of an information system?

22.4 Explain the key elements in the philosophy of the USDP.

22.5 In what ways does the participative design approach agree with object-oriented approaches, such as OPEN and the USDP?

22.6 How does the full USDP approach differ from the simplified approach followed in this book?

22.7 Name the five levels of Humphrey's model of process maturity.

22.8 Distinguish between the hard systems view and the soft systems view.

22.9 Why might a methodology based on a hard systems approach be unsuccessful in a situation where the goals of the organization are unclear?

22.10 What general advantages are claimed for using a methodology?

22.11 What might be the disadvantages of using an inappropriate methodology?

Case Study Work, Exercises and Projects

Do some research in your library or on the Internet, and collect material that describes four or more different systems development methodologies. Try to make these as different from each other as possible, for example by choosing one that is structured (e.g. SSADM), one that is object-oriented (e.g. USDP), one based on a soft systems view (e.g. SSM) and one from another tradition (e.g. ETHICS). Then use the following questions as a basis for comparison.

22.A What techniques are used by each methodology? In particular, how do they represent the process view, the data view and the temporal view of the system?

22.B To what extent does each methodology cover the full project life cycle, from project selection through to implementation and maintenance?

22.C How far do you think each methodology can be adapted to suit differing projects or circumstances?

22.D Can you find a statement that gives an underlying philosophy for each methodology? If not, is it possible to identify the intellectual tradition from which the methodology has been derived?

Further Reading

- Flynn (1998) introduces the many traditions behind today's methodologies, and discusses why there is still disagreement about the 'right' way to do systems development. Jayaratna (1994) offers an alternative framework for the evaluation of methodologies. Both of these books should probably be required reading for anyone who has the responsibility for recommending a methodology for adoption by an organization.
- For more detailed information about any specific methodology, readers should ideally consult a primary source, e.g. Jacobson et al. (1992) for Objectory, Rumbaugh et al. (1991) for OMT, Checkland and Scholes (1990) for Soft Systems Methodology, Mumford (1995) for ETHICS, Graham et al. (1998) for OPEN, Jacobson et al. (1999) for the USDP.

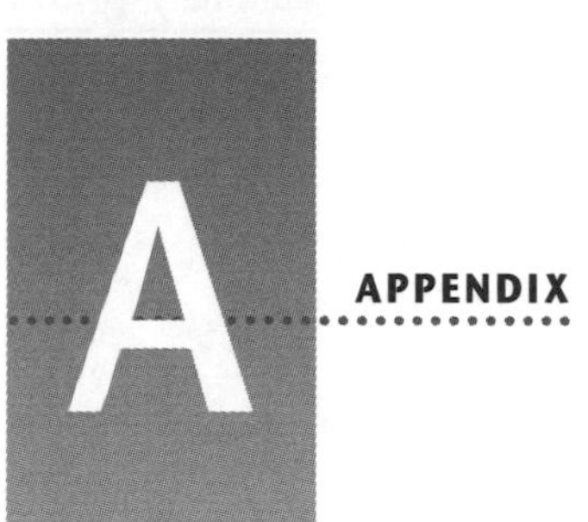

APPENDIX

Notation Summary

Use Case Diagram

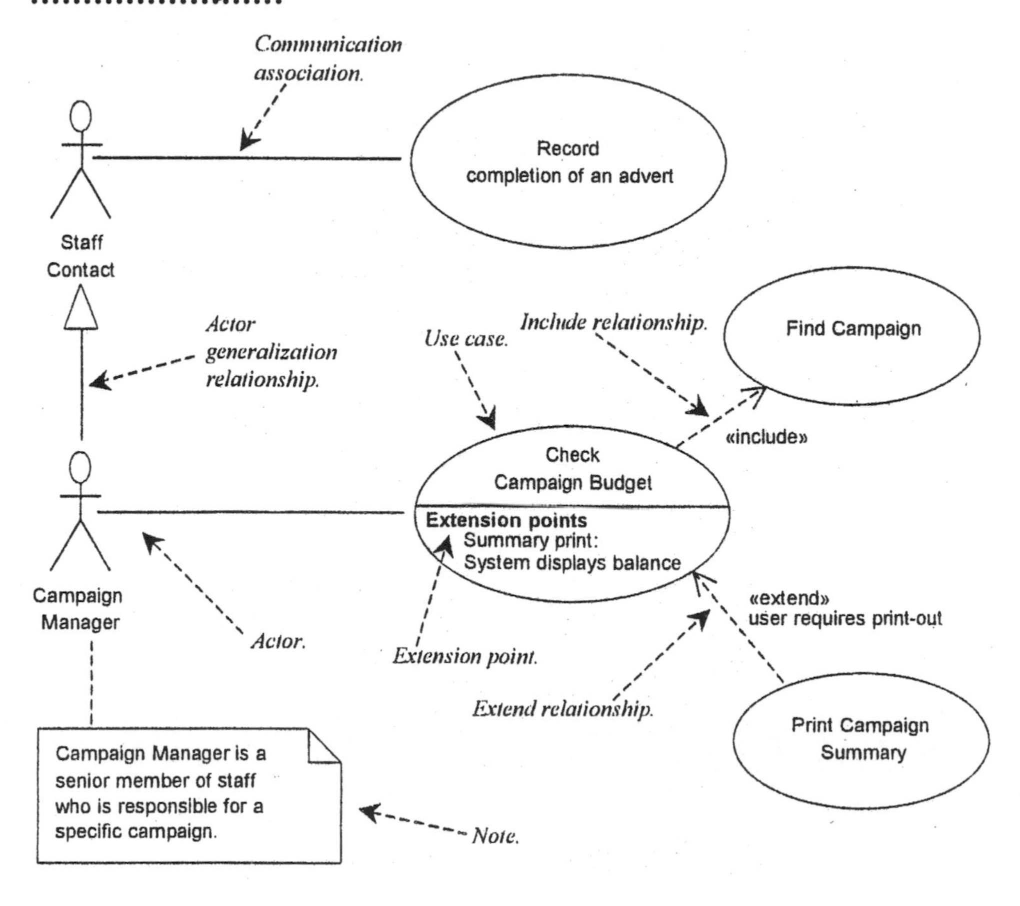

Static Structure Diagrams

Object instance notation

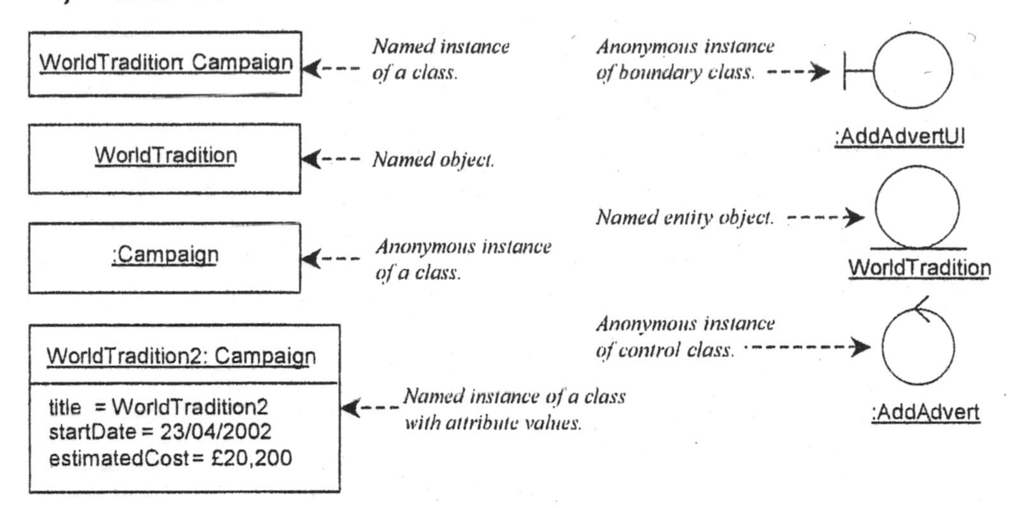

Class notation

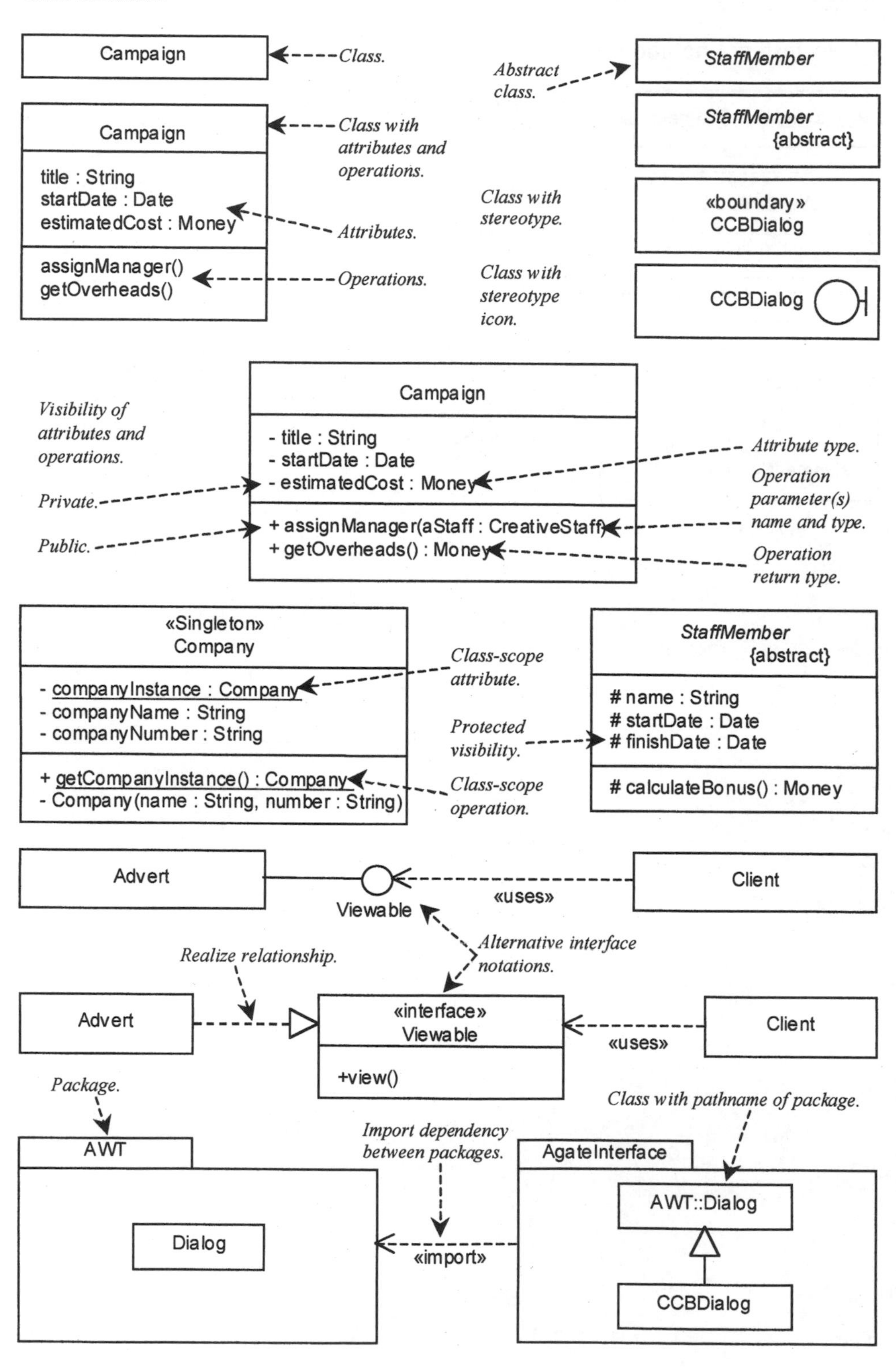

Associations

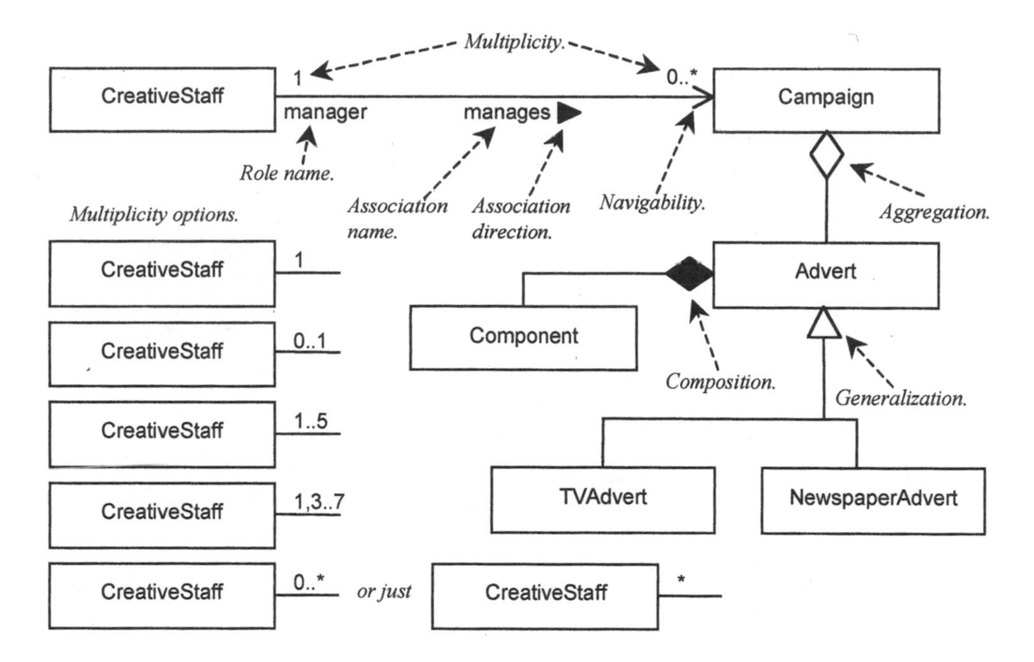

Behaviour Diagrams

Statechart diagram

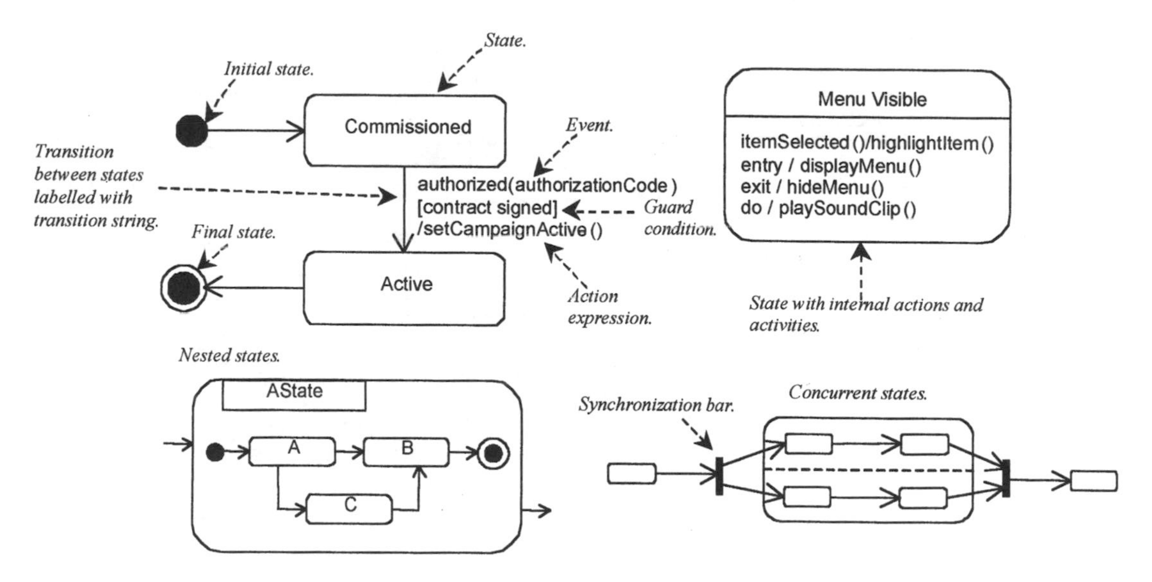

Activity diagram

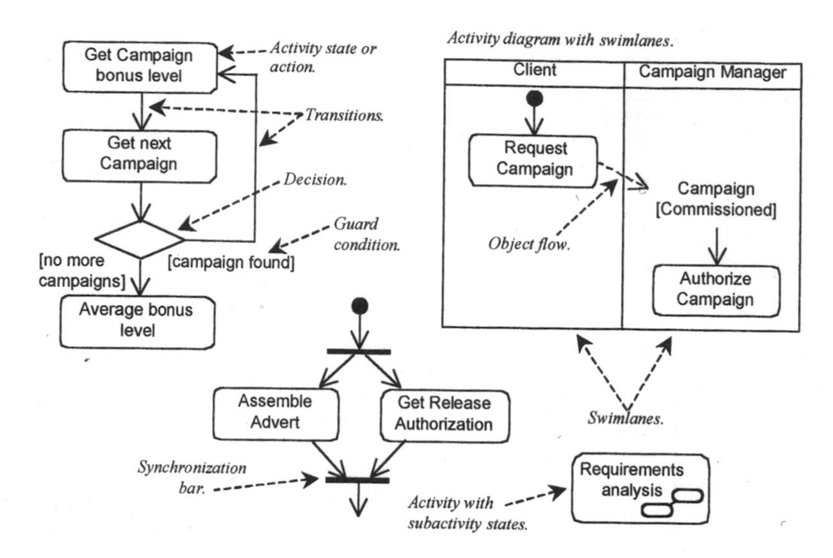

Interaction Diagrams

Sequence diagram

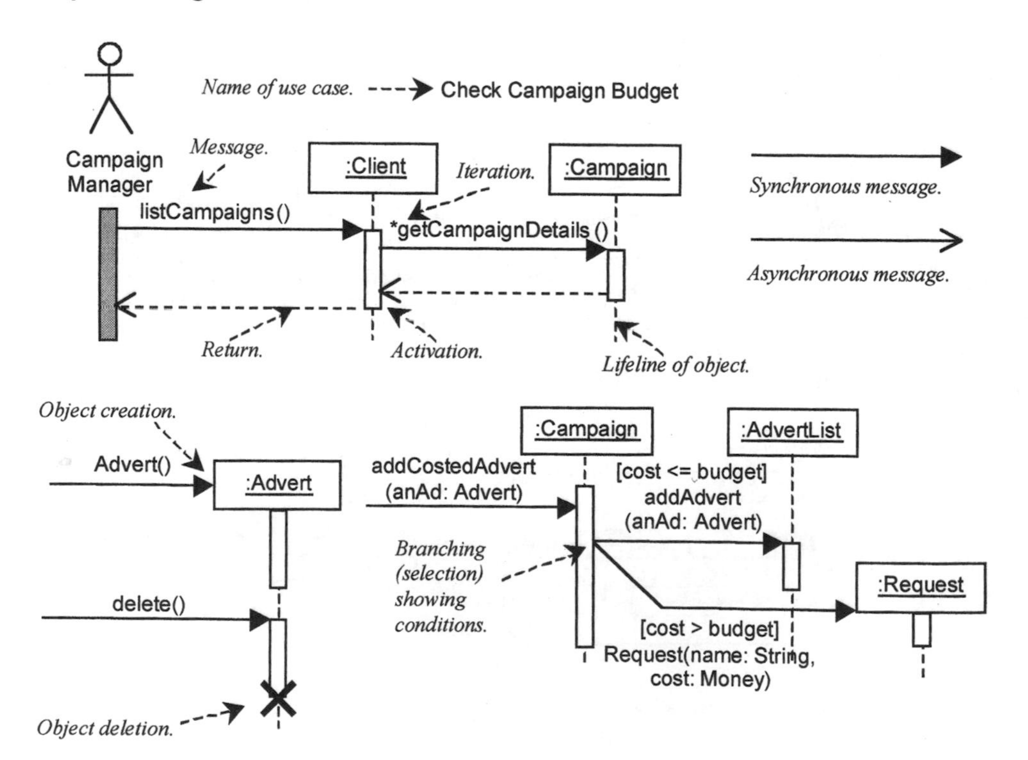

Collaboration diagram

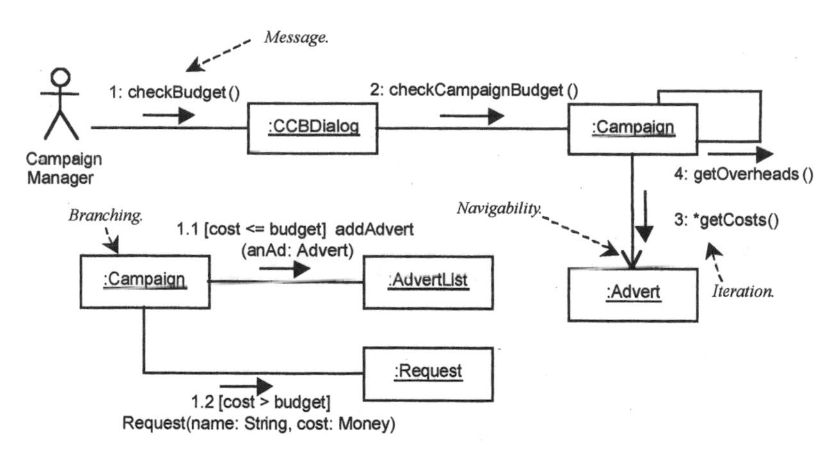

Implementation Diagrams

Component diagram

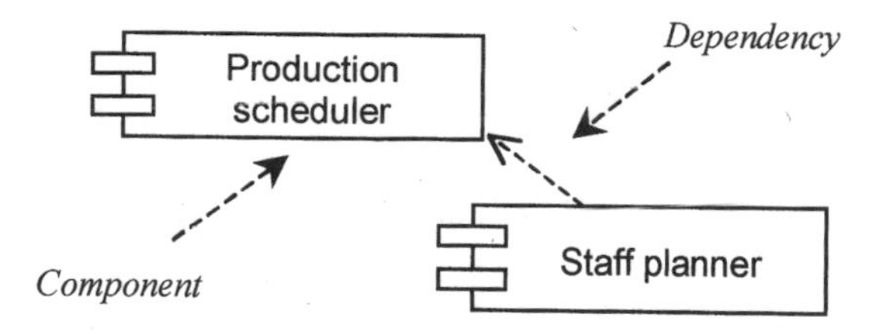

Deployment diagram

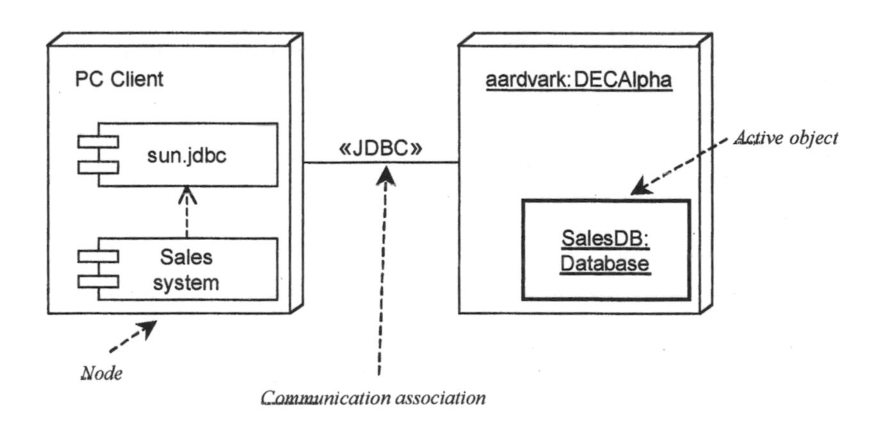

APPENDIX

Selected Solutions and Answer Pointers

In this section we give solutions to a selection of the review questions in the chapters, and also pointers on possible approaches to some of the end-of-chapter case study work, exercises and projects.

Answers to Selected Review Questions

1.3 Even if something isn't a system (or might not be one) thinking of it as one still gives useful insights.

1.5 Feedback is sampling one or more outputs of a system for comparison to a control value. Feed-forward is sampling a system input, usually before it enters the system. The control value may be an output, input or an internal measure of system performance.

1.7 A management support system provides information that helps managers to make decisions. Most use feedback or feed-forward to monitor the performance of that part of the organization for which the manager is responsible.

1.9 Business goals and strategy are defined first. An IS strategy identifies applications that can help to meet business goals, and an IT strategy identifies IT needed to develop and run the applications. Each informs its predecessor about what can realistically be achieved. The process is iterative.

1.10 One of the simplest definitions says that information is data with a structure and a meaning derived from the context in which it is used.

2.1 They differ in their view of the problems because their view of the meaning and purpose of IS development also differs (see also Review Question 1.6).

2.3 The simplest definition is 'fitness for purpose'. Since it can be hard to identify the purpose, a better alternative is 'meeting all user requirements, both stated and implied'.

2.6 The system may address irrelevant problems. It may not fit the way that people work. It may be unsuitable for its environment. It may be out of date before delivery. Political difficulties may lead to delay or cancellation.

2.7 A stakeholder has an interest in a project because they are (or will be) affected by its progress or by its results.

3.1 Advantages of the traditional waterfall life cycle include:

- teams with specialized skills can be assigned to tasks in particular phases,
- progress can be evaluated at the end of each phase,
- attendant risk can be controlled and managed.

3.2 Disadvantages of the traditional waterfall life cycle include:

- real projects rarely follow a simple sequential life cycle,
- iterations are almost inevitable,
- the elapsed time between inception and delivery is frequently too long,
- it is unresponsive to changes in the technology or requirements.

3.5 Prototyping is not necessarily concerned with delivery a working system whereas an incremental approach delivers a working system in successive increments. Note that in the Unified Software Development Process an increment may be any life cycle product.

3.7 Syntactic correctness is concerned with using the notation (e.g. UML) correctly, consistency relates to producing a set of models or diagrams that are consistent with each other and completeness refers to producing models that are completely defined.

3.8 The term requirements traceability refers to the capability of tracking each requirement to all the systems development deliverables, from analysis models to program code, that relate to it.

3.9 A diagram may be syntactically correct, consistent with other diagrams and models, and complete but it may not relate accurately or completely to the user requirements, the most important criterion for any diagram or model.

4.2 Semantics is the study of meaning. In object-oriented development it is used to denote the meaning that an element has for the user. The semantics of an object include its purpose, description, relationships and behaviour seen from a user perspective.

4.3 Other parts of a system only see an object's interface (services it can perform and operation signatures). Internal details including data are hidden and can only be accessed by a message that contains a valid signature.

4.4 Polymorphism means that when one message is sent to objects of different types, each has an appropriate, but different, implementation for its response. The object that sends the message need not know which type of object is addressed. One way of implementing polymorphism is through inheritance and overriding.

4.6 A subclass inherits all characteristics of its superclass and other ancestors (some may be overridden, but are still technically inherited). Each subclass is different from its ancestors in at least one way.

5.2 Icons, two-dimensional elements, paths and strings.

5.4 To promote communication between team members in a project. To communicate over time to other people who will work on the system. To communicate good practice and experience.

5.7 A rectangle with rounded ends.

5.8 Transitions.

5.10 The start state (a filled black circle) and the final state (a filled black circle within another circle).

6.1 Examples of functional requirements are: the need for a process to be run that allocates staff to lines based on their skills and experience, and holidays and sick leave; printing out an allocation list; amending the allocation list. Examples of non-functional requirements include: printing the allocation list by 12.00 noon; the need to handle 200 operatives' details.

6.4 Use cases are produced to model the functionality of the system from the users' point of view and to show which users will communicate with the system. They show the scope of the system.

6.6 An essential use case documents the interaction between user and system in a way that is free of technological and implementation details, and a real use case describes the concrete detail of a use case in terms of its design.

7.2 A purely graphical model does not have the precision to cover every aspect of system behaviour and structure. A purely textual model would be too large, too complex and extremely difficult to understand and to maintain.

7.3 An attribute is a characteristic of a class (every person has a height). An attribute value is a characteristic of an instance (this author is 1.75 m tall).

7.4 An element's stability is the relative infrequency of change in its description.

7.6 Multiplicity denotes the range of values for the number of objects that can be linked to a single object by a specific association. It is a constraint because it limits the behaviour of a system. If a client can have only one staff contact, it should be impossible to link a second.

7.10 A link is a connection between two objects. 'Changing' a link (say by substituting another object at one end) is equivalent to destroying the link and creating a new one. (Think about two objects tied with a single length of string. In a substitution, there is a moment when neither one nor the other is connected—unless you tie on the second before untying the first, but an object link cannot do this.)

7.12 A collaboration diagram shows only those classes that collaborate to provide the functionality of a particular use case (or operation); the links that arc shown are those that are required for that purpose. A class diagram typically shows all the classes in a particular package and all the associations between them.

8.1 Use of a component saves time and work. A friend of one of the authors once said 'Have you ever wondered how much it would cost to make your own light bulb?'

8.3 Objects are well encapsulated, and object structures can be designed this way. The hierarchic nature of generalization abstracts out the more general features of a class. Hierarchic organization of models helps the developer to find components easily when they are needed. Composition encapsulates whole structures within a composite object.

8.4 A component of a composition cannot be shared with another composition. The component has a coincident lifetime with the composition (although a component can be explicitly detached before the composition is destroyed).

8.5 This is a basis for polymorphism. The superclass operation defines the signature, but each subclass has a different method that implements the behaviour (see Chapter 10).

8.6 An abstract class has no instances and exists only as a superclass in a hierarchy. It provides a generalized basis for concrete subclasses that do have instances.

8.12 An antipattern documents unsuccessful attempts at solving problems and suggests how the failed solution may be adapted to solve the problem successfully.

9.1 Collaboration diagrams discourage both using a large number of messages between two objects and having too many parameters for each message as these are clumsy to represent on the diagram.

9.2 Small self-contained classes are easier to develop, test and maintain.

9.3 Sequence diagrams have a time dimension (normally vertically down the page) while collaboration diagrams do not. Collaborations show the links between objects, which are not shown on sequence diagrams.

9.5 An object lifeline represents the existence of an object during an interaction represented in a sequence diagram.

9.6 The focus of control indicates which operation is executing at a particular stage in an interaction represented in a sequence diagram.

9.8 Sequence numbers are written in a nested style in a collaboration diagram to represent nested procedural calls.

10.1 They confirm the user's view of the logical behaviour of a model. They also specify what the designer and programmer must produce to meet the users' requirements.

10.2 Decision tables are particularly suited to representing decisions with complex multiple input conditions and complex multiple outcomes, where the precise sequence of steps is either not significant or is not known.

10.4 An algorithm defines the step-by-step behaviour of an operation. A non-algorithmic approach defines only inputs and results.

10.5 Non-algorithmic methods of operation specification emphasize encapsulation.

10.9 OCL expressions have:

- a context within which the expression is valid (for example, a specified class);
- a property within the context to which the expression applies (for example, an attribute of the specified class);
- an operation that is applied to the property (for example, a mathematical expression that tests the value of the attribute).

11.2 A guard condition is evaluated when a particular event occurs and only if the condition is true does the associated transition fire.

11.3 All the guard conditions from a state should be mutually exclusive so that for each set of circumstances there is only one valid transition from a state. If they are not mutually exclusive more than one transition may be valid and the behaviour of the statechart is indeterminate.

11.4 An object can be in concurrent states if it has a complex composite state with concurrent substates.

11.5 Nested states are substates of a composite state and describe the life cycle of an object in detail. Nested states need not be concurrent but may be. Concurrent nested states are substates that an object may occupy at the same time.

11.6 An action is considered to be instantaneous (its duration is actually determined by the processing environment) while an activity persists for the duration of a state, for example (its duration is dependent upon the occurrence of events that may cause a state change).

11.7 A statechart describes the dynamic behaviour and associated state changes of instances of a class.

12.3 User wants report etc.—analysis. Selection of business objects etc.—logical design. Size of paper etc.—physical design.

12.5 Seamlessness means that the same model (class diagram) is used and successively refined throughout the project.

12.8 We elaborate user interface and application control classes, we add mechanisms to support data management. The class diagram is also updated with the types and visibility of attributes and operations and to show how associations are designed.

12.10 Functional, efficient, economical, reliable, secure, flexible, general, buildable, manageable, maintainable, usable, reusable.

13.1 Open layered architectures are more difficult to maintain because each layer may communicate with all lower layers hence increasing the degree of coupling in the architecture. A change to one layer may ripple to many layers.

13.2 A closed layer architecture may require more processing as messages have to be passed through intervening layers.

13.4 The main differences between the MVC and the layered architecture include the update propagation mechanism and the separation of the presentation layer into the View and Controller components in the MVC.

13.6 A broker decouples sub-systems by acting as an intermediate messaging-passing component through which all messages are passed. As a result a sub-system is aware of the broker and not directly in communication with the other sub-systems. This makes it easier to move the sub-systems to distributed computers.

13.7 User requirements may dictate concurrent behaviour that cannot be supported using facilities such as multi-tasking but requires the use of a custom-built scheduler component.

14.1 Private, public or protected visibility.

14.2 Attributes should be designated private to enforce encapsulation.

14.3 The application of coupling and cohesion produce coherent, focused classes that are loosely coupled and more amenable to reuse.

14.5 Collection classes can be used to hold the object identifiers of the linked objects at the many end of an association. Collection classes provide collection specific behaviour for manipulating the collection.

14.6 A collection class of object identifiers should be included in a class if it is not used by another class and it does not increase the complexity of the class unduly.

14.8 Normalization is useful when decomposing complex objects or when implementing the system with a relational database management system.

15.2 The main aspects of changeability are maintainability, extensibility, restructuring and portability.

15.3 The class constructor in the Singleton pattern is private so that it can only be accessed by the class-scope `instance()` method. This ensures that the Singleton class has total control over its own instantiation.

15.4 The Singleton pattern ensures that only one instance of a class exists and provides system wide access to that instance.

15.7 A pattern catalogue is a group of largely unrelated patterns, which may be used together or independently. A pattern language is a group of related patterns that relate to a particular problem domain.

16.3 Dialogue metaphor describes interaction in terms of conversation between user and system involving different kinds of communication. Direct manipulation metaphor represents objects of interest to the user as objects on the screen that they can manipulate through the use of the mouse. Dialogue follows sequence determined by system. Direct manipulation is event-driven, and the user can determine sequence of events.

16.6 User may hit Return key without thinking and delete the Client in error.

16.9 Possible advantages: structured—aid management of projects, apply standards that aid communication, force consideration of all aspects of HCI design; ethnographic—analyst gets detailed understanding of context of system, active user involvement, social and political factors taken into account; scenario-based—help to think through possible alternative routes in use cases, can be used to justify design decisions, valuable for testing programs. Possible disadvantages: structured—can be bureaucratic; ethnographic—can be time-consuming; scenario-based—generates large volume of documentation.

17.2 A horizontal prototype deals with only one layer of the system architecture, usually the user interface. A vertical prototype takes one sub-system and develops it through each layer.

17.6 We use statechart diagrams to model the lifetime of instances of business classes.

17.8 A list of states, for each state the valid events that can cause a transition from that state, the state that each transition leads to, and any operations associated with the transition into the new state.

17.11 Java `EventListener` only handles changes to interface objects. MVC deals with changes to Model objects. The Java `Observer` and `Observable` interfaces provide MVC mechanisms.

18.3 Tagged data, with a tag for the class of each object and the attribute of each value within each object. This way, it is possible to reconstruct any object from the data in the file without having to hard code the structure of every possible complex object. (This is the approach used by the Java `ObjectOutputStream` and `ObjectInputStream` classes, and by XML and SOAP.)

18.6 1. Remove repeating groups. Ensure all row column intersections contain only atomic values.

2. Make sure every attribute is dependent on the whole primary key. Create a separate table for part-key dependencies.

3. Ensure every attribute is dependent on the primary key and not on another non-key attribute. Create a separate table for non-key dependencies.

18.8 Object Manipulation Language and Object Definition Language.

19.3 Package diagrams show the logical grouping of classes in a system, whereas component diagrams show the physical components of a system. During implementation, package diagrams can be used to show the grouping of physical components into sub-systems; component diagrams can be combined with deployment diagrams to show the physical location of components of the system.

19.6 Possible tests would be to test validation of date, test validation of start time, check validation of job number etc. More detailed validation of time could be to check it is within a certain amount of time of current time—if not, then a warning should be displayed.

19.7 Review of cost benefit analysis. Summary of functional requirements met and amended. Review of achievement of non-functional requirements. Assessment of user satisfaction. Problems and issues with the system. Extract of quantitative data for future planning. Identification of candidate components for reuse. Possible future developments. Actions required. (See Section 19.9.2 for the detail.)

19.9 Because analysts or designers will have the wider view of the system and can ensure that changes fit in and do not have a detrimental impact on other sub-systems.

20.1 Saving time and money in developing the components, and saving time and money in testing the components.

20.3 'A component is a type, class or any other work product that has been specifically engineered to be reusable.' (Jacobson et al., 1997)

20.5 It considers a component to be an executable unit of code rather than a type or class. It specifies that a component should have an interface, and that it should be capable of being connected together with other components via its interface.

21.1 The slack time for an activity is the difference between its latest and earliest start times.

21.2 A milestone is on the critical path if its earliest start time and its latest start time are the same.

21.3 The Gantt chart can be used to show the partial overlap of activities, it is a useful aid for resource smoothing and it can represent progress on the project activities.

21.7 The underlying principles of XP are communication, simplicity, feedback and courage.

21.9 Object wrappers provide existing non-object-oriented systems with an object-oriented style of interface hence making integration easier.

22.1 A methodology is essentially a set of principles. A method is an instantiation of the principles in a given situation.

22.2 A task is something you do in a particular project. Tasks have products. A technique specifies how to carry out a task. A task might be 'Analyse the requirements for a use case'. One technique for doing this would be the UML collaboration diagram.

22.4 The USDP is use-case driven, architecture-centric and iterative.

22.5 PD approaches also typically emphasize:

- a scenario-based analysis of requirements that resembles use-case modelling;
- the extensive use of prototypes and mock-ups;
- an experimental and iterative life cycle;
- the active participation of users.

22.8 Hard systems methodologies assume that system objectives can be clearly identified and defined. This can lead to solving the wrong problem or developing a wonderful system that nobody needs. Or the project may sink in quicksand and end by being cancelled.

22.10 The problems already mentioned for Review Question 22.8 apply here too. Also disproportionate cost and delay if the methodology is too complex; quality problems if its techniques do not provide adequate models of the application; user rejection if it does not encourage sufficient participation.

Answer Pointers for Selected Case Study Work, Exercises and Projects

1.B Some main sub-systems are: on-line sales, retail shops, supplies, deliveries, systems support and accounts, and more. Some of the control mechanisms involve supplier re-orders, the product catalogue, network performance and security. Most have some human activities and automated support. One example of feedback includes on-line shoppers, watching the progress of their orders. The market researcher uses feed-forward (what attracts customers to web pages).

1.C Main business aim, say: 'To establish FoodCo as an independent branded name supplying a range of high quality food products to consumers'. Subsidiary aims: diversification of customer base; achievement of international recognition and sales. Each will be translated into measurable objectives, for example as a basis for the selection of information systems development projects.

2.C Stakeholders should include many of the following. Good if you listed them all, excellent if you thought of some not shown below (provided you can justify their inclusion).

Patients and potential patients, patients' relatives, ambulance drivers, paramedics, control room operators, accident and emergency staff, supervisors of professional stakeholders, managers who control affected budgets, policy level managers, taxpayers (or purchasers of medical insurance policies if this is how the system is funded), general medical practice staff, politicians (particularly if the service is publicly funded), members of the general public (who make the emergency calls), other road users.

3.A An incremental development can be justified for the following reasons:

- Useful increments can be delivered quite quickly. For example, staff management and material tracking could be implemented initially.
- Users can gain experience with the systems.
- Risk is minimized.

Requirements can be refined in light of the initial increments.

4.A The human activity system referred to is the application domain for the proposed IS. Other systems might include the project team, the analyst's department, the business planning system and the wider (political and cultural) system of the organization. Formal and informal structures of communication and relationships are the interfaces. Other installed software and hardware systems may be important (consider the discussion in Box B1.2).

4.D You should have equivalents for most of the following. The names are not significant at this stage, nor is an exact match in the way that you have grouped concepts together.

`Factory, Product, ProductRange, PackedProduct, SaladPack, VegetablePack, CookedProduct, Sauce, Pickle, SandwichTopping, Ingredient, Customer, Supermarket, Brand, Farm, Supplier, Employee, Consumer.`

5.A Some kinds of information systems can be used to model the real world in order to try out ideas. For example decision support systems typically model some aspect of a business and allow staff and managers to ask 'What if?' questions: 'What would happen to demand for a product if the price was increased by 10%?' or 'If we targeted a particular area with a mailshot, what kind of response to our product could we expect, based on what we know about the population of that area?' However, a customer in an information system is not a model of the customer, it is a set of data values that describe attributes of the customer. Also some things in information systems are the real-world objects. An invoice in a sales order processing system is the real invoice; it is not a model. In object-oriented systems, there is sometimes a belief that the operations of objects are things that those objects do to themselves. (Rumbaugh et al. (1992) suggest operations for a `Bicycle` class, like `move` and `repair`.) Typically the operations of objects are actually operations that we want the system to carry out on those objects, and we package them in the class as a way of organizing the design of the software system.

5.B Designing cars, designing aircraft (models to use in wind tunnels), architecture and town planning, packaging design for products.

6.B Here are some of the use cases that should be in the diagram with the actor in brackets. `Check Staff Availability (Production Planner)`, `Enter Details of Staff Illness (Production Planner)`, `Print Availability Lists (Production Planner)`. There is a need for some means of entering details of staff holidays. The decision about who does this will affect the scope of the system. It could be done by the staff themselves and authorized on-line by the factory manager, or this process could be done on paper and only authorized holidays entered by the production planners.

7.A The following are sample descriptions for two of the use cases.

Record employee leaving the line

Normally employees are recorded as leaving the line when they clock off at the end of a working shift. Although there are breaks in the operation of the line during a shift these are not normally recorded as employees leaving the line. Date, time and location are recorded.

Stop run

When the production line stops for a routine reason, e.g. for a break, to restock or to reload equipment, the time the run stopped is recorded and a reason is recorded. The line supervisor or chargehand can do this.

7.B For the use case realization for `Record employee leaving the line`, you should have a collaboration involving `Employee`, `Supervisor`, `Production Line`, `ProductionLineRun`, and `EmployeeAbsence`, as well as a boundary class and a control class.

8.C Possible subclasses include `TelevisionAdvert`, `RadioAdvert`, `Magazine Advert`, `PosterAdvert`, `LeafletAdvert`. We could introduce another layer of hierarchy by grouping `NewspaperAdvert` and `MagazineAdvert` under `PrintMediaAdvert`, and `TelevisionAdvert` and `RadioAdvert` under `BroadcastMediaAdvert`. (You may have chosen equally valid alternative names.)

8.E Some generalization and composition can be justified by the inclusion of the following classes. `Operative`, `RoutineBreak`, `AbsenceRecord`.

9.A The sequence diagrams should be derivable and consistent with the collaboration diagrams produced during use case realization when you answered Exercise 7.B. However, you will be adding more detail in terms of message signatures and message types.

9.B Variations in allocations of responsibility will depend upon how much responsibility the control class has and how much is devolved to the entity classes or the boundary classes. At one extreme the control class orchestrates all the functionality of the use case, at the other the control class delegates the complete control of the use case to one of the entity classes. A good design will lie between these extremes.

10.A One of the more complex (and therefore one that is well worth trying) would be `ProductionLineRun.start()`. Preconditions for this operation should be suggested by your use case description.

10.D Most decision tables can be converted easily into Structured English with either case or nested-if. For very simple tables (two outcomes) if-then-else may be enough.

11.A The events that affect `ProductionLine` include `start run`, `end run`, `detect problem`, `pause run`. The possible states for `ProductionLine` include `Idle`, `Running`, `ProblemInterrupted` and `Paused`.

12.A Examples include sequence of entry, branching points (where the user has a choice), repetition of entries (can the user enter more than one holiday at the same time?), commands that the user might need to use while entering data (but not whether they use a menu, function keys or control keys).

12.D In Windows there are many standards, for example: the use of function keys, particularly in combination with Alt and Ctrl keys; the standards for the appearance of menus, for example, menu entries followed by dots ('...') when the menu entry leads to a dialogue box; the positioning of certain buttons in dialogue boxes ('OK' and 'Cancel').

13.A The procedure should include:

- determine criteria for sub-division,
- partition system according to these criteria,
- review sub-systems to minimize message passing and maximize potential for reuse,
- specify interfaces of each sub-system.

13.B The FoodCo Production Control System could contain the sub-systems Employee Management, Product Control and Production Line Management.

14.A All attributes should designate private and operations public. Choose data types that reflect the domains from which the attribute values are selected.

14.B The one-way associations are `Line-LineFault`. The two-way associations are `Supervisor-ProductionLine` and `Line-LineRun`.

15.B The `ProductionLine` class could use the state pattern with the state subclasses `Idle`, `Running`, `ProblemInterrupted`, `Waiting`. This use of the state pattern reduces the complexity of the `ProductionLine` class but may increase the storage and processing requirements for the application as a whole.

16.A See answer pointer to 12.D.

16.C Something along these lines . . .

First, Rik runs off the three availability lists to show who is available for work the following week. He then starts with operatives who are available all week. For each operative, he views their record on screen, looking at their skills and experience, the line they are currently working on, and how long they have been on that line. He allocates each operative in turn to a line and a session in one of the factories.

This does not provide any detail of actual interaction with the system.

17.A If you are expecting to develop for a windowing environment, you will need a dialogue window as a minimum. Depending on how you handle looking up information, for example a list of valid reasons for stopping the line, you may need separate windows in which to display these look-ups. However, in the factory environment you may want to use a simpler device with an LCD screen for display and a sealed keyboard with dedicated keys. A PC with a mouse may be unnecessary and unsuitable in a messy environment. However, we are assuming a windowing environment for the other exercises in this chapter.

18.B You should end up with the following tables (or similar names): `SalesOrder`, `OrderLine`, `Customer` and `Product`.

18.E This will be similar to Figure 18.23, with a `ClientBroker` class to handle the operation to find each `Client`.

19.B Many libraries now use a web browser to access catalogue services. If this is the case, then your deployment diagram will include the client machines (PCs, Apple Macs or workstations), the web server, and probably another machine running the library software. Library staff may access the system from simple terminals for use cases to issue and return books. They will connect directly to the machine running the software, not via the web server. (The actual configuration will depend on your particular system.)

19.E Issues to consider are as follows. Is the manual organized around the tasks a user carries out? Has it got an index? Can you find the terms in the index that you, the user, know, or does it use computer jargon? Does it show screen shots? Are they the same as actual screens or windows in the version you are using? (You should be able to think of other criteria.)

19.F Possible inclusions for bug reports. User name, telephone no., building, room, address etc. Date and time bug occurred. Type of machine on which bug occurred. Operating system of machine. Software in which bug occurred. Other software running at the same time. Program/window/function being used at time of bug. Any error messages displayed for the user. What the user expected to happen. What actually happened. What the user did (key strokes, mouse clicks on buttons or menus etc.) immediately beforehand.

20.C You could either include the encryption package within the security package or have it as a separate package. The Security Core Classes will need its services and will have a dependency on it. Will it need any kind of user interface classes, for example, if it requires setting up with some kind of parameters? If it does, it will presumably need somewhere to store these parameters, in which case it will also need data storage services.

21.B Possible pilot systems include Sales Order Processing or Bar Code Production. Although these are important systems their current operation is sufficient for the company and hence new replacement systems although useful, are not critical. Of the two, the Bar Code Production system is preferable as the pilot project as it is smaller.

Glossary

Abstract class: a class that has no instances; a superclass that acts only as a generalized template for its instantiated subclasses.

Abstract operation: an operation that is not implemented in the class in which it appears (usually an abstract superclass), but will be implemented in a subclass.

Abstraction: a simplified representation that contains only those features that are relevant for a particular task; the act of separating out the general or reusable parts of an element of a system from its particular implementation.

Activation: the execution of an operation, represented in interaction sequence diagrams as a long thin rectangle.

Activity: an activity is some behaviour that may persist for the duration of a state.

Activity diagram: a variation of a statechart diagram that focuses on a flow of activity driven by internal processing within an object rather than by events that are external to it. In an activity diagram most (or all) states are action states (also called *activities*), each of which represents the execution of an operation.

Actor: an actor is an external entity of any form that interacts with the system. Actors may be physical devices, humans or information systems.

Adornment: an element attached to another model element, for example a stereotype icon or a constraint.

Aggregation: a whole–part association between two or more objects, where one represents the whole and the others parts of that whole.

Algorithm: a description of the internal logic of a process or decision in terms of a sequence of smaller steps.

Analysis class stereotype: one of three specialised kinds of class (boundary, control and entity classes (*q.v.*)) that feature in analysis class diagrams. The separation of concerns that these represent forms the basis of the architecture recommended for most models developed following USDP guidelines (*cf.* stereotype).

Antipattern: documents unsuccessful attempts at providing solutions to certain recurring problems but includes reworked solutions that are effective.

Association: a logical connection, usually between different classes although in some circumstances a class can have an association with itself. An association describes possible links between objects, and may correspond either to logical relationships in the application domain or to message paths in software.

Association class: a class that is modelled in order to provide a location for attributes or operations that properly belong to an association between other classes.

Association instance: another name for a link (*q.v.*).

Attribute: an element of the data structure that, together with operations, defines a class. Describes some property of instances of the class.

Boundary class: a stereotyped class that provides an interface to users or other systems.

Business rule: see *enterprise rule.*

Capta: data that has been selected for processing due to its relevance to a particular purpose.

Cardinality: the number of elements in a set; contrast with *multiplicity* (*q.v.*).

Class: a descriptor for a collection of objects that are logically similar in terms of their behaviour and the structure of their data.

Class diagram: a UML diagram that shows classes with their attributes and operations, together with the associations between classes.

Class Responsibility Collaboration (CRC): CRC cards provide a technique for exploring the possible ways of allocating responsibilities to classes and the collaborations that are necessary to fulfil the responsibilities

Class-scope: a class-scope attribute occurs only once and is attached to the class (not to any individual object. A class-scope operation is accessed through the class (i.e. prefixed with the class name) not through an object. Model elements that are of class scope are underlined in class diagrams.

Cohesion: a measure of the degree to which an element of a model contributes to a single purpose.

Collaboration: the structure and links between a group of instances that participate in a behaviour. The behaviour can be that of an operation or a use case (or any other behavioural classifier in UML).

Collaboration diagram: a collaboration diagram shows an interaction between objects and the context of the interaction in terms of the links between the objects.

Collection class: provides collection-specific behaviour to maintain a collection. Used when designing associations with a many multiplicity to hold collections of object identifiers.

Common Object Request Broker Architecture (CORBA): a mechanism to support the construction of systems in which objects, possibly written in different languages, reside on different machines and are able to interact by message passing.

Component: an executable software module with a well-defined interface and identity.

Component diagram: a diagram that shows the organization of and dependencies among components. One of two UML implementation diagrams (*q.v.*).

Composition: a strong form of aggregation with a lifetime dependency between each part and the whole. No part can belong to more than one composition at a time, and if the composite whole is deleted its parts are deleted with it.

Concrete class: a class that may have instances.

Concurrent states: if an object may be in two or more states at the same time, then these states are concurrent states.

Constructor operation: an operation that creates new instances of a class.

Context (in OCL—q.v.): the domain within which an OCL expression is valid, for example, a class.

Context (of a pattern): the circumstances in which a particular problem occurs.

Contract: a black box description of a service (of a class or sub-system) that specifies the results of the service and the conditions under which it will be provided.

Control class: a stereotyped class that controls the interaction between boundary classes and entity classes.

Coupling: relates to the degree of interconnectedness between design components and is reflected by the number of links and the degree of interaction an object has with other objects.

Critical Path Analysis (CPA): a diagrammatic technique for analysing the dependencies between project tasks and determining those tasks that must be completed on time if the project itself is to be completed on time.

Data: raw facts, not yet identified as relevant to any particular purpose.

Dependency: a relationship between two model elements, such that a change in one element may require a change in the dependent element.

Deployment diagram: A diagram that shows the run-time configuration of processing nodes (*q.v.*) and the components, processes and objects that are located on them. One of two UML implementation diagrams (*q.v.*).

Design constraint: a constraint that limits the design options that may be used. Common design constraints include cost and data storage requirements.

Destructor operation: an operation that destroys instances of a class.

Domain model: an analysis class model that is independent of any particular use cases or applications, and that typically contains only entity objects. A domain model may serve as a basis for the analysis and design of components that can be reused in more than one software system.

Encapsulation: hiding internal details of a sub-system (typically a class) from the view of other sub-systems, so that each can be maintained or modified without affecting the operation of other parts of the system.

Enterprise (or business) rule: a statement that expresses business constraints on the multiplicity of an association; for example, an order is placed by exactly one customer.

Entity class: a stereotyped class that represents objects in the business domain model.

Event: an occurrence that is of significance to the information system.

Exception: a mechanism for handling errors in object-oriented languages.

Extend relationship: a relationship between use cases where one use case extends or adds new actions to another. Written as a stereotype: «extend».

Extreme Programming (XP): an approach to systems development that focuses on producing the simplest coding solution for application requirements. It uses pair programming, where program code is always written by two developers working at the same workstation.

Forces (of a pattern): the particular issues that must be addressed in resolving a problem.

Functional requirement: a requirement that specifies a part of the functionality required by the user.

Generalization: the abstraction of common features among elements (for example, classes) by the creation of a hierarchy of more general elements (for example, superclasses) that encapsulate the common features.

Guard condition: a Boolean expression associated with a transition that is evaluated at the time the event fires. The transition only takes place if the condition is true. A guard condition is a function that may involve parameters of the triggering event and also attributes and links of the object that owns the statechart.

Implementation diagram: a generic term for the two UML diagrams used in modelling the implementation of a system.

Include relationship: a relationship between use cases where one use case includes the actions described in another use case. Written as a stereotype: «include».

Incremental development: involves some initial analysis to scope the problem and identify the major requirements. The requirements are then reviewed and those that deliver most benefit to the client become the focus of the first increment of development and delivery. The installation of the first increment provides valuable feedback to the development team and informs the development of the second increment and so on.

Information: facts that have been selected as relevant to a purpose and then organized or processed in such a way that they have meaning for that purpose.

Inheritance: the mechanism by which object-oriented programming languages implement a relationship of generalization and specialization between classes. A subclass automatically acquires features of its superclasses.

Instance: a single object, usually called an instance in the context of its membership of a particular class or type (also object instance).

Instance diagram: a UML diagram similar in form to a class diagram, but that contains object instances instead of classes, links instead of asociations and may show attribute values (also known as an object diagram).

Instance value (of an attribute): the value of an attribute that is taken by a particular object at a particular time.

Integrity constraint: a constraint that has to be enforced to ensure that the information system holds data that is mutually consistent and is manipulated correctly. Referential integrity ensures that an object identifier in one object actually refers to an object that exists. Dependency constraints ensure that attribute dependencies, values are maintained consistently, where the value of one attribute is calculated from other attributes, are maintained consistently. Domain integrity ensures that attributes only hold permissible values.

Interaction: defines the message passing between objects within the context of a collaboration to achieve a particular behaviour.

Interaction diagram: an umbrella term for sequence diagrams and collaboration diagrams.

Interface: that part of the boundary between two interacting systems across which they communicate; the set of all signatures for the public operations of a class or package.

Interface class: a system interacts with its actors via its interface or boundary classes

Invariant: an aspect of a UML model expressed as a formal statement that must always remain true. For example, the value of a derived attribute `totalCost` may need always to be equal to the total of all `cost` attribute values.

Knowledge: a complex structure of information, usually one that allows its possessor to decide how to behave in particular situations.

Legacy system: any computerized information system, that has probably been used for some time, that was built with older technologies (maybe using different development approaches at different times) and that, most importantly, continues to deliver benefit to the organization.

Life cycle (of a project): the phases through which a development project passes from the inception of the idea to completion of the product and its eventual decommissioning.

Lifeline: a lifeline is a vertical dashed line that represents the existence of an object on an interaction sequence diagrams. An object symbol containing the object's name is placed at the top of a lifeline.

Link: a connection between objects; an instance of an association.

Message: a request to an object that it provide some specified service, either an action that it can carry out or some information that it can provide.

Message passing: a metaphor for the way that objects interact in an object-oriented system by sending each other messages that request services, or request or supply information. Since objects interact only through the messages they exchange, their internal details can remain hidden from each other.

Method: the implementation of an operation.

Methodology: comprises an approach to software development (e.g. object-orientation), a series of techniques and notations (e.g. the Unified Modelling Language—UML) that support the approach, a life cycle model (e.g. spiral incremental) to structure the development process, and a unifying set of procedures and philosophy.

Modular construction: an approach that aims to build systems that are easy to maintain, modify or extend. Modular construction relies on modules that are essentially decoupled sub-systems, with their internal details encapsulated.

Multiplicity: a constraint that specifies the range of permitted *cardinalities* (*q.v.*), for example in an association role or in a composite class. An association may have a multiplicity of 1..5; a particular instance of that association may have a cardinality of 3.

Node: A physical computational resource used by a system at run-time, typically having processing capability and memory.

Non-functional requirement: a requirement that relates to system features such as performance, maintainability and portability.

Normalization: a technique that groups attributes based upon functional dependencies according to several rules to produce normalized data structures that are largely redundancy free.

Object: a single thing or concept, either in a model of an application domain or in a software system, that can be represented as an encapsulation of state, behaviour and identity; a member of a class that defines a set of similar objects.

Object constraint language (OCL): a formal language that supplements the graphical notations of UML. OCL is generally used to give precise definitions for operation logic, or for properties such as invariants (q.v.).

Object diagram: see *instance diagram*.

Operation: an aspect of the behaviour that defines a class; an element of the services that are provided by a class; a specification of an element of system functionality that will be implemented as a method of an object.

Operation (in OCL—q.v.): usually an arithmetic, set or type operator, such as '+', 'size' or 'isEmpty', that is applied to the property (q.v.) in an OCL expression.

Operation signature: determined by the operation's name, the number and type of its parameters and the type of the return value if any. Polymorphically redefined operations have the same signature.

Package: a mechanism for grouping UML elements, usually classes, into groups. Packages can be nested within other packages.

Pattern: a pattern is an abstract solution to a commonly occurring problem in a given context.

Polymorphism: the ability of different methods to implement the same operation, and thus to respond to the same message in different ways that are appropriate to their class. For example, objects of different subclasses in an inheritance hierarchy may respond differently to the same message, yet with a common meaning to their responses.

Post-condition: part of an operation specification; those conditions that must be true before the operation can execute.

Pre-condition: part of an operation specification; those conditions that must be true after the operation has executed—in other words the valid results of the operation.

Primary operation: an operation to create or destroy an instance of a class, or to get or set the value of an attribute.

Property: a feature or characteristic of a UML element, usually one for which there is no specific UML notation.

Property (in OCL—q.v.): that specific element of the context (q.v.) to which an OCL expression applies, for example an attribute of a class.

Prototype: a prototype is a system or partially complete system that is built quickly to explore some aspect of the system requirements. It is not intended as the final working system.

Query operation: an operation that returns information but causes no change of state within a model or a software system.

Realize relationship: a relationship between two classes where one class offers the interface of the other but does not necessarily have the same structure of the other. Commonly used to show that a class supports an interface. Written as a stereotype: «realize».

Refactoring: restructuring and simplifying programme code so that duplication is removed and flexibility is enhanced.

Relation: a group of related data items organized in columns and rows, also known as a table.

Repository: the part of a CASE tool environment that handles the storage of models, including diagrams, specifications and definitions.

Responsibility: a high level description of behaviour a class exhibits. It reflects the knowledge or information that is available to that class, either stored within its own attributes or requested via collaboration with other classes, and also the services that it can offer to other objects.

Sequence diagram: or interaction sequence diagram, shows an interaction between objects arranged in a time sequence. Sequence diagrams can be drawn at different levels of detail and also to meet different purposes at several stages in the development life cycle.

Service: a useful function (or set of functionality) that is carried out by an object (or a sub-system) when requested to do so by another object.

Signal: an asynchronous communication between objects that may have parameters.

Software architecture: describes the sub-systems and components of a software system and the relationships between the components.

Specialization: the other face of generalization; an element (for example, a class) is said to be specialized when it has a set of characteristics that uniquely distinguish it from other elements. Distinguishes subclasses from their superclass.

Stakeholders: anyone who is affected by the information system. Stakeholders not only include users and development team members, but also resource managers and the quality assurance team, for example.

State: the state of an object is determined by values of some of its attributes and the presence or absence of certain links with other objects. It reflects a particular condition for the object and normally persists for a period of time until a transition to another state is triggered by an event. Instantaneous 'flow-through' states are allowed in UML 1.4.

Stereotype: a specialized UML modelling element. The stereotype name is contained within matched guillemets «...». For example, an interface package is a stereotype of a package.

Stimulus: an interaction between two objects that conveys information with an expectation of some action.

Subclass: a specialized class that acquires general features from its ancestor superclasses in a generalization hierarchy, but that also adds one or more specialized characteristics of its own.

Sub-system: a part of a system that can be regarded as a system in its own right.

Superclass: a generalized class that is an abstraction of the common characteristics of its subclasses in a generalization hierarchy.

Synchronizing operation: an operation that ensures that those attribute values which are dependent upon each other (e.g. may be calculated from each other) have consistent values.

System: an abstraction of a complex interacting set of elements, for which it is possible to identify a boundary, an environment, inputs and outputs, a control mechanism and some process or transformation that the system achieves.

Table: group of related data items organized in columns and rows. Used to store data in relational databases.

Task: a specific activity or step in a project.

Technique: a method for carrying out a project task.

Transaction: an elementary exchange, say of an item of capta (*q.v.*) or of a unit of value.

Transition: the movement from one state or activity to another, triggered by an event. A transition may start and end at the same state.

Type: a stereotype of class that is distinct from an implementation class; a type is defined by its attributes and operations but, unlike an implementation class, may not contain any methods. Classes that represent the concepts of the application domain are in fact types. An object may change its type dynamically during system execution, and may thus appear at different times to belong to different classes.

Usability requirement: user requirement that describes criteria by which the ease of use of the system can be judged.

Use case: describes, from a user's perspective, a behaviourally related set of transactions that are normally performed together to produce some value for the user. Use cases can be represented graphically in a use case diagram, each use case being described in the data dictionary. Use cases may be modelled at varying degrees of abstraction, essential use cases, the most abstract, are technologically and implementation independent whereas real use cases describe how the use case actually operates in a particular environment.

Use case realization: a set of model elements that show the internal behaviour of the software that corresponds to the use case—usually a collaboration.

User requirement: something that users require a software system to do (functional requirement); alternatively, a standard for the performance of a system (non-functional requirement).

User story: in Extreme Programming requirements are captured as user stories. A user story is very similar to a use case.

Visibility: UML modelling elements (e.g. attributes or operations) may be designated with different levels of accessibility or visibility. Public visibility means that the element is directly accessible by any class; private visibility means that the element may only be used by the class that it belongs to; protected visibility means that the element may only be used by either the class that includes it or a subclass of that class; and package visibility means that an element is visible to objects in the package.

Wrapper: or object wrapper, used to integrate object-oriented and non-object-oriented systems by encapsulating the non-object-oriented system with an object-oriented style of interface.

Bibliography

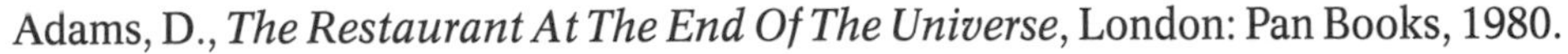

Adams, D., *The Restaurant At The End Of The Universe*, London: Pan Books, 1980.

Alexander, C., Ishikawa, S., Silverstein, M., Jacobson, M., Fiksdahl-King, I. and Angel, S., *A Pattern Language: Towns, Buildings, Construction*, New York: Oxford University Press, 1977.

Allison, B., O'Sullivan, T., Owen, A., Rice, J., Rothwell, A. and Saunders, C., *Research Skills for Students*, London: Kogan Page, 1996.

Allen, C. P., *Effective Structured Techniques: From Strategy to CASE*, Hemel Hempstead, UK: Prentice-Hall, 1991.

Allen, P. and Frost, S., *Component-Based Development for Enterprise Systems: Applying the SELECT Perspective™*, Cambridge: Cambridge University Press; SIGS Books, 1998.

Alur, D., Crupi, J. and Malks, D., *Core J2EE Patterns: Best Practices and Design Strategies*, Upper Saddle River, NJ: Prentice Hall and Sun Microsystems Press.

ANSI, *American National Standard for Information Systems: Database Language SQL*, ANSI X3,135-1986. New York: ANSI, 1986.

Apple Computer Inc., *Macintosh Human Interface Guidelines*, Reading, MA: Addison-Wesley, 1996.

Avison, D., and Fitzgerald, G., *Information Systems Development*, Oxford: Blackwell Scientific, 1988.

Barker, C., 'London Ambulance Service gets IT right', *Computing*, 25 June, 1998.

Beck, K., *Extreme Programming Explained: Embrace Change*, Reading, MA: Addison-Wesley, 2000.

Beck, K. and Cunningham, W., 'A laboratory for teaching object-oriented thinking', *Proceedings of OOPSLA '89*, pp.1–6, 1989.

Bell, D., *The Coming Of Post-Industrial Society*, New York: Basic Books, 1973. Cited in Webster (1995).

Bellin, D. and Simone, S. S., *The CRC Card Book*, Reading, MA: Addison-Wesley, 1997.

Bennett, S., Skelton, J. and Lunn, K., *Schaum's Outline of UML*, Maidenhead: McGraw-Hill, 2001.

Beveridge, T., 'Java Documentation' in *Java Unleashed*, Indianapolis: Sams.net Publishing, 1996.

Blaha, M. and Premerlani, W., *Object-Oriented Modeling and Design for Database Applications*, Upper Saddle River, NJ: Prentice-Hall, 1998.

Boehm, B. W., *Software Engineering Economics*, Englewood Cliffs, NJ: Prentice-Hall, 1981.

Boehm, B. W., 'A Spiral Model of Software Development and Enhancement', in Thayer, R. H. (Ed.), *Tutorial: Software Engineering Project Management*, Los Alamitos, CA: IEEE Computer Society Press, 1988.

Booch, G., *Object-Oriented Analysis and Design with Applications* (2nd Ed.), Menlo Park, CA: Benjamin/Cummings, 1994.

Booch, G., Rumbaugh, J. and Jacobson, I., *The Unified Modeling Language User Guide*, Reading, MA: Addison-Wesley; ACM Press, 1999.

Booth, P., *An Introduction to Human-Computer Interaction*, Hove: Lawrence Erlbaum Associates, 1989.

Brown, K. and Whitenack, B. G., 'Crossing Chasms: A Pattern Language for Object-RDBMS Integration' in Vlissides, J. M., Coplien, J. O. and Kerth, N. L. (Eds.) *Pattern Languages of Programme Design 2*, pp.227–238, Reading, MA: Addison-Wesley, 1996.

Brown, W. J., Malveau, R. C., McCormick, H. W. and Mowbray, T. J., *AntiPatterns: Refactoring Software, Architectures and Projects in Crisis*, New York, NY: John Wiley, 1998.

Browne, D., *STUDIO: Structured User-interface Design for Interaction Optimization*, Hemel Hempstead: Prentice-Hall, 1994.

Budgen, D., *Software Design*, Reading, MA: Addison-Wesley, 1994.

Buschmann, F., Meunier, R., Rohnert, H., Sommerlad, P. and Stal, M., *Pattern Oriented Software Architecture Volume 1*, Chichester: John Wiley, 1996.

Bødker, S., Gronbaek, K., and Kyng, M., 'Cooperative design: Techniques and experiences from the Scandinavian scene', in Schuler, D. and Namioka, A. (Eds.), *Participatory Design: Principles and Practices*, Hillsdale, NJ: Erlbaum, 1993.

Capers Jones, T., *Programming Productivity*, New York, NY: McGraw-Hill, 1986. Cited in Yourdon, 1989.

Carlow International Inc., *Human-Computer Interface Guidelines*, prepared for the Goddard Space Flight Center, Falls Church, VA: Carlow International Inc., 1992.

Carroll, J. M. (Ed.), *Scenario-Based Design: Envisioning Work and Technology in System Development*, New York: John Wiley, 1995.

Cattell, R. G. G. and Barry, D. K. (Eds.) *The Object Database Standard: ODMG 2.0*, San Francisco: Morgan Kaufmann, 1997.

CCTA, *SSADM Foundation: Business systems development with SSADM*, London: The Stationery Office, 2000.

Checkland, P., *Systems Thinking, Systems Practice*, Chichester: John Wiley, 1981.

Checkland, P., 'Information Systems and Systems Thinking: Time To Unite?', *International Journal of Information Management*, **8**, pp.234–243, 1988.

Checkland, P., unpublished presentation given at *Systems for Sustainability: People, Organisations and Environments*, 5th International Conference of the United Kingdom Systems Society, Milton Keynes: The Open University, July 1997.

Checkland, P. and Holwell, S., *Information, Systems and Information Systems: Making Sense Of The Field*, Chichester: John Wiley, 1998.

Checkland, P. and Scholes, J., *Soft Systems Methodology In Action*, Chichester: John Wiley, 1990.

Coad, P. and Yourdon, E., *Object-Oriented Analysis* (2nd Ed.), Englewood Cliffs, NJ: Yourdon Press; Prentice-Hall, 1990.

Coad, P. and Yourdon, E., *Object-Oriented Design*, Englewood Cliffs, NJ: Yourdon Press; Prentice-Hall, 1991.

Coad, P. with North, D. and Mayfield, M., *Object Models: Strategies, Patterns and Applications* (2nd Ed.), Upper Saddle River, NJ: Yourdon Press; Prentice-Hall, 1997.

Cockburn, A., *Writing Effective Use Cases*, Reading, MA: Addison-Wesley, 2000.

Codd, E. J., 'A Relational Model for Large Shared Data Banks', *Communications of the ACM*, **13**, (6), pp.377–387, 1970. (Also **26**, (1), pp.64–69, 1983)

Coleman, D., Arnold, P., Bodoff, S., Dollin, C., Gilchrist, H., Hayes, F. and Jeremaes, P., *Object-Oriented Development: The Fusion Method*, Englewood Cliffs, New Jersey:

Prentice-Hall International, 1994.
Collins, T., 'Banking's Big Brother option', *Computer Weekly*, 19 November, 1998a.
Collins, T., 'MPs lambast NHS for double project fiasco', *Computer Weekly*, 12 December, 1998b.
Collins, T., 'Lords inquiry to follow up Chinook campaign', *Computer Weekly*, 8 March, 2001.
Connor, D., *Information System Specification And Design Road Map*, Englewood Cliffs, NJ: Prentice-Hall, 1985.
Constantine, L., 'The Case for Essential Use Cases', *Object Magazine*, May 1997.
Cook, S. and Daniels, J., *Designing Object Systems: Object-Oriented Modelling With Syntropy*, Hemel Hempstead: Prentice-Hall, 1994.
Coplien J. O., *Advanced C++: Programming Styles and Idioms*, Reading, MA: Addison-Wesley, 1992.
Coplien, J., O., 'A Generative Development Process Pattern Language' in Coplien, J. O. and Schmidt, D. C., (Eds.), *Pattern Languages of Program Design.* Reading MA: Addison-Wesley, 1995.
Coplien, J. O. and Schmidt, D. C. (Eds.), *Pattern Languages of Program Design*, Reading, MA: Addison-Wesley, 1995.
Coplien, J. O., *Software Patterns*, New York: SIGS Books, 1996.
Cunningham, W., 'The CHECKS Pattern Language of Information Integrity' in Coplien, J. O. and Schmidt, D. C., (Eds.), *Pattern Languages of Program Design.* Reading, MA: Addison-Wesley, 1995.
Daniels, J., unpublished presentation given at *Migrating to Object Technology*, TattOO '95, Leicester: De Montfort University, January 1995.
Date, C. J., *An Introduction to Database Systems* (6th Ed.), Reading, MA: Addison-Wesley, 1995.
Date, C. J. and Darwen, H., *Foundation for Object/Relational Databases: The Third Manifesto*, Reading, MA: Addison-Wesley, 1998.
de Champeaux, D., *Object-Oriented Development Process and Metrics*, Upper Saddle River, NJ: Prentice Hall, 1997.
Deitel, H. M. and Deitel, P. J., *Java: How to Program*, Upper Saddle River, NJ: Prentice-Hall, 1997.
DeMarco, T., *Structured Analysis and System Specification*, Upper Saddle River, NJ: Yourdon Press; Prentice-Hall, 1979.
DeMarco, T., *Controlling Software Projects*, Englewood Cliffs, NJ: Yourdon Press, 1982.
Douglass B. P., *Real-Time UML Second Edition: Developing Efficient Objects for Embedded Systems*, Reading, MA: Addison-Wesley; ACM Press, 1999.
Dix, A., Finlay, J., Abowd, G. and Beale, R., *Human–Computer Interaction*, Hemel Hempstead: Prentice-Hall, 1998.
Drummond, H., 'The politics of risk: trials and tribulations of the Taurus project', *Journal of Information Technology*, **11**, pp.347–357, 1996.
Eaglestone, B. and Ridley, M., *Object Databases: An Introduction*, Maidenhead: McGraw-Hill, 1998.
Ellwood, W., 'Seduced by Technology', *New Internationalist*, December 1996.
Eriksson, H-E. and Penker, M., *UML Toolkit*, NewYork: John Wiley, 1998.
Fidler, C. and Rogerson, S., *Strategic Management Support Systems*, London: Pitman, 1996.
Finkelstein, C., *An Introduction to Information Engineering*, Wokingham: Addison-Wesley, 1989.
Flowers, S., 'IS Project Risk—The Role of Management in Project Failure',

Proceedings of BIT '97, Manchester Metropolitan University, November 1997.

Flynn, D., *Information Systems Requirements: Determination And Analysis*, (2nd Ed.), Maidenhead: McGraw-Hill, 1998.

Fowler, M. with Scott, K., *UML Distilled: Applying The Standard Object Modeling Language*, Reading, MA: Addison-Wesley, 1997a.

Fowler, M., *Analysis Patterns: Reusable Object Models*, Reading, MA: Addison-Wesley, 1997b.

Gabriel, R., *Patterns of Software: Tales from the Software Community*, New York: Oxford University Press, 1996.

Gaines, B. R. and Shaw, M. L. G., 'Dialogue engineering' in Sime, M. E. and Coombs, M. J. (Eds.), *Designing for human-computer communication*, London: Academic Press, 1983.

Gamma, E., Helm, R., Johnson, R. and Vlissides, J.M., *Design Patterns: Elements of Reusable Object-Oriented Software*, Reading, MA: Addison-Wesley, 1995.

Gane, C. and Sarson, T., *Structured Systems Analysis: Tools and Techniques*, Englewood Cliffs, NJ: Prentice-Hall, 1978.

Gilb, T., *Principles of Software Engineering Management*, Wokingham: Addison-Wesley, 1988.

Goldsmith, S., *A Practical Guide to Real-Time Systems Development*, Hemel Hempstead: Prentice Hall, 1993.

Goodland, M. with Slater, C., *SSADM Version 4: A Practical Approach*, Maidenhead: McGraw-Hill, 1995.

Goss, E.P., Buelow, V. and Ramchandani, H., 'The Real Impact of a Point-of-Care Computer System on ICU Nurses', *Journal of Systems Management*, pp.43–47, January/February 1995.

Graham, I., Henderson-Sellers, B. and Younessi, H., *The OPEN Process Specification*, Harlow: Addison-Wesley, 1998.

Graham, I., *Object Oriented Methods*, Wokingham: Addison-Wesley, 1993.

Graham, I., *Migrating to Object Technology*, Wokingham: Addison-Wesley, 1995.

Greenbaum, J. and Kyng, M., *Design at Work: Cooperative Design of Computer Systems*, Hillsdale, NJ: Lawrence Erlbaum Associates, 1991.

Hammersley, M. and Atkinson, P., *Ethnography: Principles in Practice*, (2nd Ed.), London: Routledge, 1995.

Harel, D., 'Statecharts: a visual formalism for complex systems', *Science of Computer Programming*, **8**, pp.231–274, 1987.

Harel, D., 'On visual formalisms', *Communications of the ACM*, **31**, (5), pp.514–530, 1988.

Harel, D. and Politi, M., *Modeling Reactive Systems with Statecharts: The STATEMATE Approach*, New York, NY: McGraw-Hill, 1998.

Hart, A., *Knowledge Acquisition for Expert Systems*, London: Kogan Page, 1989. (New edition 1997 published by Chapman & Hall.)

Hatley, D. J. and Pirbhai, I. A., *Strategies for Real-Time Systems Design*, New York: Dorset House, 1987.

Hay, D., *Data Model Patterns Conventions of Thought*, New York: Dorset House, 1996.

Henderson-Sellers, B. and Unhelkar, B, *Open Modeling with UML*, Wokingham: Addison Wesley Longman, 2000.

Hicks, M., J., *Problem Solving in Business and Management*, London: Chapman & Hall, 1991.

Hopkins, T. and Horan, B., *Smalltalk: an introduction to application development using VisualWorks*, Hemel Hempstead: Prentice-Hall, 1995.

Horrocks, I., *Constructing the User Interface with Statecharts*, Harlow: Addison-Wesley, 1999.

Howe, D.R., *Data Analysis for Data Base Design*, (3rd Ed.), Oxford: Butterworth-Heinemann, 2001

Humphrey, W. S., *Managing The Software Process*, Wokingham: Addison-Wesley, 1989.

Hyde, M., 'Boo rises from the ashes', *Guardian*, 3 November, 2000.

IBM, *IBM San Francisco Base*, Rochester, MN: IBM, 1998.

Ince, D., *ISO 9001 And Software Quality Assurance*, Maidenhead: McGraw-Hill, 1994.

Jackson, M., *Principles of Program Design*, London: Academic Press, 1975.

Jacobson, I., Booch, G., and Rumbaugh, J., *The Unified Software Development Process*, Reading, MA: Addison-Wesley; ACM Press, 1999.

Jacobson, I., Christerson, M., Jonsson, P. and Övergaard, G., *Object-Oriented Software Engineering: A Use Case Driven Approach*, Wokingham: Addison-Wesley, 1992.

Jacobson, I., Ericsson, M. and Jacobson, A., *The Object Advantage: Business Process Reengineering with Object Technology*, New York, NY: ACM Press, 1995.

Jacobson, I., Griss, M. and Jonsson, P., *Software Reuse: Architecture, Process and Organization for Business Success*, Harlow: Addison-Wesley, 1997.

Jayaratna, N., *Understanding and Evaluating Methodologies; NIMSAD: A Systemic Framework*, Maidenhead: McGraw-Hill, 1994.

Koestler, A., *The Ghost In The Machine*, London: Hutchinson, 1967.

Kyng, M., 'Making Representations Work', *Communications of the ACM*, **38**, 9, pp 46–55, September 1995.

Larman, C., *Applying UML and Patterns: An Introduction to Object-Oriented Analysis and Design*, Upper Saddle River, NJ: Prentice-Hall, 1998.

Lim, W. C., 'Effects of reuse on quality, productivity and economics', *IEEE Software*, **11**, (5), pp.23–30, 1994.

Loomis, M. E. S., *Object Databases: The Essentials*, Reading, MA: Addison-Wesley, 1995.

Lorenz, M. and Kidd, J., *Object-Oriented Software Metrics: A Practical Guide*, Englewood Cliffs, NJ: Prentice-Hall, 1994.

Maguire, M., *RESPECT User Requirements Framework Handbook*, European Usability Support Centres, 1997.

Martin, D., Riehle, D. and Buschmann, F. (Eds.), *Pattern Languages of Program Design 3*, Reading, MA: Addison-Wesley, 1998.

McBride, N., 'Business Use Of The Internet: Strategic Decision Or Another Bandwagon?', *European Management Journal*, **15**, pp.58–67, 1997.

Melton, A., (Ed.), *Software Measurement*, London: Thompson Computer Press, 1995.

Meyer, B., *Object Success: A Manager's Guide to Object Orientation, its Impact on the Corporation, and its Use for Reengineering the Software Process,* Hemel Hempstead: Prentice-Hall, 1995.

Meyer, B., *Object-Oriented Software Construction* (2nd Ed.), Upper Saddle River, NJ: Prentice-Hall PTR, 1997.

Meyer, B., *Object-Oriented Software Construction*, Hemel Hempstead: Prentice-Hall International, 1988.

Meyer, B., "Design by Contract", in *Advances in Object-Oriented Software Engineering*, Mandrioli, D. and Meyer, B. (Eds), London: Prentice-Hall, 1991.

Microsoft Corporation, *The Windows Interface Guidelines for Software Design*, Redmond, WA: Microsoft Press, 1997.

Muller, P-A., *Instant UML*, Birmingham: Wrox Press, 1997.

Mumford, E., *Effective Systems Design And Requirements Analysis*, Basingstoke: Macmillan, 1995.

Mumford, E., *Systems Design: Ethical Tools for Ethical Change*, Basingstoke: Macmillan, 1996.

Newkirk J. and Martin R., *Extreme Programming in Practice*, Upper Saddle River, NJ: Addison-Wesley, 2001.

OMG, *The Common Object Request Broker Architecture and Specification; Revision 2.0*, Framingham, MA: Object Management Group Inc, 1995.

OMG, *OMG Unified Modeling Language Specification—Version 1.4*, OMG, 2001.

Oppenheim, A. N., *Questionnaire Design, Interviewing and Attitude Measurement*, Continuum International, 2000.

Orfali, R. and Harkey, D., *Client/Server Programming with JAVA and CORBA* (2nd Ed.), New York, NY: John Wiley, 1998.

Page-Jones, M., *The Practical Guide to Structured Systems Design* (2nd Ed.), Englewood Cliffs, NJ: Prentice-Hall, 1988.

Peters, T., *Thriving on Chaos: Handbook for a Management Revolution*, London: Macmillan, 1988.

Philips, T., 'A Slicker System For The City', *The Guardian On-Line*, 3 April, 1997.

Pollice G., *Using the Rational Unified Process for Small Projects: Expanding Upon eXtreme Programming*, Rational Software Corporation, 2001 (at www.rational.com).

Porter, M., *Competitive Strategy*, New York: Free Press, 1985.

Preece, J. with Rogers, Y., Sharp, H., Benyon, D., Holland, S. and Carey, T., *Human–Computer Interaction*, Wokingham: Addison-Wesley, 1994.

Pressman, R. and Ince, D. (Editor), *Software Engineering: A Practitioner's Approach*, European Adaptation, London: McGraw-Hill, 2000.

Riehle, D. and Zullighoven, H., 'Understanding and Using Patterns in Software Development', *Theory and Practice of Object Systems*, **2**, (1), pp.3-13, 1996.

Rifkin, J., *The End Of Work*, New York: Putnam Publishers, 1995. Cited in Ellwood, W., 'Seduced by Technology', *New Internationalist*, December 1996.

Roberts, G., *The Users' Role in Systems Development*, Chichester: John Wiley; Central Computer and Telecommunications Agency, 1989.

Robinson, B. and Prior, M., *Systems Analysis Techniques*, London: International Thomson Publishing, 1995

Rosenberg, D, with Scott, K., *Use Case Driven Object Modeling with UML: a Practical Approach*, Upper Saddle River, NJ: Addison-Wesley, 1999.

Rosson, M. B. and Carroll, J. M., 'Narrowing the Specification-Implementation Gap in Scenario-Based Design', in Carroll, J. M. (Ed.), *Scenario-Based Design: Envisioning Work and Technology in System Development*, New York: John Wiley, 1995.

Rumbaugh, J., 'Models through the development process', *Journal of Object-Oriented Programming*, May 1997.

Rumbaugh, J., Blaha M., Premerlani, W., Eddy, F. and Lorensen, W., *Object-Oriented Modeling and Design*, Englewood Cliffs, NJ: Prentice-Hall International, 1991.

Rumbaugh, J., Jacobson, I. and Booch, G., *The Unified Modeling Language Reference Manual*, Reading, MA: Addison-Wesley; ACM Press, 1999.

Sachs, P., 'Transforming Work: Collaboration, Learning and Design', *Communications of the ACM*, **38**, (9), pp.36–44, 1995.

Sanders, J. and Curran, E., *Software quality: a framework for success in software development and support*, Wokingham: Addison-Wesley, 1994.

Sauer, C., *Why Information Systems Fail: A Case Study Approach*, Maidenhead: McGraw-Hill, 1993.

Schmidt, D., Stal, M., Rohnert, H., and Buschmann, F., *Pattern Oriented Software Architecture: Patterns for Concurrent and Networked Objects: Volume 2*, Chichester: John Wiley, 2000.

Schneider, K., 'Bug delays £25m court case system', *Computer Weekly*, 22 December, 1997.

Selic, B., Gullekson, G., and Ward, P. T., *Real-Time Object-Oriented Modeling*, New York, NY: John Wiley, 1994.

Senn, J. A., *Analysis & Design of Information Systems*, (2nd Ed.), New York: McGraw-Hill, 1989.

Shackel, B., 'Human factors and usability', in Preece, J. and Keller, L. (Eds.), *Human–Computer Interaction: Selected Readings*, Hemel Hempstead: Prentice-Hall, 1990.

Shlaer, S. and Mellor, S., *Object-Oriented Systems Analysis: Modeling the World in Data*, Englewood Cliffs, NJ: Prentice-Hall, 1988.

Shlaer, S. and Mellor, S., *Object Lifecycles: Modeling the World in States*, Englewood Cliffs, NJ: Prentice-Hall, 1992.

Shneiderman, B., *Designing the User Interface* (3rd Ed.), Reading, MA: Addison-Wesley, 1997.

Silberschatz, A., Korth, H. F. and Sudarshan, S., *Database System Concepts* (3rd Ed.), New York: McGraw-Hill, 1996.

Skidmore, S., *Introducing Systems Analysis* (2nd Ed.), Oxford: NCC Blackwell, 1994.

Skidmore, S., Mills, G. and Farmer, R., *SSADM Models & Methods*, Manchester: NCC Blackwell, 1994.

Sommerville, I., *Software Engineering* (4th Ed.), Reading, MA: Addison-Wesley, 1992.

Soni, D., Nord, R. and Hofmeister, C., 'Software Architecture in Industrial Applications', in *Proceedings of the 17th International Conference on Software Engineering*, pp.196–207, Seattle, WA: ACM Press, 1995.

Standish Group, *Chaos Report*, Standish Consulting Group, 1995. Cited in Flowers, S., 'IS Project Risk—The Role of Management in Project Failure', *Proceedings of BIT '97*, Manchester Metropolitan University, November 1997.

Stapleton, J., *Dynamic Systems Development Method*, Harlow: Addison-Wesley, 1997.

Symons, V., 'Evaluation of information systems: IS development in the processing company', *Journal of Information Technology*, **5**, pp.194–204, 1990.

Tanenbaum, A. S., *Modern Operating Systems*, Englewood Cliffs, NJ: Prentice Hall, 1992.

Texel, P., and Williams, C., B., *Use Case Combined with Booch/OMT/UML: Processes and Products*, Upper Saddle River, NJ: Prentice Hall, 1997.

The Concise Oxford Dictionary (9th Ed.), Oxford: OUP, 1995.

Timmers, P., *Electronic Commerce*, Chichester: Wiley, 2000.

Turban, E. and Aronson, J. E., *Decision Support Systems and Intelligent Systems* (6th Ed.), Upper Saddle River, NJ: Prentice Hall, 2001.

Vlissides, J.M., Coplien, J. O. and Kerth, N., (Eds.), *Pattern Languages of Program Design 2*, Reading, MA: Addison-Wesley, 1996.

Wall Street Journal, 22 September, 1998, reprinted in *The Guardian Editor*, 26 September, 1998.

Ward, P. and Mellor, S., *Structured Development for Real-Time Systems Vol 1:*

Introduction & Tools, Englewood Cliffs, NJ: Yourdon Computing Press, 1985.

Ward, P. and Mellor, S., *Structured Development for Real-Time Systems Vol 2: Essential Modelling Techniques*, Englewood Cliffs, NJ: Yourdon Computing Press, 1985.

Ward, P. and Mellor, S., *Structured Development for Real-Time Systems Vol 3: Implementation Modelling Techniques*, Englewood Cliffs, NJ: Yourdon Computing Press, 1986.

Webster, F., *Theories Of The Information Society*, London: Routledge, 1995.

Weir, C. and Daniels, J., 'Software Architecture Document', in *Proceedings of Object Technology 98*, Oxford, 1998.

Whiteside, J., Bennett, J. and Holtzblatt, K., 'Usability engineering: our experience and evolution', in Helander, M. (Ed.), *Handbook of Human-Computer Interaction,* Amsterdam: North-Holland, 1988.

Whitten, J., Bentley, L. and Barlow, V., *Systems Analysis and Design Methods* (3rd Ed.), Irwin, 1994.

Willcocks, L. and Lester, S., 'Any Way Out Of The Labyrinth? Information Technology Productivity Revisited', *The Future Of Information Systems: UK Academy For Information Systems—First Conference*, Cranfield University, April 1996.

Wirfs-Brock, R., Wilkerson, B. and Wiener, L., *Designing Object-Oriented Software,* Englewood Cliffs, NJ: Prentice-Hall International, 1990.

Yourdon, E. and Constantine, L. L., *Structured Design: Fundamentals of a Discipline of Computer Program and Systems Design*, Englewood Cliffs, NJ, USA: Yourdon Press; Prentice-Hall, 1979

Yourdon, E., *Modern Structured Analysis*, Englewood Cliffs, NJ: Prentice-Hall, 1989.

Yourdon, E., *Object-Oriented Systems Design: An Integrated Approach*, Englewood Cliffs, NJ: Prentice-Hall International, 1994.

Yourdon, E., *Structured Walkthroughs*, Englewood Cliffs, NJ: Yourdon Press, 1985

Zuboff, S., *In The Age Of The Smart Machine: The Future Of Work And Power*, Oxford: Heinemann, 1988.

Index

D

T